Electronic Fundamentals and Applications

PRENTICE-HALL ELECTRICAL ENGINEERING SERIES

JOHN D. RYDER

Professor of Electrical Engineering
Michigan State University

ELECTRONIC FUNDAMENTALS AND APPLICATIONS

FOURTH EDITION

LONDON
PITMAN PAPERBACKS

Fourth edition 1970

SIR ISAAC PITMAN AND SONS LTD.
Pitman House, Parker Street, Kingsway, London, W.C.2
P.O, Box 6038, Portal Street, Nairobi, Kenya

SIR ISAAC PITMAN (AUST.) PTY. LTD.
Pitman House, Bouverie Street, Carlton, Victoria 3053, Australia

PITMAN PUBLISHING COMPANY S.A. LTD.
P.O. Box 9898, Johannesburg, S. Africa

PITMAN PUBLISHING CORPORATION
6 East 43rd Street, New York, N.Y. 10017, U.S.A.

SIR ISAAC PITMAN (CANADA) LTD.
Pitman House, 381–383 Church Street, Toronto, 3, Canada

THE COPP CLARK PUBLISHING COMPANY
517 Wellington Street, Toronto, 2B, Canada

ISBN: 0 273 31491 2

(*Cased edition:* 273 31488 2)

Printed in Great Britain by
William Clowes and Sons, Limited, London and Beccles

GO–(T.1336)

PREFACE TO THE FOURTH EDITION

In the face of the ever-widening scope of electronics research and development, it is possible for a textbook to cover the field only by stressing fundamentals; only such an approach can integrate the whole subject. This book emphasizes the basic application, the analytic method, the fundamental process; it gives less attention to the details of individual electronic devices. This preparation seems desirable from the systems point of view, since many of these basic processes are also fundamental in non-electronic systems. Oscillation, modulation, frequency conversion, feedback, and logic are examples of processes which have many applications.

Solid-state devices have now been adopted in most areas of communications, electronics and computation. In choice of material and treatment, this revision reflects that trend. Many of the phenomena which underly both solid-state and vacuum devices have been treated jointly, and parallelism in application has been stressed wherever appropriate. The physical aspects of solid-state devices are more thoroughly covered than in the previous edition, although the major stress is still on basic understanding of principles on which applications may be based. Chapters on pulse communications and logic circuits have been added, and the discussion of tuned amplifiers has been considerably expanded.

J. D. Ryder

CONTENTS

10 LINEAR SMALL-SIGNAL RC AMPLIFIERS 241

11 WIDE BAND AND FREQUENCY SELECTIVE AMPLIFIERS 271

17 AM AND FM DEMODULATION 443

18 WAVE-SHAPING AND SWITCHING 469

19 LOGICAL SWITCHING 502

Electronic Fundamentals and Applications

1

MOTION OF CHARGED PARTICLES IN FIELDS

Controlled movement of electric charge through vacuum, gas, or semiconductor materials is a primary attribute of electronic devices. Early forms employed electrons or gas ions as the moving charges, but the development of the semiconductor family of devices has introduced another conduction mode, that associated with *holes* as positive charge carriers. The semiconductor adds complication, in that the charge motion is through a crystal lattice, and subject to interatomic forces, rather than being in free space as is electron movement in the vacuum tube. Besides the ballistic properties of electrons drifting in electric or magnetic fields, the new devices call for study of charge motion resulting from diffusion of charge in a concentration gradient.

Thus knowledge of atomic structure as well as the ballistic properties of atoms and charges in electric and magnetic fields becomes basic to the understanding of all forms of electronic devices.

1-1 EARLY HISTORY

Modern electronic science is considered as beginning in 1883 with the discovery, by Edison, that an electric current would pass between a metal plate and the heated filament in the vacuum of one of his lamps, when the plate was positive to the filament. J. A. Fleming applied this Edison effect to the rectification or detection of radio signals about 1897, and in 1906 de Forest added the grid or control element to the Fleming two-element *diode*, yielding the three-element or *triode* detector for radio signals. Further development by Arnold, Langumir, and others led to its successful operation as an amplifier.

Various anomalies in the conduction and rectification properties of certain metal oxides and sulfides were reported during the nineteenth century, and natural crystals of galena, silicon, and iron pyrites were used as detecting devices for radio signals in the early part of this century. These devices were eclipsed by the triode vacuum tube, but returned to prominence as detectors during the radar development of World War II, because of better performance at the higher frequencies. This application was followed by research on semiconductor and crystalline materials, the achievement of the transistor or control action in a solid crystal in 1948 by Bardeen and Brattain, and the junction transistor by Shockley a little later. Additional research has produced a theory of conduction for semiconductor materials which provides an understanding of solid-state diode and transistor design and performance.

1-2 THE ATOM

The atom, as the basic building block of the chemical elements, was little understood in the early twentieth century, until Niels Bohr proposed a dynamic atom model much like the solar system. Negative charges were assumed in motion in precise orbits about a central nucleus of positive charge, the whole being electrically neutral. Bohr's atomic theory successfully explained line spectra, X rays, and the conduction of electric current through gases.

Other observed phenomena were not well explained by the original theory, and these exceptions have required modification of the original Bohr atom model. Thus the negative charges have been assigned wave as well as corpuscular properties, energy levels have been substituted for the orbits, probability functions have replaced certainty, and the mathematical formulation of wave mechanics has developed. The result is a better concept of the atom, and of its function in electric conduction in gases, liquids, and the solid state.

1-3 THE ELECTRON, THE HOLE

The Edison effect showed that the current passing was due to a procession of negative particles; that is, they were attracted to a positive electrode. J. J. Thomson observed similar small charges in his research on "cathode rays" (circa 1897), and G. Johnstone Stoney proposed that they be named *electrons*. No smaller charge has ever been observed, and the electron is now accepted as the fundamental unit of electrical charge.

Although the physical form of the electron is not determinable, it is often assumed as a spherical particle, but it is known to act as a wave as well. No satisfactory explanation for this duality of corpuscular and wave properties is yet available. The concept employed in a particular discussion is usually that which leads to the most satisfactory explanation of the observed phenomena.

Thomson's early experiments showed that all electrons have the same ratio of charge to mass (e/m), now believed to have the value 1.759×10^{11} coulombs per kilogram. Millikan, in 1910, measured the negative charge on the individual electron, a value now taken as 1.602×10^{-19} coulomb. It is apparent that the mass of the electron must be 9.106×10^{-31} kilogram.

Solid-state theory has also supplied us with the concept of the *hole*. This is represented by an electron site in a crystal lattice from which the electron is missing; the hole thus has a positive charge of magnitude equal to that of the electronic charge. This subject is further discussed in Chapter 2.

1-4 NEUTRONS, PROTONS, PHOTONS, IONS

As now postulated, atoms are made up of a central nucleus of positive charge and mass attributed to *protons* and *neutrons*, with the nucleus surrounded by electrons moving in orbits of fixed energy level.

The proton carries a positive charge of magnitude equal to that of the electron, and has a mass approximately equal to that of the hydrogen atom H^1, whose nucleus consists of only one proton. This makes the mass of the proton about 1849 times as great as that of the electron.

The number of protons in the nucleus is equal to the atomic number of the element, and, for a normal atom, this positive charge is balanced by an equal negative charge carried by the electrons in the orbits.

The neutron carries zero electrical charge, but has a mass 0.08 per cent larger than that of the proton. Neutrons are present in varying numbers in the nuclei of all atoms except H^1. Atoms of elements having identical proton numbers or positive charge but with differing numbers of neutrons are called *isotopes*.

The *photon* is a bundle of radiant energy, and may be considered as having an equivalent radiation mass as determined from its energy by use of the Einstein relation

$$E = mc^2 \qquad \text{joules}$$

where m is in kilograms, and c is the velocity of light in free space $= 3 \times 10^8$ meters per second. The photon must often be given a dual character as a wave or as a particle, to satisfactorily explain certain experimental observations. In this it parallels the viewpoint we discussed concerning the electron.

The frequency of the radiant energy determines the amount of energy carried by a photon. That is, the *quantum* of energy associated with a photon is given by

$$W = hf \qquad \text{joules}$$

where h is Planck's constant and f is the frequency of the radiation in hertz. Light, heat, X rays and radio waves are all examples of photon radiation.

A positive *ion* also appears as an atom which has lost one or more of its external and more loosely held electrons, the positive charge of the nucleus then overbalancing the negative charges of the remaining electrons. A negative ion is found as an atom which has acquired one or more excess electrons.

1-5 METRIC SYSTEM MAGNITUDES

Since the m.k.s.a. system of metric units is now universal in electrical engineering, it is advisable to also adopt the principle of identifying the major orders of magnitude by distinctive prefixes, which is a characteristic of the metric system. The prefixes will be found useful in designating small subdivisions of large basic units, or when dealing with large numbers of basic small units.

The prefixes of the metric system are:

Multiple	Prefix	Symbol
10^{12}	tera	T
10^{9}	giga	G
10^{6}	mega	M
10^{3}	kilo	k

Multiple	Prefix	Symbol
10^{-2}	centi	c
10^{-3}	milli	m
10^{-6}	micro	μ
10^{-9}	nano	n
10^{-12}	pico	p
10^{-15}	atto	a

These designations have the support of the International Electrotechnical Commission and of the Institute of Electrical and Electronics Engineers.

In electrical work, where frequency has been stated as cycles per second, an obviously incorrect usage has been the world "cycles" standing alone. To correct this situation and also to conform to international practice, a recent IEEE standard (1965) has replaced cycles per second with the unit designation *hertz* (Hz). We then have kilohertz (kHz) and megahertz (MHz) for designations of frequencies previously expressed as kilocycles per second and megacycles per second.

The electrical profession is fortunate in having a truly international set of units and symbols.

1-6 MOTION OF CHARGE IN AN ELECTRIC FIELD

Because conduction occurs by movement of charge, it is desirable to understand the motion of individual charges under forces imposed by electric fields. In such an analysis, it is usual to assume that the *charge density*, or number of charges per unit volume, is so low that the repulsion effects between charges may be neglected. It is also assumed that the charge masses are so small that the gravitational forces are negligible compared with the forces impressed by the fields present.

The *field intensity* in an electric field is $\mathscr{E}$ volts per meter, defined as the force acting per unit of positive charge. That is,

$$\mathscr{E} = \frac{\mathbf{f}}{q} \qquad \text{V/m} \quad (1\text{-}1)$$

and thus

$$\mathbf{f} = q\mathscr{E} \qquad \text{N} \quad (1\text{-}2)$$

where q is in coulombs. The positive direction of the intensity $\mathscr{E}$ is then the direction of the force on the positive charge or from positive to negative electrode.

An electron has a negative charge of 1.60×10^{-19} coulomb, a quantity which will be known as $-e$. The force on an electron in a field of intensity $\mathscr{E}$ is then

$$\mathbf{f}_e = -e\mathscr{E} \qquad (1\text{-}3)$$

and $\mathbf{f}_e$ is opposite to the defined force on a positive charge, or is directed toward the positive electrode.

Due to this force the electron will be accelerated by the electric field since $\mathbf{f} = m\, dv/dt$, so

$$\mathbf{a}_e = \frac{d\mathbf{v}_e}{dt} = -\frac{e\mathscr{E}}{m} \qquad \text{m/s}^2 \quad (1\text{-}4)$$

This acceleration is directed along the lines of field flux and negatively, or towards the positive electrode.

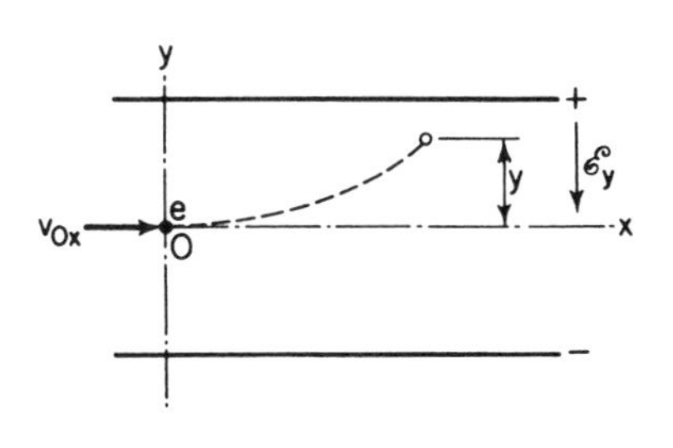

Figure 1-1. Path of a moving electron in a uniform electric field.

To further illustrate the motion of a charge, consider an electron released at a point $P(x_0, y_0)$ with an initial velocity v_{0x}, wholly in the $+x$ direction, as in Fig. 1-1. Acting in the region is a uniform electric field of magnitude $\mathscr{E}_y$ and directed along $-y$. The field is created by a potential which is positive towards $+y$. The equations of motion of the electron are therefore

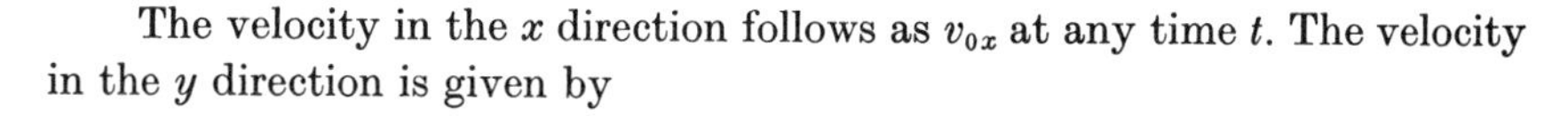

$$a_x = \frac{dv_x}{dt} = 0, \qquad a_y = \frac{dv_y}{dt} = -\frac{e\mathscr{E}_y}{m}, \qquad a_z = \frac{dv_z}{dt} = 0 \qquad (1\text{-}5)$$

The velocity in the x direction follows as v_{0x} at any time t. The velocity in the y direction is given by

$$v_y = -\int \frac{e\mathscr{E}_y}{m}\,dt + C_1$$

If $\mathscr{E}_y$ is not a time function, then

$$v_y = -\frac{e\mathscr{E}_y t}{m} \tag{1-6}$$

since $C_1 = v_y = 0$ at $t = 0$. By the defined conditions $v_z = 0$.

The resultant velocity at time t is

$$v = \sqrt{v_x^2 + v_y^2} \tag{1-7}$$

The x position at time t is directly given by

$$x = v_x t + x_0 \tag{1-8}$$

and the y position follows by integration of Eq. 1-6 as

$$y = \frac{-e\mathscr{E}_y t^2}{2m} + y_0 \tag{1-9}$$

Thus the probable position of the electron may be determined as a function of time.

1-7 ENERGY INTERCHANGE WITH A MOVING CHARGE

The work done on a charge of mass m, in moving between two points A and B in an electric field of intensity $\mathscr{E}$, is equal to the energy acquired by the charge. The work done on a charge by the force f is predicted by

$$E = \int_A^B f\cos\theta\,ds = \int_A^B q\mathscr{E}\cos\theta\,ds \tag{1-10}$$

where θ is the angle between the positive direction of $\mathbf{f}$ and the direction of movement ds.

For the case in which the charge has the value $-e$, and the movement ds is opposite to the field intensity or is towards the positive electrode, the angle θ is zero. The kinetic energy acquired by the electron is then

$$E = \frac{m(v^2 - v_0^2)}{2} = -e\int_A^B \mathscr{E}\,ds \tag{1-11}$$

The integral of the field intensity is the negative of the potential V between A and B, so that

$$E = \frac{m(v^2 - v_0^2)}{2} = eV \qquad \text{joules} \tag{1-12}$$

for the electron.

If v_0 is zero, then the final velocity reached by an electron in rising through a potential V is

$$v = \sqrt{\frac{2eV}{m}} \qquad \text{m/s} \tag{1-13}$$

Since the work integral is independent of the path in a conservative field, the path or time of movement does not affect this result. However, because of relativistic variation of mass (discussed in Section 1-11), the applicable value of v is limited.

1-8 THE ELECTRON VOLT

Equation 1-12 allows definition of an *electron volt*, eV, as the energy acquired by an electron in rising through a potential of one volt. As an energy unit

$$1 \text{ eV} = 1 \times 1.60 \times 10^{-19} = 1.60 \times 10^{-19} \qquad \text{joule}$$

Rising through a potential of 500 V, an electron acquires an energy of 500 eV; this numerical equivalence has made the electron volt a popular energy unit.

The electron volt is much used in stating the energy of high-speed particles in high-energy nuclear physics. The abbreviations MeV and BeV are used, meaning "million electron volts" and "billion electron volts," respectively.

1-9 ELECTRIC CURRENT

An electric current is conceived as movement of electric charge. Benjamin Franklin in 1747 was one of the first to propose that an electric current be looked upon as an electric fluid passing from positive to negative terminal in a metallic circuit external to the source. The definition of "positive" and "negative" was made in an arbitrary manner in that early day, and was based upon an observed difference in the charges held by electrified glass and hard rubber rods. The techniques of electrical science have been built upon this simple assumption, and years of usage have made it a universal convention.

The Franklin assumption of the direction of a positive current implies a movement of positive charges, whereas conduction may occur with either positive or negative charges in various media. Thus a movement of negative electrons from A to B must be taken as a negative current, or a conventional current from B to A.

A current is defined as charge passing a point per second. It is often convenient to measure a current in terms of *charge density* ρ, moving with a velocity v past P, the point of observation. The charge passing through a unit area per second is the *current density* J, and

$$J = \rho v \qquad \text{A/m}^2 \quad (1\text{-}14)$$

For a charge of unit value q, with n charges per cubic meter,

$$J = nqv \qquad \text{A/m}^2 \quad (1\text{-}15)$$

For a charge of electrons

$$J = -nev \qquad \text{A/m}^2 \quad (1\text{-}16)$$

the negative sign indicating that the direction of the conventional current is opposite to the velocity vector of the electrons.

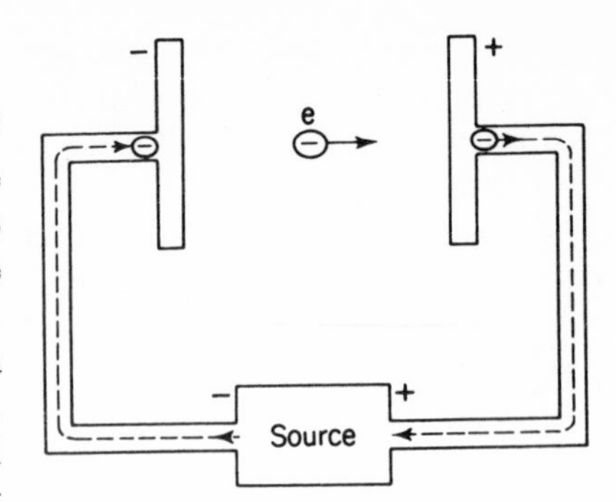

Figure 1-2. Current induced in a circuit by electron motion.

During the movement of an electron in space between electrodes, as in Fig. 1-2, the electron is accelerated and increases its kinetic energy. The energy converted must be supplied by the system. The value of current may be found from consideration of the rate of energy transfer from the source. The battery of potential V contributes energy equal to the change in kinetic energy of the charge, or

$$E = qv = \frac{m(v^2 - v_0^2)}{2} \tag{1-17}$$

Taking the time derivative

$$V\frac{dq}{dt} = -Vi = mv\frac{dv}{dt} = mva \tag{1-18}$$

For the electron $a = -e\mathscr{E}/m$, and $v = at$, so that

$$i = \frac{e^2\mathscr{E}^2 t}{m^2 V} \tag{1-19}$$

This is the current at time t, due to the transit of one electron in a field intensity $\mathscr{E}$, with an accelerating potential V. Figure 1-3 plots the current represented by one electron traveling between two plane electrodes separated 1 cm and accelerated by 100 V.

During its motion in the field the moving electron induces a charge on the positive electrode and an oppositely changing charge on the negative electrode. The free electrons in the copper circuit conductors are given a slight movement in the direction of the negative electrode. This movement of induced charge represents a conventional current out of the positive battery terminal. Thus the source supplies energy to the electron.

The electron forces a charge redistribution in the circuit and when it strikes the electrode the forces are neutralized, and current and energy interchange stop. Even if the electron never reaches the electrode, a current has existed and an energy transfer has taken place. This reasoning is supported by the basic law of current continuity.

It is equally possible to transfer energy *from* an electron to an external circuit by projecting the electron into a retarding field, or causing its dece-

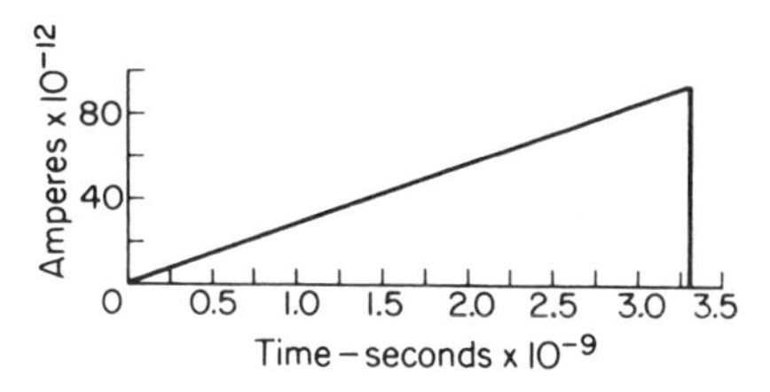

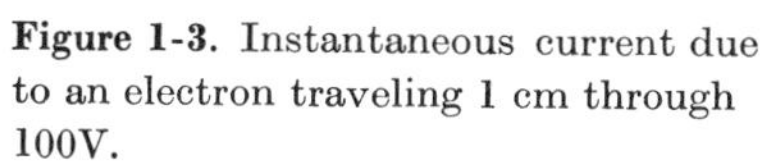

Figure 1-3. Instantaneous current due to an electron traveling 1 cm through 100V.

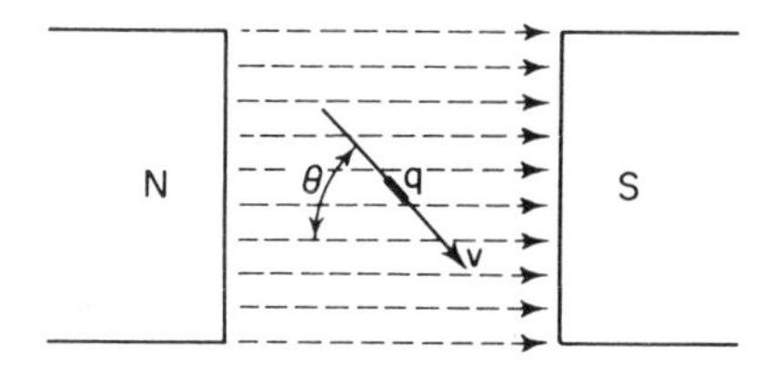

Figure 1-4. Force on a moving electron in a magnetic field.

leration. The induced current then is directed so as to transfer energy to the source.

On impact with an electrode, the kinetic energy of the moving charge is converted to heat and other radiant energy. When the energy is very great, some of the radiation may occur in the X-ray region.

1-10 CHARGE IN A MAGNETIC FIELD

The force on a charge q moving at velocity $\mathbf{v}$ in a magnetic field of intensity $\mathbf{B}$, oriented as in Fig. 1-4, is

$$\mathbf{f}_m = q(\mathbf{v} \times \mathbf{B})$$

For the electron of $-e$ charge this becomes

$$\mathbf{f}_m = -e(\mathbf{v} \times \mathbf{B}) = -evB \sin \theta \tag{1-20}$$

The cross product indicates that the force $\mathbf{f}_m$ is at right angles to the plane of $\mathbf{v}$ and $\mathbf{B}$, and has a positive direction given by the motion of a right-hand screw rotated from $\mathbf{v}$ to $\mathbf{B}$. Because of the negative charge the force on the electron is oppositely directed. The electron in Fig. 1-4 is acted upon by a force perpendicular to and into the page.

The force is perpendicular to the motion of the charge and no work is done; if $\mathbf{v}$ and $\mathbf{B}$ are constant, then $\mathbf{f}_m$ produces a constant acceleration at right angles to the velocity. Such an acceleration gives circular motion, as indicated in Fig. 1-5.

In circular motion the centrifugal force is $f_c = mv^2/r$, where v is the peripheral velocity. That velocity is the component normal to the field $\mathbf{B}$, so

$$\frac{m(v \sin \theta)^2}{r} = evB \sin \theta \tag{1-21}$$

The radius r of the circular path is

$$r = \frac{mv \sin \theta}{eB} \tag{1-22}$$

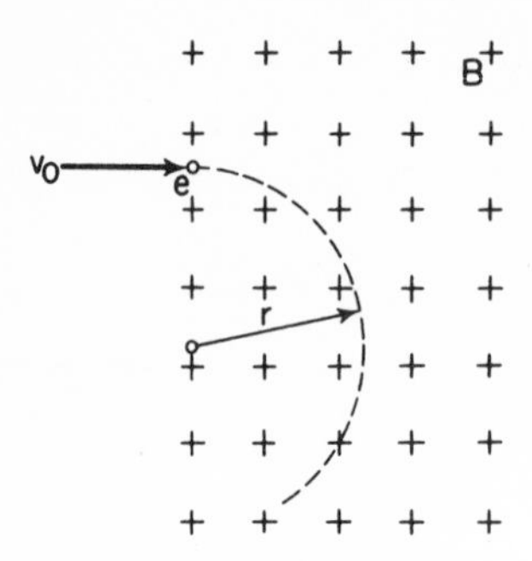

Figure 1-5. Path of an electron in a magnetic field normal to and into the page.

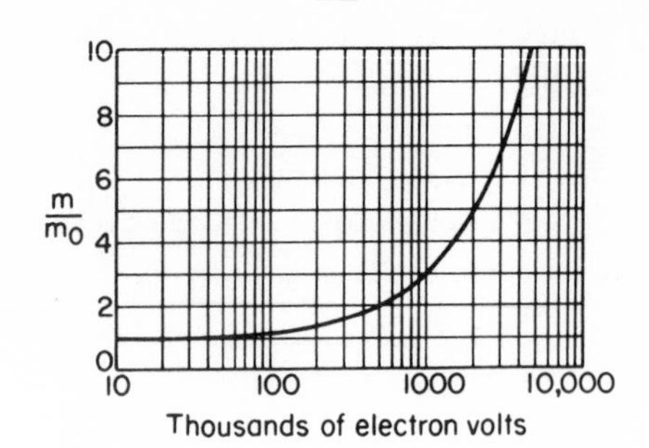

Figure 1-6. The relativistic variation of mass of an electron.

The angular velocity is

$$\omega = \frac{v \sin \theta}{r} = \frac{eB}{m} \tag{1-23}$$

A circle is completed in time T, and

$$Tv \sin \theta = 2\pi r$$

$$T = \frac{2\pi m}{eB} \tag{1-24}$$

The time of a revolution is independent of the peripheral velocity, being dependent only on B. That is, a high-velocity electron travels around a circle of large radius in the same time that a low-velocity electron takes to travel around a circle of proportionately smaller radius.

1-11 RELATIVISTIC EFFECTS ON MASS

Newton's second law states

$$f = \frac{d(mv)}{dt} = m\frac{dv}{dt} + v\frac{dm}{dv}\frac{dv}{dt} \tag{1-25}$$

If mass is assumed not to be a function of velocity, then the right-hand term goes to zero, leaving $f = m\, dv/dt$ as the usual low-velocity form of the basic law.

However, the theory of relativity requires that the masses of all bodies vary with velocity, and this has been experimentally proven. The mass of a moving body is then correctly expressed as

$$m = \frac{m_0}{\sqrt{1 - v^2/c^2}} \tag{1-26}$$

where m_0 is the *rest mass* at zero velocity. Common usage of the simpler form of expression for Newton's law is justified, since the mass is not appreciably different from the rest mass until the velocity becomes comparable to that of light at $c = 3 \times 10^8$ m/s. Figure 1-6 illustrates the variation of m as a function of particle energy in electron volts. The mass of an electron is doubled at an energy of about 500,000 eV, tripled at 1 MeV.

Applying Eq. 1-26 to Eq. 1-25, and taking dm/dv leads to

$$f = \frac{m_0}{(1 - v^2/c^2)^{3/2}} \frac{dv}{dt} \tag{1-27}$$

However,

$$W = \int_0^t f\, ds = \int_0^t fv\, dt = \int_0^v \frac{m_0 v}{(1 - v^2/c^2)^{3/2}} dv = eV$$

$$m_0 c^2 \left(\frac{1}{\sqrt{1 - v^2/c^2}} - 1 \right) = eV \tag{1-28}$$

which relates the velocity of an electron to its accelerating potential V, giving due regard to the relativistic variation of mass. From this result

$$\frac{v}{c} = \sqrt{1 - \frac{1}{[(eV/m_0 c^2) + 1]^2}} \tag{1-29}$$

and this is the velocity reached when an electron rises through a potential V, with regard for relativistic change in mass.

The expression shows that c is a limiting velocity, since v approaches c as V increases without limit. The variation of m with velocity is neglected in most engineering applications since the accelerating voltages are relatively low. However, the relativistic change in mass is important in high-energy particle accelerators, becoming an important design factor.

1-12 FOCUSING OF ELECTRON BEAMS

In an electric field in space the electrons tend to accelerate parallel to the flux lines or normal to the equipotentials. If Fig. 1-7 is looked upon as a cross section of a set of cylinders with the indicated voltage applied, the equipotentials between the cylinders A and B will lie as sketched and will constitute a lens for focusing an electron beam.

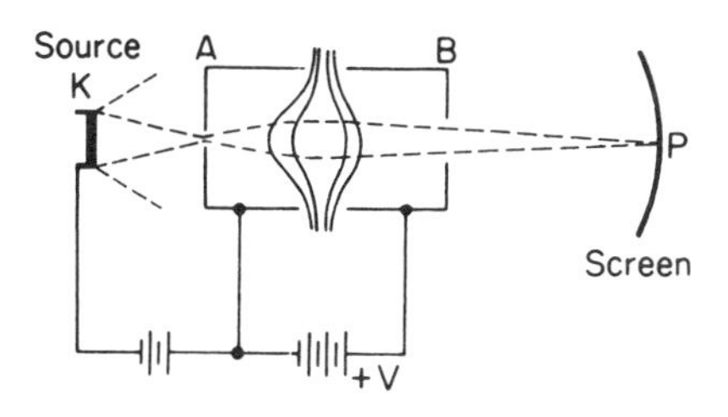

Figure 1-7. Electric-field focusing of an electron beam.

Electrons entering a small hole in the end of A will be given a small convergent force on first entering the field at the junction of the cylinders. Upon emerging through the field at the far side of the lens section the electrons will be given a small divergent force due to the oppositely curved equipotentials. However, the electrons have been accelerated due to the potential V existing between the cylinders, and the initial convergent force acts on the electrons for a longer time than the final divergent force. The result is a small inward net force for convergence of the electrons, which will cause the electrons to emerge in a bundle at B. The magnitude of the net convergent force can be controlled by variation of V, and such variation can produce an ultimate point of beam focus at P.

The equipotentials are not well-shaped near the fringes of the field and the diaphragm at B is intended to cut off improperly focused electrons passing through the edges of the field. Designing such a lens is more difficult than designing an optical lens, since electron paths are curves whereas light travels in straight lines.

Electron optics is one area in which the electron is desirably treated as having wave properties and wave length. The wave length has been found to be related to the velocity of the electron as

$$\lambda = \frac{h}{mv} \quad \text{m} \qquad (1\text{-}30)$$

where h is Planck's constant. The range of wave lengths corresponding to usual velocities in electron-optical devices is found to fall below that of visible light and extends into the X-ray region.

The electron lens suffers in sharpness if λ is not the same for all electrons in the beam, just as an optical lens is not equally sharp for all colors. Since λ is a function of accelerating voltage V and initial electron velocity, voltage V should be large to suppress the initial velocity effects.

Magnetic lenses, or combinations of magnetic and electric systems, are other possibilities for focusing electron beams. Because such lenses provide for magnification of an electron pattern formed at A, the electron microscope becomes possible. Since the wave length of the electrons is considerably less than visible or ultraviolet light, much smaller objects may be resolved than by instruments using conventional optical systems.

1-13 THE CATHODE-RAY TUBE

The cathode-ray tube, employing electron deflection by fields, is utilized as a television image tube (kinescope) and also as a device for visual study of electrical phenomena in the laboratory. Basically, a cathode-ray tube consists of an electron source, two sets of deflecting electrodes surrounding the electron beam, and a fluorescent indicating screen, as shown in Fig. 1-8. When designed for magnetic deflection of the beam, the deflection plates may be eliminated.

The electron source or heated cathode, K, emits electrons into the evacuated space, and these are accelerated by a positive potential on anode A_1. The grid G provides control of the number of electrons passing or of the

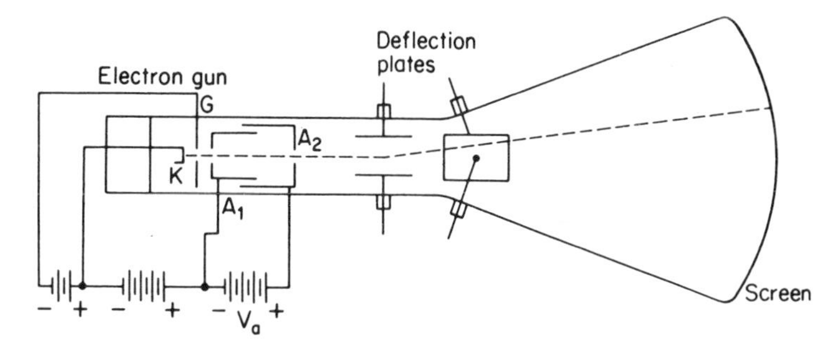

Figure 1-8. The cathode-ray tube.

beam intensity. The electron lens set up between cylinders A_1 and A_2 permits focusing of the electrons into a small pencil beam, perhaps 0.1 mm in diameter at the screen. The interior surface of the screen is thinly coated with a phosphor, a material which fluoresces or emits light when struck by high-speed electrons, thus indicating the position of the electron beam as a lighted spot.

One pair of deflecting plates is usually positioned to produce a horizontal electric field, for deflecting the electron beam horizontally or in the x direction when potentials are applied to these x-axis plates. A vertical or y deflection of the beam is obtained by application of potentials to a pair of y-axis plates, oriented to produce a vertical electric field and a vertical deflection. Simultaneous application of potentials to both x and y systems gives control of the beam and the spot of light, in both x and y coordinates. The small electron masses introduce no appreciable inertia effects when the applied deflection potentials have frequencies up to or above 10^8 hertz.

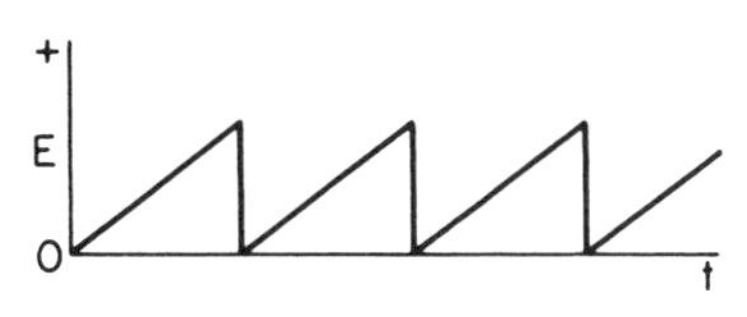

Figure 1-9. Wave form of a sawtooth sweep voltage.

To view the wave forms of time-varying electrical potentials, it is necessary that they appear plotted against time as an axis. A voltage increasing linearly with time is called a *sweep*, and takes a sawtooth form as in Fig. 1-9. If a sweep voltage is placed on the x-axis plates, and an ac voltage is applied to the y-axis electrodes, then the wave form of the ac voltage will be traced out against time, in accordance with Fig. 1-10. Repetitive sweeps, synchronized to multiple cycles of the input signal, will cause the image to be repeated and to visually stand in place.

At high spot velocities the spot movement might be thought too rapid for human sight, but due to persistence of vision in the human eye and to persistence of the image on the screen, discussed in Section 1-15, the figure appears as a solid-line plot of the varying phenomena when the repetition rate of the figure is twenty or more per second.

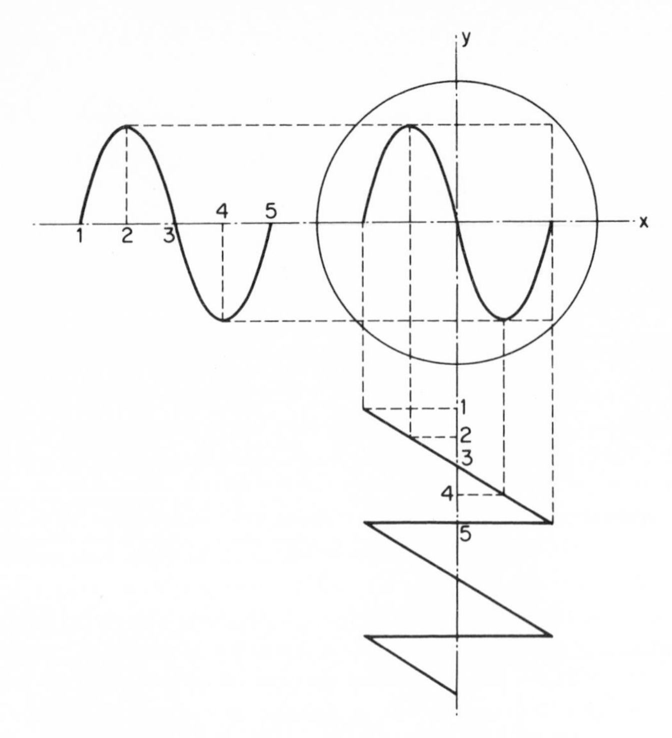

Figure 1-10. A sine wave traced as a time function by a sweep voltage.

1-14 DEFLECTION OF THE CATHODE-RAY BEAM

Deflection of the cathode-ray beam by an electric field is illustrated in Fig. 1-11. The accelerating potential V_a increases the beam velocity to v_{0x}. The time of flight in l_d, the length of the deflecting plates, may be assumed short compared to the time of a cycle of the deflection voltage V_d, or V_d can be considered constant during the flight of an electron. Fringing of the deflection fields is also neglected.

In the deflection field the path of the electron is parabolic, and from Eq. 1-9 the slope of that curve at $x = l_d$ or the slope of the tangent $0'$, P' is

$$\frac{dy}{dx} = -\frac{e\mathscr{E}_y l_d}{mv_{0x}^2} = \tan\theta \tag{1-31}$$

From the figure

$$x - 0' = \frac{y}{\tan\theta} = \frac{at^2}{2\tan\theta} \tag{1-32}$$

Since $t = l_d/v_{0x}$, then

$$x - 0' = \frac{l_d}{2} \tag{1-33}$$

and $0'$ is at the center of the deflection plates.

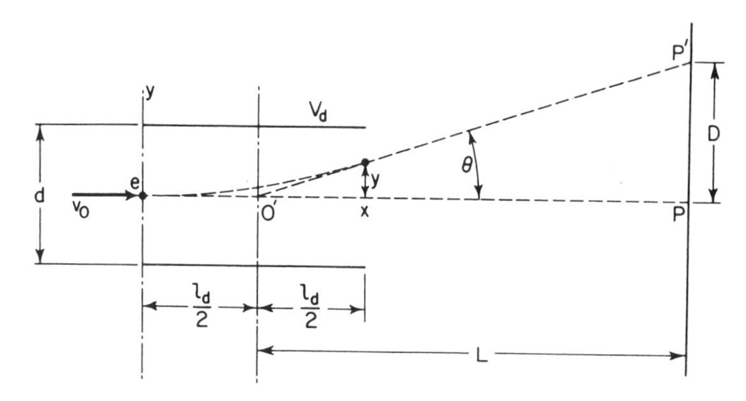

Figure 1-11. Electric-field deflection of the cathode-ray beam.

Deflection D on the screen can be found by use of Eq. 1-13 and the definition of $\mathscr{E}_y$ as

$$D = L \tan \theta = \frac{Ll_d}{2dV_a} V_d \qquad \text{m} \quad (1\text{-}34)$$

The *deflection sensitivity* is defined as the screen deflection per volt of deflecting potential V_d, or

$$S_e = \frac{Ll_d}{2dV_a} \qquad \text{m/V} \quad (1\text{-}35)$$

The length l_d and the spacing d are limited by the requirements for plate clearance for large deflections, and L is set by allowable mechanical size for the bulb. The accelerating voltage should be high for maximum spot brightness, so that a given design is a compromise of these conflicting factors. Laboratory tube sensitivities are in the range of 0.2–0.5 millimeter per volt.

Upon impact, part of the beam energy is transformed to visible light on the screen, but most of the energy will be converted as heat in the screen. A bright stationary spot can burn and damage the screen.

Deflection of the beam may also be accomplished with current coils producing a transverse magnetic field, as in Fig. 1-12. Assuming a uniform magnetic field over the distance l_m in Fig. 1-13, with no fringing of flux, the

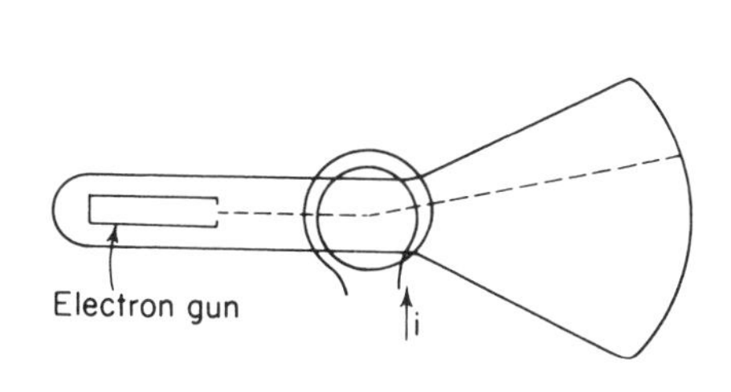

Figure 1-12. Position of coils for magnetic deflection.

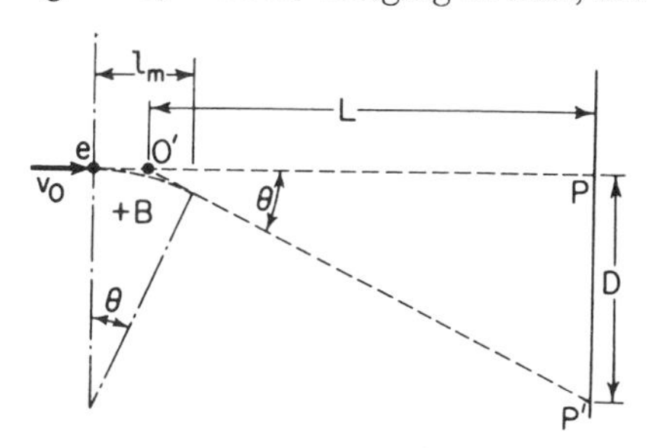

Figure 1-13. Magnetic deflection of the cathode-ray beam.

beam will be deflected in a circular path of radius $r = eB/mv_{0x}$. From the tube geometry

$$\sin\theta = \frac{l_m}{r} = \frac{eBl_m}{mv_{0x}} \tag{1-36}$$

Distance L is measured to the intersection of a tangent to the arc with the line of the original velocity v_{0x}. Point 0′ is not at the center of l_m as in the electric field case, but may be assumed as central with small error. Then $\tan\theta = D/L$ and by trigonometric identity

$$\tan\theta = \frac{eBl_m/mv_{0x}}{\sqrt{1 - (eBl_m/mv_{0x})^2}} \tag{1-37}$$

For the usual values of B and a large radius, the second term under the radical is negligible with respect to unity, and so

$$D = \sqrt{\frac{e}{m}}\,\frac{Ll_mB}{\sqrt{2V_a}} \qquad \text{m} \tag{1-38}$$

The magnetic deflection sensitivity is defined as meters deflection per weber per square meter, or

$$S_m = \sqrt{\frac{e}{m}}\,\frac{Ll_m}{\sqrt{2V_a}} \qquad \text{m/W} \tag{1-39}$$

Flux density B is proportional to coil current in air-cored coils, and it is often convenient to state the sensitivity as ampere turns required on the exciting coils. Usual sensitivities are 0.5 to 1.0 millimeter deflection per ampere turn.

Magnetic deflection sensitivity is a function of $1/\sqrt{V_a}$, in contrast to the $1/V_a$ relation for electric deflection. With magnetic deflection the loss of sensitivity is less as V_a is increased for greater spot brightness, so that it is customary to use magnetic deflection in television systems, where high spot brightness is essential.

In all tubes there remain a few ionized gas atoms. These are deflected by the field systems, but because of their great mass and the factor $\sqrt{e/m}$ in Eq. 1-38, the deflection given by the magnetic field will be very small and the charged atoms will always strike the screen near the undeflected beam position. Continued bombardment of this nature will create a brown spot on the screen. If a thin aluminum coating is placed on the inner surface of the screen, the high-velocity electrons pass through and excite the fluorescent material, but the large ions are reflected away by the atoms of the screen. The aluminum layer also serves as a reflector and returns to the screen light which would otherwise be lost in the tube interior.

1-15 FLUORESCENT SCREEN PROPERTIES

Screen materials vary in luminescent properties, and the brightness depends on beam current (or, more properly, beam power) and the size of the spot.

The phosphorescent or persistence properties cause retention of the image, and various materials are suited to particular applications. A *long-persistence* screen may retain a visible image from a few seconds up to several minutes, a *medium-persistence* screen indicates a retained image for a time in milliseconds, and a *short-persistence* material indicates a retention of the image in the microsecond range. A list of standard screen characteristics is given in Table 1-1.

Table 1-1

CATHODE-RAY TUBE FLUORESCENT SCREENS

Phosphor number	Color	Persistence	Application
P-1	Green	Medium	General oscillography
P-2	Blue-green	Long	Transient visualization
P-4	White	Short	Television picture tube
P-7	Blue-white; then yellow	Very long	Radar screens
P-11	Blue	Short	Photography
P-16	Bluish purple	Very short	Telvision scanner
P-19	Orange	Long	Radar

Removal of the charge conveyed by the beam to the screen is necessary. Fortunately, as the electrons strike the screen they cause the screen to emit other electrons, as a splash effect. The emitted electrons are gathered by a graphite coating over the interior bulb walls and connected to the second anode. The emission effect, discussed in Chapter 3 as *secondary emission*, causes sufficient charge loss from the screen to maintain its potential at a few volts positive with respect to the second anode.

1-16 THE CATHODE-RAY OSCILLOSCOPE

When assembled with power supplies, signal amplifiers, and sweep wave form generator, the cathode-ray tube becomes part of a cathode-ray oscilloscope, well suited to laboratory instrumentation.

Broad frequency response amplifiers are provided so that signal inputs in millivolts or microvolts may produce usable screen deflections. These amplifiers must transmit all necessary harmonic components of the highest frequencies present in the signals, and must not alter the phase relations. Frequently these amplifiers are also designed to transmit the dc component of the signal as well.

The sweep oscillator must provide a voltage output which is linear with time, so that single or multiple waves of a wide variety of input signals can be viewed. Circuits must also be included to start or trigger the sweep wave at a desired time in the input signal cycle, or at the beginning of a nonrepetitive transient.

Power supplies for the accelerating and focus voltages, and for supply of the amplifiers are included. Controls for focus of the spot, for intensity variation of the trace, and for moving the zero axes are also provided.

High anode potentials provide high spot brightness and must be used for the recording of fast transients. Actual anode voltages in use range from 500 to 80,000 volts, with usual laboratory equipment employing 1500 to 7000 volts; television equipment ranges up to 24,000 volts. Another solution to the need for increased brightness is furnished by use of one or more intensifier electrodes around the glass bulb between the deflection plates and the screen. These rings operate at potentials above that of the accelerating anode and further raise the beam velocity and energy. Such a tube has low V_a with respect to deflection sensitivity, but acts as one with high V_a with respect to spot brightness.

The accelerating potential source is usually connected with the tube cathode as negative and below the ground potential at which the second anode is maintained. Since the common electrode of the deflection system is connected to the second anode, and the deflection connections are handled by the user, the grounded anode is an important safety feature.

The capacitance of the deflection plates is commonly in the range of 2 to 5 pF. This reactance is usually negligible to at least 10^7 hertz, and amplifier performance will often provide the upper frequency limit. Special techniques, in which a periodic wave is sampled at repeated intervals, allow frequencies as high as 10^9 hertz to be viewed and studied on the oscilloscope screen.

PROBLEMS

1-1. Two large metal plates are separated by 0.5 cm, with the right plate being 150 V positive to the left plate. An electron with $v_0 = 10^6$ m/s to the right is released at the center of the left plate.
(a) What will be the electron's velocity upon reaching the right plate?
(b) Compute the time of flight.
(c) How much energy is conveyed to the electrode by the impact?

1-2. In the preceding problem the potential on the plates is reversed in polarity when the electron is 0.3 cm from the left plate.
(a) Which electrode is reached by the electron?
(b) Find the velocity of arrival.
(c) What is the electron's energy at the instant of potential reversal and what is the energy given up on impact? Account for the difference.

1-3. An electron with upward v_0 due to energy of 250 eV enters the field as shown in part (a), Fig. 1-14.

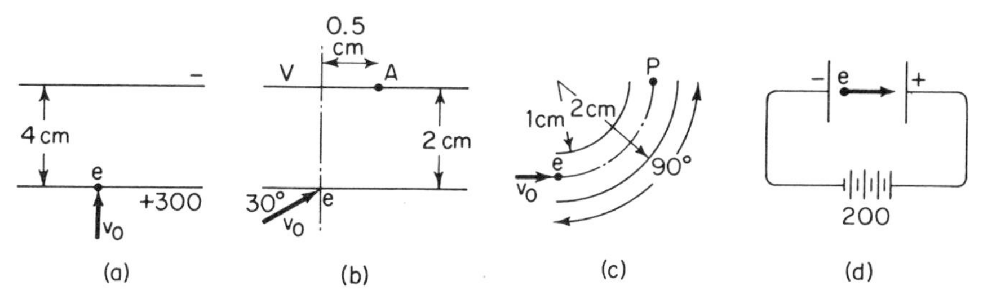

Figure 1-14.

(a) How far up does the electron travel?

(b) What is the time at which the electron strikes an electrode?

(c) Plot a curve of current in the external circuit vs. the time of flight of the electron.

1-4. An electron with energy of 75 eV enters an electric field as shown in (b), Fig. 1-14.

(a) What potential must be established between the plates to make the electron hit point A?

(b) Plot a curve of current from the source V during the flight of the electron.

1-5. In Fig. 1-14(c) an electron is released as shown, with $v_0 = 0.5 \times 10^6$ m/s and the inner electrode positive.

(a) What is the needed electric field intensity along the path midway between the electrodes to make the electron reach P?

(b) Replace the electric field with a magnetic field and specify its density and direction to make the electron reach P.

1-6. A positively charged particle having 1.60×10^{-19} coulomb and an initial velocity of 3×10^4 m/s is injected at right angles to a magnetic field of 0.10 W/m^2. Its path has a radius of 10 cm. Find the mass of the particle.

1-7. In Fig. 1-14(b) the upper plate is 100 V negative to the lower plate. The injected electron has an energy of 1000 eV. Instead of the indicated 30°, what angle of injection will give the greatest horizontal travel before return to the lower plate?

1-8. An electron with a y-directed energy of 300 eV enters a magnetic field at the origin of axes.

(a) What flux density and direction of magnetic field is required to make the electron reach the point $x = 2$ cm, $y = 0$, $z = 0$?

(b) What is the time of flight?

1-9. An electron with 10 eV energy directed at 90° to the lower of a parallel plate system, enters the field at $t = 0$. The plate separation is 1 cm, and the potential of the upper plate varies as $v = 10^9 t$, positive to the lower plate. Find the velocity, energy, and time of flight of the electron upon impact.

1-10. It is usual to assume that in copper there is one conduction, or free electron, per atom and there are 5×10^{22} atoms per cm^3 of copper. If a piece of No. 10 copper wire connects the circuits of (d), Fig. 1-14, with the plates separated 0.1 cm, then:

(a) Find the drift velocity of the electrons in the wire at the instant just before the electron strikes the positive plate.

(b) Find the velocity of the electron at the same instant. Explain the difference in these velocities.

(c) Show that the rate of energy output from the battery at that instant is equal to the rate of energy increase of the electron.

1-11. An electron with an energy in the horizontal direction of 100 eV enters the field between two horizontal plates 2 cm long and separated 2 cm. The upper plate is at +100 V. Find the direction and flux density of a magnetic field which will prevent deflection of the electron from its path between the plates. Neglect fringing of the fields.

1-12. The potential across an X-ray tube is 1 MV. Considering relativistic mass corrections, determine:

(a) If the electron starts from rest at the negative electrode, what is its velocity on arrival at the anode?

(b) What is the mass of the electron on arrival?

1-13. The x-deflection plates of a cathode-ray tube are 1.0 cm long and separated 0.4 cm. Distance L is 15 cm, and the accelerating potential is 800 V.

(a) Find the screen deflection produced by a dc voltage of 200 V.

(b) Find the screen deflection produced by a 100 V rms ac voltage.

1-14. The cathode-ray tube of Problem 1-13 has applied to the deflection plates a sinusoidal voltage of 200 V peak, and frequency of 2 GHz. An electron enters the space between the plates at voltage zero. Find the deflection of this electron on the screen, and compare with the deflection due to 200 V dc of the previous problem.

1-15. Voltages of $v_y = 100 \sin (2\pi 1000t)$ and $v_x = 120 \sin (2\pi 250t)$ are simultaneously applied to the plates of a cathode-ray tube, with deflection sensitivity for both sets of plates of 0.23 mm/V. What are the x and y coordinates of the spot at $t = 0.0005$ s? The origin is the spot position at $t = 0$.

1-16. A 40-hertz sine voltage is applied to the x plates and a 60-hertz sine voltage of equal amplitude is applied to the y plates of a cathode-ray tube. The two waves both are zero and going positive at $t = 0$. Graphically construct the complete path of the spot on the screen.

REFERENCES

1. Dumont, A. B., "Elimination of Distortion in Cathode-Ray Tubes," *Electronics*, Vol. 8, p. 16 (January 1936).

2. Coslett, V. E., *Introduction to Electron Optics*. Oxford University Press, Inc., New York, 1946.

3. Epstein, D. W., and L. Pensak, "Improved Cathode-Ray Tubes with Metal-Backed Luminescent Screens," *RCA Rev.*, Vol. 7, p. 5 (March 1946).

4. Spangenberg, K. R., *Vacuum Tubes*. McGraw-Hill Book Company, New York, 1948.

5. Harman, W. W., *Fundamentals of Electronic Motion*. McGraw-Hill Book Company, New York, 1953.

6. Haas, A., and R. W. Hallows, *The Oscilloscope at Work*. Philosophical Library, Inc., New York, 1956.

2

SOLID-STATE CONDUCTION: SEMICONDUCTOR DIODES

A *conductor* (copper, for example) has an electrical resistivity in the range from 10^{-8} to 10^{-6} ohm-meter (10^{-6} to 10^{-4} ohm-centimeter)†. Materials classed as *insulators* (quartz or polyethylene are examples) have resistivities in the range from 10^7 to above 10^{18} ohm-m (10^9 to 10^{20} ohm-cm).

Materials having resistivities intermediate between conductors and insulators are known as *semiconductors*. Of major importance at present are germanium, with a resistivity of 0.6 ohm-m (60 ohm-cm) in the very pure state, and silicon with a resistivity of 1.5×10^3 ohm-m (1.5×10^5 ohm-cm). A number of intermetallic compounds such as GaAs and PbTe are also of importance, as are some of the metallic sulfides and oxides.

In semiconductors the number of charges available for conduction is zero at 0°K, but increases with temperature. Desired conduction properties are achieved by additions of other elements to control the number and sign of the charge carriers. Conduction in semiconductors occurs both by drift under an electric field and by diffusion from one region of high charge density to another region of lower charge density.

The unilateral two-element solid-state *diode* is a basic application of controlled conduction properties in the solid state, and an understanding of its properties is an objective of this chapter.

†With the m.k.s.a. unit system, resistivity is measured in ohm-meters; conductivity is expressed in mhos per meter. However, semiconductor technology frequently employs the ohm-centimeter and mho per centimeter. Equations utilizing conductivity and resistivity must employ the meter dimension for correctness, but text references to typical values may utilize the centimeter units for consistency with past literature.

2-1 THE BOHR ATOMIC MODEL

The Bohr theory, discussed briefly in Chapter 1, supposes an atom model having a positively charged nucleus surrounded by a group of electrons, established in definite orbits. The number of protons in the nucleus is equal to N, the atomic number, and for a normal atom the number of electrons is also N, their total charge balancing the positive nuclear charge and making the atom electrically neutral. Distinctions between different elements are due to the varying number of positive charges in the nucleus.

Because of the positive charge, an electric field exists around the nucleus. It might be supposed that the centrifugal force on an electron rotating in an orbit around the nucleus is balanced by the attraction between the electron and the nucleus. However, such orbital motion implies a central acceleration, and classical electrodynamic theory requires radiation of energy from an accelerated charge. If this radiation of energy from electrons in atoms of materials had occurred since the beginning of time, the electrons would certainly have lost all energy and fallen into the nucleus; matter as we know it would no longer exist.

However, Bohr proposed that there were stable orbits in which an electron could exist and not radiate energy. He postulated that these orbits would have angular momentum quantized in steps of $nh/2\pi$ where n could have only integer values and h is Planck's constant. By use of this postulate, it is possible to calculate the energy levels and radii of the electronic orbits for a simple atom such as hydrogen, with one proton in the nucleus and one rotating electron. Then by inference, and with confirmation from spectroscopic data, it is possible to extend the theory to more complex atoms.

By Coulomb's law the force between the proton of the hydrogen nucleus and its electron is $e(-e)/4\pi\epsilon_v r^2$, and the equality of attractive and centrifugal forces allows us to write

$$f_e = \frac{-e^2}{4\pi\epsilon_v r^2} = \frac{mv^2}{r} = f_c \tag{2-1}$$

where m is the electron mass. The kinetic energy of the orbiting electron is

$$\text{K.E.} = \frac{mv^2}{2} = \frac{e^2}{8\pi\epsilon_v r} \tag{2-2}$$

The potential energy of the electron in the field of the nucleus is

$$\text{P.E.} = -\int_\infty^r f_e\,dr = \frac{-e^2}{4\pi\epsilon_v r} \tag{2-3}$$

and the total energy of the electron is

$$W = \frac{-e^2}{4\pi\epsilon_v r} + \frac{e^2}{8\pi\epsilon_v r} = \frac{-e^2}{8\pi\epsilon_v r} \qquad \text{joules} \tag{2-4}$$

The negative sign is the result of establishing the electronic energy at infinity as the reference; the electron in orbit has less total energy than an electron at infinity.

The Bohr requirement of quantized angular momentum† may now be introduced:

$$mr^2\omega = mrv = \frac{nh}{2\pi}$$

$$v = \frac{nh}{2\pi mr} \qquad (2\text{-}5)$$

Substituting in Eq. 2-2 gives

$$r = \frac{n^2h^2\epsilon_v}{\pi me^2} \qquad \text{m} \qquad (2\text{-}6)$$

as the expression for the radii of successive allowed orbits, which increase in the progression $n^2 = 1, 4, 9, \ldots$. For hydrogen with one electron, the smallest orbit can be found to have a radius of 0.53×10^{-10} meter. The energy levels of the allowed orbits are obtainable from Eq. 2-4 as

$$W = \frac{-me^4}{8\epsilon_v^2n^2h^2} \qquad \text{joules} \qquad (2\text{-}7)$$

$$= \frac{-13.6}{n^2} \qquad \text{eV}$$

Hydrogen has only one electron, but this electron may occupy any one of the many orbits, spaced outward from the neighborhood of the nucleus and related in energy as $-1/n^2$. The value for $n = 1$ is -2.17×10^{-18} joule, and this is the first orbit. It is also the most stable condition, since in general the requirement of least energy will establish the normal condition in the atom.

The more complex atoms have structures which can be inferred, in principle, from this simple example for hydrogen.

†It is possible to develop a physical basis for the Bohr assumption of quantized angular momentum by requiring that the stable orbits of the electrons be those in which the orbital circumference is equal to an integral number of electron wave lengths. Assuming circular orbits, then

$$n\lambda = 2\pi r$$

where $n = 1, 2, 3, \ldots$. It has already been stated that

$$\lambda = \frac{h}{mv}$$

Using $\omega = v/r$, these equations lead directly to the Bohr postulate requiring that the angular momentum be quantized, namely

$$mr^2\omega = \frac{nh}{2\pi}$$

2-2 THE WAVE-MECHANICAL ATOM MODEL

Schrödinger proposed a wave equation to predict the states of the electrons in the more complex atoms, and the Bohr postulate and the discrete energy levels for the electrons appeared as consequences of the solution of the Schrödinger equation. A new field of *wave mechanics* then developed, suited to particles of subatomic dimensions. Wave mechanics has supplemented and replaced the simple Bohr geometrical model of the atom. The Bohr model supplies some useful qualitative concepts, but emphasis is now placed on energy levels for the electrons rather than on orbits. Because of the wave–particle duality, the electron wave packet cannot be exactly located in space or time, and so probability functions are necessary for the prediction of electron occupancy of these definite energy levels.

The *total quantum number* n results from the solution of the wave equation. Each integer value of $n = 1, 2, 3, \ldots$ indicates a level of energy at which an electron may exist in an atom. Further study showed that there were actually a number of electrons having the same value of n, or existing in a *shell* of energy levels about the nucleus. It was found that just as the energy was quantized, so was the orbital angular momentum quantized within a given n shell. Thus arose the concept of subshells, with a second quantum number l. Slight differences in orbital energy were found when the atom was placed in a magnetic field, resulting from the orientation of the orbits. A third quantum number m_l became necessary to describe these conditions. A fourth number, m_s, entered to account for the electrons spinning on their own axes, acting as tiny magnets, with an angular momentum of $\pm\frac{1}{2}(h/2\pi)$. The spin quantum number has values of $\pm\frac{1}{2}$, to indicate the direction of spin as parallel or antiparallel to the angular momentum of the orbital motion.

The number of allowed levels described by combinations of these numbers far exceeds the number of electrons. However, *Pauli's exclusion principle* states that no two electrons in an atom can have an identical set of quantum numbers, and application of this principle reduces the possible combinations to exactly the number of levels which are found experimentally.

The set of quantum numbers comprise

$$
\begin{aligned}
n &= 1, 2, 3, \ldots \\
l &= 0, 1, 2, \ldots, n-1 \\
m_l &= -l, (-l+1), \ldots, (l-1), l \\
m_s &= \pm\tfrac{1}{2}
\end{aligned}
$$

Use of the exclusion principle then leads to logical construction of the periodic table of the elements; part of the table is presented in Table 2-1.

Table 2-1

CONSTRUCTION OF PART OF THE PERIODIC TABLE OF THE ELEMENTS

Atomic Element		Electron Configuration							
		$n = 1$	2		3			4	
Number		$l = 0$	0	1	0	1	2	0	1
1	H	1							
2	He	2							
3	Li	2	1						
4	Be	2	2						
5	B	2	2	1					
6	C	2	2	2					
7	N	2	2	3					
8	O	2	2	4					
9	F	2	2	5					
10	Ne	2	2	6					
11	Na	2	2	6	1				
12	Mg	2	2	6	2				
13	Al	2	2	6	2	1			
14	Si	2	2	6	2	2			
15	P	2	2	6	2	3			
16	S	2	2	6	2	4			
17	Cl	2	2	6	2	5			
18	Ar	2	2	6	2	6			
19	K	2	2	6	2	6		1	
20	Ca	2	2	6	2	6		2	
21	Sc	2	2	6	2	6	1	2	
22	Ti	2	2	6	2	6	2	2	
23	V	2	2	6	2	6	3	2	
24	Cr	2	2	6	2	6	5	1	
25	Mn	2	2	6	2	6	5	2	
26	Fe	2	2	6	2	6	6	2	
27	Co	2	2	6	2	6	7	2	
28	Ni	2	2	6	2	6	8	2	
29	Cu	2	2	6	2	6	10	1	
30	Zn	2	2	6	2	6	10	2	
31	Ga	2	2	6	2	6	10	2	1
32	Ge	2	2	6	2	6	10	2	2

As indicated by the total of m_l and m_s levels, a filled subshell will have $2(2l + 1)$ electrons and the number of subshells is n. Thus hydrogen has only one electron in the $n = 1$ shell, but helium has two electrons in that shell with opposite spins. Lithium will have two electrons with $n = 1$, and one electron in the $l = 0$ subshell of the $n = 2$ shell. By the $2(2l + 1)$ rule, six electrons are allowed in the $l = 1$ subshell, as for neon; eight electrons then fill the $n = 2$ shell. The remainder of the table follows, with minor deviations for the heavier and more complex elements.

The inner, lower-energy orbits are shielded by the fields of the electrons in the outside shell; the inner shells are also completely filled and therefore stable. The electrons of the outer shells are easily affected by external forces and fields, and the electrons in the outermost shell are called the *valence* electrons, since these external electrons enter into the valence bonds with a neighboring atom. The chemical properties of an element are determined by the number and arrangement of the outer shell electrons. Thus lithium, sodium, potassium, and copper are chemically similar, since they have one electron in the outer shell. Carbon, silicon, and germanium have four outer or valence electrons—this family will be of interest in the study of semiconductors. When the outer shell has all available levels filled, as for helium, neon, and argon, the element is chemically stable, and a filled shell is taken as an indication of such a condition.

These rules on electron arrangement within the atom are a result of the wave nature of the electron, and are confirmed by experimental observation of the line spectra of the elements.

2-3 STRUCTURE AND ATOMIC BONDS

As a material coalesces from a gaseous condition in which there is a random atomic location, the distances between the atoms become less and the cohesive forces increase, enabling the material to maintain its shape in the solid state. In many solids the atoms assume regular, periodic positions in a structure known as a *crystal.* A crystal may be visualized as a repetitive structure of unit cells or parallelepipeds which define fixed points in a space lattice which serve as atom locations. The form of the unit cell assumed by a solid is dependent on the magnitude and nature of the binding forces between atoms, and the atomic spacing and arrangement will adjust to minimize the electron energies.

In metals the three most common crystal forms are the *body-centered cubic,* in which a repeated cubic cell has an atom at each of the corners and at the center of the volume, the *face-centered cubic,* in which atoms are located at the corners and in the centers of all faces, and the *close-packed hexagonal,* which has atom locations at the corners of, and certain points within, a hexagonal prism. If the atoms are considered as hard spheres, and if the spheres are packed tightly together, then an atom would have six near neighbors. The face-centered cubic form and the hexagonal form represent two patterns in which the spheres may be close-packed, with two-layer or three-layer symmetry, respectively.

Metals having the close-packed face-centered form include the low-valence good conductors, such as copper, silver, gold, and platinum; the high-valence alkali metals form body-centered cubic crystals. Because of the larger interatomic distances found in this form, the interatomic binding energies of the alkali metals are small. A few metals such as tin, iron, cobalt, and manganese change crystal form above or below a critical temperature.

The body-centered cubic form is less densely packed and an atom has only four near neighbors; the centered atom approaches these four neighbors on a diagonal plane of the cubic cell.

Of particular importance in the semiconductor field is the *tetrahedral* structure formed by carbon as diamond, by germanium and silicon, by gray tin and certain other semiconductor compounds. The unit tetrahedron has one atom at each vertex, and one atom at the center of the figure, as in Fig. 2-1(a). Each bond to its four near neighbors has two electrons, one valence electron from each atom, as diagrammed in Fig. 2-1(b). These are *covalent bonds*, and each atom is surrounded by a stable shell of eight electrons.

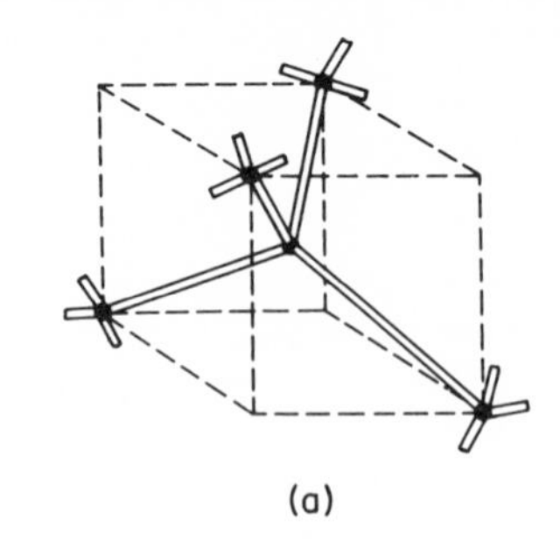

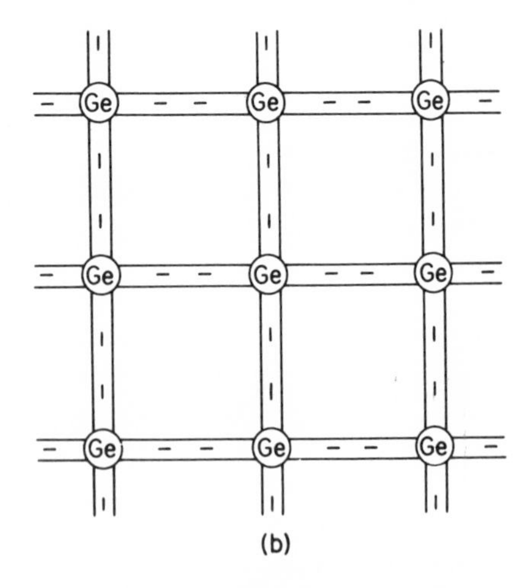

Figure 2-1. (a) Cubic diamond structure; (b) covalent bonds in germanium.

The energies binding the atoms together in solids are dependent on the type of atomic interaction present, of which *ionic*, *covalent*, and *metallic* bonds are important to our discussion.

Ionic bonds occur in compounds formed from electropositive elements of low valence and electronegative elements of high valence, as the Group I–Group VII compounds, of which NaCl is an example. It forms a face-centered cubic crystal. The sodium atom is found to have lost its one valence electron to the chlorine atom, just filling the outer shell of the latter with eight electrons. The molecule is then formed from the resultant positive sodium ion and the negative chlorine ion, and the electrostatic forces provide the binding of the molecule. As the atoms approach each other, the electrostatic forces increase, but a quantum-mechanical repulsion also develops due to crowding of the electrons. The two atoms of the molecule remain at a distance R_0 at which the total electron energies are at a minimum, or the attractive and repulsive forces are equal.

With covalent bonds, two or more atoms share electrons in a valence bond. This implies equal energy levels, but the Pauli principle cannot be violated and so the shared electrons must have opposite spins. The coupling between the spins provides binding forces which hold the molecule together. There is a repulsive force between the electrons, and the atoms remain at that spacing corresponding to an equality of the forces, or the minimum energy condition for the electrons. The useful semiconductor elements germanium and silicon, and some compounds of elements from Groups III and V, are among those forming such covalent bonds.

Metallic bonds are of an electrostatic nature and result from attraction between the fixed ions of the metal and a cloud of valence electrons which behave much as an electron gas. The atomic spacing in metal crystals may approximate 2×10^{-8} to 2.5×10^{-8} cm, and since the normal orbits of the outermost electrons may have diameters of 2×10^{-8} cm, it is apparent that the outer orbits of neighboring atoms may overlap and the electric fields from neighboring nuclei will cancel, forming the potential wells indicated in Fig. 2-2(a). A valence electron has the highest energy and is at the greatest

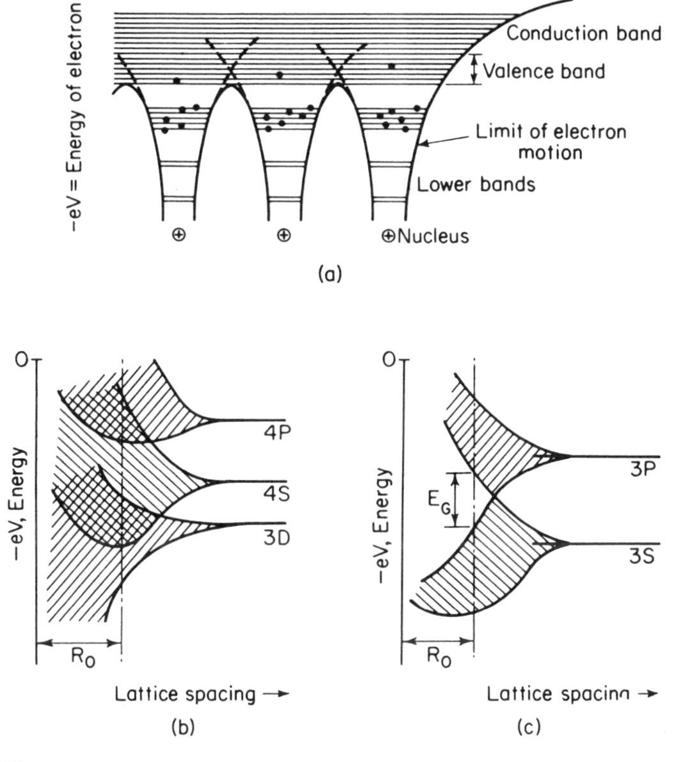

Figure 2-2. (a) Electron energy levels and potential wells in a metal at 0°K. (b) Splitting of energy levels in copper. (c) Splitting of energy levels in silicon. R_0 = normal lattice spacing.

distance from the nucleus. It is no longer under the influence of a particular nucleus and is at an energy above the top of the potential well. The valence electrons in such crystals may move so freely from the region of one atom to the next that only the boundaries of the solid limit their travel. This limit is indicated in the figure.

The inner shells of lower-energy electrons are shielded from most external influences by the fields of the outer electrons and are unaffected by forces from surrounding nuclei. Limits of motion for these lower-orbit electrons are shown as the sides of the potential wells created by the nuclei.

The valence electrons of the metal cannot all exist at the energy levels they had in the atoms when isolated. Such existence would imply many electrons with the same quantum numbers, and would violate Pauli's principle. The valence electrons are now electrons of the crystal rather than of a given atom, and must obey the Pauli exclusion principle when applied to the crystal as an atomic system.

Because of the interactions of the fields as the nuclei approach each other in the crystallization process, each of the energy levels of an atom, as shown at the right of Fig. 2-2(b), splits into n levels of energy, where n is theoretically the number of atoms in the crystal system. Each original energy level becomes a band or almost a continuum of energy levels, as the atoms near each other. At some internuclear distance the total energy of the electrons present in these bands at 0°K will be a minimum, and this condition will fix the lattice distance, shown as R_0 in Fig. 2-2(b).

In the monovalent metals there is only one valence electron per atom, but the outer shell contains two energy states (positive and negative spin). The splitting of these two original levels in all the atoms of the crystal provides $2n$ levels for n electrons. Thus at 0°K, when the electrons are in their lowest energy states, there will be unfilled higher-energy levels adjacent to the valence levels. These are indicated as a *conduction band* of levels in Fig. 2-2(a), and their role in conduction will be discussed.

2-4 *ENERGY-LEVEL VIEW OF ELECTRICAL CONDUCTION*

An applied electric field will accelerate charges and this implies an increase in the energy of the charges. But in the solid, electrons can increase their energies only by transfer to higher permitted energy levels in the band structure of Fig. 2-2(a). Thus for conduction under an electric field, *charges must make transitions to higher energy levels.* Such transfer requires, first, that unfilled higher energy levels exist in the crystal, and second, that sufficient energy can be supplied to the charges from the electric field to raise them to those unfilled levels.

These requirements are easily met in the monovalent metals and they are considered good conductors. At 0°K all electrons must be in their lowest energy levels, and Fig. 2-3(a) shows the filled levels of the valence electron energy band as solid horizontal lines; the higher unfilled levels are drawn cross-hatched. Such a diagram is an expansion of the upper portion of Fig. 2-2(a), or represents a vertical cross section of Fig. 2-2(b), along the R_0 line.

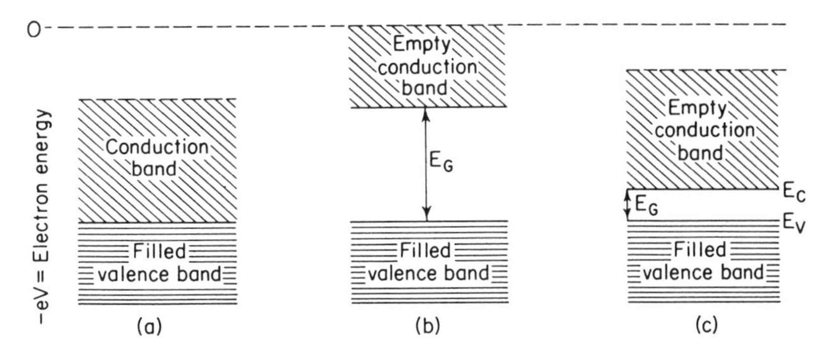

Figure 2-3. Band structure: (a) metal; (b) insulator; (c) semiconductor.

At 0°K in some nonmetals the valence electrons are found to completely fill a band of levels. These electrons cannot increase their energies because there are no unfilled higher-energy levels in the band into which transitions can be made. Therefore, it is possible to state a quantum-mechanical rule that *electrons in a filled band cannot take part in electrical conduction.*

That is, the valence electrons are found to exist inside and to fill all valence levels as shown in Fig. 2-3(b) at 0°K. A band of allowable energy levels does exist at a higher energy, but this band is empty at absolute zero. The energy which must be given to an electron to raise it out of the potential well, i.e., the energy difference between the top of the filled valence levels and the lowest level of the conduction band, is called the *gap energy*, E_G. Once raised to the conduction band, the electron is then in an unfilled band and can take part in conduction.

Another explanation of the phenomenon is illustrated by the energy level vs. lattice spacing diagram for silicon of Fig. 2-2(c). The $3S$ and $3P$ designations derive from spectroscopy, but represent, respectively, two electrons of $n = 3$, $l = 0$ and two electrons of $n = 3$, $l = 1$, shown for silicon in Table 2-1. At the crossover point to the right of the stable atomic spacing R_0, the two $3P$ states form a band with the two $3S$ states. To maintain the minimum energy condition, the two $3P$ electrons fill the two extra $3S$ levels, and the upper band levels are empty. At the normal spacing, R_0, there is a gap of energies, E_G, between the lower filled band and the upper empty band. The energies represented by this gap are said to be "unallowed" or forbidden.

If the unallowed energy gap is of considerable energy magnitude, as in diamond where $E_G = 7$ eV, there is little probability of the valence electrons acquiring such amounts of energy from usual electric field or thermal sources, and transfers to the upper band will not occur. With the lower band filled and the upper band empty, conduction cannot take place, by reason of the above quantum-mechanical rule. Materials with E_G values of a few eV or more, as diagrammed in Fig. 2-3(b), are classified as *insulators.*

In materials with gap energies of the order of 1 eV, as shown in Fig. 2-3(c), there will be no electrons in the upper band at 0°K; such materials are insulators at that temperature. As the temperature is raised, there is a probability that some electrons will receive enough energy from thermal sources to transfer across the gap. These elevated electrons are then in a partially filled energy band and can take part in electrical conduction if an electric field is applied. Materials with such intermediate gap energies are known as *semiconductors.*

When some of the valence electrons are raised to the conduction band in a semiconductor, vacant electron sites or *holes* are left in the valence band of levels. We may then think of the valence band as unfilled and these vacancies appear to take part in conduction, as will be explained in Section 2-5.

Elements in which the unallowed band of energies is nonexistent, or in which the valence and conduction bands are continuous, have already been classified as *metals* and good conductors. In metals, all the valence electrons are in an unfilled band and so the number of free charges is not changed by temperature. Therefore, the metals have positive coefficients of resistance change with temperature, because the increased thermal agitation of the atoms in the crystal interferes with charge drift, or reduces the *mobility* of the charges.

In semiconductors, the number of charges in the conduction band, and the number of vacancies in the valence band, increase with rising temperature. This increase in number of available charges reduces the electrical resistance and usually overrides the effect of reduced charge mobility caused by thermally agitated atoms. Thus semiconductor materials have negative temperature coefficients of resistance.

The Group IV elements each have four valence electrons in the outer shell. The inner shells having 2, 8 or 2, 8, 18 electrons are filled and stable. Group IV comprises carbon, silicon, germanium, tin, and lead, with gap energies as follows:

C (diamond)	7 eV
Si	1.15 eV
Ge	0.75 eV
Sn (gray)	0.1 eV
Pb	0 eV

Diamond is considered an insulator at usual temperatures and metallic tin and lead are conductors. It is the reasonable gap energy and the four electrons in covalent bonds which make germanium and silicon of interest as semiconductors.

The covalent bond may be used to provide another explanation of the gap energy of the semiconductor. At 0°K all electrons are rigidly held in these interatomic bonds. As the temperature is raised, the electrons must exist either in the bond, or must have received sufficient thermal energy to rupture the bond and become free. Lesser amounts of energy are insufficient to break a bond; such smaller amounts of energy cannot be accepted by the electron or are the forbidden energies of the gap. Therefore the energy of the unallowed gap is equal to the rupture energy of the covalent bond in the material.

The valence-bond and the energy-band viewpoints of electron behavior in solids are two ways by which we are able to consider the same problem, that of electrical conduction in semiconductors.

2-5 THE INTRINSIC SEMICONDUCTOR

As just argued, an electron in a valence bond of a semiconductor may accept an energy E_G from a thermal, electrical, or radiant energy source, may break its bond and transfer to the upper conduction levels. There it may take part in electrical conduction.

We have also introduced the concept that a departing electron leaves a vacancy or hole in the valence bond, and this is indicated schematically in Fig. 2-4. The atom is now an ion, and the hole appears as a fixed positive charge $+e$, with effective mass m_h, generally not equal to that of the electron. This difference arises because the carriers move in the field of the atomic lattice as well as in the electric field, and react more slowly.

Since a hole is attractive to electrons, it can be filled by an electron which breaks from a nearby valence bond, leaving a hole. The process may be repeated with this second hole, and so a hole progresses under the attraction of an electric field. The process occurs in a step-wise manner, that is, the move-

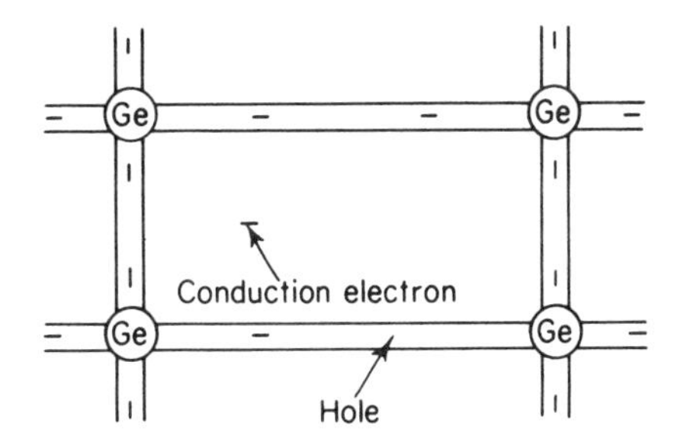

Figure 2-4. Result of formation of an electron–hole pair.

ment of a hole is actually a succession of transfers. Overlooking certain quantum-mechanical refinements, a hole can be looked upon as a positive charge, able to move through the crystal lattice. The hole velocity will depend on the rapidity with which the process of repeated recombination of holes and electrons can proceed.

The gap energy was supplied to the electron to break it free of the valence bond, and this amount of energy is released as a radiation quantum when the electron recombines with a hole. The quantum may be reabsorbed in the breaking of another bond, producing an electron–hole pair, or it may be radiated. In some materials this energy appears as visible light.

Under thermal equilibrium in a pure semiconductor the number of holes will equal the number of conduction electrons, since they are always produced as *electron–hole pairs*, every electron transferring to the conduction band leaving a vacancy or hole in the valence band. Thermal energy is the most usual source for such pair production, and the process of conduction by thermally generated mobile charges in a pure semiconductor is called *intrinsic conduction.*

When an electric field is applied, energy is available to superimpose a directed drift component on the random thermal motions of the charges. The velocities of holes and electrons drifting under an applied electric field $\mathscr{E}$ are

$$v_h = \mu_h \mathscr{E} \qquad \text{m/s} \quad (2\text{-}8)$$

$$v_e = \mu_e \mathscr{E} \qquad \text{m/s} \quad (2\text{-}9)$$

where the subscript h refers to the positive holes and the subscript e refers to the negative electrons. In a given field v_e will be negative to v_h, but since the charges are of opposite sign the currents due to both are in the same direction. They may often be considered as separate components, however.

The proportionality factor μ is known as the *mobility,* measured in meter^2 per volt-second. The mobility or freedom of charge movement is dependent on the thermal atomic agitation and also on the lattice structure.

Under equilibrium conditions in intrinsic material, new electron–hole pairs are continuously produced by thermal energies at rate $g(T)$ dependent on the material and temperature. To maintain a constant concentration of pairs, the rate of recombination $r(T)$ must be equal to $g(T)$. The rate of recombination is physically dependent on the availability of the charges (i.e., on the density of holes and on the density of electrons), and on the crystal properties as embodied in a proportionality constant K. Then

$$g(T) = r(T) = Kn_i p_i$$

where electron and hole densities per cubic meter are indicated by n and p, and intrinsic conditions by subscript i. With pair generation $n_i = p_i$, and so

$$g(T) = r(T) = Kn_i p_i = Kn_i^2$$

and

$$n_i p_i = n_i^2 \tag{2-10}$$

Since the densities n_i and p_i are an inherent property of a semiconductor at a given temperature, then the product n_i^2, or the square of the intrinsic charge density, also has a fixed value at a given temperature, dependent only on the material. This result is analogous to the constant solubility product in chemistry.

The $n_i p_i = n_i^2$ value for an intrinsic material will be shown in Section 2-8 as a material constant, dependent only on temperature. That is,

$$n_i p_i = n_i^2 = k_C k_V T^3 \epsilon^{-E_G e/kT} \tag{2-11}$$

where $k_C = 2(2\pi m_e k/h^2)^{3/2}$, $k_V = 2(2\pi m_h k/h^2)^{3/2}$, and $E_G e$ represents the energy, in joules, required to break a valence bond. The symbols m_e and m_h are the effective masses of the electrons and the holes in the material. For germanium, the product $k_C k_V$ approximates 3.1×10^{44}, and for silicon has the value 1.5×10^{45}, per m^3.

In intrinsic germanium with an energy gap of 0.75 eV, the intrinsic electron density at 300°K is 2.4×10^{19} electrons per m^3. Since germanium has about 4.4×10^{28} atoms per m^3, it is apparent that only about five atoms in every 10^{10} atoms of germanium will have broken valence bonds and can contribute to intrinsic conduction at usual temperatures. In contrast, copper has available about 10^{28} electrons per m^3 as conduction charges.

While there are few broken valence bonds at usual ambient temperatures, the number of intrinsic free charges increases rapidly with temperature. Since the resistivity is inversely related to the number of free charges, the intrinsic resistivity is given by a relation of the form

$$\rho = \rho_0 \epsilon^{E_G e/kT} \tag{2-12}$$

where ρ_0 is a constant of proportionality.

It has been indicated that electrons may be raised to the conduction band, and holes left in the valence band, by raising the temperature of a semiconductor. The band structure is then as in Fig. 2-5. Intrinsic conduc-

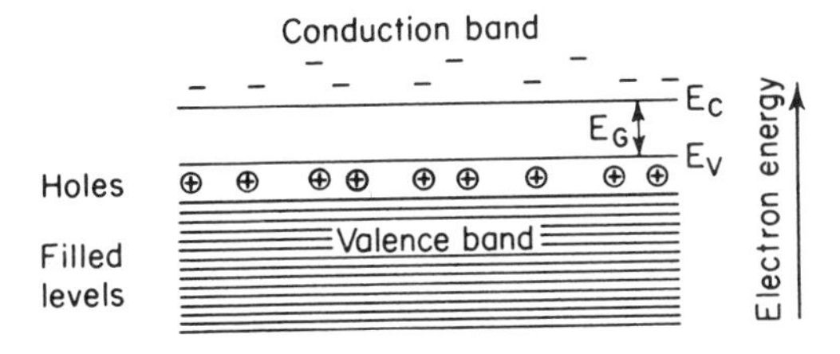

Figure 2-5. Intrinsic semiconductor band structure at elevated temperature.

tion must be suppressed in most semiconductor devices, since the intrinsic current might otherwise mask the operating currents of holes or electrons. Because of the strong dependence of n_i or p_i on temperature, it is necessary to limit the operating temperature of germanium devices to about 85°C to 100°C, and of silicon devices to 190°C to 200°C. The higher permissible temperature for silicon is a result of its larger energy gap.

2-6 *IMPURITY SEMICONDUCTORS: n AND p MATERIALS*

To provide the conduction properties needed in most semiconductor devices the purified extrinsic semiconductor is doped with controlled amounts of impurity elements chosen from Group III or Group V of the periodic table of the elements. The Group III elements have three valence electrons and include boron, aluminum, gallium, and indium. The Group V elements are phosphorus, arsenic, antimony, and bismuth, and these have five valence electrons. The impurities are added at rates of only one atom of impurity per 10^6 to 10^{10} semiconductor atoms, but these ratios are sufficient to alter the resistivity of silicon and germanium as indicated in Fig. 2-6.

In the growth process of the crystal the impurity atoms are separated by thousands of semiconductor atoms, and will not be able to control the form of the crystal lattice. Therefore an impurity atom must assume a crystal position by substituting for one of the germanium or silicon atoms. When a five-valence atom such as arsenic enters a silicon lattice and substitutes for a silicon atom, four of its five outer-shell electrons take places in the covalent bonds, as would the electrons of the silicon atom it replaces,

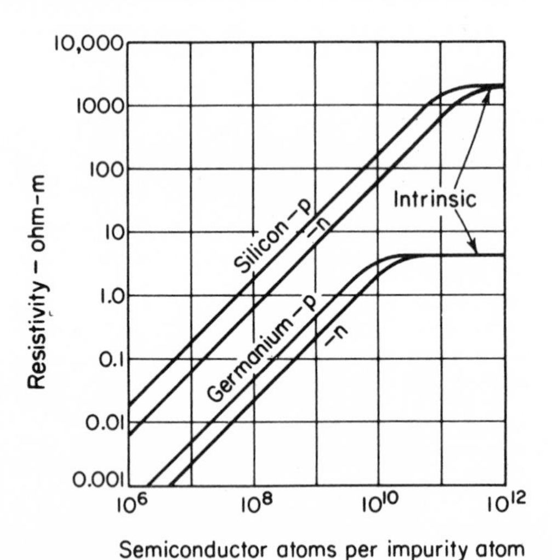

Figure 2-6. Resistivity at 25°C as a function of impurity percentage.

and one electron is left over, as in Fig. 2-7(a). The average energy of the arsenic valence electrons is higher than that of the silicon valence electrons, and at 0°K this energy level is in the unallowed gap of pure silicon and only about 0.05 eV below the bottom of the conduction band. This difference is about 0.01 eV for germanium. At temperatures of 20°K or above there is sufficient thermal energy available that this fifth electron can transfer to the silicon conduction band, where it is free to move under the influence of an electric field and contribute to a current. The arsenic atoms are then positively ionized, but must remain bound as positive charges in the lattice. The energy levels are indicated in Fig. 2-8(a).

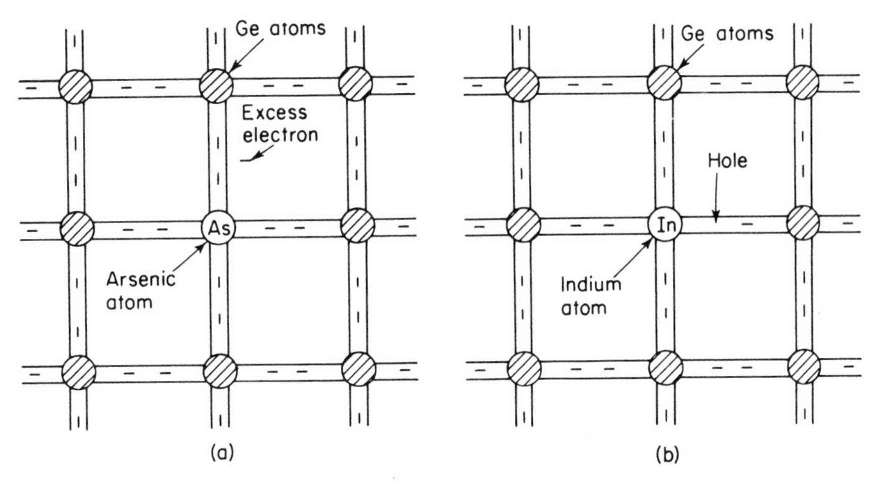

Figure 2-7. (a) Representation of germanium with arsenic as an n impurity; (b) with indium as a p impurity.

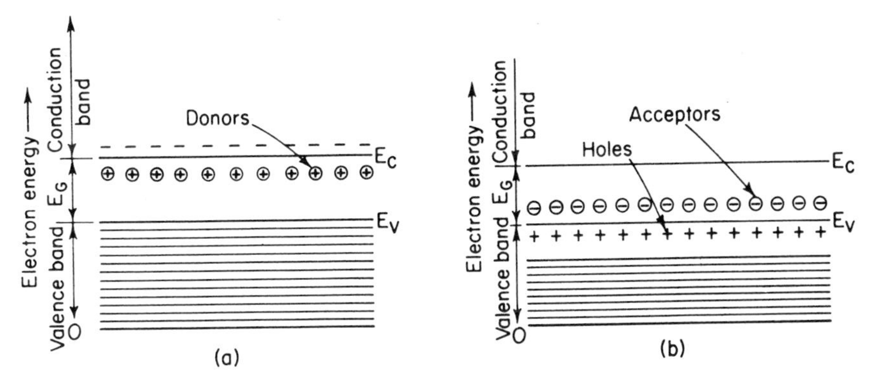

Figure 2-8. (a) Ionized donors and their electrons in the conduction band; (b) acceptors and the holes created in the valence band.

The Group V impurities which provide an extra electron for the conduction band, without simultaneously creating a hole, are *donor* impurities because each atom donates an electron. Conduction in materials with Group V impurities will therefore be predominantly by negative electrons as majority carriers and the material is said to be of n type.

When a Group III atom enters a silicon lattice and substitutes for a silicon atom, its three valence electrons enter the covalent bonds with adjacent silicon atoms, but an electron deficiency is left in the fourth bond. The Group III materials have some available energy levels about 0.01 eV above E_V, the energy level at the top of the valence band in germanium; in silicon these levels are from 0.046 to 0.16 eV into the energy gap. These are very narrow bands, almost discrete states, since the density of impurity atoms is so low that their nuclei are not appreciably influenced by any other nearby impurity atoms. At room temperature the thermal agitation energy, kT, approximates 0.025 eV. There is sufficient thermal energy present that electrons can be stolen from the neighboring silicon atoms to fill the higher-energy impurity atom levels. Holes are thereby created in the valence band of the semiconductor, and so conduction can occur due to valence band empty levels, as indicated in Fig. 2-8(b).

The Group III impurity is known as an *acceptor*, since it accepts bound electrons from the valence band. Each acceptor atom becomes negatively ionized, due to its extra electron, but the ion represents a fixed charge in the structure. Since the acceptors add holes in the valence band without adding conduction electrons, the conduction will be mainly by positive charges and the material is then said to be *p-type*. Holes are the majority carriers in p material.

In both n and p materials the electrons which create the ions have acquired higher energy, but the energy needed is much less than the energy needed to create electron–hole pairs. Thus it is possible to establish conditions for conduction by holes in p material, and conduction by electrons in n material, at temperatures much below those at which intrinsic conduction by holes and electrons occurs. Throughout the above processes the crystal as a whole remains electrically neutral, since no charge has been added or removed.

Figure 2-9 illustrates the manner in which the mobile charge concentrations vary as a temperature function. In the region of low temperature ($T < 200°K$) the sample of n-impurity silicon has only partially ionized donors, and the number of charges increases with temperature. In the constant charge region ($200° < T < 400°$) all donors are ionized, and the number of charges in the conduction band is constant. Above 400°K the number of charges thermally generated becomes appreciable, and the conduction process eventually becomes completely intrinsic.

Because of the ease with which donor electrons can be thermally excited

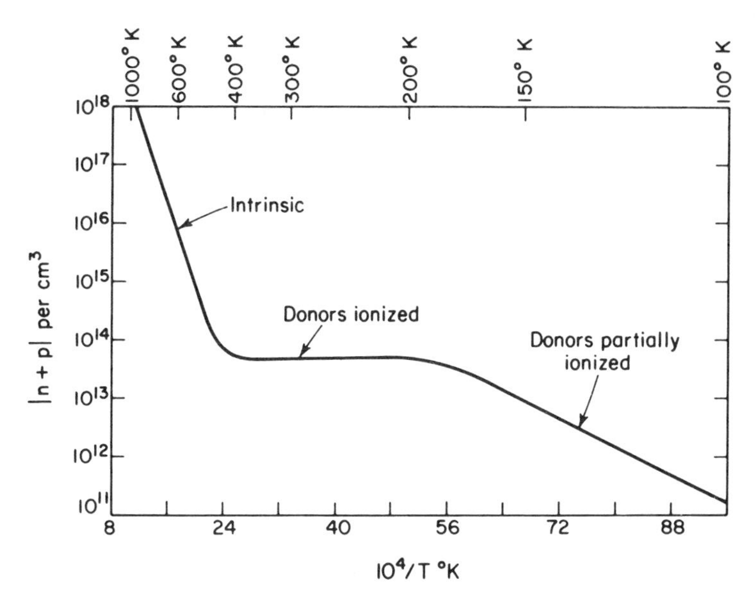

Figure 2-9. Variation in carrier density in n silicon.

to the conduction band of an n semiconductor, it is reasonable to assume that the density of conduction electrons is equal to N_D, the density of impurity atoms. That is, the donors can be assumed totally ionized at usual ambient temperatures. Likewise, the acceptor states may be assumed completely filled in a p semiconductor at usual temperatures, and so the density of holes in the valence band is N_A, where N_A is the number of acceptor atoms per unit volume.

The concentration of impurity atoms is so low compared to the number of semiconductor atoms that the thermal generation and recombination process is not appreciably altered by doping. It was previously stated that recombination was proportional to the concentrations of the two types of charges present, or

$$n_p p_p = n_i^2$$

where the subscripts indicate a p material. Because the concentration of the majority charges is increased over its intrinsic value by doping, it is found that the concentration of the minority charges is reduced, to maintain the concentration product, n_i^2, constant.

The amount of this reduction may be found by use of an equation expressing the overall charge neutrality of the material as

$$p + N_D = n + N_A \tag{2-13}$$

where the doping densities are N_D and N_A, and all impurity atoms are assumed to be ionized.

The left side of the equation states that the total positive charge is due to holes in the valence band, p, plus the positive charge associated with the ionized donor atoms, N_D. The right side of the equation shows the conduction band electronic charge, n, plus the negative charge represented in the filled acceptor states, N_A.

Using

$$np = n_i^2$$

leads to

$$n = \frac{n_i^2}{p}; \qquad p = \frac{n_i^2}{n}$$

and simultaneous solution with Eq. 2-13 yields

$$p = \frac{N_A - N_D}{2} \pm \sqrt{\frac{(N_A - N_D)^2}{4} + n_i^2}$$

$$n = \frac{N_D - N_A}{2} \pm \sqrt{\frac{(N_D - N_A)^2}{4} + n_i^2}$$

For the intrinsic case, in which $N_A = N_D = 0$, these equations lead to $n = p = n_i$, as they should.

For an n semiconductor, in which $N_D \gg n_i$ and $N_A \cong 0$, then the hole density is

$$p_n = \frac{-N_D + N_D\sqrt{1 + (4n_i^2/N_D^2)}}{2}$$

The second term under the radical is less than unity; the radical may be expanded and higher-order terms dropped, allowing the hole density in n material to be stated as

$$p_n \cong -\frac{N_D}{2} + \frac{N_D}{2}\left(1 + \frac{2n_i^2}{N_D^2}\right) = \frac{n_i^2}{N_D} \tag{2-14}$$

The second relation leads to the density of negative charges in the n material as

$$n_n \cong \frac{N_D}{2} + \frac{N_D}{2}\left(1 + \frac{2n_i^2}{N_D^2}\right) = N_D + \frac{n_i^2}{N_D} \cong N_D \tag{2-15}$$

thus determining the charge densities in the n impurity material.

Repeating, for a p impurity semiconductor in which $N_A \gg n_i$, and $N_D \cong 0$, we can state the charge densities in p material as

$$n_p \cong \frac{n_i^2}{N_A} \tag{2-16}$$

$$p_p \cong N_A + \frac{n_i^2}{N_A} \cong N_A \tag{2-17}$$

The density of majority carriers thus approximates the impurity atom density at usual ambient temperatures, while the recombination process limits the minority carrier density to a level substantially below the intrinsic n_i or p_i level. This reduction in minority carrier density may be illustrated by an example with germanium, which at 300°K has $n_i \cong 2.4 \times 10^{19}$ charges per m^3. The material is made extrinsic with an indium p impurity at the rate of one indium atom per 4×10^8 germanium atoms; donor density is assumed to be zero. Since there are 4.4×10^{28} germanium atoms per m^3, the acceptor density $N_A = (4.4 \times 10^{28})/(4 \times 10^8) = 1.10 \times 10^{20}/m^3$. With $n_i^2 = 5.76 \times 10^{38}$, then

$$p_p \cong N_A + \frac{n_i^2}{N_A} = 1.10 \times 10^{20} + \frac{5.76 \times 10^{38}}{1.10 \times 10^{20}} = 1.15 \times 10^{20}/m^3 \cong N_A$$

$$n_p \cong \frac{n_i^2}{N_A} = \frac{5.76 \times 10^{38}}{1.10 \times 10^{20}} = 5.25 \times 10^{18}/m^3$$

Because of the addition of a p impurity, the concentration of minority charges or electrons is only about one-fifth of the value it would have in intrinsic material.

Table 2-2 summarizes data on the characteristics of germanium and silicon, as the most important semiconductor materials.

2-7 THE FERMI-DIRAC ENERGY DISTRIBUTION IN METALS

In the preceding discussions the presence of charges in the conduction band or of holes in the valence band has been predicated upon the attainment of the needed thermal energies by the valence electrons. Since the electron density is so great and the individual motions are random, the problem of determining the energy levels occupied by electrons at a given temperature can only be given a statistical solution. That is, as the electrons receive or lose energy in interactions with the atoms, random energies having a statistical mean value will be transferred in each interaction, and an average distance or *mean free path* will be traversed by the electrons in a *mean free time* between interactions.

The occupancy of a given energy state is predicted not only by the division of energy among the electrons, but the occupancy must conform with the Pauli exclusion principle as well. It has been found that the predictions of the *Fermi-Dirac energy distribution* are supported by physical observation. The number of electrons occupying energy states between E and $E + dE$ in a unit volume is given by

$$N(E)\,dE = Z(E)F(E)\,dE \tag{2-18}$$

where

$$Z(E) = CE^{1/2}$$

$$C = 4\pi(2m_e^*)^{3/2}h^{-3}$$

where m_e^* is the effective mass of an electron in the semiconductor. The particle mass is dependent on the wave nature of the electron and its freedom of movement in the lattice; the value may differ slightly from the free space mass, m. Representative values are given in Table 2-2.

The function $Z(E)$ predicts the number of available energy states in the incremental energy range E to $E + dE$. The *Fermi function* $F(E)$ expresses the probable fraction of those states which are occupied at temperature T. The Fermi function is

$$F(E) = \frac{1}{1 + \epsilon^{(E - E_F)e/kT}} \tag{2-19}$$

where E_F is called the Fermi energy level and is here given in electron volts. The term kT/e has dimensions of volts, and is often referred to as the *voltage equivalent of temperature*.

At 0°K the Fermi function has values

$$F(E) = 1 \quad \text{for} \quad E < E_F$$

Table 2-2

CHARACTERISTICS OF GERMANIUM AND SILICON†

	Germanium	Silicon
Atomic number	32	14
Atomic weight	72.6	28.08
Melting point, °C	937	1420
Atoms/m^3	4.4×10^{28}	5.0×10^{28}
Dielectric constant	15.8	11.7
Intrinsic resistivity, ohm-m, at 300°K	0.45	2400
Gap energy, E_G, eV 0°K	0.75	1.15
μ_e, m^2/V-s, 300°K	0.39	0.13
μ_h, m^2/V-s, 300°K	0.19	0.05
D_e, m^2/s, 300°K	0.01	0.0031
D_h, m^2/s, 300°K	0.0045	0.0013
n_i, m^{-3}, 300°K	2.4×10^{19}	1.5×10^{16}
τ_e, μs	100–1000	50–500
τ_h, μs	100–1000	50–500
L_e, m	0.001–0.003	0.0004–0.001
L_h, m	0.0007–0.002	0.0002–0.0006
Lattice constant a, m	5.66×10^{-10}	5.43×10^{-10}
Donor levels‡	$E_C - 0.01$ eV	$E_C - 0.05$ eV
Acceptor levels	$E_V + 0.01$ eV	$E_V + 0.046$ eV
m_e^*/m, mass ratio in lattice§	0.55	1.1
m_h^*/m_h, mass ratio in lattice	0.37	0.59

†Data primarily from E. M. Conwell, *Proc. IRE*, Vol. 40, p. 1327 (1952); also Vol. 46, p. 1281 (1958).
‡These values will vary for different impurities.
§Mass of electron in the lattice/mass of electron in space.

$$F(E) = \tfrac{1}{2} \quad \text{for} \quad E = E_F$$
$$F(E) = 0 \quad \text{for} \quad E > E_F$$

These results indicate that at absolute zero all energy states below E_F have a probability of occupancy of 1, or these states are filled; energy states above E_F have zero probability of occupancy and are empty. At the Fermi level, $E = E_F$, the probability of occupancy is $\frac{1}{2}$. Physically, since all levels below E_F are filled at 0°K, it can be stated that *the Fermi level represents the energy of the highest occupied state at 0°K.*

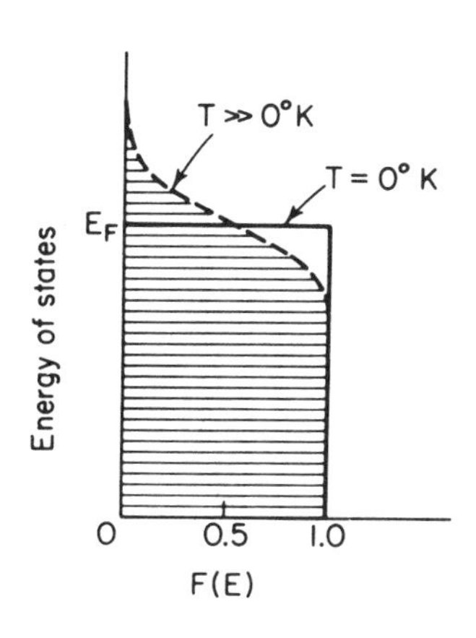

Figure 2-10. The Fermi distribution at 0°K and at temperature T.

The Fermi function for $T = 0$°K is plotted as the solid curve in Fig. 2-10. The occupancy of states cuts off sharply at E_F at that temperature. The dashed curve of the figure represents the Fermi function for some higher T. For energies much below E_F the denominator of Eq. 2-19 approximates unity. Because of the exponential term, the change from a probability of one for filled levels to a probability of zero occurs for energies which differ from E_F by only a few multiples of kT/e. For instance, with E only $5kT/e$ above E_F, a state would be filled in only 0.7 per cent of the possibilities, yet $5kT/e$ represents an energy change of only 0.125 eV at 300°K.

For E at least a few kT/e above E_F, we may say that

$$\epsilon^{(E-E_F)e/kT} \gg 1$$

and Eq. 2-19 reduces to the Boltzmann form

$$F(E) = \frac{1}{\epsilon^{(E-E_F)e/kT}} \tag{2-20}$$

Integration of Eq. 2-18 over all the free electrons of a metal will give a value for N, the known number of free electrons per unit volume. From this the value of E_F at 0°K can be found. Thus for $E < E_F$, or for $F(E) = 1$,

$$N = C \int_0^{E_F} E^{1/2}\, dE$$

$$E_F = \frac{h^2}{2m_e^*}\left(\frac{3N}{8\pi}\right)^{2/3} = 3.64 \times 10^{-19} N^{2/3} \tag{2-21}$$

For copper, where there is one valence electron per atom and approximately 10^{29} atoms per m³, $N \cong 10^{29}$ valence electrons per m³ and the value of E_F is 7 eV at 0°K. The Fermi level is thus determined by the density of valence or free electrons in metals, and is also found to be a minor function of temperature. For most metals E_F lies in the range from 2 to 10 eV.

2-8 THE FERMI LEVEL IN INTRINSIC SEMICONDUCTORS

In an intrinsic semiconductor at 0°K, all the states in the valence band are filled and all the states in the conduction band are empty. The Fermi level with a probability of occupancy of $\frac{1}{2}$ must then be located somewhere between E_V, the energy at the top of the valence band, and E_C, that at the bottom of the conduction band. Thus the Fermi level will lie in the forbidden gap.

In an intrinsic semiconductor at a temperature well above 0°K there will be thermal electrons in the conduction band and an equal number of holes in the valence band. Shockley has shown that the number of energy states in the conduction band varies as

$$Z(E) = C(E - E_C)^{1/2} \tag{2-22}$$

at the energy levels near the bottom of the conduction band. However, the expression is valid for our purposes even when the integration is extended to ∞, since it can be anticipated that the thermal electrons will most probably be in the lower-energy levels of the conduction band.

It has been shown that E_F will lie in the forbidden gap, and probably will be centered or remote from E_C. Then using the simplified Fermi form of Eq. 2-20 and Eq. 2-22 the density of thermal electrons in the conduction band is

$$n = C \int_{E_C}^{\infty} (E - E_C)^{1/2}\, \epsilon^{-(E-E_F)e/kT}\, dE$$

If a substitution of $x = (E - E_C)e/kT$ is made, it is possible to write the expression as

$$n = C\left(\frac{kT}{e}\right)^{3/2} \epsilon^{-(E_C-E_F)e/kT} \int_0^{\infty} x^{1/2}\epsilon^{-x}\, dx$$

which has a solution†

$$n = N_C\epsilon^{-(E_C-E_F)e/kT} \tag{2-23}$$

where

$$N_C = 2\left(\frac{2\pi m_e^* k}{h^2}\right)^{3/2} T^{3/2}$$

$$= 5 \times 10^{25}\ \text{m}^{-3} \quad \text{for germanium at 300°K.}$$

The result of Eq. 2-23 states the density of electrons in the conduction band as a temperature function.

There will be an equal density of holes in the valence band of the intrinsic semiconductor. In Eq. 2-18 the Fermi function $F(E)$ was used to give the fraction of states occupied. It follows that the thermal holes at the top of the

†B. O. Pierce, "A Short Table of Integrals," No. 496.

valence band have a probability of nonoccupancy of $[1 - F(E)]$. The density of holes or nonoccupied states near the top of the valence band may then be written as

$$p = \int_0^{E_V} Z(E)[1 - F(E)]\, dE \tag{2-24}$$

The zero limit refers to the energy at the bottom of the valence band. Since the holes have been created by the most energetic electrons, it is reasonable that they lie near the top of the valence band, and Shockley has shown that the number of energy states there is given by

$$Z(E) = C(E_V - E)^{1/2}$$

If we again assume that E_F lies in the central region of the energy gap, then E_F will be large with respect to any E in the valence band, and we can neglect the exponential in the denominator of $[1 - F(E)]$, giving

$$1 - F(E) \cong \epsilon^{(E-E_F)e/kT} \tag{2-25}$$

The density of holes in the valence band can then be written as

$$p = C\int_0^{E_V} (E_V - E)^{1/2}\epsilon^{(E-E_F)e/kT}\, dE \tag{2-26}$$

By a method analogous to that used for the conduction band, it is possible to arrive at an expression for p, the density of holes in the valence band, as

$$p = N_V\epsilon^{-(E_F-E_V)e/kT} \tag{2-27}$$

where

$$N_V = 2\left(\frac{2\pi m_h^* k}{h^2}\right)^{3/2} T^{3/2}$$

where m_h^* is the effective hole mass, usually near the value of m_e. The value of N_V is then approximately that of N_C.

Because $n_i = p_i = n = p$ in intrinsic semiconductors, it follows from Eqs. 2-23 and 2-27 that

$$\begin{aligned} E_F &= \frac{E_C + E_V}{2} + \frac{kT}{2e}\ln\frac{N_V}{N_C} \\ &= \frac{E_C + E_V}{2} + \frac{3kT}{4e}\ln\frac{m_h^*}{m_e^*} \end{aligned} \tag{2-28}$$

If m_e^* is assumed equal to m_h^*, then the Fermi level in intrinsic material will lie at the middle of the forbidden energy gap. Since m_h^* may at times be slightly larger than m_e^*, the Fermi level will increase slightly with temperature. The location of the Fermi level in an instrinsic semiconductor is shown in Fig. 2-11.

Since

$$N_C N_V = 4\left(\frac{2\pi k}{h^2}\right)^3 (m_e^* m_h^*)^{3/2} T^3$$

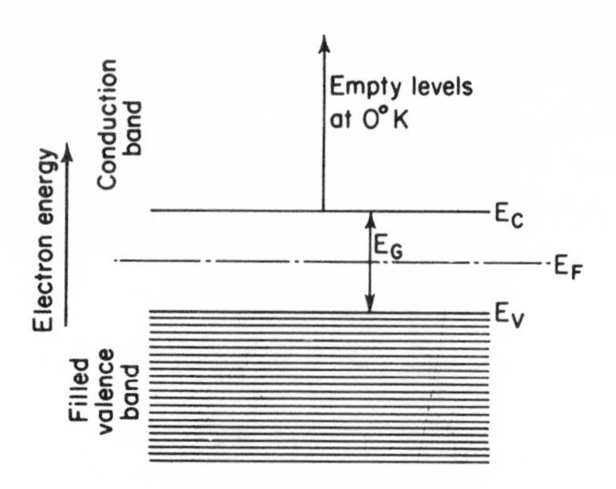

Figure 2-11. Location of the Fermi level at 0°K in an intrinsic semiconductor.

then the product

$$np = n_i^2 = N_C N_V \epsilon^{-(E_C - E_V)e/kT} \tag{2-29}$$

supports the previous development which showed n_i^2 a constant for a given material and temperature. The relation is independent of the Fermi level. Using the gap energy of 0.75 eV for germanium, and with $N_C \cong N_V = 5 \times 10^{25}$, the intrinsic carrier density in germanium at 300°K is $n_i = p_i = 2.4 \times 10^{19}\ \text{m}^{-3}$.

2-9 THE FERMI LEVEL IN *n* AND *p* SEMICONDUCTORS

An n semiconductor will have N_D impurity atoms per m³, and the percentage of these ionized or activated will be designated as n_d. The energy level of the donors is E_d, where E_d is below but near the bottom of the conduction band.

At $T = 0$°K the valence band is filled with electrons, all the donor levels are filled, the donors are not ionized, and the conduction band is empty. Therefore the Fermi level must lie below E_C and above E_d. It is reasonable to assume E_{Fn} midway between the donor level and E_C, by analogy with the results of the previous section.

At higher temperatures, but still below a temperature at which intrinsic charges are present in appreciable numbers, the easy access to the conduction band from the donor level will cause most donor electrons to transfer, leaving most donor atoms activated. While the Fermi probability for occupancy of a conduction level is less than the probability of occupancy of a donor level, there are so many conduction levels and comparatively so few donor levels that an electron activated from a donor has almost no probability of finding another donor atom, and must then remain free in the conduction band. It follows that $n_d \cong N_D$, where N_D is the doping density of the material.

If E_{Fn} is remote from E_C by several kT/e as before, the situation parallels that which leads to Eq. 2-23, and the density of electrons in the conduction band is given by that expression. If E_{Fn} is also remote from E_d by at least a few kT/e, then the density of empty donor states follows from Eq. 2-25 as

$$n_d [1 - F(E)] = N_D[1 - F(E)] = N_D\epsilon^{(E_d - E_{Fn})e/kT} \tag{2-30}$$

Since the density of conduction-band electrons must be equal to the density of empty donor levels, then

$$N_D\epsilon^{(E_d - E_{Fn})e/kT} = N_C\epsilon^{-(E_C - E_{Fn})e/kT}$$

and so, in the n semiconductor,

$$E_{F_n} = \frac{E_d + E_C}{2} - \frac{kT}{2e} \ln \frac{N_C}{N_D} \tag{2-31}$$

At $T = 0°\text{K}$ the Fermi level will lie halfway between the donor level and the bottom of the conduction band. As T increases, the Fermi level will drop. The condition at 0°K is illustrated in Fig. 2-12(a).

At some temperature above absolute zero, but below that at which intrinsic conduction is appreciable, it may be assumed that all acceptor levels of a p semiconductor are filled, leaving nonoccupied holes in the valence band. If the density of acceptor atoms is N_A, then methods similar to the above will show that the Fermi level in a p semiconductor is

$$E_{F_p} = \frac{E_a + E_V}{2} + \frac{kT}{2e} \ln \frac{N_V}{N_A} \tag{2-32}$$

where E_a is the acceptor energy level. Thus at 0°K the Fermi level will be

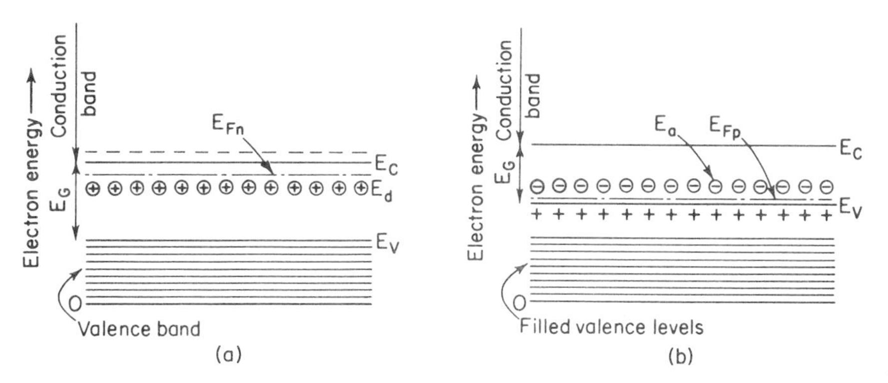

Figure 2-12. Location of Fermi levels in (a) n material; (b) p material.

halfway between the top of the valence band and the acceptor level in a p semiconductor, as shown in Fig. 2-12(b).

As the temperature goes up, the Fermi level in an n material falls in the energy gap, and the Fermi level of a p material rises. At a sufficiently high temperature the Fermi level in either material will reach the half-gap level with the conditions of intrinsic conduction. These changes are illustrated in Fig. 2-13.

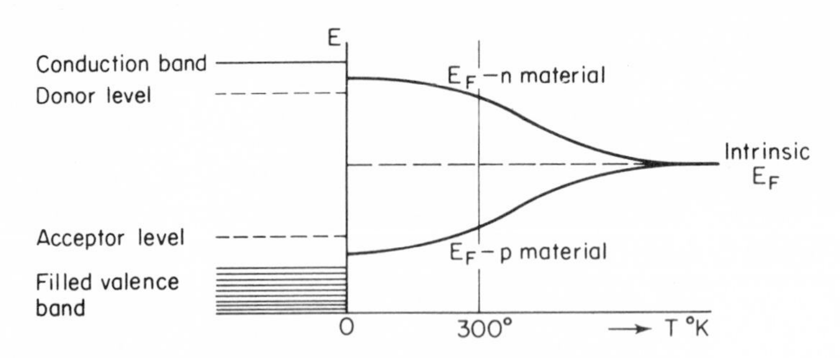

Figure 2-13. Variation of the Fermi level in n and p materials as a temperature function. Electron and hole masses are assumed to be equal.

2-10 CONDUCTION BY DRIFT OF CHARGE

Directed charge motion and conduction, in semiconductors, occurs through two mechanisms: (1) charge drift under an applied electric field, and (2) diffusion of charge from a region of high charge density to a region of lower density. The process of conduction by drift of charge will be considered in this section.

With only thermal energies acting, the conduction electrons and holes move within the crystal in completely random directions, but with statistically determinable mean free paths and mean free times between collisions with atoms. When an electric field is applied to the material, a component of velocity in the direction of the field attraction is added to the random thermal motions. The movement of each charge consists of a series of random displacements, but with a mean and directed velocity component owing to the electric field.

The current density due to electron motion across a plane in a region has been stated in Section 1-9 as

$$J_e = -nev_e$$

where n is the electron density per m^3, and v_e is the net mean free velocity, the random thermal components cancelling across the plane of measurement. A similar relation for holes is

$$J_h = pev_h$$

where p and v_h are density and velocity values for the holes. Since v_e is oppositely directed to v_h, then the total current density through a reference plane due to both electrons and holes is

$$J = e(nv_e + pv_h) \qquad \text{A/m}^2 \quad (2\text{-}33)$$

We have already defined charge mobility μ as the constant of proportionality between the acting field intensity and the mean velocity, taken as the quotient of the mean free path and the mean free time. That is,

$$\mu_e = \frac{v_e}{\mathscr{E}}, \qquad \mu_h = \frac{v_h}{\mathscr{E}} \qquad (2\text{-}34)$$

where μ has units of m^2 per volt-s. The total current is

$$J = e\mathscr{E}(n\mu_e + p\mu_h) \qquad \text{A/m}^2 \quad (2\text{-}35)$$

Rearrangement of the above leads to

$$\sigma = \frac{J}{\mathscr{E}} = e(n\mu_e + p\mu_h) \qquad \text{mho-m} \quad (2\text{-}36)$$

for the *conductivity*, or current density per unit of field intensity.

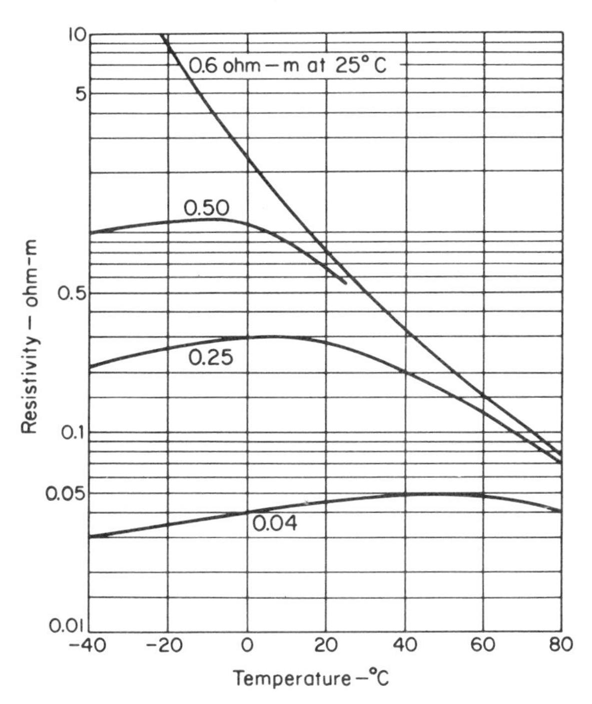

Figure 2-14. Variation of resistivity of germanium with temperature and purity.

For metals the value of n is very large and not a temperature function in the usual ranges, and p is zero. Temperature rise and the resultant increased agitation of the atoms reduces the mean free path and mean free time between electron collisions with atoms; path and time do not fall identically, however. As a result the mean velocity and the mobility both fall with rising temperature, causing the slowly rising resistivity of metals as a function of temperature.

In doped semiconductors there are few free charges at very low temperatures and the resistivity is high. As the temperature rises, the donor and acceptor levels contribute free charges and the resistivity falls until all their levels are activated, at possibly $-40°C$. The mobility then decreases somewhat with rising temperature due to increased lattice agitation, and the resistivity rises slowly. Above about 50°C in germanium and 150°C in silicon, additional carriers enter by thermal generation and the resistivity then falls rapidly. The resistivity curves appear as in Fig. 2-14, with the curve for 60 ohm-cm (0.60 ohm-m) material representing pure intrinsic germanium, where the fall in resistivity is almost entirely due to the increase in number of charge carriers by thermal-pair generation with rising temperature.

2-11 CONDUCTION BY CHARGE DIFFUSION

In a homogeneous field-free region of a semiconductor in equilibrium, the charge movement is due to the thermal agitation energies, and the net passage of charge across any reference plane is zero. However, materials may be produced in which the density of free electrons or free holes may vary with distance. In a medium with an electron density gradient, there will be more electrons in the high-density region with velocity components directed toward the low-density region than there are electrons in the low-density region with velocity components directed toward the high electron-density region. A net movement of charge will then occur by thermal agitation, and the process is known as *diffusion*. Similar reasoning could be applied to a region with a hole density gradient.

A crystal of such properties might be grown from a p-doped melt, with its characteristics changed by addition of sufficient n impurity as to gradually override the p characteristics. A charge density gradient is thus created, as indicated in Fig. 2-15.

For a study of the diffusion process we may again consider hole flow and electron flow as independent. Figure 2-16(a) represents a region with a density gradient of holes, with the hole density decreasing from left to right. Only one-dimensional charge movement will be considered, to simplify the analysis. Distances Δx are laid off at each side of a reference plane x, with Δx equal to the mean free paths of the respective charges, or the mean distance which is travelled per mean free time Δt. It will be assumed that the

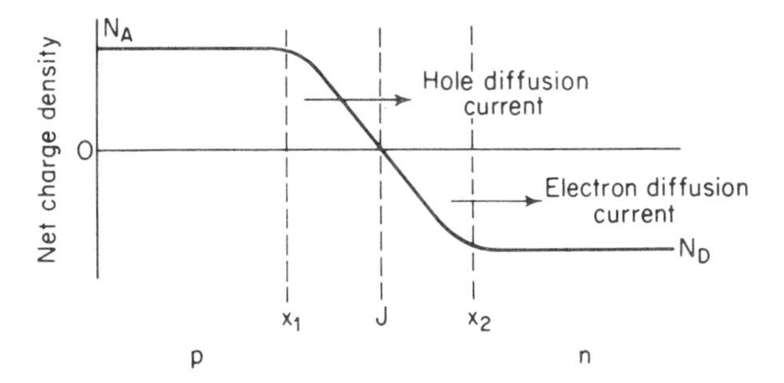

Figure 2-15. Development of charge density gradient in nonhomogeneous material.

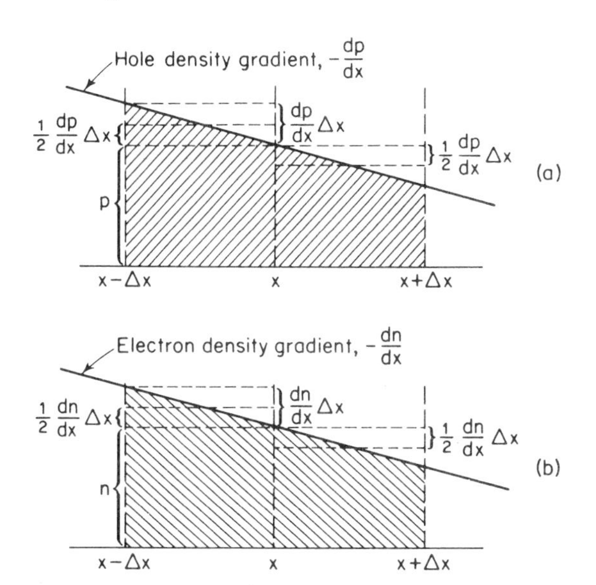

Figure 2-16. Density gradient of holes (unit cross section). Below, density gradient of electrons.

recombination rate is small, i.e., the lifetime of a charge in the material is long with respect to Δt; it is also assumed that thermal-pair generation is negligible. Because of these assumptions the number of free charges is constant during the period of analysis.

Since no external field acts, the charges will continue their random thermal motions. Because of the selection of Δx as the mean free path and the one-dimensional motion, one-half of the x-directed charges in the region $x - \Delta x$ to x will move to the right across the x reference plane in time Δt, the other half of the charges, having negative x velocity components, will move to the left across the $x - \Delta x$ plane and out of the region. Likewise, one-half of the carriers in the region x to $x + \Delta x$ will cross the plane at x

in the same time while moving to the left, the other half of the charges travelling to the right and out of the region.

The density of holes at x is taken as p, and the density gradient of holes in the $+x$ direction is $-dp/dx$. The figure shows that the average hole density in the region of width Δx between $x - \Delta x$ and x is

$$p - \frac{1}{2}\Delta x\frac{dp}{dx} \tag{2-37}$$

the negative sign arising from the negative gradient. The number of holes in the volume under discussion, of width Δx and unit cross section, is

$$\Delta x\left(p - \frac{1}{2}\Delta x\frac{dp}{dx}\right)$$

One-half of these holes will pass to the right across the reference plane in time Δt, the hole flow to the right per unit time being

$$P_{\text{right}} = \frac{1}{2}\frac{\Delta x}{\Delta t}\left(p - \frac{1}{2}\Delta x\frac{dp}{dx}\right) \tag{2-38}$$

By similar reasoning, the number of holes in the volume of width x to $x + \Delta x$ and of unit cross section is

$$\Delta x\left(p + \frac{1}{2}\Delta x\frac{dp}{dx}\right)$$

and one-half of these holes will pass to the left across the reference plane in time Δt. The rate of hole flow to the left is then

$$P_{\text{left}} = \frac{1}{2}\frac{\Delta x}{\Delta t}\left(p + \frac{1}{2}\Delta x\frac{dp}{dx}\right) \tag{2-39}$$

Since dp/dx is negative, the rate of charge flow across the plane to the right exceeds the rate of charge flow to the left; the net hole flow rate to the right is

$$\text{net } P_{\text{right}} = -\frac{1}{2}\frac{\Delta x}{\Delta t}\Delta x\frac{dp}{dx} \tag{2-40}$$

Thus there is a flow of charge and a current wherever a charge density gradient exists.

A *hole diffusion constant* D_h can be defined as

$$D_h = \frac{1}{2}\frac{\Delta x}{\Delta t}\Delta x = \frac{\Delta v\,\Delta x}{2} \qquad \text{m}^2/\text{s} \tag{2-41}$$

where Δv is the mean thermal velocity and Δx is the mean free path.

The charge flow across the reference plane per unit time is the diffusion current density, and for holes can be written from Eq. 2-40 by including $+e$, the charge per hole. Then

$$J_h = -eD_h\frac{dp}{dx} \tag{2-42}$$

representing a conventional current in the direction of the decreasing hole density, or to the right.

By similar reasoning, using n material and electrons in Fig. 2-16(b), an expression for the current density due to electron diffusion is

$$J_e = eD_e\frac{dn}{dx} \tag{2-43}$$

which represents a conventional current to the left.

The charge density gradients cause both kinds of charges to move in the direction of decreasing concentration, but the derived electric currents are in opposite directions because of the opposite charge signs. Thus a hole diffusion current will pass in the direction of decreasing hole density, a conventional current due to electron diffusion will be directed toward increasing electron density. For a material with a transition gradient between p and n materials, as in Fig. 2-15, both current components move to the right.

In semiconductors the drift and diffusion actions may be simultaneously present. Equation 2-35 may be combined with Eqs. 2-42 and 2-43 to yield expressions for total electron and hole current densities as

$$J_h = e\mu_h p\mathscr{E} - eD_h\frac{dp}{dx} \qquad \text{A/m}^2 \tag{2-44}$$

$$J_e = e\mu_e n\mathscr{E} + eD_e\frac{dn}{dx} \qquad \text{A/m}^2 \tag{2-45}$$

Diffusion current requires a charge inhomogeneity or the presence of a space gradient of charge density. Drift current is a function of an electric field, whereas diffusion may occur in regions free of electric fields, a condition which exists in many semiconductor devices. In each equation, the first term on the right is a drift component, the second term appears because of diffusion.

Diffusion causes a charge redistribution which leads to the rise of a potential across a junction of materials with a charge inhomogeneity. We may then expect that a drift current will be established due to the potential, and, with the n material positive to the p side, the potential across the junction is just sufficient to create a drift current equal and opposite to the diffusion current, resulting in a net zero current across the isolated junction.

2-12 MEASUREMENT OF CHARGE MOBILITY

Mobility μ, being defined as velocity per unit electric field intensity, can be measured by a transit-time experiment in the circuit of Fig. 2-17.

A rod of n germanium has a current I applied, allowing measurement of the voltage V over the distance A_2, and determination of the field intensity as $\mathscr{E} = V/A_2$. Two additional probes serve as emitter and collector and are accurately separated by the distance A_1. A microsecond rectangular voltage

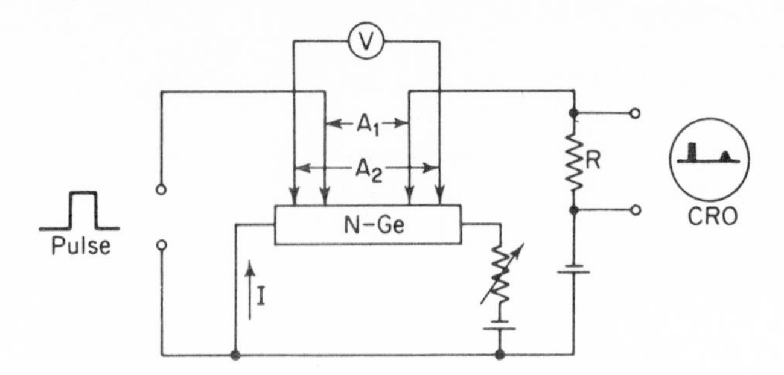

Figure 2-17. Measurement of mobility. (After Haynes–Shockley)

pulse is applied to the emitter probe, to introduce holes at the emitter, at a pulse repetition rate of 100–500 per second. The current from the output probe or collector produces a voltage across R which is applied to a cathode-ray oscilloscope whose sweep is triggered by the input pulse. The input pulse then appears on the screen at the start of the sweep.

Under the action of the electric field the emitted holes drift down the bar and reach the collector probe t seconds after the injection pulse. The drifting pulse of holes is acted upon by the diffusion process, and the received pulse will not be an image of the input-pulse shape, but will be smeared on the time axis. However, use of the known sweep-time calibration, and measurements taken between the center of the applied pulse and the peak of the received signal will allow calculation of t. The velocity v follows because A_1 is known. Mobility for the holes follows as

$$\mu_h = \frac{v_h}{\mathscr{E}}$$

If a rod of p characteristics is used, with reversed voltages, then electron mobility μ_e can be determined.

2-13 LIFETIMES OF THE CHARGES

Light energy applied to a bar of n material can break valence bonds and create electron–hole pairs; it is necessary, however, that the photons carry energy greater than the gap energy; for germanium this is possible with frequencies in the visible or near infrared.

Let a bar of n material be irradiated with high-intensity light for a pulse of microsecond order. Just after the pulse there will be an excess of electron–hole pairs over the number normally present at thermal equilibrium. Because the bar is of n characteristics, it will have present a large number of electrons, and the addition of more electrons from the radiant pair-generation will not appreciably change the conductivity due to electrons. However, the number of holes produced by the radiation is very greatly in excess of the thermal-equilibrium hole density, so that the hole conductivity is greatly increased.

After the flash of light the excess carriers recombine until the equilibrium density is once more reached, and the hole conductivity will decay at an exponential rate, or

$$p = p_0 \epsilon^{-t/\tau_h} \tag{2-46}$$

where the time constant τ_h defines the *mean lifetime* of the holes in the n material. Similar definitions apply to the lifetime of electrons in p material.

Lifetimes depend on temperature and impurity concentration and are usually in the range from 1 to 1000 μs. Lifetime in the bulk crystal may be considerably longer than the surface lifetime which the illumination method essentially measures. Crystal faults and surface imperfections and dislocations appear to aid in the mechanical energy transfer which is involved in recombination of a hole and an electron, and so the recombination process is accelerated on surfaces. Emission of the electron and hole energy by electromagnetic means is encountered, the emission frequencies including those used to create the pairs in the first place. Such radiation is a process of potential importance as a light source.

If p_0 is the hole density produced by the light flash, then the diffusion length is the distance to which the charges diffuse before their density is reduced to p_0/ϵ by the process of recombination. The diffusion length is given the symbol L_h or L_e, and is

$$L = \sqrt{D\tau} \tag{2-47}$$

For germanium, L_h is in the range of 0.007 to 0.2 cm. Since the charges may go through many collisions with atoms before recombination conditions occur, the diffusion length may be many times that of the mean free path.

It is normally desired that recombination be negligible in the charge flow through the base element of a transistor, and so the base width must be kept small with respect to the diffusion length of the charges in the material.

2-14 THE *pn* JUNCTION IN EQUILIBRIUM

A *junction* between p and n materials can be formed in a single crystal of semiconductor by growing a crystal from an n melt, and changing the melt characteristics by addition of sufficient p impurity to override the effects of the n impurity; the free electrons from the n impurity fill holes in some of the p material and are immobilized by the atoms, but additional acceptors are available to form a p material. The *region of transition* between the n and p characteristic can be quite narrow, in the range from 10^{-4} to 10^{-6} cm. The theoretical *junction plane* is assumed to lie where the density of donors and acceptors is equal. Other methods of junction formation will be discussed later.

Due to the charge inhomogeneity introduced by the junction, some of the electrons supplied by the five valence donor atoms on the n side will diffuse across the junction plane to the region of low electron density in p. This action leaves ionized donor atoms which create a positive space charge in the n region near the junction. The electrons which diffused to p will fill valence levels created by the acceptor atoms; these electrons are immobilized by the fixed valence-three acceptors and create a negative space charge in the p region near the junction. Figure 2-18 pictures such a *charge depletion region* as essentially cleared of majority free charges but with the fixed atomic ions remaining. Because of the charge separation, an electric field and potential V_B develops across the junction under equilibrium conditions, or with the junction externally isolated. The effect of the field is restricted to the narrow transition region because of the high dielectric constant of the semiconductor.

Energy-level diagrams for the materials before junction formation are shown in Fig. 2-19(a). The impurities are so sparsely distributed even in heavily doped materials that the impurity atoms do not alter the basic energy levels. Thus the energies at the band edges, E_C and E_V, are the same in either n or p for a given semiconductor. However, after formation of the

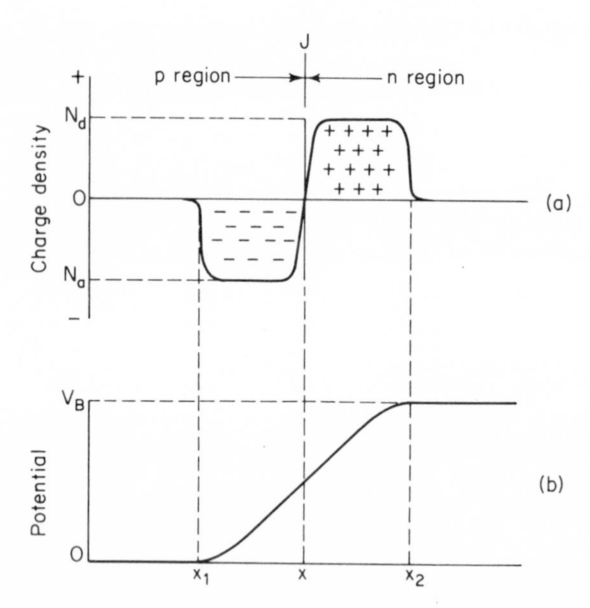

Figure 2-18. (a) Development of space-charge transition region. (b) Theoretical change in diffusion potential in an isolated junction.

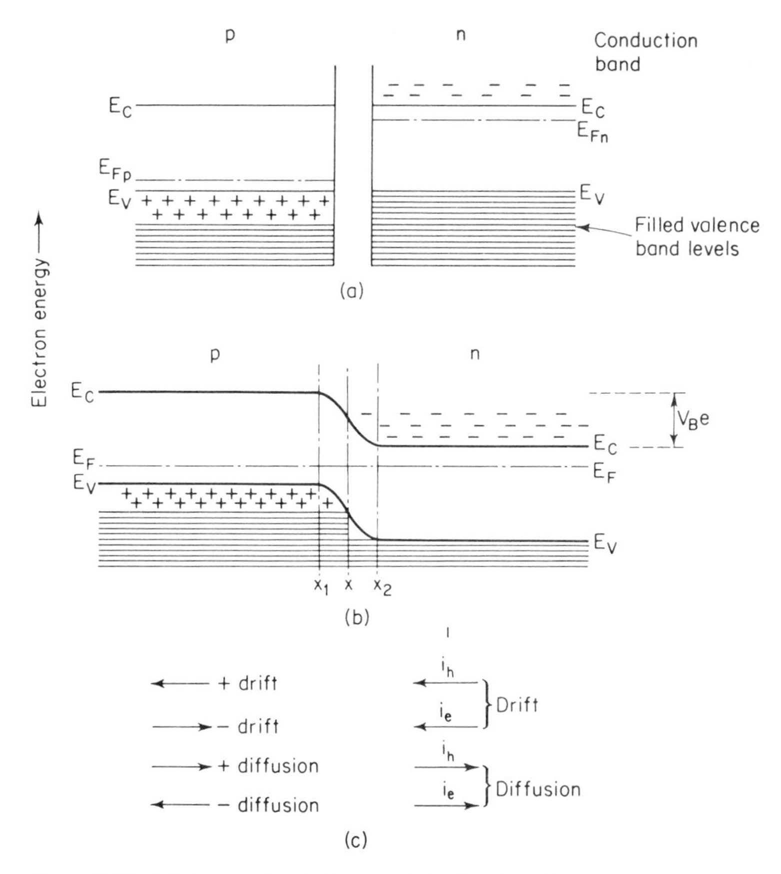

Figure 2-19. (a) Energy levels in n and p before junction formation. (b) The junction in equilibrium. (c) Detailed charge flow and current balance at equilibrium.

junction, the Fermi level is constant throughout a system at equilibrium, as in Fig. 2-19(b). The energetic electrons have adjusted the energies of the two materials to produce a physical system in which the total energy is a minimum. The situation is analogous to the manner in which the water levels adjust in connecting tanks.

At formation of the junction, a transient charge flow occurs as described, and the internal voltage V_B develops. An equilibrium is reached at that potential which maintains a drift current just equal to the diffusion current due to the concentration gradients; the net junction current must be zero. But a drift component to p is composed of p-directed holes and n-directed electrons; a diffusion current toward n is composed of n-directed holes and p-directed electrons, as indicated in Fig. 2-19(c). Equality of drift and diffusion components can occur only if equality is true for each carrier; that is, from Fig. 2-19(c):

$$i_{h\ \text{drift}} + i_{e\ \text{drift}} = i_{h\ \text{diff}} + i_{e\ \text{diff}}$$

$$i_{h\ \text{drift}} - i_{h\ \text{diff}} = i_{e\ \text{diff}} - i_{e\ \text{drift}}$$

Equilibrium forbids charge generation, so that each side of the above must be identically zero, or the diffusion and drift currents must balance for each carrier separately. The drift of holes from n to p due to $+V_B$ on n will equal the diffusion movement of holes from p to n; the drift of electrons from p to n due to V_B will cancel the diffusion of electrons from n to p.

The region over which the material transition occurs may be assumed as 10^{-6} m for a typical junction. With $V_B = 0.1$ V, the field intensity tending to produce a drift current is then 100,000 V/m. Using constants from Table 2-3 with p germanium of 1 mho/m conductivity, n germanium of 0.01 mho/m, the possible hole diffusion current can be calculated from $eD_h\, dp/dx$ as approximating 26×10^7 A/m². Actually, of course, this current is just inhibited by the barrier potential.

The band edge energies and the Fermi levels are not changed by formation of the junction, but the resultant transient flow of energetic charges and the alignment of the Fermi levels across the junction forces a relative shift in band edge energies. The magnitude of this shift is a measure of the energy differences of the Fermi levels; it indicates the relative energy carried from one region into the other by the transient charge movement at the time of junction formation. In the materials outside the transition region the normal carrier concentrations exist, and there is no indication of the presence of the junction. This is illustrated in Fig. 2-19(b).

Equation 2-23 is an expression for the density of conduction band electrons in a semiconductor under equilibrium, as determined by the energies of the conduction band edge and the Fermi level of the material. Outside the

Table 2-3

CHARACTERISTICS OF TYPICAL p AND n MATERIALS (GERMANIUM)†

Densities per m³ at 300°K

Conductivity $(\text{ohm-m})^{-1}$	p 1	p 0.01	n 1	n 0.01
p_p	3.68×10^{23}	3.68×10^{21}		
n_p	1.70×10^{15}	1.70×10^{17}		
n_i	2.4×10^{19}	2.4×10^{19}	2.4×10^{19}	2.4×10^{19}
n_n			1.75×10^{23}	1.75×10^{21}
p_n			3.57×10^{15}	3.57×10^{17}
For intrinsic level reference:				
E_F, V	+0.240	+0.125	−0.221	−0.106

†After R.D. Middlebrook, *An Introduction to Junction Transistor Theory*, John Wiley & Sons, Inc., New York, 1957.

junction face in the n material, or to the right of x_2 in Fig. 2-19(b), the electron density is

$$n_n = N_C \epsilon^{-(E_{Cn} - E_{Fn})e/kT} \tag{2-48}$$

where $E_{Cn} - E_{Fn}$ is the relative energy difference.

On the p side of the junction region, or to the left of x_1, the electrons will be due to thermal-pair generation, and the electron density is

$$n_i = N_C \epsilon^{-(E_{Cp} - E_{Fp})e/kT} \tag{2-49}$$

where $E_{Cp} - E_{Fp}$ is the relative energy difference of these levels in the p material. The ratio of the electron densities on the two sides of the transition region then is

$$\frac{n_n}{n_i} = \frac{\epsilon^{-(E_{Cn} - E_{Fn})e/kT}}{\epsilon^{-(E_{Cp} - E_{Fp})e/kT}}$$

While the energies E_{Cn} and E_{Cp} appear to have been changed by the formation of the junction, it was the alignment of the Fermi levels which caused the shift. Therefore it is convenient to say $E_{Cp} = E_{Cn}$ after junction formation, as well as before, and to measure the shift in energies as a function of the Fermi levels. Then

$$\frac{n_n}{n_i} = \epsilon^{(E_{Fn} - E_{Fp})e/kT} \tag{2-50}$$

A similar relation, but with a negative exponent, could be written for holes, starting with Eq. 2-27.

The alignment of the Fermi levels across the junction has created a potential difference, as well as current components which balance when the junction is in equilibrium. Using this requirement for the electron currents, or setting $J_e = 0$ in Eq. 2-45, leads to

$$-\frac{\mu_e}{D_e} \mathscr{E}\, dx = \frac{dn}{n}$$

Considering only one-dimensional charge flow, the above may be integrated from x_1 to x_2 of Fig. 2-18, across the width of the transition region, giving

$$-\frac{\mu_e}{D_e} \int_{x_1}^{x_2} \mathscr{E}\, dx = \int_{n_p}^{n_n} \frac{dn}{n}$$

where the limits on electron densities are those in the materials outside the bounds of the transition region. The integration of $-\mathscr{E}$ from x_1 to x_2 leads to the barrier potential V_B, as a rise from p to n. The result of the integration is

$$\frac{n_n}{n_p} = \epsilon^{V_B(\mu_e/D_e)} \tag{2-51}$$

thus relating the electron density at the junction face in the n region to the density at the face in the p region, as a function of the barrier potential.

Using a similar process for the hole current of Eq. 2-44, there results

$$\frac{p_p}{p_n} = \epsilon^{V_B(\mu_h/D_h)} \tag{2-52}$$

These expressions, comparing equilibrium densities in the semiconductor on the two sides of a junction region in terms of the barrier potential, are called the *Boltzmann equations*, and are basic relations of a junction.

Equations 2-50 and 2-51 are expressions for the same charge density ratio across the junction. Two conclusions may be drawn by comparison of the exponents. First, the potential is related to the difference of the Fermi levels of the n and p regions, as

$$V_B = E_{Fn} - E_{Fp} \tag{2-53}$$

and this establishes the magnitude of the barrier potential in terms of the physical factors causing it. Secondly, the mobility and diffusion constants are related as $\mu_e/D_e = e/kT$ or $\mu_h/D_h = e/kT$. With the use of appropriate subscripts implied, the general expression is

$$D = \frac{kT}{e}\mu \tag{2-54}$$

and this is known as the *Einstein relation*. The mobility and the diffusion constant are related to the fundamental charge energies.

The Einstein relation when employed in Eqs. 2-51 and 2-52 leads to useful forms of the Boltzmann equations, as

$$\frac{n_n}{n_p} = \epsilon^{V_Be/kT} \tag{2-55}$$

$$\frac{p_p}{p_n} = \epsilon^{V_Be/kT} \tag{2-56}$$

It may also be shown that the barrier potential is related to the characteristics of the materials as

$$V_B = \frac{kT}{e}\ln\frac{N_AN_D}{n_i^2}$$

It is seen that the impurity additions are a mechanism for producing two materials of differing electrical characteristics within a single crystal of semiconductor. The impurity-produced n and p majority densities create different Fermi levels and thus a potential difference at the n and p junction, in the equilibrium state.

2-15 CURRENTS THROUGH THE JUNCTION

The conductivity of the junction semiconductor crystal is determined by the properties of the material at or near the junction; elsewhere the crystal is assumed to be a good conductor.

In Fig. 2-19(b) for the junction in equilibrium, the n region is positive to p. Electrons tend to run downhill or to n; higher electron energy is upward in the diagram. Holes tend to move upward to p, and higher hole energy is downward. The electrons in n and the holes in p have a random distribution of thermal energy and random velocities. With application of a small *forward voltage* V, positive to p as in Fig. 2-20(a), the transition region is partially

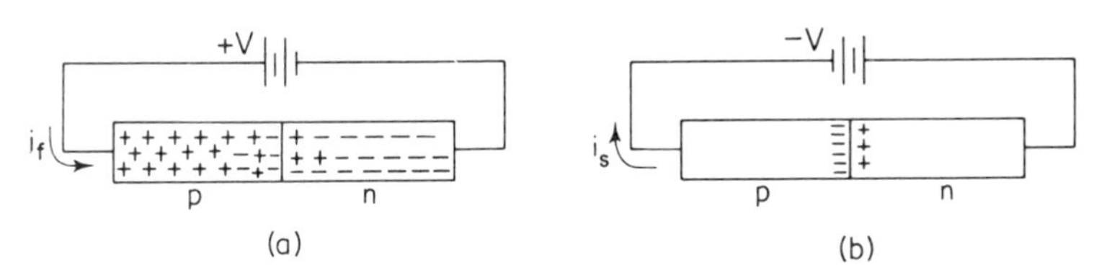

Figure 2-20. (a) Junction under forward bias. (b) Depletion region under reverse bias.

resupplied with mobile charges or is decreased in width. The equilibrium of the isolated junction is upset and the requirement of equality between diffusion and drift current is removed. The consequent reduction in barrier potential by applying V in opposition to V_B is indicated in Fig. 2-21(a), the junction potential then being $V_B - V$. The upward jog in E_F is equal to the effect of the applied voltage V.

A great number of charges now have sufficient energy to override the reduced barrier height, and so the number of holes crossing the energy barrier is increased. Figure 2-22(a) illustrates the variation in hole and electron flows across the junction with forward bias, in a junction with a highly doped p region and a lightly doped n region. Starting in p, the highest energy holes will have sufficient energy to cross over the barrier and appear on the n side of the transition region. These holes are then minority charges in n, and will increase the hole density at x_2 considerably above normal in n. This increased hole density provides the necessary charge density gradient for the diffusion process, and the holes will diffuse to the right.

As the holes diffuse away from the junction, they encounter electrons in the n material and recombine. Due to this recombination, the hole current decays exponentially with distance from the junction, ultimately declining to the density of thermally generated holes in n.

To supply the charges for recombination, there is a flow of electrons from right to left in n, which gradually decreases due to recombination with the diffusing holes. This is a drift component of majority carriers being swept to the junction by the potential, whereas the holes move by diffusion across the potential barrier. The total current must, obviously, be constant in value everywhere.

Some of the higher-energy electrons at the transition will override the

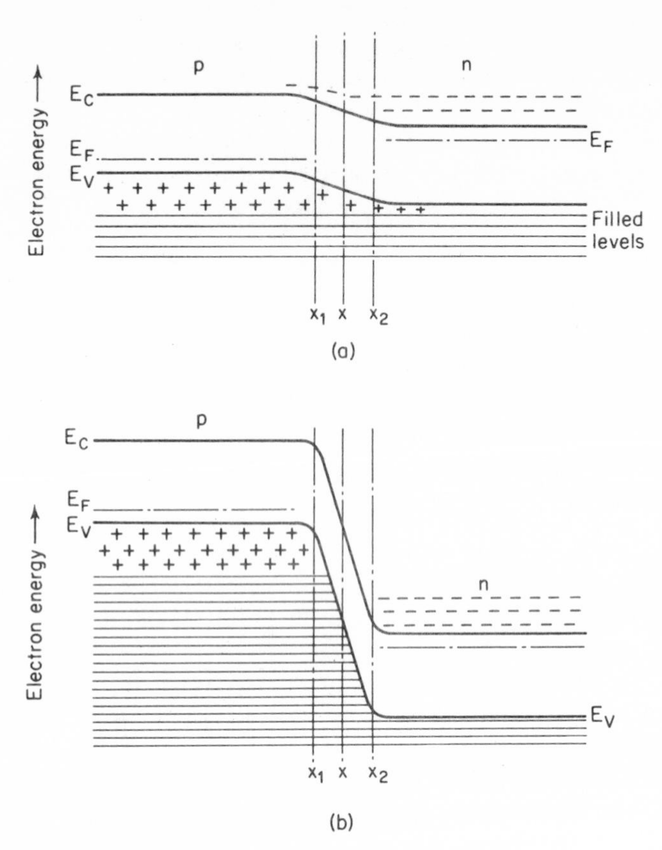

Figure 2-21. (a) Energy levels with forward bias; (b) energy levels with reverse bias.

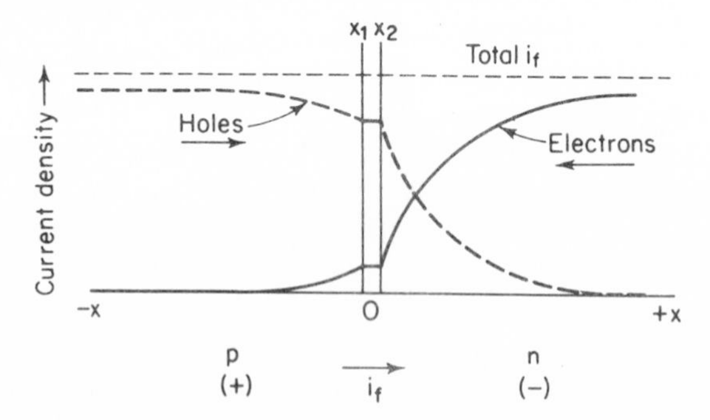

Figure 2-22. Electron and hole currents in a forward-biased junction.

barrier and appear as an electron density at x_1. Electron diffusion to the left occurs in p, with recombination causing the diffusion component to decay exponentially toward the equilibrium density on the p side, and reducing the hole density there by recombination.

In the p region the density of holes is normally that of the acceptors, namely N_A. In the n region the electron density is substantially that of the donors, N_D. Both of these majority charge densities are large with respect to the intrinsic values. Thus when a forward bias is applied to a junction, the majority densities do not appreciably change, and it is the minority densities of holes on the n side and electrons on the p side which are increased by significant factors over their equilibrium levels.

The minority holes injected into the n region over the potential barrier control the phenomena at the barrier. They may be so numerous as to cause a positive space charge and accelerate electrons into the depletion region. The equilibrium value of the pn concentration product may be exceeded, and the recombination rate increases as the concentration product moves back to the equilibrium value. The result is a high conductivity and low voltage drop in the lightly doped and high-resistance n region, and large forward currents move for low values of applied voltage.

The supply of holes in the p region being equal to the impurity density means that the forward current is not subject to appreciable variation with temperature; it is limited by the velocity distribution of the carriers and by the effective barrier height. While control of the phenomena at the junction is by minority carriers, it must still be recognized that at the semiconductor extremities the current i_f injected into p as a hole current emerges from the terminal on n as a flow of electrons, or conduction remote from the junction is by majority charges.

Under a condition of reverse bias, or with the p region negative to n, the space charge regions of Fig. 2-18(a) are thickened. The reverse bias adds to V_B, and makes the energy step at the junction higher, as in Fig. 2-21(b). There will then be few charges with sufficient energy to ride over the barrier. In fact, at 300°K the value of $kT/e = 0.026$ V, and for reverse voltages only a little greater than this, the diffusion process is completely inhibited, and the drift current predominates. However, the drift current is limited to a low value because of the shortage of mobile charges in the thickened depletion layer. The reverse current is almost entirely composed of carriers produced by thermal-pair generation in and near the charge-depleted regions. Holes so generated are swept to the negative terminal on p, and electrons generated are attracted to the positive terminal on n.

Because all the charges move "downhill" on the energy diagram and because thermal-pair generation of charges is a constant at a given temperature, the drift current under reverse junction potential (called the *reverse saturation current*) is independent of applied voltage. Since the rate of charge generation in the depletion region depends strongly upon temperature, the reverse saturation current is also strongly sensitive to temperature. However, the reverse current is small at usual ambient temperatures, because of the low rate of pair generation.

If $p_p \gg n_n$, then nearly all the current across the junction is carried by holes, as indicated by Fig. 2-22(a). The p region injects holes into the n region with forward bias. If $n_n \gg p_p$, then the electrons constitute the major current component at the junction, and the n region injects electrons into the p region with forward bias. The material with the greatest impurity density, or the lowest resistivity, then controls the situation. In a typical case with a p to n ratio of 100:1, more than 99 per cent of the forward junction carriers will be holes. Choice of material and processing method must be a compromise, since a low resistance is desired for large current-carrying capacity, but high-resistance material must be used for large reverse voltage ratings.

2-16 THE DIODE EQUATION

It is now possible to predict the current which will pass through a pn junction with an applied potential. The voltage across the junction will be small and comparable to kT/e in the forward direction, or with $+V$ applied to p. In fact, the forward junction potential will rarely exceed a few tenths of a volt.

The Boltzmann equation for holes, Eq. 2-56, may be rearranged as

$$p_n = p_p \epsilon^{-V_B e/kT}$$

in the equilibrium case. The hole density p_p is that due to acceptors in the p region just to the left of the junction in Fig. 2-19(b). Also p_n is the hole density in the n region just to the right of the junction region.

With a forward voltage V applied, with V opposing V_B, the hole density on the right of the junction becomes

$$p_n + \Delta p_n = p_p \epsilon^{-(V_B - V)e/kT} = p_p\, \epsilon^{-V_B e/kT}\, \epsilon^{Ve/kT} \tag{2-57}$$

The change Δp_n results because more holes now have sufficient energy to cross the reduced potential barrier and appear on the face of the transition region in n. These holes then diffuse to the right in the n region, and constitute the hole current of Fig. 2-22.

The increase in hole density at the right of the junction is caused by application of V and, through subtraction of Eq. 2-56 from 2-57, this increase in hole density is

$$\Delta p_n = p_p \epsilon^{-V_B e/kT}(\epsilon^{Ve/kT} - 1) \tag{2-58}$$

Multiplication by the junction area, A, and by the particle charge and velocity, leads to a product of the form of pev_h, or a current. Thus the hole current at the right face of the junction is

$$i_h = Ap_p e v_h \epsilon^{-V_B e/kT}(\epsilon^{Ve/kT} - 1)$$
$$= B_h(\epsilon^{Ve/kT} - 1) \tag{2-59}$$

By similar methods, starting with the Boltzmann relation of Eq. 2-51 for electrons, it is possible to write the electron current in p, at the left face of the junction, as

$$i_e = B_e(\epsilon^{Ve/kT} - 1) \tag{2-60}$$

The total current through the junction is then

$$i = i_h + i_e = (B_h + B_e)(\epsilon^{Ve/kT} - 1) \tag{2-61}$$

The constants may be evaluated by noting that $i = -(B_h + B_e)$ for large negative values of V. However, we have already reasoned physically that the current for large negative voltages will be the reverse saturation current due to thermal charges swept out of the depletion region. Therefore

$$i = i_s(\epsilon^{Ve/kT} - 1) \tag{2-62}$$

is *the diode equation*.

In the forward direction ($+V$), and with V large with respect to kT/e (perhaps $V = +0.130$ V, whereas $kT/e = 0.026$ V at 300°K), then

$$i \cong i_s \epsilon^{Ve/kT} \tag{2-63}$$

indicating that the forward current increases exponentially with large values of V. In the reverse potential direction ($-V$), the exponential term of Eq. 2-62 quickly becomes negligible with respect to unity, and

$$i \cong -i_s$$

Thus the volt–ampere curve of Fig. 2-23 is predicted, and agrees well with actual diode performance, when the voltage is measured directly across the junction.

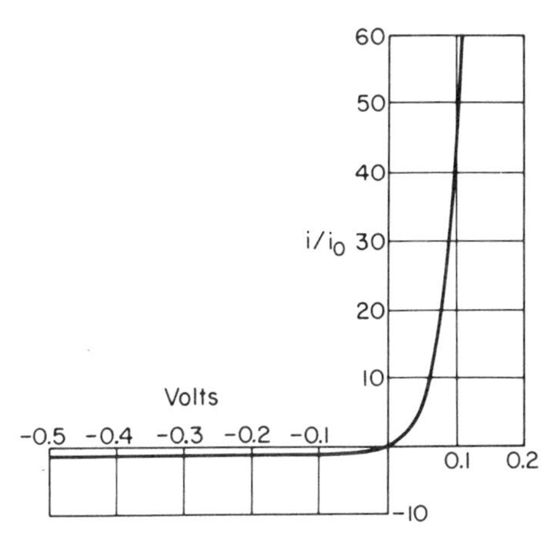

Figure 2-23. Theoretical volt–ampere curve for a junction diode.

With reverse junction bias, only the reverse saturation current i_s remains. This current is almost entirely due to thermal generation in the junction region, with the charges riding "downhill" on the energy diagram—thus all those generated are swept out, or the current is saturated.

It is obvious that

$$i_s = K\epsilon^{-V_B e/kT} \tag{2-64}$$

Since V_B is related to the energy difference between the Fermi levels in the n and p regions, then large V_B values will be associated with large E_G materials. This explains the important role of silicon devices, the larger energy gap of silicon leading to significantly smaller reverse saturation currents.

The junction diode is a nonlinear and essentially unilateral device, and therefore is a rectifier. The nonlinear forward characteristic is a result of the exponential distribution of energy among the charges, each increment of reduction in barrier potential permitting increased numbers of charges to pass over the barrier.

The reverse diode current is actually made up of two components, the saturation current i_s, and a leakage component. The latter is due to surface leakage across the junction and is a function of voltage. Thus the total reverse current increases slightly with increasing voltages. Another departure from the ideal theory occurs at some high reverse voltage at which the reverse current rises suddenly. This is called the *breakdown potential*, and applications make use of the reverse-biased diode as a voltage regulator; the physical actions which occur will be discussed later in detail.

2-17 *PHYSICAL SIGNIFICANCE OF i_s*

Under the reverse potential situation illustrated by Figs. 2-21(b) and 2-24, even the highest energy holes in p cannot overcome the barrier; the electrons in n face a similar situation and thus the majority carriers are essentially blocked from diffusion across the junction.

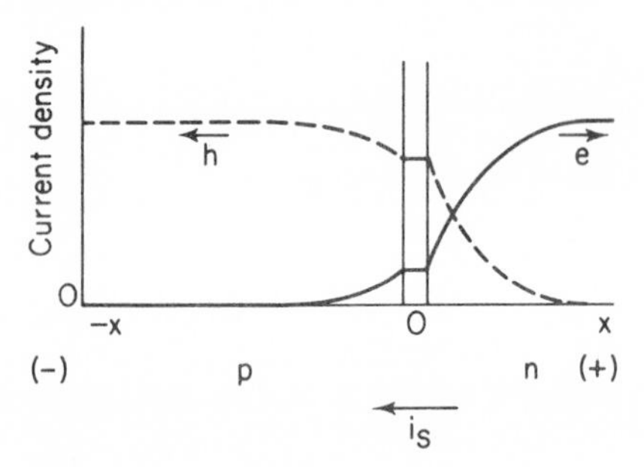

Figure 2-24. Current densities in the reverse-biased junction.

Minority carriers are generated by thermal energies throughout the materials. However, electrons generated in the p material, or holes generated in the n material, remote from the junction, will be likely to recombine with the majority carriers and never reach the junction. Therefore, charges contributing to the reverse current i_s will be largely those thermally generated near the junction, and generally within the transition region.

Holes appearing on the n face of the junction are swept into p by the junction bias, the junction appears as a sink, and $p = 0$ at the junction surface at $x = 0$. A similar action results for electrons thermally generated in p in the transition region or near the junction. This density condition on the junction faces establishes the necessary density gradient for minority diffusion toward the barrier, or for holes in n and electrons in p.

The current density at the junction is a result of thermal generation, recombination, and diffusion of charge. Consider p_0 as the equilibrium density of holes in n, remote from the reverse-biased junction. By reason of Eq. 2-46 and the definition of hole lifetime, we have

$$p = p_0 \epsilon^{-t/\tau_h}$$

as an expression of the hole density caused by recombination of charge. The recombination rate is

$$\frac{dp}{dx} = -\frac{p_0}{\tau_h}\epsilon^{-t/\tau_h} = \frac{-p}{\tau_h} \tag{2-65}$$

Under the equilibrium conditions existing far from the junction, $p = p_0$, and the rate of thermal generation must equal the rate of recombination. The rate of generation will therefore be p_0/τ_h, and this rate will be the same throughout the material for a given temperature. The rate of recombination, however, is proportional to the hole density existing, and while it is $-p_0/\tau_h$ in remote areas in equilibrium, it will be $-p/\tau_h$ in the region near the junction where the equilibrium has been upset by the presence of the junction and its potentials.

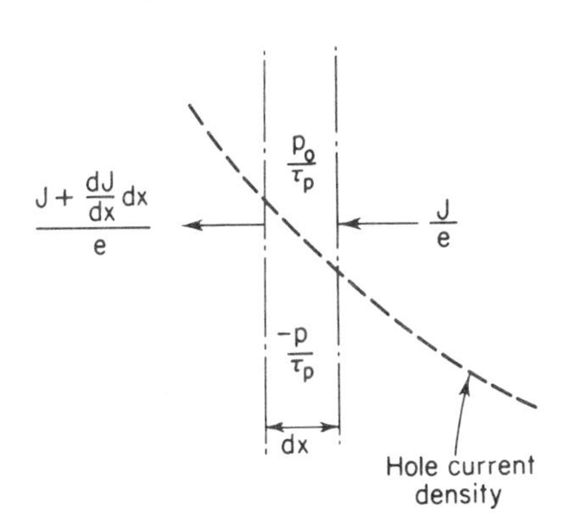

Figure 2-25. Factors affecting hole-current density in n.

Figure 2-25 shows the factors affecting the hole density in a region of thickness dx, located in n very close to the junction. Entering from the right is a diffusion flow of holes J/e, where J is the diffusion current density. Leaving the left face of the elemental region is a diffusion flow of holes given by

$$-\frac{J + (dJ/dx)\,dx}{e}$$

Per unit time there is a thermally generated density of $(p_0/\tau_h)\,dx$, and lost by recombination is a density of $-(p/\tau_h)\,dx$.

Combining these gains and losses of holes under steady-state assumptions leads to

$$\frac{J}{e} - \frac{J + (dJ/dx)\,dx}{e} + \frac{p_0}{\tau_h}dx - \frac{p}{\tau_h}dx = 0 \tag{2-66}$$

$$\frac{dJ}{dx} = \frac{e}{\tau_h}(p_0 - p) \tag{2-67}$$

Diffusion hole density is established by Eq. 2-42 as

$$J_h = -eD_h\frac{dp}{dx} \tag{2-68}$$

and so

$$\frac{d^2p}{dx^2} = -\frac{p_0 - p}{\tau_h D_h} = \frac{p - p_0}{L_h^2} \tag{2-69}$$

the last substitution following by reason of the definition of the diffusion length L_h in Eq. 2-47.

Solution of the above equation, with $p = 0$ at $x = 0$ at the junction, due to the sweeping action of the reverse potential, and $p = p_0$ at $x = \infty$, gives

$$p = p_0(1 - \epsilon^{-x/L}) \tag{2-70}$$

for the hole density in the transition region. Taking the derivative and substituting in Eq. 2-68 for the hole diffusion density gives

$$J_{sh} = \frac{eD_h p_0}{L_h}$$

where the constants are the values *for holes in the n material.*

A similar derivation could be carried out for the electrons diffusing on the p side of the transition, and the total reverse saturation current, for a junction of area A, can be written as

$$i_s = Ae\left(\frac{D_h p_0}{L_h} + \frac{D_e n_0}{L_e}\right) \tag{2-71}$$

$$= A\frac{kT}{e}\left(\frac{\sigma_h}{L_h} + \frac{\sigma_e}{L_e}\right) \tag{2-72}$$

Thus the reverse saturation current is related to the physical properties of the semiconductor materials, and to the temperature. The p notation refers to hole densities, and the n notation to the density of electrons; h and e refer to hole and electron properties.

Using a reference temperature T_0, and a current i_0 as the value of i_s at that temperature, it is possible to write

$$\frac{i_s}{i_0} = \left(\frac{T}{T_0}\right)^3 \epsilon^{(E_G e/k)\ (T - T_0/TT_0)} \tag{2-73}$$

with impurity conduction dominant.

Evaluating $E_G e/k$ with $E_G = 0.75$ eV for germanium, then

$$\frac{i_s}{i_0} = \left(\frac{T}{T_0}\right)^3 \epsilon^{8700(T - T_0/TT_0)} \tag{2-74}$$

Using $E_G = 1.15$ eV for silicon,

$$\frac{i_s}{i_0} = \left(\frac{T}{T_0}\right)^3 \epsilon^{13300(T - T_0/TT_0)} \tag{2-75}$$

The curves of Fig. 2-26 are plotted from these relations. The sensitivity to temperature is less than is indicated by the curves, since the total reverse current also includes surface leakage currents which are not temperature sensitive.

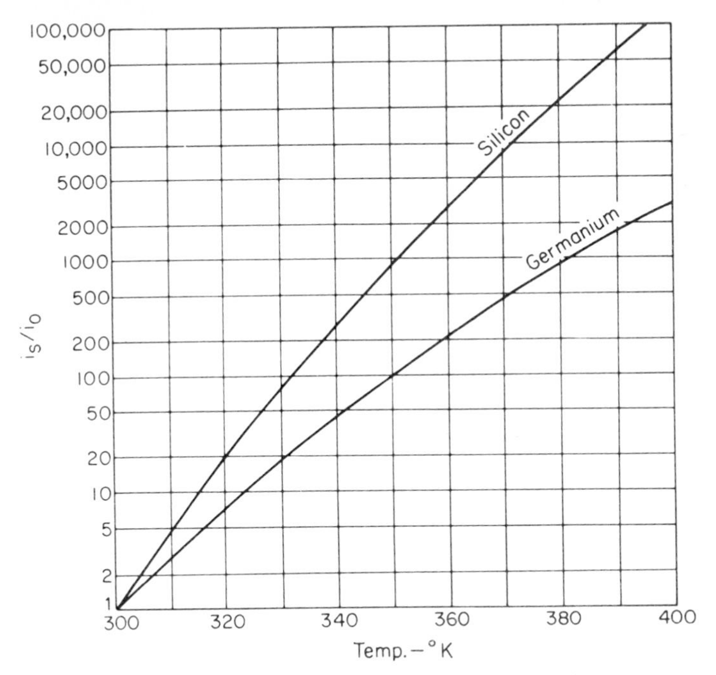

Figure 2-26. Theoretical variation of reverse saturation current.

2-18 THE DIODE FORWARD RESISTANCE

The diode resistance in the forward direction is low in value, due to the blocking effect of the barrier potential as well as the bulk resistance of the material. The barrier effect represents an effective resistance, and the bulk resistance is desirably negligible, although this ideal cannot always be reached.

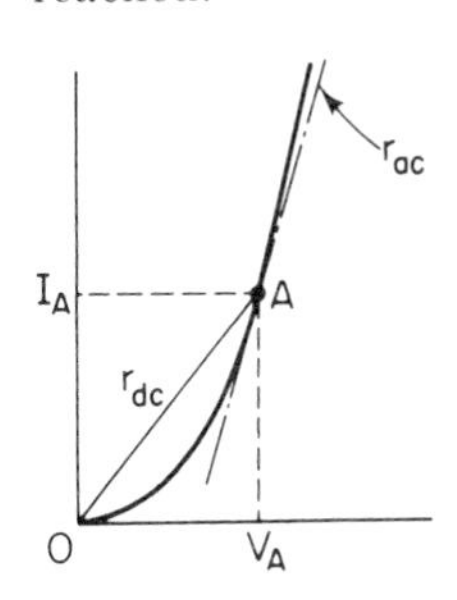

Figure 2-27. Resistance of the diode.

Using the forward portion of the diode volt–ampere curve of Fig. 2-27, with $+V$ to p, the dc or *static resistance* at a current I_A can be derived from the volt–ampere relation as

$$R_{dc} = \frac{V_A}{I_A} = \frac{V_A}{i_s(\epsilon^{V_A e/kT} - 1)} \quad \text{ohms} \qquad (2\text{-}76)$$

For ac or variation of current around an operating point, it is the slope of the curve that represents the *dynamic resistance*

$$\frac{1}{r_{ac}} = \frac{di_A}{dv_A} = i_s \frac{e}{kT} \epsilon^{v_A e/kT} = \frac{e}{kT}(i_A + i_s)$$

At 27°C (300°K) this becomes

$$r_{ac} = \frac{0.026}{i_A + i_s} \cong \frac{0.026}{i_A} \qquad \text{ohms} \quad (2\text{-}77)$$

where the current is in amperes.

2-19 *CAPACITANCE EFFECTS IN THE REVERSE-BIASED JUNCTION*

Poisson's equation relates charge density and potential, and may be developed by considering an infinitesimal cube of sides dx, dy, dz located at a point x in a region containing space charge of density ρ, as in Fig. 2-28. The charge enclosed in the cube is

$$dq = \rho \, dx \, dy \, dz$$

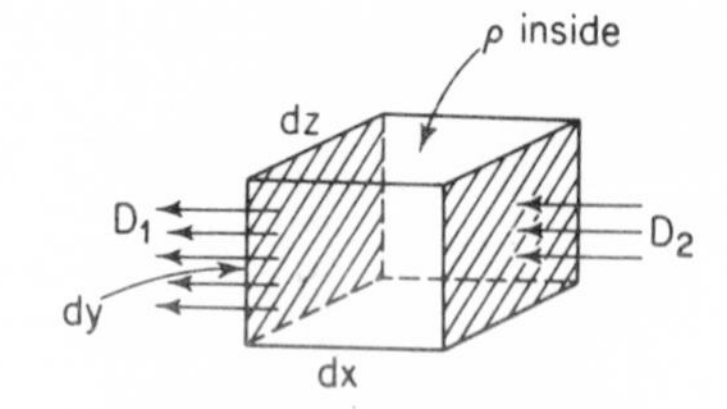

Figure 2-28. Development of Poisson's equation.

Assuming one-dimensional or x-directed flux, and defining inward flux as negative, then with D_2 as the flux density at the right face, the inward flux is $-D_2 \, dy \, dz$. With D_1 as the flux density at the left face, the leaving flux is $D_1 \, dy \, dz$. By definition, the net outward flux, $(D_1 - D_2) \, dy \, dz$, is equal to the enclosed charge, or

$$(D_1 - D_2) \, dy \, dz = \rho \, dx \, dy \, dz \qquad (2\text{-}78)$$

Dividing by dx,

$$\frac{(D_1 - D_2)}{dx} = \frac{dD}{dx} = \rho \qquad (2\text{-}79)$$

since $(D_1 - D_2)/dx$ is the rate of change of flux density in the x direction.

By definition, $D = \epsilon \mathscr{E} = -\epsilon \, dV_x/dx$, where V_x is the potential at point x, and $\epsilon = \epsilon_v \epsilon_r$. Therefore

$$\frac{d^2 V_x}{dx^2} = -\frac{\rho}{\epsilon} \qquad (2\text{-}80)$$

is *Poisson's equation*, in one dimension.

We will now apply Poisson's equation to determine the capacitance represented by the transition region of a reverse-biased junction.

The transition region in a *pn* junction was described in Section 2-14. Upset of the equilibrium condition by application of a reverse potential increases the effect of the barrier potential and causes an enlargement or widening of the transition region. With the resultant separation of charge, the transition region acts as a capacitance.

While previous discussion implied that the characteristics of the semiconductor change from fully p to fully n in a short transition zone, it is possible to manufacture transitions with impurity gradients varying from the *gradual junction* of a grown crystal, to an *abrupt* or step form resulting from alloying a p-forming impurity onto an n-germanium wafer. Some of the p impurity diffuses shallowly into the germanium, converting it to p form; the remainder of the impurity serves as a contact. The concentration N_D of impurity atoms in the n material is relatively constant up to the limit of diffusion of the p material, as illustrated in Fig. 2-29(a). The p region will have the higher conductivity, or $N_A \gg N_D$.

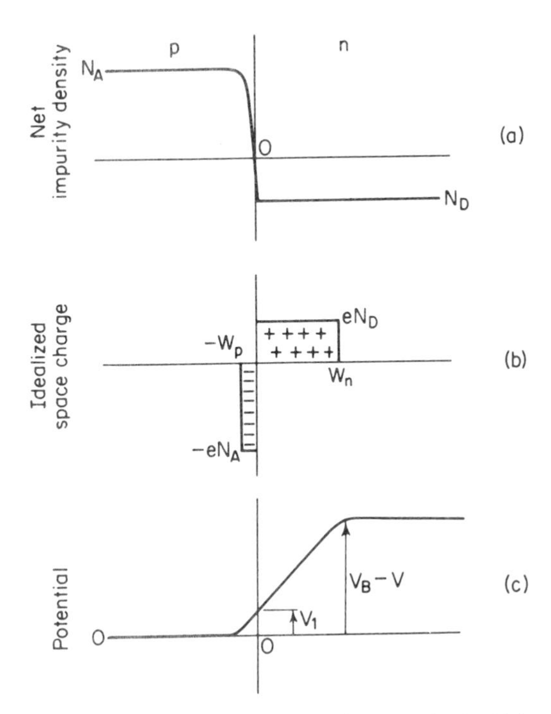

Figure 2-29. Variation of space charge and potential through the abrupt junction.

The distribution of the space charge in an abrupt junction may be idealized to the form indicated in Fig. 2-29(b), to provide a charge variation with distance which is adaptable to mathematical analysis. If the area of the transition region is large in extent, the field flux will be directed along x and Eq. 2-80 will apply. For the p side of the junction

$$\frac{d^2 V_p}{dx^2} = -\frac{\rho}{\epsilon} = \frac{eN_A}{\epsilon}, \qquad x < 0 \tag{2-81}$$

and for the n side

$$\frac{d^2 V_n}{dx^2} = -\frac{eN_D}{\epsilon}, \qquad x > 0 \tag{2-82}$$

A double integration of Eq. 2-81 and evaluation of the first constant at $x = -W_p$, where $dV_p/dx = 0$, leads to

$$V_p = \frac{eN_A}{\epsilon}\left(\frac{x^2}{2} + W_p x\right) + C_p \tag{2-83}$$

Similarly, for Eq. 2-82, with $dV_n/dx = 0$ at $x = W_n$, we have

$$V_n = \frac{eN_D}{\epsilon}\left(-\frac{x^2}{2} + W_n x\right) + C_n \tag{2-84}$$

The voltage is continuous at $x = 0$, or $V_p = V_n = V_1 = C_p = C_n$. At $x = -W_p$ the potential V_p is zero, and

$$V_1 = \frac{eN_A}{2\epsilon} W_p^2 \tag{2-85}$$

The total charge must be zero for neutrality, and it is possible to write

$$N_A W_p = N_D W_n \tag{2-86}$$

The potential across the junction is $V_B - V$, where V will be negative for reverse bias. Using the above neutrality condition, and writing $V_p = V_B - V$ at $x = W_n$, there results

$$V_B - V = \frac{e}{2\epsilon}(N_D W_n^2 + N_A W_p^2) = \frac{e}{2\epsilon} N_A W_p^2\left(1 + \frac{N_A}{N_D}\right) \tag{2-87}$$

The effective widths of the transition regions are

$$W_p = \left[\frac{2\epsilon(V_B - V)}{eN_A(1 + N_A/N_D)}\right]^{1/2} \tag{2-88}$$

$$W_n = \left[\frac{2\epsilon(V_B - V)}{eN_D(1 + N_D/N_A)}\right]^{1/2} \tag{2-89}$$

As stated, $N_A \gg N_D$ for a useful alloyed junction, and by Eq. 2-86 the transition region extends largely into the n or high-resistivity material. The p side of the junction is of lower resistivity and very thin, so it is reasonable to assume that the potential appears primarily across the n transition region.

The dynamic or incremental capacitance represented by the reverse-biased transition region is found as the change of charge per unit voltage change, or $C_T = \partial Q/\partial V$. Using $eN_D = \sigma_e/\mu_e$ from Eq. 2-36 and with $N_D/N_A \ll 1$, Eq. 2-89 becomes

$$W_n = \left[\frac{2\epsilon\mu_e(V_B - V)}{\sigma_e}\right]^{1/2} \tag{2-90}$$

The charge on the n side is $Q = eN_D W_n = \sigma_e W_n/\mu_e$. The capacitance of the abrupt junction under reverse bias is then

$$C_T = \frac{\partial Q}{\partial W}\frac{\partial W}{\partial(V_B - V)} = \epsilon\left[\frac{\sigma_e}{2\epsilon\mu_e(V_B - V)}\right]^{1/2} \quad \text{F/m}^2 \tag{2-91}$$

This result is similar to the expression for a parallel plate capacitor in which $C_T = \epsilon_v \epsilon_r A/W$ farads, except that W is a voltage function.

The transition width W_n varies as the square root of the applied voltage, an increased voltage sweeping out more charge and making the dielectric layer thicker. Dependence is logically on the n resistivity because of the assumptions, and because of the thinness of the p region. Typical values of C_T of an abrupt junction will be found in the range from 30 to 100 picofarads.

For a *gradual junction* the distribution of impurities may be as indicated in Fig. 2-30(a) with the space charge variation idealized to the situation indicated in (c). Assuming all donors and acceptors activated, the space charge density can be expressed as

$$\rho = ax, \qquad \frac{-W}{2} < x < \frac{W}{2} \tag{2-92}$$

where the variation rate of the impurity concentration is

$$a = -\frac{e(N_A + N_D)}{W_T}$$

Substituting for ρ in Poisson's equation gives

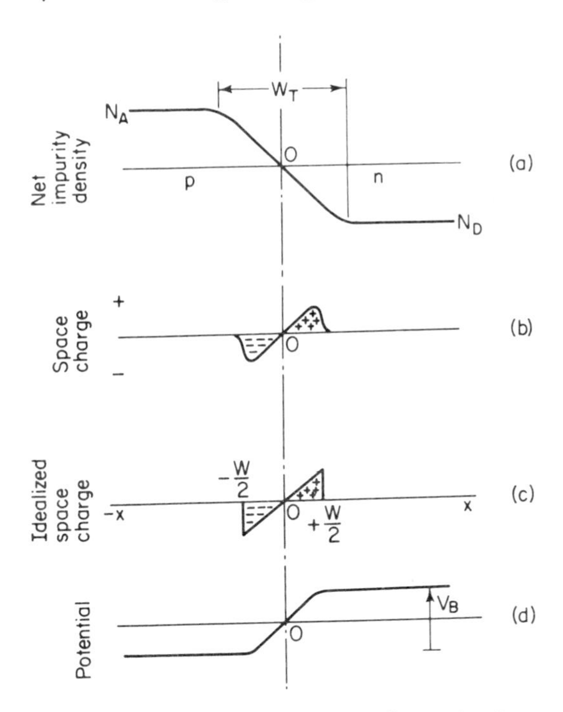

Figure 2-30. Gradual junction: (a) charge density distribution; (b) space charge distribution; (c) assumed space charge distribution; (d) potential variation.

$$\frac{d^2V_x}{dx^2} = -\frac{ax}{\epsilon}, \qquad \frac{-W}{2} < x < \frac{W}{2} \tag{2-93}$$

as the applicable expression.

With $dV_x/dx = 0$ at $x = \pm W/2$, and $V_x = 0$ at $x = 0$, a double integration yields an expression

$$V_x = \frac{ax}{2\epsilon}\left(-\frac{x^2}{3} + \frac{W^2}{4}\right)$$

and this variation appears in Fig. 2-30(d). Evaluation of the potential at $x = \pm W/2$ and subtraction leads to an expression for the total potential as $V_B - V = aW^3/12\epsilon$, from which the width of the transition region is

$$W = \left[\frac{12\epsilon(V_B - V)}{a}\right]^{1/3} \tag{2-94}$$

The maximum charge density occurs at $x = \pm W/2$, where $\rho_{\max} = aW/2$. The charge on one side of the junction is $Q = aW^2/8$, or

$$Q = \frac{a^{1/3}}{8}[12\epsilon(V_B - V)]^{2/3}$$

The dynamic transition region capacitance for the gradual junction is then

$$C_T = \frac{\partial Q}{\partial(V_B - V)} = \epsilon\left[\frac{a}{12\epsilon(V_B - V)}\right]^{1/3} \qquad \text{F/m}^2 \tag{2-95}$$

The capacitance varies as the inverse cube root of the applied voltage rather than the square root, as for the abrupt junction. The transition region is symmetrical around the junction plane.

The capacitance equation might also be witten

$$C_T = \epsilon\left[\frac{e(N_A + N_D)}{12\epsilon W_T(V_B - V)}\right]^{1/3} = \epsilon\left[\frac{(\sigma_e/\mu_e) + (\sigma_h/\mu_h)}{12\epsilon W_T(V_B - V)}\right]^{1/3} \tag{2-96}$$

and again this is the form expected for a parallel-plate capacitor, but with W as a voltage function. The capacitance can be controlled by the charge density gradient a and a lower capacitance achieved than for the abrupt junction. Note that $\epsilon = \epsilon_v\epsilon_r$ throughout.

A typical value for C_T with a graded junction lies in the range from 2 to 50 picofarads.

2-20 THE DIFFUSION OR STORAGE CAPACITANCE

With changes in forward voltage, there must be an incremental change in the hole density at the n face of the junction, and a similar electron density change at the p face. This change in charge density present is equivalent to a *storage* or *diffusion capacitance*, C_D, since the amount of stored charge changes with applied voltage, as in a capacitor.

As might be expected, the capacitance C_D is related to the diffusion length and the diffusion constant. That is,

$$C_D = \frac{\partial Q}{\partial V} = \frac{L_e^2}{D_e r_e} + \frac{L_h^2}{D_h r_h} = \frac{\tau_e}{r_e} + \frac{\tau_h}{r_h} \tag{2-97}$$

The resistances r_e and r_h are the electron and hole components of the diode ac resistance of Eq. 2-77.

The diffusion capacitance may have values from ten to several hundred picofarads, and this serves to limit the frequency range of the junction diode rectifier. The capacitance will also be of importance in high-frequency applications of the transistor.

In the use of the diode in switching circuits the emf is abruptly changed from forward to reverse, and the diffusion capacitance must completely discharge. The passage of displacement current after the potential has reversed often degrades pulse or step wave forms of current, and much development has gone into special switching diodes in which this storage effect is minimized.

The diffusion capacitance is effective with forward potential and is created by the minority carriers; conversely, the transition capacitance is effective with reverse bias and is due to majority charges.

2-21 THE ZENER DIODE

At some value of reverse voltage, about −140 V for the diode of Fig. 2-31, the theoretical diode voltage–current relation fails and the reverse current increases rapidly. This is known as *breakdown* of the diode. The voltage at which breakdown occurs is stable, and with its low internal dynamic resistance a properly designed diode can be used as a voltage regulator. The device is then known as a *Zener diode*, after the man who conducted early studies on the phenomenon.

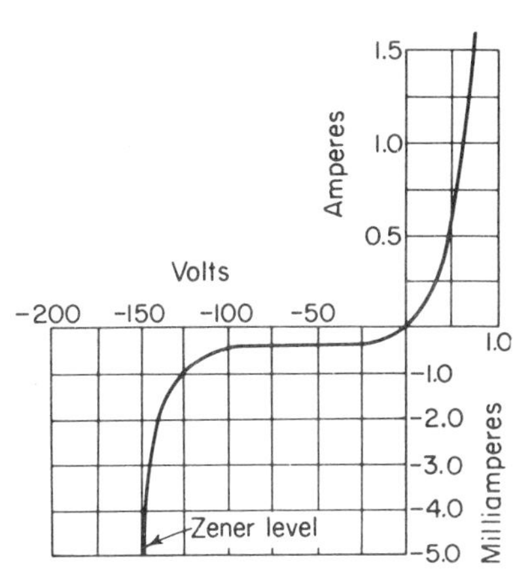

Figure 2-31. Junction diode characteristic, showing Zener level.

Choice of material and doping permits design of units in which the breakdown occurs at reverse voltages from about three volts to several hundred volts. The breakdown phenomenon is reversible and not damaging, if the power loss is limited and a safe operating temperature maintained.

The original Zener theory proposed that the sudden rise of current at a critical voltage was due to rupture of covalent bonds in atoms in the transition region, because of the high field intensity created by the sum of the barrier potential and the reverse voltage. For silicon units, field intensities of 1.5×10^6 V/cm would be required, and with depletion layers of 10^{-4} to 10^{-5} cm, these fields could be achieved at potentials of 15 to 150 V.

However, in usual junctions the Zener theory seems less likely as an explanation than does that of *avalanche breakdown*. It is postulated that the electrons and holes of the reverse saturation current are accelerated to high drift velocities by the reverse potential across the diode. At a critical field level the moving electrons acquire sufficient energy to break valence bonds upon collision with the fixed atoms; after each such collision there exists a new electron–hole pair. These charges are also accelerated and create more pairs upon collision, and the current rapidly goes into an avalanche condition. The critical requirement is that the field intensity be high enough to permit a charge to acquire the needed energy in a distance less than one mean free path.

The barrier height, being a function of the Fermi levels and therefore of the impurity concentrations, is constant, and therefore the voltage across the transition region increases directly with applied voltage V. The transition width increases as either $V^{1/2}$ or $V^{1/3}$. Therefore, the field intensity across the transition layer increases as $V^{1/2}$ or $V^{2/3}$, and can supply the basic mechanism for avalanche breakdown.

Another physical mechanism appears in heavily doped regulator diodes designed to operate at 5 volts or less. Electrons on the p side of the junction may have energies equal to empty conduction-band levels on the n side, and some of these electrons apparently pass directly through the barrier, as a result of their wave properties. The *tunneling* action occurs at a definite voltage and so the devices are suited to regulator use in this voltage range. This phenomenon will be discussed further in the next section.

Figure 2-32 shows an enlarged section of the characteristic of several diode regulators. Since the temperature coefficients of the tunneling and avalanche processes are opposite in effect, it is possible to find a region around the five volt level at which the effects balance and produce a diode with a zero temperature coefficient of voltage change.

Regulators of Zener type may be used as series elements to provide a fixed voltage drop, or they may be employed in a shunt circuit as in Fig. 2-33. The operating current will usually approximate 20 per cent of the maximum rating, $I_{\max}$. The value of R can then be found as

$$R = \frac{E - E_d}{I_L + 0.2\, I_{\max}} \tag{2-98}$$

The total of the diode and load currents remains constant, and the voltage is maintained at E_d.

The dynamic internal resistance of the diode is that shown to changing

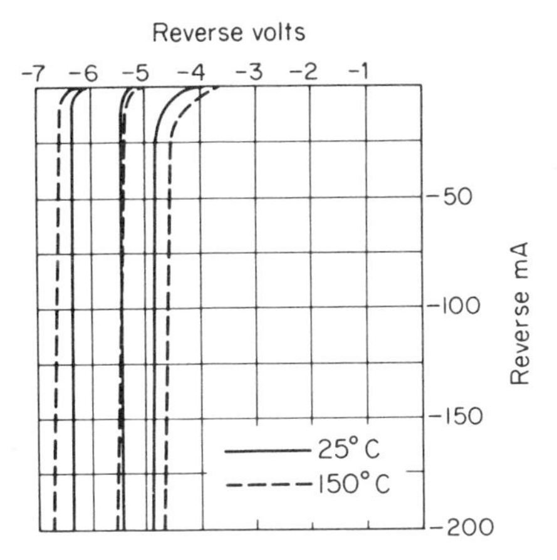

Figure 2-32. Temperature variation of the Zener regulating level.

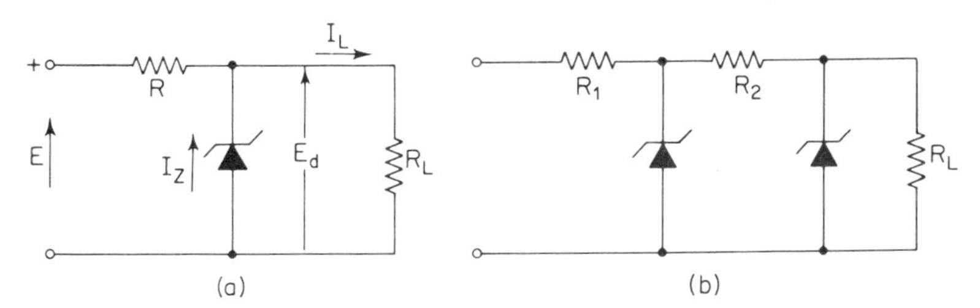

Figure 2-33. Zener diode regulator circuits.

currents, and will be low over the near-vertical portion of the diode curve. Diodes regulating at five volts may have dynamic resistances of an ohm or less, which may increase to 100 ohms or more for units designed to regulate at a level of several hundred volts.

2-22 THE TUNNEL DIODE

In 1957 Esaki discovered that certain heavily doped diodes displayed tunneling effects in the forward direction. With abrupt junctions and depletion regions only 10^{-6} to 10^{-7} cm thick, the volt–ampere curves at low voltages can take the form of Fig. 2-34. The curve shows a region of negative internal resistance, or decreasing voltage for an increasing current, from about 0.1 to 0.2 volts, and this provides a useful circuit property.

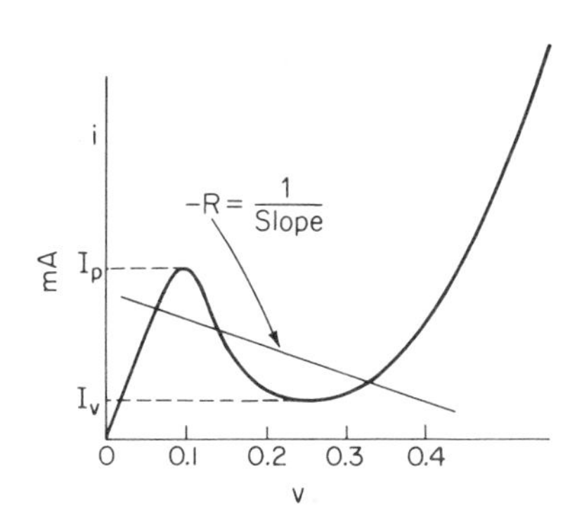

Figure 2-34. Volt–ampere curve of a tunnel diode.

Because of the heavy doping of the materials the Fermi level for the n material will move into the conduction band above E_C, and the Fermi level for the p material will move into the valence band below E_V. The energy level situation for an unbiased juction is then as in Fig. 2-35(a).

Charge tunneling through the barrier requires the transfer of a charge from an occupied level across the gap to an open level at equal energy. At (a) there are electrons in the conduction band in n but these appear opposite filled levels in p and tunneling is small. As a forward voltage is applied, the n levels are raised relative to the p levels, as in Fig. 2-35(b), and conduction-band electrons face valence-band holes across the barrier. Tunneling occurs and the current increases to the peak value, I_p. As further forward voltage is applied, the filled electron levels gradually move above the open valence levels and the current decreases due to reduced tunneling. This decrease of current with increase of voltage is characteristic of a negative resistance. The valley current I_v is then reached and beyond I_v the energy levels become as in a normal forward-biased junction; the charges diffuse across the junction and the current–voltage relation follows the usual diode law.

The peak current I_p and the valley current I_v are stable operating points and their ratio may reach 15:1 in some diodes. In switching between these states, the time of transition may be only a few nanoseconds, due to the quan-

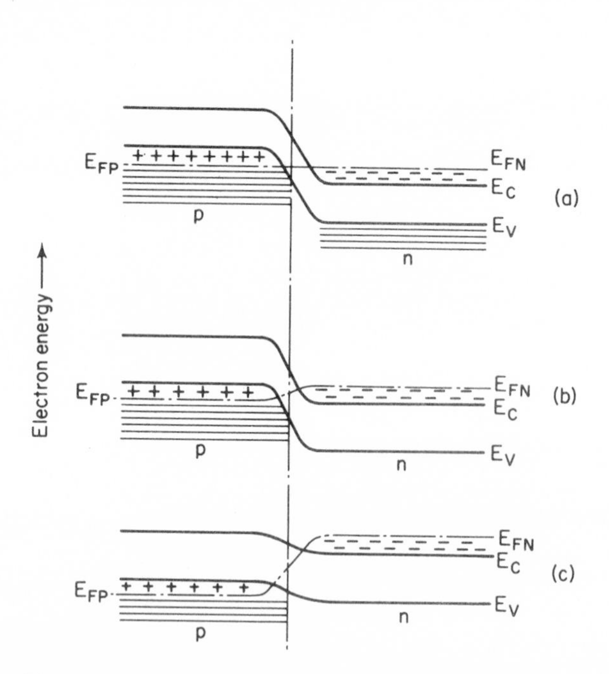

Figure 2-35. (a) Zero bias; (b) small forward bias; (c) large forward bias.

tum-mechanical or wave nature of the electron transfers through the barrier.

The negative resistance portion of the characteristic adds another circuit element to the usal R, L, C combinations. Even though limited by low voltage and high capacitance, the diodes are useful as oscillators and amplifiers.

2-23 THE POINT-CONTACT DIODE

Early diodes were of point-contact form, a tiny wire "cat-whisker" contacting the surface of a natural semiconductor such as galena or iron pyrites. These devices were notably unstable, both mechanically and electrically, and modern forms appear as in Fig. 2-36. The semiconductor wafer is only a few millimeters square and a fraction of a millimeter thick, and the contact point is one or two mils in diameter, of aluminum or gold-gallium alloy.

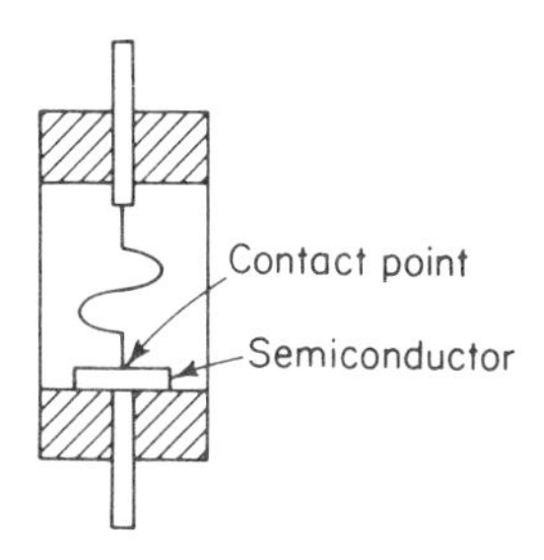

Figure 2-36. Point-contact diode.

The semiconductor is of n material and, during processing, a high current pulse is passed between contact and wafer, welding the wire in place. The welding operation forms a small p region around the contact, and the theory of operation follows that given for a junction unit.

The point-contact diode has a low capacitance and is suited to operation at frequencies of 10 gigahertz or more. The general shape of the volt–ampere curve is the same as for a junction unit, but currents and power ratings are much reduced. The point-contact diode has considerably lower reverse current and therefore its *rectification ratio*, or ratio of forward to reverse current at a given applied voltage, may be as high as 10^7. The use of gold in the contact is dictated by a desire for increased switching speed, since gold tends to promote recombination and reduces storage time effects.

2-24 THE HALL EFFECT

Since conduction theory was scant, observation by Hall in 1879 of phenomena due to an apparent flow of charge opposite to that present in conventional currents could be reported only as a conduction anomaly. Now known as the *Hall effect* and considered due to hole movement, the experiment provides a means of measurement of charge mobility, and is applied in several devices for the measurement of magnetic flux density.

Figure 2-37 illustrates a conductor in which holes are assumed to move in the x direction, representing a current density $J_x = I_x/yz$. A uniform magnetic field of density B_y is introduced perpendicular to the conductor

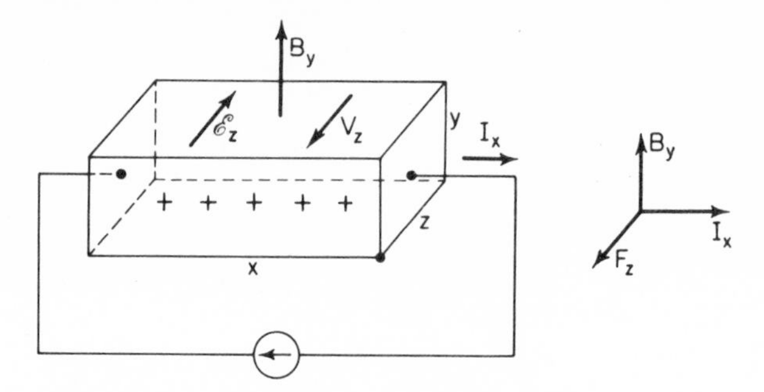

Figure 2-37. Hall effect on a current of holes.

and current. Hole movement through the magnetic field produces a force on the holes

$$\mathbf{f} = \mathbf{v} \times \mathbf{B}$$

directed toward the near face, and creating a surface charge of holes as indicated. An electric field $\mathscr{E}_z$ is then established by this charge, and equilibrium is reached when the electric field force on the charges balances the magnetic force. That is,

$$q\mathscr{E} = q(\mathbf{v} \times \mathbf{B}) \tag{2-99}$$

In the case diagrammed, the electric field is directed on $-z$ and

$$\mathscr{E}_z = -v_x B_y \tag{2-100}$$

The field establishes a potential across the conductor as

$$V_z = v_x B_y z \tag{2-101}$$

A current composed of electrons would reverse the polarity of the potential.

At equilibrium the ratio of the electric field intensity per unit current density to the magnetic field intensity is defined as the Hall coefficient, R_H, That is,

$$R_H = \frac{\mathscr{E}_z}{J_x B_y} = \frac{1}{nq} \tag{2-102}$$

after use of Eq. 2-100. Thus the charge density can be determined.

Since the current I_x and voltage V_z can be readily measured, it is possible to express R_H in a more convenient form as

$$R_H = \frac{V_z}{z}\frac{yz}{I_x B_y} = \frac{V_z y}{I_x B_y} \tag{2-103}$$

The conductivity is $\sigma = nq\mu$, and the charge mobility in the sample can be written as

$$\mu = \sigma R_H = \frac{J_x}{\mathscr{E}_x} \frac{V_z y}{I_x B_y} = \frac{x V_z}{z V_x B_y} \tag{2-104}$$

Thus the sign of the carriers, their density, and their mobility are measurable. It is evident that the Hall effect allows measurement of the conduction characteristics due to the majority carriers, and if both types of carriers are present in comparable numbers, the experiment will fail.

2-25 LARGE-AREA RECTIFIERS

Older forms of semiconductor rectifiers employ the barrier-layer properties of metallic contacts to cuprous oxide, selenium, and copper sulfide. These devices consist of a semiconductor bonded onto a metal base. The forward direction of current is with the semiconductor positive; this implies that the electron flow is from the metal, and the operation can be explained if the semiconductor is considered of p properties.

In the copper-oxide rectifier, a layer of cuprous oxide is formed on a copper base by oxidation at high temperature. Each disk, as shown in Fig. 2-38, is limited to about 8 V in the reverse direction, and units are stacked

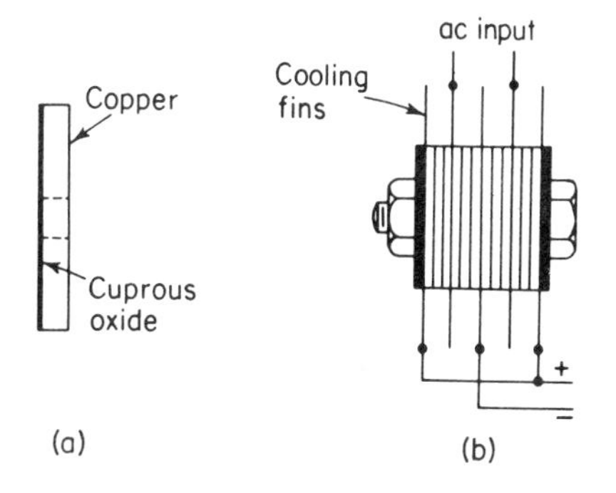

Figure 2-38. (a) Single copper–cuprous-oxide disk; (b) stacked rectifier unit.

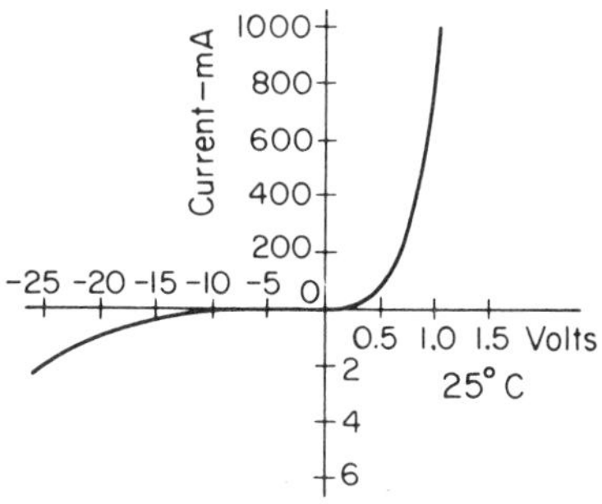

Figure 2-39. Selenium volt–ampere characterisic.

Table 2-4

RECTIFIER PROPERTIES

Material	Current Density, A/cm^2	Maximum Temperature, °C	Reverse Volts, Maximum
CuO	0.04	45	8
Selenium	0.10	45	26
Mg-CuS	6	35	3.5

in series on an insulated rod for higher voltages. Possible current is dependent on disk area, and disks may be paralleled for higher currents. The rise of reverse current limits the operating temperature to about 45°C.

Selenium deposited on an iron or aluminum base also provides a useful rectifier and a typical volt–ampere curve appears in Fig. 2-39. The plates will withstand about 26 V reverse emf, and may be operated up to 100°C.

Properties of the several types are summarized in Table 2-4.

2-26 OTHER SEMICONDUCTOR MATERIALS

It has been emphasized that each atom of silicon or germanium has four valence electrons which form covalent bonds with four neighbor atoms, so that each atom is at the center of a tetrahedron. These factors: valence electron to atom ratio of four, the covalent bonds, and the tetrahedral structure, are found to serve as criteria in the search for additional semiconducting materials.

A gallium atom (Group III) has three valence electrons, arsenic (Group V) has five valence electrons, and these elements form gallium arsenide, GaAs. This material has a four-to-one valence electron to atom ratio, forms tetrahedral crystals, and is a valuable semiconductor. Other III-V group compounds can be formed and many are semiconductors; characteristics for some are given in Table 2-5. It is possible to extend the search by applying similar criteria to compounds involving the II-VI group elements, or to multiple combinations employing atoms from three or more groups, many occurring in nature as minerals. Such multiple compounds include Cu_3AsS_4, $AgAsSe_2$, and $AgSbTe_2$.

Table 2-5

INTERMETALLIC COMPOUNDS

Material	E_G (eV)	μ_e (m^2/V-s)	μ_h (m^2/V-s)
CdS	2.4	0.02	. . .
AlSb	1.52	>0.04	>0.04
GaAs	1.35	0.40	0.025
InP	1.25	0.34	0.065
GaSb	0.68	0.40	0.07
InAs	0.35	3.00	0.02
PbTe	0.27	0.12	0.048
PbSe	0.22	0.12	0.087
InSb	0.18	7.70	0.12
For comparison:			
Si	1.15	0.12	0.05
Ge	0.75	0.39	0.19

Replacement of an atom by a heavier atom of the same group will reduce the energy gap, the melting point, and the thermal conductivity, and will increase the mobility of the charges.

After a new compound has been prepared and found to be stable, measurements of thermoelectric power, thermal conductivity, the Hall coefficient, and the rectification properties contribute information toward possible application. Materials for thermoelectric cooling, high-temperature diodes or transistors, and solar cells are especially in demand today.

PROBLEMS

2-1. Compute and plot the Fermi function $F(E)$ for copper at 50°K, if $N \cong 10^{29}$ valence electrons per m^3.

2-2. A sample of germanium has intrinsic resistivity of 0.10 ohm-m at 85°C. If the mobility of the electrons is 0.36 m^2/V-s and that of holes is 0.17 m^2-s, compute the number of germanium atoms per m^3 having broken valence bonds at this temperature.

2-3. A germanium wafer of cross section 3 mm × 1 mm has electron and hole mobility as in Problem 2-2. If there is one ionized antimony atom per 10^7 germanium atoms, compute the conductivity of the material, neglecting intrinsic conduction.

2-4. A silicon wafer with p resistivity of 100 ohm-m due to gallium has dimensions of 3 mm × 0.5 mm. With electron mobility of 0.12 m^2/V-sec, hole mobility of 0.04 m^2/V-sec, compute the number of silicon atoms present per ionized gallium atom.

2-5. Phosphorus is added to a pure silicon sample in the amount of 10^{13} atoms per cm^3. If all donor atoms are activated, what is the resistivity at 20°C?

2-6. Repeat Problem 2-5 for germanium.

2-7. Why does silicon absorb light of wave length shorter than a certain critical value, and become transparent for light of longer wave length? What is the critical wave length for silicon? For germanium?

2-8. Determine the donor concentration in n germanium having a resistivity of 0.015 ohm-m, $\mu_e = 0.36$ m^2/V-s. Repeat for P germanium of equal resistivity, $\mu_h = 0.17$ m^2/V-s.

2-9. Compute the rectification ratio (ratio of forward to reverse current at the same voltage) for a germanium pn junction at 0.15 V, and $i_s = 25 \times 10^{-6}$ A; $T = 300$°K.

2-10. Plot the forward and reverse volt–ampere curves for the junction of Problem 2-9, over the range of 2 V forward to 10 V reverse.

2-11. With germanium having the characteristics of Table 2-2, calculate the intrinsic resistivity at 350°K.

2-12. Intrinsic germanium has $n_i = 2.4 \times 10^{19}$ carriers/m^3 at 300°K. Find the number of grams of indium which must be added per kilogram of germanium to reduce the resitivity to 0.04 ohm-m, assuming complete activation.

2-13. With germanium of 0.54 ohm-m intrinsic resistivity, an impurity concentration of one antimony atom per 10^8 germanium atoms is added. Find the Fermi energy level after the addition.

2-14. A germanium diode has an area of 1 mm^2. The n region has a resistivity of 0.05 ohm-m, a lifetime of 50 μs, and a length of 1.5 mm. The conductivity of the p region is high. Under this assumption, compute the reverse saturation current at 300°K.

2-15. For the diode of Problem 2-14, compute the forward voltage drop at a current of 10 A. Also compute the static resistance at this current.

2-16. The diode of Problem 2-14 has a junction of abrupt form. Calculate the thickness of the barrier region, and determine the junction capacitance at -10 V, -100 V.

2-17. Germanium has one arsenic atom added per 10^8 germanium atoms. How many holes and electrons are present for conduction, at 300°K?

2-18. A silicon *pn* junction has an area of 0.01 cm^2, with $N_A = 10^{25}$/m^3 in the p region, $N_D = 10^{22}$/m^3 in the n region. The gradual junction doping extends over a distance $W_T = 2 \times 10^{-4}$ cm. Since avalanching occurs with electric fields of 4.5×10^7 V/cm, what will be the avalanche breakdown voltage of this diode?

REFERENCES

1. Seitz, F., *The Physics of Metals*. McGraw-Hill Book Company, New York, 1943.

2. Shockley, W., *Electrons and Holes in Semiconductors*. D. Van Nostrand Co., Inc., Princeton, N. J., 1950.

3. Hall, R. N., "Germanium Rectifier Characteristics," *Phys. Rev.*, Vol. 83, p. 228 (July 1951).

4. Conwell, E. M., "Properties of Silicon and Germanium," *Proc. IRE*, Vol. 40, p. 1327 (1952); also Vol. 46, p. 1281 (1958).

5. Shockley, W., "Transistor Electronics: Imperfections, Unipolar, and Analog Transistors," *Proc. IRE*, Vol. 40, p. 1303 (1952).

6. Breckenridge, R. G., "Semiconducting Intermetallic Compounds," *Phys. Rev.*, Vol. 90, p. 488 (May 1953).

7. Middlebrook, R. D., *An Introduction to Junction Transistor Theory*. John Wiley & Sons, Inc., New York, 1957.

8. Shive, J. N., *The Properties, Physics, and Design of Semiconductor Devices*. D. Van Nostrand Co., Inc., Princeton, N. J., 1957.

9. Dekker, A. J., *Solid-State Physics.* Prentice-Hall, Inc., Englewood Cliffs, N. J., 1957.

10. Dewitt, D., and A. L. Rosoff, *Transistor Electronics.* McGraw-Hill Book Company, New York, 1957.

11. Van der Ziel, A., *Solid-State Physical Electronics.* Prentice-Hall, Inc., Englewood Cliffs, N. J., 1957.

12. Esaki, L., "New Phenomena in Narrow Ge PN Junctions," *Phys. Rev.*, Vol. 109, p. 603 (1958).

13. Nussbaum, A., *Semiconductor Device Physics.* Prentice-Hall, Inc., Englewood Cliffs, N. J., 1962.

14. Adler, R. B., A. C. Smith, and R. L. Longini, *Introduction to Semiconductor Physics* (SEEC, Vol. 1). John Wiley & Sons, Inc., New York, 1964.

3

ELECTRON EMISSION: THE VACUUM TUBE

The valence electrons of a metal possess sufficient energy to permit thermal movements within the crystal lattice of the conductor, but additional field forces exist at the surface of a metal which restrict the charge movements to the metal interior. These forces at the surface must be overcome by the electrons, if they are to pass through the surface. The process by which electrons are freed of the restrictive surface forces and emerge into the surrounding space is called *electron emission*, and this process is basic to most vacuum and gaseous forms of electron devices.

All forms of electron emission appear to involve the same basic phenomena—the provision of additional energy to overcome the surface forces. The various forms of emission are differentiated, however, by designation of the means by which the additional energy is supplied. If the additional energy is supplied by thermal means, the process is known as *thermionic emission*; use of radiant energy or light produces *photoelectric emission*; application of electric fields sufficient to directly extract the electrons produces *high-field* or *autoelectronic emission*; and emission caused by mechanical bombardment with a primary beam of electrons or other particles is known as *secondary emission*.

3-1 SURFACE FORCES; THE WORK FUNCTION

Because of crystal growth irregularities or surface treatment, there will be a surface roughness of the order of atomic diameters. The surface atoms will have unfilled valence bonds because of the absence of neighbor atoms. To be emitted, an electron must therefore have sufficient energy to exceed these indeterminate near-surface forces. These forces will be a function of surface condition, deposited layers of metals, or adsorbed gases.

At greater distances from the surface the forces due to individual surface atoms will compensate, and the force field existing becomes due to attraction from the *image* or induced surface charge. When the electron leaves the surface, a positive charge is induced on the surface, as in Fig. 3-1(a). Since the metal surface must be an equipotential in the field, there is no difference in field distribution if the surface charge is assumed concentrated at $+e$, the image, as in Fig. 3-1(b). Since the field at the electron is unchanged, the force on the electron is also unchanged, and Fig. 3-1(b) can be used to compute the force existing.

From Coulomb's law the force between the electron $(-e)$, and its image $(+e)$ is

$$f_x = \frac{(+e)(-e)}{4\pi\epsilon_v(2x)^2} = \frac{-e^2}{16\pi\epsilon_v x^2} \tag{3-1}$$

where $\epsilon_v = 10^7/4\pi c^2$, the permittivity of space. The negative force is to the left, or in the negative x direction. It should be remembered that the surface is assumed plane, and this is only true at distances large with respect to atomic diameters.

If the electron is removed to an infinite distance, or emitted, the work done against f_x must be

$$E = -\int_x^\infty f_x\,dx = \frac{e^2}{16\pi\epsilon_v x} \qquad \text{joules} \tag{3-2}$$

indicating that the work is an inverse function of distance, as shown by Fig. 3-2.

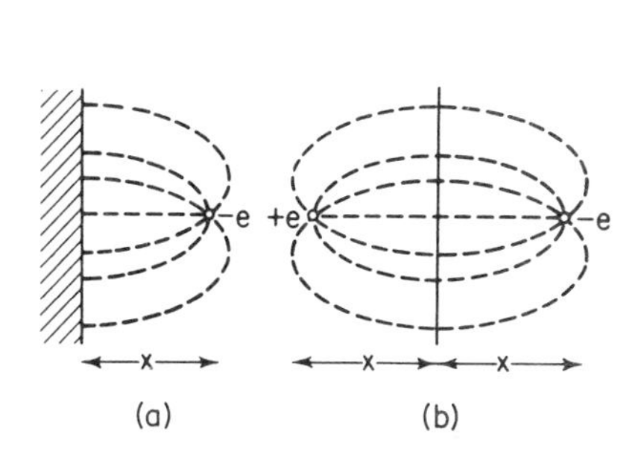

Figure 3-1. The image principle.

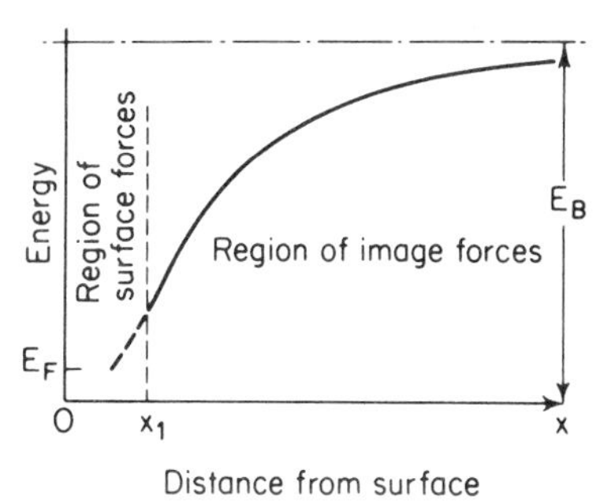

Figure 3-2. Energy conditions near the surface.

The work done by the electron, or the energy required for the electron to overcome the forces at the surface and pass into space, is called the surface barrier energy, E_B.

To be emitted, an electron must possess a kinetic energy at least equal to this sum of the energy required to overcome the near-surface forces, and the energy needed to overcome the far-from-surface image forces. Any excess

Table 3-1

VALUES OF THE EMISSION CONSTANTS†

Element	A_0 (A/m²/deg²)	b_0 (°K⁻¹)	E_W (eV)	Melting Point (°K)
Calcium	60.2×10^4	37,100	3.2	1083
Carbon	60.2×10^4	54,400	4.7	....
Cesium	16.2×10^4	21,000	1.81	299
Copper	65×10^4	47,600	4.1	1356
Molybdenum	60.2×10^4	49,900	4.3	2895
Nickel	26.8×10^4	58,000	5.0	1725
Tantalum	60.2×10^4	47,600	4.1	3123
Thorium	60.2×10^4	39,400	3.4	2118
Tungsten	60.2×10^4	52,400	4.52	3643

†J. A. Becker, *Rev. Mod. Phys.*, Vol. 7, p. 95 (1935).

of kinetic energy over that needed to override this barrier energy E_B remains with the electron as kinetic energy after emission.

The Fermi-Dirac distribution predicts that many electrons do possess an energy near the value of E_F, even at absolute zero. Thus it is only necessary to supply additional energy E_W, where

$$E_W = E_B - E_F \tag{3-3}$$

to secure the emission of an electron. The energy E_W is known as the *work function* of a surface, and is the energy needed by an electron at absolute zero to override the surface forces and cause emission.

The work function depends on the material, its surface regularity, adsorbed films, and surface impurities. The work function of a given material may vary from point to point over the surface, as a result of variations in the near-surface forces. Values of work function given in Table 3-1 are believed representative, but surface conditions enter into the determination, and the work function becomes dependent on experimental procedures.

3-2 *THE DUSHMAN EQUATION FOR THERMIONIC EMISSION*

The energies possessed by the valence electrons of a metal result in randomly directed velocities. If an electron is to be emitted, its velocity must have a sufficiently large component toward the surface to overcome the surface force. Consideration of electrons with such outward-directed velocity components leads to an electron energy distribution like that of Fig. 3-3 for absolute zero

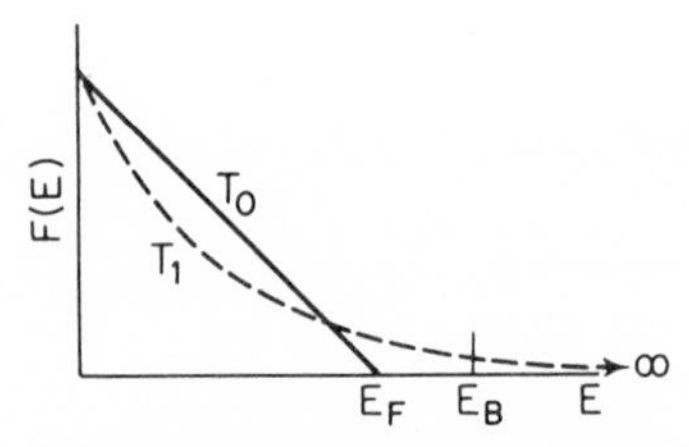

Figure 3-3. Energy distribution among electrons having surface-directed velocity components.

T_0, and a higher temperature T_1. The curve for T_1 approaches the axis asymptotically, and so there will be some electrons with energies exceeding E_B. These are the electrons capable of passing through the surface and being emitted.

To find the current emitted per unit surface area, it is necessary to determine the equation for the number of electrons having an energy E at a given temperature T from thermodynamic considerations, integrate this equation from E_B to infinity to obtain the number of electrons capable of passing the surface, and multiply by the charge and the average velocity of the electrons. An equation of the form

$$J = A_0 T^2 \epsilon^{-(E_B - E_F)e/kT} \qquad \text{A/m}^2 \quad (3\text{-}4)$$

is the result. The term $E_B - E_F$ is recognizable as the work function E_W. The equation is more commonly written in the form

$$i = A_0 S T^2 \epsilon^{-b_0/T} \qquad \text{A} \quad (3\text{-}5)$$

where

T = temperature of the source, °K;

$b_0 = eE_W/k = 11{,}600E_W$;

S = surface area, m^2; and

$A_0 = 4\pi mek^2/h^3$, $A/m^2/deg^2$.

An equation with the exponent of T as $\frac{1}{2}$ was developed by Richardson in 1914. However, Eq. 3-5 is believed to have a better theoretical base. It was developed by S. Dushman in 1923 and is known as Dushman's equation. Since the exponential term provides the major effect, experiment cannot show which exponent of T is correct.

Theory indicates that A_0 should be a universal constant of value 120×10^4 $A/m^2/deg^2$, but experimental values do not agree in all cases. This is explained as due to the reflection of many electrons by surface atoms, and thus the emission constant is lower than the theoretical value.

If Eq. 3-5 is written as

$$\log \frac{i}{T^2} = \log A_0 S - \frac{0.4343}{b_0/T} \qquad (3\text{-}6)$$

and the left-hand side plotted against $1/T$, the result should be a straight line with intercept of $\log A_0S$ and a slope of $-0.4343b_0$. By such a plot of experimental data, values of A_0 and b_0 may be determined.

Because of the asymptotic form of the curve for T_1 in Fig. 3-3, some electrons have kinetic energies after emission. The mean of these energies will be given by kT/e, which has a value of 0.24 eV for a metal at 2800°K. Thus initial velocities will be small compared to the velocities reached by the electrons in electron tubes, where they are accelerated by potentials of hundreds of volts, and the initial velocities are usually neglected.

3-3 PHOTOEMISSION

Light, while having wave properties, is also conceived as consisting of bundles of energy called *photons*. The energy carried by each photon is related to the frequency of the light or radiant energy as

$$W = hf \qquad \text{joules} \quad (3\text{-}7)$$

where h is Planck's constant, and f is in hertz.

Photoelectric emission differs from thermionic emission only in that the work-function energy is supplied by light or radiant energy, through bombardment with photons. To cause emission of an electron upon impact of a photon with an atom, the photon energy hf must at least equal the work-function energy eE_W; any excess photon energy appears as kinetic energy of the emitted electron. Einstein stated this in equation form as

$$hf = eE_W + \frac{mv^2}{2} \quad (3\text{-}8)$$

After the impact and energy absorption, the photon disappears.

The Einstein equation, based on the corpuscular theory of light, is experimentally supported and satisfactorily explains the phenomena of photoemission. The concept of the photon as both a wave and a particle is similar to the dual concept employed for the electron. The unification of these two concepts has not yet been achieved; each concept is employed in those cases where it most adequately explains the phenomena.

Since eE_W is a constant for a given emitting surface, then at some frequency f_0, where

$$f_0 = \frac{eE_W}{h} \qquad \text{Hz} \quad (3\text{-}9)$$

the emission will cease. This is a *threshold frequency*, no lower frequency having sufficient energy per photon to cause emission.

The work functions of various materials may be determined from their threshold frequencies, and with care in surface cleanliness, good checks can be obtained with values derived by thermionic methods. This agreement supports the belief that the basic emission phenomena are similar.

Materials of small work function serve best as photoelectric cathode materials; cesium and rubidium are frequently used.

3-4 SECONDARY EMISSION

To produce *secondary emission*, the surface of a conductor or nonconductor is bombarded by a primary beam of electrons or other particles, such as gas ions. The secondary electrons are emitted because of the mechanical energy contributed to the surface atoms. These secondary electrons will be attracted to a higher potential electrode and represent a charge movement away from the bombarded surface, in contrast to the primary beam which constitutes a charge movement to the surface.

The ratio of the average number of secondary electrons emitted to the number of primary electrons, or equivalent charges, striking the surface per unit time, is called the *secondary-emission coefficient* δ. On some surfaces, δ can have a value as high as 10. In Fig. 3-4 the curve of δ is seen to rise with increased primary-beam energy, until a maximum is reached at about 400 to 600 eV. At these energies the primary beam penetrates deeply, perhaps to 100 atom diameters, and further increases in depth of penetration result only in more electrons being recaptured by other atoms on their way to the surface.

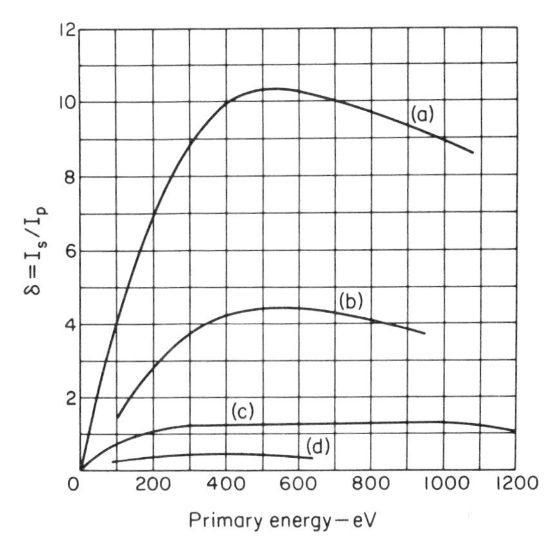

Figure 3-4. Secondary emission ratio: (a) Cs on Ag; (b) Na on Ag; (c) nickel; (d) carbonized nickel.

Any material is capable of secondary emission, but the high-emission materials are of low work function. Surface condition and adsorbed films may change the emission factor considerably.

3-5 THE SCHOTTKY EFFECT AND HIGH-FIELD EMISSION

A positive potential placed on an electrode near an electron-emitting surface will aid the emission process. The resultant electric field accelerates the electrons and opposes the image forces near the cathode surface. The outward-directed energy given the electron is eV, where V is the potential at any distance x; this is illustrated as opposite to the surface barrier energy in Fig. 3-5.

At a distance x_0 from the surface the field force on the electron is equal to the image force (from Eq. 3-1), so

$$e\mathscr{E} = \frac{e^2}{16\pi\epsilon_v x_0^2} \tag{3-10}$$

The reduction in the barrier energy ΔW is equal to twice the field-derived energy at x_0, or $\Delta W = 2e\mathscr{E}x_0$. Solving Eq. 3-10 for x_0, it follows that

$$\Delta W = \frac{e^{3/2}}{2}\sqrt{\frac{\mathscr{E}}{\pi\epsilon_v}} \tag{3-11}$$

Suppose a potential of 1000 V is placed on an electrode parallel to and at a distance of 1 cm from the emitting surface. The field intensity $\mathscr{E}$ is 100,000 V/m, and the value of ΔW due to the field is found to be 0.012 eV. While this is a very small reduction in the apparent work function, it can still produce a measurable increase in the emitted current.

This is called the *Schottky effect*, named after the man who explained the lowering of the work function by an applied electric field. As a result of this effect, the temperature saturation current predicted by the Dushman equation is to a small extent a function of the applied electric field.

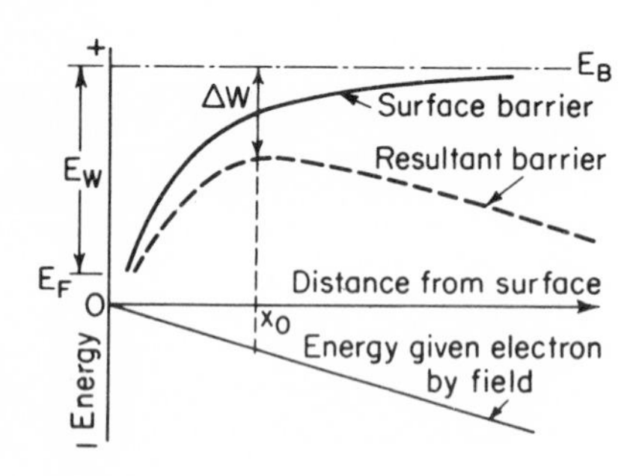

Figure 3-5. Lowering of the surface barrier by an electric field.

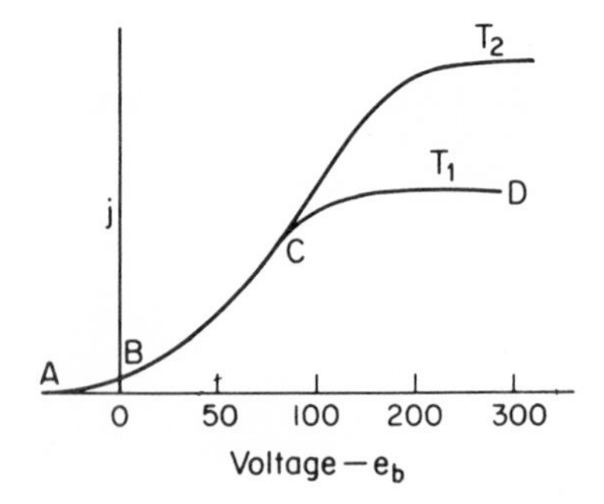

Figure 3-6. Volt–ampere curve for a vacuum diode.

If the applied accelerating field is greatly increased, then at some large field intensity and resultant small x_0, the surface barrier region becomes so thin that it is possible for electrons to tunnel through the barrier. As a result, emission currents may reach very high values at fields approximating 10^8 V/m. This result is called *high-field emission*.

3-6 VOLT-AMPERE CURVE OF THE THERMIONIC DIODE

A basic application of thermionic emission occurs in the thermionic *diode* or two-element vacuum rectifier. Such a device consists of a thermionic emitting surface which is known as the *cathode*, since it is the negative element, and a positive plate or *anode* to collect the emitted electrons.

With the emitting cathode heated to temperature T_1, a voltage is applied between anode and cathode with anode positive. A volt–ampere curve relating potential across and current through the device will appear as in Fig. 3-6. The curve is seen to consist of three regions. The toe from A to B indicates that some electrons have sufficient kinetic energy upon emission to move against a small retarding field and reach a negative collector. From B to C the current rises with positive anode voltage but is less than the possible emission as predicted by the Dushman equation. In the region from C to D all the current predicted by the Dushman equation reaches the anode. No further emission increase is possible at temperature T_1, and the current is *temperature-saturated*. However, if the temperature of the cathode surface is raised to T_2, a higher current is emitted and reaches the anode, and a new value of saturation current is reached.

The slight increase in current with increasing voltage in the saturation region is that predicted by the Schottky effect.

3-7 THE SPACE-CHARGE EQUATION

An explanation is needed for the region B-C of the volt–ampere curve, where the current received by the anode is less than the predicted electron emission value. In a diode without a positive anode potential, electrons are emitted and fill the space around the cathode with negative charge. The electrons at the cathode surface are then faced with a repelling electric field due to this negative charge. Equilibrium will be reached at that negative space charge which provides a repelling field at the cathode surface just sufficient to prevent further electrons from leaving the cathode.

When the anode is made positive to the cathode, electrons are drawn from the space cloud to the anode, reducing the charge in the space and therefore the repelling field at the surface. Equilibrium will be reestablished at some reduced value of space-charge density and resultant surface field at which the number of electrons able to leave the cathode is equal to the number being withdrawn by the anode.

If the emission leaving the cathode is momentarily greater than that of the current reaching the anode, the charge in the space grows and an increased repelling field at the cathode reduces the number of electrons leaving the cathode. If the number of electrons withdrawn by the anode exceeds the current leaving the cathode, then the space charge decreases and more electrons are permitted to leave the cathode and replenish the space charge.

The space-charge cloud of electrons acts as a velocity filter, permitting only electrons with velocities or energies above a certain minimum to leave the cathode region.

If a potential is applied between the parallel plates of Fig. 3-7, the potential distribution would follow the straight line at (a). If the cathode is heated,

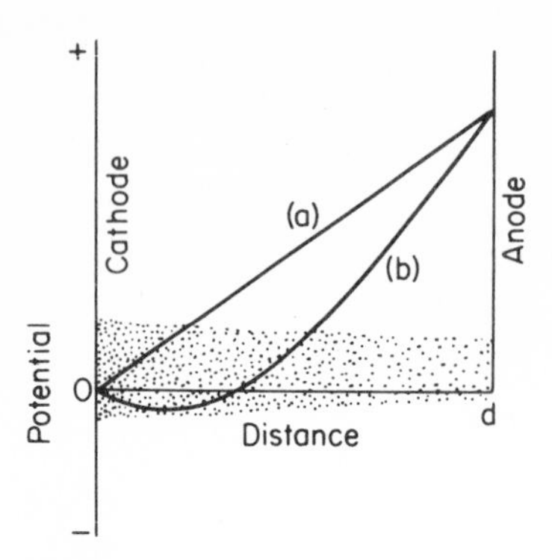

Figure 3-7. Potential variation in a diode: (a) no space charge; (b) with space charge.

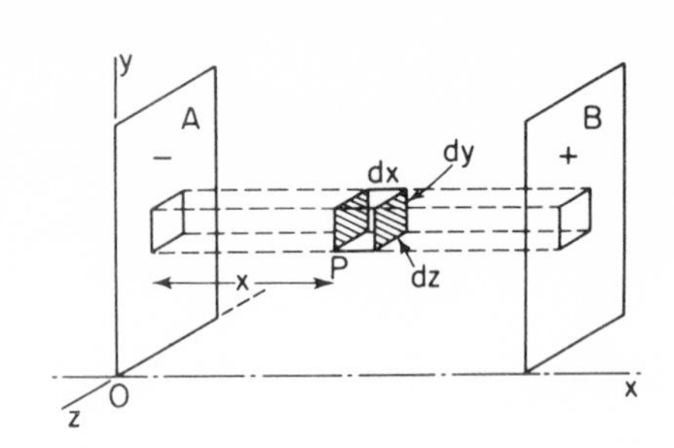

Figure 3-8. Analysis of the parallel-plane vacuum diode.

electrons are emitted and enter the space; this effectively lowers the potential at all points and the potential distribution becomes that of (b). The negative gradient or repelling field at the cathode is apparent as the negative slope of the curve. As the potential minimum of (b) is raised or lowered by changes in electron density or in anode potential, the repelling field at the cathode surface will vary and cause changes in the emitted current. The potential minimum is usually very close to the cathode surface, and is referred to as the *virtual cathode*, since it constitutes a dense cloud of electrons which may be regarded as the source of electrons which move to the anode.

The parallel-plane diode may also be represented by Fig. 3-8, wherein the cathode at the left is heated to a high temperature, sufficient that the current will never be saturated. The right-hand plate is cold and a positive potential e_b is applied with respect to the cathode. The electrodes are in vacuum, sufficiently good that collisions of electrons with gas atoms will be negligible.

The electrode areas are assumed large so that the electric field will be one-dimensional and everywhere perpendicular to the electrodes. Electrons emitted from A are attracted by the potential on B, and represent a current density J in the space. If ρ is the charge density at P and v is the velocity of charge movement toward B, then for electrons

$$J = -\rho v \qquad \text{A/m}^2 \qquad (3\text{-}12)$$

The charge density is related to the space potential by Poisson's equation, developed in Section 2-19, as

$$\frac{d^2V}{dx^2} = -\frac{\rho}{\epsilon_v} \qquad (3\text{-}13)$$

Neglecting emission energies as small, the velocity of the electrons at P is a function of the potential

$$\frac{mv^2}{2} = eV \qquad (3\text{-}14)$$

Combining the three equations above leads to

$$\frac{d^2V}{dx^2} = \frac{J}{\epsilon_v\sqrt{2eV/m}} \tag{3-15}$$

Integrating Eq. 3-15, and neglecting the first constant as small, leads to a solution

$$J = \frac{4\epsilon_v}{9}\sqrt{\frac{2e}{m}}\,\frac{V^{3/2}}{x^2} \qquad \text{A/m}^2 \tag{3-16}$$

since $V = 0$ at $x = 0$. The particular values of interest are those at the anode, or $V = e_b$ and $x = d$, so the current density at the anode is

$$J = 2.34 \times 10^{-6}\frac{e_b^{3/2}}{d^2} \qquad \text{A/m}^2 \tag{3-17}$$

where d is the anode-cathode separation in meters.

This result is known as the *three-halves power law* or the *Langmuir-Child equation*, after the two investigators who independently developed it. The equation predicts the value of current for the region *B-C* of Fig. 3-6 as proportional to the three-halves power of the applied voltage, and explains how the space charge limits the current to a value less than the possible emission at the lower voltages. The Dushman equation predicts the current that *may* be emitted at a temperature T; the Langmuir-Child law establishes the current *actually* emitted and collected by the anode. Most vacuum-tube operation is in the region where the current is limited by the space charge.

Langmuir also obtained an expression for the space-charge-limited current of a diode with long concentric cylindrical electrodes. This is

$$i = 14.7 \times 10^{-6}\frac{le_b^{3/2}}{r_a\beta^2} \qquad \text{A} \tag{3-18}$$

where l is the length of the cylinders and r_a is the anode radius. This equation is for total current and not current density.

The factor β is a function of log r_a/r_k, where r_k is the cathode radius. This function appears as a result of integration of the field equations in cylindrical coordinates. Values of β^2 are tabulated in Table 3-2; for $r_a/r_k > 10$ the term β^2 may be assumed as unity with small error.

Table 3-2

VALUES OF THE CONSTANT β^2†

r_a/r_k	β^2	r_a/r_k	β^2
1.00	0.000	20.0	1.072
3.00	0.517	50.0	1.094
5.00	0.767	100.0	1.078
10.00	0.978	∞	1.00

†From I., Langmuir and K. Blodgett, *Phys. Rev.*, Vol. 22, p. 347 (1923).

The density J from Eq. 3-16, in terms of V and x, may be equated to that of Eq. 3-17, since the current density is equal at all points in the space for the parallel-plane case. Solution for V gives the potential at any distance x as

$$V = e_b\left(\frac{x}{d}\right)^{4/3} \qquad \text{V} \quad (3\text{-}19)$$

This potential distribution does not indicate the potential minimum due to the electron concentration near the cathode. However, had the constant been determined in the integration of Eq. 3-15, the distribution would have the predicted form of (b) in Fig. 3-7.

Deviations from plane and cylindrical shapes are usual, and values of the exponent of e_b are found in the range from 1.3 to 1.8. This neither invalidates the basic theory nor changes the fact that the vacuum diode is a nonlinear circuit element.

If the emission velocities are neglected, then an electron will arrive at the anode with an energy given by

$$E = ee_b \qquad \text{joules}$$

Upon impact with the anode, this energy is converted into heat. At voltages above about 50,000 a minor amount of the radiation also appears in the X-ray region.

In most vacuum-tube designs the power loss must be removed by radiation, which requires appropriate anode area and material to withstand a considerable temperature rise. In small tubes the design limits the temperature rise to that at which the anodes do not show color. In higher-power radiation-cooled tubes the anodes are of molybdenum, tantalum, or graphite to operate at much higher temperatures. Another class of tubes employs water or forced air for conduction cooling.

3-8 *THERMIONIC CATHODES*

Cathodes for thermionic emission are of two types: *directly heated* (or filamentary) and *indirectly heated* (or equipotential). These types are illustrated in Fig. 3-9.

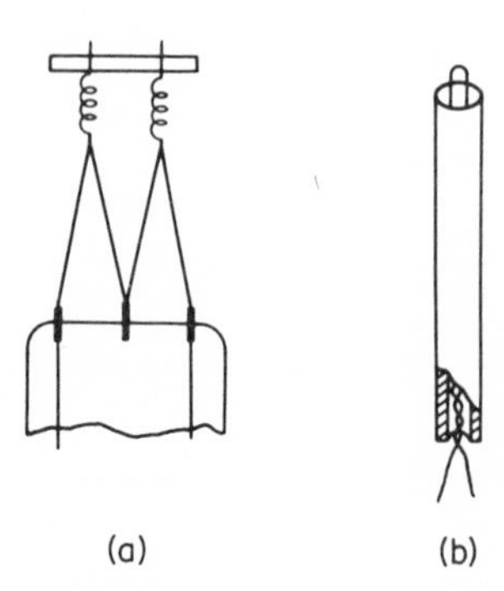

Figure 3-9. Cathodes: (a) filamentary; (b) indirectly heated.

In the filamentary construction, a wire is heated by passage of an electric current and electron emission takes place directly from the wire surface. The indirectly heated form employs a tungsten-wire heater coated with a refractory insulator and inserted in a thin, hollow nickel cyclinder. The outside surface of the cylinder is coated with a layer of emitting material. Passage of current through the heater element raises the cylinder surface to emitting temperature. Owing to the heat storage in the cylinder and heater, cathodes of this type require a short time to reach emitting temperature, but this thermal inertia eliminates cyclic variations in temperature and emission, which occur with the filamentary construction when the cathode is heated with alternating current.

Since the heating current passes directly through the filamentary cathode, there is a voltage drop along the wire and the entire cathode is not at the same potential. This causes a variation in emission along the wire. The effect is eliminated by the indirectly heated construction, and the latter type is often referred to as being *equipotential.*

The list of materials suited to thermionic emission is fairly limited. Desirable qualities are low work function, high melting point, and a high vapor pressure. The latter is desirable to reduce the rate of evaporation of the metal at high temperatures. Many low work-function materials, such as cesium, have low melting points and are not useful. The list is limited, therefore, to tungsten, thoriated tungsten, and oxide coatings on a nickel or alloy base.

Tungsten, with a melting point of 3643°K, may be operated to 2500° to 2600°K without excessive evaporation. This high temperature overcomes the disadvantage of the relatively high work function. However, because of the large heating power required to maintain this operating temperature, tungsten is rarely used as a direct emitter today.

Thoriated tungsten, or tungsten containing a few per cent of thorium oxide, gives improved emission characteristics. In operation, some of the thorium oxide is reduced to thorium, which partially covers the tungsten surface as a monoatomic layer of thorium atoms. The increased emission is caused by the attraction of the thorium atoms toward the tungsten, creating a subsurface electric field favorable to electron transit through the surface. The monoatomic emitting layer is the result of a delicate balance between diffusion of thorium atoms to the surface and evaporation of thorium atoms from the surface. An operating temperature of about 1900°K insures equilibrium conditions.

Oxide-coated cathodes are almost always employed with indirect heating and are formed by deposition of barium and strontium carbonates on a nickel base. During processing, these compounds break down to the oxides and to pure barium and strontium. The pure metals increase the conductivity of the oxide layer and provide an internal field favorable to electron transit, as for thoriated tungsten. These cathodes operate at 900° to 1100°K.

Cathodes may be compared by use of the *relative emission efficiency*, defined as milliamperes of emission per watt of heating power. This figure, and other emission constants believed representative, appear in Table 3-3.

Table 3-3

CHARACTERISTICS OF CATHODE MATERIALS

Material	A_0	b_0	E_W (eV)	Emission Efficiency (mA/W)	Operating Temperature (°K)
Tungsten	60.2×10^4	52,400	4.52	4—20	2500—2600
Thorium on tungsten	3.0×10^4	30,500	2.6	50—100	1900—2000
Oxide coatings	0.01×10^4	11,600	1.0	100—1000	900—1200

Over 95 per cent of all vacuum tubes currently employ oxide coatings for their cathodes, due to the small heating-power requirements and the high emission efficiency.

PROBLEMS

3-1. An electron in tungsten has an energy equal to E_F at 0°K. It is given an additional energy directed outward of 12×10^{-19} joule. If it is able to leave the tungsten surface, what will be its velocity after emission?

3-2. A certain tube has the following emission data taken:

I, A	T°K	I, A	T°K
0.000382	1900	0.0694	2300
0.00169	2000	0.1952	2400
0.00665	2100	0.5037	2500
0.0219	2200	1.169	2600
		2.75	2700

Determine by graphical means the values of the emission coefficients A_0 and b_0, if the emitting area is 1.8 cm^2.

3-3. A cesium photoemissive surface has a work function of 1.81 V. Find the threshold wave length for this surface; also the maximum emission velocity if electrons are emitted when this surface is struck by light of 5300 Å wave length.

3-4. (a) Calculate the energy carried by photons of red light of $\lambda = 6439$ Å; yellow light of $\lambda = 5890$ Å; and ultraviolet light of $\lambda = 3302$ Å and 2537 Å.

(b) Each of the photons of (a) strikes a sodium surface of 1.9 V work function. If electrons are emitted, find their velocities.

3-5. Threshold wave lengths are measured for platinum at 2570 Å; potassium, 7000 Å; cadimum, 3140 Å; and magnesium, 3430 Å. Compute the work functions.

3-6. A flash of light from the ultraviolet line of mercury at 2537 Å, lasting 0.1 s, strikes a sodium surface of 1.5 cm^2 in a phototube. The surface has a work function of 1.9 V, and the light has a power density of 0.5 W/cm². If one-tenth the photons cause the emission of an electron, find the current flowing during the flash.

3-7. If secondary electrons are emitted with average energy of 10 eV and δ for a particular surface is 9, find the mA emission per input watt, if the primary beam has fallen through 300 V and the surface work function is 0.85 eV.

3-8. (a) To what value must the work function of a tungsten surface be changed to raise the current density to 20,000 A/cm² at a temperature of 2600°K?

(b) What value of field intensity could provide this reduction of work function?

3-9. A large oxide-coated cathode is to have an emission current of 100 A when operated at 1050°K.

(a) Determine the surface area needed.

(b) Find the surface area of tungsten needed if operated at the same temperature as the oxide-coated cathode.

3-10. At 2400°K a tungsten filament is observed to increase its emission periodically by amounts up to 5 per cent. What change in value of work function E_W is required to explain this change?

3-11. (a) For a vacuum diode with parallel electrodes, spaced 0.5 cm and with $E_b = 250$ V, plot the variation of potential across the space under the assumption of space-charge conditions and negligible emission velocities of the electrons.

(b) Make a similar plot of charge density, and of velocity of the electrons. What relationship between v and ρ is indicated? Why is this true?

3-12. A parallel-plate diode with spacing of 0.20 cm has an anode-cathode voltage of 100 V. The cathode is capable of emitting an infinite number of electrons.

(a) How much heat in W/m² will the anode have to dissipate?

(b) What will be the power dissipated per unit area if the anode-cathode voltage is doubled?

(c) If the anode surface emissivity is 0.4, find the anode temperature for both (a) and (b). Assume that the surrounding temperature is 20°C and that the heat is radiated from only one side of the anode.

3-13. A cylindrical-plate diode, with a thoriated tungsten filament of length 2.5 cm and diameter 0.04 cm, has a filament temperature of 1950°K. The anode diameter is 2 cm.

If the current through the tube is to be one-eighth of temperature saturation value, what anode-cathode voltage should be used?

3-14. A certain vacuum diode on test shows that 25 V is required to cause 180 mA anode current.

(a) How much voltage is required if the anode current is to be increased to 325 mA, assuming sufficient emission?

(b) What is the maximum current possible without causing the anode loss to exceed 25 W?

(c) If the anode area is 4 cm^2 and emissivity 0.6, what will be the anode temperature rise above a 20°C ambient, for part (b)?

3-15. A vacuum diode is to be built with cylindrical anode and thoriated-tungsten filament, to give a saturation current of 250 mA at an anode voltage of 105 V. The anode must be large enough to radiate the input under this condition at a rate of 1.3 W/cm^2. Find the diameter and length of the anode if $\beta^2 = 1.0$.

REFERENCES

1. Langmuir, I., "The Effect of Space Charge and Residual Gases on Thermionic Currents in High Vacuum," *Phys. Rev.*, Vol. 2, p. 40, (1913).
2. Dushman, S., "Electron Emission from Metals as Functions of Temperature," *Phys. Rev.*, Vol. 21, p. 623 (1923).
3. Langmuir, I., "The Electron Emission from Thoriated-Tungsten Filaments," *Phys. Rev.*, Vol. 22, p. 623 (1923).
4. Wooldridge, D. E., "Theory of Secondary Emission," *Phys. Rev.*, Vol. 56, p. 562 (1939).
5. Glover, A. M., "A Review of the Development of Sensitive Phototubes," *Proc. IRE*, Vol. 29, p. 413 (1941).
6. Spangenberg, K. R., *Vacuum Tubes*. McGraw-Hill Book Company, New York, 1948.
7. Van der Ziel, A., *Solid-State Physical Electronics*. Prentice-Hall, Inc., Englewood Cliffs, N. J., 1957.

4

SINGLE-PHASE RECTIFIER CIRCUITS AND FILTERS

Most electronic equipment is supplied with direct current from the ac lines through the use of diodes as rectifiers. This chapter will consider such rectifier circuits with single-phase supply; circuits with polyphase supply and suited to high power output, will be treated in a later chapter.

4-1 THE IDEAL DIODE

The volt-ampere relation for a semiconductor diode is predicted by the exponential relation of Eq. 2-62, that for the vacuum diode by the three-halves power law in Eq. 3-17. Curves for practical devices of both types appear in Figs. 4-1 and 4-2.

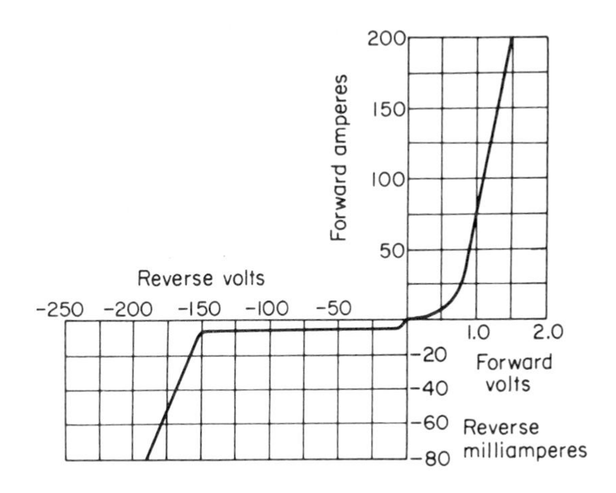

Figure 4-1. Characteristic of a 1N412A silicon diode.

The internal forward drop of a semiconductor diode will usually be less than a volt. The several hundred ohm forward resistance of the vacuum diode, and the small resistance of the semiconductor diode, compared to usual loads of several thousand ohms, make the consideration of forward diode resistance usually unnecessary in rectifier circuit analysis. That is, we assume the diode as lossless or *ideal* in the forward direction.

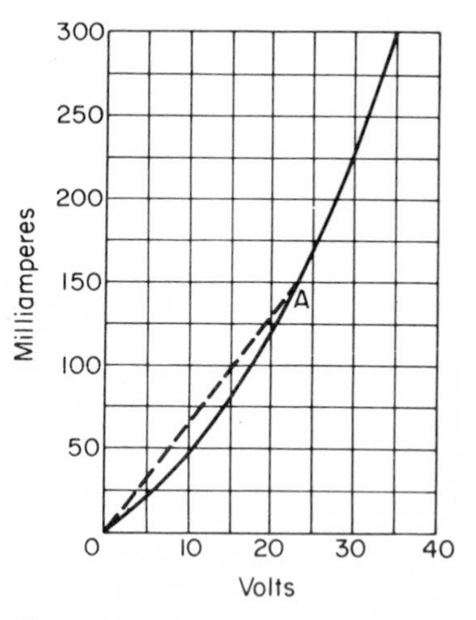

Figure 4-2. Characteristic of 5Z3 vacuum diode.

The vacuum diode appears as an open circuit in the reverse direction, and the semiconductor diode reverse resistance is usually so large that it may also be considered as infinite. Thus the volt-ampere characteristic of an ideal diode may be viewed as made of two straight lines, as in Fig. 4-3. Such a volt-ampere characteristic consisting of straight-line segments is said to be *piece-wise linear*. For the ideal diode, positive anode voltages will place the operation along the vertical line segment and the device will be linear; negative anode voltages place operation on the horizontal segment and the current is zero.

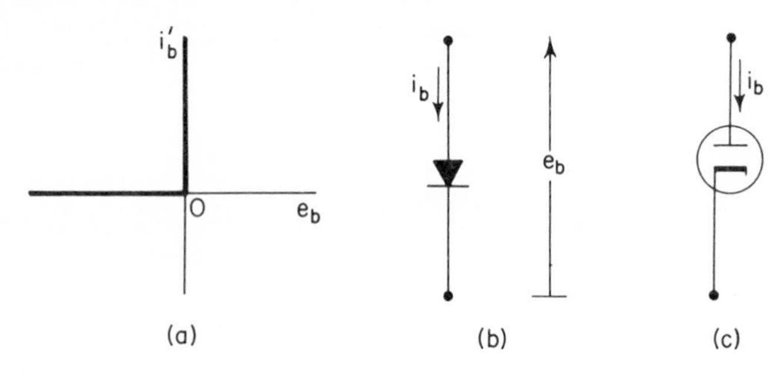

Figure 4-3. (a) Volt–ampere curve for ideal diode; (b) diode symbol with e_b indicated for forward conduction; (c) vacuum diode.

4-2 *THE HALF-WAVE RECTIFIER CIRCUIT*

A *half-wave rectifier circuit* is shown in Fig. 4-4. While not directly of great importance, the half-wave circuit is a basic building block for more complex rectifier circuits.

With the diode assumed ideal, a sinusoidal voltage e applied to the two

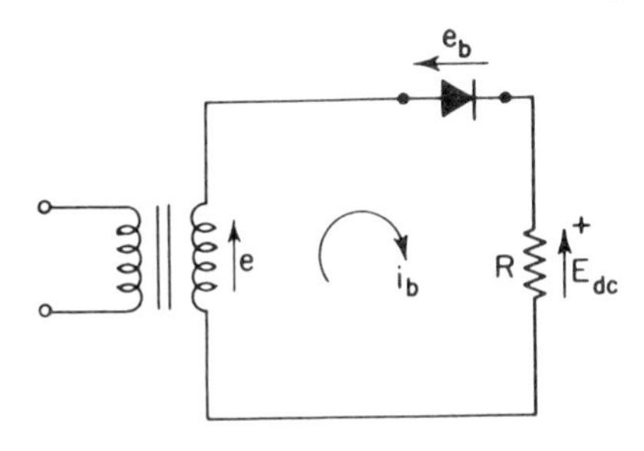

Figure 4-4. Half-wave diode rectifier circuit.

line segments of the diode volt-ampere curve yields values for the currents on the two half cycles as

$$i_b = \frac{E_m}{R} \sin \omega t, \qquad 0 \leq \sin \omega t \leq 1 \tag{4-1}$$

$$i_b = 0, \qquad -1 \leq \sin \omega t \leq 0 \tag{4-2}$$

The wave form of the current through the diode and resistance load is that shown in Fig. 4-5. The diode acts as a voltage-actuated switch, connecting the ac source to the load in the period in which the diode anode is positive to cathode, and disconnecting source and load when the diode voltage reverses.

A Fourier analysis of the half-sinusoid voltage pulses appearing across the resistance load yields

$$e = \frac{E_m}{\pi} + \frac{E_m}{2} \sin \omega t - \frac{2E_m}{3\pi} \cos 2\omega t - \frac{2E_m}{15\pi} \cos 4\omega t \cdots \tag{4-3}$$

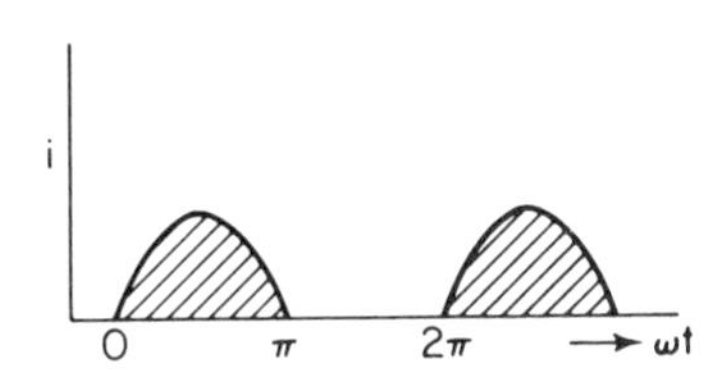

Figure 4-5. Current wave form in resistance load—half-wave circuit.

Thus a diode is a frequency converter in which an input frequency is changed to a large number of output frequencies. In rectification, the dc or zero frequency component is the desired output; thus the dc load voltage appears as E_m/π, and the dc or average value of load current is I_m/π, where $I_m = E_m/R$.

The dc power dissipated in the load is that component which would be effective in charging a battery or in an electrolytic process. From Eq. 4-3 it appears that currents of harmonic frequencies are also present; they dissipate power in the load resistance and reduce the power efficiency if dc power is considered as the useful output. Because of the magnitude of the ac components of the load voltage, or the irregularities of the wave form of Fig. 4-5, the half-wave rectifier circuit does not provide "smooth" or ripple-free direct current, and the output is not directly useful for the supply of much electronic equipment.

4-3 FULL-WAVE RECTIFIER CIRCUIT

It can be reasoned that if the load of the half-wave rectifier circuit were supplied with current during the inactive half cycle, the efficiency might be raised and smoother dc obtained. The full-wave circuit of Fig. 4-6 accomplishes

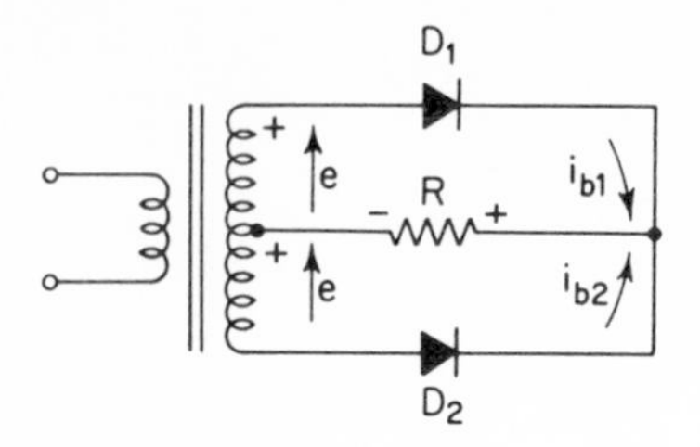

Figure 4-6. Full-wave circuit.

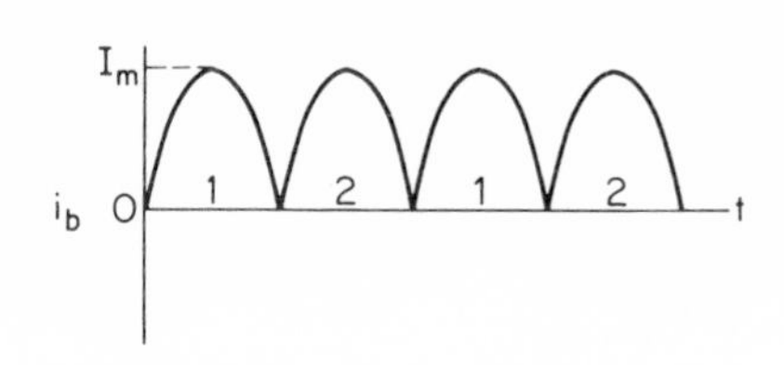

Figure 4-7. Current wave form in resistance load—full-wave circuit.

this by connecting a second half-wave circuit, supplied out of phase, to the common load resistance. The circuit is widely employed and its performance will be analyzed in detail.

Since conduction occurs when a diode anode is positive to cathode, the diodes will conduct on opposite half-cycles, resulting in a load current of half-sine pulses, as in Fig. 4-7. With ideal diodes, the currents are

$$\left.\begin{aligned} i_{b1} &= \frac{E_m}{R}\sin\omega t \\ i_{b2} &= 0 \end{aligned}\right\} \qquad 0 \le \sin\omega t \le 1 \tag{4-4}$$

$$\left.\begin{aligned} i_{b1} &= 0 \\ i_{b2} &= -\frac{E_m}{R}\sin\omega t \end{aligned}\right\} \qquad -1 \le \sin\omega t \le 0 \tag{4-5}$$

Each diode then operates independently and under exactly the same conditions as in the half-wave circuit, only the load currents being combined. A dc ammeter in series with the load R will measure the dc or average current as

$$I_{\mathrm{dc}} = \frac{1}{2\pi}\int_0^{2\pi} i_b\,d\omega t = \frac{1}{2\pi}\int_0^{\pi}\frac{E_m}{R}\sin\omega t\,d\omega t + \frac{1}{2\pi}\int_{\pi}^{2\pi}\frac{(-E_m)}{R}\sin\omega t\,d\omega t \tag{4-6}$$

$$= \frac{2E_m}{\pi R} = \frac{2I_m}{\pi} \tag{4-7}$$

and defining the peak current I_m as well. The dc load voltage is then $E_{\mathrm{dc}} = 2E_m/\pi$, or twice the voltage of the half-wave circuit.

The total ac power input to the circuit is $P = I_{\mathrm{rms}}{}^2R$, neglecting the transformer losses. The current I_{rms} can be computed from the wave form of Fig. 4-7 and the conventional effective current definition as

$$I_{\mathrm{rms}} = \sqrt{\frac{1}{2\pi}\int_0^{2\pi} i_b^2\,d\omega t} = \sqrt{\frac{2}{2\pi}\int_0^{\pi} I_m^2\sin^2\omega t\,d\omega t} = \frac{I_m}{\sqrt{2}} \tag{4-8}$$

and thus the power input is

$$P = \frac{I_m^2R}{2} = \frac{E_m^2}{2R} \tag{4-9}$$

The voltage wave form applied to the load, causing the current wave of Fig. 4-7, may be analyzed as

$$e = \frac{2E_m}{\pi} - \frac{4E_m}{3\pi}\cos 2\omega t - \frac{4E_m}{15\pi}\cos 4\omega t - \frac{4E_m}{35\pi}\cos 6\omega t \cdots \tag{4-10}$$

The lowest frequency alternating component present is twice the supply frequency, $\omega/2\pi$, which makes removal of the ac harmonics easy with filter circuits. Filters make the output current sufficiently smooth for most uses, by preventing the ac components from passing through the load resistance. They will be discussed at a later point in this chapter.

The peak voltage appearing across a diode during the reverse half-cycle is called the *peak reverse voltage*. During the interval in which D_1 is nonconducting, the instantaneous voltages in the outside loop of the circuit are

$$e_{b1} = -2E_m \sin \omega t \tag{4-11}$$

The maximum value of e_{b1} occurs when $\omega t = 3\pi/2$, and thus the peak reverse voltage for the full-wave circuit is $2E_m$, or the peak of the total transformer voltage. In the half-wave circuit the peak reverse voltage is only E_m. This peak reverse voltage must be limited to the diode rating to prevent voltage breakdown. A second basic diode limit is the *peak current* which is carried by the diode, and this is I_m. The easier filtering and the doubled output voltage makes the full-wave circuit more desirable than the half-wave circuit.

4-4 *RIPPLE FACTOR*

In many applications, a residual pulsation in the output direct current is not desirable. A measure of the purity of the dc output is called the *ripple factor*, γ, defined as the ratio of two current (or voltage) components:

$$\gamma = \text{ripple factor} = \frac{\text{effective value of ac components}}{\text{average, or dc component}}$$

The effective value of the total load current is

$$I_{\text{rms}} = \sqrt{I_{\text{dc}}^2 + I_{\text{ac}}^2} \tag{4-12}$$

from which I_{ac} can be found, and the ripple factor written as

$$\gamma = \sqrt{(I_{\text{rms}}/I_{\text{dc}})^2 - 1} \tag{4-13}$$

Use of I_{dc} and I_{rms} for the half-wave rectifier circuit gives a value of 1.21 or 121 per cent for the ripple factor with resistive load. A similar computation gives the ripple factor for the full-wave circuit with resistive load as 0.48, or 48 per cent. Neither of these figures is sufficiently low for direct use of the outputs with much electronic equipment where $\gamma \leq 0.001$ may be required. Filter circuits thus are necessary.

4-5 *THE BRIDGE RECTIFIER CIRCUIT*

For obvious reasons the circuit of Fig. 4-8 is called a *bridge rectifier circuit.* Its performance is that of a full-wave circuit, but without the need for a center connection on the supply transformer.

In the half-cycle in which the transformer voltage e has the polarity indicated, diodes D_1 and D_2 conduct as shown by the arrows A. Diodes D_3 and D_4 are then reverse-biased and nonconducting. In the following half-cycle, D_3 and D_4 conduct, causing a second load current pulse in the same direction as the pulse from the first half-cycle. Full-wave current is obtained in the load, with two diodes conducting simultaneously and in series on each half-cycle.

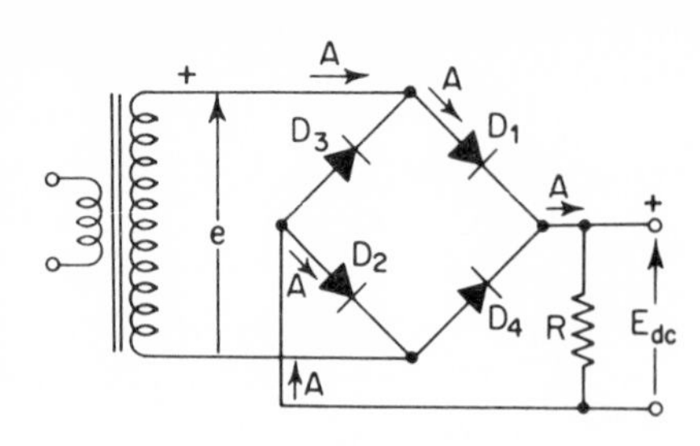

Figure 4-8. The bridge rectifier circuit.

The same dc voltage can be obtained from a transformer in a bridge circuit as from another transformer with twice the secondary turns in a center-tap circuit. Current passes on both half-cycles in the bridge-circuit transformer winding, but only on half the cycle in each winding of the center-tap circuit. For equal output the current rating of a transformer secondary in the bridge circuit must be about 40 per cent larger than for the center-tap circuit transformer. With half the turns and only a 40 per cent larger current rating, the bridge circuit uses its transformer more efficiently.

Inspection of the circuit shows that the peak reverse voltage across a nonconducting diode is equal to the peak of the transformer voltage. For a given dc output voltage, the bridge circuit requires a diode of only half the peak reverse rating as is required by the center-tap circuit.

Vacuum diodes require three separate and insulated heater-circuit sources with the bridge circuit. This is, obviously, a disadvantage. Because of the flexibility of the semiconductor diode, the bridge circuit is now receiving wide application with those devices, however.

4-6 *VOLTAGE-MULTIPLYING RECTIFIER CIRCUITS*

By the avoidance of heater circuitry, the semiconductor diode also makes the circuits of Fig. 4-9 attractive. With a symmetrical ac wave applied, the output dc voltage of *a* will approximate the peak-to-peak value of the input voltage, and the output of the circuit of *b* will approximate twice the input voltage peak-to-peak value. Higher multiples of the input voltage can be obtained, although at reduced efficiency.

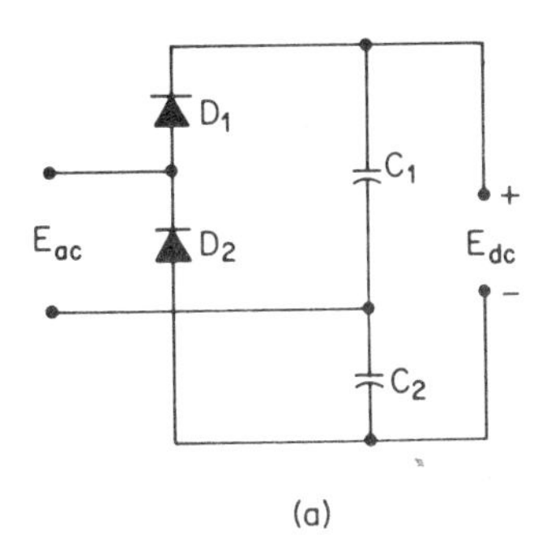

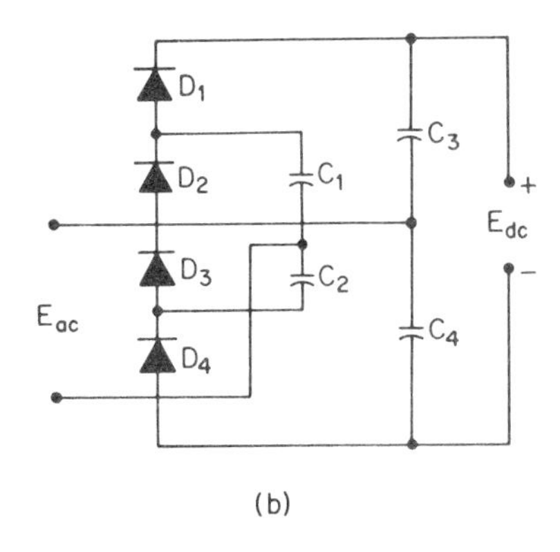

Figure 4-9. (a) Voltage-doubler circuit; (b) voltage-quadrupler circuit.

In the *voltage doubler* of Fig. 4-9(a), with the upper ac terminal assumed positive, diode D_1 conducts and charges capacitor C_1 to the peak of the supply voltage. In the next half-cycle diode D_2 conducts and charges capacitor C_2 to the peak of that half-cycle. Capacitors C_1 and C_2 are in series for the output circuit and produce a voltage which, at no load, equals the sum of the positive and negative peaks of the applied ac wave, or twice the peak value of a symmetrical wave.

The diodes conduct high peak currents for short time intervals and C_1 and C_2 must be of high capacitance to maintain the voltage over the nonconducting intervals. For high ωCR values the circuit provides both good regulation and low ripple; the dynamic regulation for surge currents, as encountered with Class B amplifiers, is especially good. Further filtering may be added if the ripple produced by the capacitance is not sufficiently low.

4-7 THE SHUNT-CAPACITOR FILTER

Filter circuits are employed to reduce rectifier output ripple, either bypassing the alternating output components around the load by a shunt capacitor, or limiting their magnitude to a low value in the load by a series inductor. Combinations of capacitance and inductance may be found to be more efficient than either element alone.

The shunt-capacitor filter of Fig. 4-10 is a simple and efficient form. The capacitance is so chosen that $X_c \ll R$, and the alternating currents find a low-reactance shunt in C. Only a small alternating current component passes in R, producing a small ripple voltage.

The capacitor alters the conditions under which the diode operates. When the diode output voltage is increasing, the capacitor stores energy by charging to the peak of the input cycle, as indicated in (b), Fig. 4-10. With falling source voltage, the diode disconnects the source and load at the instant when the source voltage starts to fall faster than the capacitor voltage

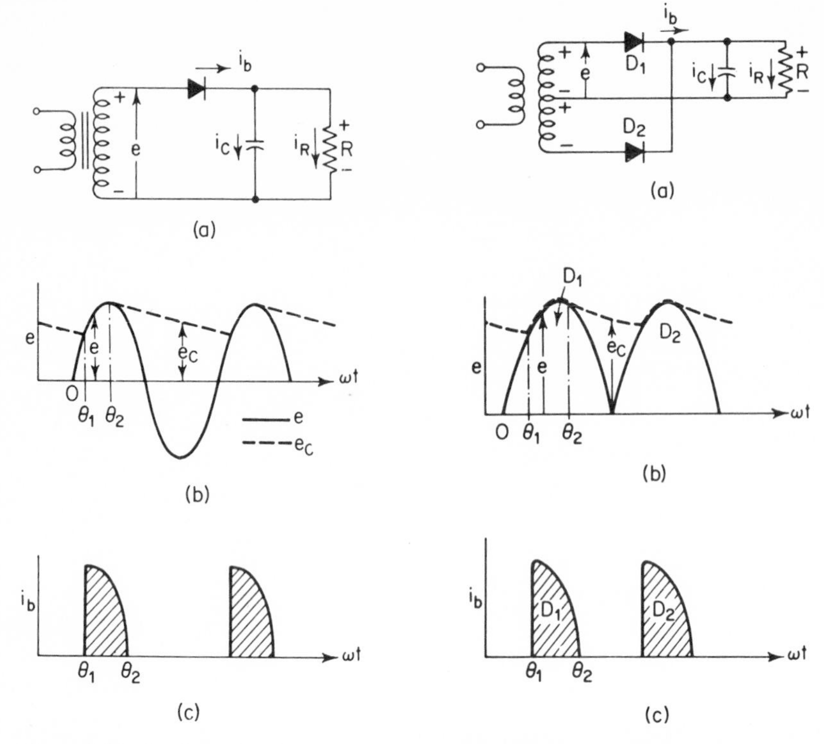

Figure 4-10. (a) Capacitor filter and half-wave rectifier circuit;
(b) voltage wave forms;
(c) diode current wave forms.

Figure 4-11. (a) Capacitor filter with full-wave circuit;
(b) circuit voltages;
(c) diode current waves.

can fall, as determined by the time constant of C and the load. The capacitor continues to maintain the load voltage at a higher value and lower ripple than if the capacitor were not present.

The diode delivers a charging pulse of current with each cycle, and then disconnects source and load. The circuit is excited by repeated current transients of the form of Fig. 4-10(c), and the source voltage series of Eq. 4-3 does not apply because of the discontinuous current pulses.

The full-wave circuit of Fig. 4-11 is more frequently employed and will be analyzed here. During the charging interval the voltage e_C across the capacitor is equal to the supply voltage

$$e_C = e = E_m \sin \omega t, \qquad \theta_1 \leq \omega t \leq \theta_2 \tag{4-14}$$

neglecting diode resistance and transformer reactance. During this charging interval

$$i_b = i_C + i_R = C\frac{de_C}{dt} + \frac{E_m}{R}\sin \omega t$$

Then

$$i_b = E_m\left(\omega C \cos \omega t + \frac{1}{R} \sin \omega t\right), \qquad \theta_1 \leq \omega t \leq \theta_2 \tag{4-15}$$

At θ_2, when diode conduction ends, or $i_b = 0$, then $i_R = -i_C$ and the capacitor discharges through R. Then

$$\omega C \cos \theta_2 = -\frac{1}{R} \sin \theta_2$$

from which

$$\theta_2 = \tan^{-1}(-\omega CR) \tag{4-16}$$

This equation introduces the dimensionless parameter ωCR which will be found useful in describing filter circuit performance.

Since θ_2 lies in the second quadrant, Eq. 4-15 may be written as

$$i_b = \frac{E_m}{R}\sqrt{1 + \omega^2C^2R^2} \sin(\omega t + \varphi)$$

where

$$\varphi = \tan^{-1}(\omega CR) = \pi - \tan^{-1}(-\omega CR)$$

Then

$$i_b = \frac{E_m}{R}\sqrt{1 + \omega^2C^2R^2} \sin(\theta_2 - \omega t), \qquad \theta_1 \leq \omega t \leq \theta_2 \tag{4-17}$$

is a complete expression for the current pulse through the diode.

At θ_2 the voltage output from the diode is falling faster than the capacitor can discharge through R. The diode then has a positive cathode and ceases conduction, disconnecting the source from the load. For the period between $\omega t = \theta_2$ and $\omega t = \pi + \theta_1$ it can be seen that $i_R = -i_C$, $e_C/R = -C\, de_C/dt$, and

$$\frac{de_C}{dt} + \frac{e_C}{RC} = 0 \tag{4-18}$$

becomes the circuit equation; its solution is

$$e_C = A\epsilon^{-t/RC}$$

At $\omega t = \theta_2$, $e_C = E_m \sin \theta_2$, and so

$$e_C = E_m \sin \theta_2 \epsilon^{-(\omega t - \theta_2)/\omega CR}, \qquad \theta_2 \leq \omega t \leq (\pi + \theta_1) \tag{4-19}$$

becomes the expression for the capacitor voltage during the interval when the capacitor is supplying the load current.

For large ωCR (large C or large R), the exponential term approaches unity, and the voltage curve in the interval shows only a small slope; thus the ripple will be small. For smaller ωCR (R smaller, or with heavier load current) the exponential term has a greater variation and the voltage during the interval will drop below its value at θ_2. Thus the *ripple with a capacitor filter increases with increase in load current.*

At $\omega t = \pi + \theta_1$, the capacitor voltage is again equal to the supply through the other diode, or $e_C = E_m |\sin(\pi + \theta_1)| = E_m \sin \theta_1$. Substituting for e_C in Eq. 4-19 leads to

$$\sin \theta_1 = \sin \theta_2 \epsilon^{-(\theta_1 - \theta_2 + \pi)/\omega CR} \tag{4-20}$$

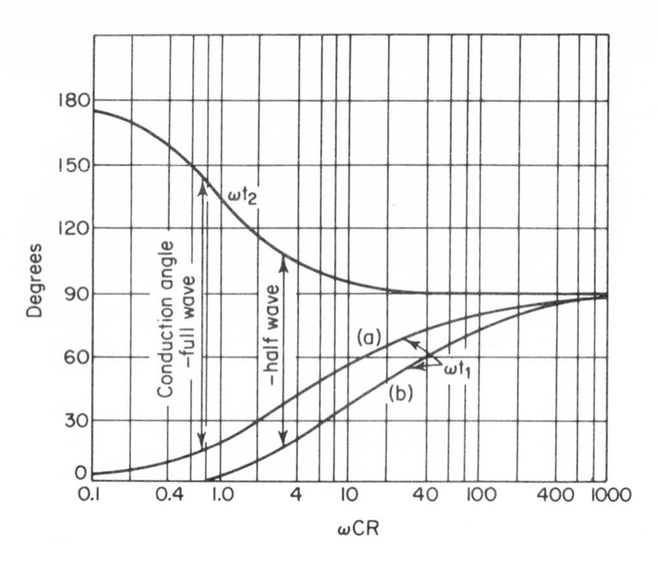

Figure 4-12. Conduction angles for the shunt-capacitor filter: (a) full-wave; (b) half-wave.

This is a transcendental equation which can be solved by computer for values of θ_1 and θ_2 as functions of ωCR. Results are plotted in Fig. 4-12 for both the half-wave and full-wave circuits.

Having values of θ_1 and θ_2, typical diode-current wave forms can be plotted from Eq. 4-17 and appear in Fig. 4-13. Since θ_2 decreases and θ_1 increases with increase in ωCR, the conduction angle, $\theta_2 - \theta_1$, becomes very small for large values of capacitance and given R values. The peak amplitude of the current also increases with ωCR, its value being determined at $\omega t = \theta_1$ from Eq. 4-15 as

$$i_{\text{peak}} = E_m\left(\omega C \cos \theta_1 + \frac{1}{R} \sin \theta_1\right) \tag{4-21}$$

for $1/\omega C \ll R$.

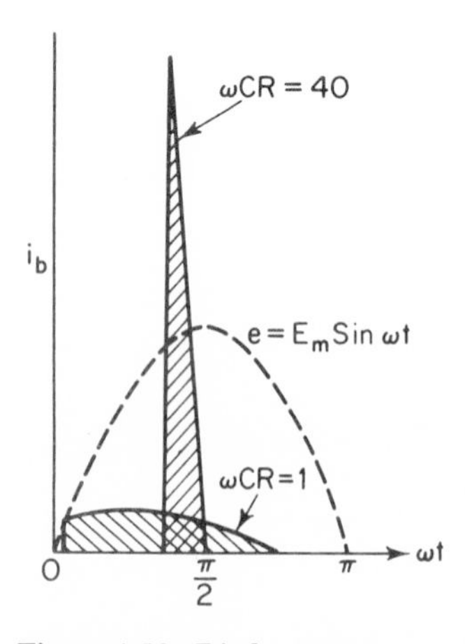

Figure 4-13. Diode current waves in half-wave circuit with capacitor filter.

The direct voltage across the load may be obtained by averaging the voltages given by Eqs. 4-14 and 4-19 and summing, noting that there are two such current intervals per cycle. This operation leads to

$$E_{dc} = \frac{E_m}{\pi}\sqrt{1 + \omega^2C^2R^2}[1 - \cos(\theta_2 - \theta_1)] \tag{4-22}$$

The ratio E_{dc}/E_m is plotted in Fig. 4-14 as a function of the parameter ωCR. For low values of this parameter the ratio approaches $2/\pi$ as the value for a purely resistive load. For good voltage regulation with varying loads, sufficient capacitance should be used to ensure operation on the upper plateau of the curve, or with $\omega CR > 20$.

If the output voltage wave form is approximated by the solid curve of Fig. 4-15, simplified and reasonably accurate expressions can be derived for the ripple, γ, and E_{dc}. The wave form is assumed made up of straight lines with a peak E_m and a peak-to-peak ripple magnitude of E_R. The average or dc value is

$$E_{dc} = E_m - \frac{E_R}{2} \tag{4-23}$$

During the capacitor discharge interval, θ_2 to $\pi + \theta_1$, the capacitor is assumed to lose charge at a constant rate or constant current, $dq/dt = I_{dc}$.

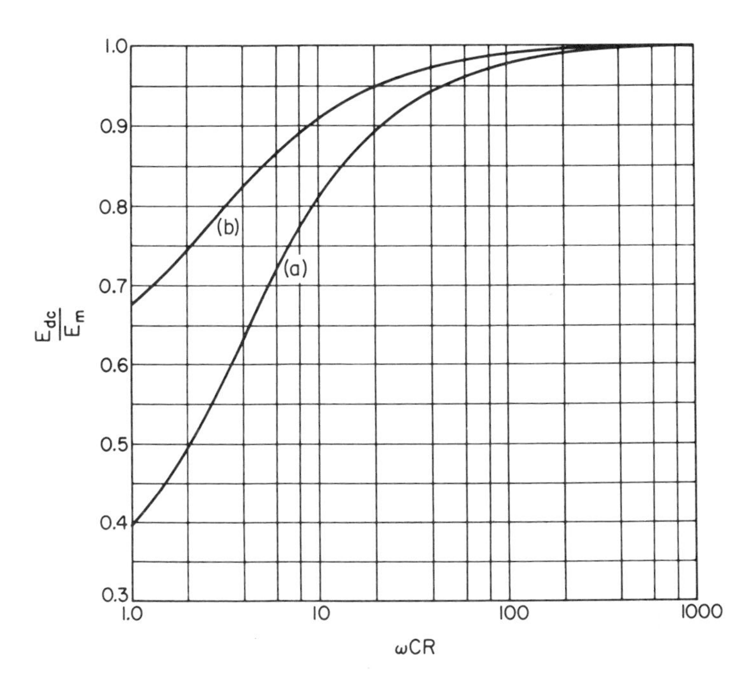

Figure 4-14. Variation of E_{dc}/E_m for capacitor filter: (a) half-wave input; (b) full-wave input.

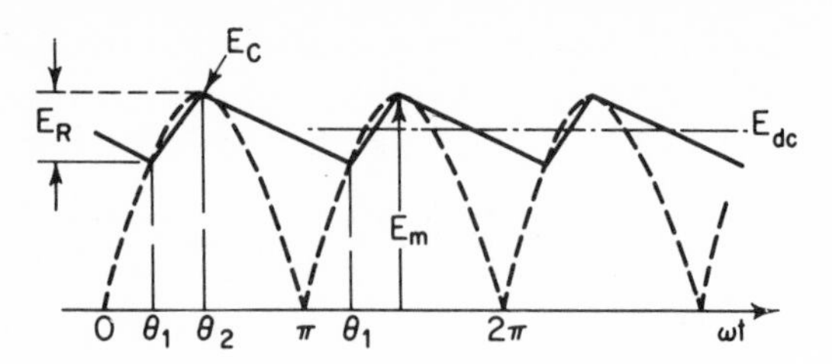

Figure 4-15. Output wave approximation, full-wave circuit.

This assumption introduces an anomaly, since if the load current is constant the load voltage cannot vary as shown; the assumption leads to useful results when the ripple is small, however. The rate of loss of capacitor potential then is

$$\frac{de_C}{dt} = \frac{E_R}{\pi - (\theta_2 - \theta_1)} = \frac{1}{\omega C}\frac{dq}{dt}$$

so that

$$E_R = \frac{[\pi - (\theta_2 - \theta_1)]I_{dc}}{\omega C} \tag{4-24}$$

where ω applies to the supply frequency, as before.

By usual methods it can be found that the effective value of the ripple component is

$$E_{ac} = \frac{E_R}{2\sqrt{3}} = \frac{[\pi - (\theta_2 - \theta_1)]I_{dc}}{2\sqrt{3}\,\omega C} \tag{4-25}$$

and the ripple factor can be written, using $E_{dc} = I_{dc}R$, as

$$\gamma = \frac{E_{ac}}{I_{dc}R} = \frac{\pi - (\theta_2 - \theta_1)}{2\sqrt{3}\,\omega CR} \tag{4-26}$$

for the full-wave rectifier with capacitor filter.

With semiconductor diodes, it becomes possible to use filter designs with $\omega CR > 100$. Vacuum diodes could not safely supply the high peak currents called for by such ωCR values and short conduction angles, but semiconductor diodes have larger allowable ratios of peak to average currents. Using electrolytic capacitors of 40 to 100 μF it is possible to obtain excellent regulation for varying or dynamic loads along with low ripple values as indicated in Fig. 4-16.

Reference to Fig. 4-12 shows that at $\omega CR = 100$, the conduction angle, $\theta_2 - \theta_1$, is less than 0.2 radian or 10°. Neglect of this small angle with respect to π radians in the numerator of Eq. 4-26 introduces only small error in the computation of the ripple. Then

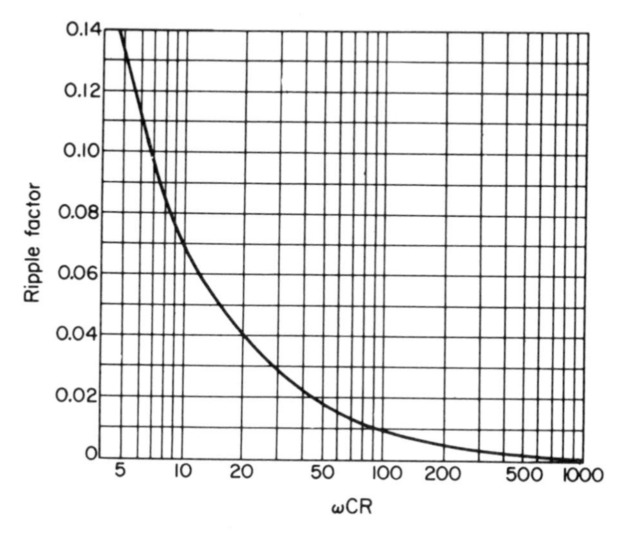

Figure 4-16. Ripple factor vs. ωCR, full-wave circuit.

$$\gamma \cong \frac{1}{4\sqrt{3}\, fCR} \tag{4-27}$$

Using Eqs. 4-23 and 4-24, a simplified dc voltage expression can be obtained as

$$E_{dc} = \frac{E_m}{1 + \dfrac{\pi - (\theta_2 - \theta_1)}{2\omega CR}} \cong \frac{E_m}{1 + \dfrac{1}{4fCR}} \tag{4-28}$$

The final expression follows because of the neglect of the conduction angle with respect to π.

For design purposes it is possible to write

$$E_{rms} = \frac{E_{dc}}{\sqrt{2}}\left[1 + \frac{\pi - (\theta_2 - \theta_1)}{2\omega CR}\right] \cong \frac{E_{dc}}{\sqrt{2}}\left[1 + \frac{1}{4fCR}\right] \tag{4-29}$$

The latter relation may be used to determine the approximate secondary voltage, E_{rms}, of one-half of the supply transformer for a specified E_{dc} output. For a specified ripple, Eq. 4-27 gives the needed capacitor value as

$$C \geqq \frac{1}{4\sqrt{3}\, fR\gamma} \tag{4-30}$$

When the voltage drop in a vacuum diode must be included in the design, the methods of Section 4-13 should be employed.

Because of frequent application of the voltage-doubler circuit and its

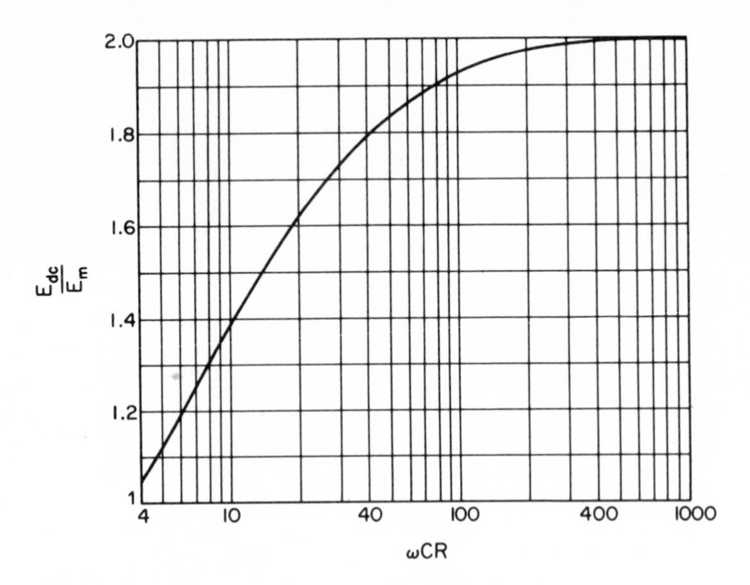

Figure 4-17. Output voltage vs. ωCR for the voltage doubler.

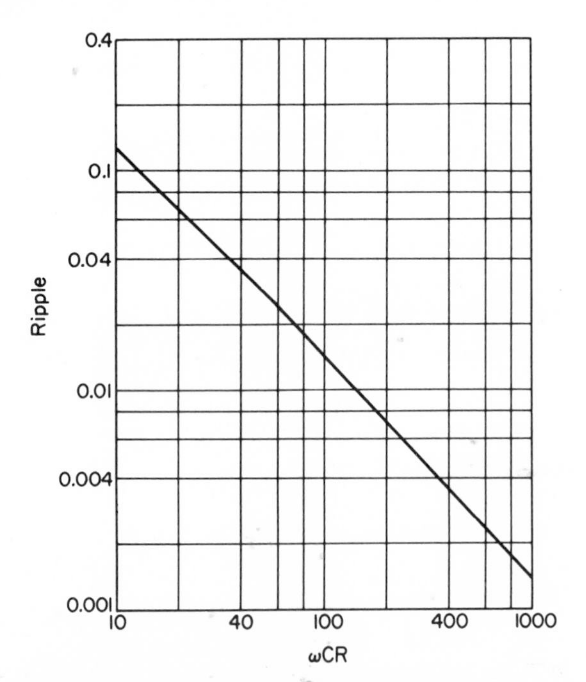

Figure 4-18. Ripple factor vs. ωCR for the voltage doubler.

use of the shunt-capacitor filter, performance curves for E_{dc}/E_m and ripple are included as Figs. 4-17 and 4-18. These can be derived by methods of analysis similar to those used in this section.

4-8 THE SERIES-INDUCTOR FILTER

An inductor in series with the resistive rectifier load presents a high impedance to the alternating components in the diode output, reducing the amplitude of all with respect to the dc current, and lowering the ripple. Magnetic energy is stored when the current tends to rise above the average value, and returned to the circuit when the current tends to fall below the average value. A series-inductor filter is shown connected to a half-wave diode in Fig. 4-19(a).

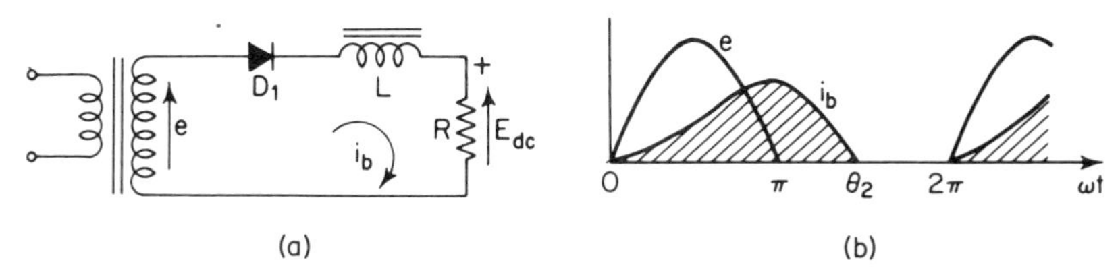

Figure 4-19. (a) Series-inductor filter, half-wave circuit; (b) current and voltage wave forms.

The equation for the circuit is

$$\frac{di_b}{dt} + \frac{i_b R}{L} = \frac{E_m \sin \omega t}{L}$$

and a complete solution can be obtained as

$$i_b = \frac{E_m}{R\sqrt{1 + \omega^2 L^2/R^2}}[\sin (\omega t - \alpha) + \epsilon^{-R\omega t/\omega L} \sin \alpha] \tag{4-31}$$

with $\alpha = \tan^{-1} \omega L/R$, and noting that $i_b = 0$ when $\omega t = 0$. The current pulses of Fig. 4-19(b) are of this form. Because of the discontinuous conduction the half-wave circuit produces a low average voltage, and is rarely used with an inductor filter.

However, the conduction angle of a diode exceeds 180°, and with full-wave supply the conduction angle of the first diode overlaps the starting angle of the second diode and there is a continuous current in the load. Neglecting the effects of the transformer reactance, the current appears to commutate between the diodes at voltage zero, the diode having an anode most positive to its cathode being conducting. The load current then appears as in Fig. 4-20(b).

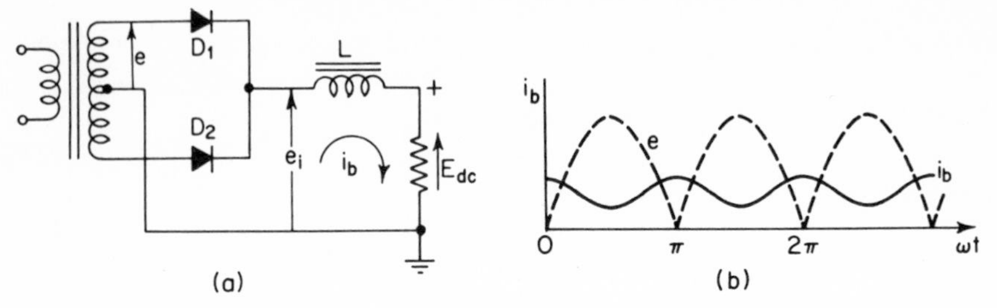

Figure 4-20. (a) Inductor filter and full-wave circuit; (b) load currents.

Analysis is simplified by reason of the continuous current, since the voltage wave applied to the input of the filter is that of the full-wave rectifier circuit, given in Eq. 4-10 as

$$e_t = \frac{2E_m}{\pi} - \frac{4E_m}{\pi}\left(\tfrac{1}{3}\cos 2\omega t + \tfrac{1}{15}\cos 4\omega t + \tfrac{1}{35}\cos 6\omega t \ldots\right)$$

There is no average voltage across an inductance, and so the dc voltage $2E_m/\pi$ will appear across the inductor resistance R_c and the load R. The dc load quantities then are

$$E_{dc} = \frac{2E_m}{\pi}\left(\frac{R}{R + R_c}\right), \qquad I_{dc} = \frac{2E_m}{\pi}\left(\frac{i}{R + R_c}\right) \tag{4-32}$$

The choke or inductor resistance should be small for good efficiency, and the R_c of the choke will hereafter be neglected.

At the second harmonic of the supply the input impedance of the filter and load is

$$Z_2 = R\sqrt{1 + \frac{4\omega^2 L^2}{R^2}} \tag{4-33}$$

At the next higher, or fourth harmonic

$$Z_4 = R\sqrt{1 + \frac{16\omega^2 L^2}{R^2}} \tag{4-34}$$

If $\omega^2 L^2/R^2 \gg 1$, as is usual, then $|Z_4/Z_2| = 2$. From the series for the applied voltage it can be seen that $E_2/E_4 = 5$, so that $I_2/I_4 \cong 10$. Thus the fourth and higher harmonics are small with respect to the second in producing ripple components in the load. Small error will be introduced if the second harmonic is used alone in computing the ripple.

Then

$$I_2 = \frac{4E_m}{3\sqrt{2}\pi}\,\frac{1}{R\sqrt{1 + 4\omega^2 L^2/R^2}}, \qquad I_{dc} = \frac{2E_m}{\pi R}$$

and the ripple will be

$$\gamma = \frac{\sqrt{2}}{3\sqrt{1 + 4\omega^2 L^2/R^2}} \tag{4-35}$$

It has been assumed that $\omega^2 L^2/R^2 \gg 1$, so

$$\gamma = \frac{0.236}{\omega L/R} \tag{4-36}$$

This result shows that *the ripple will decrease as the load current rises*, or as R decreases. This is in contrast to the capacitor filter in which the ripple increases with load current. However, the capacitor filter produces a higher dc voltage for a given E_m, with greater peak current demands.

4-9 THE LC FILTER

To provide lower ripple factors than are possible with the circuits already discussed, shunt capacitance and series inductance may be combined in the *LC* filter of Fig. 4-21. Information gained from the analysis of the capacitor filter helps to explain the circuit action at light loads, and at heavy loads analysis by the method of the previous section is possible.

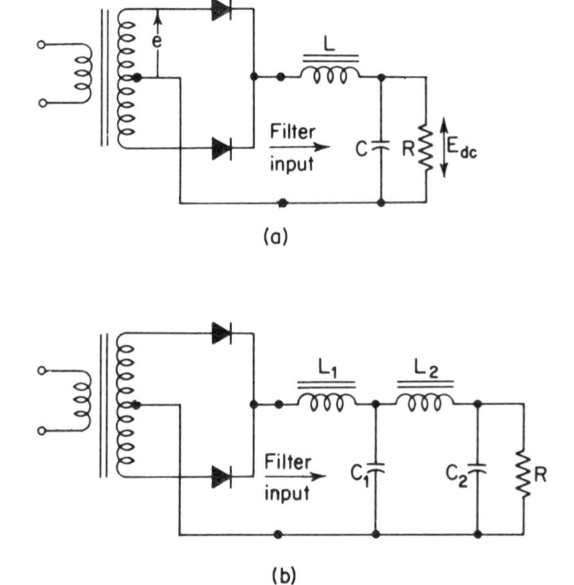

Figure 4-21. *LC* filters.

For light loads when $R \cong \infty$, the capacitor charges to the peak of the applied voltage, and the dc voltage will be E_m. As a small current is taken by the load, the diode switches the source onto the filter for an instant, and the capacitor charges to the peak again on each cycle; the dc voltage is slightly lower owing to the average capacitor voltage being below the peak of the wave. The current is so low that the small energy stored in the inductor

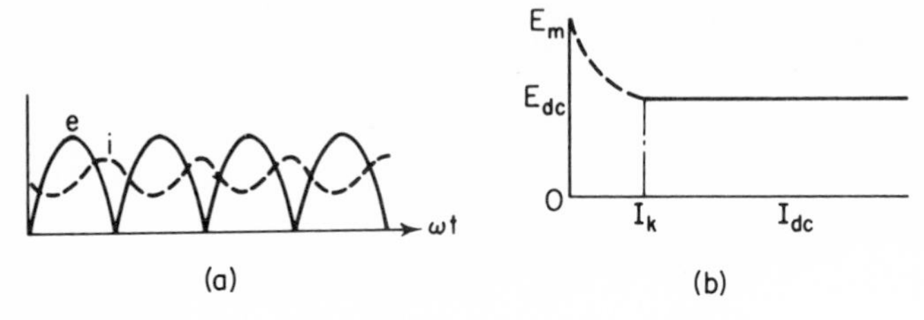

Figure 4-22. (a) *LC* filter current and voltage; (b) E_{dc} vs. I_{dc}.

has no effect; the action is much the same as with the full-wave capacitor filter. The voltage follows the dashed portion of the load curve in Fig. 4-22(b), decreasing with increasing current.

For still larger load currents, the conduction angle lengthens owing to the discharge of the capacitor and the presence of the inductor. At a load current I_k the conduction angle of each diode becomes 180°, and further increases in load current now build up the inductive stored energy so that the current in the inductor and load is never zero. In effect, the usual full-wave rectifier output of sinusoidal half-wave voltages, as at Fig. 4-22(a), is applied to the filter input. That is, with continuous load current above I_k, an alternating voltage of known harmonic content from Eq. 4-10, as

$$e = \frac{2E_m}{\pi} - \frac{4E_m}{\pi}\left(\tfrac{1}{3}\cos 2\omega t + \tfrac{1}{15}\cos 4\omega t + \tfrac{1}{35}\cos 6\omega t + \cdots\right)$$

is applied to an impedance, and ordinary circuit analysis methods apply.

The inductor L has a high reactance at the second harmonic frequency; it will have double this value of reactance at the fourth harmonic. Thus the fourth harmonic current will be only 10 per cent of the second harmonic value, as was previously shown. The reactance of C at the fourth harmonic is one-half of its reactance at the second harmonic, so that a load voltage produced by fourth-harmonic current will be only five per cent of the second-harmonic voltage. Therefore, currents of harmonic frequency higher than the second will be neglected with small error in the analysis which follows.

At the second harmonic, $2f$, the input impedance of the filter of Fig. 4-21(a) is

$$Z_2 = 2j\omega L - \frac{jR/2\omega C}{R - j/2\omega C}$$

and if the capacitive reactance is small with respect to R, then

$$|Z_2| = \frac{4\omega^2 LC - 1}{2\omega C} \tag{4-37}$$

The effective second harmonic voltage, when applied to Z_2, leads to a current

$$I_2 = \frac{8\omega C E_m}{3\pi\sqrt{2}\,(4\omega^2 LC - 1)} \tag{4-38}$$

Since in filters X_C is always small with respect to R, then the second-harmonic current passes predominantly through C, or $I_2 \cong I_C$. If I_C and I_R are the capacitor and load currents, it follows that $|I_R/I_C| = 1/2\omega CR$. Then

$$|I_R| = \frac{4E_m}{3\pi\sqrt{2}\,R(4\omega^2 LC - 1)} \tag{4-39}$$

From the Fourier series, $E_{dc} = 2E_m/\pi$, and $I_{dc} = E_{dc}/R$, so the ripple in the load is

$$\gamma = \frac{\sqrt{2}}{3(4\omega^2 LC - 1)} = \frac{0.47}{4\omega^2 LC - 1} \tag{4-40}$$

Ripple factor is plotted as a function of $\omega^2 LC$ in Fig. 4-23. As long as current conduction is continuous in the inductance L, *the ripple factor is independent of load current*. This feature permits the design of filters having specified ripple factors that can operate over a considerable current range. The ripple factors obtainable with LC filters are smaller than are possible with practical components in the circuits previously discussed. Obviously, for small ripple, the condition of $4\omega^2 LC = 1$ should be avoided.

A second section of filter may be added, as in Fig. 4-21(b), to achieve still further ripple reduction. By an analysis similar to that just undertaken, it can be shown that the ripple factor then is

$$\gamma = \frac{0.47}{(4\omega^2 L_1 C_1 - 1)(4\omega^2 L_2 C_2 - 1)} \tag{4-41}$$

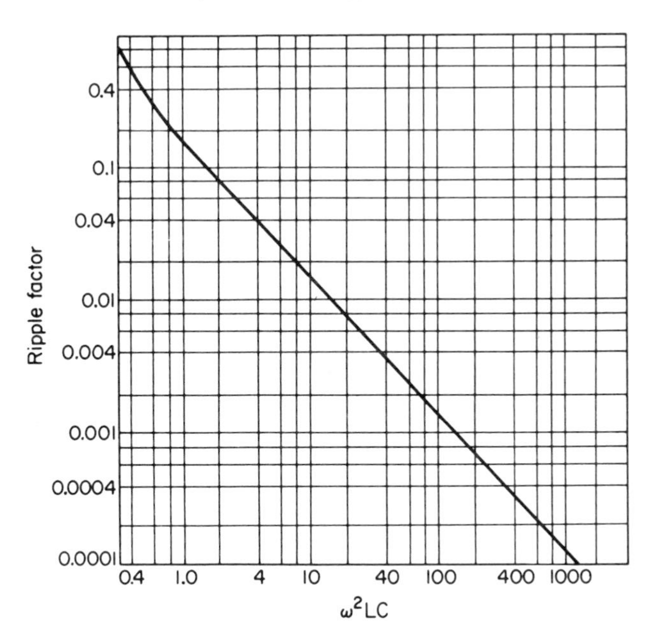

Figure 4-23. Ripple factor vs. $\omega^2 LC$ for LC filter with continuous current.

The filter could be extended to any number of sections, each reducing the ripple by the factor $1/(4\omega^2LC - 1)$.

To avoid the rise in voltage for currents less than I_k, a *bleeder* resistor may be placed in shunt with the load. The value of this resistor is such that it draws a current of approximately I_k A, and the rectifier load current cannot then drop below I_k. This insures continuous conduction in the inductor, and also improves the voltage regulation of the circuit for varying load.

The inductors or *chokes* will have inductances of the order of 5 to 30 henrys, with air gaps in the magnetic structures to reduce magnetic saturation by the dc load component. The capacitors will be of the order of 2 to 100 or more microfarads, of oil-filled paper or electrolytic types. While paper capacitors are usual for high voltages, it is now becoming common to employ series connections of cheaper electrolytic capacitors. It is then desirable to connect a high resistance across each capacitor to insure equal sharing of the applied dc voltage.

4-10 THE CRITICAL VALUE OF THE INPUT INDUCTOR

The value of I_k is that load current at which the diode output current just becomes continuous. However, the current from the diodes has two components, the dc value and the ac value, of which the second harmonic is of major interest. The dc component is determined by the load resistance, and the ac component by the filter input impedance; these two impedances must be in proper relation if the current is to be continuous. Then for $i > 0$, the condition of Fig. 4-24 becomes limiting and the dc component must exceed the second harmonic peak current or

$$\frac{2E_m}{\pi R} \geq \frac{4E_m}{3\pi Z_2} \tag{4-42}$$

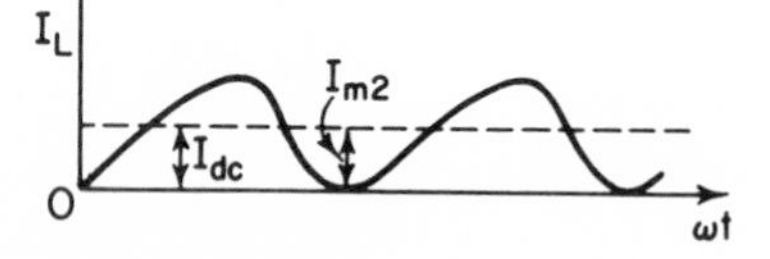

Figure 4-24. Current in inductor when $I_{dc} = I_k$.

The limiting condition comes from the equality as

$$Z_2 = \tfrac{2}{3}R$$

By reason of previous assumptions $Z_2 \cong 2\omega L$, and so the minimum value of the input inductor needed to maintain continuous current, will be

$$L_k \geq \frac{R}{3\omega} \cong \frac{R}{1100} \qquad \text{H} \tag{4-43}$$

for a supply at 60 hertz. The value of I_k will then be

$$I_k = \frac{2E_m}{3\pi\omega L} = \frac{0.212E_m}{\omega L} \qquad \text{A} \tag{4-44}$$

and the bleeder resistor should have a value $R_B = E_{dc}/I_k$.

The value of I_k can be reduced, allowing use of a higher value bleeder with less power wastage, by raising the ac input impedance of the filter, Z_2. A larger inductance implies higher cost and greater space, but the problem can be solved by use of a *swinging choke*. Such an inductor takes advantage of the variation of inductance of an iron-core reactor due to saturation by the dc current, and is designed to have an inductance at full dc load just large enough for the required filtering. At lighter current loads, the value of inductance rises, increasing the ac circuit impedance and reducing the value of I_k at which discontinuous current begins in the inductor.

The use of a swinging choke improves the dynamic regulation of the output voltage. At heavy load currents the inductance is small and load current surges can be supplied by drawing on the input transformer. If L were constant at a large value, its current could not readily change; a surge could then only be supplied by the charge available in the output capacitor and the voltage would fall with current output.

4-11 THE π FILTER

Placing a capacitor across the input terminals of the LC filter forms the π filter or capacitor-input filter of Fig. 4-25. The addition of the input capacitor with full-wave diode rectification results in a light-load (high R) voltage higher than obtainable from the L-section filter. At heavy currents (lower R) the average voltage on C_1 falls much below E_m.

Capacitor C_1 takes a charging pulse and charges to E_m before being disconnected from the source by a diode. The capacitor then discharges through L and R until a second diode again connects the source. The discharge current of C_1 is difficult to evaluate, but for an approximate analysis it may be assumed that the current from C_1 is constant at I_{dc}. This seems reasonable since L will tend to maintain constant current.

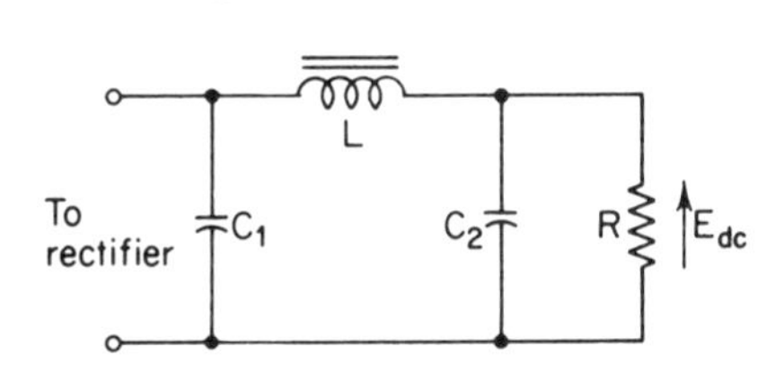

Figure 4-25. The π filter.

Following the method of Section 4-9, the wave form across C_1 may be approximated by Fig. 4-26. For small ripple $\omega C_1 R$ is large and $\theta_2 - \theta_1$, the angle of conduction, is small. This angle may then be neglected with respect to π in Eq. 4-24, and the ripple voltage is

$$E_R = \frac{\pi I_{dc}}{\omega C_1} \tag{4-45}$$

The dc output voltage is given by use of Eq. 4-23 as

$$E_{dc} = E_m - I_{dc}\left(\frac{\pi}{2\omega C_1} + R_c\right) = E_m\left(1 - \frac{\pi}{2\omega C_1 R} - \frac{R_c}{R}\right) \tag{4-46}$$

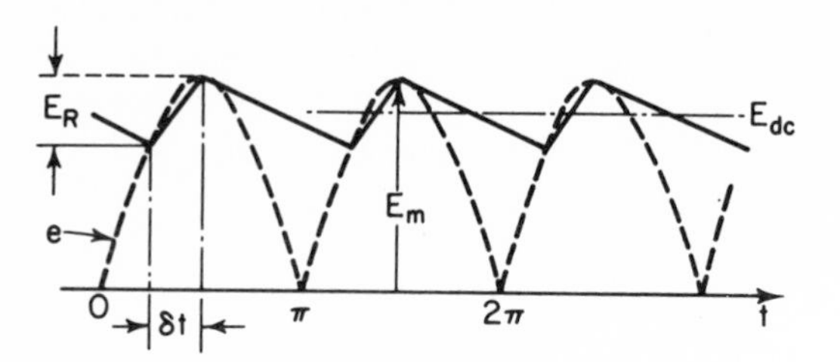

Figure 4-26. Approximate input wave across C_1.

where R_c is the resistance of the filter choke. The value of dc voltage is thus a function of C_1.

The triangular ripple wave form of Fig. 4-26 has a Fourier series given by

$$e_R = -\frac{E_R}{\pi}\sin 2\omega t + \frac{E_R}{2\pi}\sin 4\omega t - \frac{E_R}{3\pi}\sin 6\omega t \cdots \tag{4-47}$$

This wave form is applied across L and C_2 in series if $1/2\omega C_2 \ll R$, where R represents the load and bleeder in parallel. The second harmonic is predominant and leads to a voltage across C_2 of

$$E_2 = \frac{E_R}{\pi\sqrt{2}}\left(\frac{1}{4\omega^2 LC_2 - 1}\right)$$

Use of E_R from Eq. 4-45 allows the ripple to be written as

$$\gamma = \frac{E_2}{E_{dc}} = \frac{1}{\sqrt{2}\,\omega C_1 R(4\omega^2 LC_2 - 1)} \tag{4-48}$$

For the capacitive input filter, *the ripple increases with load current.* The value of output voltage increases with C_1 by reason of Eq. 4-46, and the ripple is a function of both C_1 and C_2.

The input capacitor provides a means of obtaining a higher voltage from a given transformer than is possible with the LC filter, and materially reduces the ripple. The penalty is poorer dynamic regulation than is obtained from the shunt capacitor circuit, and poorer static regulation than exists with the LC circuit.

4-12 THE RESISTIVE FILTER

Because of cost and space requirements, the inductor of the LC filter is occasionally replaced with a resistor, as in Fig. 4-27. This increases the internal voltage drop, but this may not be serious in designs for small steady-current loads, as in cathode-ray oscilloscope supplies. When used with a full-wave rectifier, the previous sections showed that the second-harmonic voltage is the major source of ripple, and at the input to the filter has an rms value

$$E_2 = \frac{4E_m}{3\sqrt{2}\,\pi}$$

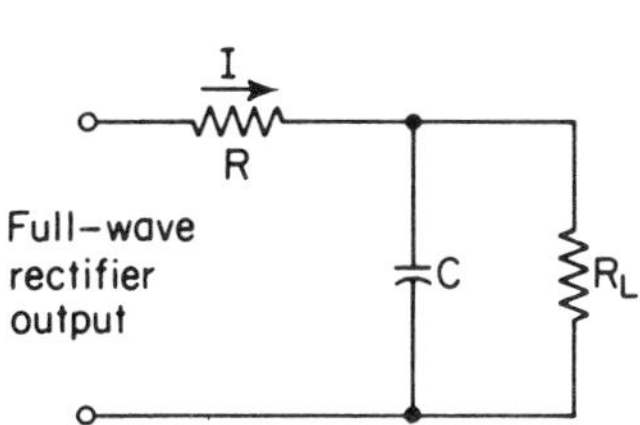

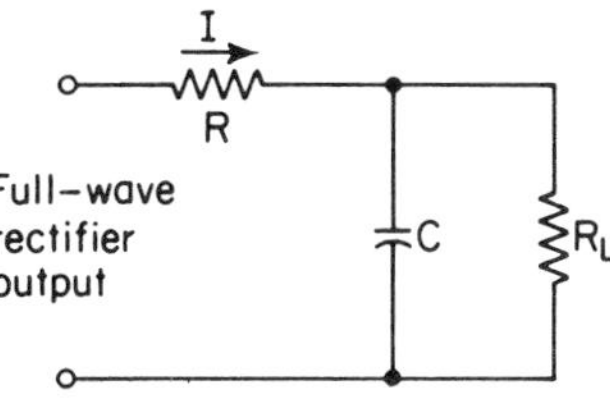

Figure 4-27. The RC filter.

Figure 4-28. Filtering and voltage adjusting with an RC filter.

If $1/2\omega C \ll R_L$, then the second-harmonic current is

$$I_2 \cong \frac{4E_m}{3\sqrt{2}\,\pi R}$$

and this produces a ripple voltage across C of

$$E_C = \frac{\sqrt{2}\,E_m}{3\pi\omega CR} \tag{4-49}$$

The load dc voltage is

$$E_{\text{dc}} = \frac{2E_m}{\pi}\left(\frac{R_L}{R + R_L}\right) \tag{4-50}$$

and so the ripple is

$$\gamma = \frac{1 + R/R_L}{3\sqrt{2}\,\omega CR} \tag{4-51}$$

Large values of C provide small ripple. The circuit is often used as a voltage-dropping section, to provide additional filtering as well as a lowered voltage, as in Fig. 4-28.

4-13 OPERATION CHARACTERISTICS FOR VACUUM DIODES

Circuits employing semiconductor diodes perform in a manner closely approaching the theoretical, and the relations derived are suited to design. However, the design of rectifier circuits utilizing vacuum diodes is often carried out by use of *operation characteristics*, illustrated in Figs. 4-29 and 4-30, in order to include the effect of the diode internal resistance. The characteristics are plots of performance of actual diode types, in terms of dc output voltage available from various filter circuits.

The curves of Fig. 4-29, for a capacitor-input filter with $C_1 = 10\ \mu$F, show the rise to E_m at low currents, and the rather poor voltage regulation with varying load. Using values selected from the chart, it is possible to subtract the voltage drop in the choke and arrive at probable dc output voltages.

Figure 4-30 is of similar nature for a choke-input or L-section filter. The better voltage regulation is apparent, as well as the tendency for the

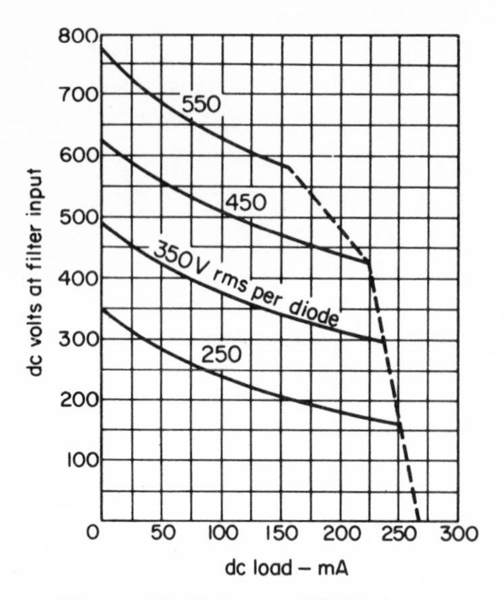

Figure 4-29. Full-wave 5U4G tube, capacitor input, $C = 10\ \mu$F.

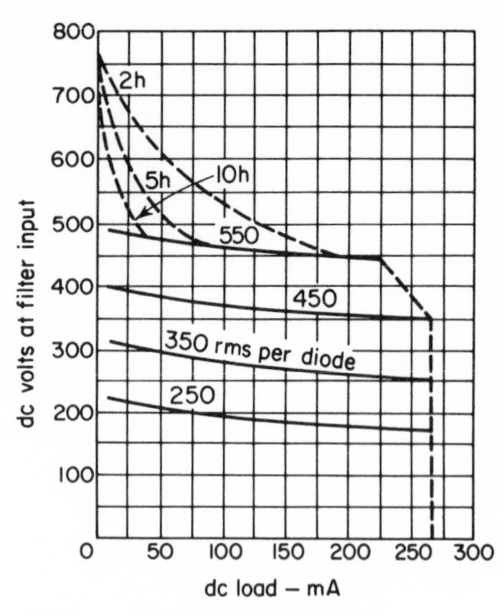

Figure 4-30. Full-wave 5U4G tube, inductor input, $C = 16\ \mu$F.

circuit to become a capacitor-input form at light currents, where the effect of L disappears.

Charts should be employed for each specific type of vacuum diode employed.

4-14 CIRCUIT SELECTION AND OPERATION

The maximum reverse voltage rating of a vacuum diode is fixed by the insulation breakdown limits, while the semiconductor diode is limited by the effects of reverse avalanche breakdown.

The peak forward current is also a limitation. It is related to the maximum available emission in the vacuum diode, or to the maximum instantaneous heat loss in the semiconductor. The average current is a limit, being fixed by the allowable average heat loss in either device.

Input filter chokes have been utilized in the past to protect the vacuum diodes from high current peaks, caused by filter capacitor charging, but the choke input filter gives a lower dc output voltage from a given transformer. Shunt-capacitor filters were avoided because of the peak currents required. The availability of small silicon diodes, in which peak-to-average current ratios as high as fifteen are common, and the use of electrolytic capacitors in series for high voltages, have changed the design concept. The input filter choke with its cost and weight may be eliminated in some designs. Since the dc output voltage approximates E_m in shunt-capacitor filters, there is then no need for expensive bleeders to prevent the output voltage from rising at low currents.

Shunt-capacitor filters have excellent dynamic regulation characteristics with varying loads, as are met with Class B amplifiers. This excellent regulation is illustrated for high ωCR values in Fig. 4-14.

Silicon diodes may be used in series for high reverse voltages. Each diode should then be shunted with a resistor of about 100,000 ohms, to equalize the diode reverse voltages. Similarly, electrolytic capacitors in series should be individually shunted by resistance, since their dc resistance may vary and cause unequal distribution of the applied dc voltage. These resistors also serve to discharge the capacitors when the power is disconnected.

Operating temperatures are important. The actual temperature of the junction is critical, and high ambient temperatures may dictate the reduction of electrical load. A typical derating curve for various ambient temperatures is shown in Fig. 4-31. Heat sinks or conductive cooling may be required in some applications.

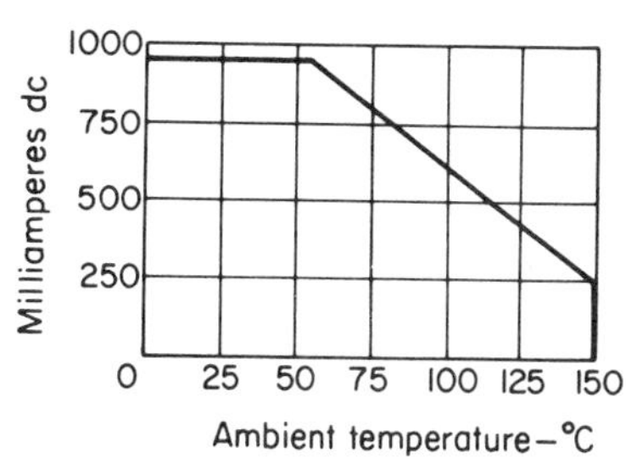

Figure 4-31. Derating curve for a small silicon diode.

PROBLEMS

4-1. A diode carries a current given by $i = 60t$. The wave starts at $t = 0$ and falls to zero at $t = \frac{1}{60}$ s, after which it repeats indefinitely. Find the dc voltage across a 50-ohm load, the total power loss in the load, and the ripple factor.

4-2. A half-wave ideal rectifier uses an applied voltage of 250 rms and 60 hertz, and the load is 1750 ohms. Find:

(a) I_{dc}.
(b) E_{dc}.
(c) I_{rms} in the load.
(d) The load power loss.

4-3. The silicon diode of Fig. 4-1 is used to charge a 12-V battery from a 36-V rms source in a half-wave circuit. If the charging rate is to be 50 A, find the value of series resistance needed, assuming the battery resistance is 0.07 ohm.

4-4. A diode is rated $I_m = 675$ mA, average $I = 350$ mA. Two of these units are used in a full-wave circuit with a transformer supplying 210 V rms each side of the center-tap. Neglect diode resistance and find:

(a) The value of load resistance which must be present to give the greatest dc power output without exceeding any diode rating.
(b) Specify the dc load voltage and current for (a).

4-5. Find the value of load R permissible for maximum dc power output from a full-wave rectifier using diodes rated P.I.V. $= 400$ V, $I_m = 5$ A, $I_{av} =$ 750 mA.

4-6. A full-wave rectifier is to supply a load of 3000 ohms at 300 V, dc. Find:
(a) E_{rms} of the transformer required.
(b) Peak diode current.
(c) Power output of the transformer.

4-7. Using the method which led to Eq. 4-26, develop a similar expression for the ripple in the output of a half-wave circuit with shunt-capacitor filter.

4-8. Determine the ac transformer voltage needed to supply a half-wave rectifier circuit to charge a 6-V battery of resistance 0.1 ohm at a charging rate of 3 A.

4-9. A transformer with an 800-V center-tapped secondary winding supplies a full-wave rectifier circuit. The load is 1000 ohms, and a shunt capacitor of 60 μF is used. Find:
(a) The dc load voltage.
(b) The peak current demand on a diode.
(c) The ripple factor.

4-10. Prove that the rms value of the triangular ripple wave of Fig. 4-15 is given by $E_R/2\sqrt{3}$.

4-11. Utilizing the methods of the shunt-capacitor filter analysis, find an expression for the ripple factor of the voltage-doubling rectifier, in terms of ωCR, assuming $C_1 = C_2 = C$.

4-12. Plot a diode current wave form for the shunt-capacitor circuit and full-wave rectification, with $\omega CR = 40$, $E_m = 270$ V, $I_{dc} = 300$ mA.

4-13. A shunt-capacitor filter supplies a resistance from a full-wave rectifier circuit. The following readings are taken: $E_{dc} = 425$ V, $I_{dc} = 227$ mA, $I_{rms} = 235$ mA in the load. Find the existing ripple and size of capacitor being used.

4-14. With a full-wave rectifier and LC filter, the applied rms voltage per side of the transformer is 300, $f = 60$ hertz, $L = 10$ H, $C = 4$ μF. Find E_{dc}, the ripple factor, and the maximum value that R may have without the dc voltage rising excessively.

4-15. With a full-wave circuit supplying a load with 0.2 A at 300 V, find the per cent ripple to be expected when a 5-henry choke or a 10-μF capacitor are used as elements in:
(a) Shunt-capacitor filter.
(b) Series-inductor filter.
(c) L-section filter.

4-16. A π filter uses two 20-μF capacitors and a 20-henry inductor. With a load of 3000 ohms, and a transformer supplying 250 V to center-tap at 60 hertz, determine the dc output voltage and the ripple, using full-wave rectification.

4-17. Design an *LC* filter for a full-wave rectifier circuit supplied at 60 hertz. It is to have a 3000-ohm load at 300 V E_{dc}, and a ripple factor of 0.005. Use a critical value for the choke.

4-18. A π-section filter is to supply a 3000 ohm load with a current of 100 mA. Design the circuit for a ripple of 0.005, and specify the transformer voltages.

4-19. A full-wave rectifier uses the 5U4G tube of Figs. 4-29 and 4-30. For a load requiring 300 V dc and 175 mA, find the ac transformer voltages for a capacitor-input filter with $C_1 = 10\ \mu F$, and for an inductor-input filter with output $C = 16\ \mu F$. Choke inductance and resistance are 10 henrys and 250 ohms.

REFERENCES

1. Waidelich, D. L., "Diode Rectifying Circuits with Capacitance Filters," *Trans. AIEE*, Vol. 60, p. 1161 (1941).
2. Waidelich, D. L., "The Full-Wave Voltage-Doubling Rectifier Circuit," *Proc. IRE*, Vol. 29, p. 554 (1941).
3. Schade, O. H., "Analysis of Rectifier Operation," *Proc. IRE*, Vol. 31, p. 341 (1943).
4. Waidelich, D. L., "Analysis of Full-Wave Rectifier and Capacitive Input Filter," *Electronics*, Vol. 20, p. 120 (September 1947).
5. Henkels, H. W., "Germanium and Silicon Rectifiers," *Proc. IRE*, Vol. 46, p. 1086 (1958).
6. *Silicon Rectifier Handbook*. Motorola, Inc., Phoenix, Ariz., 1966.

5

TWO-PORT NETWORKS

The linear electrical network of greatest interest to our studies connects a two-terminal source to a two-terminal load and so is referred to as a four-terminal or more properly a *two-port network*. The characteristics of transistor or tube may be linearized in a piece-wise sense, and the properties of two-port networks may then be applied to the transistor and tube.

This chapter provides a review of such network material, and may be omitted by the student with a satisfactory network theory background.

5-1 THE TWO-PORT NETWORK

Two-port networks may be studied as a class, without regard to the exact internal circuitry, and a general two-port network with defined currents and voltages appears in Fig. 5-1. Currents are assumed positive inward so that specification of input or output terminals does not alter an analysis.

It is the terminal quantities v_1, i_1, v_2, i_2 by which the network responds to external forcing functions, and specification of these quantities is equivalent to specification of the performance of the network. A network analysis, dependent on terminal quantities, applies whether a linear network is active or passive. It is also possible to apply network relations to devices that do not have linear volt–ampere curves, by assumption of linearity of the device in the range over which the device is operated. Linearity is illustrated in Fig. 5-2(a) and piece-wise linearity from A to B is shown in Fig. 5-2(c).

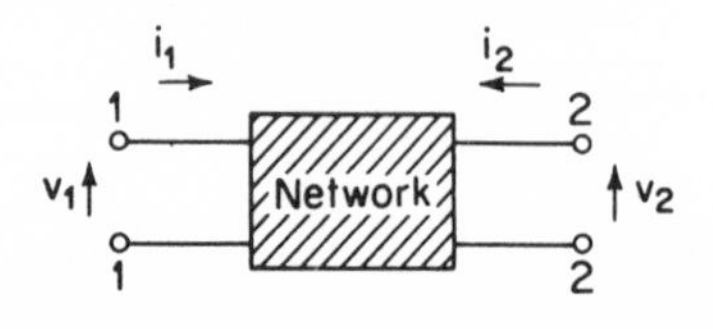

Figure 5-1. Two-port network.

Any pair of the network terminal variables v_1, i_1, v_2, i_2 may be arbitrarily chosen as independent variables, leading to two equa-

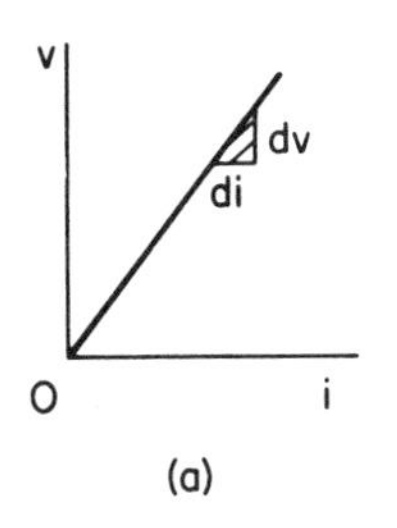

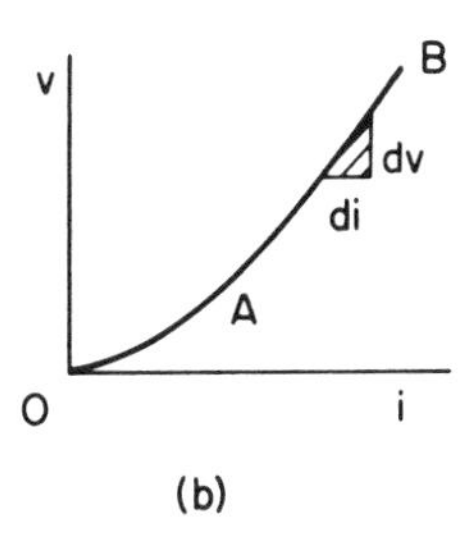

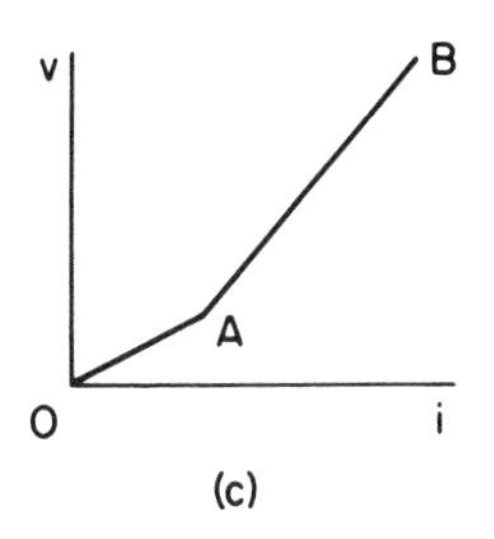

Figure 5-2. (a) Volt–ampere curve for a linear device; (b) nonlinear device; (c) piecewise linear approximation.

tions which may be solved for the other two dependent variables. Choice of three of the six possible independent variable pairs as i_1, and i_2, v_1 and v_2, and v_2 and i_1, leads to three sets of circuit parameter definitions which have been found useful in electronic analysis.

5-2 *OPEN-CIRCUIT IMPEDANCE PARAMETERS*

Choosing i_1 and i_2 as the independent variables implies the following general functional relations for the network of Fig. 5-1:

$$v_1 = f_1(i_1, i_2)$$

$$v_2 = f_2(i_1, i_2)$$

The circuits of interest are to be operated with changing or ac currents, and the effect of varying voltages or currents is to move along the slope of the volt–ampere curve. These changes may be mathematically specified by taking the total differentials as

$$dv_1 = \frac{\partial v_1}{\partial i_1} di_1 + \frac{\partial v_1}{\partial i_2} di_2 \tag{5-1}$$

$$dv_2 = \frac{\partial v_2}{\partial i_1} di_1 + \frac{\partial v_2}{\partial i_2} di_2 \tag{5-2}$$

If ac voltages are employed, and operation is over only the quasi-linear region of the appropriate volt–ampere characteristics, then *the partial derivatives become constants with dimensions of impedance.* For sufficiently small sinusoidal signals the above equations may be written as

$$\mathbf{V}_1 = \mathbf{z}_i\mathbf{I}_1 + \mathbf{z}_r\mathbf{I}_2 \tag{5-3}$$

$$\mathbf{V}_2 = \mathbf{z}_f\mathbf{I}_1 + \mathbf{z}_o\mathbf{I}_2 \tag{5-4}$$

which in matrix form become

$$\begin{bmatrix} \mathbf{V}_1 \\ \mathbf{V}_2 \end{bmatrix} = \begin{bmatrix} \mathbf{z}_i & \mathbf{z}_r \\ \mathbf{z}_f & \mathbf{z}_o \end{bmatrix} \begin{bmatrix} \mathbf{I}_1 \\ \mathbf{I}_2 \end{bmatrix} \qquad \begin{matrix} (5\text{-}3) \\ (5\text{-}4) \end{matrix}$$

An advantage of the use of matrix notation is here demonstrated, in that it separates and identifies the independent variables $\mathbf{I}_1$ and $\mathbf{I}_2$, the circuit function, and the dependent variables $\mathbf{V}_1$ and $\mathbf{V}_2$ on the left side.

The **z** impedances may also be written

$$\mathbf{z}_i = \mathbf{z}_{11}, \qquad \mathbf{z}_f = \mathbf{z}_{21}$$
$$\mathbf{z}_r = \mathbf{z}_{12}, \qquad \mathbf{z}_o = \mathbf{z}_{22}$$

but the letter subscripts seem preferred in electronic study and will be used here.

While the $\mathbf{z}$ parameters have been defined as partial derivatives, or as the slopes of the volt-ampere relations, they may be correlated with actual measurements made on the network with assigned terminations. If the network is open-circuited at the 2,2 terminals, then $\mathbf{I}_2 = 0$, and Eq. 5-3 leads to $\mathbf{z}_i = \mathbf{V}_1/\mathbf{I}_1$. By use of the same termination in Eq. 5-4, $\mathbf{z}_f = \mathbf{V}_2/\mathbf{I}_1$. If the network is open-circuited at the 1,1 terminals, then $\mathbf{I}_1 = 0$, and the equations yield $\mathbf{z}_r = \mathbf{V}_1/\mathbf{I}_2$, and $\mathbf{z}_o = \mathbf{V}_2/\mathbf{I}_2$. These ratios can be measured on the network, using ac signals.

The $\mathbf{z}$ impedances have become known as the *open-circuit impedance parameters* for the two-port network, and their definitions may be collected as:

$$\left.\begin{aligned}
\mathbf{z}_i &= \mathbf{z}_{11} = \frac{\partial v_1}{\partial i_1} = \frac{\mathbf{V}_1}{\mathbf{I}_1} = \text{open-circuit input impedance, } \mathbf{I}_2 = 0\\
\mathbf{z}_r &= \mathbf{z}_{12} = \frac{\partial v_1}{\partial i_2} = \frac{\mathbf{V}_1}{\mathbf{I}_2} = \text{open-circuit reverse transfer impedance, } \mathbf{I}_1 = 0\\
\mathbf{z}_f &= \mathbf{z}_{21} = \frac{\partial v_2}{\partial i_1} = \frac{\mathbf{V}_2}{\mathbf{I}_1} = \text{open-circuit forward transfer impedance, } \mathbf{I}_2 = 0\\
\mathbf{z}_o &= \mathbf{z}_{22} = \frac{\partial v_2}{\partial i_2} = \frac{\mathbf{V}_2}{\mathbf{I}_2} = \text{open-circuit output impedance, } \mathbf{I}_1 = 0
\end{aligned}\right\} \tag{5-5}$$

Measurements of this nature can be made upon any linear, or piece-wise linear network over the operating range, whether active or passive; these definitions and others like them are extremely fundamental.

5-3 SHORT-CIRCUIT ADMITTANCE PARAMETERS

If we choose v_1 and v_2 as the independent variables of the two-port network, then the general functional relations become

$$i_1 = f_3(v_1, v_2)$$
$$i_2 = f_4(v_1, v_2)$$

and upon taking the total differentials as before:

$$di_1 = \frac{\partial i_1}{\partial v_1} dv_1 + \frac{\partial i_1}{\partial v_2} dv_2 \tag{5-6}$$

$$di_2 = \frac{\partial i_2}{\partial v_1} dv_1 + \frac{\partial i_2}{\partial v_2} dv_2 \tag{5-7}$$

If sinusoidal currents are employed and operation is over a range in which the slopes of the volt–ampere curves or the partial derivatives are constant, then

$$\mathbf{I}_1 = \mathbf{y}_i\mathbf{V}_1 + \mathbf{y}_r\mathbf{V}_2 \tag{5-8}$$

$$\mathbf{I}_2 = \mathbf{y}_f\mathbf{V}_1 + \mathbf{y}_o\mathbf{V}_2 \tag{5-9}$$

or

$$\begin{bmatrix}\mathbf{I}_1\\ \mathbf{I}_2\end{bmatrix} = \begin{bmatrix}\mathbf{y}_i & \mathbf{y}_r\\ \mathbf{y}_f & \mathbf{y}_o\end{bmatrix}\begin{bmatrix}\mathbf{V}_1\\ \mathbf{V}_2\end{bmatrix} \qquad \begin{matrix}(5\text{-}8)\\(5\text{-}9)\end{matrix}$$

The **y** parameters may be defined from measurements on the two-port network, or from the above equations, by using short-circuit terminations. If the 2,2 terminals be short-circuited, then $\mathbf{V}_2 = 0$, and Eq. 5-8 yields $\mathbf{y}_i = \mathbf{I}_1/\mathbf{V}_1$, while Eq. 5-9 gives $\mathbf{y}_f = \mathbf{I}_2/\mathbf{V}_1$. If the 1,1 terminals be short-circuited, appropriate measurements may be made, or since $\mathbf{V}_1 = 0$, then Eqs. 5-8 and 5-9 give $\mathbf{y}_r = \mathbf{I}_1/\mathbf{V}_2$ and $\mathbf{y}_o = \mathbf{I}_2/\mathbf{V}_2$.

These admittances have become known as the *short-circuit admittance parameters* for the two-port network, and their definitions can be summarized as:

$$\left.\begin{aligned}
\mathbf{y}_i &= \mathbf{y}_{11} = \frac{\partial i_1}{\partial v_1} = \frac{\mathbf{I}_1}{\mathbf{V}_1} = \text{short-circuit input admittance, } \mathbf{V}_2 = 0\\
\mathbf{y}_r &= \mathbf{y}_{12} = \frac{\partial i_1}{\partial v_2} = \frac{\mathbf{I}_1}{\mathbf{V}_2} = \text{short-circuit reverse transfer admittance, } \mathbf{V}_1 = 0\\
\mathbf{y}_f &= \mathbf{y}_{21} = \frac{\partial i_2}{\partial v_1} = \frac{\mathbf{I}_2}{\mathbf{V}_1} = \text{short-circuit forward transfer admittance, } \mathbf{V}_2 = 0\\
\mathbf{y}_o &= \mathbf{y}_{22} = \frac{\partial i_2}{\partial v_2} = \frac{\mathbf{I}_2}{\mathbf{V}_2} = \text{short-circuit output admittance, } \mathbf{V}_1 = 0
\end{aligned}\right\} \tag{5-10}$$

5-4 HYDRID OR *h* PARAMETERS

It has been found that the third choice of i_1 and v_2 as independent variables leads to a set of network parameters of considerable value in transistor circuit analysis. If this choice of variables is made,

$$v_1 = f_5(i_1, v_2)$$

$$i_2 = f_6(i_1, v_2)$$

Taking the total differentials,

$$dv_1 = \frac{\partial v_1}{\partial i_1}di_1 + \frac{\partial v_1}{\partial v_2}dv_2 \tag{5-11}$$

$$di_2 = \frac{\partial i_2}{\partial i_1}di_1 + \frac{\partial i_2}{\partial v_2}dv_2 \tag{5-12}$$

Again, for small sinusoidal signals limited to the quasi-linear region the partial derivatives become constants and

$$\mathbf{V}_1 = \mathbf{h}_i\mathbf{I}_1 + \mathbf{h}_r\mathbf{V}_2 \tag{5-13}$$

$$\mathbf{I}_2 = \mathbf{h}_f\mathbf{I}_1 + \mathbf{h}_o\mathbf{V}_2 \tag{5-14}$$

or

$$\begin{bmatrix}\mathbf{V}_1\\ \mathbf{I}_2\end{bmatrix} = \begin{bmatrix}\mathbf{h}_i & \mathbf{h}_r\\ \mathbf{h}_f & \mathbf{h}_o\end{bmatrix}\begin{bmatrix}\mathbf{I}_1\\ \mathbf{V}_2\end{bmatrix} \qquad (5\text{-}13),\ (5\text{-}14)$$

The $\mathbf{h}$ coefficients are known as the *hybrid parameters*, since both open- and short-circuit terminations are used in defining them for the general network. By appropriate choice of open-circuit ($\mathbf{I}_1 = 0$) and short-circuit ($\mathbf{V}_2 = 0$) conditions as applied to Eqs. 5-13 and 5-14, it is possible to arrive at definitions for the small signal ac values of the $\mathbf{h}$ parameters as

$$\left.\begin{aligned}
\mathbf{h}_i &= \mathbf{h}_{11} = \frac{\partial v_1}{\partial i_1} = \frac{\mathbf{V}_1}{\mathbf{I}_1} = \text{short-circuit input impedance, } \mathbf{V}_2 = 0\\
\mathbf{h}_r &= \mathbf{h}_{12} = \frac{\partial v_1}{\partial v_2} = \frac{\mathbf{V}_1}{\mathbf{V}_2} = \text{open-circuit reverse voltage ratio, } \mathbf{I}_1 = 0\\
\mathbf{h}_f &= \mathbf{h}_{21} = \frac{\partial i_2}{\partial i_1} = \frac{\mathbf{I}_2}{\mathbf{I}_1} = \text{short-circuit forward current ratio, } \mathbf{V}_2 = 0\\
\mathbf{h}_o &= \mathbf{h}_{22} = \frac{\partial i_2}{\partial v_2} = \frac{\mathbf{I}_2}{\mathbf{V}_2} = \text{open-circuit output admittance, } \mathbf{I}_1 = 0
\end{aligned}\right\} \tag{5-15}$$

While all these defined parameters have meaning as slopes of graphical characteristics, their magnitudes are usually obtained by measurement, using small signals in appropriately terminated circuits. A particular advantage of the hybrid parameters is the accuracy with which they can be measured in circuits of the type encountered with electronic devices.

Relations which pemit the $\mathbf{z}$ parameters to be calculated from $\mathbf{h}$ measurements are often needed, and can be obtained by solution of Eqs. 5-3, 5-4, 5-13, and 5-14, leading to

$$\mathbf{h}_i = \mathbf{z}_i - \frac{\mathbf{z}_r\mathbf{z}_f}{\mathbf{z}_o}, \qquad \mathbf{h}_f = -\frac{\mathbf{z}_f}{\mathbf{z}_o} \tag{5-16}$$

$$\mathbf{h}_r = \frac{\mathbf{z}_r}{\mathbf{z}_o}, \qquad \mathbf{h}_o = \frac{1}{\mathbf{z}_o}$$

and the inverse relations

$$\mathbf{z}_i = \mathbf{h}_i - \frac{\mathbf{h}_r\mathbf{h}_f}{\mathbf{h}_o}, \qquad \mathbf{z}_f = -\frac{\mathbf{h}_f}{\mathbf{h}_o} \tag{5-17}$$

$$\mathbf{z}_r = \frac{\mathbf{h}_r}{\mathbf{h}_o}, \qquad \mathbf{z}_o = \frac{1}{\mathbf{h}_o}$$

5-5 DEFINED IMPEDANCES OF THE TWO-PORT NETWORK

As a general circuit parameter, the *input impedance* at the 1,1 terminals of the loaded two-port network of Fig. 5-3 may be found in terms of the $\mathbf{z}$ parameters as $\mathbf{Z}_i$. From Eq. 5-3

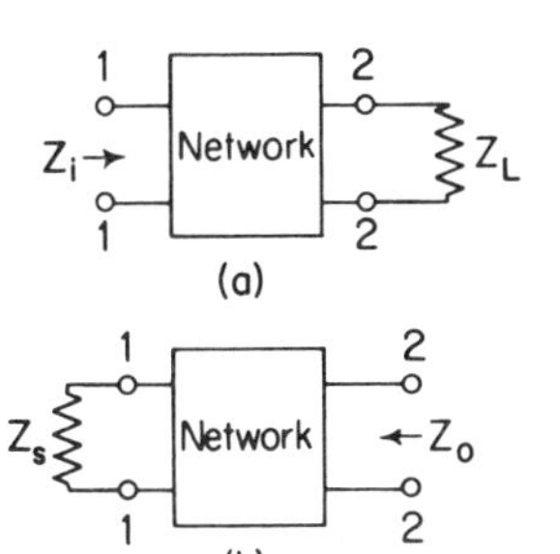

Figure 5-3. Input and output impedances.

$$\mathbf{Z}_i = \frac{\mathbf{V}_1}{\mathbf{I}_1} = \mathbf{z}_i + \mathbf{z}_r \frac{\mathbf{I}_2}{\mathbf{I}_1} \tag{5-18}$$

using the upper-case $\mathbf{Z}$ to indicate an external circuit value.

With a load $\mathbf{Z}_L$ on the 2,2 terminals, $\mathbf{V}_2/\mathbf{I}_2 = -\mathbf{Z}_L$, the negative sign appearing because of the defined direction of $\mathbf{I}_2$. Then from Eq. 5-4

$$\frac{\mathbf{I}_2}{\mathbf{I}_1} = \frac{-\mathbf{z}_f}{\mathbf{z}_o + \mathbf{Z}_L} \tag{5-19}$$

and this result will be useful later. Using this loaded current ratio in Eq. 5-18, the input impedance becomes

$$\mathbf{Z}_i = \mathbf{z}_i - \frac{\mathbf{z}_r \mathbf{z}_f}{\mathbf{z}_o + \mathbf{Z}_L} \tag{5-20}$$

By definition $\mathbf{z}_i$ is the open-circuit input impedance. Under load its value is modified by the coupled-in effect of the output circuit and load, with $\mathbf{z}_r \mathbf{z}_f$ as coupling members.

The *output impedance* looking back into the 2,2 terminals can be found as

$$\mathbf{Z}_o = \frac{\mathbf{V}_2}{\mathbf{I}_2} = \mathbf{z}_f \frac{\mathbf{I}_1}{\mathbf{I}_2} + \mathbf{z}_o \tag{5-21}$$

With source impedance $\mathbf{Z}_s$ across the terminals, $\mathbf{V}_1/\mathbf{I}_1 = \mathbf{Z}_s$. Then from Eq. 5-3,

$$\frac{\mathbf{I}_1}{\mathbf{I}_2} = \frac{-\mathbf{z}_r}{\mathbf{z}_i + \mathbf{Z}_s} \tag{5-22}$$

Therefore the network output impedance is

$$\mathbf{Z}_o = \mathbf{z}_o - \frac{\mathbf{z}_r \mathbf{z}_f}{\mathbf{z}_i + \mathbf{Z}_s} \tag{5-23}$$

The open-circuit output impedance is $\mathbf{z}_o$ by definition. Under conditions of drive by a generator of impedance $\mathbf{Z}_s$, the value of $\mathbf{z}_o$ is modified by the coupled-in effect of the input circuit.

The *transfer impedance* $\mathbf{Z}_T$ is defined as the ratio of voltage applied in one mesh to the current produced in a second mesh, all other emf's being removed, so that

$$\mathbf{Z}_{T12} = \frac{\mathbf{V}_1}{\mathbf{I}_2} = \mathbf{z}_r - \frac{\mathbf{z}_i(\mathbf{z}_o + \mathbf{Z}_L)}{\mathbf{z}_f} \tag{5-24}$$

which is a useful relation.

5-6 EQUIVALENT CIRCUITS FOR ACTIVE NETWORKS

A T or π network may be developed as equivalent to any linear, bilateral, passive network, and it is possible to extend the method to active circuits, such as those employing transistors or vacuum tubes, in which $\mathbf{z}_r \neq \mathbf{z}_f$, but where neither $\mathbf{z}_r$ nor $\mathbf{z}_f$ are zero. If either of the reverse or forward transfer parameters is zero, then the network is unilateral.

Again choosing $\mathbf{I}_1$ and $\mathbf{I}_2$ as the independent variables, along with the $\mathbf{z}$ open-circuit parameters, the general two-port relations are

$$\begin{bmatrix} \mathbf{V}_1 \\ \mathbf{V}_2 \end{bmatrix} = \begin{bmatrix} \mathbf{z}_i & \mathbf{z}_r \\ \mathbf{z}_f & \mathbf{z}_o \end{bmatrix} \begin{bmatrix} \mathbf{I}_1 \\ \mathbf{I}_2 \end{bmatrix} \qquad \begin{matrix} (5\text{-}25) \\ (5\text{-}26) \end{matrix}$$

This pair of equations is represented by the equivalent circuit of Fig. 5-4(a), as can be shown by comparing the circuit equations with the diagram. The circuit is complicated by the presence of two transfer generators, and a more convenient one-generator form may be obtained by algebraic manipulation of the equations.

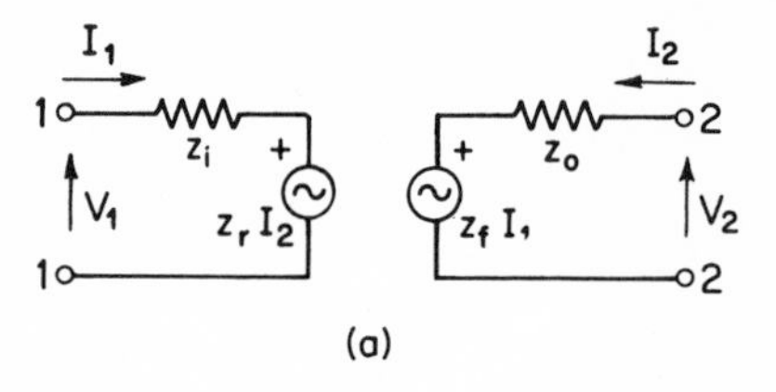

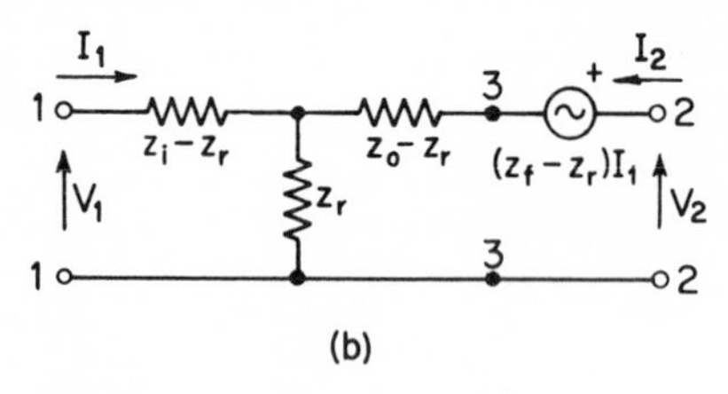

Figure 5-4. Active equivalent networks.

Addition and subtraction of $\mathbf{z}_r\mathbf{I}_1$ to Eq. 5-26 allow the equations to be written as

$$\begin{bmatrix} \mathbf{V}_1 \\ \mathbf{V}_2 - (\mathbf{z}_f - \mathbf{z}_r)\mathbf{I}_1 \end{bmatrix} = \begin{bmatrix} \mathbf{z}_i & \mathbf{z}_r \\ \mathbf{z}_r & \mathbf{z}_o \end{bmatrix} \begin{bmatrix} \mathbf{I}_1 \\ \mathbf{I}_2 \end{bmatrix} \qquad \begin{matrix} (5\text{-}27) \\ (5\text{-}28) \end{matrix}$$

and these equations, still equivalent to Eqs. 5-25 and 5-26, can be represented by the circuit of Fig. 5-4(b). The circuit matrix is that of a passive T circuit which appears between the 1,1 and 3,3 terminals. The addition of the generator $(\mathbf{z}_f - \mathbf{z}_r)\mathbf{I}_1$ in the output circuit provides the active circuit element. This equivalent will be found useful in transistor circuit analysis.

A modified circuit form appears in Fig. 5-5, where the transfer generator has been transformed to a current source. This source, bridged across $\mathbf{z}_o - \mathbf{z}_r$, must supply a current of

$$\frac{\mathbf{z}_f - \mathbf{z}_r}{\mathbf{z}_o - \mathbf{z}_r}\mathbf{I}_1$$

and with output open the generator introduces into the circuit a potential of

$$\mathbf{E} = (\mathbf{z}_f - \mathbf{z}_r)\mathbf{I}_1$$

as did the voltage-source generator of Fig. 5-4(b).

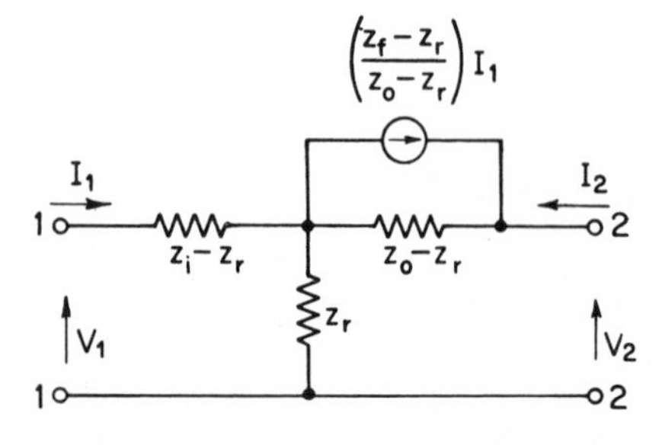

Figure 5-5. Current-source two-port network.

Equivalent circuits for the $\mathbf{y}$ parameters may also be found. Taking the equations

$$\begin{bmatrix}\mathbf{I}_1\\ \mathbf{I}_2\end{bmatrix} = \begin{bmatrix}\mathbf{y}_i & \mathbf{y}_r\\ \mathbf{y}_f & \mathbf{y}_o\end{bmatrix}\begin{bmatrix}\mathbf{V}_1\\ \mathbf{V}_2\end{bmatrix}$$

the two-generator circuit of Fig. 5-6(a) can be drawn. By adding and subtracting $\mathbf{y}_r\mathbf{V}_1$ in the second equation it is possible to write

$$\begin{bmatrix}\mathbf{I}_1\\ \mathbf{I}_2 - (\mathbf{y}_f - \mathbf{y}_r)\mathbf{V}_1\end{bmatrix} = \begin{bmatrix}\mathbf{y}_i & \mathbf{y}_r\\ \mathbf{y}_r & \mathbf{y}_o\end{bmatrix}\begin{bmatrix}\mathbf{V}_1\\ \mathbf{V}_2\end{bmatrix} \qquad \begin{matrix}(5\text{-}29)\\ (5\text{-}30)\end{matrix}$$

and these equations are represented by the circuit of Fig. 5-6(b). Again a passive network appears represented by a symmetric impedance matrix, to which is added an active element, the transfer generator.

The hybrid Eqs. 5-13 and 5-14 lead to an equivalent circuit as shown in Fig. 5-7.

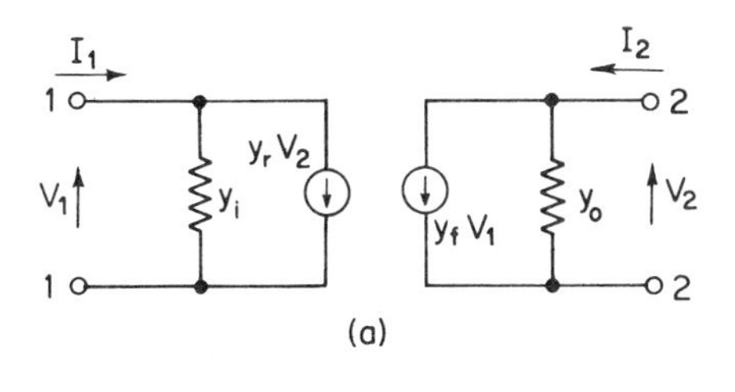

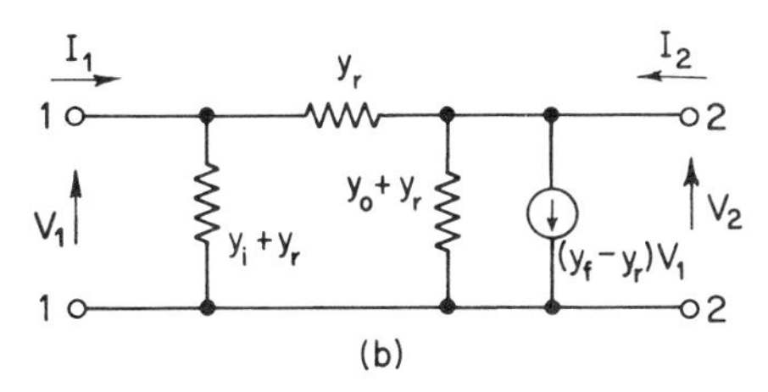

Figure 5-6. Equivalent circuits using **y** parameters.

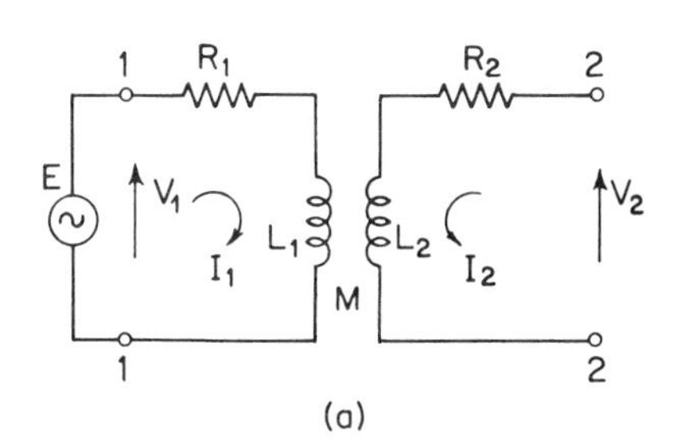

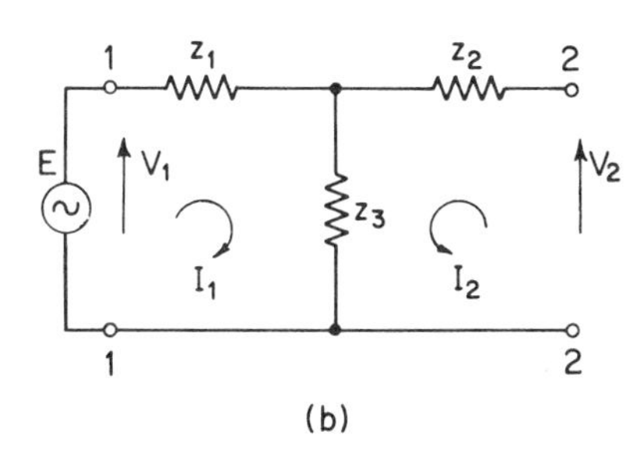

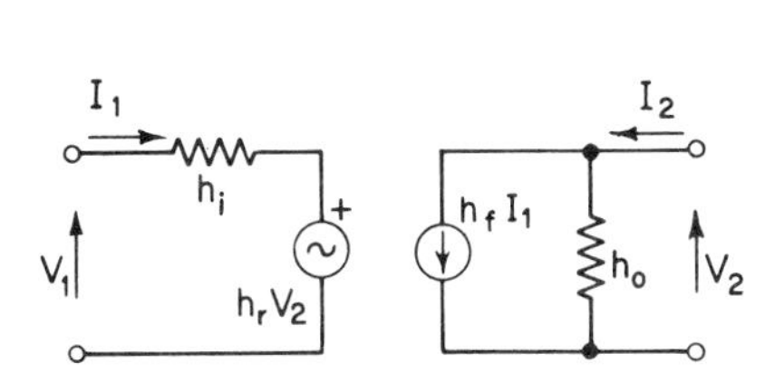

Figure 5-7. The hybrid parameter equivalent circuit.

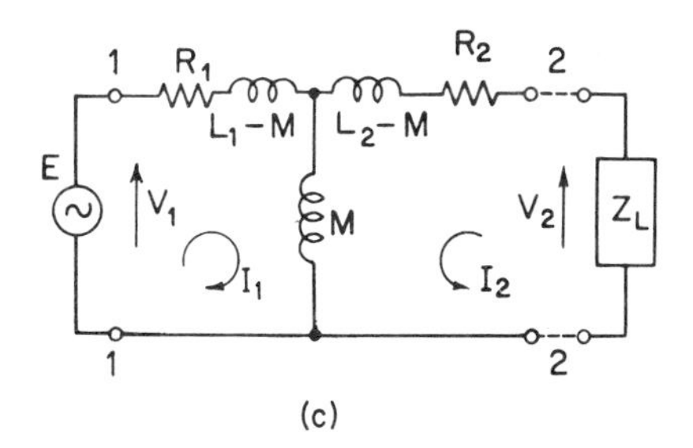

Figure 5-8. Inductive coupling and its equivalent circuit.

5-7 INDUCTIVELY COUPLED CIRCUITS

An inductively coupled circuit, as shown in Fig. 5-8(a), is frequently employed as a coupling element between successive amplifier tubes or transistors. The magnetic flux path is assumed predominantly air, or made of magnetic material such as powdered iron with substantially constant permeability.

The mutual inductance present then determines a *coefficient of coupling* k, where

$$k = \frac{M}{\sqrt{L_1 L_2}} \tag{5-31}$$

The leakage flux may be large and the value of k small but under control of the designer, who can choose numbers of turns, dimensions, and physical coil location in fixing k.

For convenience in circuit analysis, it is desirable to replace the inductively coupled circuit of Fig. 5-8(a), with a T network which is equivalent over a specified frequency range. By reference to (a) and using the definitions of Section 5-2, it can be found that

$$\mathbf{z}_i = R_1 + j\omega L_1$$

$$\mathbf{z}_o = R_2 + j\omega L_2$$

$$\mathbf{z}_f = \mathbf{z}_r = j\omega M$$

Similar application of the $\mathbf{z}$ parameter definitions to the T network of Fig. 5-8(b) leads to

$$\mathbf{z}_i = \mathbf{z}_1 + \mathbf{z}_3$$

$$\mathbf{z}_o = \mathbf{z}_2 + \mathbf{z}_3$$

$$\mathbf{z}_f = \mathbf{z}_r = \mathbf{z}_3$$

If the T network of Fig. 5-8(b) is to be equivalent to the inductively coupled circuit, then these measured parameters must be equal to force external voltage and current equality. It follows that

$$\mathbf{z}_1 = \mathbf{z}_i - \mathbf{z}_r = R_1 + j\omega(L_1 - M) \tag{5-32}$$

$$\mathbf{z}_2 = \mathbf{z}_o - \mathbf{z}_r = R_2 + j\omega(L_2 - M) \tag{5-33}$$

$$\mathbf{z}_3 = \mathbf{z}_f = \mathbf{z}_r = j\omega M \tag{5-34}$$

These quantities may be used to replace the elements of the T network, resulting in the circuit in Fig. 5-8(c). This circuit is then equivalent to the inductively coupled circuit of Fig. 5-8(a).

5-8 THE DECIBEL AND VU IN SYSTEM MEASUREMENT

The measurement of the output of many active circuits or amplifiers is made by use of a unit of *power ratio* known as the *decibel*. The gain of a cascaded amplifier can be found from the individual stage gains as

$$A = A_1 \times A_2 \times A_3 \times \dots \qquad (5\text{-}35)$$

However, if the gains are expressed as powers of 10, then

$$A = 10^a \times 10^b \times 10^c \times \dots = 10^{a+b+c+\dots}$$

or

$$\log_{10} A = a + b + c + \dots$$

Use of a logarithmic unit allows gains or losses to be directly added or subtracted. The telephone industry proposed a logarithmic unit, named the *bel* for Alexander Graham Bell and defined as the logarithm of a power ratio, or

$$\text{number of bels} = \log_{10} \frac{P_2}{P_1} \qquad (5\text{-}36)$$

It has been found that a unit one-tenth as large is more convenient, since such a unit approximates the power change required to produce a just detectable change in the intensity of a sound. This smaller unit is called the *decibel*, abbreviated dB, defined as

$$\text{number of dB} = 10 \log_{10} \frac{P_2}{P_1} \qquad (5\text{-}37)$$

If the output of an amplifier under one condition is 3.5 W and under a second condition is 7 W, then

$$\text{number of dB} = 10 \log_{10} \frac{7.0}{3.5} = 3.0$$

A change of power of 2 to 1 has resulted in a change of 3.0 dB in power level. A negative sign in the result would indicate that a power loss has taken place.

Although the decibel is a power ratio and not an absolute power measurement, it can be used for absolute measurements if a certain reference, or zero, level for P_1 is adopted beforehand and stated. A variety of various reference values have been used; one which has become common is 0.001 W. Consequently, the amplifier above with 7 W output is

$$10 \log_{10} \frac{7}{0.001} = 38.4 \text{ dB}$$

above zero level. The amplifier under the 3.5 W output condition would then have an output level of 35.4 dB above zero level, and this would be stated as +35.4 dB above 0.001 W reference.

Power is $P = E^2/R$, and the output and input powers of an amplifier may be compared as

$$\text{dB gain} = 10 \log_{10} \frac{E_2^2/R_2}{E_1^2/R_1}$$

If $R_1 = R_2$, or the resistors in which the power is measured are equal, then

$$\text{dB gain} = 20 \log_{10} \frac{E_2}{E_1} \qquad (5\text{-}38)$$

Even though the resistances in which the input and output powers are measured are not equal, and the usage is technically incorrect, it has still become common to state amplifier gains by use of Eq. 5-38. In such cases, it is advisable to use the notation dBV, to indicate that it is more correctly a voltage ratio rather than a power ratio which is being expressed.

In the broadcast field when 1 mW is employed as the zero reference, it is common practice to state absolute power in terms of *volume units*, or VU, where 10 dB above 0.001 W equals 10 VU. In other words,

$$\text{number of VU} = 10 \log_{10} \frac{P}{0.001} \tag{5-39}$$

where P is the amount of power measured. The term VU applies only when the monitoring is done with a meter of standardized damping characteristics.

Experiment shows that the ear hears sound intensities on a proportional, or logarithmic scale, and not on a linear one. Since the output of many amplifiers is ultimately converted to sound and received by the human ear, it is important to have a power unit consistent with properties of the ear, and the decibel is such a unit. Therefore the use of the decibel unit is justified on a psychological as well as convenience basis.

PROBLEMS

5-1. Calculate the **z** parameters for the circuit of (a), Fig. 5-9.

5-2. Determine the **h** parameters for the circuit of (a), Fig. 5-9.

5-3. Calculate the **y** and **h** parameters for the circuit of (b), Fig. 5-9.

5-4. If the T network of (c), Fig. 5-9, is to be equivalent to the π network of (a), compute the values of the T elements.

5-5. Write the general circuit equations for (b), Fig. 5-9, using $\mathbf{V}_1$ and $\mathbf{V}_2$ as the independent variables.

5-6. If $\mathbf{V}_2$ and $\mathbf{I}_2$ are selected as independent variables for a two-port network, write the definitions of the appropriate circuit parameters which may be designated b, with appropriate subscripts.

5-7. Find expressions for $\mathbf{z}_1$, $\mathbf{z}_2$, and $\mathbf{z}_3$ which will make the T network of (c) equivalent to that of (b), Fig. 5-9.

5-8. A microphone has an output at -54 dB below a 1-mW reference. The microphone supplies an amplifier of 1000-

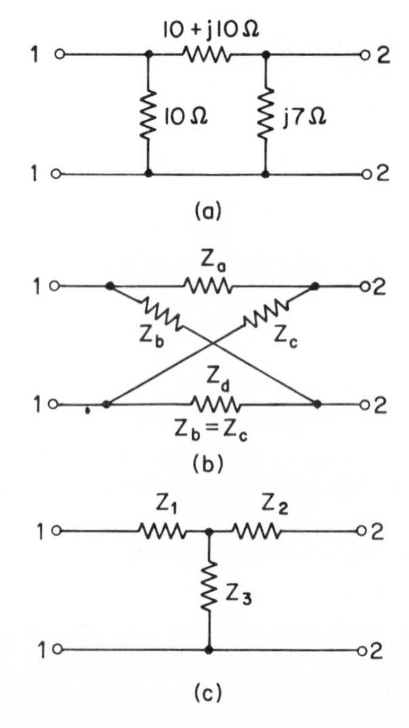

Figure 5-9.

ohm input resistance, which is to have an output level of +37 dB into a 100-ohm load.
(a) What is the output voltage of the microphone?
(b) Find the amplifier gain in decibels.
(c) Compute the amplifier power output, load current, and voltage.
(d) The output power is to be reduced to 60 per cent of original value. What will be the reduction in dB output?

5-9. Determine the input, output, and transfer impedances for the circuit of Fig. 5-9 (a).

5-10. Derive the transfer admittance for the circuit of Fig. 5-9 (b).

5-11. A transistor having $h_i = 28$ ohms, $h_r = 8 \times 10^{-4}$, $h_f = 0.98$, and $h_o = 0.6$ μmho, is used in a circuit equivalent to Fig. 5-7. If the load across which $\mathbf{V}_2$ is measured is 10,000 ohms, find the output voltage $\mathbf{V}_2$ if the input current is 0.0015 A.

5-12. The loss on a radio-frequency transmission line is 1.8 dB per 100 ft. How long can the line be and still have a power efficiency of 80 per cent?

5-13. A generator with power output of 5 mW supplies an amplifier which feeds a 500-ohm load. If the load power level is to be +20 VU, what must be the dB gain in the amplifier?

5-14. A radio receiver has an input impedance which is 70 ohms resistive. The antenna signal applied to this input is 50 μV. The electrical output to the loud-speaker is to be 1.8 W. Find:
(a) The input dB level, 1-mW reference.
(b) Receiver dB gain.
(c) Output power level, 1-mW reference.

REFERENCES

1. Guillemin, E. A., *Communication Networks*, Vol. 2. John Wiley & Sons, Inc., New York, 1935.
2. Peterson, L. C., "Equivalent Circuits of Linear Four-Terminal Networks," *Bell Syst. Tech. Jour.*, Vol. 27, p. 593 (1948).
3. Brown, J. S., and F. D. Bennett, "The Application of Matrices to Vacuum-Tube Circuits," *Proc. IRE*, Vol. 36, p. 844 (1948).
4. LeCorbeiller, P., *Matrix Analysis of Electric Networks*. John Wiley & Sons, Inc., New York, 1950.
5. Shekel, K., "Matrix Representation of Transistor Circuits," *Proc. IRE*, Vol. 40, p. 1493 (1952).
6. Ryder, J. D., *Networks, Lines, and Fields*, 2nd ed. Prentice-Hall, Inc., Englewood Cliffs, N. J., 1955.
7. Bellman, R., *Introduction to Matrix Analysis*. McGraw-Hill Book Company, New York, 1960.
8. Frame, J. S., "Matrix Functions and Applications," *IEEE Spectrum*, Vol. 1, (1964), No. 3; page 208; No. 4, page 102; No. 5, page 100; No. 6, page 123; No. 7, page 103.

6

THE TRANSISTOR: PHYSICAL VIEWPOINT

In 1948 Bardeen and Brattain found that the current through the forward-biased point contact of a semiconductor diode could control a current to a reverse-biased third electrode, mounted very close to the normal contact point. The resultant three-electrode semiconductor device was named the *transistor*, as a contraction of the words "transfer resistor." Extensive and continuing research has developed a considerable knowledge of solid-state conduction phenomena, and to the evolution of an array of solid-state devices having control, conversion, amplifying, and switching properties.

6-1 THE JUNCTION TRANSISTOR

The original discovery of controlled solid-state conduction was made with the point-contact construction, but further work by Shockley and the original team led to the growth of three-element semiconductor crystals with a thin *p* region between two *n* layers, forming two opposed diode junctions, as illustrated in Fig. 6-1. The techniques employed in such manufacture will be discussed later, but they have been developed to yield either *pnp* or *npn* units. Operation of the two types is broadly parallel, but with reversed biases and charge carriers, and we will usually base our discussion on the *pnp* construction.

The first junction is biased in the forward direction or with *p* positive for the *pnp* form, and the normal forward diode current law applies. Under the action of the forward bias the junction barrier is lowered, and holes diffuse across the barrier and into the central *n* region. Because of the physical action involved, the external electrode of the forward-biased junction is called the *emitter* (*p* for the *pnp* unit; *n* for the *npn* type).

The potential across the central *n* region is uniform. The width W is

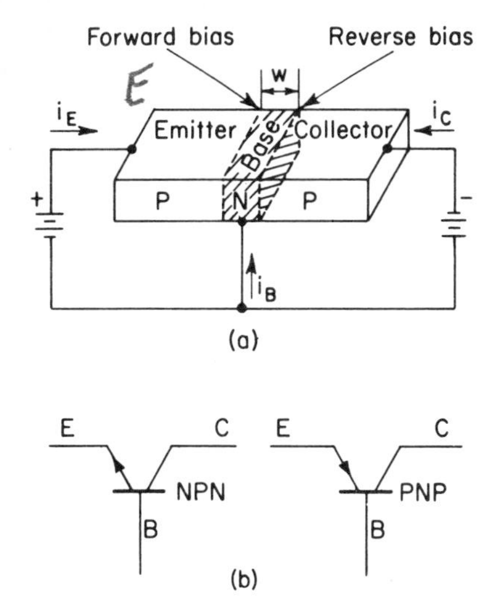

Figure 6-1. (a) Junction *pnp* transistor; (b) circuit symbols.

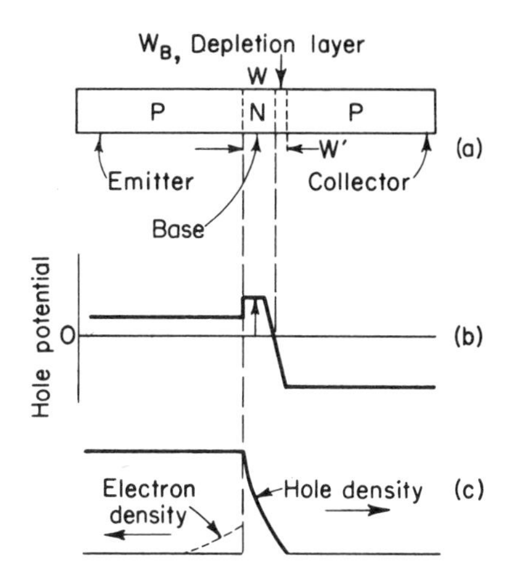

Figure 6-2. (a) *pnp* transistor; (b) hole potential distribution; (c) hole and electron densities.

designed to be very small with respect to the distance a hole can diffuse in a charge lifetime, and so most of the diffusing holes from the emitter reach the further n to p junction without recombination. This junction is reverse-biased, or with a negative potential on p, and so the holes enter the transition region in the direction of easy hole flow and are swept into the terminal on p, as indicated in Fig. 6-2(b). This p portion of the reverse-biased junction is called the *collector* (n for the *npn* unit), since it collects the charges after the transit of the structure.

The central n region is called the *base*. This was the semiconductor wafer of the point-contact unit in which solid-state control was first investigated; thus "base" as a name has historical support.

In addition to the diffusion of holes across the emitter-base junction there will be diffusion of electrons from the n base to the p emitter. However, the doping of the p emitter region is made much greater than that of the n base region, and so the density of holes diffusing into the base will be much greater than the electron density diffusing to the emitter. This ratio of impurity concentrations insures that the current across the emitter-base junction will be predominantly due to holes in the *pnp* unit. These carriers can reach the collector and contribute to the collector current; electrons crossing from base to emitter cannot do so.

A portion of the collector current will be derived from the reverse saturation current of the reverse-biased collector junction, originating from the thermal generation of holes. This leakage current, known in the transistor field as I_{CO} or I_{CBO}, is present regardless of the presence or absence of current from the emitter. Thus with $i_E = 0$, the collector current is I_{CBO}. Operating temperatures must be limited to insure that this current will be small with respect to the main working current i_E.

As the hole current diffusing into the base is varied by the applied junction potential in accordance with the diode equation, the resultant collector current will correspondingly vary. Control of the collector current is thus achieved by means of an *input current* to the emitter.

The relation between the emitter current i_E and the collector current i_C, in the static circuit condition of Fig. 6-1(a), and at a temperature at which I_{CBO} is negligible, is

$$\alpha_{FB} = \left| \frac{i_C}{i_E} \right| \tag{6-1}$$

Because recombination in the base and the electron current from the base are kept small by design, this ratio usually falls in the range from 0.90 to 0.999. By convention, the assumed positive direction of currents in Fig. 6-1(a) is taken as into the transistor.

As shown in Section 2-18, the resistance of the input forward-biased diode circuit is low, usually a few hundred ohms. The output circuit between collector and base will have an internal resistance from 10,000 ohms to several megohms, because of the reverse bias. The input power level is low because of the low input resistance, while the output power level is much higher, since the same current exists in a circuit of high impedance. Thus power and voltage amplification are possible in the transistor, making it a useful control device.

The above description of the operation of a *pnp* transistor could be readily adapted to a discussion of the *npn* construction by reversing the bias potentials and considering the normal current to be made up of electrons.

6-2 THE INTERNAL EFFICIENCY FACTORS

The ratio of Eq. 6-1 is physically determined as the result of three factors. The first of these is the emitter efficiency, γ, defined as the proportion of the total emitter current carried by holes at the emitter junction (or electrons for *npn*). For highly-doped emitter regions this factor can be made close to unity.

If p_0 is the hole density at the base side of the emitter-base junction, then the hole current entering the base is

$$i_h = ev_h p_0$$

As shown by Fig. 6-2(c), the hole density declines from p_0 at the emitter junction to a density of zero at the transition region face of the collector junction, where the holes are rapidly swept into the collector. The amount of hole diffusion is proportional to the initial gradient or slope of the density variation. The base width W is small with respect to the diffusion length L_h. The velocity v_h is that of normal equilibrium diffusion, L_h/τ_h, increased by a factor which may be approximated as L_h/W to reflect the increase in gradient or slope of the density distribution created by the relative thinness of the base. Thus

$$i_h \cong \frac{eL_h^2 p_0}{\tau_h W} = \frac{eD_h p_0}{W} \tag{6-2}$$

It can also be shown that the small electron current from the base, on the left of the base-emitter junction, is

$$i_e \cong \frac{eD_e n_0}{L_e} \tag{6-3}$$

The diffusing densities p_0 and n_0 are proportional to the respective hole density in the emitter, p_e, and the electron density in the base, n_b, so that

$$\frac{n_0}{p_0} = \frac{n_b}{p_e} \tag{6-4}$$

From the definition, the emitter efficiency is then

$$\gamma = \frac{I_h}{I_h + I_e} = \frac{1}{1 + I_e/I_h} \cong 1 - \frac{I_e}{I_h} \tag{6-5}$$

the latter expression following since I_e/I_h is small by selection of the impurity ratio for base and emitter. Substitution of Eqs. 6-2, 6-3, 6-4 allows Eq. 6-5 to be written as

$$\gamma = 1 - \frac{D_e}{D_h}\frac{W}{L_e}\frac{n_b}{p_e} \tag{6-6}$$

The ratio of the diffusion constants will usually exceed unity, but W/L_e and n_b/p_e are made small by design selection. Thus the emitter efficiency is normally very close to unity.

The second factor influencing α_{FB} is the *transport factor*, usually given as B, and defined as the ratio of the current received by the collector to the current reaching the base. With $W \ll L_h$, the transport factor can be written as

$$B \cong 1 - \frac{1}{2}\left(\frac{W}{L_h}\right)^2 \tag{6-7}$$

and this factor is close to unity.

The *collector efficiency* δ also enters in the determination of α_{FB}. This factor is defined as the ratio of hole current crossing the base-collector junction to the total collector current; δ is also called the *collector multiplication factor*. At low collector voltages δ is found to be near unity. At higher voltages, usually above the range employed in amplification, there may be generation of secondary carriers by ionization of atoms through collision with

6

highly accelerated charges in the collector. This process may cause α_{FB} to rise above unity under some high-voltage conditions. Instability can then result in certain circuits, and the effect is sometimes deliberately encouraged in switching devices to achieve higher speeds of operation.

Under steady-state conditions the current to the base feeds the recombination needs of the base. With a change in base potential, there must be a change in charge stored in the base, and a current enters. The base current, of electrons in this case, has a steady-state component to supply the small recombination losses, and a transient component to furnish the stored charge with potential change. Thus a capacitance effect is present.

The base current must also supply the charges injected across the barrier into the emitter. This diffusion into the emitter does not contribute to the collector current and so is a loss; it is usually very small with respect to the main emitter current of holes. The base current can be stated from the defined currents by use of

$$i_C + i_E + i_B = 0$$

Then

$$i_C = -\alpha_{FB} i_E$$

$$i_B = -i_E(1 - \alpha_{FB}) \tag{6-8}$$

Conduction across the base has been assumed to occur by diffusion. However, at high rates of hole injection some additional electron charge may be required in the base to maintain charge neutrality. The result is a shift of electrons and a gradient of electron density in the base that creates a small drift field, aiding hole conduction in the base; the value of α_{FB} therefore rises. At extremely high hole densities, the electron density required for charge neutrality causes more of the emitter junction current to be carried by electrons, and this reduces the emitter efficiency and α_{FB}. This is indicated by the presence of the n_b term in Eq. 6-6.

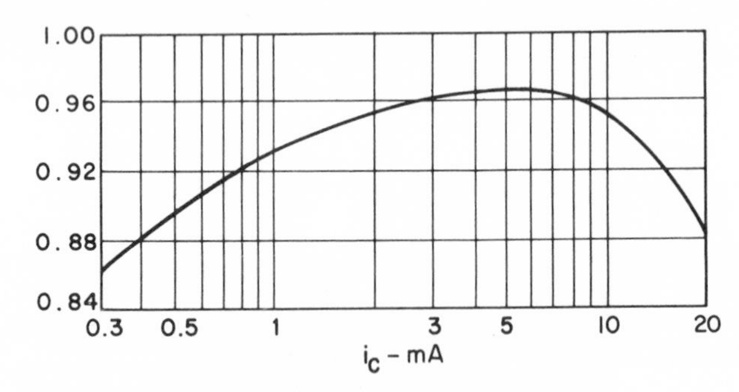

Figure 6-3. Variation of static α with current for a silicon power transistor at 25°C.

The overall effect of the various efficiency factors on α_{FB} is illustrated by the curve of Fig. 6-3 for a silicon transistor at 25°C.

6-3 THE DYNAMIC OR SMALL-SIGNAL DEFINITION OF α

The transistor is one of a number of nonlinear electrical circuit elements. Such devices are difficult to analyze for performance unless the operating characteristics are assumed to be piece-wise linear, or linear within specified ranges of the variables. This may require that the applied signals be restricted in magnitude so as to produce variations only over portions of the curved operating characteristics which can then be assumed linear.

It is the change produced by varying signals that is of interest, and thus it is the slope of the characteristic rather than the static value which is utilized. Along the curve the ratio of a change in collector current Δi_C to a change in emitter current Δi_E, with collector-base voltage held constant ($\Delta v_{CB} = 0$), is defined in the limit as the *collector-emitter short-circuit current amplification factor*, or

$$\alpha_{fb} = -h_{fb} = \left.\frac{\partial i_C}{\partial i_E}\right]_{v_{CB}=\text{constant}} \tag{6-9}$$

The factor α_{fb} may be expected to lie in the range from 0.90 to 0.99 for usual transistors. It applies for small signals.

A second amplification factor, called the *collector-base short-circuit current amplification factor* is defined as the ratio of a change in collector current Δi_C produced by a change in base current Δi_B, with v_{CE} constant. That is,

$$\beta = h_{fe} = \left.\frac{\partial i_C}{\partial i_B}\right]_{v_{CE}=\text{constant}} \tag{6-10}$$

The base current largely results from recombination losses of a few per cent of the collector current so that β or h_{fe} may reach values as high as 200.

From the definition of Eq. 6-9

$$-\Delta i_C = \alpha_{fb}\,\Delta i_E$$

and since

$$\Delta i_B = -\Delta i_E - \Delta i_C = -\Delta i_E(1 - \alpha_{fb})$$

then

$$\beta = h_{fe} = \frac{\alpha_{fb}}{1 - \alpha_{fb}} \tag{6-11}$$

and this is a useful relation.

6-4 BASE-WIDENING EFFECTS

The base width W is defined as the distance between the emitter junction and the inner edge of the collector transition region. The thickness of the transition region will increase with increasing collector voltage (in its normal reversed polarity) and the effective base width W will then decrease with collector voltage, as the transition region subtracts from the base region.

With a narrower base there will be less opportunity for recombination, and so the transport factor B, given by Eq. 6-7, will increase. As a secondary effect, the diffusion gradient facing the holes at the emitter-base junction will increase slightly with decreased base width and the emitter efficiency will increase. Thus α_{fb} becomes a function of collector voltage and its varying component.

6-5 THE COLLECTOR CONDUCTANCE

With hole density equal to zero at the collector-base junction, the collector receives all the charges available and thus the current should be independent of applied voltage. The collector appears to operate in saturation, or with an infinite collector internal resistance. However, the change in W with changing collector voltage introduces a small signal variation of current with voltage, and a finite collector conductance is present. This may be viewed as

$$g_{cb} = \left.\frac{\partial i_C}{\partial v_C}\right]_{i_E=\text{constant}} = \frac{\partial(\alpha i_E)}{\partial v_C} = i_E \frac{\partial \alpha}{\partial W}\frac{\partial W}{\partial v_C} \tag{6-12}$$

The effective base width W is related to the transition layer width W_B according to Fig. 6-2

$$W = W' - W_B$$

and W_B for an abrupt junction has been obtained as

$$W_B = \left(\frac{2\epsilon\mu_e v_C}{\sigma_e}\right)^{1/2} \tag{6-13}$$

where $\epsilon = \epsilon_v\epsilon_r$. Then

$$\frac{\partial W}{\partial v_C} = -\frac{1}{2}\left(\frac{2\epsilon\mu_e}{\sigma_e}\right)^{1/2} v_C^{-1/2} \tag{6-14}$$

If the reverse voltage is maintained low in the usual region for amplification, then multiplication may be neglected, and the effect of v_C on base width is reflected in B, the transport factor, by a reduction in base recombination. Then $\gamma = \delta \cong 1$, and from Eq. 6-7

$$\frac{\partial B}{\partial W} = -\frac{W}{L_h^2} \tag{6-15}$$

Equation 6-12 for the collector conductance can then be assembled as

$$g_{cb} = \frac{i_E}{2v_C^{1/2}L_h^2}\left(\frac{2\epsilon\mu_e}{\sigma_e}\right)^{1/2}\left[W' - \left(\frac{2\epsilon\mu_e}{\sigma_e}\right)^{1/2} v_C^{1/2}\right] \tag{6-16}$$

The expression is valid only below the multiplication range of collector voltage. It does indicate the factors affecting the collector conductance in the usual small-signal ranges of amplifier use. The conductance will increase with i_E.

6-6 THE BASE RESISTANCE, r_b AND r_b'

The base current passes from an external base terminal to the active base region. Its path may be considered composed of two portions in series: that which is due to the material lying between the junctions and representing the resistance faced by the current in traveling to the junctions from some central point of true base potential, and a second part due to the resistance between the external contact and the central point of true potential. The latter component, r_{bb}, is dependent on unit design and the distribution of current paths to the base material, and is relatively constant. The internal resistance between the junctions, r_b', is a function of signal and collector potential, since the effective base width varies with the thickness of the collector depletion region. This latter component is related to input current magnitude.

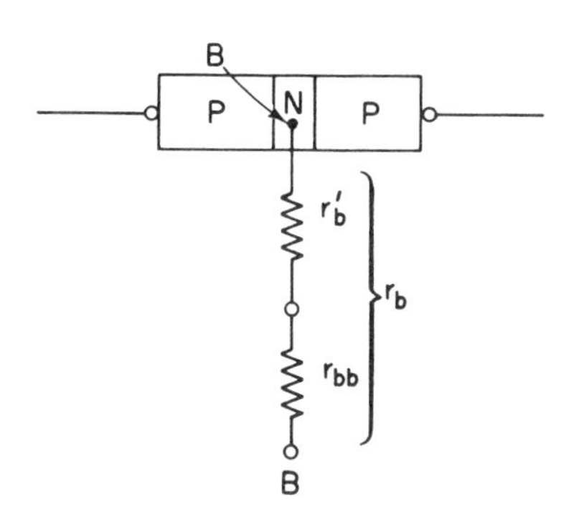

Figure 6-4. Base resistances defined.

The effective circuit is shown in Fig. 6-4. The base resistance $r_b = r_b' + r_{bb}$ is kept as small as possible by design, because its circuit effects are usually undesirable. Inability to reach the actual point of true potential of the base creates difficulties in some circuits.

To avoid the problem of parameter variation with signal, the base resistance is assumed constant as a first approximation in analysis.

6-7 EMITTER RESISTANCE

In Section 2-18 we determined the dynamic or ac resistance of a forward-biased diode for small variations of current. This is also the situation existing in the emitter circuit of a transistor, so the small-signal emitter resistance is

$$r_e = \frac{kT}{e}\frac{1}{i_E} = \frac{0.026}{i_E} \quad \text{at } 300°\text{K} \qquad \text{ohms} \quad (6\text{-}17)$$

with collector voltage held constant, or $\Delta v_C = 0$. Thus the small-signal input resistance is a reciprocal function of emitter current.

A second-order effect appears due to the presence of the collector in the transistor. The decrease of base width with increases in collector voltage reflects changes in the hole-density gradient at the emitter-base junction, and small signal variations in collector voltage can produce small changes in emitter current. There is effectively introduced into the emitter circuit a potential which depends on the collector voltage; this represents a feedback of a signal voltage from the collector to the emitter circuit.

An approximate relation for the feedback conductance can be found, using relations already developed. Let

$$g_{ec} = \left.\frac{\partial i_E}{\partial v_C}\right]_{v_E=\text{constant}} = \frac{\partial i_E}{\partial W}\frac{\partial W}{\partial v_C} \tag{6-18}$$

Assuming the emitter current to be entirely composed of holes, or $\gamma = 1$, then Eq. 6-2 gives

$$i_E = \frac{eD_h p_0}{W}$$

and

$$\frac{\partial i_E}{\partial W} = -\frac{eD_h p_0}{W^2} = -\frac{i_E}{W} \tag{6-19}$$

Using W from Eq. 6-13 and $\partial W/\partial v_C$ from Eq. 6-14, it is possible to write

$$g_{ec} = \frac{i_E}{2v_C^{1/2}}\left(\frac{2\epsilon\mu_e}{\sigma_n}\right)^{1/2}\left[\frac{1}{W' - \left(\dfrac{2\epsilon\mu_e}{\sigma_n}\right)^{1/2} v_C^{1/2}}\right] \tag{6-20}$$

This relation provides an approximation to the feedback conductance in the range in which v_C is below the multiplication level. Note that $\epsilon = \epsilon_v\epsilon_r$.

6-8 VOLTAGE BREAKDOWN

The collector junction of a transistor has definite maximum limits on applied voltage. The phenomena experienced at excess voltage levels may be Zener breakdown, avalanche multiplication, or punch-through.

Zener breakdown has already been discussed in connection with the regulating diode, and is due to the breaking of covalent bonds by a high internal electric field in the narrow depletion region.

Avalanche multiplication has also been discussed, and is the result of ionization of semiconductor atoms by collision with holes accelerated by the collector-base voltage. Such collisions can create an additional free hole and an electron, and these in turn may gain sufficient energy from the field to ionize again, in a chain reaction. A large increase in collector current can occur, as illustrated above A in Fig. 6-5. In general, the use of higher resistivity materials for the collector region will give higher avalanching voltages.

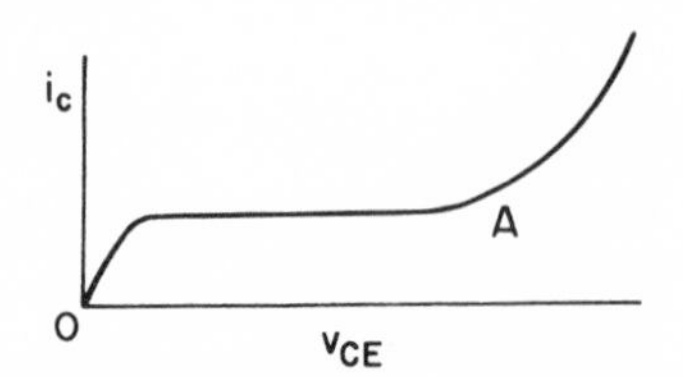

Figure 6-5. Effect of avalanche multiplication starting at A.

An empirical relation to predict the magnitude of multiplication effects on the collector efficiency is

$$\delta = \frac{1}{1 - (v_{CB}/v_A)^m} \tag{6-21}$$

where v_A is the avalanche breakdown voltage. The exponent m for most transistors approximates 3. If v_A for a given transistor is 100 volts, then a collector voltage of 9 volts yields a collector efficiency of 1.0007. Raising the collector voltage to 50 volts increases δ to 1.143, and this certainly cannot be ignored in its effect on α_{FB}.

Punch-through is the name given when the depletion layer thickens, extends all the way through the base, and reaches the emitter region at high collector voltages. This action short-circuits the base and the transistor, and large currents are possible. However, no permanent damage is done to the transistor if other circuit elements safely limit the current magnitude.

In units with thin low-conductivity bases, punch-through usually occurs at a lower voltage than does avalanching. With thick high conductivity bases, avalanching will occur at a voltage below the punch-through level.

Low base resistance is desirable for circuit reasons and is often obtained by using lower resistance base material. Thus choice of materials must be a compromise.

6-9 THE LEAKAGE CURRENT, I_{CBO}

The presence of the reverse saturation or leakage current I_{CBO} through the collector-base junction was mentioned in Section 6-2. This current is measured as a collector-base current with the emitter open, thus the subscript CBO. The current consists of thermally-generated electron-hole pairs from the base region, surface leakage current, and thermal pairs from the depletion region which extends into the collector. Since the extent of this depletion region depends on voltage, the latter component of leakage current is voltage dependent.

The current I_{CBO} represents an undesired component through the collector-base junction, since it cannot be controlled by base or emitter current. It contributes to the total current as

$$i_C = -\alpha_{FB} i_E + I_{CBO} \tag{6-22}$$

the negative sign resulting from the defined current directions in Fig. 6-1.

Usually I_{CBO} will be small and can be neglected. However, as was i_s in the diode, the leakage current I_{CBO} is a strong function of temperature. The major limiting factor on high temperature transistor operation is the magnitude reached by I_{CBO} in comparison to i_C; silicon is usually operable to 200°C, but germanium only to about 85°C.

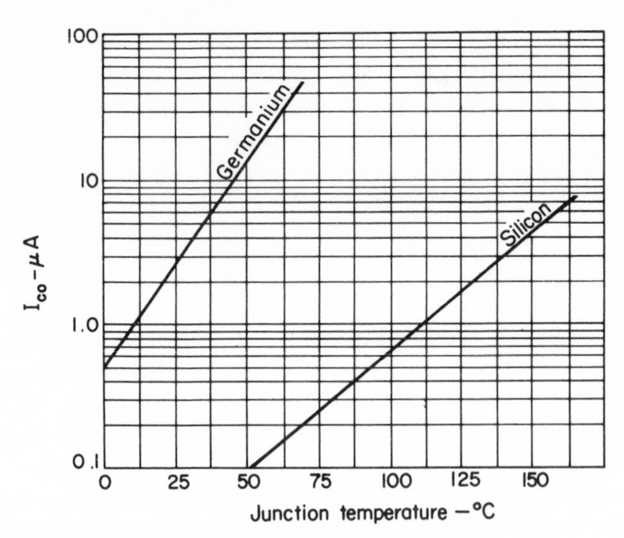

Figure 6-6. Temperature dependence of I_{CO}.

From the theoretical rate of pair generation, I_{CBO} in germanium should double for every 8°C rise in temperature; for silicon this step should be 6°C. However, because some of the leakage current is not temperature dependent, the usual rule of thumb is to expect I_{CBO} to double about every 10°C for germanium, and about every 18°C for silicon. Individual types may be considered to vary as in Fig. 6-6.

The silicon reverse saturation current at room temperature may be only one per cent of the germanium leakage for comparable units; thus silicon can operate at higher temperatures before I_{CBO} becomes large enough to seriously affect the control current.

6-10 A PHYSICAL EQUIVALENT CIRCUIT FOR THE TRANSISTOR

It is now possible to assemble the elements of the transistor into a circuit model which will reflect the physical phenomena discussed in this chapter. Such a model appears in Fig. 6-7.

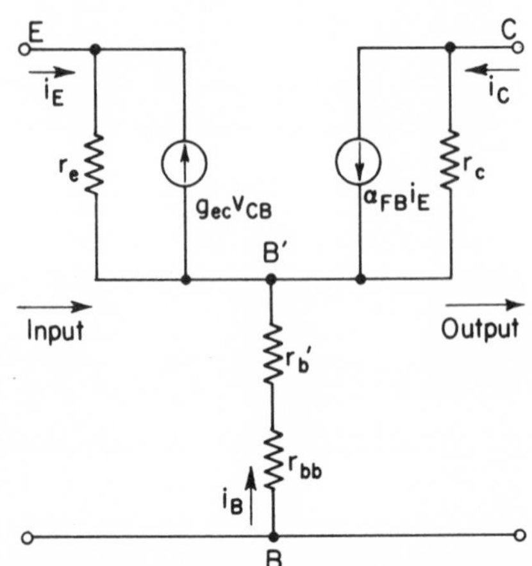

Figure 6-7. Physical equivalent circuit for a transistor.

The input circuit has a resistance r_e, due to the forward-biased diode. To this is added $r_b = r_b' + r_{bb}$, since this resistance is common to the input and output circuits. Also appearing is a current generator $g_{ec}v_{CB}$ which introduces a feedback current from the output circuit to the input. This feedback is due to the reaction of the collector voltage on the emitter efficiency, as the base width changes with collector voltage change. In many transistors this feedback is small enough that it can be neglected.

The output circuit combines the feed-forward current generator $\alpha_{fb}\, i_e$ with the collector conductance g_c, which appears as a large resistance r_c.

All the above applies only at low frequencies, since the various capacitive reactances associated with the transistor have not yet been introduced. However, it is possible to carry on a great deal of circuit analysis from the simplified low-frequency viewpoint.

6-11 THE EBERS–MOLL EQUIVALENT CIRCUIT

Another model based on physical reasoning, and of considerable value in the prediction of large-signal switching operation, is that contributed by Ebers and Moll. From the previous discussion, the currents across the two junctions, as controlled by the emitter-base voltage, are

$$i_{EF} = I_{ES}(\epsilon^{ev_{EB}/kT} - 1) \tag{6-23}$$

$$i_{CF} = -\alpha_F I_{ES}(\epsilon^{ev_{EB}/kT} - 1) \tag{6-24}$$

the assigned directions of i_E and i_C being inward as before. The currents in the reverse direction, under control of the collector-base junction voltage, are

$$i_{CR} = I_{CS}(\epsilon^{ev_{CB}/kT} - 1) \tag{6-25}$$

$$i_{ER} = -\alpha_R I_{CS}(\epsilon^{ev_{CB}/kT} - 1) \tag{6-26}$$

The reverse short-circuit current amplification factor α_R is usually much less than unity; the term corresponds to the feedback element. These equations may be combined as

$$i_E = I_{ES}(\epsilon^{ev_{EB}/kT} - 1) - \alpha_R I_{CS}(\epsilon^{ev_{CB}/kT} - 1) \tag{6-27}$$

$$i_C = -\alpha_F I_{ES}(\epsilon^{ev_{EB}/kT} - 1) + I_{CS}(\epsilon^{ev_{CB}/kT} - 1) \tag{6-28}$$

The first equation states that the emitter current i_E has two components, one that corresponds to the usual forward-biased diode current, and a second return or feedback component as a function of v_{CB}. The result may be interpreted as the diode and reverse current generator connected to the emitter terminal of Fig. 6-8(a).

The second equation states that the collector current also has two components, one being the result of forward carrier transport across the base, controlled by the emitter-base voltage, and the second a diode current controlled by the collector-base voltage. This equation is represented by the forward current generator $\alpha_F I_F$ and the parallel diode of the output circuit.

As written, these equations apply for both types of transistor. For understanding, a voltage will be positive when it appears across a diode in the normal forward direction. The voltage term will be negative when it appears across a diode biased in the reverse direction. The equations may be better understood if applied to a *pnp* unit as an example. Then the diode and the current I_F of the input circuit provide the usual forward hole current across the emitter-base junction with v_{EB} positive, with the feedback current provided by the dependent current generator $\alpha_R I_R$, computed with v_{CB} negative. In the output circuit the normal forward hole curent is given by the current generator $\alpha_F I_F$, and the reverse saturation current of the junction is given by the diode current expression, with v_{CB} negative. Then I_{ES} and I_{CS} are the respective reverse saturation currents, taken with $v_{CB} = 0$ and $v_{EB} = 0$, respectively.

By defining new saturation currents

$$I_{EO} = I_{ES}(1 - \alpha_F\alpha_R)$$

$$I_{CO} = I_{CS}(1 - \alpha_F\alpha_R)$$

Eqs. 6-27 and 6-28 can be rewritten in terms of terminal currents as independent variables, instead of the less useful diode currents. Then

$$i_E = I_{EO}(\epsilon^{ev_{EB}/kT} - 1) - \alpha_R i_C \qquad (6\text{-}29)$$

$$i_C = -\alpha_F i_E - I_{CO}(\epsilon^{ev_{CB}/kT} - 1) \qquad (6\text{-}30)$$

and these equations are represented by the circuit model in Fig. 6-8(b). The reverse saturation current now fits the regular definition for that quantity, and is measured with the emitter circuit open, or $i_E = 0$, and V_{CB} negative.

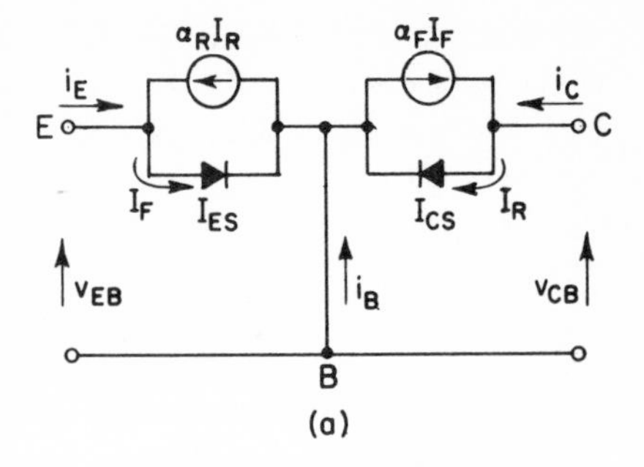

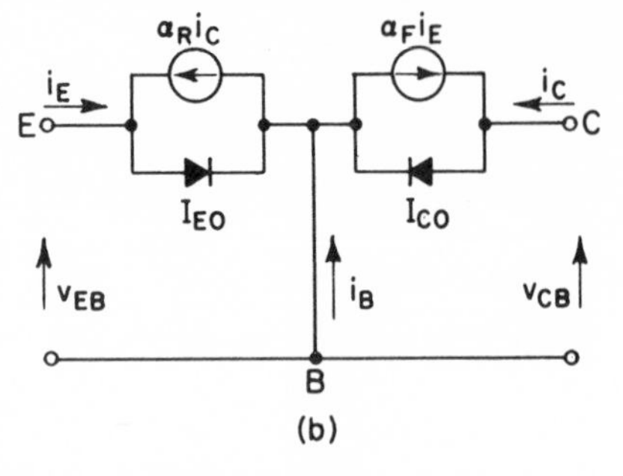

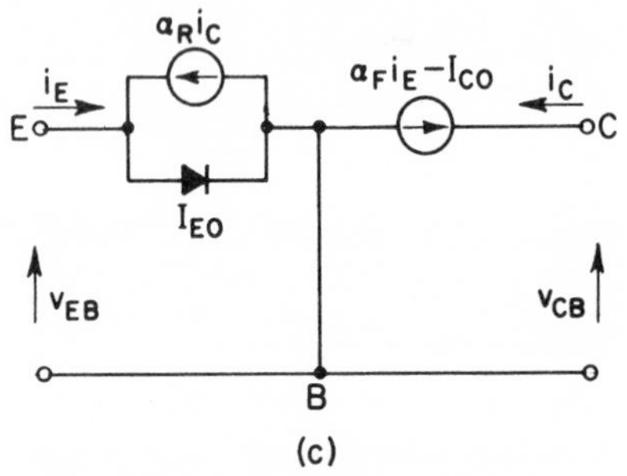

Figure 6-8. The Ebers–Moll model: (a) complete; (b) in terms of diode currents; (c) in the normal operating region.

Various conditions of operation can be classified: *cutoff*, or with both junctions reverse-biased and, therefore, the current generators removed from Fig. 6-8(b); the *saturation state* or with both junctions forward-biased; and the *normal state*. Under the normal condition the emitter junction is forward-biased and the collector-base junction is reverse-biased. For reverse voltages large with respect to a few kT/e, the collector current exponential

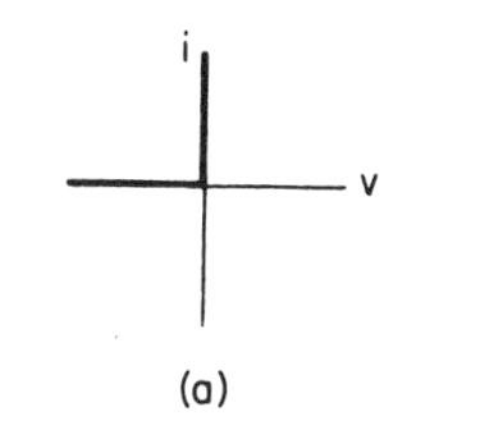

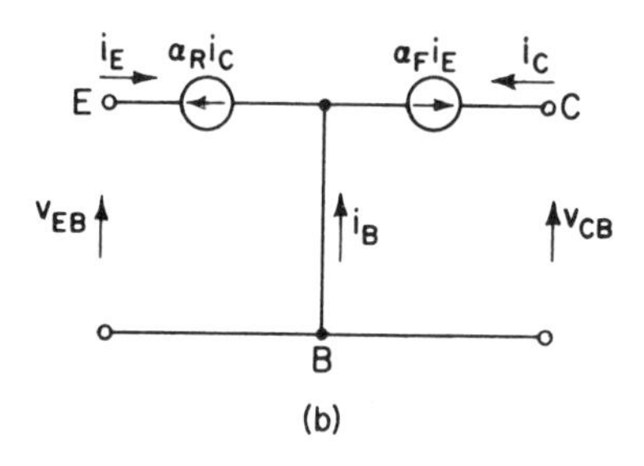

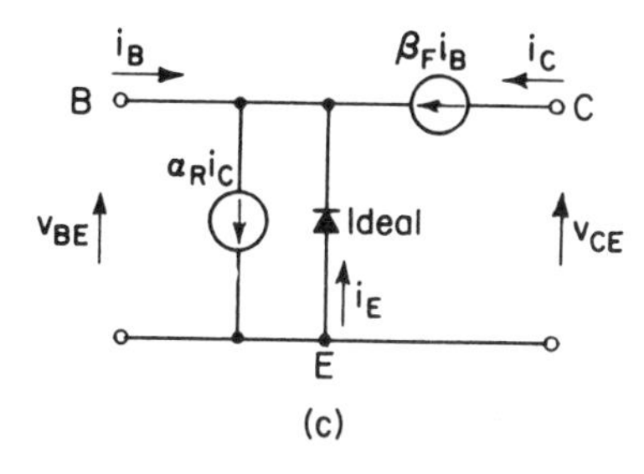

Figure 6-9. (a) Ideal diode volt–ampere characteristic; (b) common-base transistor equivalent circuit; (c) common-emitter equivalent circuit.

reduces to zero, and the circuit equations become

$$i_E = I_{EO}\epsilon^{ev_{EB}/kT} - \alpha_R i_C \qquad (6\text{-}31)$$

$$i_C = -\alpha_F i_E + I_{CO} \qquad (6\text{-}32)$$

These equations are represented by the circuit of Fig. 6-8(c). In some cases the feedback generator α_R in the input circuit can be considered negligible in effect and dropped from the circuit.

It is possible to still further simplify the equivalent circuit of Fig. 6-8, and this step provides another circuit model, suited to use with large changes of input and output voltage. As before, a diode may be idealized to the piece-wise linear volt–ampere curve of Fig. 6-9(a); this step neglects the forward resistance of the input diode and the reverse leakage current of the output diode.

Thus I_{CO} and I_{EO} are assumed zero, and the circuit model reduces to that of Fig. 6-9(b) for the base common to both input and output, and to Fig. 6-9(c) for the emitter as the common element. Predicted volt–ampere curves for these cases take the form of Fig. 6-10, and comparison with experimental performance curves of the next chapter will show them to be of similar form.

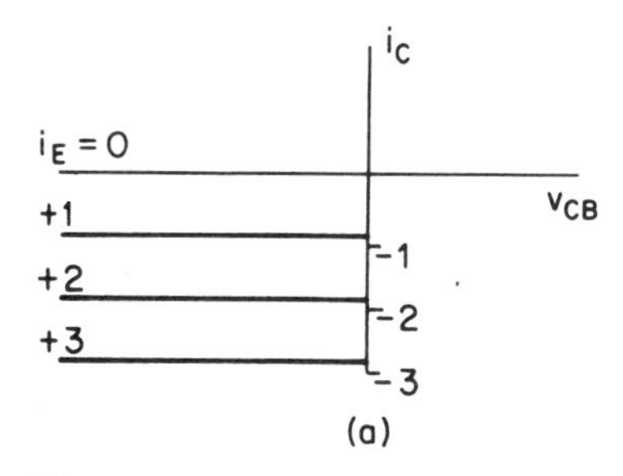

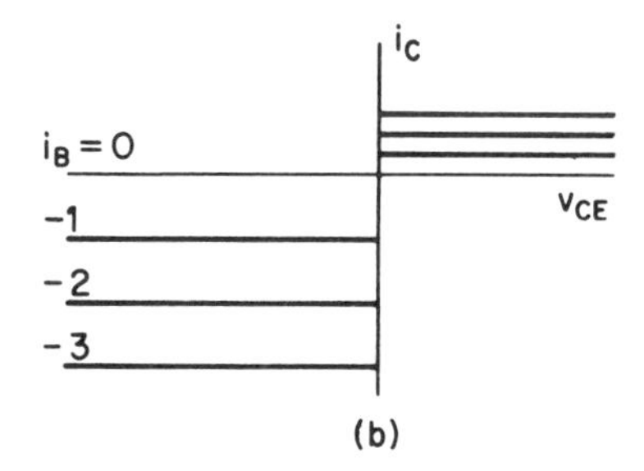

Figure 6-10. (a) Ideal common-base curves; (b) same for common-emitter circuit.

6-12 THE FIELD-EFFECT TRANSISTOR

A basically different device, but employing solid-state principles, is the *field-effect transistor* (FET) proposed by Shockley in 1948 and now of importance. The device is *unipolar*, employing only the free majority carriers in the conducting region, whereas the usual transistor uses both positive and negative carriers and is said to be *bipolar*. The majority carriers are restricted to a channel and control of the current is by application of an electric field.

In the theoretical device of Fig. 6-11(a), a current is established in the n bar by an applied voltage with negative terminal supplying electrons to the *source* electrode, and the *drain* collecting these electrons at the other end of the bar. Material of p type could be used as well, with conduction then by holes. Formed on each side of the channel at the midpoint are p layers and junctions. Suppose 4 V applied between source and drain, and the p *gate* electrodes connected to the source. At the midpoint of the bar, the potential is approximately +2 V with reference to the source. The gate being connected to the source is at −2 V to the bar and is back-biased. Thus a depletion region is formed in the bar under the gate electrodes. This depletion region reduces the cross section of the channel available for source-drain conduction, raises the resistance of the channel, and reduces the drain current i_D. The extent to which the depletion region extends into the channel can be controlled by an additional voltage applied between gate and source. As the gate-source voltage nears a value called the *pinch-off voltage*, V_p, the channel nears closure and the drain

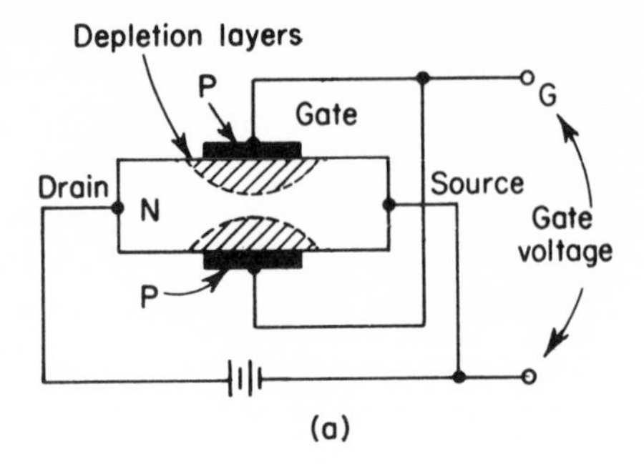

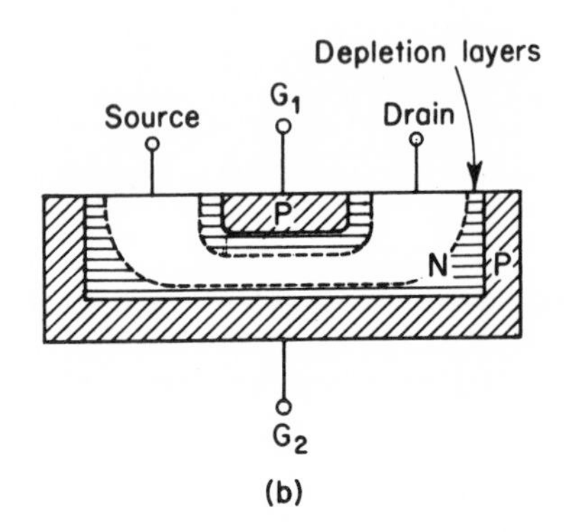

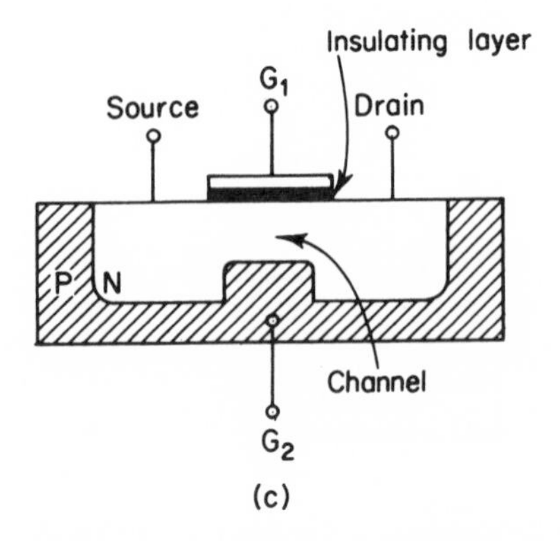

Figure 6-11. (a) Theoretical FET; (b) a JFET; (c) an IGFET.

current becomes relatively independent of drain-source voltage, as in Fig. 6-12. Thus the drain current is controllable by gate-source voltage and we have a *junction field-effect transistor* or JFET, as in Fig. 6-11(b).

Since the source-drain potential is distributed along the bar, the end of the gate electrode nearer the drain will be more negative or have a greater reverse bias than will the end near the source. The depletion region will be thicker on the drain end.

A second form is the *insulated-gate field-effect transistor*, or IGFET, of Fig. 6-11(c). The p substrate wafer forms a pn junction with the channel, and the upper gate electrode is insulated by a very thin silicon-dioxide layer on the surface of the silicon bar. The upper gate electrode represents one plate of a capacitor, the channel being the other electrode. With a positive voltage placed on the gate, G_1, a negative charge is induced in the channel and this attracts holes from the p substrate, creating a positive space charge in the

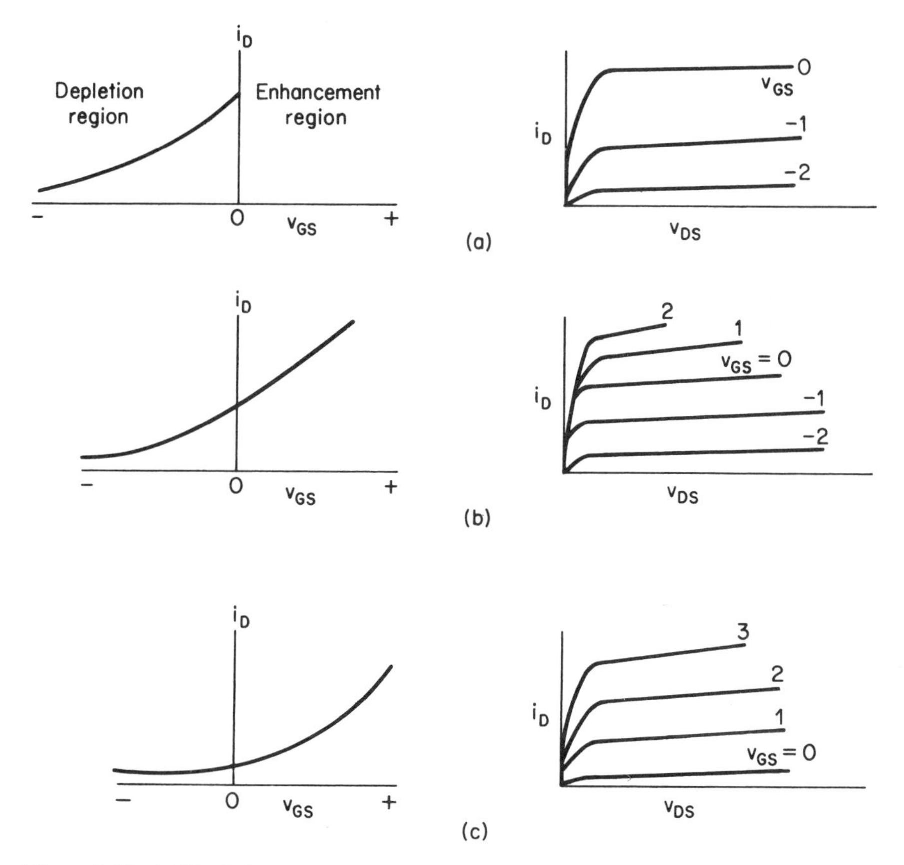

Figure 6-12. (a) Depletion-mode FET; (b) depletion-enhancement mode; (c) enhancement mode.

channel and increasing the conduction of electrons and the drain current. A negative voltage on G_1 has the opposite effect on the drain current. Since there is no conductive path between gate G_1 and the channel in the IGFET, its input impedance may be some hundreds of megohms in parallel with a small capacitance.

A variation is the MOSFET, or *metal-oxide semiconductor* FET, illustrated in Fig. 6-13. The oxide insulator is very thin, of the order of 0.01 to 0.5×10^{-6} m. A thin n channel is laid down between source and drain electrodes on a p substrate under the insulated gate. A field is set up in the oxide and generates a depletion region under the gate. This depletion region penetrates more deeply with increased negative potential until pinch-off occurs in the n channel. The MOSFET has high-frequency capabilities because of the low gate capacitance.

As just discussed, field-effect transistors can be designed for operation under two basic modes—*depletion* of carriers in the channel or *enhancement* of carriers. In the depletion mode as illustrated in Fig. 6-12(a), the carriers in the channel are decreased by an appropriate change in the gate-source voltage. That is, there is considerable current at zero gate voltage, and this is reduced by applying a reverse voltage to the gate. For the enhancement mode of (c), the drain current for zero gate-source voltage is very low, and this current is enhanced by forward gate voltages. A third type of IGFET has characteristics illustrated at (b), and combines both enhancement and depletion modes and has considerable drain current at zero gate-source voltage. Such a mode of operation has both forward and reverse input characteristics as shown.

The transfer characteristic, as the relation between input voltage v_{GS} and output current i_D, can be written by an algebraic equation as

$$i_D = I_{DSS}\left(\frac{v_{SG}}{V_p} - 1\right)^2 \tag{6-33}$$

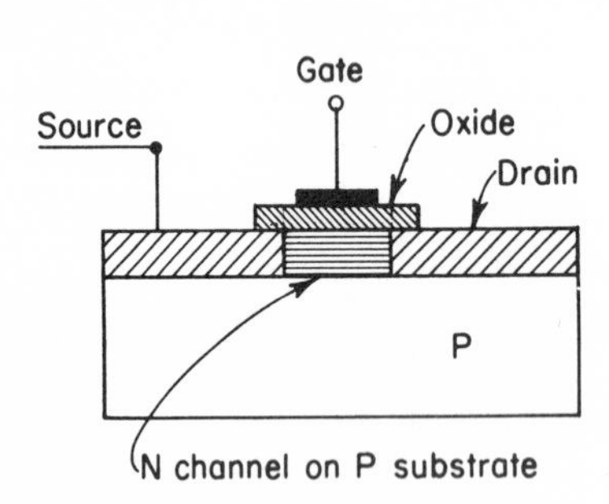

Figure 6-13. A metal-oxide semiconductor FET.

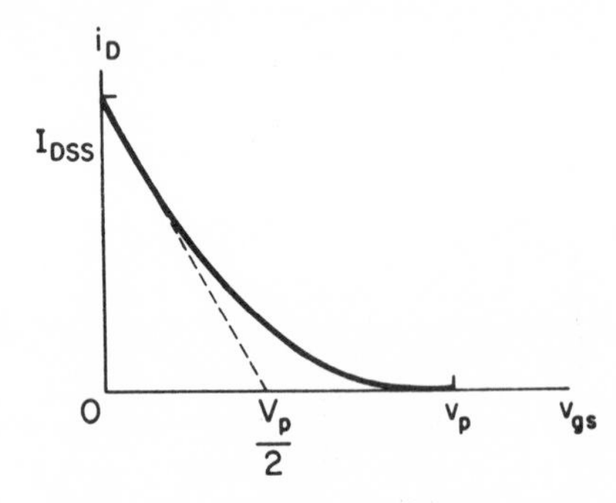

Figure 6-14. Transfer curve for an FET.

The current I_{DSS} is measured through the channel at $v_{GS} = 0$. Voltage V_p is the pinch-off voltage. This equation indicates that the FET is a perfect square-law device, and this has important implications for application. A parabola for this equation is drawn in Fig. 6-14.

The slope of the parabola is defined as the small-signal *transfer conductance* between input and output, or

$$g_m = \left.\frac{\partial i_D}{\partial v_{GS}}\right]_{v_{DS}=\text{constant}} \tag{6-34}$$

Measurement of the pinch-off voltage V_p is difficult because of the asymptotic toe of the curve, but can be found by taking the derivative of Eq. 6-33 as

$$\frac{\partial i_D}{\partial v_{GS}} = \frac{I_{DSS}}{V_p/2} \tag{6-35}$$

Thus the tangent to the curve at $i_D = I_{DSS}$ has the slope of the hypotenuse of a triangle of sides I_{DSS} and $V_p/2$; such a tangent is dashed in the figure. The intersection with the v_{GS} axis therefore is $V_p/2$.

Figure 6-15 illustrates one circuit model for the insulated-gate field effect transistor. The input resistance can range up to hundreds of megohms, and the output may be a few megohms paralleling a capacitance of a few picofarads. The device has many potential applications due to its high input impedance.

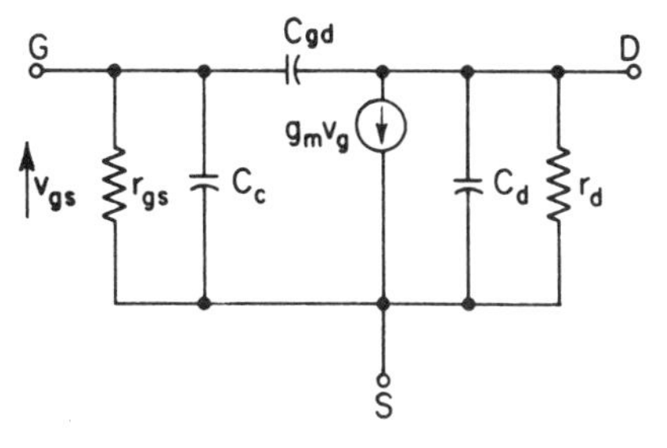

Figure 6-15. FET equivalent circuit.

6-13 THE UNIJUNCTION TRANSISTOR (UJT)

Another solid-state device is the *unijunction transistor* (UJT) illustrated in Fig. 6-16. A bar of n silicon has end contacts as bases B_1 and B_2, and a p emitter-junction located on the side of the bar. Between bases B_1 and B_2 the bar represents a resistance of 5000 to 10,000 ohms. Base B_2 is maintained positive at a potential V_{BB} with respect to B_1, and the emitter will be at some potential ηV_{BB}, due to its location on the bar. If the potential v_E applied to the emitter is less than ηV_{BB}, then the emitter is reverse-biased and only leakage current is present through the emitter junction. If v_E is more positive than ηV_{BB}, then the emitter is forward-biased and a large emitter current exists. This emitter current consists largely of holes injected into the n bar and moving to B_1. These holes call for an equal increase in electrons in the region, and the conductivity of the emitter-B_1 portion of the bar rises and the

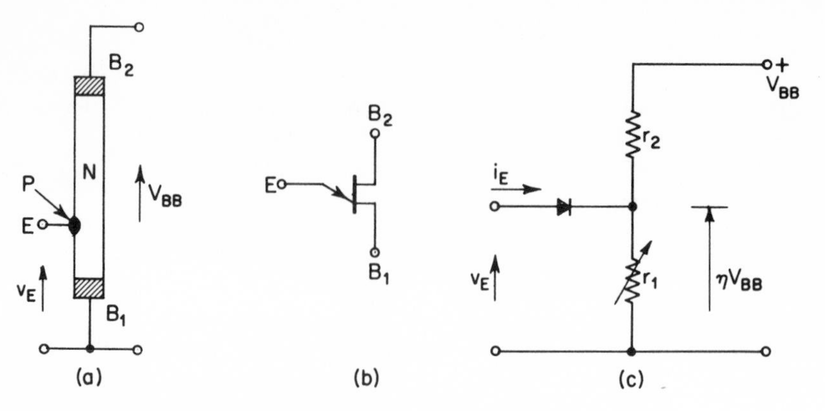

Figure 6-16. (a) Unijunction transistor; (b) circuit symbol; (c) equivalent circuit.

voltage drops. Simulation of the device as a variable voltage-divider in Fig. 6-16(c) is then justified.

Figure 6-17(a) shows the action of the emitter more clearly. With v_E less than ηV_{BB}, the emitter is reverse-biased and as v_E rises the current has the normal reverse-biased leakage value from A to B in the figure. At B, however, v_E reaches ηV_{BB} and the diode becomes forward-biased. This diode in forward bias has a volt-ampere curve given by the path ACD, and so the diode voltage must drop from B to the curve at C. The region B to C is one of negative resistance, or an increasing current with decreasing voltage. Further increases in v_E then cause the forward-biased diode to follow the normal diode curve, to which is added a small Ri drop in the emitter-base region, reaching point D.

The figure also shows a family of volt-ampere curves for various V_{BB} values. Each curve has a *peak point* V_p. There is also a *valley point* V_v at which the curve has its lowest value; between peak point and valley point is the region of negative resistance. Thus for $V_{BB} = 10$ V, the voltage rises to V_p at about 5.8 V and then the current rises and the voltage falls to V_v at 3.6 V and about 5 mA. To the right of the valley point the emitter-B_1 path is saturated with carriers and only a small positive resistance appears in that path.

The value of the peak voltage can be computed from

$$V_p = \eta V_{BB} + 0.7 \tag{6-36}$$

where the factor 0.7 V is an approximation to the inherent emitter-base forward drop of a silicon diode. From Fig. 6-17 it can be seen that $\eta \cong 0.6$ for the unit measured.

The unijunction transistor is useful as a voltage-operated switch or trigger in control applications.

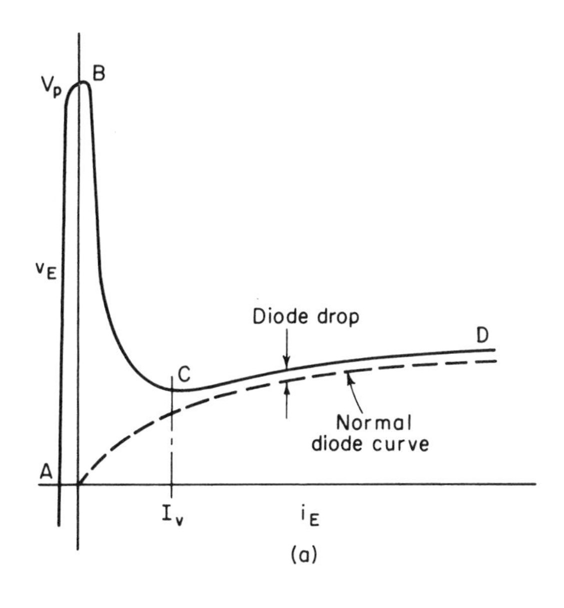

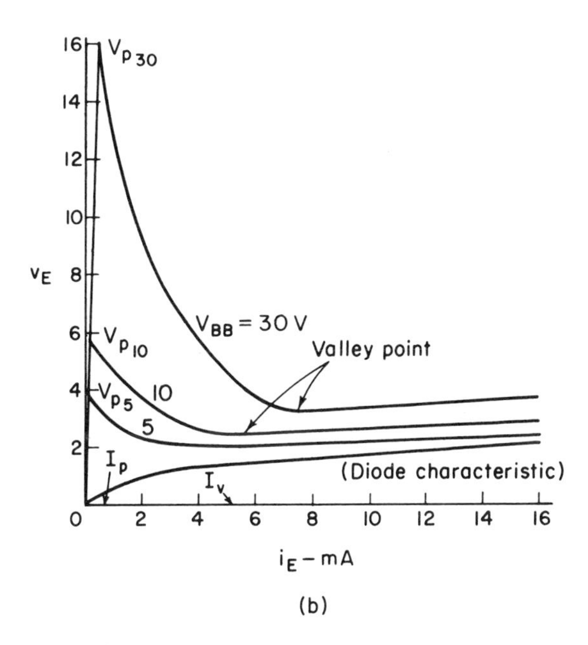

Figure 6-17. UJT emitter characteristic; (b) curve family for the UJT.

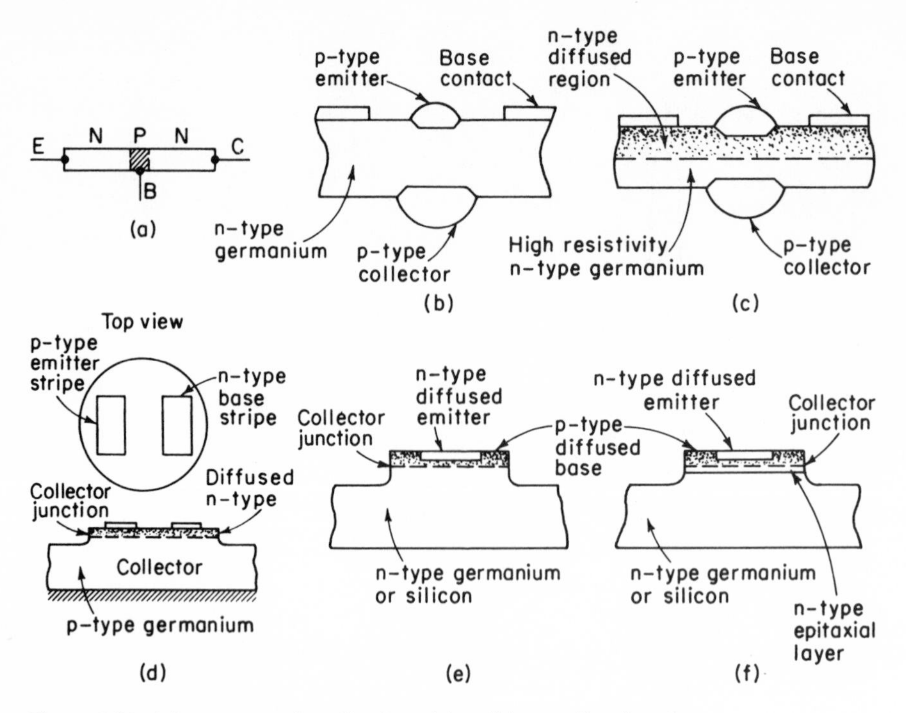

Figure 6-18. (a) *npn* grown junction transistor; (b) *pnp* alloy junction; (c) drift-field alloy transistor; (d) *pnp* mesa transistor; (e) *npn* diffused transistor; (f) epitaxial *npn* transistor.

6-14 PREPARATION OF SEMICONDUCTOR MATERIALS

Silicon is a hard, light element, appearing much like graphite, and melting at 1420°C. It is prepared, usually, by hydrogen reduction of the commercially available tetrachloride.

Germanium is a hard, dense element available in GeO_2 from which it is reduced with hydrogen. After melting above 937°C it is cast in bars of polycrystalline form and uncertain purity.

Purification of these materials to semiconductor standards is usually carried out by the process of zone melting in an inert gas atmosphere. Most impurities tend to remain dissolved in the liquid state rather than to freeze into the solid, so if a short section of an impure rod is melted and the molten zone is caused to move slowly along the rod, the impurities tend to move with the molten zone to the terminal end of the rod. By repeating this process a great number of times, the impurities can be concentrated at one end of the bar, which can then be discarded. Such heating of small sections is most easily carried out by the induction process. Germanium with resistivity above 0.6 ohm-m (60 ohm-cm) can be obtained by this purifying process. Theoretically-pure germanium should have a resistivity of about 0.65 ohm-m.

Since silicon reacts with many refractory materials, it cannot be melted in the usual laboratory crucibles. Melting of a small zone by induction heating in a vertically held rod is possible, the short molten zone being retained in place by the surface tension of the liquid silicon. The impurities can then be swept to the top of the bar as the liquid zone is moved up by transit of the heating coil. Very close temperature control is necessary to avoid flow of the molten slug, and a great many passes of the zone must be used to achieve silicon of about 1.5×10^3 ohm-m (1.5×10^5 ohm-cm) resistivity.

Solidification usually produces a polycrystalline result, and the boundaries of the crystals introduce anomalies in conduction properties. Useful and predictable semiconductors are customarily grown as large single crystals by dipping a small seed crystal into a bath of molten semiconductor, with rotation and withdrawal of the seed as the crystal grows. The seed must be properly oriented to insure growth along the desired crystal axes. An inert atmosphere is maintained over the melt.

Controlled amounts of n or p impurities are added to the melt during the growth process to produce the desired type of conduction and resistivity. Final crystals may be two to three cm in diameter and 10 to 20 cm long. These are sliced by diamond saws into wafers a fraction of a millimeter thick. The surfaces are lapped and polished, and then chemically etched to remove crystal dislocations caused by the sawing and polishing processes.

6-15 JUNCTION MANUFACTURE

An early method of producing grown pn junctions for diode or transistor use was by successive doping of the melt from which the single crystal was being grown. Starting with an n melt, an impurity of p nature is introduced at an appropriate instant in sufficient amount to override the initial doping. After a short interval another n impurity is added to override the p addition. The second n layer is, of necessity, of lower resistivity than the first n layer, and is referred to as $n+$ material. To produce more than one junction per crystal requires successive cycles of impurity introduction, each causing a further reduction of resistivity. The method is therefore useful only when the resistivity requirements are not critical.

Rate-grown junctions employ a variation of impurity concentration which depends on the rate at which the crystal is grown. As an example, the concentration of n-forming antimony increases with crystal-pulling speed, whereas the concentration of p-forming boron is essentially independent of rate of crystal growth. By cycling the growth rate, alternate n and p regions may be grown from a melt containing both antimony and boron impurities. Regions only a few thousandths of a millimeter in thickness may be grown, giving many junctions per crystal. The crystal is subsequently cut into layers, each containing a pn junction.

Alloyed junctions are produced by fusing onto the silicon or germanium base wafer a pellet of the desired impurity element. If the base is *n* silicon, then the *p* impurity pellet may be of indium, and this alloys with the silicon at the interface, creating there a thin *p*-silicon layer as a transistor emitter or diode anode; the remainder of the pellet serves as an ohmic contact. For a transistor the process may be repeated on the other side of the wafer with a pellet which will yield the desired collector characteristics. Time and temperature are controlled to produce the desired depth of penetration, and diodes of abrupt change in junction characteristics are so fabricated.

Epitaxial growth processes are used to form new layers of crystal material on a chosen crystal base; *epitaxy* is defined as oriented intergrowth of two solid materials, wherein the epitaxially grown material follows the crystal structure and orientation of the substrate. Growth occurs usually from an atmosphere containing silicon and the desired impurity in gaseous form, at an elevated temperature and for a precise time. Therefore the impurity concentration in the newly grown material can differ from that of the substrate; the resistivity and resistivity gradient are both controllable, and layers of micron thickness are obtainable.

Use of high resistivity collector material raises the breakdown voltage, but is undesirable because of the resultant increase in saturation resistance of a transistor. Epitaxial growth makes possible the deposition of a high-resistivity thin collector layer on a low-resistivity silicon substrate, giving both high breakdown voltage and low saturation resistance. The method provides the designer of transistors with increased flexibility in use of materials.

The *diffusion process* is widely employed in transistor and diode manufacture because of the ease of accurate production control. A substrate semiconductor wafer of desired resistivity is coated with a layer of doping element by deposition from a gaseous atmosphere. Exposure to a high temperature for a controlled time follows, causing the impurity atoms to migrate to the desired depth in the substrate. If an *n* silicon wafer had been chosen with boron as the diffusing impurity, then the surface of the wafer would have become *p* material to the desired depth; there would be a junction between the *p* and *n* materials at that depth.

For diodes, numbers of large wafers can be processed in one operation, yielding good uniformity of characteristics. The wafers are later diced into small diode elements.

The diffusion process leads to several basic forms of transistor which will be further discussed in the next section. The transistors which result from all these manufacturing methods are diagrammed in Fig. 6-18.

6-16 MESA AND PLANAR TRANSISTORS

The process of diffusion can be limited to desired and very small electrode areas in transistor manufacture by use of photographically applied chemical surface masks. The electrode areas can be accurately placed, and hundreds of identical transistors produced on one silicon wafer cut from a 2 or 3-cm diameter crystal.

The *mesa transistor*, so called because its elements are raised above the surrounding surface as a mesa in the desert, starts with a low-resistivity *n* silicon substrate. To obtain higher breakdown voltage a high resistivity *n* layer is epitaxially grown on this wafer, to a thickness of about ten microns. Two diffusion processes follow, carried out through masks, so that only selected areas of the *n* material are affected. A base area is first created by diffusing with a *p* impurity, allowing this impurity to reach a depth of about three microns by control of time and temperature. A new mask is applied, restricting the area of the base element which will be affected, and a strongly *n* impurity is diffused over a small emitter area, but with diffusion depth controlled to two microns. The result is a two micron thick emitter region of *n* properties, and an underlying one micron base region of *p* properties,

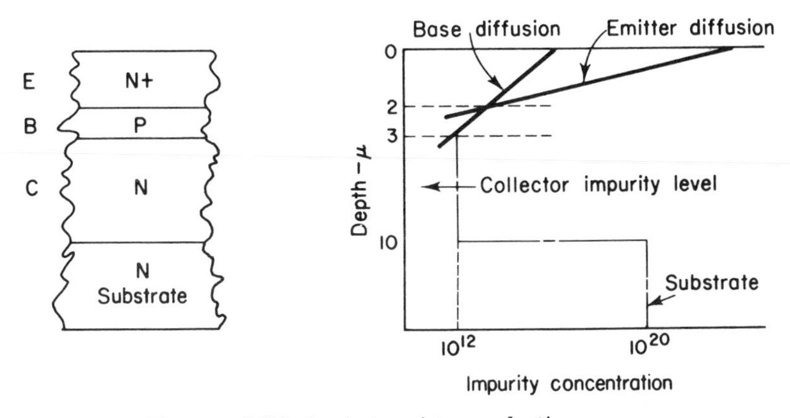

Figure 6-19. The use of diffusion in transistor production.

with seven microns of high-resistivity collector material remaining. These processes are illustrated in Fig. 6-19.

Further masks are applied and undesired and inactive areas of material are etched away, leaving the transistor isolated as a mesa, indicated in Fig. 6-20(c). Such a transistor is useful at high frequencies, having a low base resistance and a collector capacitance reduced to one to three picofarads.

The *planar transistor* is produced in somewhat similar fashion. A base is formed by diffusion of a *p* impurity

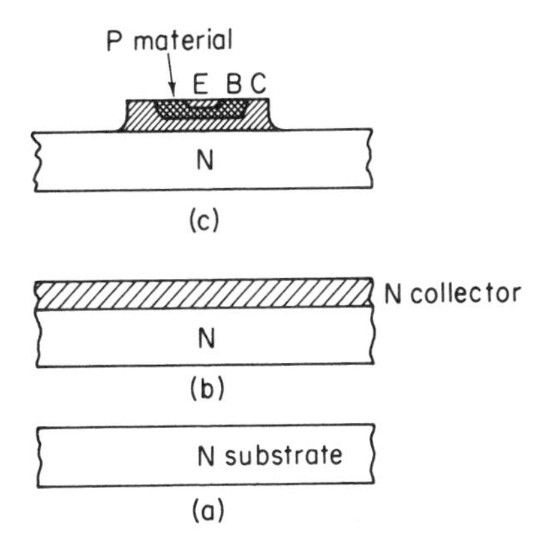

Figure 6-20. The epitaxial process.

through a mask to the desired depth in the collector *n* silicon wafer. Through another mask, an *n* impurity is diffused into a small area of the base electrode to form an emitter.

The upper surface of the transistor is left plane, and a layer of silicon dioxide is formed thereon. This layer seems especially beneficial in reflecting back into the material the holes and electrons which reach the surface, or in reducing the surface conduction anomalies that otherwise would exist where the collector-base and base-emitter junctions reach the surface. The oxide also serves as an effective and chemically inert protective coating. The silicon dioxide can be formed by several chemical reactions, one of which involves the reduction of the tetrachloride with hydrogen in the presence of carbon dioxide, as

$$SiCl_4 + 2H_2 + 2CO_2 \rightleftarrows SiO_2 + 2CO + 4HCl$$

Openings are then etched through still another mask through which a metal may be deposited as connections for the base and emitter.

6-17 INTEGRATED CIRCUITS

Etched circuits, in which circuit wiring is formed on an insulating circuit board by photo-masking and selective etching away of a thin copper sheet cemented to the board, constituted one of the first advances in mechanizing electronic circuit production. The capacitors, resistors, inductors, and active devices have their leads inserted in holes in the board and soldering is completed in one operation, by passing the wiring surface over a solder bath. Automation of assembly, elimination of human wiring errors, and greater reproducibility of circuit performance from unit to unit through precise lead arrangement are the advantages obtained. Smaller size follows as the result of miniature components designed and produced for circuit board mounting.

The next step was to "print" or deposit resistance material and metallic capacitor electrodes directly by stencil application (later by metal sputtering or evaporation) onto glass or ceramic chips. Wiring was added with silver paint applied by silk screen techniques. The first such circuits were *RC* coupling and filter networks; active devices are mounted as in the etched circuits and the result is called a *hybrid* of thin film components and attached active devices.

As the magnitude of electronic systems increased, especially in the computer field, it has been necessary to build complex systems without having the production problems increase to an equal order of complexity. To this need has been added a desire for greater reliability, and for miniaturization by the placement of great numbers of components per unit volume. A response to these needs has been found in the developing field of monolithic or *integrated silicon circuits*. Therein the techniques of semiconductor diffusion, epitaxial growth, and the ideas of the planar transistor are being applied to the production of complete electronic circuits, of high order of complexity, within or on a single wafer of silicon. Materials are deposited or diffused in layers often only a few microns thick, and are restricted to desired areas of the wafer by photographically applied masks. Both passive and active elements are constructed within the silicon wafer, and thin-film metallic interconnections are applied over the sealing silicon dioxide surface layer.

6-18 MASKING

The photographic masks used in the integrated circuit processes are similar to negatives, and are produced as much-reduced photographs of large-scale drawings. In another method of making the photographic master, a two-layer plastic is used. The underlying layer is translucent, and where the upper red layer is cut and removed, light will pass and expose the mask negative. By using large master sheets, the accuracy of line width can be made very great in the photographically reduced mask. Processing of a given circuit type may require a considerable number of such masks, all accurately registered on each other.

As a first step in photo-processing of a circuit, a p silicon wafer may be given a coat of oxide. It is then coated with a photoresist lacquer, and the disk rotated rapidly to distribute the lacquer in a thin uniform layer. The first mask is placed over the wafer and the photoresist exposed to ultraviolet light. Where the light did not strike, the resist may be chemically dissolved, exposing the silicon dioxide layer. This can then be etched away, leaving the silicon wafer open to processing. Where the light struck, the polymerized photoresist lacquer remains and this is unaffected by the etching acid.

Etching operations can be precisely controlled by timing, and the final limit on minimum width of line which may be cut in a layer such as the silicon dioxide is fixed by film resolution and light diffraction, and is in the neighborhood of 10 microns.

Because of the small size of each complex circuit, several hundred circuits may be simultaneously photographed onto a substrate wafer of five to six cm^2 area.

6-19 THE SILICON INTEGRATED CIRCUIT

Use of the techniques of planar transistor production and photographic masking leads to development of active and passive circuit elements on a silicon wafer, as indicated in Fig. 6-21. Starting with the p silicon wafer of (a), an n impurity is deposited through a mask onto the surface areas over A, B, C. This impurity is then diffused to a desired depth by control of time and temperature, forming the n conductivity pockets shown in (b). The pn diodes

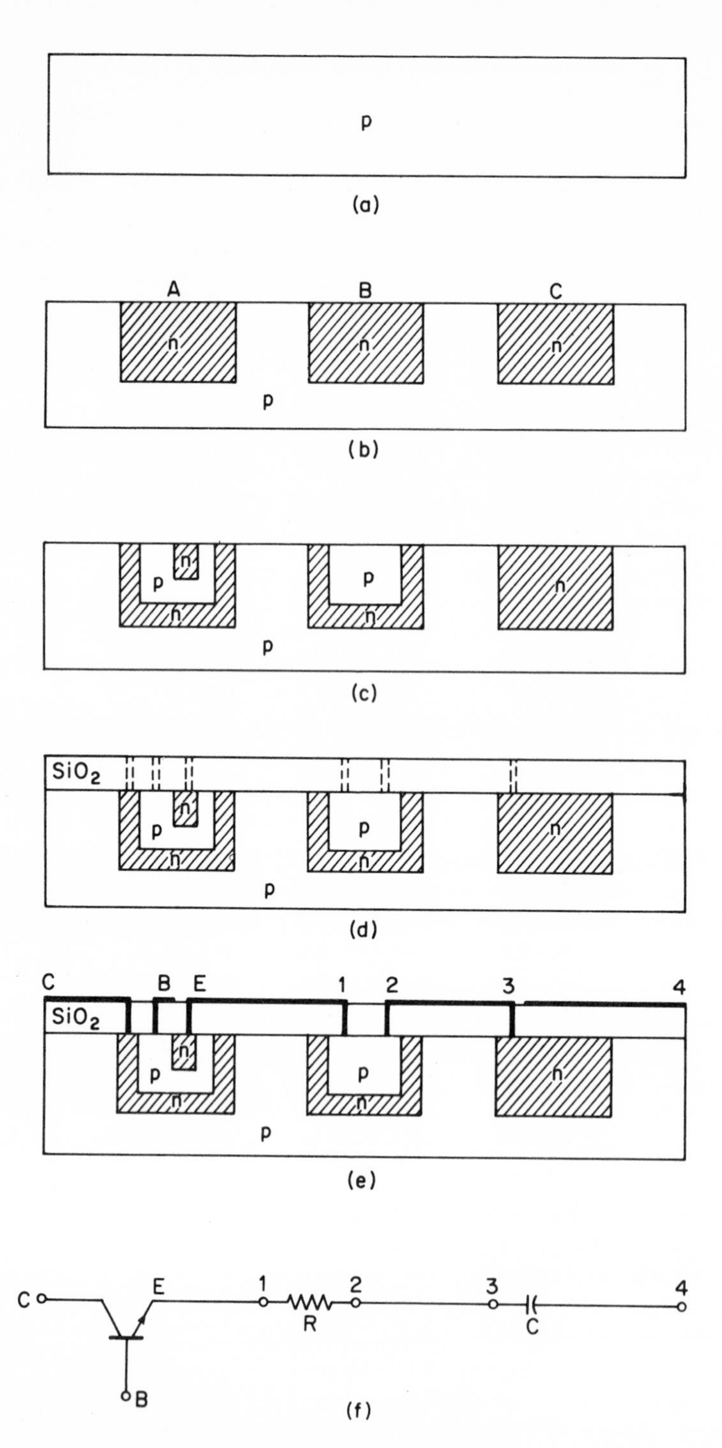

Figure 6-21. Steps in integrated circuit production.

formed with the substrate will be back-biased, and will provide isolation of the circuit elements from the substrate.

Deposition of a p impurity onto the desired areas of A and B, and subsequent diffusion into the n material, will produce the p volumes in (c). In the next step an n emitter is likewise masked and diffused into the p material of A.

The wafer is then coated with an isolating silicon dioxide layer, followed by appropriate masking and etching of holes for connections to the elements, as indicated at (d). Metallization is performed through another mask on top of the oxide, resulting in the circuit at (e).

The electrical result is shown at (f). At A, an npn transistor has been formed. At B, the p material is used as a resistor, with its value controlled by the area converted to p material. At C, the silicon dioxide is used as the dielectric of a capacitor, with the area of the electrode being that of the n layer, or that of the metallic connector.

Thus, with only the basic processes equivalent to those required for fabrication of transistors, it is possible to make complex circuits of active and passive elements. The layout of the circuit and the values of the resistors and the capacitors are functions of masking areas; the "tooling" for production is relatively simple and subject to rapid change. Two hundred identical circuits can be produced as easily on one chip as can one.

Components formed by integrated circuit techniques will not be entirely similar to their discrete component counterparts. Diodes may be directly formed, or may employ the base-emitter or collector-base junctions produced by the diffusions incident to transistor formation. These will differ somewhat in forward drop and switching time. Transistors will have a larger than normal collector-to-ground capacitance due to the capacitance introduced through the np isolation diode. The leakage current in this diode will approximate 10–100 nanoamperes. A silicon dioxide layer may also be formed to serve as isolation between the circuit elements and the substrate wafer.

The diffused resistor is associated with a distributed capacitance, and is analogous to an infinite RC ladder network with a cutoff frequency, determined by RC_s. The resistance is proportional to R_sL/W where R_s is the resistivity of the diffused region in ohms per m^2. The capacitance shunting the resistor is proportional to the resistor area LW, and so the shunt capacitance is

$$C_s = kW^2\left(\frac{R}{R_s}\right)$$

Thus a narrow resistor is indicated for a high cutoff frequency. Variations in resistance occur through variation in the diffusion characteristics and in variation of line width. The latter is controllable in the etching operation.

The use of silicon dioxide as a capacitor dielectric makes possible capacitors up to about 100 pF. A reverse-biased junction may also be employed as a capacitor, but such a capacitance is voltage sensitive.

Low resistance materials, such as gold or aluminum, are deposited to serve as interconnecting leads.

6-20 ADVANTAGES

Integrated circuits have problems of yield in production, since a fault produced by a dust particle can destroy the usefulness of a complete circuit, and scratches may result in a broken connection. However, the reduced complexity in connection, the increased reliability, and the two-dimensional nature of the components which minimizes stray coupling, all contribute to major advantages for the integrated circuit.

Another advantage is now emerging in circuit design. When using discrete components, the active device has been regarded as a high-cost item; its use was avoided when a desired circuit function could be achieved through use of additional passive components. However, in the integrated circuit, the cost is proportional to circuit area, and the cost of the processing operations is independent of the number of components present. The cost of passive elements is often higher than that of an active device, and so it becomes economical to design circuits which use additional transistors and diodes. These may perform or simulate circuit functions which are relegated to passive elements in discrete component circuitry. The development of integrated circuits therefore creates opportunity for new circuit designs, using diodes and transistors more often than resistors and capacitors, and this may eventually lead to circuit forms differing from those of the past.

At present the differential amplifier lends itself to such production in the field of analog signal circuits, while computer logic and switching circuits, already employing more transistors and diodes than passive components, are examples of integrated circuit applications in the field of digital signals.

PROBLEMS

6-1. A *pnp* transistor has $\alpha_F = 0.94$, $\alpha_R = 0.60$, $I_{CO} = 30\ \mu\text{A}$, $I_{EO} = 45\ \mu\text{A}$. With the transistor operating at cutoff or with $v_{EB} = -5$ V, $v_{CB} = -5$ V, find the three transistor currents.

6-2. With the base common to input and output circuits, as in Fig. 6-1, it is found that $i_B = 150\ \mu\text{A}$, $i_C = 1.27$ mA. Determine h_{FB} and h_{FE}.

6-3. A germanium transistor has $I_{CO} = 15\ \mu\text{A}$ at 25°C. What will be the probable leakage current at 85°C?

6-4. A silicon transistor has $\Delta i_C = 1.80$ mA for $\Delta i_E = 1.89$ mA. What change in i_B will produce an equivalent change in i_C?

6-5. An abrupt junction germanium *pnp* transistor has base width $W' = 0.0025$ cm, base conductivity $= 4$ mho/cm, $\mu_e = 0.39$ m^2/V-s, $D_e/L_e = 2 \times 10^3$ cm/s. Find the value of the effective base width W at -9 V for v_{CE}.

REFERENCES

1. Bardeen, J., and W. H. Brattain, "The Transistor, a Semiconductor Triode," *Phys. Rev.*, Vol. 74, p. 230 (1948).
2. Shockley, W., *Electrons and Holes in Semiconductors*. D. Van Nostrand Co., Inc., Princeton, N. J., 1950.
3. *Proceedings of the IRE*, special issue on the transistor. Vol. 40, p. 1289 *et seq*. (1952).
4. Keck, P. H., and M. J. E. Golay, "Crystallization of Silicon from a Floating Liquid Zone," *Phys. Rev.*, Vol. 89, p. 297 (1953).
5. Ebers, J. J., and J. L. Moll, "Large Signal Behavior of Junction Transistors," *Proc. IRE*, Vol. 42, p. 1761 (1954).
6. Pfann, W. G., "Continuous Multistage Separation by Zone-Melting," *Jour. Metals*, Vol. 7, p. 297 (1955).
7. Middlebrook, R. D., *Introduction to Junction Transistor Theory*. John Wiley & Sons, Inc., New York, 1957.
8. Pritchard, R. L., "Advances in the Understanding of the *P-N* Junction Triode," *Proc. IRE*, Vol. 46, p. 1130 (1958).
9. *Proceedings of the IRE*, Vol. 46, p. 947 *et seq*. (1958).
10. Conwell, E. M., "Properties of Silicon and Germanium," I, *Proc. IRE*, Vol. 40, p. 1327 (1952) and II, *Proc. IRE*, Vol. 46, p. 1281 (1958).
11. Shive, J. B., *The Properties, Physics, and Design of Semiconductor Devices*. D. Van Nostrand Co., Inc., Princeton, N. J., 1959.
12. Searle, C. L., A. R. Boothroyd, *et al.*, *Elementary Circuit Properties of Transistors* (SEEC, Vol. 3). John Wiley & Sons, Inc., New York, 1964.
13. Wallmark, J. T., and H. Johnson, *Field-Effect Transistors*. Prentice-Hall, Inc., Englewood Cliffs, N. J., 1966.

7

THE TRANSISTOR: BASIC AMPLIFIERS

Nonlinear circuit devices are usually modelled by equivalent circuits which simulate the device performance within a particular quasi-linear operating region. Selection is usually made of the model giving the simplest analysis with sufficient accuracy for the purpose.

The previous chapter assembled several transistor models from known physical circuit elements: diodes, resistors, and current generators. The accuracy of each representation depends on the fidelity with which these elements approximate the actual performance, and on the complexity tolerated in the model. The result was intended to have a physical relation to the device under study.

A second approach is taken here by use of the network theory reviewed in Chapter 5, in which device performance will be described by systems of algebraic equations, and with parameter determination by device measurements. Because of the linear performance of the transistor with small signals applied, the methods to be studied in this chapter are widely favored for linear amplifier design.

7-1 TRANSISTOR CURRENT AND VOLTAGE NOTATION

Standard notation for the transistor has already been used and will be further employed. The following rules are graphically indicated in Fig. 7-1:

Variable	Assigned letter	Subscript	Example
Total instantaneous quantity	Lower case	Upper case	i_C
Instantaneous value of varying component	Lower case	Lower case	i_c
dc or non-time-varying	Upper case	Upper case	I_C
ac rms quantity	Upper case	Lower case	I_c
Supply or source	Upper case	Upper case, repeated	V_{CC}

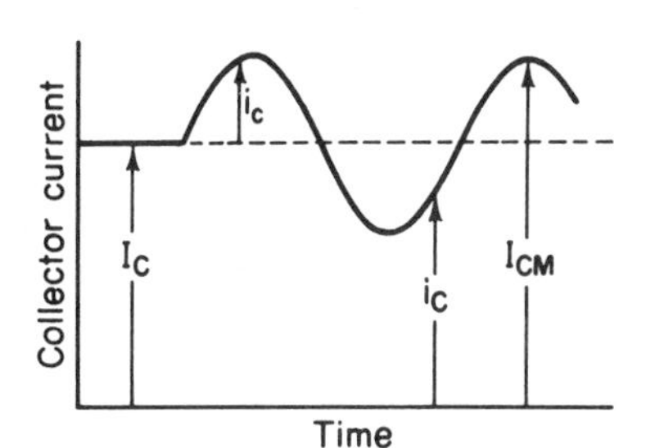

Figure 7-1. Transistor current notation.

The subscripts c or C, e or E, b or B are used to refer to collector, emitter, and base quantities, respectively, and a second letter subscript is used when needed, to indicate the common or reference electrode.

Internal transistor parameters are lower case, r, y, z, h; external impedances are indicated by capital letters.

For model circuit parameters, both numerical and letter subscripts are employed but letter subscripts will be used here; examples are i, r, f and o as first subscripts to designate the particular parameter, and its impedance matrix position. A second subscript is added to indicate the common electrode of the circuit or that used in measurement of the parameter. Such combinations as z_{ib}, y_{oe}, h_{fe} result.

7-2 *VOLT-AMPERE RELATIONS OF THE TRANSISTOR*

Volt-ampere characteristics are a fundamental means of describing the performance of a circuit device, and their introduction here for the transistor is intended to provide a bridge between the physical device and its models of the previous chapter, and the algebraic models to be studied here.

A complete forward and reverse characteristic for a *pnp* transistor is presented in Fig. 7-2. The collector current is plotted against v_{CE} as the collector voltage, with i_B, the base current, as a parameter. In the first quadrant there appears the usual forward characteristics of the collector junction as a diode, and this is not a transistor operating region. Operation therefore must be with the collector reverse-biased and in the third quadrant, where a family of i_C, v_{CE} curves is found.

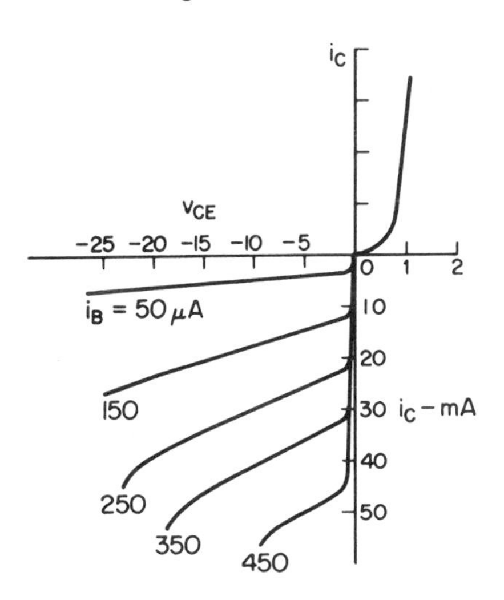

Figure 7-2. Four-quadrant common-emitter curves of a *pnp* transistor.

These *output characteristics* are expanded in Fig. 7-3, and since v_{CE} is measured to the emitter, this element becomes common to both input and output circuits, implying a circuit as in Fig. 7-4. The slopes of the output curves must represent $1/r_c$, where r_c is the output resistance of the transistor. That its value is large is indicated by the low slope of the curves; that it

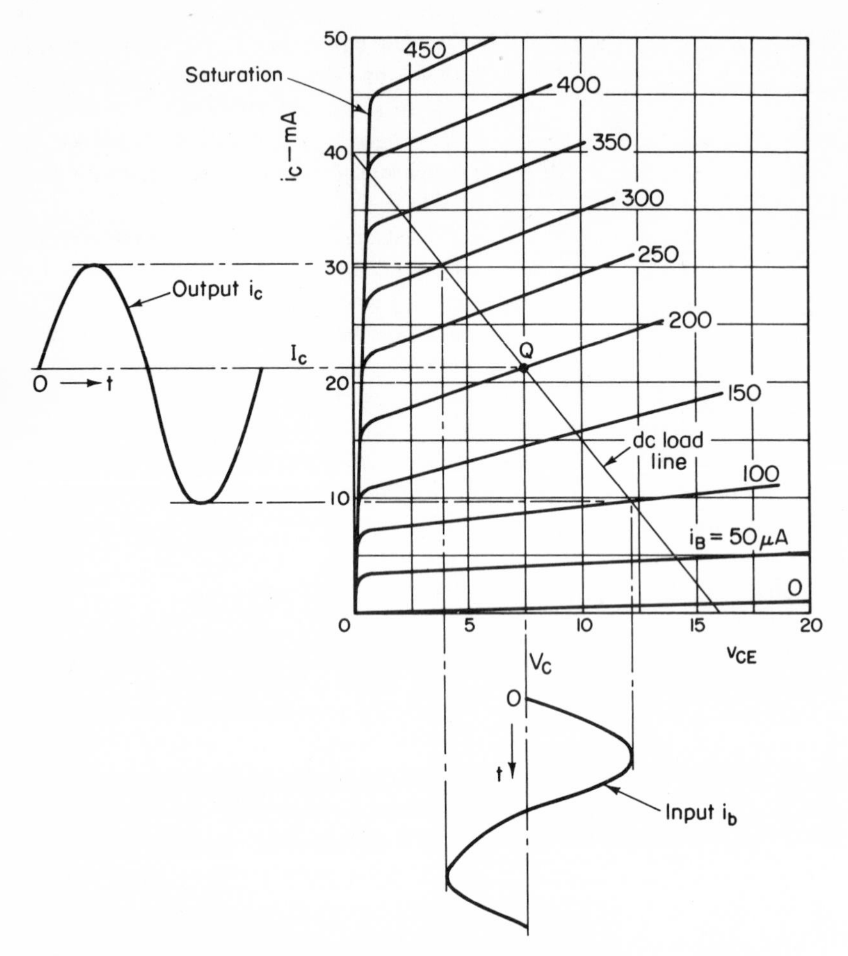

Figure 7-3. Output curves of a silicon transistor and load line, at 25°C.

decreases with increase in i_C is shown by the increased slope at the higher i_C values.

At very small values of collector voltage, the curves merge, indicating that the base current then has no control over the collector current. The collector-base junction is found to be forward-biased, and its voltage is that of a forward-biased diode. The transistor is said to be in *saturation*. The transistor acts as a constant resistance as indicated by the saturation line on the curve, and its resistance

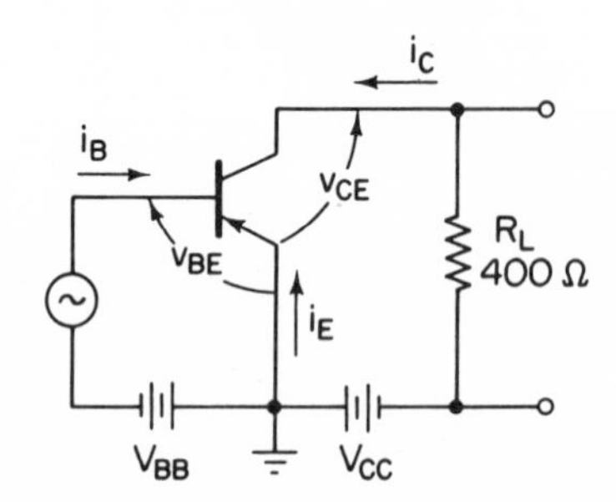

Figure 7-4. Common-emitter amplifier.

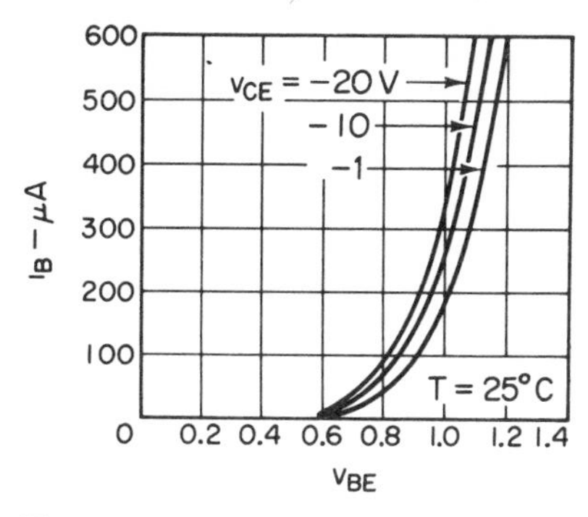

Figure 7-5. Input curves for the transistor of Figure 7-3.

is known as $R_{CE(\text{SAT})}$. This value is largely due to the bulk resistivity of the collector material.

The curve for $i_B = 0$ represents the condition of collector cutoff, close to the v_{CE} axis. The vertical spacing between the curves is a measure of h_{fe}, as defined in Eq. 6-10.

The *input characteristics* of Fig. 7-5 provide input circuit information, with the slope giving the reciprocal of the input resistance. For many purposes it is reasonable to assume v_{BE} constant at a value in the range from 0.7 V to 1.0 V for a silicon transistor, with small error; for germanium transistors this range might be from 0.2 V to 0.5 V.

7-3 *THE LOAD LINE AND DYNAMIC TRANSFER CURVE*

The characteristic curves relating i_C and v_{CE} of the transistor can be functionally expressed as

$$v_{CE} = f(i_B, i_C) \tag{7-1}$$

The current-voltage relations in the collector circuit of Fig. 7-4 are

$$v_{CE} = V_{CC} - i_C R_L \tag{7-2}$$

The transistor and load are in series, and these equations involve the same current. Simultaneous solution of Eqs. 7-1 and 7-2 would give a value for the transistor current, but this is not possible because of the functional form of Eq. 7-1.

However, V_{CC} is a constant and Eq. 7-2 is linear and of the form

$$y = b + mx$$

with x intercept V_{CC}, slope of $-1/R_L$, and y intercept $i_C = V_{CC}/R_L$. A plot of such a line, superimposed on the output characteristics of the transistor, defines all possible combinations of current and voltage which may exist in the series circuit of transistor and load, and provides a solution for Eqs. 7-1 and 7-2. In Fig. 7-4 the load may be taken as 400 ohms and $V_{CC} = 16$ V, so that the y intercept $= 16/400 = 0.04$ A, and the x intercept $= V_{CC} = 16$ V. A straight line drawn between these intercepts will have the proper slope and is the *dc load line* for the chosen load, as in Fig. 7-3.

If alternating current inputs are employed, and the output is to be a linear representation of the input, a central *operating* or *quiescent point*, Q, must be chosen on the load line as an origin, so that positive and negative swings of input may occur about it in a linear region of the curves. A Q point is determined by choice of I_B, and $I_B = 200\ \mu\text{A}$ gives a central location for the resting point in Fig. 7-3.

It can then be determined that at the Q point, $v_{CE} = 7.5$ V across the transistor, the voltage drop across the load is 8.5 V, and $I_C = 21.2$ mA.

Assuming an input ac current wave of 100 μA peak for the base current, an input i_B wave can be plotted, and carried over to the i_C axis to give an output wave. However, this information can be better obtained from a *dynamic transfer curve*. This dynamic transfer curve directly relates input and output currents for the transistor and load chosen, and is plotted in Fig. 7-6 with data taken from the intersections of the load line and the i_B curves of Fig. 7-3. In this case the selection yields a quasi-linear relation, but the departure from linearity can be neglected for sufficiently small signal variations.

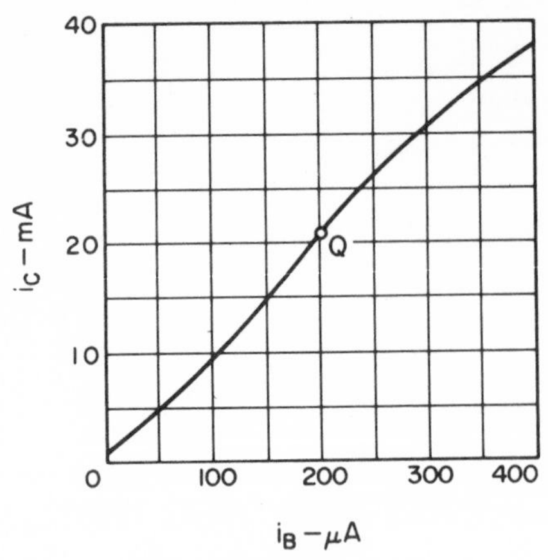

Figure 7-6. Dynamic transfer curve for the load line of Figure 7-3.

The dynamic transfer curve will have considerable use in the analysis of amplifiers with large signals, where the nonlinearity cannot be overlooked.

7-4 *THE IDEAL AMPLIFIER*

The transistor is a current input device; that is, it is responsive to variations in the flow of charges over the emitter-base barrier. In an ideal form, the transistor can be described by the circuit of Fig. 7-7(a), in which the input impedance is zero to maximize the input current, and the shunt output impedance is infinite to deliver all the output current to the load circuit. The output current would be $i_2 = h_f i_1$, where h_f is the current gain. The input-output relation is linear, or h_f is a constant.

The ideal voltage amplifier of Fig. 7-7(b) can serve as a model for the vacuum triode. The input impedance is infinite to maximize the input voltage from the source, and the series output impedance is desirably zero to

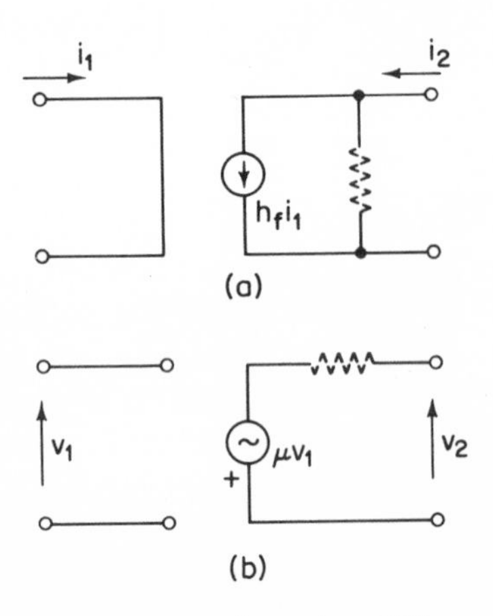

Figure 7-7. (a) Current amplifier; (b) voltage amplifier.

deliver all of the developed voltage to the load. The load voltage is then given by μv_1, and the input-output relation is linear, or μ is a constant.

No transistor or tube is perfect, however. They do have finite and nonzero impedances, they are not inherently linear, and they and their associated circuits have parasitic reactances which lead to varying performance over wide frequency ranges. The designer works to overcome the practical deficiencies of these devices.

7-5 *THE TRANSISTOR AS A SMALL-SIGNAL AMPLIFIER*

When viewed as a control generator in the active circuit models of Chapter 5, the transistor must have two ports or two pairs of terminals. However, the transistor has three internal elements, and so one of the internal elements must be chosen as common to the input and output circuits. There are then three possible circuit configurations, as shown in Fig. 7-8, with diverse characteristics and operational properties.

To maintain operation within a linear region of the transistor characteristics, the input signals must be small. The small-signal currents and voltages are defined in the figure, with all currents inward at the terminals, as is conventional.

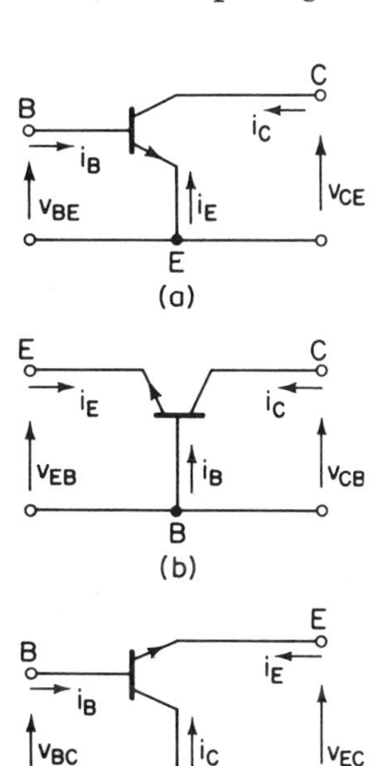

Figure 7-8. (a) Common-emitter connection; (b) common-base connection; (c) common-collector connection.

As discussed in Chapter 5, three choices of independent voltage and current variables are of use in network study of transistor and tube. These selections lead to the r (z), y, and h sets of parameters for the active device. On an historical basis the use of r parameters and the common-base circuit was first, but at present the h parameters and the common-emitter circuit are of greatest importance and will be treated first here.

The values of the h parameters differ in each of the three basic transistor circuits; they must be measured for the particular basic circuit form under study. That is, common-base h_{-b} parameters are used for common-base circuit performance, h_{-e} and h_{-c} parameters are used for common-emitter and common-collector performance calculations, respectively. The performance equations are independent of circuit form.

The r (z) parameters have the same value for a given transistor and operating point, regardless of the circuit form used, but the performance equations differ with each basic circuit. The r parameters are not susceptible to easy direct measurement, and their values usually are computed from the more readily measured h parameters.

7-6 THE h-PARAMETER EQUIVALENT CIRCUIT

We now assume that the operating point has been placed in a linear region of the transistor characteristics, and we are going to work with signal-created variations around that point. Instead of parameters relating total current and voltage values, our concern is now with incremental changes, and the slopes of the characteristic curves measured at the Q point define the parameters of interest.

The intent to use the h parameters dictates the choice of i_1 and v_2 as independent network variables, according to Section 5-4. With small-signal sinusoidal variations applied to the two ports or two terminal pairs of a transistor, viewed as the "black box" of unknown interior circuit as in Fig. 7-9(a), basic current and voltage measurements may be made. From those measurements it is possible to develop by algebraic means a circuit model of an internal network which will operate under the assigned constraints as does the actual transistor under those conditions. The result is a mathematically derived model, which has value because of its compatibility with other circuit elements in circuit analysis. For the present, the frequency of operation will be considered sufficiently low that the internal transistor capacitive reactances can be ignored.

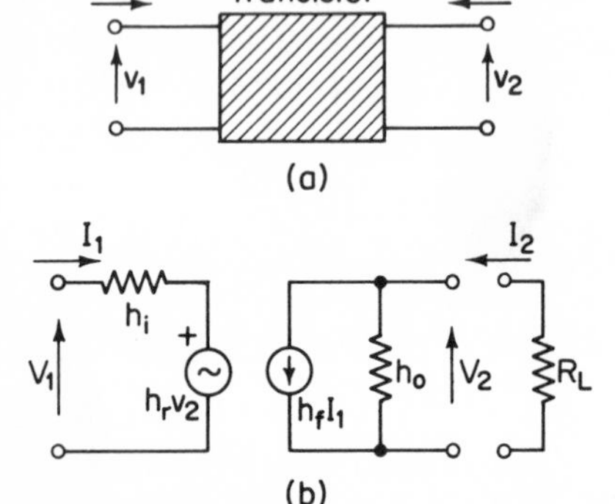

Figure 7-9. (a) Transistor as a two-port "black box." (b) The h-parameter equivalent circuit.

With the input restricted to small-signal variations, the usual network relations with h parameters, as developed in Eqs. 5-13 and 5-14, may be used as a starting point in discussion of the performance of our black box network:

$$\mathbf{V}_1 = h_i\mathbf{I}_1 + h_r\mathbf{V}_2 \tag{7-3}$$

$$\mathbf{I}_2 = h_f\mathbf{I}_1 + h_o\mathbf{V}_2 \tag{7-4}$$

or

$$\begin{bmatrix} \mathbf{V}_1 \\ \mathbf{I}_2 \end{bmatrix} = \begin{bmatrix} h_i & h_r \\ h_f & h_o \end{bmatrix} \begin{bmatrix} \mathbf{I}_1 \\ \mathbf{V}_2 \end{bmatrix} \tag{7-5}$$

It should be noted that h_i is an impedance, h_o an admittance, and h_r and h_f are dimensionless ratios.

Equations 7-3 and 7-4 can be represented by a circuit model as at Fig. 7-9(b). The first equation is of mesh form, and the first mesh includes a resistor and a transfer voltage source. The second equation is of nodal form and the mesh includes a conductance and a controlled current source. The current source represents the forward transfer generator, with an output current proportional to the input or controlling current. The output current from the generator divides between the conductance h_o and the load R_L.

Since the transistor is not unilateral, a portion of the output voltage is fed back to the input circuit and appears as the reverse transfer generator $h_r\mathbf{V}_2$. Thus the actual transistor is not ideal as a current source, having a low input resistance, a high output resistance, and a feedback element. A normal objective is to minimize the effects of these elements in order to approach ideal performance.

The h parameters are determined by ac measurements made on the network, with specified terminations. It follows from Section 5-4 that

$$\left.\begin{aligned} h_i &= \frac{\mathbf{V}_1}{\mathbf{I}_1} = \text{short-circuit input impedance,} && (\mathbf{V}_2 = 0) \\ h_r &= \frac{\mathbf{V}_1}{\mathbf{V}_2} = \text{open-circuit reverse voltage ratio,} && (\mathbf{I}_1 = 0) \\ h_f &= \frac{\mathbf{I}_2}{\mathbf{I}_1} = \text{short-circuit forward current ratio,} && (\mathbf{V}_2 = 0) \\ h_o &= \frac{\mathbf{I}_2}{\mathbf{V}_2} = \text{open-circuit output admittance,} && (\mathbf{I}_1 = 0) \end{aligned}\right\} \quad (7\text{-}6)$$

The input parameters, h_i and h_f, are measured under an output ac short circuit. An accurate ac short circuit is relatively easy to obtain in the high impedance collector circuit, and this eliminates errors which might appear due to shunt leakage and capacitance. The output parameters, h_o and h_r, are measured with the input circuit open; this permits accuracy since open circuits are accurately obtained in the usual low impedance input circuits. Accuracy and ease of measurement are major reasons for the choice of hybrid parameters in the transistor network model.

Parameter h_i is the reciprocal slope of the input characteristic; h_o is the slope of the output curve. The parameters h_r and h_f are the slopes of appropriately presented curves or represent the spacing between the input and output curves at constant current and constant voltage.

7-7 *TRANSISTOR AMPLIFIER PERFORMANCE WITH THE h PARAMETERS*

A signal source of resistance R_s and a load of resistance R_L must be added to the basic hybrid-parameter equivalent circuit to simulate actual transistor amplifier performance; the result is Fig. 7-10(a). Note that we are still employing a network representation of the transistor as an arbitrary black box;

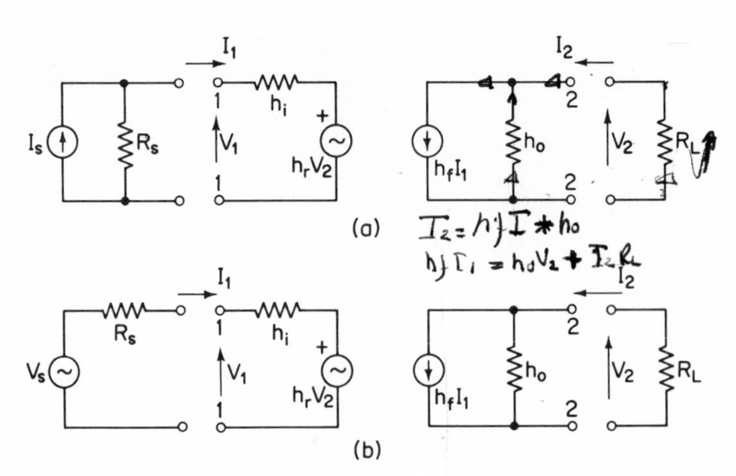

Figure 7-10. The h-parameter circuit: (a) current-source input; (b) voltage-source input.

it has not yet become necessary to specify the common element of the transistor.

Using Fig. 7-10(a), the operating equations become:

$$\mathbf{V}_1 = (\mathbf{I}_s - \mathbf{I}_1)R_s = h_i\mathbf{I}_1 + h_r\mathbf{V}_2 \tag{7-7}$$

$$0 = h_f\mathbf{I}_1 + \left(h_o + \frac{1}{R_L}\right)\mathbf{V}_2 \tag{7-8}$$

The *input impedance* at the 1,1 terminals can be obtained by solving Eq. 7-8 for $\mathbf{V}_2$ and substituting in Eq. 7-7:

$$R_i = \frac{\mathbf{V}_1}{\mathbf{I}_1} = h_i - \frac{h_r h_f R_L}{1 + h_o R_L} = \frac{h_i + R_L \Delta_h}{1 + h_o R_L} \tag{7-9}$$

where the determinant of the h matrix of Eq. 7-5 is

$$\Delta_h = h_i h_o - h_r h_f \tag{7-10}$$

The *output impedance* at the 2,2 terminals can be found by the usual method, letting $\mathbf{I}_s = 0$ and noting that $\mathbf{V}_2 = -\mathbf{I}_2 R_L$. Then

$$G_o = \frac{1}{R_o} = h_o - \frac{h_r h_f}{h_i + R_s} \tag{7-11}$$

and

$$R_o = \frac{h_i + R_s}{\Delta_h + R_s h_o} \tag{7-12}$$

Equations 7-9 and 7-11 show how the operating characteristics depend on the feedback parameter, h_r, through Δ_h. If the feedback is small, or $h_r h_f R_L \ll (1 + h_o R_L)$, then the input impedance is h_i; if $h_r h_f \ll (h_i + R_s)$, the output admittance is h_o. In such cases the feedback generator can be dropped from the circuit model.

By again noting that $\mathbf{V}_2 = -\mathbf{I}_2 R_L$, Eq. 7-8 may be used to calculate the *internal transistor current gain*, $\mathbf{A}'_i$, from terminals 1,1 to the load, as

$$\mathbf{A}'_i = \frac{\mathbf{I}_2}{\mathbf{I}_1} = h_f\left(\frac{1}{1 + h_o R_L}\right) \tag{7-13}$$

Since h_f is the short-circuit current gain of the transistor alone, the presence of the load and output admittance reduces this theoretical gain by the factor $1/(1 + h_o R_L)$. Transistor amplifier outputs approach the performance of true current sources, and in cascade each stage may approximate a current source driving the following input.

To permit calculation of the current gain, including the effect of the source resistance R_s, the equations may be solved again, leading to

$$\mathbf{A}_i = \frac{\mathbf{I}_2}{\mathbf{I}_s} = h_f\left(\frac{1}{1 + h_o R_L}\right)\left(\frac{R_s}{R_s + h_i - \dfrac{h_r h_f R_L}{1 + h_o R_L}}\right) \tag{7-14}$$

$$= \mathbf{A}'_i\left(\frac{R_s}{R_s + h_i - \dfrac{h_r h_f R_L}{1 + h_o R_L}}\right) \tag{7-15}$$

Thus the internal gain $\mathbf{A}'_i$ may be approached if $R_s \longrightarrow \infty$ as would be the case for a true current source input. Note that the final denominator term disappears if h_r is made small by transistor design.

By use of Eq. 7-9, Eq. 7-14 can be modified to give

$$\mathbf{A}_i = h_f\left(\frac{1/h_o}{1/h_o + R_L}\right)\left(\frac{R_s}{R_s + R_i}\right) \tag{7-16}$$

The first parentheses can be recognized as the current division factor between the load and the shunted output resistance $1/h_o$. The second parentheses show current division between the source resistance and the transistor input resistance. Obviously, to approach the ideal current source performance given by h_f, the value of h_o should be small (r_o large) and R_s should be large.

The *voltage gain* from input to output is also of importance. This performance factor can be found if the input source be changed to one of voltage form, as in Fig. 7-10(b). The network equations are modified by using $\mathbf{V}_1 = \mathbf{V}_s - R_s \mathbf{I}_1$, leading to

$$\mathbf{A}_v = \frac{\mathbf{V}_2}{\mathbf{V}_s} = \frac{-h_f R_L}{(h_i + R_s)(1 + h_o R_L) - h_r h_f R_L} \tag{7-17}$$

The negative sign indicates a phase reversal of the output voltage with respect to the input voltage. The equation again shows the effect of h_r, and indicates the possibility of gain instability ($\mathbf{A}_v \longrightarrow \infty$), if h_r were ever to become large. Much device and circuit work has been done to reduce the effect of this parameter in amplifiers.

The *internal voltage gain*, $\mathbf{A}'_v$, from terminals 1,1 to the output can be written from the above by making $R_s = 0$, so that

$$\mathbf{A}_v' = \frac{\mathbf{V}_2}{\mathbf{V}_1} = \frac{-h_f R_L}{h_i(1 + h_o R_L) - h_r h_f R_L} \tag{7-18}$$

This gain can be maximized with large R_L.

Since power must be supplied to the finite input resistance of the transistor, it is of interest to determine how much the circuit increases this power, as the *operating power gain*. Because of the different ways of computing voltage and current gains, the power gain can also be calculated in various ways. However, for our purpose, it does not seem proper to include the power dissipated in R_s as part of the input power; more properly that power might be considered as a loss in the preceding circuit. Then the operating power gain will be given by the product $|A_i' A_v'|$, or

$$\text{operating P.G.} = |A_i' A_v'| = \left(\frac{h_f}{1 + h_o R_L}\right)\left[\frac{h_f R_L}{h_i(1 + h_o R_L) - h_r h_f R_L}\right] \tag{7-19}$$

The result of this calculation is usually stated in decibels.

7-8 THE COMMON-EMITTER AMPLIFIER; h PARAMETERS

The common-emitter (C-E) amplifier of Fig. 7-11 results if the emitter is chosen as the common element between input and output circuits. The currents then become $I_1 = I_b$, $I_2 = I_c$. The circuit is widely favored since it can be designed for good voltage and current gains, as well as the highest potential power gain of the three basic circuits. It also has output and input impedances which are more nearly of the same order than is true for the other circuit forms. Therefore, common-emitter circuits in cascade, without matching of input and output impedances with transformers, give better power transfer than is possible with the other basic circuits. This is an economic advantage.

To illustrate the magnitude of the performance which may be expected, we select a transistor having the following parameters as measured in the C-E circuit, indicated by the use of e as the second subscript:

$$h_{ie} = 900 \text{ ohms}, \qquad h_{fe} = 24$$

$$h_{re} = 7.1 \times 10^{-4}, \qquad h_{oe} = 22 \times 10^{-6} \text{ mho}$$

and add a source of resistance $R_s = 2000$ ohms, and a load $R_L = 10{,}000$ ohms. From the equations of the preceding section:

$$h_{ie}h_{oe} = 900 \times 22 \times 10^{-6} = 0.0198$$

$$h_{re}h_{fe} = 7.1 \times 10^{-4} \times 24 = 0.0170$$

$$\Delta_{he} = 0.0198 - 0.0170 = 0.0028$$

$$1 + h_o R_L = 1 + (22 \times 10^{-6} \times 10^4) = 1.22$$

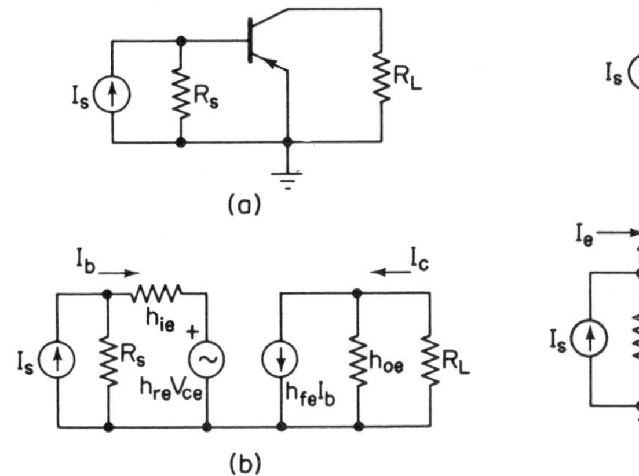

Figure 7-11. The common-emitter amplifier.

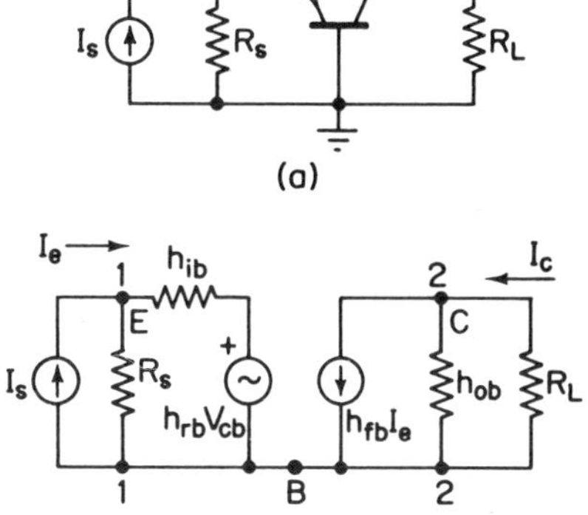

Figure 7-12. The common-base amplifier.

Then

$$R_{ie} = 900 - \frac{170}{1.22} = 756 \text{ ohms}$$

$$R_{oe} = \frac{900 + 2000}{0.0028 + 0.044} = 62{,}000 \text{ ohms}$$

$$A'_{ie} = \frac{24}{1.22} = 19.7$$

$$A_{ie} = \frac{19.7 \times 2000}{2000 + 900 - 170/1.22} = 14.3$$

$$A'_{ve} = \frac{-24 \times 10^4}{(900 \times 1.22) - 1.70} = -258$$

$$A_{ve} = \frac{-24 \times 10^4}{(2000 + 900) \times 1.22 - 170} = -71.2$$

Operating P.G. $= A'_{ie}A'_{ve} = 5070$, equivalent to a gain of 37.1 dB.

7-9 THE COMMON-BASE AMPLIFIER; h PARAMETERS

The common-base (C-B) amplifier of Fig. 7-12 has useful properties as an impedance transformer which also provides power gain. It has a low input resistance and a high output resistance and serves to couple a low-resistance source to a high-resistance load. To illustrate, we use the same transistor as in the preceding example, with the currents designated as $I_1 = I_e$, $I_2 = I_c$. Measured with the base as the common element, the parameters are:

$$h_{ib} = 36 \text{ ohms}, \qquad h_{fb} = -\alpha_{fb} = -0.96$$

$$h_{rb} = 0.84 \times 10^{-4}, \qquad h_{ob} = 0.88 \times 10^{-6} \text{ mho}$$

With $R_s = 2000$ ohms, $R_L = 10{,}000$ ohms, then

$$h_{ib}h_{ob} = 0.315 \times 10^{-4}$$

$$h_{rb}h_{fb} = -0.80 \times 10^{-4}$$

$$\Delta_{hb} = 1.12 \times 10^{-4}$$

$$1 + h_{ob}R_L = 1.0085 \cong 1$$

Note that $h_{ob}R_L \ll 1$; this will be a usual condition with normal loads in the common-base circuit. Then

$$R_{ib} = 36 + 1.12 \cong 37 \text{ ohms}$$

$$R_{ob} = \frac{36 + 2000}{(17.6 + 1.12)10^{-4}} = 1.09 \times 10^6 \text{ ohms}$$

$$A'_{ib} = -0.96$$

$$A_{ib} = -0.96 \frac{2000}{2000 + 36 + 0.8} = -0.94$$

$$A'_{vb} = \frac{0.96 \times 10^4}{36 + 0.8} = 261$$

$$A_{vb} = \frac{0.96 \times 10^4}{2000 + 36 + 0.8} = 4.7$$

Operating P.G. = 250, which is equivalent to 24 dB.

There is no voltage phase reversal in this circuit. The current gain is near but less than unity, approximating α_{fb}, but the major advantage of the circuit is in its impedance-transforming property. The output impedance is higher than that of the C-E circuit by a considerable factor, and the input resistance is lower than the C-E circuit. The input current essentially appears in the output circuit, but the impedance is only 37 ohms while the output resistance is over 1 megohm. The circuit acts as a pump forcing water from a low-pressure system into a high-pressure system, and there is an obvious power gain.

7-10 THE COMMON-COLLECTOR AMPLIFIER; h PARAMETERS

This circuit, shown in Fig. 7-13, is also called an *emitter follower,* since the output voltage from emitter to collector tends to equal or follow the changes in input voltage applied between base and collector. The circuit has a high input resistance and a low output resistance, thus reversing the impedance transforming properties of the common-base circuit; it also approximates unity voltage gain.

Our particular transistor will have common-collector h parameters as follows:

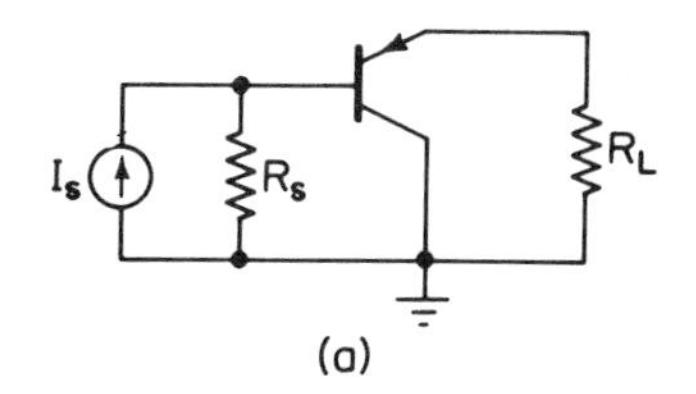

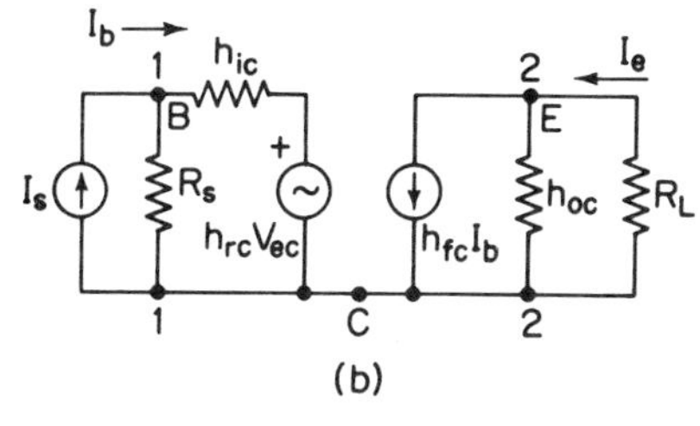

Figure 7-13. The common-collector amplifier.

$$h_{ic} = 900 \text{ ohms}, \quad h_{fc} = -25$$

$$h_{rc} \cong 1, \quad h_{oc} = 22 \times 10^{-6} \text{ mho}$$

We will choose source and load resistances to give a better impedance match; that is, make $R_s = 10{,}000$ ohms and $R_L = 500$ ohms. Then

$$h_{ic}h_{oc} = 0.0198$$

$$h_{rc}h_{fc} = -25$$

$$\Delta_{hc} = 25$$

$$1 + h_{oc}R_L = 1.011 \cong 1$$

It will usually be possible to assume that $h_{oc}R_L \ll 1$.

Using the common equations of Section 7-7, then

$$R_{ic} = 900 + (25 \times 500) = 13{,}400 \text{ ohms}$$

$$R_{oc} = \frac{900 + 10{,}000}{25 + 0.22} = 432 \text{ ohms}$$

$$A'_{ic} = -25$$

$$A_{ic} = -25\left(\frac{10{,}000}{10{,}900 + 12{,}500}\right) = -10.7$$

$$A'_{vc} = \frac{25 \times 500}{900 + 12{,}500} = 0.933$$

$$A_{vc} = \frac{25 \times 500}{10{,}900 + 12{,}500} = 0.535$$

Operating P.G. $= 23.3$, equivalent to 13.7 dB gain.

The circuit finds considerable use as a means of matching high impedance sources to lower impedance loads.

7-11 *MATCHING OF TRANSISTOR AMPLIFIER IMPEDANCES IN CASCADE*

The calculations of the preceding sections have indicated that the transistor gain is sensitive to the input source resistance and the output load. Maximum power gain would be achieved if R_s and R_i were matched, and if the output resistance R_o were also matched to the load R_L. In cascaded stages the source resistance will be the output resistance of the preceding amplifier, and the load may be the input resistance of a following stage. Since R_o is a function of R_s, and R_i is a function of R_L, the optimum conditions for power gain are not readily evident.

With matched input conditions $R_s = R_i$ and the source will supply power to the transistor input in the amount

$$P_{in} = \frac{E_s^2}{4R_s} = I_i^2 R_i \tag{7-20}$$

Output power is $I_2^2 R_L$ and will be maximum if $R_L = R_o$. The power gain then would be

$$\text{available power gain} = \frac{I_2^2 R_L}{I_i^2 R_i} = A_i'^2 \frac{R_L}{R_i} \tag{7-21}$$

With input matching it is necessary that

$$R_s = R_i = \frac{h_i + R_L \Delta_h}{1 + h_o R_L}$$

For output matching it is necessary that

$$R_L = R_o = \frac{h_i + R_s}{\Delta_h + h_o R_s}$$

Solving simultaneously leads to

$$R_s = \sqrt{\frac{h_i \Delta_h}{h_o}}, \qquad R_L = \sqrt{\frac{h_i}{h_o \Delta_h}} \tag{7-22}$$

These expressions relate source and load to the transistor parameters for simultaneous matched input and output conditions, or for *maximum available power gain*. Using A_i' from Eq. 7-13 in Eq. 7-21, it is possible to write the M.A.P.G. as

$$\text{M.A.P.G.} = \frac{h_f^2}{(\sqrt{\Delta_h} + \sqrt{h_o h_i})^2} \tag{7-23}$$

Since this expression involves only transistor parameters, it is useful as a theoretical figure of merit for transistor comparison. The perfection with which amplifiers are matched is measured by comparison of the actual power gain with the M.A.P.G. figure.

In usual amplifiers the M.A.P.G. value can be achieved only with use of impedance-matching transformers between amplifier stages. The use or nonuse of transformers is dependent on cost, space, and the operating frequency; with the common-emitter circuit it is often cheaper to accept a mismatch without transformers and then obtain the needed power gain through additional amplification.

For low power-level amplifiers in which current gain is the greatest consideration, power matching will not be used. For such cases $R_i \ll R_s$, so as not to lose input current through the source shunt resistance, and $R_L \ll R_o$ so as to channel most of the output current into the load, which may often be the input circuit of the following amplifier stage. These are both desirable conditions to make the amplifiers more nearly approach the performance of ideal current generators.

7-12 THE T CIRCUIT MODEL FOR THE C-B AMPLIFIER

A T equivalent circuit using the impedance or z parameters is employed in the analysis of certain circuits. In contrast with the h parameters, the z parameters are invariant from circuit to circuit, with differing circuit equations being used for the three basic circuits. An advantage then arises, namely, the possibility of comparing the performances of the three circuits and developing an understanding of their relative values of impedance and gain. In the following, it will again be assumed that the frequency is low enough that transistor reactances can be ignored; the z parameters then become a set of r parameters.

To obtain transistor relations utilizing the r parameters, it is necessary to select $\mathbf{I}_1$ and $\mathbf{I}_2$ as the independent variables, as in Section 5-2. Equations are obtained for the transistor as

$$\begin{bmatrix} \mathbf{V}_{eb} \\ \mathbf{V}_{cb} \end{bmatrix} = \begin{bmatrix} r_{ib} & r_{rb} \\ r_{fb} & r_{ob} \end{bmatrix} \begin{bmatrix} \mathbf{I}_e \\ \mathbf{I}_c \end{bmatrix} \qquad \begin{matrix} (7\text{-}24) \\ (7\text{-}25) \end{matrix}$$

for the common-base circuit of Fig. 7-14. The definitions of the r parameters are those of Eq. 5-5.

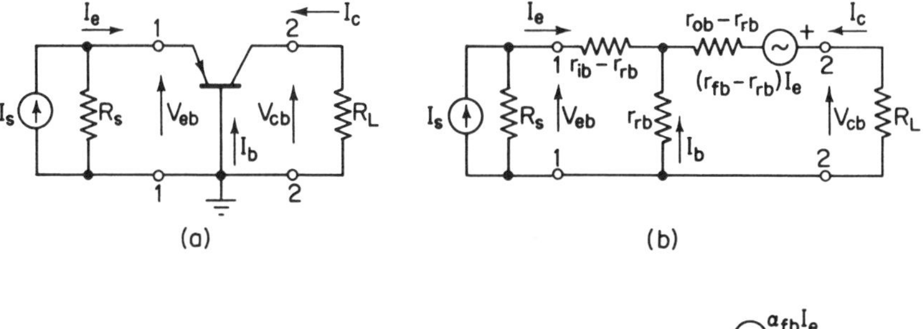

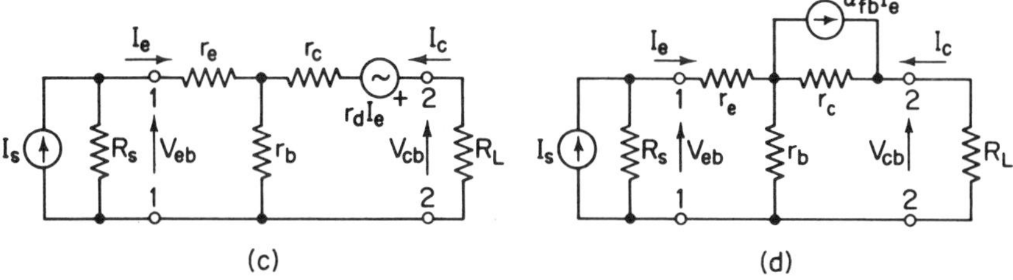

Figure 7-14. (a) Common-base circuit; (b) r-parameter equivalent circuit; (c) T circuit; (d) current-source circuit.

These equations may be algebraically manipulated by the addition and subtraction of $r_{rb}I_c$ to Eq. 7-25, to yield

$$\begin{bmatrix} \mathbf{V}_{eb} \\ \mathbf{V}_{cb} - (r_{fb} - r_{rb})\mathbf{I}_e \end{bmatrix} = \begin{bmatrix} r_{ib} & r_{rb} \\ r_{rb} & r_{ob} \end{bmatrix} \begin{bmatrix} \mathbf{I}_e \\ \mathbf{I}_c \end{bmatrix} \quad \begin{matrix} (7\text{-}26) \\ (7\text{-}27) \end{matrix}$$

and from these equations a circuit can be drawn as in Fig. 7-14(b), mathematically equivalent to the common-base amplifier.

A physical interpretation can be provided for the parameters by relating the emitter, base, and collector regions of the transistor with this circuit, through the definitions:

$$r_e = r_{ib} - r_{rb} = \text{emitter resistance}$$

$$r_b = r_{rb} \qquad = \text{base resistance} = r_b' + r_{bb}$$

$$r_c = r_{ob} - r_{rb} = \text{collector resistance}$$

$$r_d = r_{fb} - r_{rb} = \text{transfer resistance}$$

Equations 7-26 and 7-27 then become

$$\begin{bmatrix} \mathbf{V}_{eb} \\ \mathbf{V}_{cb} \end{bmatrix} = \begin{bmatrix} r_e + r_b & r_b \\ r_b + r_d & r_b + r_c \end{bmatrix} \begin{bmatrix} \mathbf{I}_e \\ \mathbf{I}_c \end{bmatrix} \quad \begin{matrix} (7\text{-}28) \\ (7\text{-}29) \end{matrix}$$

which apply to the circuit of Fig. 7-14(c).

A second circuit form is that at d, where the voltage source is replaced with an equivalent current source by use of Norton's theorem. The resultant circuit correlates well with the physically-derived model of Fig. 6-8.

The circuits of Fig. 7-14(c) and (d) show a forward transfer generator under control of the input current. Feedback to the input is given by the $r_b\mathbf{I}_c$ voltage drop. Each parameter has physical meaning, r_e being related to the previously-derived resistance of the emitter-base diode, r_b to the base resistance $r_b' + r_{bb}$, and r_c to the resistance of the collector junction under the normal reverse bias.

The transfer resistance is related to the current gain α_{fb}. For short-circuit termination $V_{cb} = 0$, and Eq. 7-29 gives

$$\left|\frac{I_c}{I_e}\right| = \frac{r_{fb}}{r_{ob}} = \frac{r_b + r_d}{r_b + r_c} = \alpha_{fb} \qquad (7\text{-}30)$$

Usually r_b is of the order of 100 ohms, while r_c is of megohm order, or $r_b \ll r_c$. Then

$$r_d \cong \alpha_{fb} r_c \qquad (7\text{-}31)$$

Since α_{fb} is near unity, then r_d is close to the r_c value. At high frequencies, where r_c appears shunted by a capacitance, this relation fails.

The relation

$$\beta = h_{fe} = \frac{\alpha_{fb}}{1 - \alpha_{fb}}$$

was obtained in Eq. 6-11. Frequent conversions will be made using relations which may be derived as

$$\alpha_{fb} = \frac{h_{fe}}{1 + h_{fe}}, \qquad 1 - \alpha_{fb} = \frac{1}{1 + h_{fe}} \tag{7-32}$$

thus relating the current gains of the common-base and common-emitter circuits.

The *input resistance* R_{ib} can be found as

$$R_{ib} = r_i - \frac{r_r r_f}{r_o + R_L} = r_e + r_b - \frac{r_b(r_b + r_d)}{r_b + r_c + R_L} \tag{7-33}$$

To maximize the current gain, the load R_L will usually be small with respect to r_c. Also $r_b \ll r_c$, so that

$$R_{ib} \cong r_e + r_b(1 - \alpha_{fb}) = r_e + \frac{r_b}{1 + h_{fe}} \tag{7-34}$$

The second term will usually be small and the input resistance will differ little from r_e.

The *output resistance* at the 2,2 terminals follows as

$$\begin{aligned} R_{ob} &= r_o - \frac{r_r r_f}{r_i + R_s} = r_b + r_c - \frac{r_b(r_b + r_d)}{r_e + r_b + R_s} \\ &\cong \frac{r_c}{1 + h_{fe}}\left[1 + \frac{h_{fe}(r_e + R_s)}{r_e + r_b + R_s}\right] \end{aligned} \tag{7-35}$$

after making the usual magnitude assumptions. The output resistance is high, yielding the step-up in impedances previously demonstrated with the h parameters for this circuit.

The *current gain* can be obtained by use of $\mathbf{V}_{cb} = -\mathbf{I}_c R_L$ in Eq. 7-29, so that

$$A'_{ib} = \frac{\mathbf{I}_c}{\mathbf{I}_e} = -\frac{r_b + r_d}{r_b + r_c + R_L} \cong -\alpha_{fb} = -\frac{h_{fe}}{1 + h_{fe}} \tag{7-36}$$

The current gain is less than but near unity.

The *voltage gain* $\mathbf{V}_{cb}/\mathbf{V}_s$ is complicated but can be reduced to

$$\mathbf{A}_{vb} = \frac{\mathbf{V}_{cb}}{\mathbf{V}_s} \cong \frac{\alpha_{fb} R_L}{r_e + r_b(1 - \alpha_{fb}) + R_s} = \frac{h_{fe}}{1 + h_{fe}}\left(\frac{R_L}{r_e + R_s + \dfrac{r_b}{1 + h_{fe}}}\right) \tag{7-37}$$

The *internal voltage gain* $\mathbf{A}'_{vb}$ can be written by placing $R_s = 0$, and so the operating power gain $|A'_{ib}A'_{vb}|$ from 1,1 to the load is

$$\text{operating P.G.} \cong \frac{\alpha_{fb}^2 R_L}{r_e + r_b(1 - \alpha_{fb})} = \frac{h_{fe}}{1 + h_{fe}}\left[\frac{h_{fe} R_L}{r_b + r_e(1 + h_{fe})}\right] \tag{7-38}$$

7-13 *T EQUIVALENT CIRCUITS FOR* C-E *AND* C-C *AMPLIFIERS*

The elements of the common-base (C-B) circuit can be rearranged into the form of the common-emitter (C-E) amplifier as in Fig. 7-15. The input signal polarity is reversed with respect to emitter and base, giving a reversed-phase output.

Circuit equations for Fig. 7-15(b) are

$$\mathbf{V}_{be} = (r_b + r_e)\mathbf{I}_b + r_e\mathbf{I}_c \tag{7-39}$$

$$\mathbf{V}_{ce} = r_e\mathbf{I}_b + r_d\mathbf{I}_e + (r_e + r_c)\mathbf{I}_c \tag{7-40}$$

It is more convenient to have the output of the forward transfer generator made a function of input current $\mathbf{I}_b$; this can be achieved by noting that $\mathbf{I}_e = -(\mathbf{I}_b + \mathbf{I}_c)$, and so

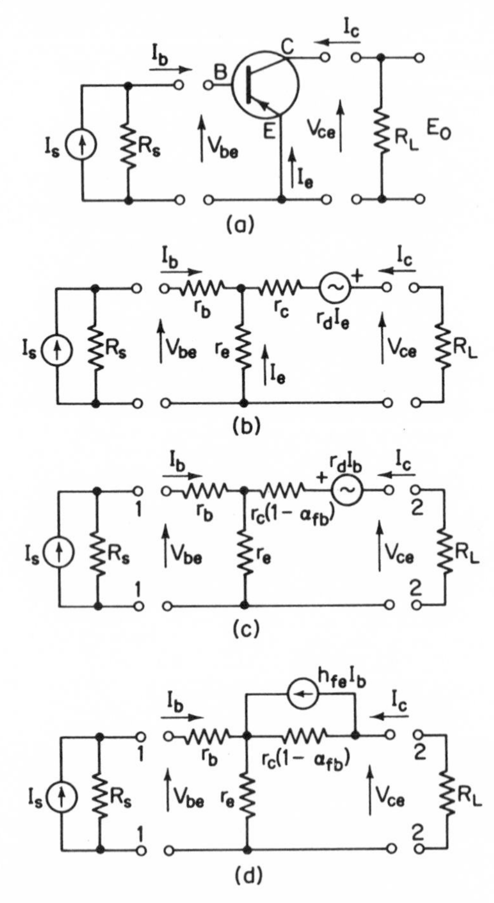

Figure 7-15. Evolution of the common-emitter circuit.

$$\mathbf{V}_{ce} = (r_e - r_d)\mathbf{I}_b + [r_c(1 - \alpha_{fb}) + r_e]\mathbf{I}_c$$

Using the assumptions that r_b and r_e are small with respect to r_d and r_c and that $R_L \ll r_c$, the performance relations for the C-E circuit are

$$R_{ie} \cong r_b + \frac{r_e}{1 - \alpha_{fb}} = r_b + r_e(1 + h_{fe}) \tag{7-41}$$

$$R_{oe} \cong \frac{r_c}{1 + h_{fe}}\left(1 + \frac{h_{fe}r_e}{r_b + r_e + R_s}\right) \tag{7-42}$$

$$\mathbf{A}'_{ie} = \frac{\mathbf{I}_c}{\mathbf{I}_b} = \frac{\alpha_{fb}r_c - r_e}{r_e + r_c(1 - \alpha_{fb}) + R_L} \cong h_{fe} \tag{7-43}$$

$$\begin{aligned}\mathbf{A}'_{ve} = \frac{\mathbf{V}_{ce}}{\mathbf{V}_{be}} &= \frac{-(\alpha_{fb}r_c - r_e)R_L}{(r_e + r_b)[r_c(1 - \alpha_{fb}) + R_L] + r_e(\alpha_{fb}r_c + r_b)} \\ &\cong \frac{-h_{fe}}{1 + h_{fe}}\left(\frac{R_L}{r_e + \dfrac{r_b}{1 + h_{fe}}}\right)\end{aligned} \tag{7-44}$$

$$\text{operating P.G.} \cong \frac{h_{fe}^2 R_L}{r_b + r_e(1 + h_{fe})} \tag{7-45}$$

The common-collector (C-C) circuit follows by rearrangement of the elements of the C-E circuit, being shown in Fig. 7-16(b). The network relations are then

$$\begin{bmatrix}\mathbf{V}_{bc} \\ \mathbf{V}_{ec}\end{bmatrix} = \begin{bmatrix} r_b + r_c & r_c(1 - \alpha_{fb}) \\ r_c & r_e + r_c(1 - \alpha_{fb})\end{bmatrix} \begin{bmatrix}\mathbf{I}_b \\ \mathbf{I}_e\end{bmatrix} \tag{7-46, 7-47}$$

Making the usual magnitude assumptions, we obtain the performance relations:

$$R_{ic} \cong \frac{(1 + h_{fe})R_L}{1 + (1 + h_{fe})R_L/r_c} \cong (1 + h_{fe})R_L \tag{7-48}$$

$$R_{oc} \cong r_e + \frac{r_b + R_s}{1 + h_{fe}} \tag{7-49}$$

$$\mathbf{A}'_{ic} = \frac{\mathbf{I}_e}{\mathbf{I}_b} \cong \frac{-r_c}{r_e + r_c(1 - \alpha_{fb}) + R_L} \cong -(1 + h_{fe}) \tag{7-50}$$

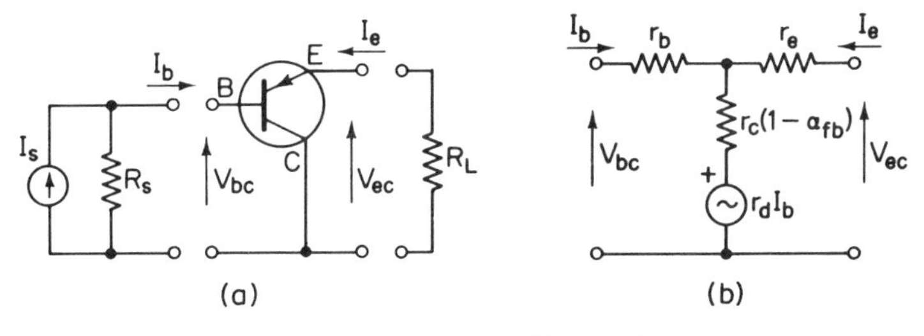

Figure 7-16. Evolution of the common-collector circuit.

$$\mathbf{A}'_{vc} = \frac{\mathbf{V}_{ec}}{\mathbf{V}_{bc}} \cong \frac{1}{1 + (1 - \alpha_{fb})r_b/R_L} \cong 1 \tag{7-51}$$

$$\text{operating P.G.} \cong 1 + h_{fe} \tag{7-52}$$

The input and output resistance expressions merit further discussion. From Eq. 7-48, the input resistance is approximately that of the load multiplied by the current gain, and from Eq. 7-49, with R_s large as is usual, the output resistance is approximately that of the input circuit divided by the current gain.

It should again be noted that the internal T impedance parameters r_e, r_b, r_c, r_d for a given transistor, are invariant at a given Q point in the three basic amplifier circuits. The equivalent circuits and the performance relations change with the circuit, in contrast to the h parameters, whose values change in invariant performance relations for the basic circuits.

7-14 *COMPARISON OF AMPLIFIER PERFORMANCE*

Since the T circuit performance relations change in form between the three basic amplifiers, it is possible to use the r relations to develop some comparisons between the basic circuits.

As was pointed out in the h-parameter analysis, the common-emitter circuit is widely applied because it has high current gain approximating h_{fe}, the highest available power gain, and input and output resistances near enough to the same order that cascading without impedance-matching transformers does not give severe losses in gain. Since R_{ie} is a function of $(1 + h_{fe})$ and R_{oe} is a function of $1/(1 + h_{fe})$, the choice of a transistor with high h_{fe} increases the input resistance and reduces the output resistance, making a match more possible in cascaded amplifiers.

In the common-base (C-B) circuit, the input current to the emitter is not amplified, and the current gain is less than unity, given by α_{fb}. The input must supply the emitter current, and since this is large the input resistance must be quite low, considerably lower than for the C-E circuit. The output resistance is higher than that of the C-E circuit by approximately the factor $(1 + h_{fe})$, and the voltage gain across this high output resistance is also high.

The common-collector (C-C) circuit has a voltage gain slightly less than unity, since the output voltage across the emitter load can never exceed the input voltage on the base because the emitter-base junction would then be back-biased. The emitter current would then decrease and the emitter voltage would decrease, restoring the normal forward junction voltage relation. The input being to the base, the current is amplified to emitter level and the current gain is high. The input to the base implies a high input resistance, approximating $(1 + h_{fe})R_L$, and with output limited in voltage level and at a high current, the C-C stage has a low output resistance, of the order of $R_s/(1 + h_{fe})$. The C-C circuit is largely suited to use as a buffer amplifier, to couple high resistance sources to a lower level of resistance.

The usual range of performance of these amplifiers is summarized in Table 7-1, and in Figs. 7-17 to 7-20. In general, the C-E circuit is useful because of its high gain and its reasonable impedance levels. The usual appli-

Table 7-1

APPROXIMATE TRANSISTOR AMPLIFIER RELATIONS†

	C-E	C-B	C-C
R_i	$r_b + (1 + h_{fe})r_e$	$r_e + \dfrac{r_b}{1 + h_{fe}}$	$(1 + h_{fe})R_L$
R_i, range	200–1500	30–1000	10^3–5×10^5
R_o	$\left(\dfrac{r_c}{1 + h_{fe}}\right)\left(1 + \dfrac{h_{fe}r_e}{r_b + r_e + R_s}\right)$	$\left(\dfrac{r_c}{1 + h_{fe}}\right)\left[1 + \dfrac{h_{fe}(r_e + R_s)}{r_e + r_b + R_s}\right]$	$r_e + \dfrac{r_b + R_s}{1 + h_{fe}}$
R_o, range	5×10^3–10^5	10^5–10^6	10^2–10^4
$\mathbf{A}_i'$	h_{fe}	$\dfrac{-h_{fe}}{1 + h_{fe}}$	$-(1 + h_{fe})$
$\mathbf{A}_v'$	$\dfrac{-h_{fe}R_L}{r_b + r_e(1 + h_{fe})}$	$\dfrac{h_{fe}R_L}{r_b + r_e(1 + h_{fe})}$	1
Power gain	$h_{fe}\left[\dfrac{h_{fe}R_L}{r_b + r_e(1 + h_{fe})}\right]$	$\left(\dfrac{h_{fe}}{1 + h_{fe}}\right)\left[\dfrac{h_{fe}R_L}{r_b + r_e(1 + h_{fe})}\right]$	$1 + h_{fe}$
Power gain, range in dB	30–40	15–30	12–16

†Assuming $(r_e + r_b) \ll R_L \ll r_c$.

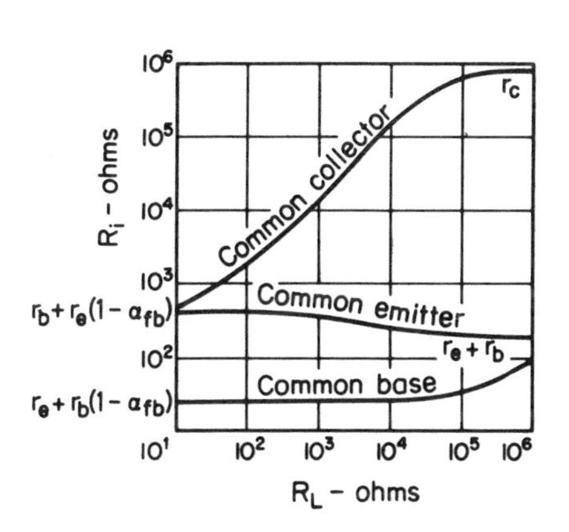

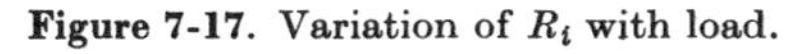
Figure 7-17. Variation of R_i with load.

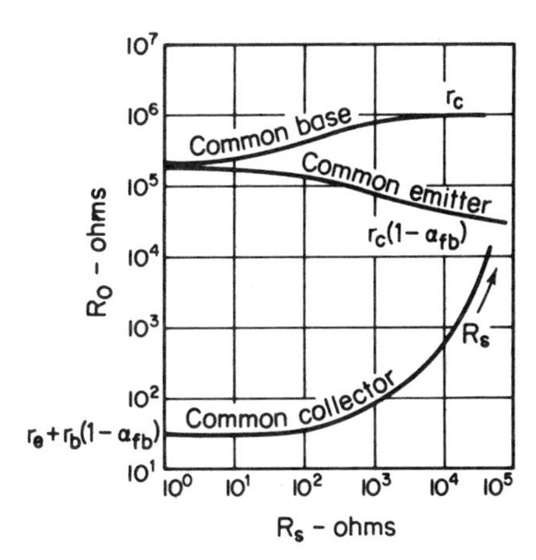

Figure 7-18. Variation of R_o with input source resistance.

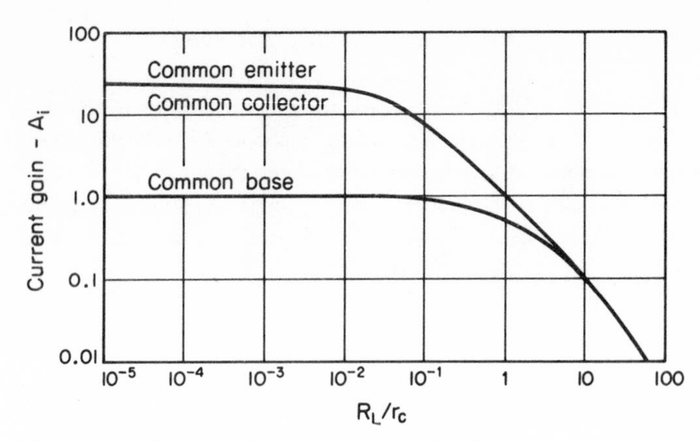

Figure 7-19. Current gain as a function of R_L/r_c.

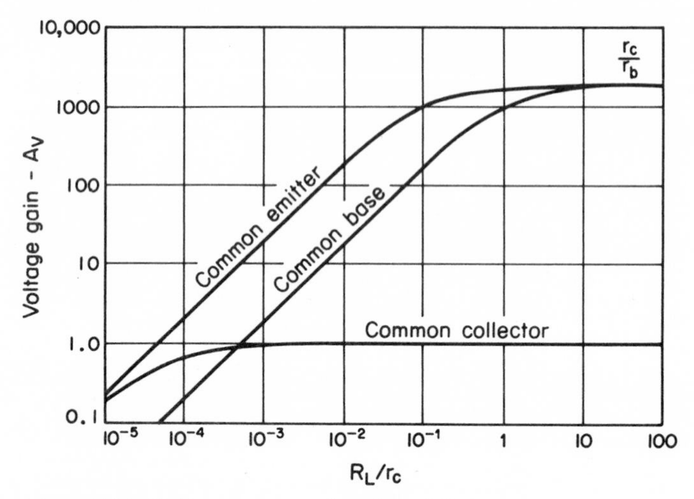

Figure 7-20. Voltage gain as a function of R_L/r_c.

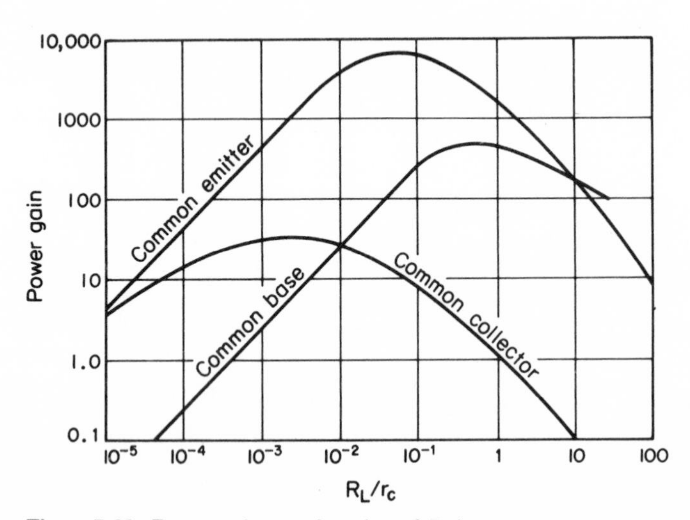

Figure 7-21. Power gain as a function of R_L/r_c.

cations for the C-B and C-C circuits employ their impedance-transforming properties.

Power gain is often a major consideration, and for its study the curves of Fig. 7-21 are provided.

7-15 RELATIONS FOR CONVERSION OF THE PARAMETERS

Transistor manufacturers usually publish the h parameters of a transistor type, although the information is sometimes available only in the C-E form. Useful values of the r parameters must often be computed. By applying the h parameter definitions to the r parameter equivalent circuits, and solving the multiple equations resulting, a set of conversion equations can be derived. The rigorous forms of many of these equations are difficult to use, and so approximate forms are given, as obtained by application of the usual magnitude approximations on the parameters. Then:

C-B *and* C-C *Parameters as Functions of* C-E *Parameters*

$$h_{ib} \cong \frac{h_{ie}}{1 + h_{fe}}, \qquad h_{ic} = h_{ie}$$

$$h_{ob} \cong \frac{h_{oe}}{1 + h_{fe}}, \qquad h_{oc} = h_{oe}$$

$$h_{fb} = \frac{-h_{fe}}{1 + h_{fe}}, \qquad h_{fe} = -(1 + h_{fe})$$

$$h_{rb} \cong \frac{\Delta_{he} - h_{re}}{1 + h_{fe}}, \qquad h_{rc} \cong 1$$

C-E *and* C-C *Parameters as Functions of* C-B *Parameters*

$$h_{ie} \cong \frac{h_{ib}}{1 + h_{fb}}, \qquad h_{ic} \cong \frac{h_{ib}}{1 + h_{fb}}$$

$$h_{oe} \cong \frac{h_{ob}}{1 + h_{fb}}, \qquad h_{oc} \cong \frac{h_{ob}}{1 + h_{fb}}$$

$$h_{fe} \cong \frac{-h_{fb}}{1 + h_{fb}}, \qquad h_{fc} \cong \frac{-1}{1 + h_{fb}} = \frac{-1}{1 - \alpha_{fb}}$$

$$h_{re} \cong \frac{\Delta_{hb} - h_{rb}}{1 + h_{fb}}, \qquad h_{rc} \cong 1$$

r Parameters as Functions of h Parameters

$$r_e = \frac{h_{re}}{h_{oe}}, \qquad r_e = h_{ib} - \frac{h_{rb}}{h_{ob}}(1 + h_{fb})$$

$$r_b = h_{ie} - \frac{h_{re}}{h_{oe}}(1 + h_{fe}), \qquad r_b = \frac{h_{rb}}{h_{ob}}$$

$$r_c = \frac{1 + h_{fe}}{h_{oe}}, \qquad r_c = \frac{1 - h_{rb}}{h_{ob}}$$

$$\alpha_{fb} = \frac{h_{fe}}{1 + h_{fe}}, \qquad \alpha_{fb} = -h_{fb}$$

h Parameters as Functions of r Parameters

$$h_{ie} \cong r_b + \frac{r_e}{1 - \alpha_{fb}}, \qquad h_{fe} = \frac{\alpha_{fb}}{1 - \alpha_{fb}} \equiv \beta$$

$$h_{oe} \cong \frac{1}{r_c(1 - \alpha_{fb})}, \qquad h_{re} \cong \frac{r_e}{r_c(1 - \alpha_{fb})}$$

The above relations were used to compute the h parameters for the transistor used in the examples of Sections 7-8, 7-9, and 7-10. These parameters were

$$\begin{array}{lll}
h_{ie} = 900 \text{ ohms}, & h_{ib} = 36 \text{ ohms}, & h_{ic} = 900 \text{ ohms} \\
h_{fe} = 24, & h_{fb} = -0.96, & h_{fc} = -25 \\
h_{re} = 7.1 \times 10^{-4}, & h_{rb} = 0.84 \times 10^{-4}, & h_{rc} \cong 1 \\
h_{oe} = 22 \times 10^{-6} \text{ mho}, & h_{ob} = 0.88 \times 10^{-6} \text{ mho}, & h_{oc} = 22 \times 10^{-6} \text{ mho} \\
\Delta_{he} = 28 \times 10^{-4}, & \Delta_{hb} = 1.12 \times 10^{-4}, & \Delta_{he} = 25
\end{array}$$

from which the T equivalent parameters would be

$$r_e = 32 \text{ ohms}, \qquad r_c = 1.12 \times 10^6 \text{ ohms}$$

$$r_b = 100 \text{ ohms}, \qquad r_d = 1.08 \times 10^6 \text{ ohms}$$

$$\alpha_{fb} = 0.96$$

7-16 *THE TRANSCONDUCTANCE MODEL FOR THE TRANSISTOR AMPLIFIER*

While current gain has been emphasized in the previous circuit models of this chapter, there are also needs for analysis in terms of voltage gain, particularly when comparing a transistor amplifier with a vacuum tube amplifier, which is a voltage-operated device. For this purpose a simple equivalent circuit may be developed, as the common-emitter circuit shown in Fig. 7-22. To this circuit we will later add the internal transistor capacitances, for study of the response of the amplifier over extended frequency ranges.

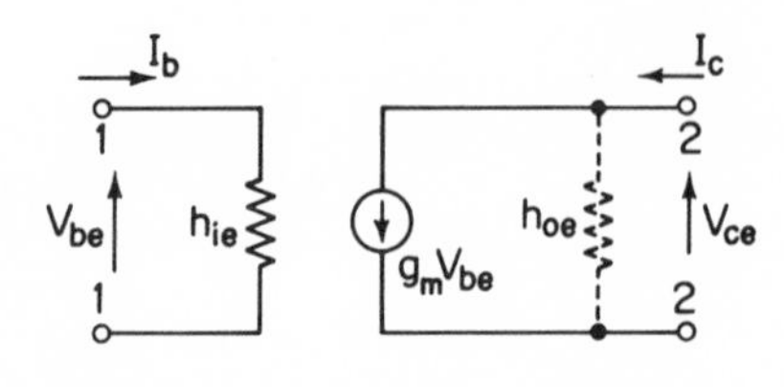

Figure 7-22. The transconductance model.

The reverse transfer generator is usually small enough to be neglected and the input circuit then contains the resistance h_{ie} only. The input current is

$$\mathbf{I}_b \cong \frac{\mathbf{V}_{be}}{h_{ie}}$$

and

$$\mathbf{I}_c = h_{fe}\mathbf{I}_b \cong \frac{h_{fe}}{h_{ie}}\mathbf{V}_{be} \tag{7-53}$$

By the basic definitions

$$\dot{i}_C = \frac{\partial i_C/\partial i_B}{\partial v_{BE}/\partial i_B} = \frac{\partial i_C}{\partial v_{BE}} = g_m \tag{7-54}$$

The parameter g_m is the transfer conductance or the *transconductance*. Reverting to sinusoidal variation, the above derivative indicates that

$$g_m \cong \frac{h_{fe}}{h_{ie}} \tag{7-55}$$

The transconductance is the slope of the transfer characteristic, representing current change in the collector circuit, for unit change in base-emitter voltage.

The transconductance can also be given a physical interpretation. The collector current can be written for the forward-biased emitter junction as

$$i_C = \alpha_{FB} I_{CO} \epsilon^{ev_{BE}/kT} \tag{7-56}$$

for $v_{BE} \gg kT/e$. Taking the derivative of this expression

$$\frac{\partial i_C}{\partial v_{BE}} = g_m = \frac{e}{kT}|i_C| = \frac{11{,}600}{T}|i_C| \qquad \text{mho} \tag{7-57}$$

This expression shows that the transconductance is a function of i_C and is an inverse function of T. For $T = 300°\text{K}$, the expression becomes $g_m = 38{,}500$ μmho/mA, and this is a theoretical limit. The expression may be converted, allowing g_m to be more clearly related to i_C, as

$$g_m \cong \frac{i_C \text{ in mA}}{26}$$

and so the theoretical value of g_m can be directly predicted from the i_C value.

The actual g_m is reduced from this value since v_{BE} should be measured from the internal point of true base potential. The external v_{BE} is greater, and g_m less, by reason of the resistance drop in r_{bb}.

Since $r_o = 1/h_o$ is generally quite high for junction transistors, and the load values are usually of smaller magnitude to maximize the current gain, then small error is introduced by eliminating h_o from the equivalent circuit, as indicated.

It is apparent from the circuit of this transconductance model that

$$R_i = h_{ie}, \qquad A_i = h_{fe}$$
$$R_o = \infty, \qquad A_v = g_m R_L$$

7-17 COMPOSITE AMPLIFIERS

The usual means of obtaining gains greater than are possible with one transistor is to use several transistors in serial connection or in cascade. The cascade arrangement need not be built of identical circuits, and some interesting circuit properties can be developed by cascading the varying input and output impedances of the basic circuits.

One such circuit is the *emitter-coupled amplifier* of Fig. 7-23. The low output impedance of the C-C stage employing transistor Q_1 is well matched to the low input impedance of the following C-B stage using Q_2. The coupling resistor R_E, of a few thousand ohms, is usually large with respect to R_{i2} of Q_2. Thus R_{i2} is the load on the first or C-C stage, and its input resistance is

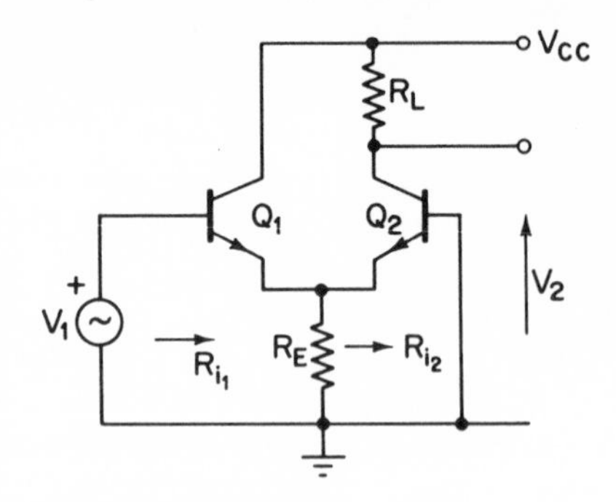

Figure 7-23. Emitter-coupled composite amplifier.

$$R_{i1} \cong (1 + h_{fe})R_L = (1 + h_{fe})\left(\frac{r_{b2}}{1 + h_{fe}}\right) \cong r_{b2} \tag{7-58}$$

The current gain of this composite amplifier is that of a C-C circuit, with load R_{i2}, followed by a C-B amplifier with a load R_L. Then

$$\text{overall } \mathbf{A}_i \cong -(1 + h_{fe})\left(\frac{h_{fe2}}{1 + h_{fe2}}\right) \tag{7-59}$$

which is equal to h_{fe}, if the transistors are identical. This is the theoretical gain of a C-E amplifier.

Use of the circuit provides an output resistance approximating r_{c2} and an input resistance near r_{b2}. It mimimizes the effect of transistor internal capacities which are difficult to overcome in the usual single transistor form of C-E amplifier. The circuit gives good response at the higher frequencies, and also may serve as a differential amplifier with inputs to both bases.

The composite amplifier of Fig. 7-24 (a) is known as the *Darlington circuit*. It behaves much as a single transistor of composite characteristics. The analysis for current gain is indicated in the figure, from which

$$\text{overall } \mathbf{A}_i = \frac{\mathbf{I}_2}{\mathbf{I}_b} = (1 + h_{fe1})(1 + h_{fe2}) \cong h_{fe1} + h_{fe2} + h_{fe1}h_{fe2} \tag{7-60}$$

which is approximately the product of the h_{fe} values.

Since the input resistance of Q_2 is that of a C-C stage, or

$$R_{i2} \cong (1 + h_{fe})R_L$$

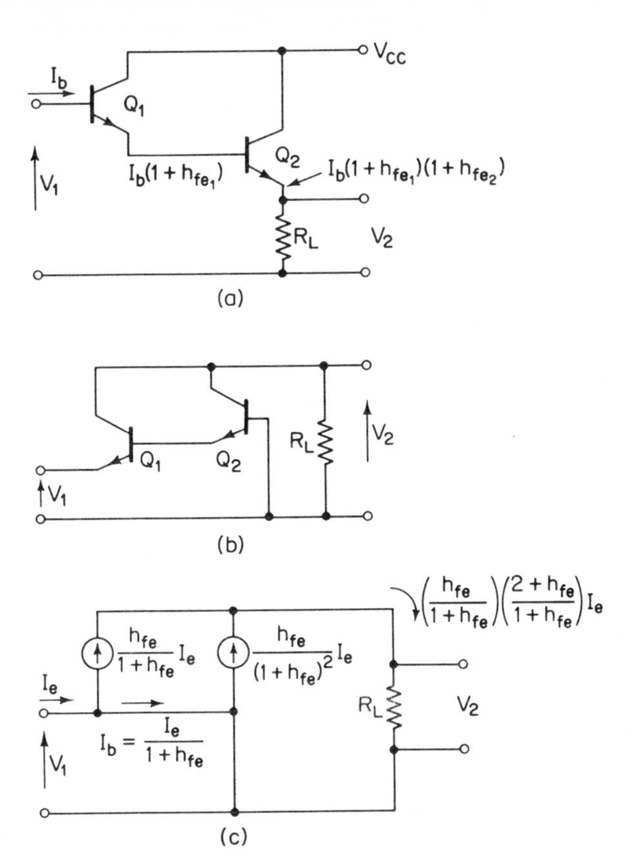

Figure 7-24. The Darlington composite amplifier: (a) C-C form; (b) C-B form; (c) equivalent for (b).

and this is the load on the first stage, then the input resistance of the composite transistor is

$$R_{i1} \cong (1 + h_{fe})^2 R_L \tag{7-61}$$

assuming equal transistors. This result is $(1 + h_{fe})$ times that of the emitter follower. However, it is usually necessary to supply the base bias current separately through a resistive network, and this network will shunt and lower the effective input resistance.

The common-collector form of the Darlington amplifier provides a major increase in current gain and input resistance with almost no increase in circuit complexity over that of a single transistor in the emitter-follower circuit.

A similar C-B circuit, giving a high value of composite h_{fe} is that of Fig. 7-24(b). The analysis follows directly from the equivalent circuit at (c). The base current of the first transistor Q_1 is $\mathbf{I}_e/(1 + h_{fe})$ and this is amplified by the second C-B stage using an identical transistor at Q_2. The current through the load is the sum of the collector currents of Q_1 and Q_2, or

$$\mathbf{I}_2 = \left(\frac{h_{fe}}{1 + h_{fe}} + \frac{h_{fe}}{(1 + h_{fe})^2}\right)\mathbf{I}_e$$

from which

$$\text{overall } \mathbf{A}_i = \frac{\mathbf{I}_2}{\mathbf{I}_e} = \left(\frac{h_{fe}}{1 + h_{fe}}\right)\left(\frac{2 + h_{fe}}{1 + h_{fe}}\right) \cong 1 \tag{7-62}$$

The composite transistor thus has a current gain very close to unity. For the transistor of the previous examples, with $h_{fe} = 24$, $\alpha_{fb} = 0.96$, the composite α_{fb} becomes 0.9984 and the composite h_{fe} reaches the value of 624.

This high value of h_{fe} is obtained with only the addition of one transistor to the usual C-C amplifier, and the circuit is of considerable value.

PROBLEMS

7-1. The transistor of Fig. 7-3 is used in the circuit of Fig. 7-4 with a load of 500 ohms and $V_{CC} = 20$ V. Select a reasonable Q point, specify I_B and I_C, and determine the maximum and minimum currents for a sinusoidal input signal of 140 μA rms.

7-2. The transistor of Fig. 7-25 is used with a Q point at $I_B = 200\ \mu$A and a collector supply of 30 V in the circuit of Fig. 7-4. Determine the ac output voltage, if the input resistance R_i is 3000 ohms, the input voltage is 0.21 V rms, and the load is 5000 ohms.

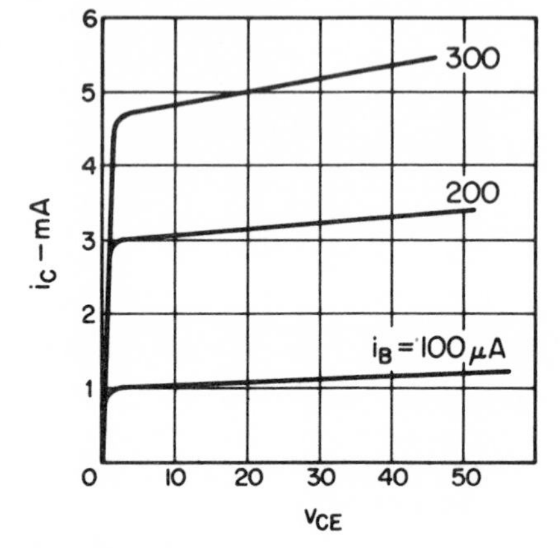

Figure 7-25.

7-3. Determine R_i and R_o, $\mathbf{A}_i$ and $\mathbf{A}_v$, for a transistor with $h_{ie} = 1250$ ohms, $h_{re} = 1.2 \times 10^{-4}$, $h_{fe} = 24$, $h_{oe} = 5 \times 10^{-6}$ mho, $R_s = 400$ ohms, and $R_L = 1000$ ohms in the C-E circuit.

7-4. The h constants of an *npn* transistor are: $h_{ib} = 35$ ohms, $h_{rb} = 1.3 \times 10^{-4}$, $h_{fb} = -0.94$, $h_{ob} = 1.2\ \mu$mho. Find the power gain possible when this transistor is used in a C-B circuit with $R_L = 20{,}000$ ohms, $R_s = 250$ ohms.

7-5. Design a C-B circuit to employ the transistor of Problem 7-3, to give the maximum available power gain. State the gain in dB.

7-6. Repeat Problem 7-5, with the transistor of Problem 7-3 in a C-E circuit.

7-7. Using the transistor of Problem 7-4, plot the variation of R_i as a function of R_L, and R_o as a function of R_s, in the common-base circuit.

7-8. For the transistor of Problem 7-4, plot the variation of power gain against R_L, as it would occur in each of the three basic circuits.

7-9. Determine the value of maximum available power gain possible with the transistor of Problem 7-3 in each of the three basic circuits.

7-10. Determine the input and output resistances, the overall current and voltage gains, and the operating power gain for a transistor having $r_b = 30$ ohms, $r_e = 400$ ohms, $r_c = 750{,}000$ ohms, $\alpha_{fb} = 0.95$, with $R_L = 10{,}000$ ohms, $R_s = 400$ ohms, in the C-B circuit.

7-11. Repeat Problem 7-9 for the C-E circuit.

7-12. A transistor, having $r_e = 25$ ohms, $r_b = 400$ ohms, $r_c = 10^6$ ohms, $r_d = 0.95 \times 10^6$ ohms, delivers an output of 6 V across a 1500-ohm resistive load, in the common-emitter circuit. What is the ac input current?

7-13. The constants of a transistor are $\alpha_{fb} = 0.96$, $r_c = 10^6$ ohms, $r_e = 400$ ohms, $r_b = 37$ ohms. Determine the current gain expected if two of these transistors are connected in cascade, with a second-stage load of 1000 ohms. Assume the input of the second stage is connected across a load of 1000 ohms of the first stage. The common-base circuit is used.

7-14. A transistor has the following parameters: $h_{ib} = 40$ ohms, $h_{rb} = 2.0 \times 10^{-4}$, $h_{fb} = -0.98$, $h_{ob} = 0.50$ μmho. Determine the approximate parameters for use in the C-E and C-C circuits.

7-15. A transistor has $r_e = 30$ ohms, $r_b = 100$ ohms, $r_c = 1.2 \times 10^6$ ohms, and $\alpha_{fb} = 0.98$.

(a) Compute the h_{-e}, and the h_{-b} parameters.

(b) Determine the M.A.P.G. possible with this transistor in the C-E circuit.

(c) What load should be used?

7-16. A C-E Darlington amplifier uses two transistors having the characteristics given in Problem 7-14. Find the input resistance and current gain.

7-17. Develop the transconductance common-emitter equivalent circuit for the transistor of Problem 7-15.

REFERENCES

1. *IRE Standards on Letter Symbols for Semiconductor Devices, Proc. IRE*, Vol. 44, p. 934 (1956).

2. Pritchard, R. L., "Electronic Network Representation of Transistors—A Survey," *IRE Trans. on Network Theory*, CT-3 (March, 1956).

3. Middlebrook, R. D., *An Introduction to Junction Transistor Theory*. John Wiley & Sons, Inc., New York, 1957.

4. Linvill, J. G., and J. F. Gibbons, *Transistors and Active Circuits*. McGraw-Hill Book Company, New York, 1961.

5. Joyce, M. V., and K. K. Clarke, *Transistor Circuit Analysis*. Addison-Wesley Publ. Co., Reading, Mass., 1961.

6. Searle, C. L., *et al.*, *Elementary Circuit Properties of Transistors* (SEEC, Vol. 3). John Wiley & Sons, Inc., New York, 1964.

7. Angelo, E. J., Jr., *Electronic Circuits*. McGraw-Hill Book Company, New York, 1964.

8. *Transistor Manual*, 7th ed. General Electric Co., Syracuse, N. Y., 1964.

9. Glasford, G. M., *Linear Analysis of Electronic Circuits*. Addison-Wesley Publ. Co., Reading, Mass., 1965.

10. Thornton, R. D., J. G. Linvill, *et al.*, *Handbook of Transistor Circuits and Measurements* (SEEC, Vol. 7). John Wiley & Sons, Inc., New York, 1966.

8

THE TRANSISTOR: DC BIAS AND STABILITY

The transistor currents are functions of temperature because of the thermally-related value of I_{CBO}. With I_{CBO} changing as much as 100 per cent for 10°C rise in temperature in germanium, or for 18°C rise in silicon, the Q point currents and voltages are not stable with temperature variation. In some circuits there is danger of a current increase sufficient to destroy the transistor through overheating.

There is also much variation in parameters from transistor to transistor, and replacement transistors may not operate at the Q point intended in the circuit design.

Thus stabilization of the operating point is a subject of importance. For this study we will return to the dc current view of the transistor of Chapter 6.

8-1 VARIATION OF THE PARAMETERS WITH CURRENT

For design information on small-signal transistors, the parameters are often specified at 25°C and with a collector-emitter voltage of 5 V magnitude. These parameters depend, however, on the operating point, and the variation of the parameters of a typical silicon unit is shown in Figs. 8-1 and 8-2.

It has been predicted that

$$r_e = \frac{kT/e}{i_E} = \frac{26}{i_E \text{ in mA}}$$

and the curve of Fig. 8-1(b) shows such a relation. That r_c decreases with i_E has also been predicted, as a function of the increased collector efficiency created by the development of a drift field in the base at high currents.

The variations of the h parameters in Fig. 8-2 can be similarly accounted

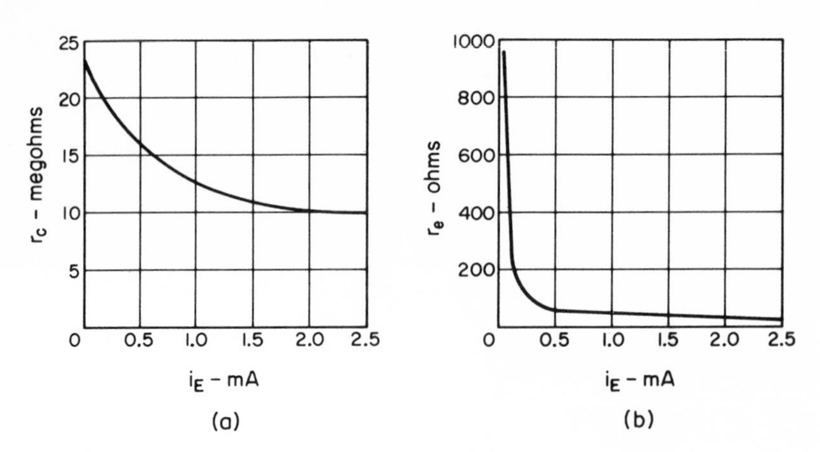

Figure 8-1. Variation of r parameters with current for a small transistor.

for by physical reasoning. The h_{re} change is a result of base narrowing, increasing the feedback into the emitter at higher v_{CE} values. The increases in h_{fe} are the results of base narrowing and reduced recombination at the higher voltages.

8-2 VARIATION OF THE PARAMETERS WITH TEMPERATURE

The effect of junction temperature on the internal transistor h parameters is illustrated in Fig. 8-3. Factors producing these changes include

1. The temperature coefficient of the v_{BE} voltage.
2. The variation of h_{FE} with temperature.
3. The increase of I_{CBO} with temperature.

It may be reasoned that the increase of average charge energy with temperature (kT/e) will make it easier for the charges to diffuse over the emitter-base barrier, and therefore less applied forward potential will be needed for a given current. This reasoning is confirmed by the curves of Fig. 8-4, relating v_{BE} and i_E for various temperatures.

To maintain a constant value of $i_E = 3$ mA, the base-emitter voltage must change -0.017 V for a temperature increase from 0° to 70°C, for the unit of the curves. The temperature coefficient of base-emitter voltage therefore approximates -2.5 mV per °C. This value is found equally suitable for silicon and germanium, although the offset voltage at zero i_E will approximate 0.5 to 0.7 V for silicon, instead of the indicated 0.2 V for germanium.

Variations of h_{FE} and h_{fe} for a typical unit are plotted in Fig. 8-5. The effect of temperature on h_{fe} or α_{fb} is complex, because the current gain is a result of the emitter efficiency, the transport factor, and the collector efficiency. At low temperatures the value of h_{fe} is low because of reduced

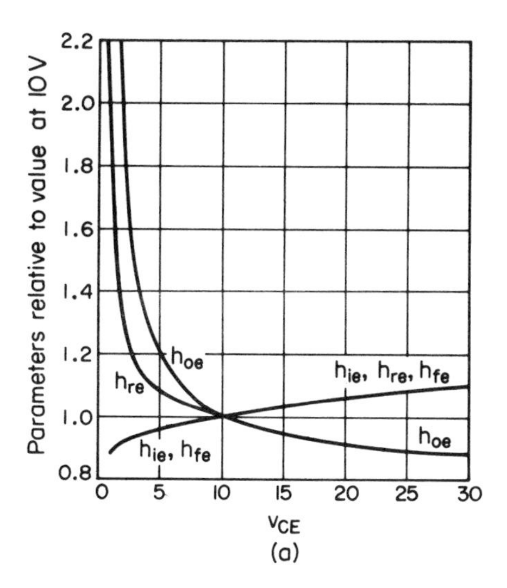

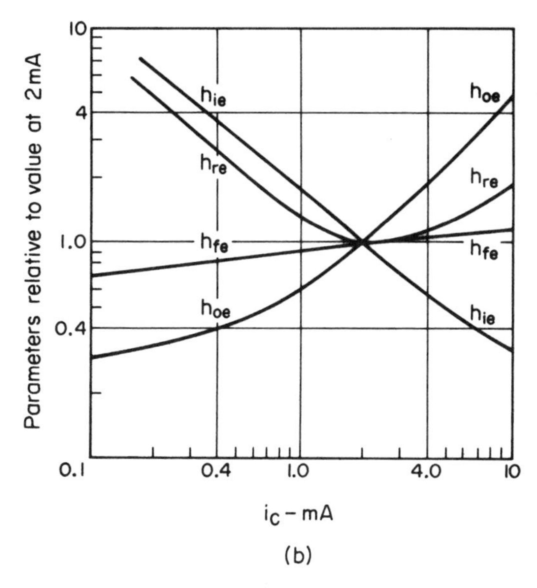

Figure 8-2. Variation of h parameters with voltage and current—silicon transistor.

charge energies; it is also low at high currents because the emitter efficiency falls. However, units can be designed to have an almost constant h_{fe} value over the normal operating current range, and in such a range the static parameter h_{FE} approximates the small-signal parameter in value.

The base resistance $r_b' + r_{bb}$ is almost entirely a bulk resistance parameter, and it decreases slightly with increased temperature. Equation 6-17

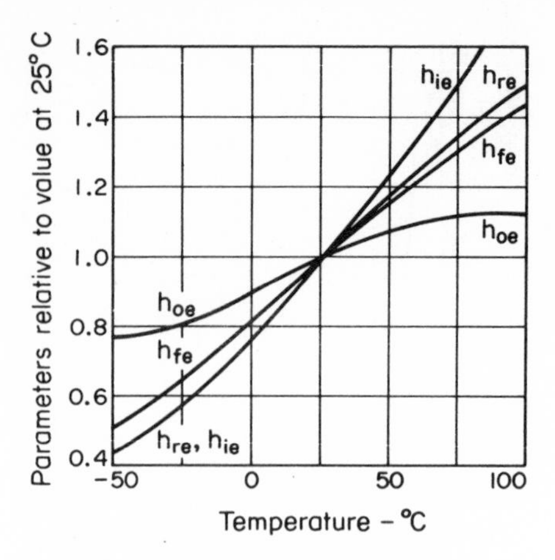

Figure 8-3. Variation of h parameters with temperature; $v_{CE} = 5$ V, $i_C = 2$ mA.

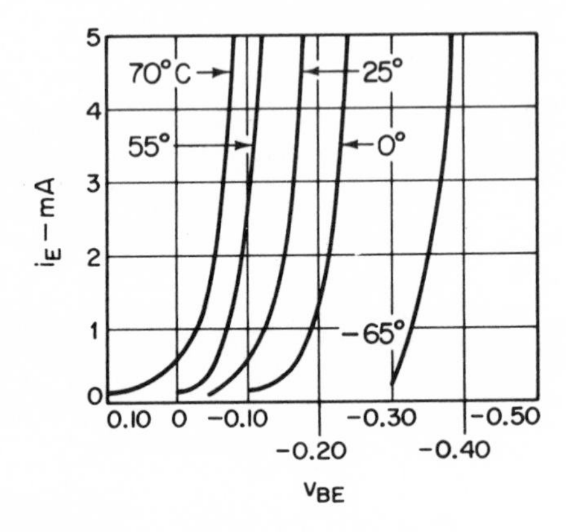

Figure 8-4. Variation of v_{BE} of a germanium transistor with temperature.

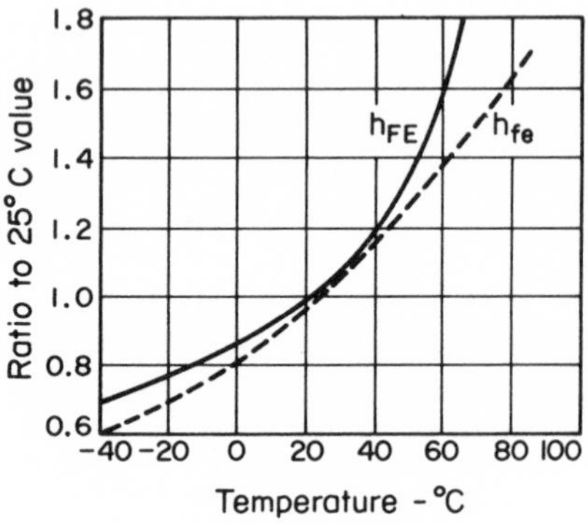

Figure 8-5. Variation of h_{FE} and h_{fe} with temperature in a germanium diode.

shows the emitter resistance to be an inverse function of temperature at constant i_E.

In the lower ambient ranges, it is the temperature coefficient of v_{BE} which predominates in causing a shift of the operating point. At high operating temperatures, it is the change in I_{CBO} which causes the greatest variation in Q point values.

8-3 THE BIAS PROBLEM

For a linear amplifier, the Q point selection must place the operation of the transistor in the center of a linear region of the dynamic transfer curve of Section 7-3. The amplifier is then able to give symmetrical ac outputs.

The problem of bias is that of stabilizing the currents at the Q point so that temperature changes or dc parameter changes introduced by replacement transistors do not shift the operating currents into a region of nonlinear performance.

From the linear diode model of Fig. 6-8(c), the total collector current is

$$i_C = -\alpha_{FB} i_E + I_{CBO} \tag{8-1}$$

The large-signal value of the current amplification factor is

$$\alpha_{FB} = \frac{i_C - I_{CBO}}{i_E} \tag{8-2}$$

when I_{CBO} is not negligible. By reason of the defined currents,

$$i_E + i_B + i_C = 0 \tag{8-3}$$

and the collector current can also be written as

$$i_C = \frac{\alpha_{FB}}{1 - \alpha_{FB}} i_B + \frac{I_{CBO}}{1 - \alpha_{FB}} = h_{FE} i_B + (1 + h_{FE}) I_{CBO} \tag{8-4}$$

Equations 8-1 and 8-4 show that the leakage current may be considered as an input signal to the transistor, and its effect on the collector current depends on whether I_{CBO} enters as an emitter current component or as a base current component.

Equation 8-4 indicates that the common-emitter output curves of Fig. 8-6 will be functions of temperature. As the temperature changes from T_1 to T_2, the operating point will shift along the load line as shown in the figure. If the input signal drives the transistor with a $\pm 30\ \mu$A base current variation, i.e., from the curve for $i_B = 10\ \mu$A to the curve for $i_B = 70\ \mu$A, then the shift in operating point to T_2 is about as much as can be tolerated without

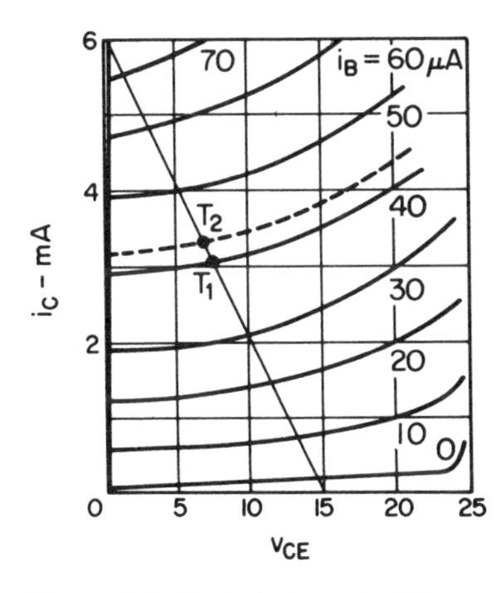

Figure 8-6. Output characteristics: silicon planar transistor.

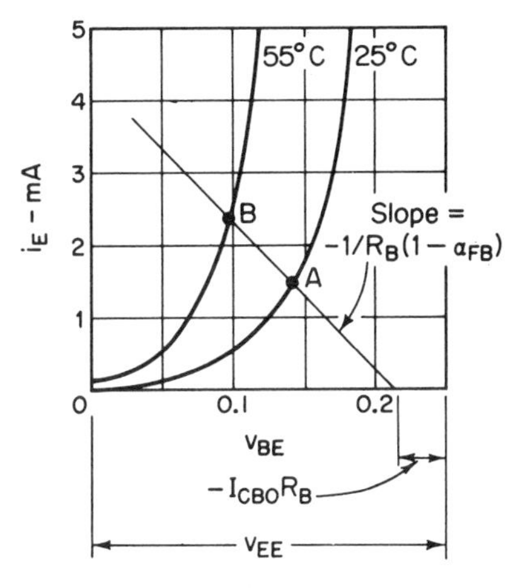

Figure 8-7. Input circuit operating conditions.

clipping of the top of the output wave. Such variation in characteristics can readily occur when changing from unit to unit of a particular transistor type. Thus constant base-current biasing does not insure a fixed location of the operating point.

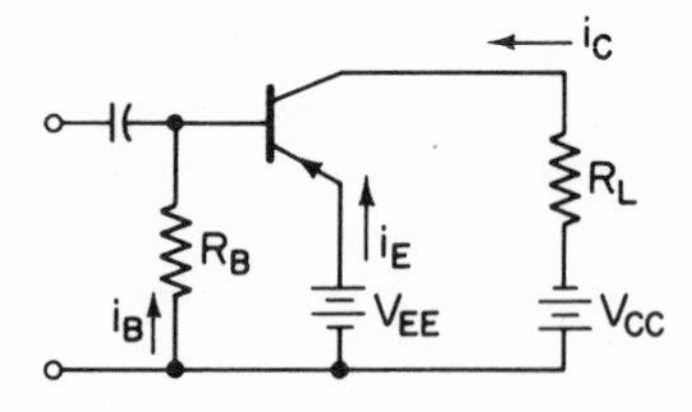

Figure 8-8. Two-battery fixed bias circuit.

The input curves of Fig. 8-7 illustrate a second biasing problem—that of shift of i_E at the operating point, when fixed emitter-base bias is used. A mesh equation may be written for the input circuit of Fig. 8-8 as

$$V_{BE} = V_{EE} + I_B R_B \tag{8-5}$$

Since $I_B = -I_E - I_C$ and $I_C = -\alpha_{FB} I_E + I_{CBO}$, it follows that

$$V_{BE} = (V_{EE} - R_B I_{CBO}) - R_B(1 - \alpha_{FB}) I_E \tag{8-6}$$

and also that the operating line of Fig. 8-7 fits this equation, using a slope of $-1/[R_B(1 - \alpha_{FB})]$. Here R_B represents the circuit resistances across which the input signal is applied.

From the input curves, the operating point at 25°C is at A, and at 55°C the operating point has shifted to B. The result is a change in I_E from 1.4 mA to 2.3 mA. Two factors cause this shift: One is the subtraction of the term $R_B I_{CBO}$ from the applied bias voltage, the second is the variation of the v_{BE} curve with temperature. Performance is improved, therefore, in circuits having low R_B values.

The shift in Q point emitter current may be considerably more damaging than indicated. The collector region is heated by the passage of the collector current, and a self-induced rise in temperature will cause I_{CBO} to rise. This increase causes the collector current and the temperature to rise once more. But this temperature rise again increases I_{CBO}, and so I_C and the temperature increase further. Without some form of limitation of the current increase, this cumulative situation can create a *thermal runaway* in the transistor and its ultimate destruction. Thermal runaway will be studied in detail in connection with power amplifiers.

8-4 THE INSTABILITY FACTOR S

The changes in output and input characteristics demonstrate that neither fixed emitter-base voltage nor fixed base current will maintain constant operating conditions for the transistor amplifier. Since it is $i_E \cong i_C$ which may cause the thermal damage to the transistor, it is logical to attempt to control i_E so as to achieve stability of the collector current within desired limits, as the leakage current changes. Such restriction is also effective in limiting Q point shifts caused by differences in characteristics of transistors upon replacement.

At elevated operating temperatures the changes in I_{CBO} override changes in v_{BE} as a cause of Q point shift. To measure the effectiveness of various circuit designs in reducing the effect of I_{CBO} on I_C, an *instability factor* S is defined as the ratio of a change in collector current to the change in reverse saturation current producing it, or

$$S = \frac{dI_C}{dI_{CBO}} \tag{8-7}$$

Ideally this ratio should be unity; that is, any change in I_C should not be greater than the change in I_{CBO}. However, practical circuits amplify I_{CBO} to some extent, S will be greater than unity, and the circuit will show some instability.

As an example of the extreme limit of instability, consider the *fixed-bias* circuit of Fig. 8-8. The equation for the collector current can be written by substitution for I_E in Eq. 8-6, giving

$$I_C = -\frac{h_{FE}}{R_B}(V_{BE} - V_{EE}) + (1 + h_{FE})I_{CBO} \tag{8-8}$$

Taking the derivative with respect to I_{CBO} leads to

$$S = 1 + h_{FE} = \frac{1}{1 - \alpha_{FB}} \tag{8-9}$$

and S may have values from 30 to 150, depending on the choice of transistor.

In some textual material S is called the *stability factor*. Since S increases as a circuit is more unstable, it seems that the choice of *instability factor* as a name more accurately describes the physical result. The instability of a circuit with change in other factors is also described by similar relations, such as dI_C/dh_{FE}, dI_C/dV_{CC}, but as variants of the more serious case these will not be discussed here.

8-5 THE INSTABILITY FACTOR WITH CURRENT-FEEDBACK BIAS

The circuit of Fig. 8-9(a) is derived from the two-battery bias circuit of Fig. 8-8 by addition of an emitter resistance R_E and rearrangement to require only one voltage source, an economic advantage. The circuit develops an emitter-bias voltage proportional to the emitter current, feeds the resultant voltage back to the input circuit for stabilization, and is referred to as *current-feedback bias*. The circuit thus measures and attempts to control I_E.

We know that V_{BE} will range only from 0.2 V to 0.7 V, depending on the choice of germanium or silicon as the transistor material, and V_{BE} will be negligible with respect to V_{CC}. A circuit relation can then be written for the input mesh as

$$V_{CC} - R_E I_E + R_B I_B = 0 \tag{8-10}$$

Also $I_E = -I_B - I_C$, and so from Eq. 8-4

$$I_C = h_{FE} I_B + (1 + h_{FE})I_{CBO} \tag{8-11}$$

Elimination of I_B leads to

$$I_C = -\left(\frac{h_{FE}}{1 + h_{FE}}\right)\left[\frac{V_{CC}}{R_E + R_B/(1 + h_{FE})}\right] + \frac{(R_E + R_B)I_{CBO}}{R_E + R_B/(1 + h_{FE})} \tag{8-12}$$

The instability of the current at the operating point as a result of change in I_{CBO} is then measured by the derivative as

$$S = \frac{R_E + R_B}{R_E + R_B/(1 + h_{FE})} = (1 + h_{FE})\frac{1 + R_B/R_E}{1 + h_{FE} + R_B/R_E} \tag{8-13}$$

Current-feedback bias thus lowers S, or improves the stability over that given by the fixed-bias circuit of Fig. 8-8. The improvement is due to the presence of R_E, and if we make $R_E \gg R_B$, the value of S approaches unity. If we make $R_E = 0$, the relation for S degenerates to that of Eq. 8-9 for the fixed-bias circuit.

A large value for R_E could give a desirably low value of S, but would cause a loss in power. It would also cause a loss in gain due to negative feedback from the output to the input, because the signal voltage across R_E subtracts from the input signal voltage. This loss of gain can be avoided by use of a capacitor of reactance $X_C \ll R_E$ at the lowest frequency, shunted across R_E. This leaves R_E effective at dc to provide the needed dc stabilization, and shunts R_E out of the circuit at signal frequencies. The cost of C must be balanced against the cost of providing additional gain elsewhere; frequently the balance indicates a decision for elimination of C.

By making $R_B = 0$, we can obtain an emitter-bias circuit of perfection, with $S = 1$; the change in I_C would be only the change in I_{CBO}. However, R_B cannot be reduced to zero since it shunts the input signal source. Therefore, solution of the instability problem requires suitable choices of R_E and R_B so as to maintain reasonable input resistances and also avoid the loss in gain and the power wastage of excessive R_E values. With R_E large, V_{CC} takes on some of the properties of a current source for supply of the collector-emitter circuit, and

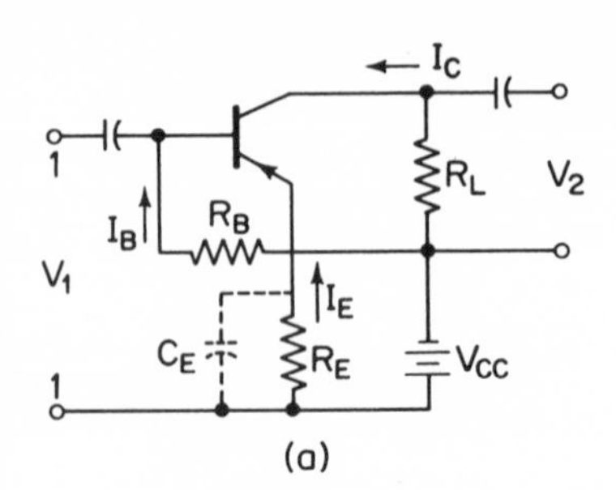

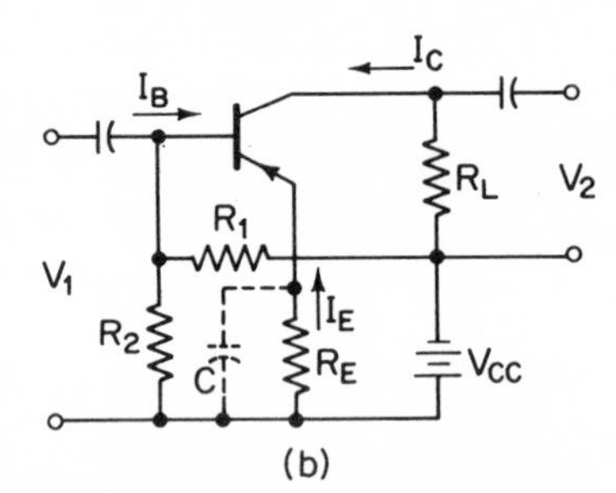

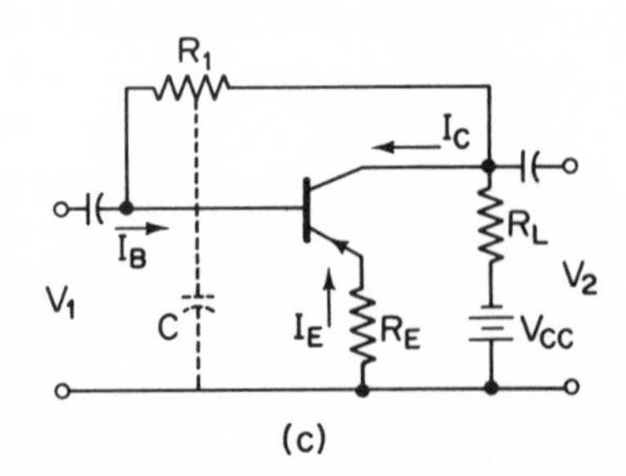

Figure 8-9. (a) One-source bias circuit; (b) emitter-base current bias; (c) voltage-feedback bias.

this would be an ideal supply for a transistor. The instability factor S is a measure of the effectiveness of the design in compromising the several competing requirements.

To illustrate the magnitude of the problem, consider a transistor with $h_{FE} = 50$, $I_C = 1$ mA, used in the circuit of Fig. 8-9(a) with $R_E = 0$. The circuit has the value of S given by Eq. 8-9, or $S = 51$. The value of I_{CBO} might be 1 μA at 25°C and 32 μA at 75°C. The change in collector current will be $(32 - 1) \times 51 = 1.581$ mA, and at the higher temperature the collector current will be 2.581 mA. If used in a hypothetical circuit with $S = 1$, the collector circuit would change only from 1.001 mA to 1.032 mA.

8-6 A FURTHER CURRENT-FEEDBACK BIAS CIRCUIT

Addition of another resistor, R_2, as in Fig. 8-9(b), to the current-feedback circuit of Fig. 8-9(a), provides the designer with greater freedom in meeting the conflicting requirements on R_B, the base bias resistance.

The parallel combination of resistors R_1 and R_2 appears as an effective value for R_B of Fig. 8-9(a), and the circuit can be reduced to that form. Equation 8-13 then applies, and so the instability factor for the circuit is

$$S = (1 + h_{FE})\left[\frac{R_1R_2/(R_1 + R_2) + R_E}{R_1R_2/(R_1 + R_2) + (1 + h_{FE})R_E}\right] \tag{8-14}$$

The ratio of R_2 to $R_1 + R_2$ determines the bias voltage for the base, but the parallel value enters the instability relation. This allows greater freedom in choice of the effective value of R_B.

8-7 VOLTAGE-FEEDBACK BIAS

The circuit of Fig. 8-9(c) returns the bias resistor R_1 to the collector rather than to V_{CC}. If I_C increases, the drop across R_L also increases. This lowers the collector-emitter voltage and reduces the bias current through R_1, tending to restore I_C toward its original value. Again neglecting V_{BE} as small, a mesh equation can be written as

$$I_ER_E - I_BR_1 - (I_C + I_B)R_L - V_{CC} = 0 \tag{8-15}$$

It is also true that

$$I_C = -\alpha_{FE}I_E + I_{CBO}$$

$$I_B = \frac{(1 - \alpha_{FB})I_C}{\alpha_{FB}} - \frac{I_{CBO}}{\alpha_{FB}}$$

Then

$$I_C = \frac{\alpha_{FB}V_{CC}}{R_E + R_L + (1 - \alpha_{FB})R_1} - \left(\frac{R_E + R_L + R_1}{R_E + R_L + (1 - \alpha_{FB})R_1}\right)I_{CBO}$$

and so

$$S = \frac{R_E + R_L + R_1}{R_E + R_L + (1 - \alpha_{FB})R_1} = \frac{1}{1 - \left(\frac{h_{FE}}{1 + h_{FE}}\right)\left(\frac{R_1}{R_E + R_L + R_1}\right)} \tag{8-16}$$

This circuit is called a *voltage-feedback bias circuit*, because it is the load voltage which is fed back through R_1 to correct the current at the operating point.

The signal voltage across R_L is also transmitted through R_1 to the base, as well as the dc component, and this reduces the gain by negative feedback. Signal transmission across R_1 can be eliminated, while retaining the dc corrective feature, by short-circuiting the alternating signal components through use of capacitor C in the diagram. To retain gain, however, the portion of R_1 between the collector and C must be large with respect to R_L. The circuit is suited to power amplifiers, where R_E must often be small because its dc power loss is detrimental to overall power efficiency.

8-8 BIAS-CIRCUIT DESIGN

The single-battery circuit of Fig. 8-9(b), with current feedback, provides latitude for meeting various performance requirements and is widely used. It will be employed here for an example of circuit design for prevention of instability and for assurance of linear operation.

Assume that the transistor is of germanium, and is described by the characteristics of Figs. 8-10 and 8-7, with $\alpha_{FB} = 0.985$, $h_{FE} = 60$, $h_{ie} = 1670$

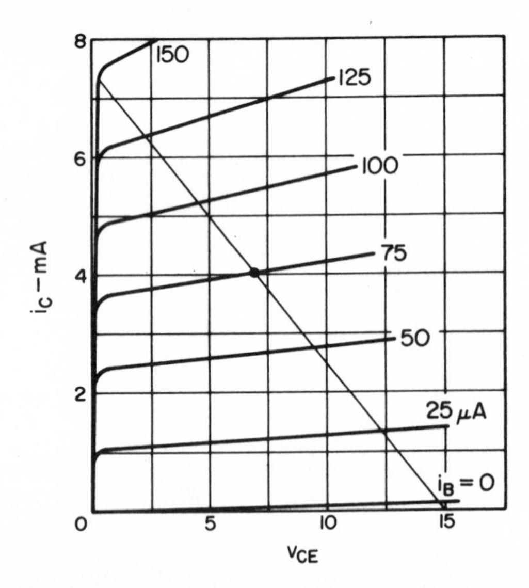

Figure 8-10. Output characteristics: germanium planar transistor.

ohms, $h_{re} = 4.5 \times 10^{-4}$, $h_{oe} = 38$ μmho. The supply voltage available is 15 V. From the output curves it is apparent that $I_E = 4$ mA and $V_{CE} = 7$ V would place the operating point in a suitably linear region, and permit symmetrical ac input voltages to be amplified. The largest possible value of $R_E + R_L$ is then fixed at $8/0.004 = 2000$ ohms, since 8 V is available for resistance drop. However, the desired high-frequency response may often limit the size of the load resistor, and this will be treated later.

It is also necessary that the resistor R_E be large enough to minimize the effect of changes in emitter-base resistance, induced by temperature change, on the dc emitter current. The value of R_E is often selected between 500 and 1000 ohms.

If the value of load resistor is set at 1500 ohms, and R_E chosen as 500 ohms, a dc load line for 2000 ohms may be drawn through the Q point at $V_{CE} = 7$ V, $I_E = 4$ mA.

Curves such as those of Fig. 6-6 can be used to find the expected change in leakage current with a specified temperature change. If I_{CBO} is 1 μA at 25 °C, this current-temperature point may be located on the graph, and a line drawn parallel to the germanium line, thus describing the leakage current change for our particular transistor. For a maximum junction temperature of 75°C, then $I_{CBO} = 30$ μA. The transistor has $h_{FE} = 60$, and if S is allowed to have its highest possible value of $1 + h_{FE}$, then the effective change in I_C would be $61 \times 30 \times 10^{-6} = 1.8$ mA, and the Q point would shift up the load line to 5.8 mA and 3.5 V at 75°C, an obviously doubtful situation for linear amplification of large signals.

A reasonable allowable shift in collector current might be 0.5 mA, and this would indicate $S \cong 17$ as a maximum for the instability factor. However, using $S = 10$ as a more conservative value, it is possible to calculate an effective R_B, the parallel value of R_1 and R_2, by Eq. 8-13. With $\alpha_{FB} = 0.985$ this yields $R_B = 5380$ ohms.

The values of R_1 and R_2 then follow from consideration of the desired base-emitter potential. From Fig. 8-7, at 25°C, V_{BE} might be taken at an average value of 0.15 V, positive to the emitter. The drop across R_E is $0.004 \times 500 = -2.0$ V, with the common-ground as reference, and thus the base is at -2.15 V to ground. The base current I_B is usually small with respect to the R_1, R_2 current, which will be $2.15/5380 = 390$ μA. Thus the voltage at the base connection on the R_1, R_2 voltage divider is -2.15 V, and R_1 and R_2 are found from

$$\frac{R_2}{R_1 + R_2} = \frac{2.15}{15} \quad \text{and} \quad \frac{R_1 R_2}{R_1 + R_2} = 5380$$

Solution leads to values of $R_1 = 37{,}600$ ohms and $R_2 = 6300$ ohms. Standard resistor values are 33,000 ohms and 6200 ohms, which are satisfactorily close to the design values. The circuit then is complete in Fig. 8-11.

With an input resistance $R_{ie} = r_b + (r_e + R_E)/(1 - \alpha_{FB}) = 880 +$

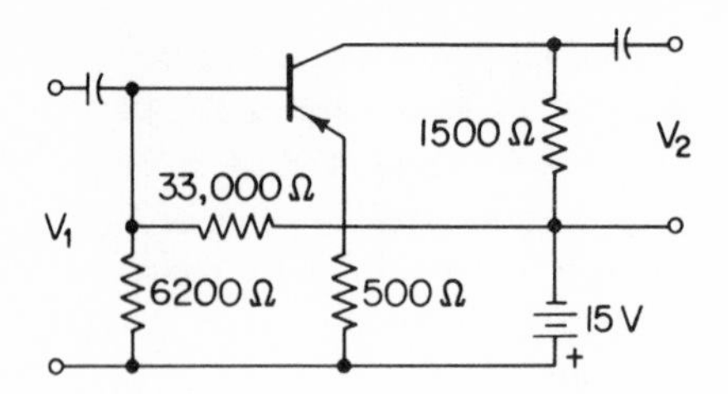

Figure 8-11. Final circuit design for $S = 10$.

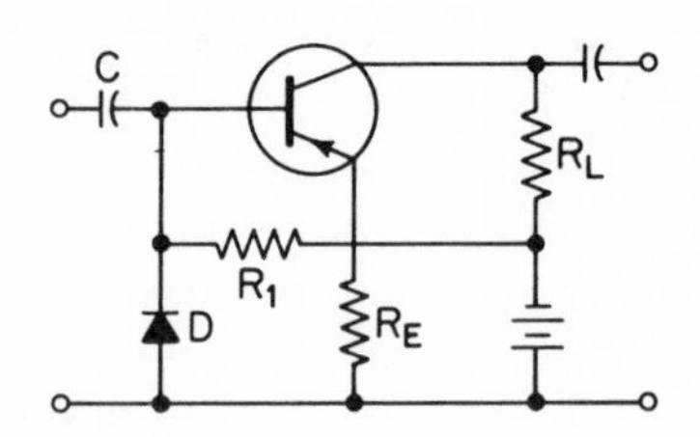

Figure 8-12. Diode temperature compensation.

$512/0.015 \cong 4300$ ohms, the effect of R_1 and R_2 in parallel is not negligible. The presence of the effective R_B value reduces the input resistance of the amplifier to 2300 ohms.

8-9 TEMPERATURE-SENSITIVE ELEMENTS IN BIAS STABILIZATION

Stability may also be obtained by the use of temperature-sensitive resistive elements as in Fig. 8-12. A diode here replaces R_2 of the circuit of Fig. 8-9(b). This diode should be selected to have a temperature dependence similar to that of the collector junction of the transistor. As the temperature rises, the value of I_C tends to increase. However, at the same time, the resistance of the diode decreases, placing a smaller voltage on the base and restoring I_C toward its original value.

PROBLEMS

Voltage magnitudes are given, it being assumed that the student will select npn *or* pnp *transistors, and make appropriate polarity changes.*

8-1. For the amplifier of Fig. 8-13, determine the expression for S, the instability factor.

8-2. A silicon transistor has $h_{FE} = 50$, $I_B = 200\ \mu A$ at the Q point, and $I_{CBO} = 1\ \mu A$ at 25°C. What will be the Q point collector current at a junction temperature of 100°C?

8-3. The leakage current in a germanium transistor is 1.5 μA at 25°C. What will the leakage current be at 85°C?

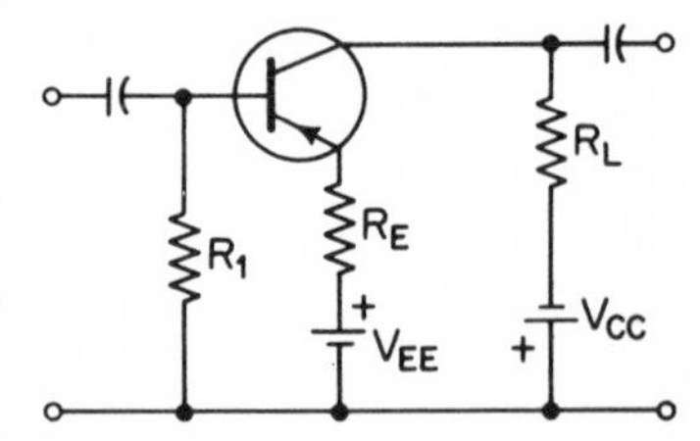

Figure 8-13.

8-4. Using a Q point at $I_C = 3$ mA with the silicon transistor of Fig. 8-10, select a load R_L and design a circuit of the form of Fig. 8-9(a). Assume the circuit must amplify signals as large as $\pm 50\ \mu A$ input, that V_{CC} available

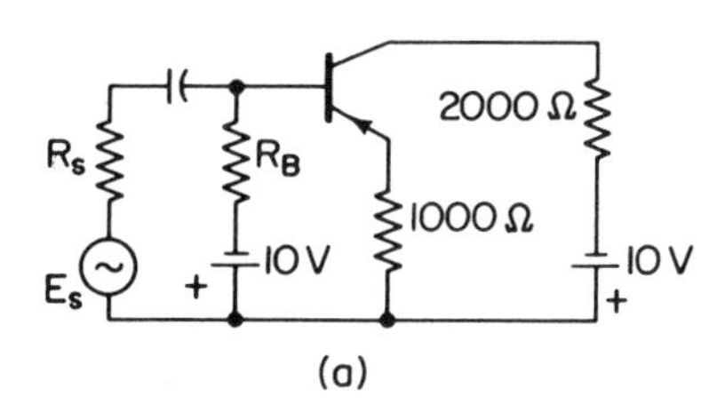

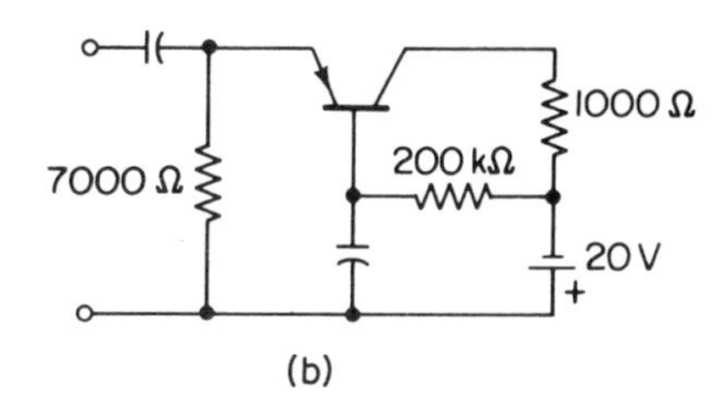

Figure 8-14.

is 12 V, that $I_{CBO} = 0.9\ \mu$A at 25°C, $h_{FE} = 45$ and that the operating junction temperature may reach 125°C. The value of S should approximate 15.

8-5. A silicon transistor having the characteristics of Fig. 8-10 is used in the circuit of Fig. 8-9(a). With $V_{EB} = 0.5$ V, $V_{CC} = 30$ V, $I_E = 3$ mA, $R_L = 6000$ ohms, $R_E = 500$ ohms, calculate R_1 and S. How much will S be changed if R_E is changed to 1000 ohms? Assume $\alpha_{FB} = 0.96$.

8-6. The circuit of Fig. 8-9(c) is used with $R_E = 0$ and $V_{CC} = 9$ V, $R_L = 2500$ ohms, $R_1 = 160{,}000$ ohms, with a transistor having $h_{FE} = 38$ and $I_{CBO} = 4\ \mu$A. Find I_C, I_B, and S.

8-7. The transistor of Fig. 8-6 is used in the circuit of Fig. 8-9(c) with $V_{CC} = 10$ V, $I_C = 2.0$ mA, $R_L = 1000$ ohms. Find values for R_1 and R_E yielding the minimum value of S.

8-8. The transistor of Fig. 8-6 is used in the circuit of Fig. 8-9(b) with $V_{CC} = 6$ V, $R_L = 2000$ ohms, $R_E = 500$ ohms, and with $I_C = 1.5$ mA. Determine R_1, R_2, for $S = 15$, with $h_{FE} = 100$, and assuming the input characteristics are those of Fig. 8-7.

8-9. In Fig. 8-14(a) determine R_B to bring I_C to 1 mA, if the transistor is germanium, and $I_{CBO} = 5\ \mu$A, $\alpha_{FB} = 0.98$, and $r_e = 30$ ohms, $r_b = 300$ ohms, $R_s = 40{,}000$ ohms.

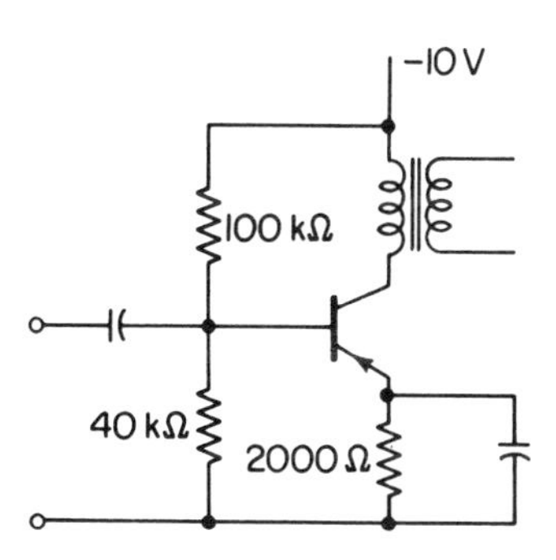

Figure 8-15.

8-10. Calculate S, and I_E for the circuit of Fig. 8-14(b), if $\alpha_{FB} = 0.99$, $I_{CBO} = 10\ \mu$A, and $V_{BE} = 0.2$ V.

8-11. The germanium transistor of Fig. 8-15 has I_{CBO} which doubles for every 10°C rise in temperature. The base-emitter voltage has a temperature coefficient of -2.4 mV/°C. Determine the relative magnitudes of the thermally induced changes of V_{EB} and the changes of I_{CBO}, as they affect the operating point. The temperature rises from 25°C. The transformer resistance is negligible.

REFERENCES

1. Shea, R. F., *Transistor Circuit Engineering*. John Wiley & Sons, Inc., New York, 1957.

2. Hunter, L. P., *Handbook of Semiconductor Electronics*, 2nd ed. McGraw-Hill Book Company, New York, 1964.

3. *Transistor Manual*, 7th ed., General Electric Co., Syracuse, N.Y., 1964.

4. Walker, R. L., *Introduction to Transistor Electronics*. Wadsworth Publ. Co., Belmont, Calif., 1966.

9

THE VACUUM TRIODE AND PENTODE

The addition of a control grid to the vacuum diode by deForest in 1906, to form the *triode*, made possible the control of power by an electric field in a vacuum. Later Hull and others added more grids to form the *tetrode* and the *pentode*, and we now have other specialized types as well.

It is intended in this chapter to lay a general foundation for the use of the triode and pentode as additional tools in electronic systems. Because of the duality between transistor and tube, the general method will follow that used in the preceding study of the transistor. It is assumed that many of the fundamental methods of analysis are familiar from their use in the preceding chapters.

9-1 THE GRID AS A CONTROL ELECTRODE

A thermionic cathode emits electrons having an energy spectrum presented in Fig. 9-1. The area under the curve to the right of 0.4 eV represents electrons able to move against a -0.4 V potential; larger fractions of the total could move against -0.3 V or -0.2 V potentials. By variation of a retarding potential on an electrode facing the cathode, the number of electrons leaving the cathode can thus be controlled. This is the principle of the *triode*, a tube basically arranged as in Fig. 9-2.

The anode is made positive to the cathode, and is attractive to electrons. A potential minimum is present near the cathode surface, due to the space charge established by the emitted electrons. If a grid of wire mesh is placed in this space-charge region, it can alter the electric field facing the electrons at the cathode surface. Being a mesh, the grid can establish this field without appreciably interfering with the electronic movement to the anode.

The potential of the grid influences the field at the cathode surface and

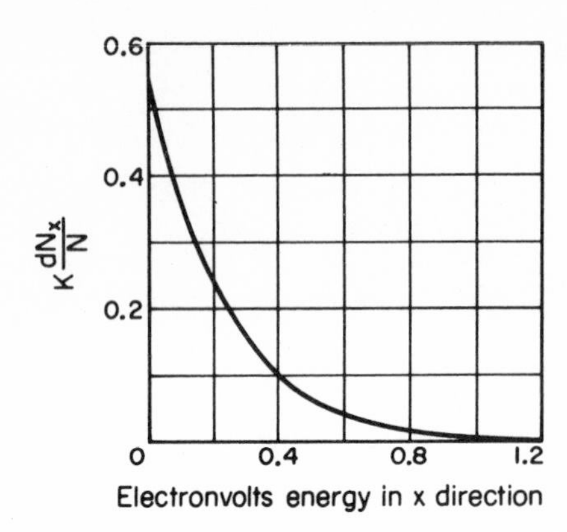

Figure 9-1. Maxwellian energy distribution: outward velocities at 2600°K.

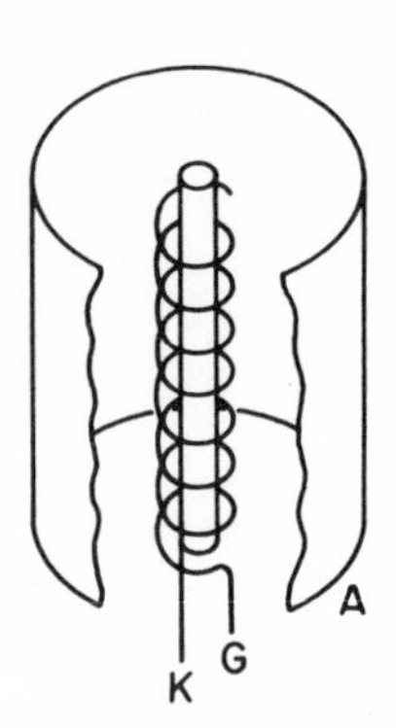

Figure 9-2. Triode structure.

adds to or nullifies the accelerating field due to the positive potential on the anode, as indicated in Fig. 9-3. With anode to cathode potential of e_b, and grid to cathode potential of e_c, it may be reasoned that the current passing between anode and cathode is related to these potentials as

$$i_b = f(e_c, e_b) \tag{9-1}$$

The grid is closer to the cathode than is the anode, and the field produced by a one volt change on the grid is more effective in varying the current than is a one volt change on the anode. The *amplification factor* μ is defined as the ratio of the effectiveness of a potential change at the grid to the same potential change at the anode. Equation 9-1 may then be modified to form the basic law of the triode

$$i_b = f(\mu e_c + e_b) \tag{9-2}$$

The grid has an appreciable area in the electron stream, but when it is negative to the cathode, few electrons will have sufficient energy to override the potential and reach the grid. Normal small-signal amplifiers are so operated and grid current is negligible. At positive grid potentials considerable grid current will be present; characteristics in the positive grid region may be nonlinear, and positive grid operation is permitted largely in power amplifiers.

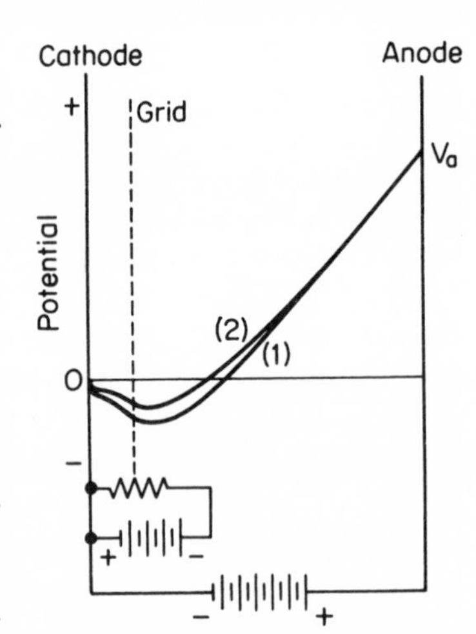

Figure 9-3. Potential variations with space charge present.

9-2 CIRCUIT NOTATION FOR THE VACUUM TUBE

As for the transistor, there is a standard circuit notation established by the Institute of Electrical and Electronics Engineers, and this will be followed here. Because of the early development of the triode, the notation reflects historical usage and lacks some of the network formalism common with the transistor. For instance, in the early days of radio broadcasting the batteries supplying filament, plate, and grid circuits were labelled A, B, and C; these letters have carried over as subscripts. Use is also made of g, k, and p as parameter subscripts on quantities related to the grid, cathode, and anode (plate) circuits.

As for the transistor, lower-case letters designate instantaneous or varying currents or voltages and upper-case letters denote rms or dc values. Some of the more common symbols are

e_c = instantaneous total grid-cathode voltage
e_g = instantaneous value of ac component of grid-cathode voltage
E_c = average or quiescent value of grid-cathode voltage
E_g = effective or rms value of ac component of grid-cathode voltage
i_b = instantaneous total anode current
i_p = instantaneous value of ac component of anode current
I_b = average or quiescent value of anode current
I_p = effective or rms value of ac component of anode current
e_b = instantaneous total anode voltage
e_p = instantaneous value of ac component of anode voltage
E_b = average or quiescent value of anode voltage
E_p = effective or rms value of ac component of anode voltage
E_{bb} = anode-circuit supply voltage
E_{cc} = grid-circuit supply voltage

Some tubes have more than one grid and numerical subscripts are used, as e_{c1}, e_{c2}, with the grid nearest the cathode considered grid 1.

A typical triode amplifier circuit appears in Fig. 9-4. The diagrams of Fig. 9-5 illustrate the following circuit relations

$$e_c = E_{cc} + e_g \tag{9-3}$$

$$i_b = I_b + i_p \tag{9-4}$$

$$E_b = E_{bb} - I_b R_L \tag{9-5}$$

$$e_b = E_b - i_p R_L \tag{9-6}$$

$$e_o = -i_p R_L = e_b - E_b \tag{9-7}$$

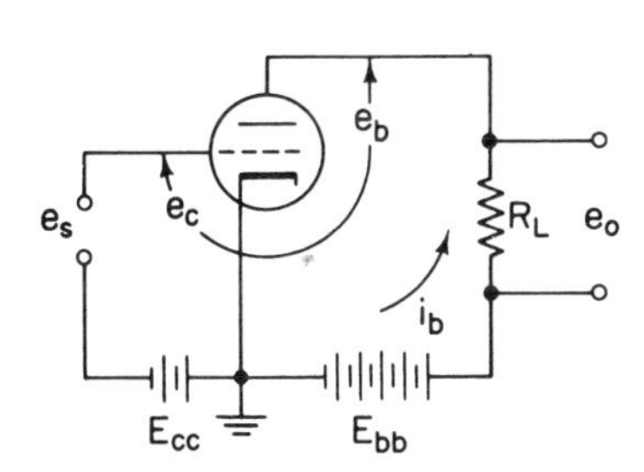

Figure 9-4. Triode amplifier circuit.

The last expression points out that the output voltage is to be taken as a rise from the common terminal. Justification for the other expressions should be evident from Fig. 9-5.

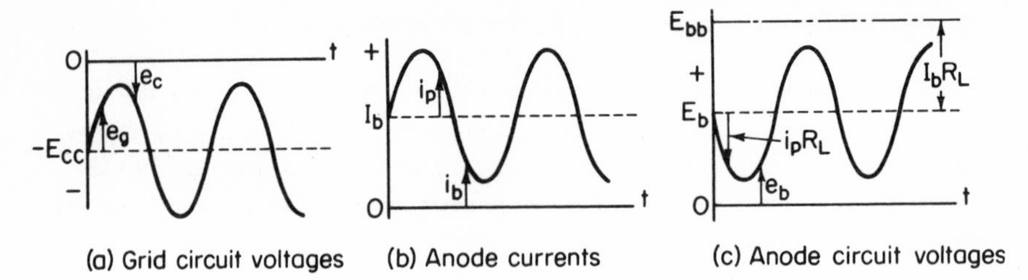

Figure 9-5. Currents and voltages of the triode.

9-3 *VOLT-AMPERE CHARACTERISTICS OF THE TRIODE*

Since the triode is a nonlinear circuit device, one means for expressing its operating characteristics is by families of volt-ampere curves. Since triode operation has been predicted by a functional relation of three variables, a surface is really indicated as the proper geometric relationship, but this can be represented by three sets of two-variable curves, with the third variable as a parameter in each case.

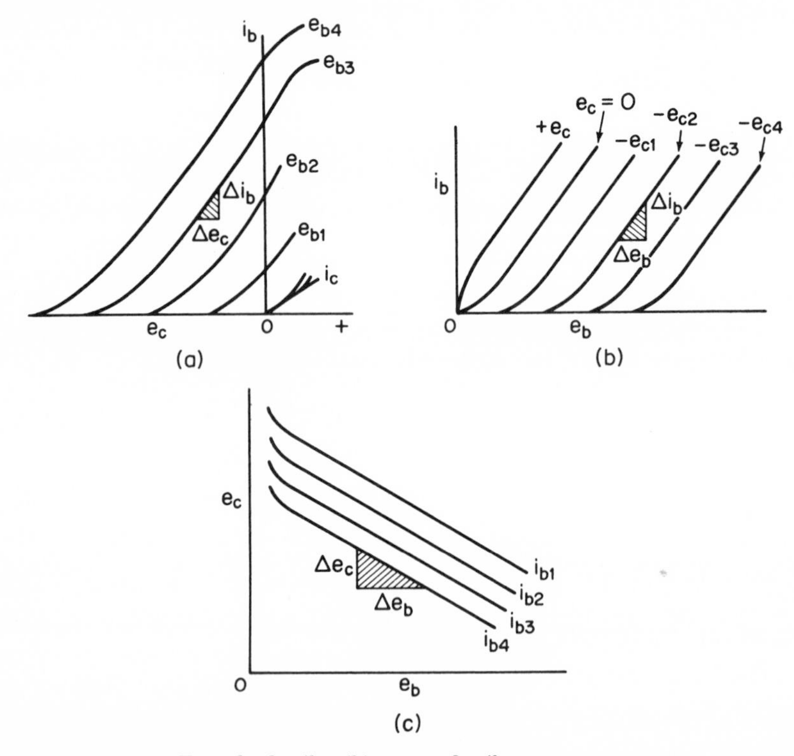

Figure 9-6. (a) Transfer family; (b) output family; (c) constant-current family.

Figure 9-6(a) relates the input voltage e_c to the output current i_b; these are the *transfer* curves. A second family in (b) relates the anode current and anode voltage; these are the *output* curves. The third family in (c) relates the input voltage and the output voltage required to maintain a constant current; therefore they constitute the *constant-current* family.

For a given anode voltage, the transfer curves indicate that a sufficiently negative grid voltage will bring the anode current to zero, and that grid voltage is called the *cutoff voltage*. It may be reasoned from the definition of μ that

$$\text{cutoff } e_c = -\frac{e_b}{\mu} \tag{9-8}$$

9-4 VACUUM-TUBE PARAMETERS

As in the case of the transistor, the vacuum tube is employed in circuits in which it is the change in current produced by a change in voltage that is of interest. The change takes place around an operating point on the curve, so that for a change of i the corresponding change in e will be found along the slope of the curve.

Referring to the transfer family of curves,

$$\left.\frac{\partial i_b}{\partial e_c}\right]_{e_b=k} = g_m \tag{9-9}$$

The partial derivative implies that e_b, the third variable, is held constant, or the change is made along the curve.

The current is in the anode circuit and the voltage is in the grid circuit, so that a transfer conductance is indicated. The derivative is given the symbol g_m, and is called the *grid-anode transconductance*. This transconductance is an important figure of merit for an amplifier, giving an indication of the change in anode current per volt change on the grid. Tubes are manufactured with g_m values ranging from a few hundred to 50,000 μmho.

The curves of Fig. 9-6(b) indicate the variation of anode current with anode potential. The reciprocal of the curve slope is a resistance, known as the *anode* or *plate resistance*. Thus

$$\frac{1}{\text{slope}} = \left.\frac{\partial e_b}{\partial i_b}\right]_{e_c=k} = r_p \tag{9-10}$$

The transconductance g_m measures the effectiveness of the grid voltage in controlling the anode current. Likewise, $g_p = 1/r_p$ measures the effectiveness of the anode voltage in changing the anode current. The parameter μ was defined as the ratio of the effect of the grid voltage to the effect of the anode voltage on the current. Therefore, by definition,

$$\mu = \frac{g_m}{g_p} = \frac{\partial i_b/\partial e_c}{\partial i_b/\partial e_b} = -\left.\frac{\partial e_b}{\partial e_c}\right]_{i_b=\text{constant}} \tag{9-11}$$

To hold i_b constant the changes of e_b and e_c must be opposite; therefore the negative sign appears. This parameter is related to the slope of the constant-current family.

Equation 9-11 is more often written as

$$\mu = g_m r_p \tag{9-12}$$

which shows that the three tube coefficients are interdependent. While in general the coefficients are variables, they may be considered constants over large regions. The amount of such variation is indicated for a small triode in Fig. 9-7.

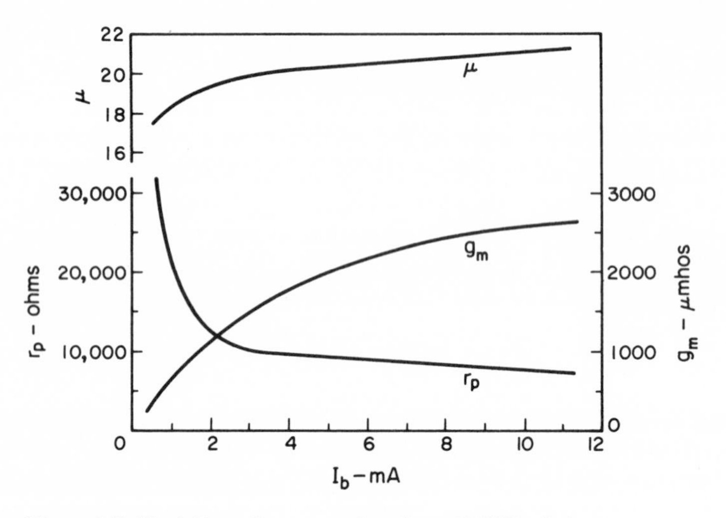

Figure 9-7. Variation of μ, g_m, and r_p for a 12AU7 triode.

The transconductance is most affected by the size of the cathode, while μ and r_p are largely affected by grid location and the spacing of the grid wire-wound helix. A closely-wound grid, or one close to the cathode will have greater effect on the cathode field and a higher μ than an open-wound grid. Means of calculating the performance of a grid are available but are so involved that design is usually by experiment.

9-5 *THE DYNAMIC CURVE*

The *static curves* of Fig. 9-6(a) are drawn without an output load. To graphically predict actual performance in a circuit, a *dynamic transfer curve* must be drawn as a plot of i_b vs. e_c, for a stated load R_L and fixed E_{bb}; such a curve appears in Fig. 9-8. The curve may be drawn by taking i_b, e_c points from the appropriate load line, drawn on the output characteristics of Fig. 9-6(b) by methods paralleling those for the transistor discussed in Section

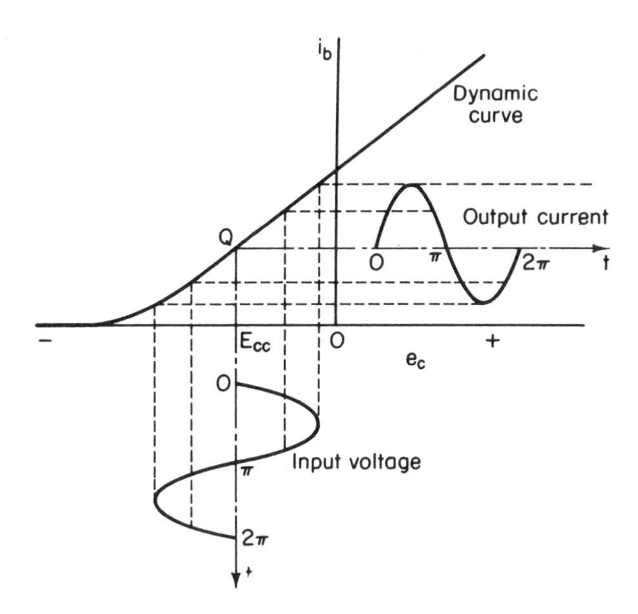

Figure 9-8. The dynamic transfer curve.

7-3. The point Q is the *operating point*, since changes of voltage or current occur about it as an origin. Choice of E_{CC}, E_{BB}, or I_B fix the Q point, and for linear operation it is usually placed near the center of the negative linear region of the dynamic characteristic.

The dynamic transfer curve allows graphical determination of the wave form of anode current for a given e_c variation, and may be used to determine wave form alteration, or distortion, if the dynamic curve is not linear. The dynamic curve may be mathematically described by a power series, and it is then used as a starting point in analysis of tube operation under large signal conditions.

9-6 EQUIVALENT CIRCUITS FOR THE VACUUM TRIODE

The triode can also be viewed as a two-port network, under the same assumptions of linearity as were required for the transistor. The variables associated with the triode are i_c, i_b, e_c, and e_b; the variable i_c is assumed zero for small-signal operation with a negative grid. Custom dictates the selection of e_c and e_b as the independent network variables, and this choice results in the use of the y parameters to describe the network, as developed in Section 5-3.

For small sinusoidal signals, Eqs. 5-8 and 5-9 state

$$I_1 = y_i V_1 + y_r V_2 \tag{9-13}$$

$$I_2 = y_f V_1 + y_o V_2 \tag{9-14}$$

For the triode

$$i_c = 0$$

$$i_b = f(e_c, e_b)$$

and the y parameters may be defined from Eq. 5-10 as

$$y_i = \frac{\partial i_1}{\partial v_1} = \frac{\partial i_c}{\partial e_c} = 0, \qquad y_r = \frac{\partial i_1}{\partial v_2} = \frac{\partial i_c}{\partial e_b} = 0$$

$$y_f = \frac{\partial i_2}{\partial v_1} = \frac{\partial i_b}{\partial e_c}, \qquad y_o = \frac{\partial i_2}{\partial v_2} = \frac{\partial i_b}{\partial e_b}$$

It is apparent by Eqs. 9-9 and 9-10 that

$$y_f = g_m, \qquad y_o = \frac{1}{r_p} \tag{9-15}$$

The use of y parameters suggests the two-generator equivalent circuit developed as Fig. 5-6(a) and redrawn as Fig. 9-9(a). With $y_i = y_r = 0$, $y_f = g_m$, $y_o = 1/r_p$, the circuit simplifies to that in Fig. 9-9(b).

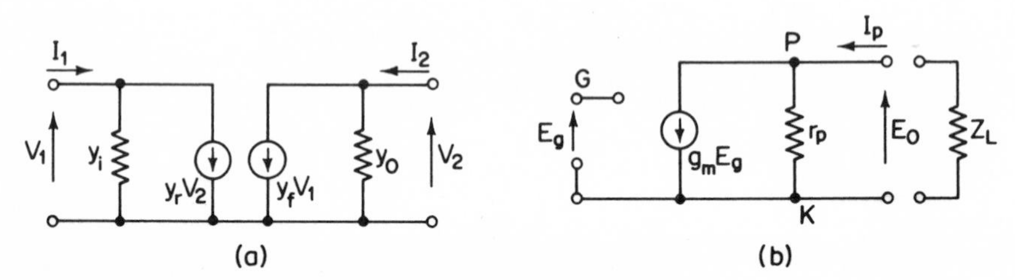

Figure 9-9. (a) Two-generator equivalent circuit with y parameters; (b) current-source equivalent circuit for the triode.

Using triode notation with $I_1 = \mathbf{I}_g$, $I_2 = \mathbf{I}_p$, $V_1 = \mathbf{E}_g$, $V_2 = \mathbf{E}_o$, it is possible to write Eq. 9-14 as

$$\mathbf{I}_p = g_m\mathbf{E}_g + \frac{\mathbf{E}_o}{r_p} \tag{9-16}$$

This equation results from application of the general y parameter equations to the triode functional relations. The equation also follows from a current summation at P, Fig. 9-9(b), and these results justify the use of Fig. 9-9(b) as the *current-source equivalent circuit* for the triode. That is, the triode can be represented by a current generator $g_m\mathbf{E}_g$, which supplies current to an internal resistance r_p and to an external load $\mathbf{Z}_L$. The output voltage is $\mathbf{E}_o$, and both input and output voltages are measured as rises from cathode reference.

Equation 9-16 may also be written as

$$-\mu\mathbf{E}_g + r_p\mathbf{I}_p = \mathbf{E}_o \tag{9-17}$$

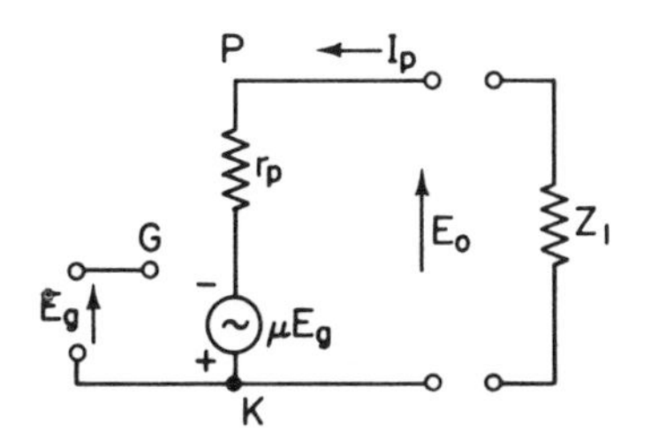

Figure 9-10. Voltage-source equivalent circuit for the triode.

and this is a voltage summation for a circuit of the form of Fig. 9-10. The voltage-source generator $\mu\mathbf{E}_g$ acts in a series circuit with an internal resistance r_p and an external load $\mathbf{Z}_L$, with a current $\mathbf{I}_p$. This circuit is known as the *voltage-source equivalent* for the triode.

In setting up the equivalent circuits and with input voltage as a rise to the grid, the transfer generators $g_m\mathbf{E}_g$ and $\mu\mathbf{E}_g$ should be made positive to the cathode, to account for the reversal of phase of the output voltage.

9-7 THE PENTODE

At frequencies of the order of a megahertz, the gain obtainable from a triode decreases and the circuits may go into oscillation. These difficulties are created because the anode, being at a higher ac potential, tends to drive a current through C_{gp} to the grid circuit of Fig. 9-11(a), in such phase as to contribute energy to the input circuit; this condition leads to oscillation or unstable gain.

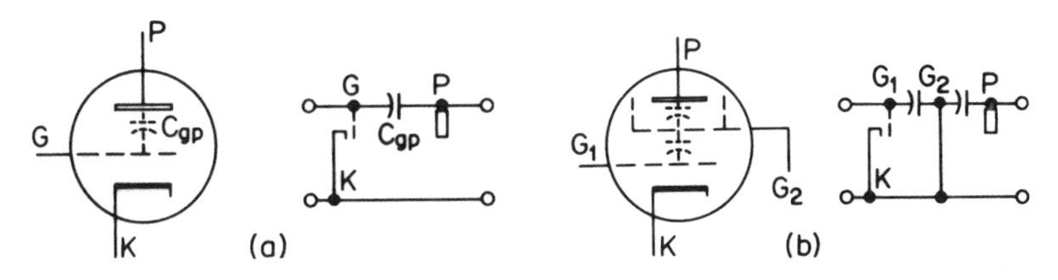

Figure 9-11. Internal capacitances: (a) triode; (b) tetrode.

However, C_{gp} can be broken into two capacitances as at Fig. 9-11(b), by a wire-mesh *screen grid*, introduced in the space between control grid and anode and maintained at cathode ac potential. Capacitive current from the anode is then short-circuited to ground, and energy flow from anode to grid becomes almost impossible. The screen grid establishes a plane of near-zero ac potential in the space, yet electrons can pass through the screen openings to the anode. The result is a *tetrode*, or four-element tube, devised by Hull about 1927.

Because of the screening effect, the anode voltage has almost zero accelerating effect at the cathode surface. To develop an anode current of electrons, a positive dc potential is applied to the screen, but the screen is still maintained at zero ac potential for shielding purposes by a capacitance of low reactance connected between screen and cathode.

The dc screen potential causes high electron velocities, and electron impact on the anode creates secondary electron emission. At low anode potentials the screen will be more positive than the anode and therefore attractive to the emitted secondary electrons. The number of secondary electrons may equal or even exceed the number of primary electrons reaching the anode, and the net anode current at low potentials may be severely reduced, causing a nonlinearity in the output curves of tetrodes for large signal swings.

To eliminate this effect of secondary emission and to further improve the screening action, Hull added a third grid about 1930, making the *pentode*. This additional grid is called the *suppressor*, placed between screen and anode and normally connected to the cathode. The resultant potential distribution at lowered anode voltage is shown in Fig. 9-12. The primary electron stream has sufficient energy to override the potential minimum at the suppressor, G_3, and to reach the anode. The low-energy secondary electrons emitted from the anode face a retarding field due to the suppressor and are driven back to the anode; thus the pentode characteristics achieve the smooth form of Fig. 9-13.

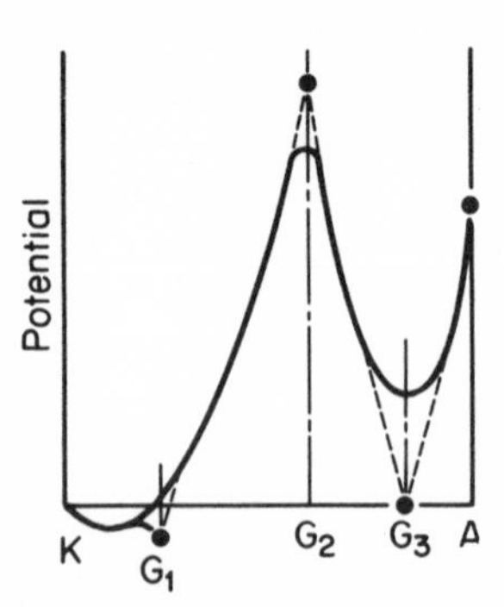

Figure 9-12. Pentode potential distribution, suppressor at cathode potential.

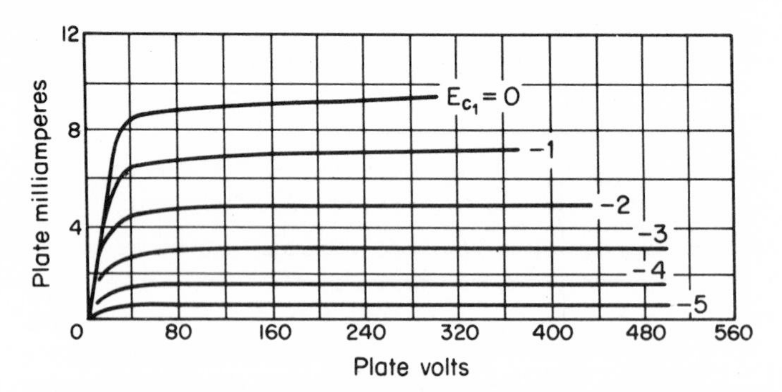

Figure 9-13. Anode characteristics of a pentode; $E_{c2} = 100$ V.

The pentode acts somewhat as a triode with the screen as anode; the real anode serves more as an electron collector than as an electron accelerator. The value of C_{gp} is reduced from about 3 pF for a triode to values approximating 0.004 pF for a pentode. The value of μ is high, reaching 2500 in some designs.

The output characteristics, by their low slope, indicate a high anode resistance of the order of one megohm, with the anode current almost independent of anode voltage. With r_p usually large with respect to Z_L, the anode current is also not dependent on load resistance. These are properties of a current source, and an equivalent circuit of that type is indicated for the pentode, as in Fig. 9-14. This is basically the same as the transconductance model for the transistor.

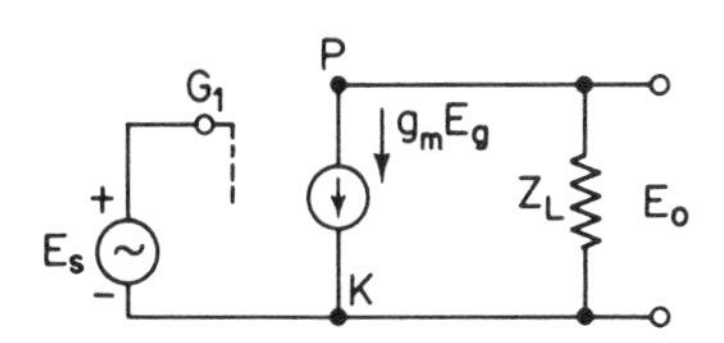

Figure 9-14. Current-source equivalent circuit for a pentode.

If the anode potential is removed, the total current will be little altered since it is the result of screen voltage acceleration. However, the current then goes entirely to the screen which has limited dissipation capability and may overheat; thus anode potential should not be removed without also removing the screen potential.

9-8 THE BEAM TUBE

Although the pentode is excellent for small-signal use, it develops minor difficulties in the amplification of large signals. The screen, located directly in the electron stream, intercepts numbers of electrons with consequent loss of power in the screen. Also the rounded shoulders of the i_b, e_b curves at low anode voltages indicate some remaining secondary emission, and cause nonlinear response for large output voltages.

By aligning the screen grid wires in the shadow of the control grid wires, the number of electrons striking the screen is reduced and the power

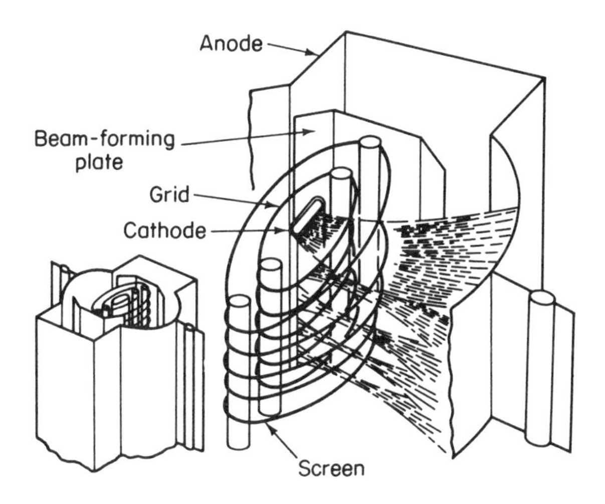

Figure 9-15. Internal structure of a beam tetrode.

efficiency is raised. The suppressor grid is also replaced with a high electron space density, the electrons being concentrated between screen and anode by beam-forming plates on each side of the cathode. The result is the *beam tetrode* of Fig. 9-15. The high electron density depresses the potential between screen and anode in a manner similar to the suppressor grid, and screening action is nearly perfect because of the uniformity of the potential plane created.

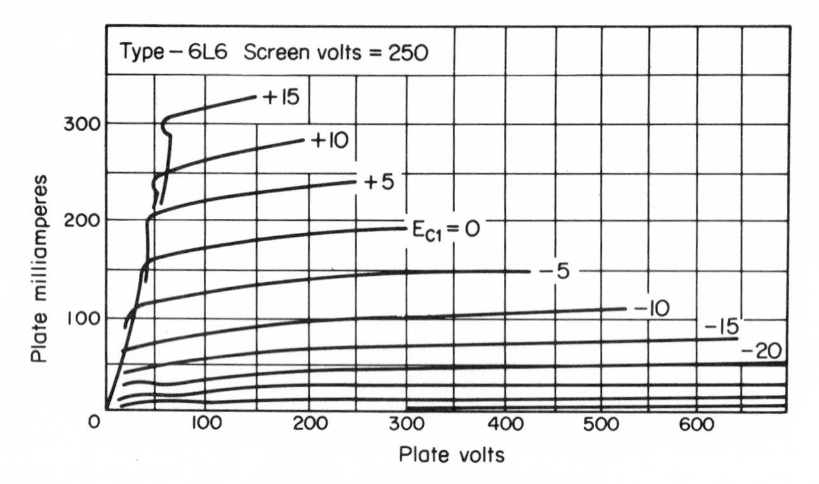

Figure 9-16. Anode characteristics of a 6L6 beam tetrode.

While technically a tetrode, the tube has pentode characteristics, as shown in Fig. 9-16. The elimination of the secondary-emission effects provides linear output curves to quite low anode voltages. The tube is therefore well suited to large-signal linear amplification; because of an increased g_m, such tubes also require smaller input voltages than are needed by pentodes to produce given power outputs.

9-9 THE VARIABLE-MU PENTODE

The gain of a pentode amplifier can be approximated as $-g_m\mathbf{Z}_L$. Since g_m is proportional to the slope of the transfer curve, the value of g_m, and therefore the gain, can be reduced by moving the operating point down the uniform-grid transfer curve of Fig. 9-17. To provide a greater range of smoothly varying g_m values and better control of the gain by bias adjustment, the *variable-mu* pentode was designed. The curve for the variable-mu grid provides a region of gradual change at low slope and low g_m, suited to easy adjustment of gain for the large signals which are handled at low gain values.

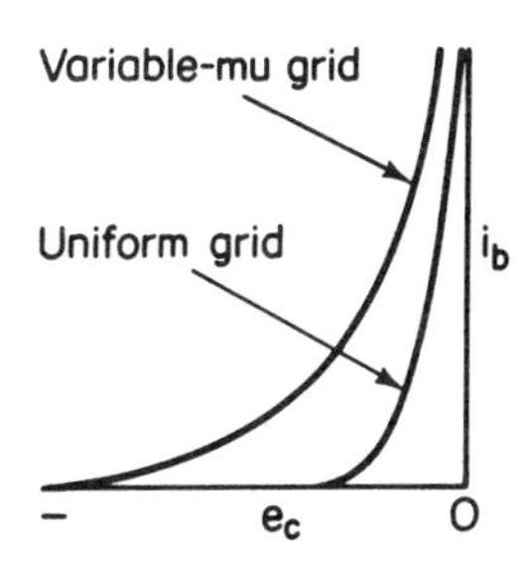

Figure 9-17. Transfer curves for constant-μ and variable-μ pentodes.

The result is obtained by winding the control-grid wires with variable pitch. With low negative bias, the value of g_m is high as an average of the whole grid. As the bias is increased negatively, the fine-pitch portions of the grid go to cutoff, and the g_m falls, being then the average of the remaining coarser-spaced portions of the grid. Either manual or automatic gain control circuits (AGC) give smooth gain change. As an example of the effect, the transconductance of a 6BA6 pentode may be 4000 μmho at -3 V E_{cc}, giving a gain of 400 with a load of 0.1 megohm. At a grid bias of -20 V the g_m is reduced to 40 μmho and the gain to four.

9-10 BASIC VACUUM-TUBE AMPLIFIER CIRCUITS

The vacuum triode has three internal elements: cathode, control grid, and anode. Multiple grid tubes, such as pentodes or hexodes, have the same basic elements to which are added extra grids for special functions.

The basic circuits represent the three possible choices of common electrode, to form a four-terminal or two-port device. The *grounded-* or *common-cathode* circuit in Fig. 9-18(a) was the earliest form and remains the most used. The *cathode-follower* of (b) is of use for coupling a high impedance to a lower impedance level, serving the same function as the emitter-follower transistor circuit. The *grounded-grid* circuit of (c) provides step-up impedance characteristics, paralleling the function of the common-base transistor circuit.

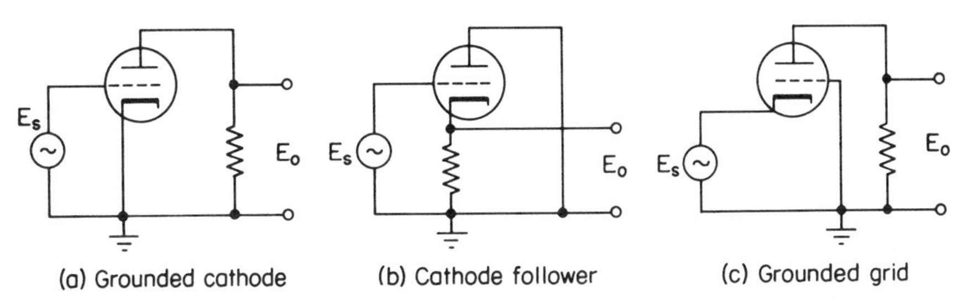

Figure 9-18. Basic amplifier circuits.

9-11 THE GROUNDED-CATHODE AMPLIFIER CIRCUIT

Usual circuit connections for the triode and pentode in the grounded-cathode circuit are shown in Fig. 9-19, with equivalent circuits drawn in Fig. 9-20.

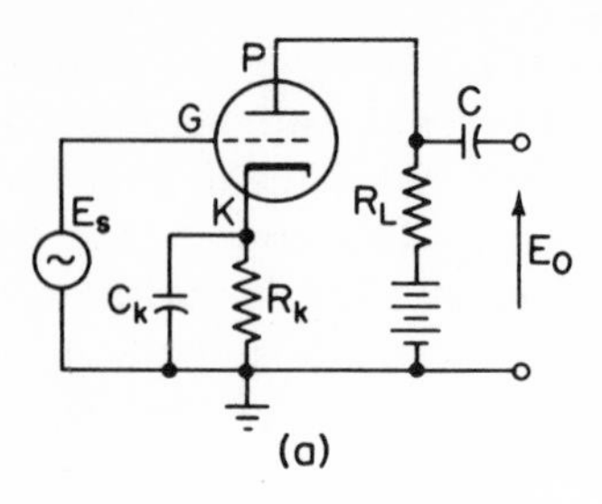

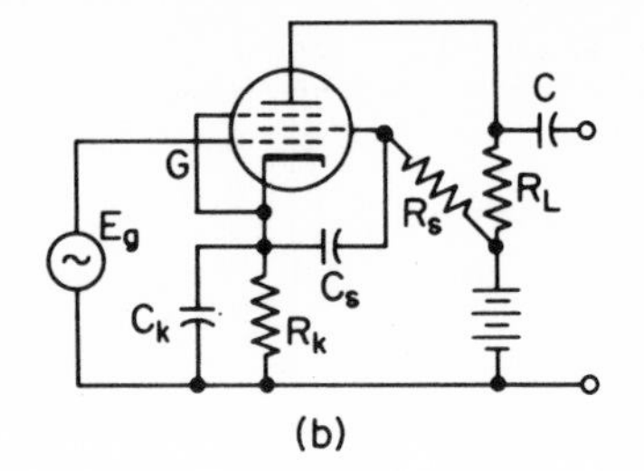

Figure 9-19. Grounded-cathode circuit: (a) triode; (b) pentode.

The functions of the cathode resistor R_k and the screen-dropping resistor R_s for the pentode will be discussed in the next section; they may be considered shunted out of the equivalent signal-frequency circuit by their bypass capacitors.

Since $\mathbf{E}_o = -R_L\mathbf{I}_p$, the output voltage can be written from Fig. 9-20(a) as

$$\mathbf{E}_o = \frac{-\mu R_L \mathbf{E}_g}{r_p + R_L} \tag{9-18}$$

Since the input current is assumed zero for the negative grid, current gain of the triode or pentode has no meaning. *Voltage gain* is defined as for the transistor, as the ratio of the output voltage rise to the input voltage rise. Then

$$\mathbf{A}_v = \frac{\mathbf{E}_o}{\mathbf{E}_g} = \frac{\mathbf{E}_o}{\mathbf{E}_s} = \frac{-\mu R_L}{r_p + R_L} = \frac{-g_m R_L}{1 + R_L/r_p} \tag{9-19}$$

The angle associated with the result is the phase shift of the amplifier; for a resistive load it is 180° in this circuit.

Figure 9-21 shows the variation of gain vs. load for a triode of $\mu = 20$. With resistive load, the dc voltage drop in large loads lowers E_b and raises r_p, and the gain does not continue to rise for very large load resistors.

Since the current-source circuit is appropriate for pentodes, its use allows the pentode gain to be written as

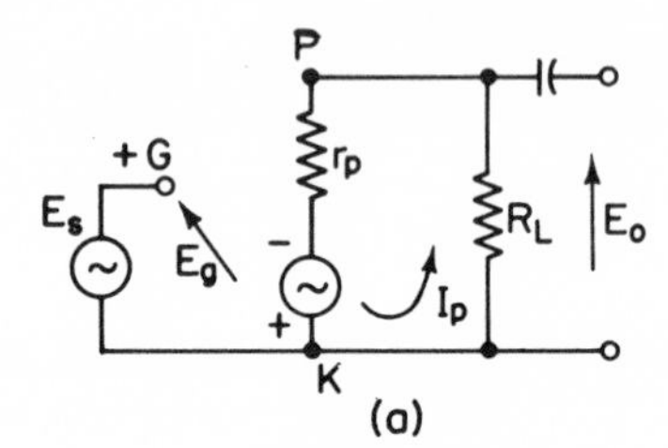

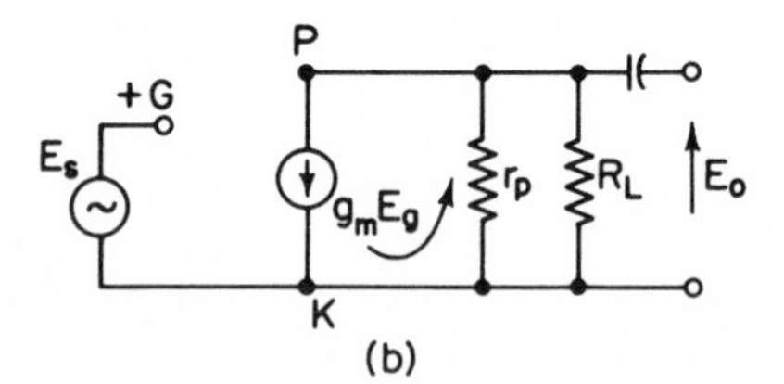

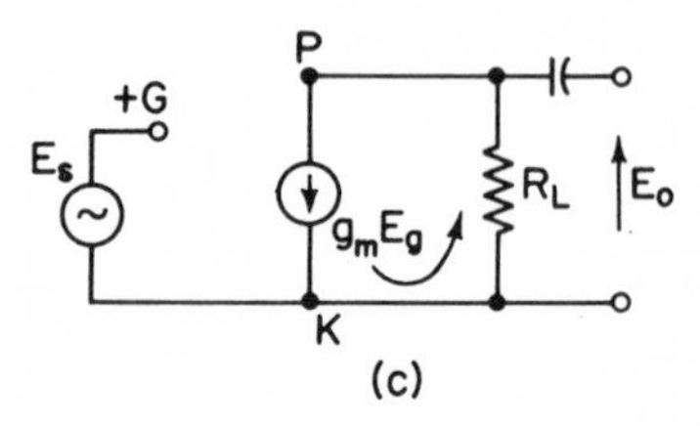

Figure 9-20. (a) Voltage-source equivalent circuit; (b) current-source equivalent circuit; (c) current-source equivalent circuit, for pentode.

$$\mathbf{E}_o = -g_m R_L \mathbf{E}_g$$

$$\mathbf{A}_v = \frac{\mathbf{E}_o}{\mathbf{E}_g} = \frac{\mathbf{E}_o}{\mathbf{E}_s} = -g_m R_L \tag{9-20}$$

The gain is proportional to g_m for both triode and pentode, and g_m becomes an important performance criterion.

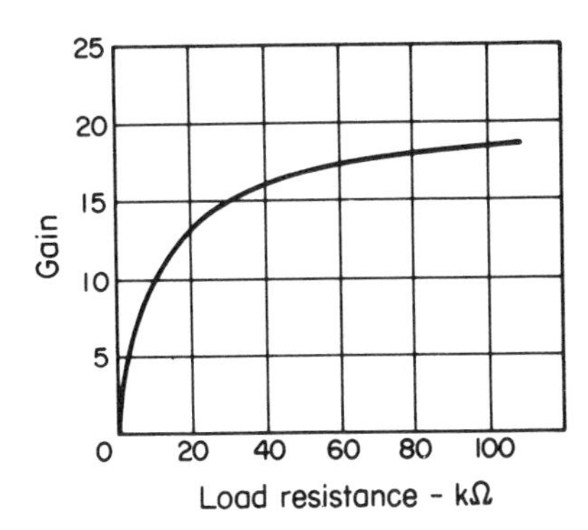

Figure 9-21. Variation of triode gain with load; $\mu = 20$, $r_p = 6500$ ohms.

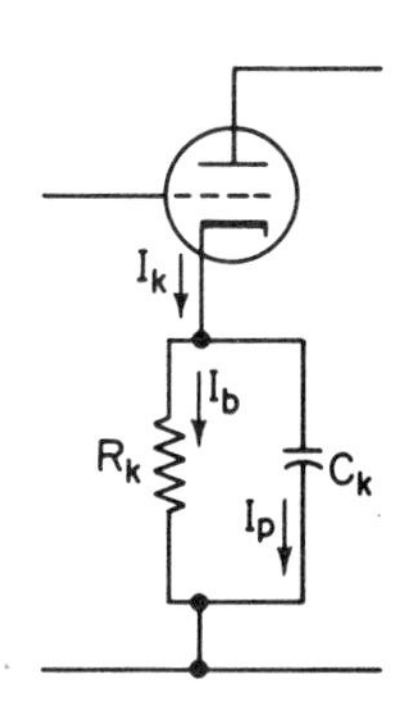

Figure 9-22. The *RC* circuit for cathode bias.

9-12 THE dc BIAS SOURCES

The grid-bias voltage of the tube can be provided by insertion of a resistor, R_k, in the cathode lead. The polarity of the voltage developed across this resistor is negative to the grid, so as to place the operating point in the linear negative portion of the transfer characteristic.

An ac signal voltage is also developed across this resistor, and the effective gain is reduced because this ac voltage subtracts from the input signal voltage. This is a form of negative feedback, to be discussed in Chapter 12. The reduction of gain can be prevented if R_k is paralleled by a capacitor C_k, of reactance small with respect to R_k. Then the ac voltage across the R_k, C_k combination is negligible, and the combination can be dropped from the ac equivalent circuit.

The dc and ac components may be assumed to divide as in Fig. 9-22, and so the value of R_k can be determined from the relations

$$E_{cc} = R_k I_b, \qquad R_k = \frac{E_{cc}}{I_b} \tag{9-21}$$

For pentodes the cathode current will also include the screen current, so that

$$R_k = \frac{E_{cc}}{(I_b + I_{c2})} \tag{9-22}$$

The capacitance C_k must be chosen as a small reactance with respect to R_k at the lowest signal frequency; this requirement is usually met if $X_c < R_k/10$. Capacitor values for audio frequencies will range from 1 to 100 μF.

The value of the screen dropping resistor for the pentode of Fig. 9-19(b) can be found through use of the difference between E_{bb} and the desired screen potential, and the screen current; that is,

$$R_s = \frac{E_{bb} - E_{c2}}{I_{c2}} \tag{9-23}$$

Capacitor C_s must offer a small reactance with respect to the screen-cathode path within the tube, and this usually requires a reactance of less than a few hundred ohms at the lowest signal frequency.

9-13 THE INPUT CIRCUIT OF THE GROUNDED-CATHODE AMPLIFIER

The triode tube has internal electrode capacitances: C_{gk} between grid and cathode; C_{gp}, between grid and anode; and C_{pk} between anode and cathode. These capacitances are of the order of 1 to 5 pF each. It might appear that the input admittance of the triode with negative grid would be wholly due to these capacitances; however, the active nature of the circuit alters the situation and the input appears as a resistance shunted by a capacitance. This apparent capacitance may exceed the actual geometric capacitance, and the phenomenon is known as the *Miller effect*.

If the reactance of C_{pk} is combined with $\mathbf{Z}_L$, the circuit becomes that of Fig. 9-23(b), and the input admittance can be found by the methods of Chapter 5. The four-terminal active network using the y parameters of Section 5-6

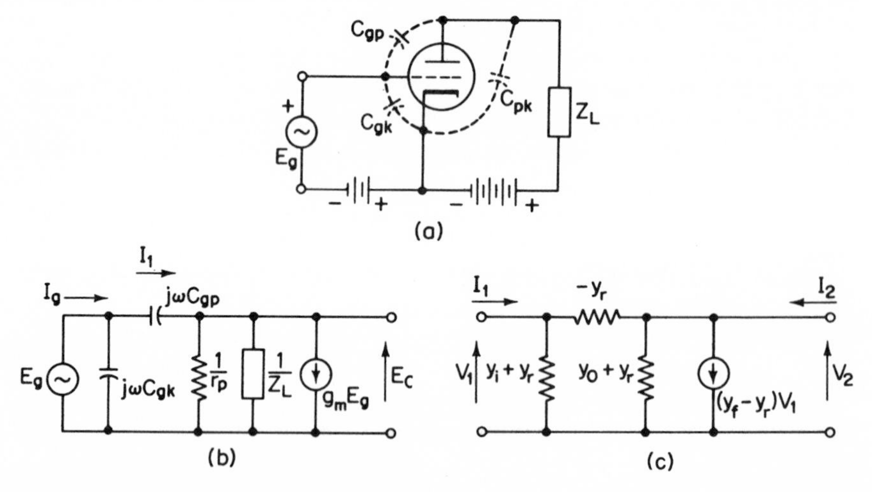

Figure 9-23. Triode internal capacitances and equivalent circuits.

is shown at (c), and can be seen to represent the triode by direct comparison of terms, as

$$y_i = j\omega(C_{gk} + C_{gp}), \qquad y_f = g_m - j\omega C_{gp}$$

$$y_r = -j\omega C_{gp}, \qquad y_o = j\omega C_{gp} + \frac{(r_p + \mathbf{Z}_L)}{r_p\mathbf{Z}_L}$$

The term $j\omega C_{gp}$ may be neglected as small in the expressions for y_f and y_o; this is equivalent to saying that the current $\mathbf{I}_1$ is negligible compared to the current due to $g_m\mathbf{E}_g$. Utilizing Eq. 5-20 for the input admittance of a network,

$$Y_i = y_i - \frac{y_r y_f}{y_o + Y_L} \tag{9-24}$$

it can be seen that

$$Y_g = j\omega\left[C_{gk} + C_{gp}\left(1 + g_m\frac{r_p\mathbf{Z}_L}{r_p + \mathbf{Z}_L}\right)\right] \tag{9-25}$$

for the grid-cathode input circuit of Fig. 9-23(b), equivalent in performance to the circuit of Fig. 9-24.

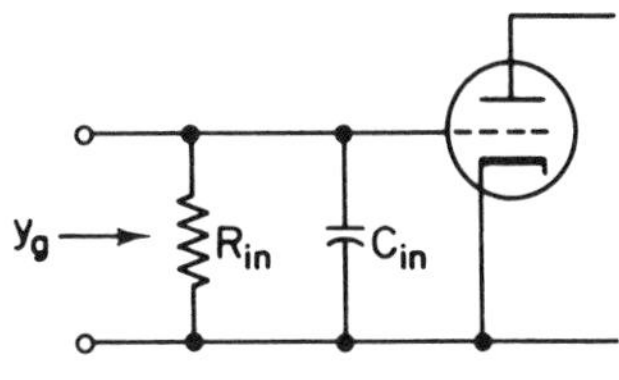

Figure 9-24. Equivalent input circuit of a triode.

The anode load will be complex, in general, as

$$\mathbf{Z}' = \frac{r_p\mathbf{Z}_L}{r_p + \mathbf{Z}_L} = R' + jX'$$

Then rewriting Eq. 9-25

$$\mathbf{Y}_g = -g_m\omega C_{gp}(\pm X') + j\omega[C_{gk} + C_{gp}(1 + g_m R')] \tag{9-26}$$

Thus the input admittance for the triode in the grounded-cathode circuit is composed of resistance and capacitance as

$$R_{\text{in}} = -\frac{1}{g_m\omega C_{gp}(\pm X')} \tag{9-27}$$

$$C_{\text{in}} = C_{gk} + C_{gp}(1 + g_m R') \tag{9-28}$$

Three cases arise from the signs in Eq. 9-26:

1. The load reactance may be capacitive, X' is then negative and R_{in} is a positive resistance which decreases with frequency. The input capacitance is C_{in}.
2. The load may be resistive and R_{in} is infinite. The value of C_{in} will be greater than C_{gk} and may be quite large for some triodes. Because of its small C_{gp}, the pentode input capacitance will approximate C_{gk}.
3. The load reactance may be inductive and X' is positive, making R_{in} negative. The current I_{gp} then feeds power back to the input circuit from the anode circuit.

Whenever the power fed back is greater than the input circuit losses the net circuit resistance is zero or negative, and the tube becomes an oscillator. It is also found that C_{in} becomes large for triodes at radio frequencies. The pentode has a very small C_{gp} by design, and is capable of satisfactory amplification at frequencies of several hundred megahertz in the grounded-cathode circuit.

9-14 THE GROUNDED-GRID AMPLIFIER

With the grid as the common element, the circuit of Fig. 9-25 is known as a *grounded-grid amplifier*. The grounding of the grid provides shielding between input and output circuits. At radio frequencies this shielding reduces energy transfer between output and input and permits the use of triodes with resonant load circuits, without threat of capacitive feedback and oscillation, and triodes are so used for stable power gain in transmitters. Triodes also produce lower internal noise than do pentodes, and the circuit is useful as an input amplifier for radio receivers handling very weak input signals, practically at the inherent tube noise level, as discussed in Section 9-17.

Equations for the two meshes are

$$\mu \mathbf{E}_g - \mathbf{E}_s = \mathbf{I}_p(r_p + R_s + R_L) \tag{9-29}$$

$$\mathbf{E}_g + \mathbf{E}_s = -\mathbf{I}_p R_s \tag{9-30}$$

The load rise from grid to anode is $\mathbf{E}_o = -\mathbf{I}_p R_L$, and since E_s is given as a negative rise above common terminal, the gain can be written as

$$\mathbf{A}_v = \frac{\mathbf{E}_o}{\mathbf{E}_s} = \frac{(\mu + 1)R_L}{r_p + (\mu + 1)R_s + R_L} \tag{9-31}$$

The output of the amplifier is in phase with the input for resistive load. Since $\mathbf{E}_s$ is in series with the transfer generator, the input voltage adds to that of the generator, and the tube has an apparent amplification factor of $\mu + 1$.

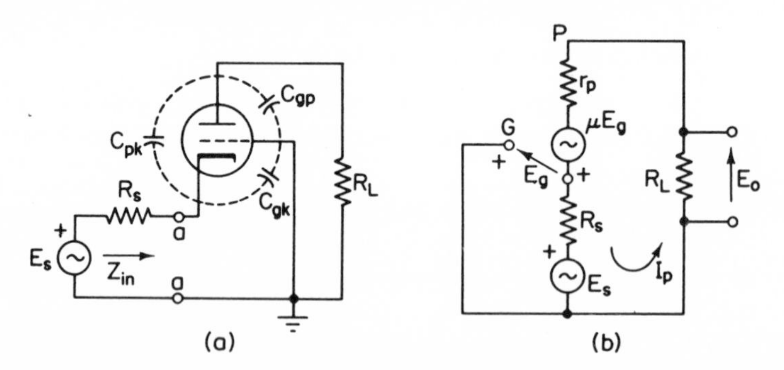

Figure 9-25. (a) Grounded-grid amplifier; (b) equivalent circuit, neglecting internal capacitances.

From Eq. 9-31 the effective output impedance can be found as

$$R_o = r_p + (\mu + 1)R_s \tag{9-32}$$

The input impedance at terminals a, a is

$$R_i = \frac{\mathbf{E}_g}{\mathbf{I}_p} = \frac{r_p + R_L}{\mu + 1} \tag{9-33}$$

The input impedance appears as the resistance of the output circuit divided by $\mu + 1$. For a tube of $\mu = 20$, $r_p = 7700$ ohms, and with a 50,000-ohm load, the input resistance is 2750 ohms.

Since there is power input to the stage, the power gain is of importance, as

$$\text{P.G.} = \frac{E_o^2/R_L}{E_s^2(\mu + 1)/(r_p + R_L)} = |A_v| \tag{9-34}$$

from terminals a, a to the output load.

The circuit is of use in transforming from a source of low impedance, such as an antenna, to a load of high impedance, in a fashion similar to the common-base transistor circuit.

9-15 THE CATHODE FOLLOWER

An amplifier with the anode as the common terminal is called a *cathode follower*. The name is logical because the cathode potential varies with and is almost equal to the input voltage. The circuit and its equivalent are shown in Fig. 9-26. In (b) the bias resistances are replaced by their parallel equivalent R_g.

Neglecting the tube capacitances, the mesh equations are

$$\mathbf{E}_g = \mathbf{E}_s - \mathbf{I}_p R_k$$

$$\mu\mathbf{E}_g = \mathbf{I}_p(r_p + R_k)$$

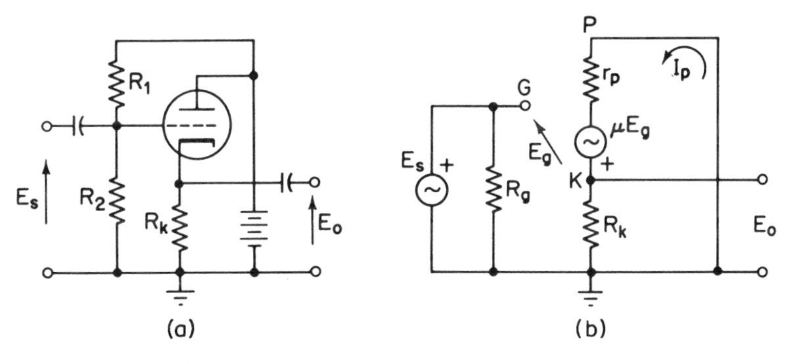

Figure 9-26. Cathode follower and its voltage-source equivalent circuit.

The voltage amplification can be found as

$$\mathbf{A}_v = \frac{g_m R_k}{1 + g_m R_k + (R_k/r_p)} = \frac{\mu R_k}{r_p + (\mu + 1)R_k} = \frac{1}{1 + (1/g_m R_k) + (1/\mu)} \tag{9-35}$$

The gain is always less than but near unity, especially as $g_m R_k$ increases. When a pentode is used, the screen bypass must connect to the cathode, since the screen must be held at cathode ac potential.

With a current $\mathbf{I}_1$ into the upper $\mathbf{E}_o$ terminal of Fig. 9-26(b)

$$\mathbf{E}_o = R_k\mathbf{I}_1 + R_k\mathbf{I}_p = -\mathbf{E}_g \tag{9-36}$$

$$\mu\mathbf{E}_g = R_k\mathbf{I}_1 + (r_p + R_k)\mathbf{I}_p \tag{9-37}$$

Solving for I_p and inserting the result in Eq. 9-36 gives the output resistance as

$$R_o = \frac{\mathbf{E}_o}{\mathbf{I}_1} = \frac{r_p R_k/(\mu + 1)}{R_k + r_p/(\mu + 1)} \tag{9-38}$$

The relation is that of two resistors in parallel, namely, R_k and the effective internal resistance of the tube, $r_p/(\mu + 1) \cong 1/g_m$.

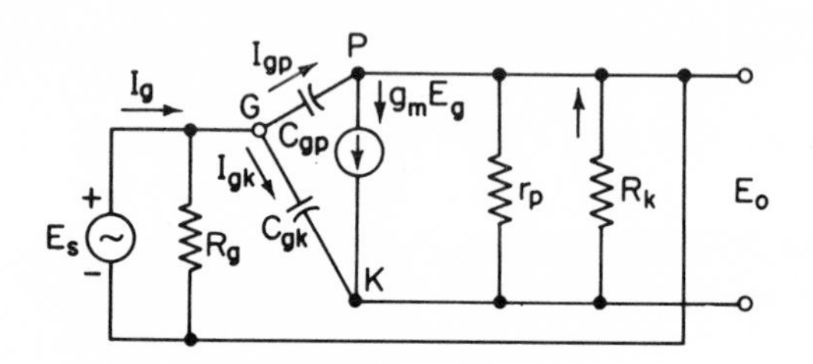

Figure 9-27. Cathode follower equivalent circuit, including capacitances.

The input admittance may be found from Fig. 9-27, with C_{pk} neglected, by writing

$$\mathbf{I}_g = \frac{\mathbf{E}_s}{R_g} + \mathbf{I}_{gp} + \mathbf{I}_{gk} = \left(\frac{1}{R_g} + j\omega C_{gp}\right)\mathbf{E}_s + j\omega C_{gk}(\mathbf{E}_s - \mathbf{E}_o) \tag{9-39}$$

Division by $\mathbf{E}_s$ gives

$$\mathbf{Y}_i = \frac{\mathbf{I}_g}{\mathbf{E}_s} = \frac{1}{R_g} + j\omega[C_{gp} + C_{gk}(1 - \mathbf{A})] \tag{9-40}$$

since $\mathbf{E}_o/\mathbf{E}_s = \mathbf{A}$. With resistive load, the gain A is real and positive, so that the input admittance is due to R_g shunted by a capacitance

$$C_{\text{in}} = C_{gp} + C_{gk}(1 - A) \tag{9-41}$$

Since the gain is near but less than unity, this capacitance is largely due to C_{gp}, and considerably less than that of the same tube in the grounded-cathode circuit.

The circuit is used as an impedance-matching device, to couple a high impedance source to a lower impedance circuit, analogous to the function of the emitter-follower with the transistor.

9-16 MODIFIED CATHODE FOLLOWER

The circuit variation of Fig. 9-28 provides a simple bias source for the cathode follower, and has added value because the input impedance is raised, an important factor in measuring equipment.

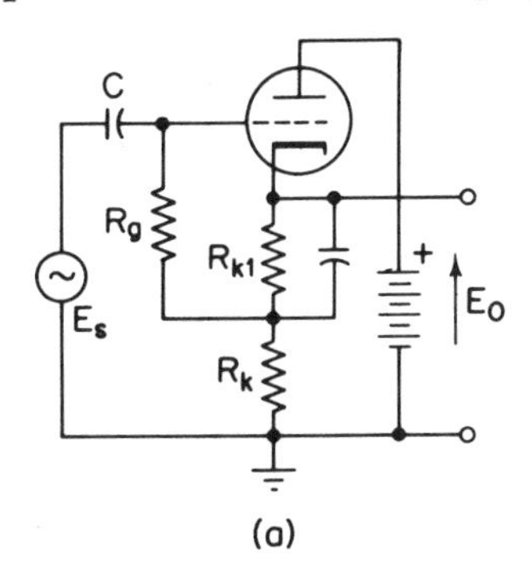

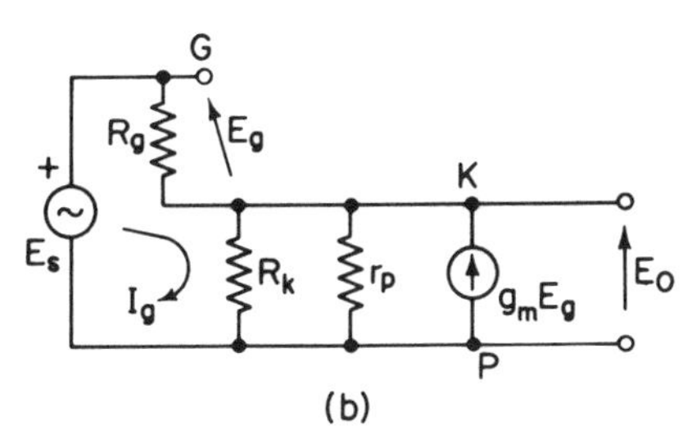

Figure 9-28. High-input-impedance cathode follower.

The resistor R_{k1} is selected to furnish the proper bias, but shunted out of the equivalent circuit with the bypass capacitor. The ac output load is then R_k. Using the circuit relations

$$\mathbf{E}_g = \mathbf{E}_s - \mathbf{E}_o = R_g\mathbf{I}_g$$

$$\mathbf{E}_o = (\mathbf{I}_g + g_m\mathbf{E}_g)\frac{R_k r_p}{R_k + r_p}$$

The gain may be found by the equation

$$\mathbf{A}_v = \frac{1}{1 + \left(\dfrac{R_g}{1 + g_m R_k}\right)\left(\dfrac{r_p + R_k}{r_p R_k}\right)} \tag{9-42}$$

and with $g_m R_k \gg 1$, the gain approaches unity.

The grid resistor R_g is connected between G and K in Fig. 9-28(b), and the voltage at its cathode end is almost equal to and follows the voltage at the grid end. Since the voltage across R_g is $\mathbf{E}_s$ in the cathode follower of Fig. 9-26, and only $\mathbf{E}_s - \mathbf{E}_o$ in Fig. 9-28, the current $\mathbf{I}_g$ is less in the latter case. The input resistance is found to be greater than R_g, as

$$R_i = \frac{\mathbf{E}_s}{\mathbf{I}_g} = R_g\left(1 + \frac{\mathbf{E}_o}{\mathbf{E}_s - \mathbf{E}_o}\right) = R_g\left(\frac{1}{1 - A}\right) \tag{9-43}$$

The circuit can be designed to have a very high R_i value.

9-17 COMPOSITE VACUUM-TUBE AMPLIFIERS

The output impedance of a cathode follower is low. The input impedance of a grounded-grid amplifier is low. A cathode follower then is suitable for exciting a grounded-grid stage; the result provides the input impedance of the cathode follower and the gain of the grounded-grid amplifier. Such a two-tube circuit is said to be *cathode-coupled*, and a circuit is shown in Fig. 9-29.

Ordinarily the two tubes are identical triodes, and

$$\mathbf{E}_s - \mathbf{E}_{g1} = R_k\mathbf{I}_1 - R_k\mathbf{I}_2$$

$$\mu\mathbf{E}_{g1} = (r_p + R_k)\mathbf{I}_1 - R_k\mathbf{I}_2$$

$$-\mu\mathbf{E}_k = \mu\mathbf{E}_{g2} = R_k(\mathbf{I}_1 - \mathbf{I}_2) - (r_p + R_L)\mathbf{I}_2$$

It is then possible to write

$$\mathbf{A}_v = \frac{\mathbf{E}_o}{\mathbf{E}_s} = \frac{\mu R_L}{r_p\left[2 + \dfrac{r_p + R_L}{(\mu + 1)R_k}\right] + R_L} \tag{9-44}$$

The cathode-coupled amplifier behaves like a grounded-cathode stage having a tube of amplification factor μ and an anode resistance greater than twice r_p of one of the triodes. Although maximum gain is theoretically obtained with R_k large, the value giving correct bias is usually satisfactory.

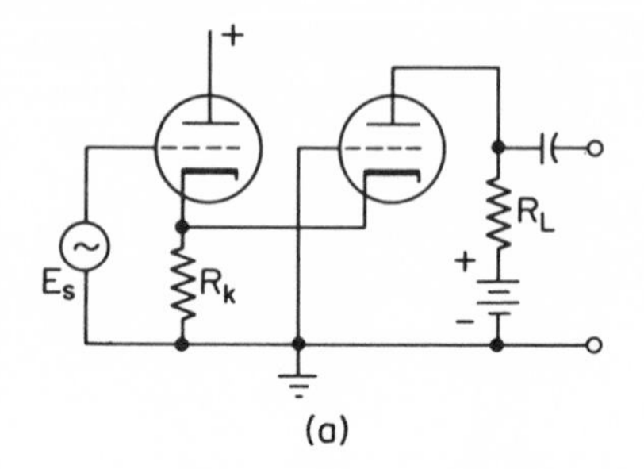

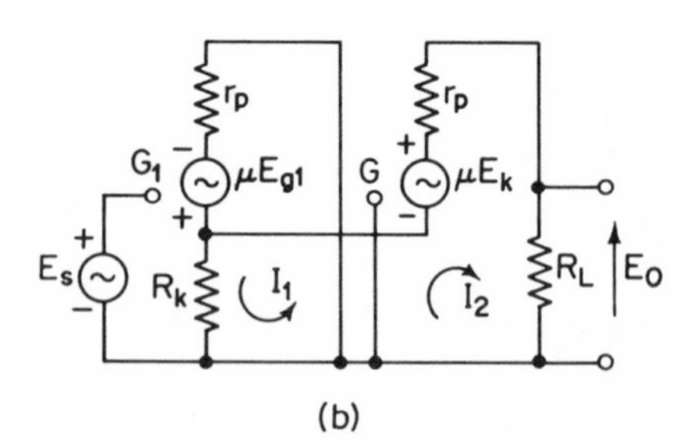

Figure 9-29. Cathode-coupled amplifier.

A composite circuit consisting of a grounded-cathode triode followed by a grounded-grid triode is often adopted for high-frequency amplification, because of several desirable characteristics. It is called a *cascode amplifier* and appears in Fig. 9-30; often it uses a dual triode.

The load on the first triode is the input resistance of the grounded-grid triode as

$$R_{i2} = \frac{r_{p2} + R_L}{\mu_2 + 1}$$

and so the overall gain can be written as the product of the gains $A_v = A_{v1}A_{v2}$, or

$$A_v = \left(\frac{-\mu_1 R_{i2}}{r_{p1} + R_{i2}}\right)\left[\frac{(\mu_2 + 1)R_L}{r_{p2} + R_L}\right] \tag{9-45}$$

For reasonably small values of R_L, it can be expected that

$$r_{p1} \gg R_{i2}$$

and so

$$A_v \cong -g_{m1}R_L \tag{9-46}$$

which is the gain of a pentode with transconductance g_{m1}.

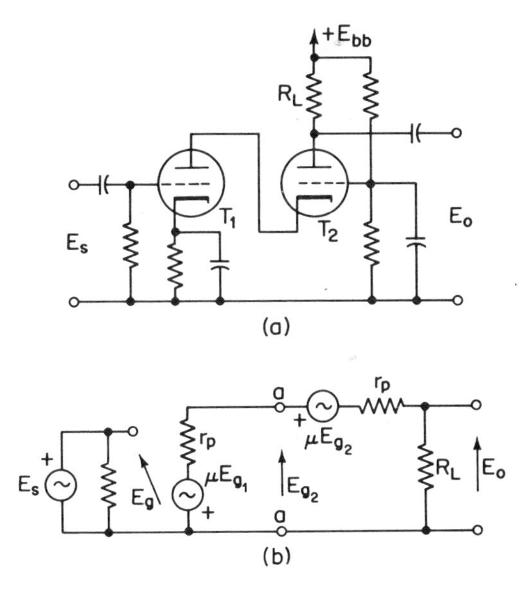

Figure 9-30. Cascode amplifier.

The gain of the grounded-cathode stage is low because of the low load, and its input capacitance is not appreciably increased due to the Miller effect. Thus its input capacitance approximates C_{gp}, and at high frequencies its loading on the source is small. Because of the low value of R_{i2}, the input capacitance of the grounded-grid tube has small effect; therefore the circuit is a useful high-frequency amplifier.

While it would appear to be replaceable by a single pentode, the circuit is found to introduce much less circuit noise than does a pentode, and this is an important factor when working with very small input signals, as will be discussed in the following chapter.

PROBLEMS

9-1. (a) From the anode characteristics of the 12AU7 triode, plot the static transfer characteristics for $E_b = 200$ V.

(b) Plot the dynamic characteristic for $E_{bb} = 200$ V, with a 25,000-ohm load.

9-2. From the following data taken on a triode find μ, g_m, and r_p, independently:

i_b, mA	e_c, V	e_b, V
8.2	0	165
6.0	0	130
6.0	−1	165
4.1	−1	130
4.1	−2	165

9-3. For a certain triode, the function of Eq. 9-2 may be expanded as

$$i_b, \text{A} = 87 \times 10^{-6}(\mu e_c + e_b) + 1.5 \times 10^{-7}(\mu e_c + e_b)^2$$

If $\mu = 15$ and the operating point is at $e_c = -7$ V and $e_b = 210$ V, find i_b, r_p, and g_m.

9-4. A triode has $g_m = 3300$ μmho, $r_p = 5100$ ohms.

(a) Find the plate-current change produced by variation of the grid voltage from -2 to -6 V, at $E_b = 140$ V.

(b) What change in plate voltage will bring the plate current back to its original value, with e_c remaining at -6 V?

9-5. For a particular triode the plate current can be found from

$$i_b, \text{mA} = 0.0042(4.2e_c + e_b)^2.$$

If $e_b = E_b = 200$ V, $E_{cc} = -10$ V, and $e_g = 3$ V peak sine wave, find the average, maximum, and minimum values of plate current.

9-6. Two triodes having the following parameters are connected in parallel:

$$\mu = 4.2, \qquad \mu = 17$$

$$g_m = 4500\ \mu\text{mho}, \qquad g_m = 1250\ \mu\text{mho}$$

Find the μ, g_m, and r_p of a single tube which would be equivalent in operation.

9-7. (a) If the dynamic grid grid transfer characteristic is expressed by an equation of the form

$$i_b = a_0 = a_1 e_g + a_2 e_g^2$$

show that a small input voltage $e_g = E \sin \omega t$ will produce an output current

$$i_b = a_0 + \frac{a_2 E^2}{2} + a_1 E \sin \omega t - \frac{a_2 E^2}{2} \cos 2\omega t.$$

(b) Find the expression for the output current if $e_g = E_1 \sin \omega t + E_3 \sin 3.5\ \omega t$.

9-8. Draw current-source equivalent circuits of (a) to (c) of Fig. 9-31.

9-9. By use of equivalent circuits find the gain or ratio E_o/E of the circuit of (b) in Fig. 9-31.

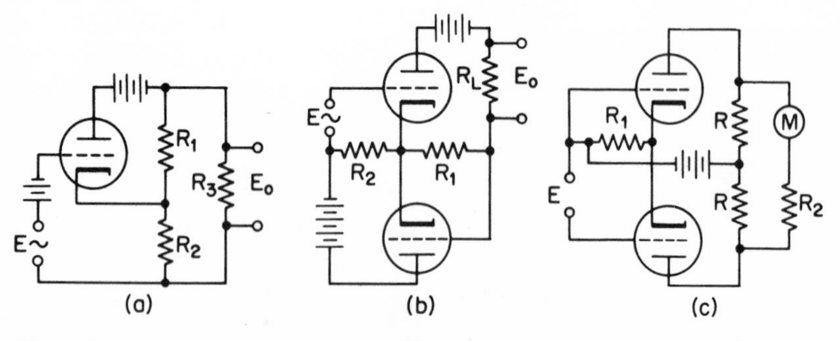

Figure 9-31.

9-10. By use of an equivalent circuit for (c) in Fig. 9-31, find the current through the meter M of resistance $R_2 = 200$ ohms, if $R = 10{,}000$ ohms, $\mu = 10$, $r_p = 7000$ ohms, $R_k = 2000$ ohms. Let $E = 0.5$ V, dc, positive to the grid, and assume identical tubes.

9-11. In (a), Fig. 9-31, let $R_1 = 10{,}000$ ohms, $R_2 = 3000$ ohms, $R_3 = 5000$ ohms, $r_p = 10{,}000$ ohms, and $g_m = 2500$ μmho. Find the ac voltage E_o, and I_p, if $E = 2.0$ V rms.

9-12. A 6BA6 pentode has the following characteristics:

$$g_m = 4400\ \mu\text{mho}, \qquad E_{c2} = 100\ \text{V}$$

$$r_p = 1.5\ \text{megohms}, \qquad I_{c2} = 4.2\ \text{mA}$$

$$I_b = 5.5\ \text{mA}, \qquad C_{gk} = 5.5\ \text{pF}$$

$$E_{bb} = 250\ \text{V}, \qquad C_{gp} = 0.0035\ \text{pF}$$

$$E_{c1} = -3\ \text{V}, \qquad C_{pk} = 5.0\ \text{pF}$$

and is used in the circuit of Fig. 9-19(b).

(a) Calculate desirable values for R_k, C_k, R_s, C_s which will be satisfactory over the frequency range 500 to 1500 kHz.

(b) With a resonant load at 1.25 MHz, of resistive impedance of 125,000 ohms, find the gain.

9-13. The 6BA6 of Problem 9-12 is to provide an output voltage of 35 V with $I_p = 0.24$ mA. Find the required rms input voltage, and the resonant load impedance.

9-14. (a) A 12AT7 triode with $\mu = 60$, $r_p = 10{,}900$ ohms, $g_m = 5500$ μmho, $C_{gp} = 1.5$ pF, $C_{gk} = 2.2$ pF, $C_{pk} = 0.5$ pF, has an anode circuit load of a 10 μH inductance, and operates at 1 megahertz. Find R_{in} and C_{in}.

(b) Repeat at a frequency of 100 megahertz. Would this tube constitute a suitable load for a radio-receiver input circuit having 5000 ohms impedance?

9-15. A 6AT6 triode has the following characteristics: $\mu = 70$, $g_m = 1200$ μ-mho, $C_{gk} = 2.3$ pF, $C_{gp} = 2.1$ pF, $C_{pk} = 1.1$ pF. With an anode load of $3000 + j3000$ ohms, plot curves of R_{in} and C_{in} over the range from 1 to 100 megahertz, using a logarithmic frequency scale.

9-16. In Fig. 9-28(a), $R_g = 10$ megohms, $R_{k1} = 2000$ ohms, $R_k = 25{,}000$ ohms, and the tube has $\mu = 20$, $r_p = 7700$ ohms, $g_m = 2600$ μmho.

(a) Find the input impedance seen by the generator, neglecting the reactance of C and the tube capacities.

(b) If $E_s = 2$ V rms, find the output voltage.

(c) What is the decibel power gain?

9-17. The grounded-grid amplifier of Fig. 9-25 uses a triode having $\mu = 70$, $r_p = 40{,}000$ ohms. With $R_s = 300$ ohms, $R_L = 25{,}000$ ohms, find the dB power gain and the input impedance R_i.

9-18. The circuit of Fig. 9-19(a) is to be used with the tube of Problem 9-16. If $E_{bb} = 250$ V, $I_b = 4.5$ mA, determine $R_k =$ for $E_{cc} = -6$ V. Find a value for $C_k =$ if the lowest frequency is to be 30 hertz.

9-19. The 6AT6 of Problem 9-15 is to have $E_{cc} = -3$ V, $I_b = 1$ mA. The value of R_k must insure a gain of 0.90 in a cathode follower. Design a circuit which will give the proper bias and gain.

9-20. In Fig. 9-29, $R_L = 500{,}00$ ohms $R_k = 1000$ ohms, and the triode of Problem 9-16 is used in each position. If $E_s = 1$ V rms, find the output voltage.

9-21. A generator has a resistance of 500 ohms and a generated voltage of 0.05 V. Using a triode in a grounded-grid circuit, with $\mu = 20$, $r_p = 7000$ ohms, design a circuit to match the generator resistance, and determine the output voltage obtainable.

REFERENCES

1. Schade, O. H., "Beam Power Tubes," *Proc. IRE,* Vol. 26, p. 137 (1938).
2. Richter, W., "Cathode-Follower Circuits," *Electronics,* Vol. 16, p. 112 (November 1943).
3. Jones, M. C., "Grounded-Grid Radio-Frequency Voltage Amplifiers," *Proc. IRE,* Vol. 32, p. 423 (1944).
4. *Coordination of Electrical Graphical Symbols.* American Standard Z32.11-1944, American Standards Association, New York.
5. *Standards on Abbreviations, Graphical Symbols, Letter Symbols, and Mathematical Signs.* Institute of Electrical and Electronics Engineers, New York, 1948.
6. Angelo, E. J., Jr., *Electronic Circuits,* 2nd ed. McGraw-Hill Book Company, New York, 1964.
7. Glasford, G. M., *Linear Analysis of Electronic Circuits.* Addison-Wesley Pub. Co., Reading, Mass., 1965.
8. "IEEE Recommended Practice for Units in Published Scientific and Technical Work." *IEEE Spectrum,* Vol. 3, p. 169 (1966).

10

LINEAR SMALL-SIGNAL RC AMPLIFIERS

A common type of amplifier handles *audio frequencies* for reproduction of speech and music with a frequency band reaching from 20 hertz to 60 kilohertz in well-designed equipment. The *video-frequency amplifier* of the television system must handle frequencies from 30 hertz to 4.5 megahertz. Such amplifiers, in which the frequency range starts near zero and extends to some thousands or millions of hertz, will be studied in this and the following chapter. Thus amplifiers are required to handle input signals with component frequencies covering wide frequency ranges; the circuit and device reactances are then effective in altering the device and load impedances. The restriction to small input signals will permit use of the equivalent circuit for either transistor or tube in the analyses.

The methods of amplifier analysis for gain, frequency response, phase response, and transient response are generally applicable to transistor or tube and will be demonstrated with either as convenient.

10-1 DISTORTION OF SIGNALS

Linear operation of an amplifier implies that the output wave form be an accurate reproduction of the wave form applied to the input circuit. This ideal is not achieved but can be approached in practical equipment; differences between output and input wave forms may appear, and such differences are known as *distortion*. Distortion may be due to operation on a nonlinear transfer curve of the active device, or it may be due to associated circuit and device reactances. To reduce distortion it is necessary to first classify it by type; then its known causes may be corrected by design changes. The several forms of distortion are:

Nonlinear distortion, produced if the transfer curve is not linear over the range used. The effect is illustrated by the curve of Fig. 10-1. The output wave differs from that at the input, and the output current contains harmonic frequencies which were not present in the sinusoidal input signal.

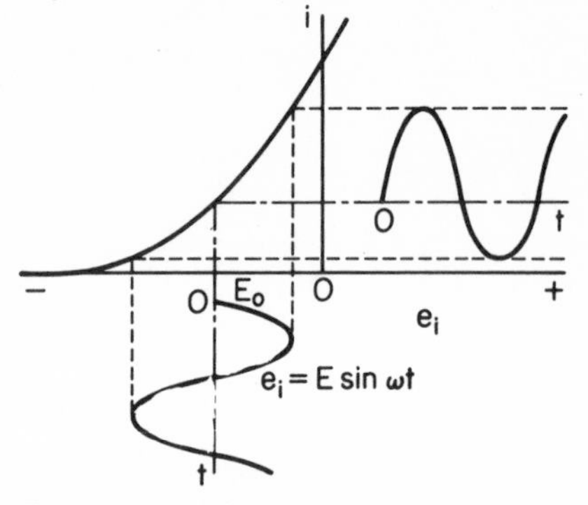

Figure 10-1. Distortion due to a nonlinear dynamic curve.

A nonlinear transfer curve also produces *intermodulation* when several frequencies are simultaneously present in the input. Additional frequencies will appear in the output as sums and differences of integral multiples of all input frequencies. These sum and difference frequencies are heard as noise in audio amplifier outputs, or give spurious signals in other circuits.

Frequency distortion is caused by unequal amplification of the component frequencies present in the spectrum of the input signal. At Fig. 10-2(a) an input wave is analyzed into a fundamental and a second harmonic. If this wave is applied to an amplifier with a gain of n times for the fundamental, and $2n$ for the second harmonic, the output wave will be that of the altered form at (b).

An ideal amplifier should give equal gain at all frequencies of interest, and frequency distortion is present if the plot of amplifier output versus frequency, for a constant sinusoidal input, is not a horizontal line. The typical curve of gain against frequency of Fig. 10-3 is normally plotted with a logarithmic frequency axis, to encompass a wide frequency range. Fre-

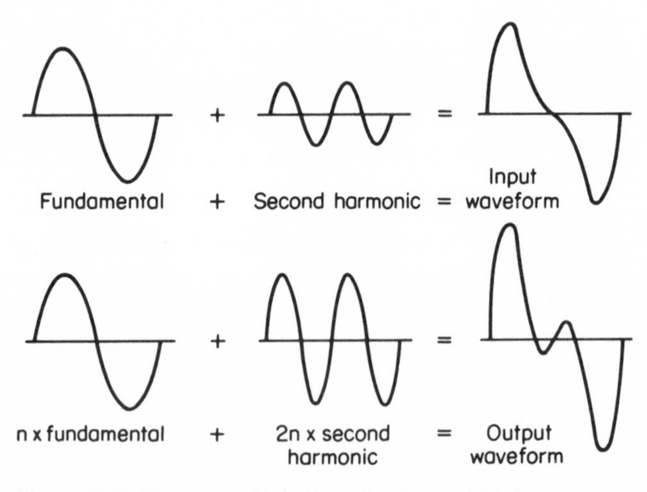

Figure 10-2. Frequency distortion: (a) above, (b) below.

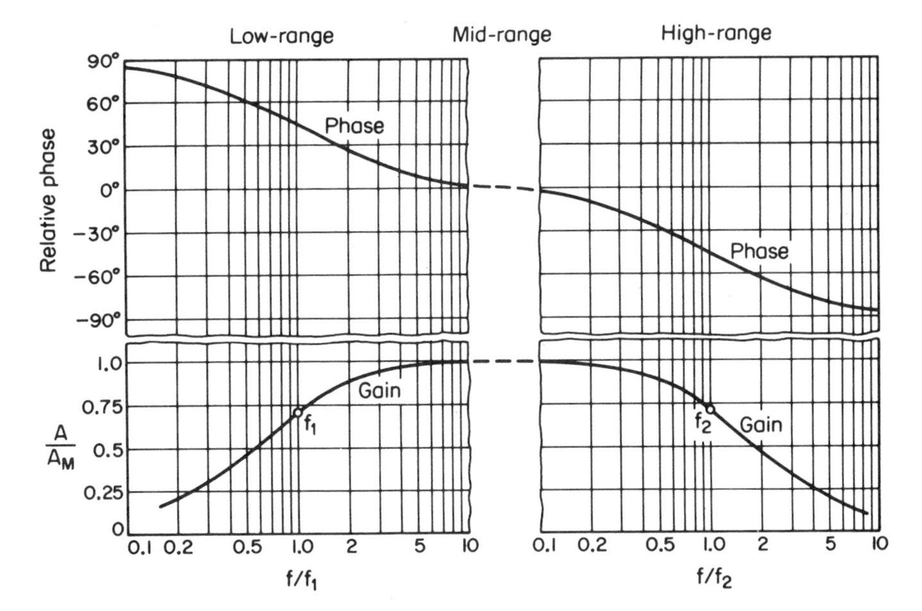

Figure 10-3. Generalized gain and phase-shift curves for *RC* amplifiers.

quency distortion is a result of reactances associated with the circuit or the active device.

Phase or delay distortion occurs if the transmission delay through the circuit is different for various frequencies. If the phase angle θ changes in proportion to frequency, i.e., if $d\theta/d\omega$ is a constant or zero there is no phase distortion. Phase distortion is created by the presence of reactive circuit and device elements. The reduction of phase distortion in television and other wide frequency band systems is an important matter.

If the operating point is located in the middle of a linear region of the transfer curve, and the signal is sufficiently small, then the nonlinear distortion should be low. Such a selection of conditions is known as *Class A operation*, and will be assumed for the linear amplifiers of this chapter.

10-2 FREQUENCY AND PHASE RESPONSE

Input signals to electronic amplifiers are often nonsinusoidal, and application of Fourier methods to such wave forms results in their representation as a broad spectrum of sine frequencies. To avoid delay and frequency distortion and resultant change in wave form of the output, the gain and phase response of an amplifier must be uniform over the input band of frequencies, but for economic reasons or technical limitations of the circuits, this ideal is rarely realized. This necessitates a trade-off between the improvement of wave form by design, and the cost to obtain that improvement.

Coupling networks are necessary to transfer the ac signal from the output of one amplifer stage to the input of the next, and to block the dc potentials. Reactances are used for these functions, and the internal capacitances of transistor and tube are also present. The variation of gain and phase with frequency is the result of the presence of these reactances in the circuit.

The use of resistance-capacitance (*RC*) coupling between two C-E transistor amplifiers is illustrated in Fig. 10-4. Resistors R_1 and R_2 are present to supply the bias for the base, capacitor C is present to block the collector potential from the base circuit of Q_2. An additional reactance is introduced by C_E, the bypass capacitor for the emitter resistor. There are also the internal transistor capacitances which do not appear implicitly in the circuit, but their effects cannot be overlooked.

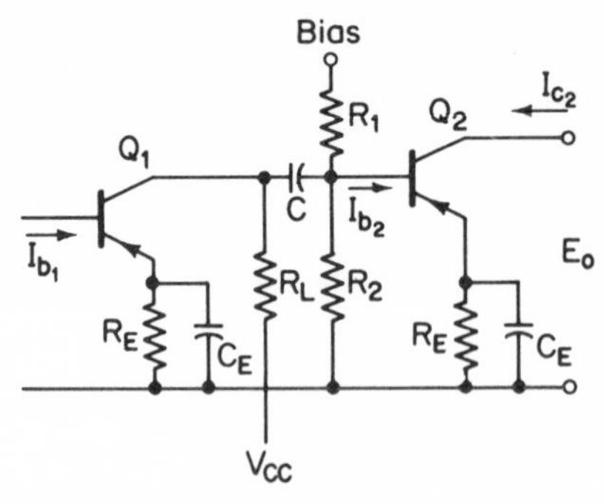

Figure 10-4. *RC* coupling between C-E stages.

Analysis of amplifier performance is usually carried out by division of the amplifier response band into three ranges, indicated in Fig. 10-3. The so-called *mid-range* of frequencies is that frequency range in which the gain and phase delay are independent of frequency, or the reactances of the bypass and coupling capacitances and the inherent internal capacitances may be considered inoperative in the circuit.

The *low range* is that in which series capacitance or shunt inductance becomes effective in modifying circuit impedances. The coupling capacitance C and the bypass capacitance C_E, or their equivalents in vacuum-tube circuits, increase in reactance with falling frequency and cause a low-frequency drop in gain.

The *high range* is that in which shunt capacitances and series inductance become effective in reducing gain; the internal device capacitances are the limiting factors. These capacitances are in shunt with load resistors of appreciable magnitude, and reduce the load value and the gain as the frequency increases. In transistors the current amplification factor $h_{fe} = \beta$ also falls at some extreme frequency, due to the time required for charge storage.

By convention, the response *band width*, *BW*, of an amplifier is defined in terms of two bounding frequencies, f_1 and f_2, at opposite ends of the mid-range, at which the voltage or current gain has fallen to $1/\sqrt{2}$ of its mid-range value. That is,

$$BW = f_2 - f_1 \qquad \text{Hz} \quad (10\text{-}1)$$

and these frequencies are indicated in Fig. 10-3. Since the power is then one-half of the mid-range value, these are also called the *half-power frequencies*, and the gain in dB will be 3 dB below the mid-range reference value. Frequencies f_1 and f_2 will be shown closely related to amplifier design parameters.

10-3 THE HYBRID-π EQUIVALENT CIRCUIT

For analysis of transistor amplifiers in the high-frequency range we must include the effect of the internal capacitances, as was done for the vacuum tube. These capacitances are included in the *hybrid-π equivalent circuit* of Fig. 10-5.

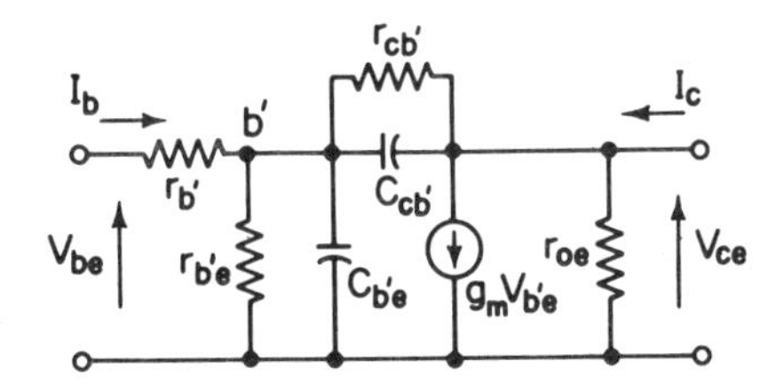

Figure 10-5. Hybrid-π equivalent circuit: $r_{b'} = 50$ ohms, $C_{b'e} = 1000$ pF, $r_{cb'} = 2$ megohms, $r_{b'e} = 1100$ ohms, $C_{cb'} = 5$ pF, $r_{oe} = 50{,}000$ ohms.

The internal capacitances and the variable portion of the base resistance actually terminate at an internal point of true base potential b'. As the input voltage is increased, there is an increase in hole density in the base, and an electron current enters the base to maintain charge neutrality. When the emitter-base voltage decreases, the stored holes and electrons return to the circuit. That is, the changes in charge create a current

$$-i_B = C_{b'e}\frac{dv_{EB}}{dt} \qquad (10\text{-}2)$$

and capacitance $C_{b'e}$ represents this effect.

The depletion layer of the collector-base reverse-biased junction represents an appreciable capacitance, and appears in the circuit as $C_{cb'}$. Resistance $r_{cb'}$ is added as a second-order effect to represent the collector current change resulting from the change in diffusion gradient as the base width changes with collector voltage. As has been defined, $g_m \cong h_{fe}/h_{ie}$.

The hybrid-π circuit provides good accuracy in the prediction of transistor performance in the high-frequency range.

10-4 A SIMPLIFIED HIGH-FREQUENCY C-E EQUIVALENT CIRCUIT

The hybrid-π circuit is more complicated than is always justified, and it can be simplified for better understanding of the effects of circuit and device parameters on performance.

Resistor $r_{b'}$ is usually small compared to R_s of the source and so will be dropped. The current transfer generator then is more useful, being controlled by the input voltage, rather than by an internal voltage. Resistor $r_{cb'}$ can be neglected as a second-order effect, and since $r_{oe} = 1/h_{oe}$ will ordinarily be large with respect to the load, it may be eliminated. For convenience, h_{ie} is shown as r_{ie}. The result is the modified high-frequency equivalent circuit of Fig. 10-6(a). Its relation to the transconductance model of Section 7-16 can be seen.

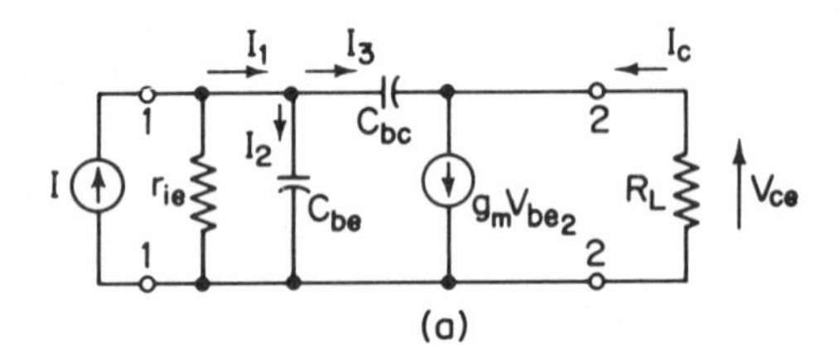

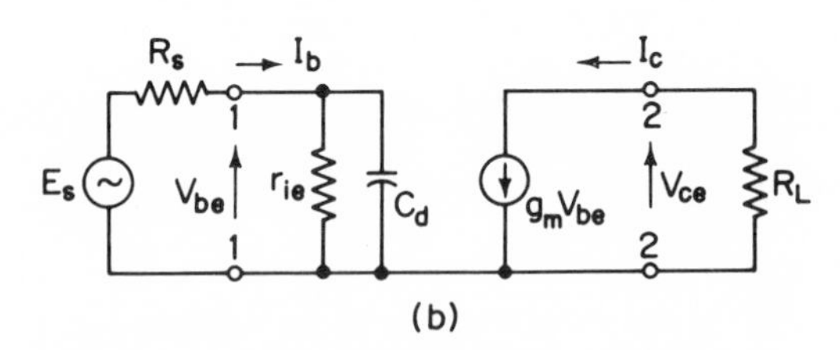

Figure 10-6. (a) Modified hybrid-π circuit; (b) further simplification of the circuit.

The input current is broken into components in Fig. 10-6(a), and their values may be written as

$$\mathbf{I}_2 = j\omega C_{be}\mathbf{V}_{be} \tag{10-3}$$

$$\mathbf{I}_3 = j\omega C_{bc}(\mathbf{V}_{be} - \mathbf{V}_{ce}) \tag{10-4}$$

The load current is $\mathbf{I}_c - \mathbf{I}_3$; however, $j\omega C_{bc}$ is small and $\mathbf{I}_3 \ll \mathbf{I}_c$. Then

$$\mathbf{V}_{ce} = -g_m\mathbf{V}_{be}R_L \tag{10-5}$$

The input current to the capacitance network is

$$\mathbf{I}_1 = j\omega\mathbf{V}_{be}[C_{be} + (1 + g_mR_L)C_{bc}]$$

This expression indicates that the input capacitance is

$$C_d = C_{be} + (1 + g_m R_L) C_{bc} \tag{10-6}$$

a result which is analogous to the Miller effect capacitance in the vacuum tube. With $g_m = 25{,}000\ \mu$mho, $R_L = 1000$ ohms, and the capacitances given in Fig. 10-5, the input capacitance of the transistor will be

$$C_d = 1000 + (1 + 25)5 = 1130 \text{ pF}$$

The result of Eq. 10-6 shows that C_{be} and C_{cb} may be replaced with a single capacitor C_d, in shunt in the circuit. The circuit of Fig. 10-6(b) is thus a useful form of representation of the transistor at higher frequencies, when in the common-emitter circuit.

10-5 THE RC-COUPLED C-E AMPLIFIER AT LOW AND MID-FREQUENCIES

We now have equivalent circuits for the transistor for use in the study of the frequency response of the common-emitter amplifier. Division of the response into low-, middle-, and high-frequency ranges leads to the circuits of Fig. 10-7 for the *RC*-coupled common-emitter amplifier.

For the initial stage of an amplifier, in Fig. 10-7(a), $\mathbf{I}_s$ and R_s will be values for the signal source. For an intermediate stage of a cascaded amplifier $\mathbf{I}_s$ and R_s will become $\mathbf{I}_{c1}$ and R_L of the preceding stage. Bias resistors R_1 and R_2 can be reduced to an equivalent R_B; this will usually be large with respect to r_{ie} by design, and can then be dropped from the circuit.

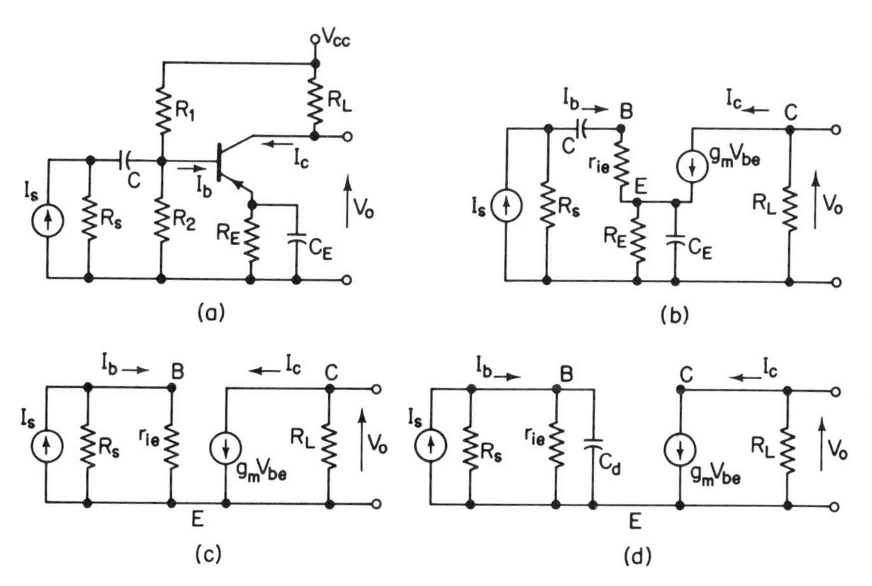

Figure 10-7. (a) *RC* amplifier; (b) low-frequency equivalent circuit; (c) mid-range equivalent; (d) high-frequency equivalent.

Capacitors C and C_E will usually have values in the microfarad range, while C_d will range from 10 to 1000 pF. At low frequencies, C and C_E will appear in the circuit as at (b), Fig. 10-7. At high frequencies, C and C_E appear as short circuits, and C_d enters as an appreciable shunting reactance, as at (d). In the mid-frequency range, C_d is an open circuit, and C and C_E are ac short circuits, thus eliminating all reactances from the equivalent circuit at (c).

For accuracy, we must recall that the input resistance of the transistor involves an approximation, because of

$$R_{ie} = \frac{\Delta_{he} + h_{ie}}{1 + h_{oe}R_L} \cong h_{ie} \equiv r_{ie} \tag{10-7}$$

The simplified form holds for most R_L values. The current generator $g_m V_{be}$ will also be referred to as $\beta_0 I_b \equiv h_{feo} I_b$. The zero subscript indicates that β is measured at a low frequency at which the internal capacitances are not effective.

The *mid-frequency current gain* can be written from the circuit of Fig. 10-7(c). Since $\mathbf{I}_b = \mathbf{I}_s R_s/(r_{ie} + R_s)$, then

$$\mathbf{A}_{iM} = \frac{\mathbf{I}_c}{\mathbf{I}_s} = \frac{\beta_0 R_s}{r_{ie} + R_s} \tag{10-8}$$

which shows that current gain is maximized at the value of β_o for $R_s \gg r_{ie}$, or for a current source as the amplifier input.

The voltage gain is not usually of such great interest in a transistor amplifier. However, when needed it may be found by use of a circuit fundamental; that is,

$$\text{voltage gain} = \frac{\mathbf{I}_c R_L}{\mathbf{I}_b r_{ie}} = \text{current gain} \frac{\text{load impedance}}{\text{input impedance}} \tag{10-9}$$

Using this relation, and noting that the amplifier input impedance is $R_s r_{ie}/(R_s + r_{ie})$, leads to

$$\mathbf{A}_{vM} = \frac{-\beta_0 R_L}{r_{ie}} = -g_m R_L \tag{10-10}$$

The latter expression follows because $\beta_0 = g_m r_{ie}$. A vacuum pentode, also behaving as a current source, has a similar gain relation.

The current through the emitter bias circuit is $\mathbf{I}_e = (1 + \beta_0)\mathbf{I}_b$, and the voltage drop to ground across the circuit, in terms of $\mathbf{I}_b$, is

$$\mathbf{V}_E = \frac{R_E(1 + \beta_0)}{1 + j\omega C_E R_E} \mathbf{I}_b \tag{10-11}$$

This defines an effective impedance for the bias circuit, and allows the writing of

$$\mathbf{Z}_E = \frac{R_E(1 + \beta_0)}{1 + j\omega\left(\dfrac{C_E}{1 + \beta_0}\right) R_E(1 + \beta_0)} \tag{10-12}$$

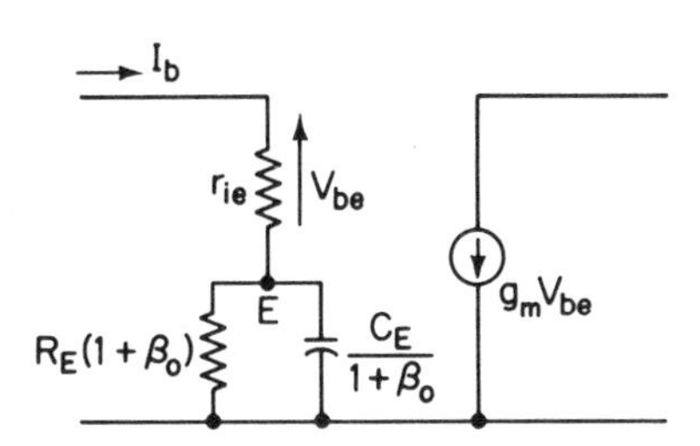

Figure 10-8. Alternative treatment of the emitter bias circuit.

which justifies the alternative circuit treatment of Fig. 10-8, where the effective values of the resistance and capacitance are $R_E(1 + \beta_0)$ and $C_E/(1 + \beta_0)$, when placed in the $\mathbf{I}_b$ circuit.

The writing of the current response for the low-frequency region is complicated because of the several RC time constants. However, a look at the time constants separately shows C_E combined with R_E in parallel with $r_{ie} + R_s$; the effective value of the resistance is usually only a few hundred or few thousand ohms. Capacitance C is combined with $R_s + r_{ie} + R_E(1 + \beta_0)$, and this resistance is many thousands of ohms. For equal time constants, and therefore equal effects on the frequency response, the ratio of C_E to C may approximate fifty or more. To increase a small capacitor C by a given factor costs little; to increase an already large capacitor may raise the cost considerably and also introduce a space problem. Therefore, the capacitance of C_E usually becomes the limiting factor at the low frequencies defined in Section 10-2. We will use this assumption of separable time constants in order to simplify the circuit analysis.

At a frequency at which the reactance of C_E becomes comparable to R_E, and at which C represents a short circuit in Fig. 10-7(b), the current I_s divides, so that

$$\mathbf{I}_b = \mathbf{I}_s \frac{R_s}{R_s + r_{ie} + \dfrac{R_E(1 + \beta_0)}{1 + j\omega C_E R_E}} \tag{10-13}$$

The collector current can be found from $\mathbf{I}_c = g_m \mathbf{V}_{be} = g_m r_{ie} \mathbf{I}_b = \beta_0 \mathbf{I}_b$, and

$$\mathbf{I}_c = \frac{\beta_0 R_s \mathbf{I}_s}{R_s + r_{ie}} \left\{ \frac{1 + j\omega C_E R_E}{1 + \dfrac{R_E(1 + \beta_0)}{R_s + r_{ie}} + j\omega C_E R_E} \right\} \tag{10-14}$$

At a lower frequency where the reactance of C begins to modify the circuit impedance, then C_E may be considered as high in reactance or open, and a further current division factor is introduced. The *low-frequency current gain* of the amplifier may then be written as

$$\mathbf{A}_{iL} = \mathbf{A}_{iM} \left\{ \frac{1 + j\omega C_E R_E}{1 + \dfrac{R_E(1 + \beta_0)}{R_s + r_{ie}} + j\omega C_E R_E} \right\} \left\{ \frac{j\omega C[R_s + r_{ie} + R_E(1 + \beta_0)]}{1 + j\omega C[R_s + r_{ie} + R_E(1 + \beta_0)]} \right\} \tag{10-15}$$

Two parameters may be defined as

$$\omega_E = \frac{1}{C_E R_E}, \qquad \omega_C = \frac{1}{C[R_s + r_{ie} + R_E(1 + \beta_0)]}$$

At a frequency where the real and reactive terms of the denominator of the first brace in Eq. 10-15 are equal, the gain will be $1/\sqrt{2}$ of its mid-range value, and this is the lower band limit. That is, from

$$\omega_1 C_E R_E = 1 + \frac{R_E(1 + \beta_0)}{R_s + r_{ie}}$$

$$\omega_1 = 2\pi f_1 = \omega_E + \frac{1 + \beta_0}{C_E(R_s + r_{ie})} \tag{10-16}$$

It can be seen that the band limit frequency f_1 is considerably above the frequency determined by the bias circuit time constant as $f_E = \omega_E/2\pi$.

By Eq. 10-16 the low-frequency limit is more a function of R_s than it is of R_E. The lowest value of f_1, for a given R_E and C_E, will occur with R_s very large, which is also desirable from the standpoint of current gain. If $R_s \to \infty$, then f_1 is brought down to f_E.

The gain expression can be simplified to

$$\mathbf{A}_{iL} = \mathbf{A}_{iM}\left\{\frac{1 + (j\omega/\omega_E)}{1 + \dfrac{R_E(1 + \beta_0)}{R_s + r_{ie}} + (j\omega/\omega_E)}\right\}\left(\frac{j\omega/\omega_C}{1 + (j\omega/\omega_C)}\right) \tag{10-17}$$

In the mid-range, $\omega \gg \omega_E$ and the gain is $\mathbf{A}_{iM}$. The frequency response is $1/\sqrt{2}$ below the mid-range value at f_1, and continues to fall below that frequency. At a still lower frequency, $f_C = \omega_C/2\pi$, the loss of gain is accelerated by the effect of C.

In design, the circuit resistances R_L, R_E, and R_B are first chosen to give an appropriate I_C value and instability factor. Then C_E is selected to place f_1 at a suitable frequency below the specified operating band. Economically a design with $\omega_E = \omega_C$ would seem to have value since neither capacitor would then be in excess. A more detailed analysis indicates that this result may be approximated if $C_E = C(1 + \beta_0)$. However, such a choice might not lead to a minimum cost situation. In fact, the cost and space requirements for C_E may be considered excessive when balanced against the additional gain its presence provides. It is often dispensed with and the gain sacrifice accepted. The gain with R_E unbypassed, can be found by letting $\omega_E = \infty$ in Eq. 10-17. Then

$$\mathbf{A}_{iL} = \mathbf{A}_{iM}\left\{\frac{1}{1 + \dfrac{R_E(1 + \beta_0)}{R_s + r_{ie}}}\right\}\left[\frac{j\omega/\omega_C}{1 + (j\omega/\omega_C)}\right] \tag{10-18}$$

The removal of the bypass introduces negative feedback, and causes a gain reduction.

The circuit of Fig. 10-9 illustrates a practical two-stage common-emitter amplifier. Using a transistor with $\beta_0 = 50$, $h_{ie} = 600$ ohms, the values of the frequency parameters for one stage are found to be

$$f_E = 3.2 \text{ Hz}, \qquad f_C = 0.56 \text{ Hz}, \qquad f_1 = 32 \text{ Hz}$$

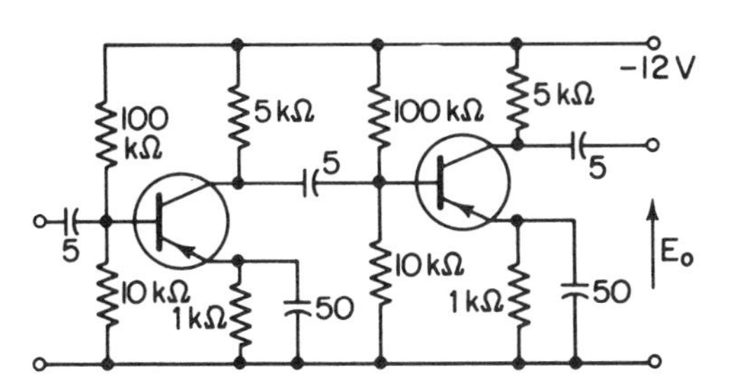

Figure 10-9. Practical audio-frequency C-E amplifier.

Thus the previous assumptions concerning the magnitudes of the important circuit parameters are supported. The band limit of 32 hertz is usually considered sufficiently low for audio-frequency performance. The calculations indicate the major effect that f_1 has in comparison to usual values of the limit frequencies f_E and f_C.

10-6 THE RC-COUPLED C-E AMPLIFIER IN THE HIGH-FREQUENCY RANGE

The circuit of Fig. 10-7(d) was derived for the common-emitter amplifier at frequencies where the reactance of C_d affects the input admittance in the high-frequency range. The current $\mathbf{I}_b$ can be written as

$$\mathbf{I}_b = \mathbf{I}_s \frac{R_s(1 + j\omega C_d r_{ie})}{R_s + r_{ie} + j\omega C_d r_{ie} R_s} \tag{10-19}$$

and

$$\mathbf{V}_{be} = \mathbf{I}_b \frac{r_{ie}}{1 + j\omega C_d r_{ie}} \tag{10-20}$$

Since $\mathbf{I}_c = g_m \mathbf{V}_{be}$, the current gain follows as

$$\mathbf{A}_{iH} = \frac{\mathbf{I}_c}{\mathbf{I}_s} = \frac{\beta_0 R_s}{R_s + r_{ie}} \left[\frac{1}{1 + j\omega C_d \dfrac{R_s r_{ie}}{R_s + r_{ie}}} \right] \tag{10-21}$$

By defining

$$\omega_2 = 2\pi f_2 = \frac{1}{C_d\left(\dfrac{R_s r_{ie}}{R_s + r_{ie}}\right)} \tag{10-22}$$

the *high-frequency gain* expression becomes

$$\mathbf{A}_{iH} = \mathbf{A}_{iM}\left[\frac{1}{1 + j(\omega/\omega_2)}\right] \tag{10-23}$$

Frequency f_2 is the upper band limit or half-power frequency, and it is an inverse function of C_d of the transistor and the input circuit resistances.

10-7 CURVES OF RESPONSE VERSUS FREQUENCY

The amplitude and phase angle of the mid-range gain are independent of frequency, and the mid-range gain is useful as a normalizing standard in the comparative study of low- and high-frequency gains, as in Eqs. 10-17 and 10-23. Ratios such as A_{iL}/A_{iM}, and A_{iH}/A_{iM} permit specification of amplifier performance in terms of circuit parameters and time constants. When written in decibel units, such relations become additive in cascaded amplifiers, and exhibit other useful properties.

For instance, consider Eq. 10-23 written in the form

$$\left|\frac{A_{iH}}{A_{iM}}\right| = \left[\frac{1}{1 + (\omega/\omega_2)^2}\right]^{1/2}$$

The ratio can be translated into decibels as

$$\text{dB}\left|\frac{A_{iH}}{A_{iM}}\right| = 20 \log \left[\frac{1}{1 + (\omega/\omega_2)^2}\right]^{1/2} = -10 \log [1 + (\omega/\omega_2)^2] \qquad (10\text{-}24)$$

In the mid-range where $\omega \ll \omega_2$, this reduces to zero dB and so the mid-range gain is the reference or zero level for gain specification in decibels.

At $\omega/\omega_2 = 1$, the equation becomes

$$\text{dB}\left|\frac{A_{iH}}{A_{iM}}\right| = -10 \log 2 = -3.01 \text{ dB}$$

and so $f = f_2$ appears again as the -3 dB or upper half-power frequency.

At values of $\omega/\omega_2 \gg 1$, the equation becomes

$$\text{dB}\left|\frac{A_{iH}}{A_{iM}}\right| = -20 \log \frac{\omega}{\omega_2} \qquad (10\text{-}25)$$

This equation is that of a straight line on a dB or log-gain vs. log-frequency graph, passing through the *corner point* at 0 dB, $\omega/\omega_2 = 1$, and falling with a slope of -20 dB per frequency decade, equivalent to -6 dB per frequency octave. Such a line is dashed in Fig. 10-10, as the asymptote which the actual gain curve follows above $\omega/\omega_2 \cong 3$.

Low-frequency gain relations for circuits with single effective time constants take the form

$$\text{dB}\left|\frac{A_{iL}}{A_{iM}}\right| = -10 \log \left[1 + \left(\frac{\omega_1}{\omega}\right)^2\right] \qquad (10\text{-}26)$$

with the half-power point or corner point at $\omega = \omega_1$. When $\omega \ll \omega_1$, the gain rises at 20 dB per frequency decade, as shown by the dashed asymptote of Fig. 10-10.

Gain-frequency curves can be easily plotted by drawing the asymptote from the corner point with the appropriate slope, locating the -3 dB point and sketching the remaining parts of the curve by use of values from Table 10-1.

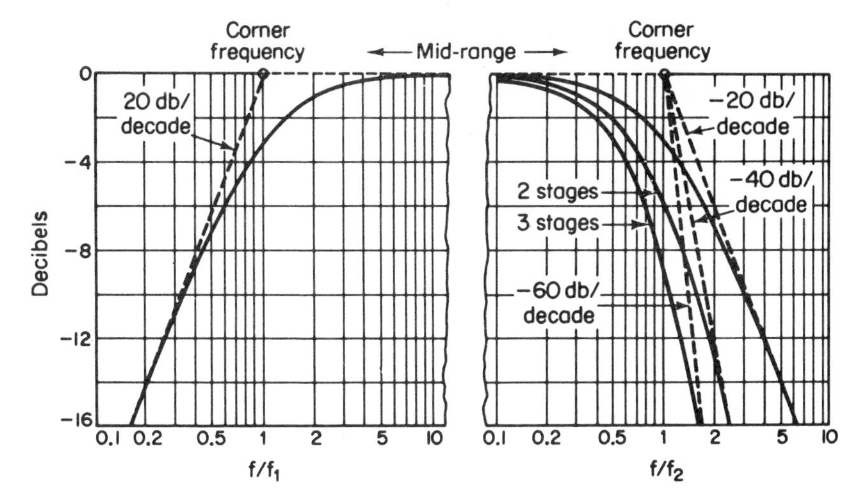

Figure 10-10. Generalized gain curves in dB.

The gain being additive, the gain of a two-stage (identical) amplifier will fall at −40 dB per decade (−12 dB per octave) at the frequency extremes. A three-stage circuit will have a gain falling at −60 dB per decade. The corner frequency gains will be −6 dB and −9 dB, as shown.

Table 10-1

GAIN-FREQUENCY DATA IN DECIBELS

ω/ω_1	Gain, dB	θ, degrees	ω/ω_2	Gain, dB	θ, degrees
0.1	−20	84.3	0.1	−0.04	− 5.8
0.3	−10.7	73.6	0.3	−0.38	−16.7
0.5	−7.0	63.4	0.5	−0.96	−26.6
0.7	−4.8	55.0	0.7	−1.7	−35.0
1.0	−3.0	45.0	1.0	−3.0	−45.0
1.5	−1.6	33.8	1.5	−5.1	−56.3
3.0	−0.42	18.4	3.0	−10.0	−71.6
5.0	−0.18	11.3	5.0	−14	−78.7
10.0	−0.04	5.8	10.0	−20	−84.3

Where the gain approaches a slope of −20 dB per decade, the phase angle of the amplifier will approach −90° at high frequencies and +90° at low frequencies. These gains and angles will be multiplied by the number of stages for n-stage amplifiers. Thus phase angle information can be approximated from the slopes of the gain curves.

Amplifiers having two time constants will have corresponding corner points. This is the case for the low-frequency response given by Eq. 10-17. The expression can be written as

$$\text{dB}\,\frac{A_{iL}}{A_{iM}} = 20\log\frac{\omega}{\omega_C} - 20\log\left[1+\left(\frac{\omega}{\omega_C}\right)^2\right]^{1/2} + 20\log\left[1+\left(\frac{\omega}{\omega_E}\right)^2\right]^{1/2} - 20\log\left[a^2+\left(\frac{\omega}{\omega_E}\right)^2\right]^{1/2} \tag{10-27}$$

If the terms are numbered 1, 2, 3, 4, the respective asymptotes can be identified in Fig. 10-11. The actual gain curve is then sketched to follow the asymptotes.

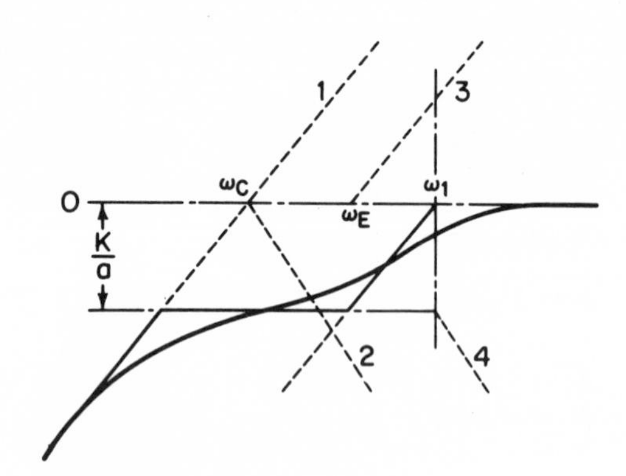

Figure 10-11. Two time-constant low-frequency responses.

10-8 FREQUENCY-LIMIT PARAMETERS OF A TRANSISTOR

Common-emitter circuit current gain falls off at high frequencies primarily because of the internal capacitances, represented by C_d. Useful transistor figures of merit can be derived from the short-circuit current gain, measured at high frequencies in the circuit of Fig. 10-12.

The input current at a high frequency is

$$\mathbf{I}_b = \left(\frac{1}{r_{ie}} + j\omega C_d\right)\mathbf{V}_{be}$$

Since $\mathbf{I}_c = g_m\mathbf{V}_{be}$, the short-circuit current gain is

$$\beta \equiv h_{fe} = \frac{\mathbf{I}_c}{\mathbf{I}_b} = \frac{g_m r_{ie}}{1 + j\omega C_d r_{ie}} \tag{10-28}$$

The value $g_m r_{ie}$ is equal to β_0, so

$$\beta = \frac{\beta_0}{1 + j\omega C_d r_{ie}} \tag{10-29}$$

With frequency $f_\beta = \omega_\beta/2\pi$, where

$$f_\beta = \frac{1}{2\pi C_d r_{ie}} \tag{10-30}$$

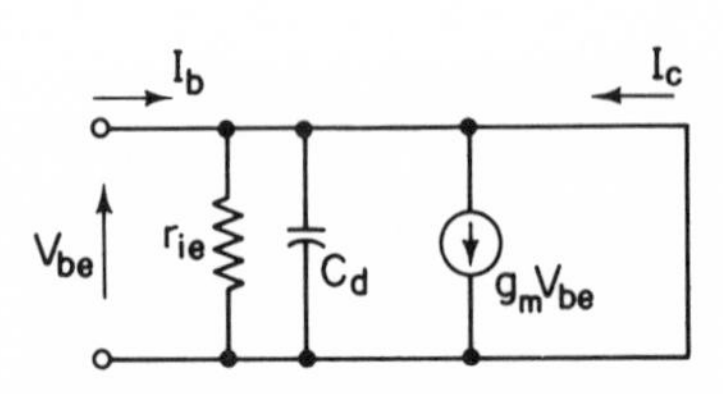

Figure 10-12. Measurement of short-circuit current gain: C-E amplifier.

the expression for β becomes

$$\beta = \frac{\beta_0}{1 + j(f/f_\beta)} \tag{10-31}$$

At $f = f_\beta$ the short-circuit current gain of the transistor *in the C-E circuit* will have fallen to $1/\sqrt{2}$ of its mid-frequency value β_0. The frequency f_β is called the β *cutoff frequency* and this serves as a high-frequency performance criterion for a given transistor, dependent only on the transistor parameters.

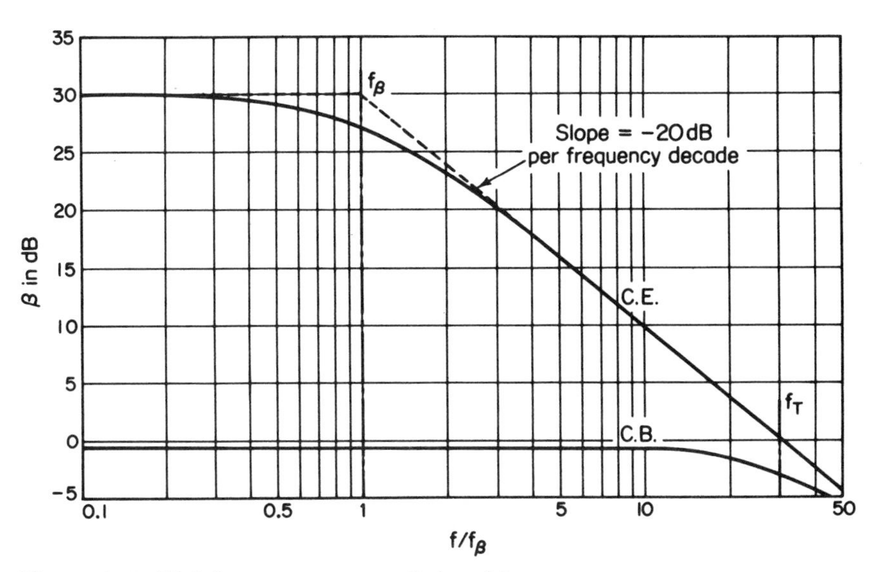

Figure 10-13. High-frequency response of a transistor.

A transistor is operable at frequencies above f_β, but with reduced gain. This is illustrated by the expression for β plotted in Fig. 10-13 in decibels, in terms of f/f_β. Using $f/f_\beta \gg 1$ in Eq. 10-31, the equation of the asymptote is

$$\beta = \frac{\beta_0}{f/f_\beta}$$

and from the corner frequency at $\beta = \beta_0, f/f_\beta = 1$, it falls with a slope of -20 dB per frequency decade. The curve illustrates the manner in which β varies with frequency above f_β.

Another performance criterion used to measure high-frequency transistor performance is f_T, the *gain-bandwidth product*. If the curve for β be extended to a higher frequency $f = f_T$ where β becomes unity or the current gain reaches zero dB, then from the asymptote equation above

$$f_T/f_\beta = \beta_0$$

and

$$f_T = \beta_0 f_\beta = \frac{\beta_0}{2\pi C_d r_{ie}} \tag{10-32}$$

Since the lower band limit for a transistor is at zero frequency, and f_β is the defined upper limit, then the transistor bandwidth is f_β. Therefore Eq. 10-32 represents the product of gain, β_0, and bandwidth f_β, and shows this dependent on transistor parameters. It indicates that for a transistor having $f_T =$ 20 MHz, a current gain of 10 is theoretically attainable with a bandwidth of 2 MHz, or a gain of 100 can be achieved with a bandwidth of 200 kHz. Another transistor type having $f_T = 5$ MHz would be expected to have proportionately poorer gain or bandwidth.

Thus f_β and f_T serve as figures of merit in the selection of a transistor for a given application; the connected circuit parameters then reduce these limit figures to the operating value of f_2 predicted by Eq. 10-22 for a given amplifier. Figures for f_β or f_T are usually given in a manufacturer's specification for a transistor type.

The first step in design of a C-E amplifier in the high-frequency range is the selection of a transistor with a suitably high value of f_β or f_T, and with a suitably small C_d. Frequency f_2 is then controlled by selection of R_s as indicated in Eq. 10-22. Lowering of R_s increases the bandwidth but lowers the current gain. Reduction of R_L also increases the bandwidth indirectly, through reduction of gain and the Miller effect capacitance introduced into C_d.

If these steps cannot be taken because of need for high current gain, then a transistor with greater f_T value must be chosen or the band-widening techniques of the next chapter must be applied.

10-9 TRANSISTOR FREQUENCY RESPONSE IN THE C-B CIRCUIT

The common-base equivalent circuit is drawn in Fig. 10-14, by rearrangement of the elements of the simplified common-emitter circuit of Fig. 10-6(a). It is of interest to compare the frequency response of a given transistor in the C-B and the C-E circuits, because of the difference in input resistance and in the location of C_{bc} in the common-base circuit.

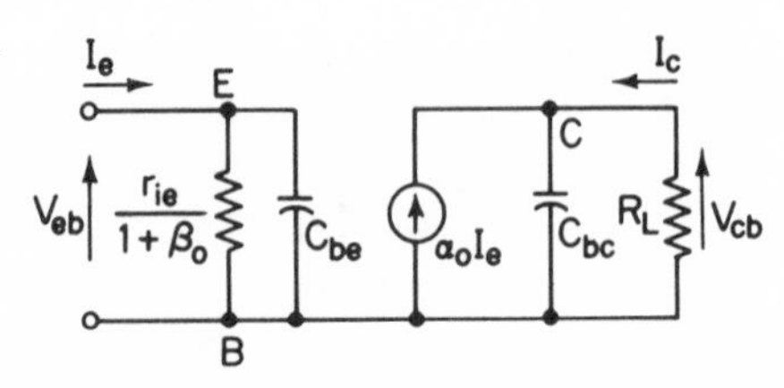

Figure 10-14. High-frequency C-B equivalent circuit.

By the usual relations $g_m V_{be} = \alpha_0 \mathbf{I}_e$ for the current source. The output capacitance is still connected to the base but now appears in shunt with the output load. Since C_{bc} is very small, capacitance C_{be} is found to control the upper frequency limit on circuit performance. Resistance h_{ib} is given in Section 7-15 as

$$h_{ib} \cong \frac{h_{ie}}{1 + \beta_0} \cong \frac{h_{ie}}{\beta_0} \tag{10-33}$$

and so $r_{ib} = h_{ib}$ is much less than the resistance across which C_{be} was connected in the common-emitter circuit. The reactance of C_{be} remains high with respect to its shunting resistance to a higher frequency than for the common-emitter circuit, and the common-base circuit gives a wider frequency response with a given transistor, than does the common-emitter circuit. However, the latter gives a higher current gain.

At mid-frequency $\mathbf{I}_c/\mathbf{I}_e = -\alpha_0$, by reason of the indicated current directions in Fig. 10-14. It can be shown from the figure that at higher frequencies

$$\alpha = \frac{\alpha_0}{1 + j\omega C_{be} r_{ib}} = -\frac{I_c}{I_e'} \tag{10-34}$$

A frequency f_α can be defined as

$$f_\alpha = \frac{\omega_\alpha}{2\pi} = \frac{1}{2\pi C_{be} r_{ib}} = \frac{1 + \beta_0}{C_{be} r_{ie}} \tag{10-35}$$

and so

$$\alpha = \frac{\alpha_0}{1 + j(f/f_\alpha)} \tag{10-36}$$

The frequency f_α is the upper band limit for the transistor short-circuit current gain in the common-base circuit. It is also called the *α cutoff frequency*.

Comparison of Eq. 10-35 and Eq. 10-32 shows that

$$f_\alpha \cong f_T = \beta_0 f_\beta \tag{10-37}$$

Thus the wider frequency response obtainable from a transistor in the common-base circuit over its response in the common-emitter circuit is demonstrated. A performance comparison also appears in Fig. 10-13.

The band limit frequency f_α can be related to transistor design by a physical equation

$$f_\alpha = \frac{2L^2}{2\pi W^2 \tau} \tag{10-38}$$

where W is the base width, L the diffusion length, and τ the charge lifetime. Thus a thin base will give a wider band response.

10-10 THE RC-COUPLED AMPLIFIER WITH TRIODE AND PENTODE

The circuit of Fig. 10-15 is an example of a triode amplifier with RC coupling. The input capacitance of the following stage is indicated as C_{in} consisting of C_{gk} plus the Miller effect capacitance.

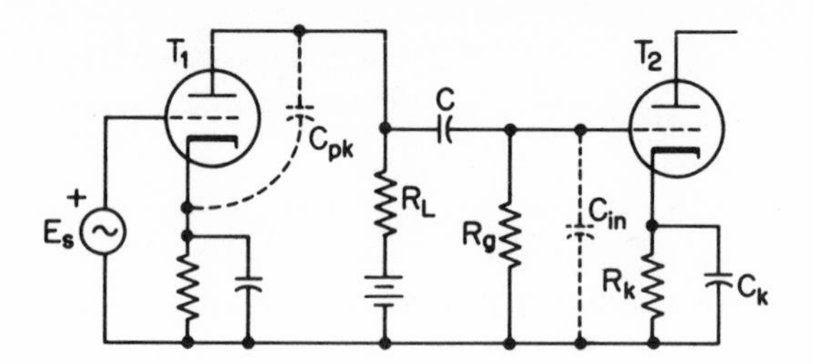

Figure 10-15. *RC* amplifier with triodes.

Following the previous method, C_k and C are effective in the low-frequency range, and the circuits of Fig. 10-16(b) follow. An effective load can be defined as

$$R'_L = \frac{r_p R_L}{r_p + R_L}$$

For pentodes this reduces simply to $R'_L = R_L$, because of the large value of r_p.

As for the transistor, the cathode-bias circuit introduces a voltage drop

$$\mathbf{V}_k = \frac{(\mu + 1)R_k\mathbf{I}_p}{1 + j\omega C_k R_k} \tag{10-39}$$

The grid leak R_g is present to provide a dc bias path to the grid and is ordinarily large with respect to R'_L, and so the C, R_g circuit provides a voltage reduction factor, but may not change appreciably the load seen by the tube. Following methods used for the transistor, the *low-frequency voltage gain* can be obtained as

$$A_{vL} = \frac{-g_m R'_L}{1 + R'_L/r_p}\left[\frac{1 + j\omega C_k R_k}{1 + \dfrac{(\mu+1)R_k}{r_p + R_k} + j\omega C_k R_k}\right]\left(\frac{j\omega C R_g}{1 + j\omega C R_g}\right) \tag{10-40}$$

Time constant parameters can be defined as

$$\omega_k = \frac{1}{C_k R_k}, \qquad \omega_C = \frac{1}{C R_g}$$

The low-frequency band limit appears when the real and reactive terms in the denominator of the first bracket are equal in magnitude. Then

$$f_1 = \frac{\omega_1}{2\pi} = f_k + \frac{\mu + 1}{2\pi C_k(r_p + R_k)} \tag{10-41}$$

This is a frequency considerably above that defined as $f_k = \omega_k/2\pi$.

The *low-frequency voltage gain* can be simplified to

$$\mathbf{A}_{vL} = \mathbf{A}_{vM}\left\{\frac{1 + j(\omega/\omega_k)}{1 + \dfrac{(\mu+1)R_k}{r_p + R_k} + j(\omega/\omega_k)}\right\}\left[\frac{j\omega/\omega_C}{1 + j(\omega/\omega_C)}\right] \tag{10-42}$$

since the mid-frequency gain can be written by making ω large in Eq. 10-40, as

$$\mathbf{A}_{vM} = \frac{-g_m R'_L}{1 + (R'_L/r_p)} \tag{10-43}$$

No assumptions have been made about the relative magnitudes of the time constants, and so the analysis holds for $\omega_k = \omega_C$. For this equal time constant case the capacitors should be selected for $C_k R_k = C R_g$. Circuit design follows by selection of a Q point, determination of R_k for the desired anode current and bias voltage, and choice of R_L to divide V_{bb} more or less equally between tube and load. Capacitor C_k can be selected to place f_1 sufficiently below the desired frequency pass band, followed by designation

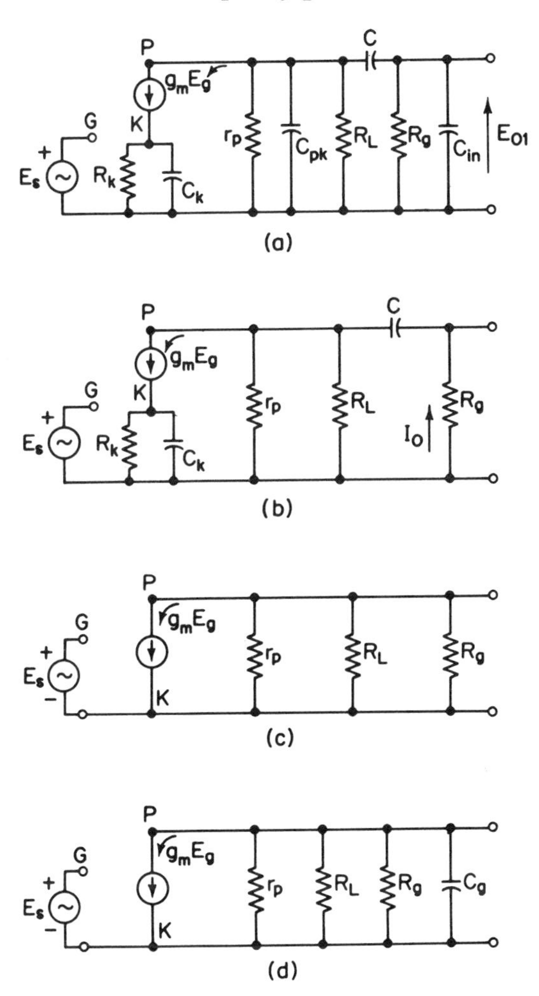

Figure 10-16. (a) Complete equivalent; (b) low-range equivalent; (c) mid-range equivalent; (d) high-range equivalent.

of C by use of the equality of time constants. For the pentode, C_s and R_s can be determined by procedures previously discussed.

In the high-frequency range, capacitances C_{pk} and C_{in} are in parallel and are called C_g as in Fig. 10-16(d). Pentodes are most usually employed in RC-coupled common-cathode amplifiers because C_g is small and the bandwidth can be greater than with the triode. Resistance r_p can be dropped as large, and the *high-frequency voltage gain* of the pentode written as

$$\mathbf{A}_{vH} = \frac{-g_m R_L}{1 + (R_L/R_g)}\left[\frac{1}{1 + j\omega C_g \dfrac{R_L R_g}{R_L + R_g}}\right] = \mathbf{A}_{vM}\left[\frac{1}{1 + j(\omega/\omega_2)}\right] \quad (10\text{-}44)$$

where

$$f_2 = \frac{\omega_2}{2\pi} = \frac{1}{2\pi C_g\left(\dfrac{R_L R_g}{R_L + R_g}\right)} \quad (10\text{-}45)$$

for the upper half-power frequency.

Response curves of the form of Fig. 10-10 may be plotted for both high- and low-frequency responses of the triode or pentode amplifier, using the corner-point techniques previously introduced. As an approximate bound, the mid-frequency gain expression of Eq. 10-43 may be considered as applying at frequencies above that at which $X_C \cong R_g/10$, and below that at which $X_{Cg} \cong 10R'_L R_g/(R'_L + R_g)$.

These expressions for the pentode and those for the common-emitter transistor amplifier are similar, since in both cases the equivalent circuit contains a transfer current generator shunted by a large internal resistance and a load. Both transistor and pentode have low-frequency band limits which are functions of circuit parameters, but the high-frequency half-power frequency is largely a function of the active device and its parasitic reactances.

10-11 EXTENSION OF THE LOW-FREQUENCY RANGE

It is possible to extend the f_2 frequency to very low values through appropriate choice of C and C_E in the RC amplifier. However, cost of C and C_E may require limitation of their magnitudes, and a circuit compensation method of extending the low-frequency range is available, as indicated in Fig. 10-17.

Resistor R_x is chosen to be large with respect to $1/\omega C_x$ down to the lowest frequency of interest, and so R_x is dropped from the equivalent circuit. With appropriate selection of C_x, the impedance of the R_1, C_x circuit can be made to rise with decreasing frequency, altering the current division ratio between $\mathbf{I}_1$ and $\mathbf{I}_2$ at such a rate as to hold $\mathbf{I}_2$ constant and independent of frequency, for the transistor case. For the pentode it is voltage $\mathbf{E}_o$ which is held constant, but the equations will be similar.

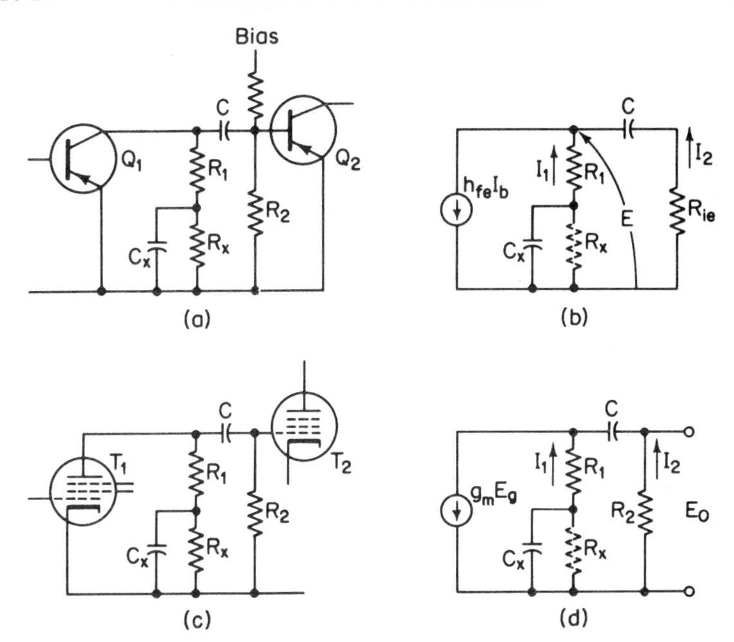

Figure 10-17. Low-frequency compensating network.

Using the current division factor, the current $\mathbf{I}_2$ can be found as

$$\mathbf{I}_2 = h_{fe}\mathbf{I}_{b1}\left[\frac{R_1 + (1/j\omega C_x)}{R_1 + (1/j\omega C_x) + R_{ie} + (1/j\omega C)}\right] \tag{10-46}$$

$$= h_{fe}\mathbf{I}_{b1}\left\{\frac{1}{1 + \dfrac{C(1 + j\omega C R_{ie})}{C_x(1 + j\omega C_x R_1)}}\right\} \tag{10-47}$$

If $\mathbf{I}_2$ is not to be a frequency function, then it is necessary that

$$1 + j\omega C R_{ie} = 1 + j\omega C_x R_1 \tag{10-48}$$

or

$$CR_{ie} = C_x R_1 \tag{10-49}$$

becomes the required circuit condition. The assumption that $R_x \gg 1/\omega C_x$ fails at some extremely low frequency, but the method does provide a means of extending f_1 to lower values than are possible with reasonable values of C in the uncompensated circuit. The equality of the time constants also produces zero phase shift, and this is of importance in some applications.

10-12 BANDWIDTH LIMITATIONS OF CASCADED AMPLIFIERS

The ratio of overall high-frequency gain to the mid-range gain for an amplifier of n identical stages may be written in dB as

$$\left|\frac{A_{\text{ov}}}{A_M}\right|, \text{dB} = -10\, n \log\left[1 + \left(\frac{\omega}{\omega_2}\right)^2\right] \tag{10-50}$$

Angular frequency ω_2 is the same for all stages.

An angular frequency $\omega_o = 2\pi f_o$, is then defined as the half-power or −3 dB frequency for the n-stage amplifier, overall, and so Eq. 10-50 is

$$-3.0 = -10\, n \log\left[1 + \left(\frac{\omega_o}{\omega_2}\right)^2\right]$$

from which the required ω_2 for each stage of the amplifier can be written in terms of the specified overall bandwidth as

$$\omega_2 = \frac{\omega_o}{(10^{0.3n} - 1)^{1/2}} = \frac{\omega_o}{(2^{1/n} - 1)^{1/2}} \tag{10-51}$$

The resultant band narrowing in multi-stage amplifiers has already been indicated in Fig. 10-10. Table 10-2 gives a specific figure for the bandwidth reduction in cascaded amplifiers. If the overall band is specified, each stage must be designed for a considerably wider bandwidth.

Table 10-2

RELATIVE CUTOFF FREQUENCIES FOR CASCADED AMPLIFIERS

ω_o = cutoff for overall amplifier, ω_2 = cutoff for one stage

n	ω_2/ω_o	n	ω_2/ω_o
1	1.00	5	2.60
2	1.56	6	2.86
3	1.96	7	3.10
4	2.30	8	3.25

The defined gain-bandwidth product for a transistor in the C-E circuit can be written from its parameters as

$$\text{transistor GBW} = f_T = \frac{g_m}{2\pi(C_{be} + C_{bc})} \tag{10-52}$$

For a pentode the gain-bandwidth product can be written by multiplying the RC-coupled amplifier gain and the f_2 frequency, giving

$$\text{pentode GBW} = \frac{g_m}{2\pi(C_{\text{in}} + C_{pk})} \cong \frac{g_m}{2\pi(C_{gk} + C_{pk})} \tag{10-53}$$

This relation is similar to that for the transistor.

As stages are cascaded, the individual stage f_2 values must be raised, and this requires reduced R_L values and lowers the stage gain. It is not surprising that there is a trade-off between gain and bandwidth and that the GBW product is independent of load value for a given transistor or pentode.

High values of gain–bandwidth product are associated with transistors or tubes having high g_m values and low internal capacitances. The designer looks first for these parameters in selecting a suitable type for a given band.

10-13 AN EQUIVALENT CIRCUIT FOR THE FIELD-EFFECT TRANSISTOR

The theory of the field-effect transistor or FET was given in Section 6-12, and a family of output characteristics is drawn in Fig. 10-18. In the region between the origin and point A, at low source-drain voltages, the current is a linear function of applied voltage, and the FET is a useful voltage-controlled variable resistor. In the "pinch-off" region to the right of A, the channel is partially closed by the gate voltage. The characteristics then parallel those of the pentode, and the device is a useful high-input impedance amplifier.

The output resistance r_d is high, but subject to control in device design in the range from about 10,000 ohms to several megohms. The output capacitance C_{ds} is about 1 pF, to which should be added another picofarad for the inherent capacitance to the case. The input resistance r_{gs}, being that of a reverse-biased diode in the FET, is of the order of 10^8 to 10^{10} ohms. It repre-

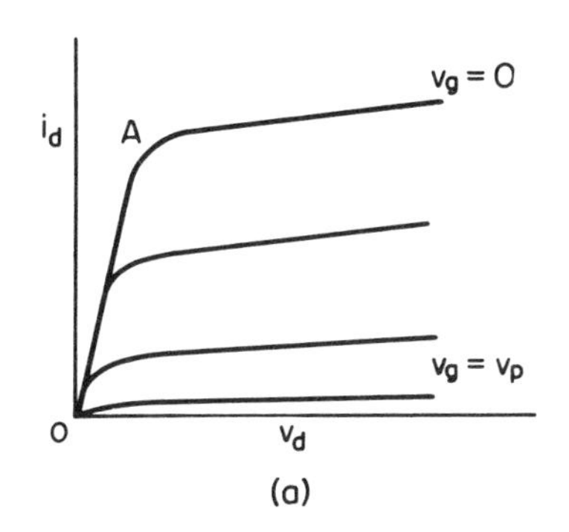

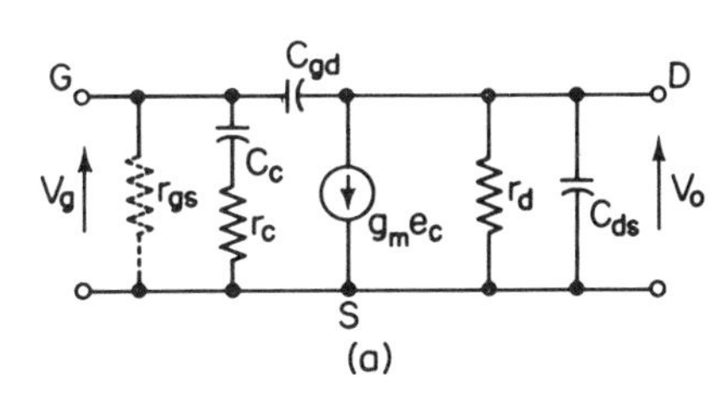

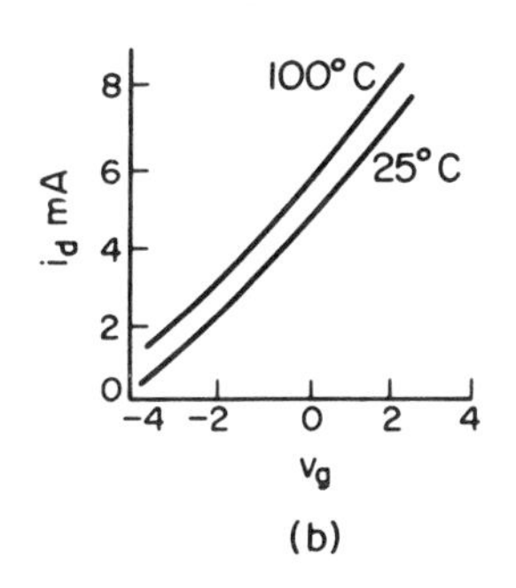

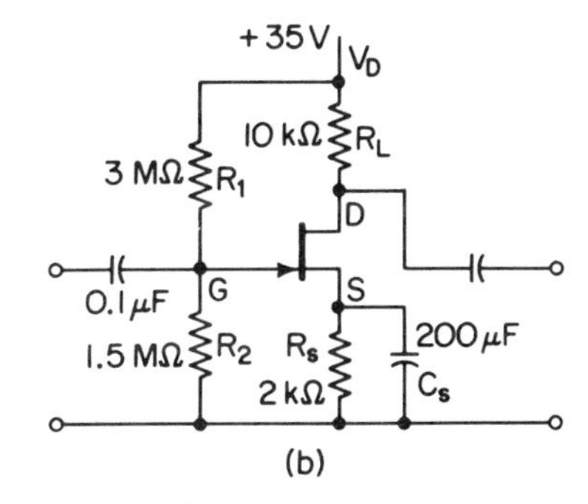

Figure 10-18. (a) MOSFET output characteristic; (b) transfer curve.

Figure 10-19. (a) MOSFET equivalent circuit; (b) grounded-source practical circuit.

sents the leakage resistance of the silicon-dioxide insulating layer in the insulated-gate (MOSFET) type of transistor and is then of the order of 10^{12} to 10^{17} ohms. The input capacitance C_c represents the effective distributed capacitance between the gate electrode and the channel. It may approximate 4 pF, and has a small loss component in series. Capacitance C_c and resistance r_c provide the high-frequency limiting factors of the device. The feedback capacitance C_{gd} is the capacitance between gate and drain; this will usually be only a small fraction of a picofarad, and can be neglected for the lower frequencies.

The device is subject to temperature variation, and this is illustrated for the transfer characteristic in Fig. 10-18(b).

10-14 OPERATION OF THE FIELD-EFFECT TRANSISTOR

The most common use of the FET is in the common-source circuit of Fig. 10-19(b). Its performance has the high input impedance and medium to high output impedance of a pentode. Neglecting the bias resistances R_1 and R_2 as large, and using the equivalent circuit of Fig. 10-19(a), the *mid-range voltage gain* can be found to be

$$\mathbf{A}_v = -\frac{g_m R_L}{1 + R_L/r_d} \tag{10-54}$$

The load may be larger or smaller than r_d. Low-frequency and high-frequency relations follow from the equations developed for the pentode.

The common-drain or *source-follower* circuit of Fig. 10-20 is similar to the cathode follower. The input impedance can be made very high, and the output impedance is low. The voltage gain is, of course, less than unity. Mid-range operating relations follow from the triode treatment, and are

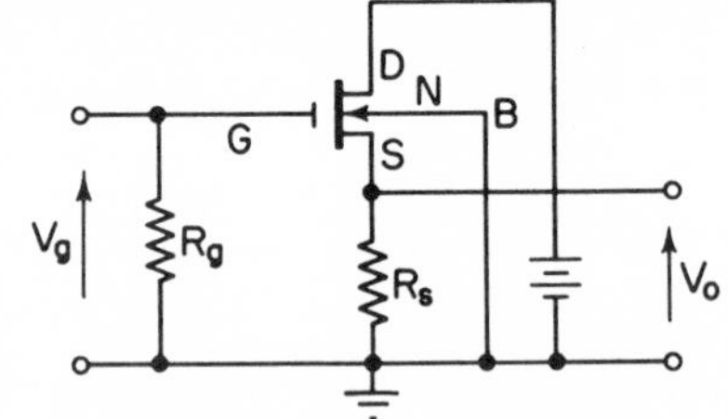

Figure 10-20. Source-follower circuit.

$$\mathbf{A}_{vM} = \frac{\mu R_s}{r_d + (\mu + 1)R_s} \simeq \frac{g_m R_s}{1 + g_m R_s} \tag{10-55}$$

The latter expression assumes $\mu \gg 1$.

The input capacitance is reduced from that of the common-source circuit, as

$$C_{\text{in}} = C_{gd} + (1 - A_{vM})C_{gs} \tag{10-56}$$

and the output resistance R_o is

$$R_o = \frac{R_s[r_d/(\mu + 1)]}{R_s + r_d/(\mu + 1)} \tag{10-57}$$

This result indicates the output load R_s shunted by the effective internal resistance $r_d/(\mu + 1)$, as was the case for the cathode follower.

With $g_m = 2500$ μmho, $R_s = 2000$ ohms, $r_d = 10{,}000$ ohms, the voltage gain is 0.83 and the output resistance is 325 ohms.

10-15 NOISE OR RANDOM CURRENTS IN AMPLIFIERS

Random variations in the current output of electronic devices are called *noise*. These fluctuations are due to the inherent particle nature of electrical conduction, and to the statistical variations in path, recombination, diffusion and other charge-flow processes. At very low signal levels the passing of a signal into an amplifier becomes analogous to the pouring of a number of marbles into a pipe, rather than to a stream of water in the usual hydraulic analogy. Other marbles already present in the pipe have random velocities and these random velocities tend to obscure the directed movements given by the input signal energies. Noise places a basic lower limit on the signal levels at which a device or an amplifier may operate, although special techniques exist for detecting a signal at very low signal-to-noise ratios, and include methods which trade increased time of transmission for more information received.

In radio reception, noise is noticed as a hiss accompanying a very weak signal; in television, noise appears as "snow" in the picture produced by a weak signal. Noise which is generated external to the system usually limits radio reception at frequencies below about 20 megahertz, but above that frequency the internally developed system noise usually limits sensitivity.

Noise has no specific frequency, being distributed over broad frequency bands. Each impact of charges or atoms produces a very short energy pulse which, if analyzed, consists of a broad distribution of frequencies. Being random in nature, this distribution has an effective power P_n, with effective current $\sqrt{\overline{i_n^2}}$ and voltage $\sqrt{\overline{v_n^2}}$. Measurements, to be meaningful, must be made over specified frequency bands on a power or voltage-squared basis, between the half-power frequencies. The power can be stated as $P_n/\Delta f$, or effective power per hertz at a center frequency. When noise is uniformly distributed over all frequencies, it is said to constitute *white noise*, analogous to white visible light which contains all frequencies or colors.

For use in comparison of the noise output of devices or circuits, a *noise figure*, N.F., is defined and usually stated in dB. This noise figure is the ratio of the signal-to-noise power P_{ss}/P_{sn} at the circuit input to the signal-to-noise power at the output of the device, P_{os}/P_{on}. That is,

$$\text{N.F.} = \frac{P_{ss}/P_{sn}}{P_{os}/P_{on}} \quad \text{dB} \qquad (10\text{-}58)$$

This figure indicates the degradation of the signal-to-noise ratio by the device under study. If the input is a voltage source with resistance R, the noise figure of an amplifier may also be stated as

$$\text{N.F.} = \frac{P_{n\,\text{out}}}{P_{n\,\text{in}}G} \tag{10-59}$$

where G is the power gain of the amplifier and P_n the noise power of the source.

Noise in electronic circuits and devices is created by several phenomena:

Thermal noise is developed by the random thermal motions and impacts of the charges and atoms in all conductors. Since it is due to thermal energies, this noise is temperature dependent. It is also found to be uniformly distributed over all frequencies presently in use. The thermal noise due to a resistance R is

$$E_{\text{noise}} = \sqrt{4kTR\,\Delta f} \tag{10-60}$$

where k is Boltzmann's constant $= 1.38 \times 10^{-23}$ joule/°K, T is temperature in °K, and Δf is the frequency band over which the power is measured.

As shown in Fig. 10-21, a thermal noise source may be considered equivalent to a noise generator of voltage E_{noise} in series with noiseless resistor R. The maximum power would be transferred to a matched load of value R, and so

$$P_{\max} = \frac{E^2_{\text{noise}}}{4R} = kT\,\Delta f \tag{10-61}$$

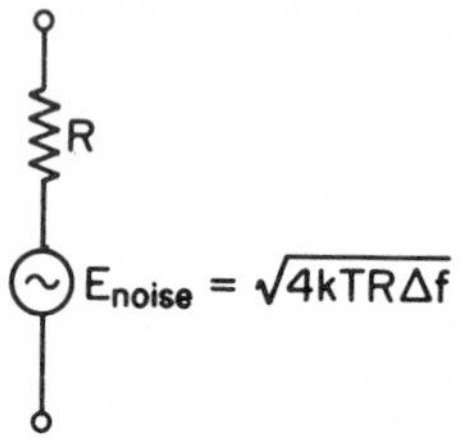

The noise power is dependent on the frequency band accepted; wide frequency response gives inherently more noise input.

Figure 10-21. Equivalent noise source for thermal noise.

For a bandwidth of 4 megahertz at 300°K ambient, the thermal noise power present in a circuit is 1.6×10^{-14} W, independent of the resistance. This power is that produced by a signal of 1.26 μV across 100 ohms.

Lowering of the operating temperature of the input circuit of a device is a means of reducing the thermal noise. This method is applied in reception of satellite and space signals, some apparatus temperatures being maintained at that of liquid helium.

Shot noise results from the random nature of charge diffusion across a junction, or of electron emission from a cathode. This noise is uniformly distributed over the frequency spectrum. The noise energy increases with the total current being transferred in the device.

Flicker or 1/f noise seems due to random variations in the emission or diffusion processes, involved with changes in surface characteristics. It seems confined largely to the region below 1000 hertz, and varies inversely with frequency. In transistors it is subject to reduction by careful processing.

In vacuum tubes the effects of flicker and shot noise are reducible by space-charge operation which gives a reservoir of electrons near the cathode, the emission is thus averaged over a period of time.

There are additional noise sources present in tubes:

Partition noise is caused by the irregular division of current between two or more positive electrodes. That is, it is arbitrary whether a particular electron goes to the screen or anode, although the time average is constant. Partition noise makes pentodes noisier than triodes.

Induced grid noise is present because of random space charge fluctuations which induce voltages on the grid. This noise is appreciable above 30 megahertz.

For comparison of tubes on a noise basis, it is usual to consider a tube as equivalent to a resistor R_{eq}, which, if inserted in series with the input of a noise-free tube, would generate the same noise current in the output as is generated by the actual tube. For triodes, this noise equivalent resistor may be approximated as

$$R_{eq} \cong \frac{2.5}{g_m} \quad \text{ohms} \quad (10\text{-}62)$$

where the noise is predominantly shot noise. For pentodes the partition noise also appears, and the equivalent resistor becomes the sum of a shot-noise resistor and a partition-noise resistor, as

$$R_{eq} \cong \frac{2.5}{g_m} + \frac{20I_s}{g_m I_k} \quad \text{ohms} \quad (10\text{-}63)$$

where I_s = screen current, and I_k = cathode current.

Typical equivalent resistance values range from 200 ohms for a 6CW4 triode of very high g_m, to as much as 1800 ohms for the 6AK5 pentode. These figures will vary between tubes of the same type, probably due to accidents of manufacture, and so can only be used as guides in tube selection.

Transistor noise figures as low as 2 or 3 dB are possible. The actual value depends on the operating point and is reduced by low current levels or by low voltages. The value of I_{CO} should be as low as possible, and r'_b should be small. The C-E and C-B circuits are generally less noisy than the C-C circuit.

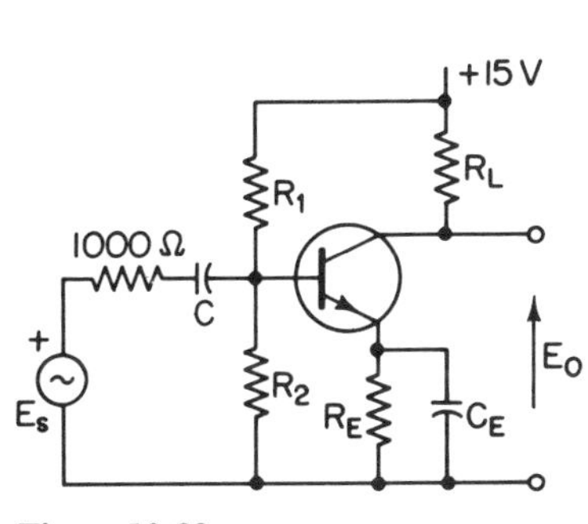

Figure 10-22.

PROBLEMS

10-1. In the circuit of Fig. 10-22, $R_L = 4000$ ohms, $R_E = 1000$ ohms, $R_1 = 60{,}000$ ohms, $R_2 = 30{,}000$ ohms, and the transistor has $h_{ie} = 1200$ ohms, $h_{re} = 2.5 \times 10^{-4}$, $h_{fe} = 100$, $h_{oe} = 50$ μmho, and $C_1 = 5$ μF. Determine C_E, if f_1 is to be 40 hertz.

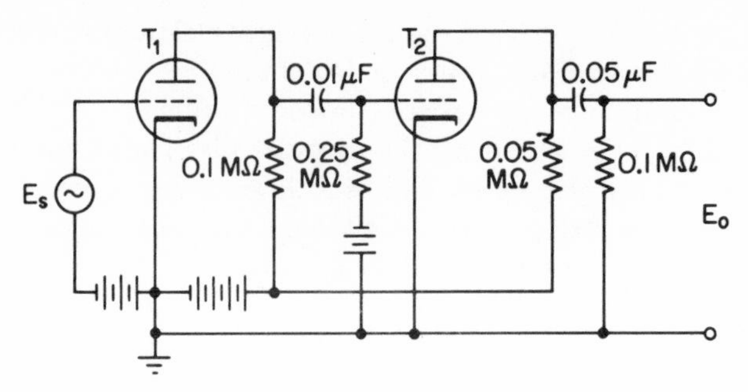

Figure 10-23.

10-2. For the first-stage circuit of Fig. 10-23, if $g_m = 2500$ μmhos, $r_p = 12{,}000$ ohms, C_{in} of the second stage $= 70$ pF, and C_{pk} of $T_1 = 10$ pF, find the bandwidth $f_2 - f_1$ in hertz.

10-3. Tube T_1 in Fig. 10-23 has $\mu = 35$, $r_p = 25{,}000$ ohms. Find the needed value of the coupling capacitor of the first stage if the voltage gain at 20 hertz is to be 80 per cent of the mid-range gain. Plot the voltage gain vs. frequency curve over the mid- and low-frequency ranges.

10-4. The circuit of Fig. 10-24 is used with two transistors of the type specified in Problem 10-1. The load of the second stage is 5000 ohms, and $R_1 = R_2 = 10{,}000$ ohms. With $C = 5$ μF, determine the overall current gain at the mid-range, and the low-frequency band limit.

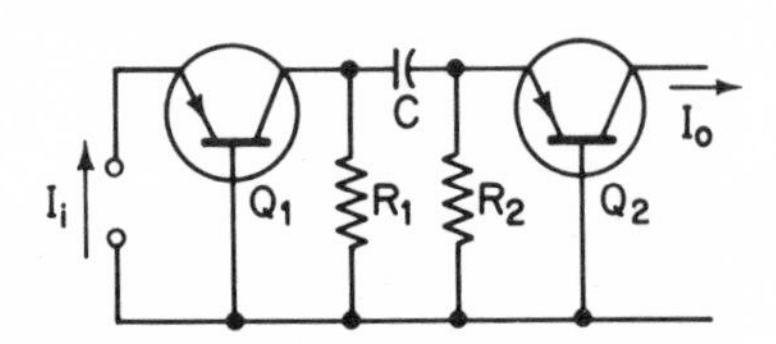

Figure 10-24.

10-5. With $\beta_0 = 80$, $C_{cb} = 1$ pF, $C_{be} = 58$ pF, $h_{ie} = 50$ ohms, find f_β and f_T of the transistor.

10-6. The transistor of Fig. 10-22 has $h_{ie} = 580$ ohms, $h_{fe} = 40$, $h_{re} = 1.5 \times 10^{-4}$, $h_{oe} = 3.0 \times 10^{-5}$ mho. With $R_L = 5000$ ohms, $R_s = 10{,}000$ ohms, and $R_E = 1000$ ohms, $C = C_E/10$, find C_E to make $f_1 = 60$ hertz, and plot A_{iL}/A_{iM} in dB in the low-frequency and mid-range regions.

10-7. Calculate the bandwidth of the amplifier of Fig. 10-24, if $C = 2$ μF, $R_1 = 5000$ ohms, $R_2 = 20{,}000$ ohms, the transistors have the characteristics of Problem 10-6, and $C_{cb} = 6$ pF, $C_{be} = 600$ pF. Assume the input source is 4000 ohms resistive, and the output load is 2000 ohms resistive.

10-8. Compute the current gain in the C-B circuit at 100,000 hertz, for a transistor having $r_e = 50$ ohms, $r_b = 400$ ohms, $r_c = 0.75$ megohms, $\alpha = 0.95$, $C_{be} = 150$ pF, and $C_{cb} = 15$ pF, with $R_L = 5000$ ohms, $R_s = 2000$ ohms.

10-9. Repeat Problem 10-8 using the C-E circuit. Explain the difference in performance, if any.

10-10. Specify the f_1 and f_2 frequencies for each stage of a four identical stage amplifier, if the overall bandwidth is to extend from 100 hertz to 450,000 hertz.

10-11. A given amplifier has an f_2 value of 30,000 hertz. At what frequency is the amplifier voltage gain down only five per cent of its mid-range value?

10-12. In the circuit of Fig. 10-4, the two transistors are identical, with $\alpha = 0.96$, $r_c = 2$ megohms, $r_b = 500$ ohms, $r_e = 40$ ohms, $R_L = 7000$ ohms, and R_1 and R_2 large, $C = 1\ \mu$F. The second stage has a load of 5000 ohms, and the input source has a resistance of 5000 ohms. With $R_E = 500$ ohms, $C_E = 50\ \mu$F, find the frequency f_1.

10-13. A transistor has mid-range $\alpha_o = 0.97$ and $f_\alpha = 0.45$ megahertz. Find the magnitude of α at 0.70 megahertz.

10-14. Each stage of a three-stage amplifier has $A_v = 30$ and an f_2 value of 400 kHz. The complete amplifier is to be modified to make the overall $f_2 =$ 600 kHz by reducing the stage loads of the pentodes used. With g_m of each pentode $= 2500\ \mu$mho, what will be the new loads and the new stage gains?

10-15. An instrument amplifier must have a current gain constant within 2.5 per cent up to 10,000 hertz. What f_2 value must the amplifier have? If composed of three identical stages, what individual stage f_2 value will be required?

10-16. Plot the high-frequency gain A_H/A_M for a six-stage amplifier with $f_2 =$ 1.0 MHz for each stage.

10-17. Three nonidentical amplifier stages are cascaded. The individual bandwidth limits are

	f_1, Hz	f_2, Hz
A	100	252,000
B	82	347,000
C	47	545,000

Determine the overall bandwidth limits.

10-18. A MOSFET amplifier has $g_m = 2000\ \mu$mho, and $r_d = 15{,}000$ ohms.
(a) Design a circuit for a source-follower to match a load of 250 ohms;
(b) What voltage gain will be obtained?

10-19. Using the transistor parameters of Problem 10-8, calculate the transistor input capacitance in the circuit of Fig. 10-22, with $R_L = 5000$ ohms.

10-20. The current-gain bandwidth product f_T for a transistor is 120×10^6. With mid-frequency $h_{fe} = 50$, $h_{ie} = 2500$ ohms, total load of 2000 ohms, determine the gain magnitude to be expected at the video amplifier band limit of 4.5 MHz.

10-21. An amplifier for a laboratory voltmeter is advertised as accurate to 95 per cent of its low-frequency gain at 100 kHz.

(a) Having two identical pentode stages, what is the stage f_2 value?

(b) A pentode having $C_{gk} = 10$ pF, $C_{pk} = 4$ pF, and $g_m = 5000$ μmhos is being considered for use in each stage of this amplifier. Will it be possible to obtain a per-stage voltage gain of 40?

REFERENCES

1. Thomas, D. E., "Transistor Cut-Off Frequency," *Proc. IRE*, Vol. 40, p. 1481 (1952).
2. Angelo, E. J., Jr., *Electronic Circuits*. McGraw-Hill Book Company, New York, 1964.
3. *Transistor Manual*, 7th ed. General Electric Company, Syracuse, N. Y., 1964.
4. Shea, R. F., *Transistor Applications*. John Wiley & Sons, Inc., New York, 1964.
5. Searle, C. L., *et al.*, *Elementary Circuit Properties of Transistors* (SEEC, Vol. 3). John Wiley & Sons, Inc., New York, 1964.
6. Glasford, G. M., *Linear Analysis of Electronic Circuits*. Addison-Wesley Pub. Co., Reading, Mass., 1965.
7. Wallmark, J. T., and H. Johnson, *Field-Effect Transistors*. Prentice-Hall, Inc., Englewood Cliffs, N. J., 1966.
8. Todd, C. D., *Junction Field-Effect Transistors*. John Wiley & Sons, Inc., New York, 1968.

11

WIDE-BAND AND FREQUENCY SELECTIVE AMPLIFIERS

Electronic devices are often called upon to amplify wave forms which include many frequencies distributed across a broad spectrum, as well as to selectively amplify bands of frequencies. The coupling circuits, in the first case, must provide uniform gain over a wide frequency spectrum, and in the second case must be designed to discriminate between wanted and unwanted frequencies.

The input wave forms represent information, as the audio frequencies of speech or music, the picture elements in television, or the components of a received pulse in radar, and alteration of amplitude or change of phase represents loss of some of the transmitted information. This may be a serious matter in the operation of communication systems.

While building on our study of the high-frequency properties of transistor or tube amplifiers, this chapter will be concerned more with frequency-responsive or frequency-selective networks than with the active devices themselves. Thus it will contribute to the understanding of communication systems, as distinct from a study of electronic devices.

11-1 BANDWIDTH REQUIREMENTS FOR VIDEO AMPLIFIERS

A type of wide-band frequency response needed for the amplification of pulse signals is indicated in Fig. 11-1, as the response of a *video amplifier*. These amplifiers are designed to have a continuous response extending from near zero frequency to as much as 10 or 20 megahertz, thus being essentially wide-band low-pass amplifiers. As used for the amplification of television signals or pulse wave forms, the lowest frequency is that of the repetition rate of the signal. This is the 30 hertz picture rate in television, or the 6 or 8 kilohertz rate required in some pulse communications. The highest frequency passed should include all useful signal components, extending to 4.5 megahertz in television and to higher frequencies for pulse amplifiers.

While the basic design principles for the video amplifier are those of Chapter 10, circuits will here be developed to overcome some of the frequency limitations imposed by the internal capacitances of transistor or tube.

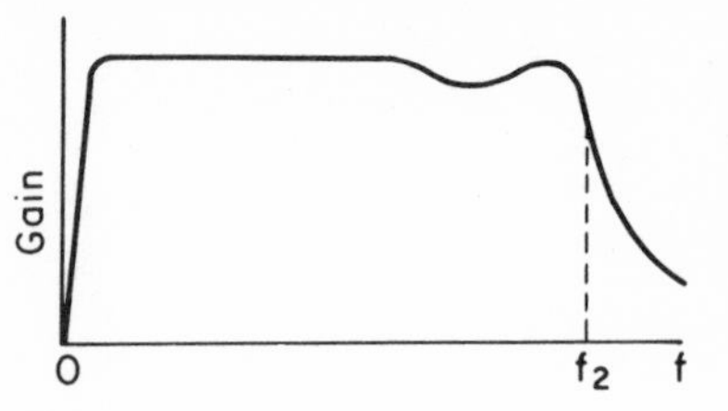

Figure 11-1. Response of a video amplifier.

Previously we analyzed such circuits in the *frequency domain*, resulting in figures for the mid-range gain, the lower and upper cutoff frequencies, a knowledge of phase shifts, and a graph of the frequency response. In this chapter it will at times be more suitable to study the circuit in the *time domain*, since video amplifiers usually have pulse inputs, and it is the transient response to such nonsinusoidal inputs in which we are interested. Therefore, the analysis will lead to performance figures for the rise time of the wave front, the overshoot of the wave in reaching its maximum value, and the sag of the top of the pulse. These results from time-domain analysis will be related to the performance indices derived in the frequency domain in Chapter 10.

11-2 THE STEP FUNCTION AND PULSE RESPONSE

The ideal pulse wave form consists of a unit *step function* at the initiation of the pulse, and a negative step function at the end of the pulse. When analyzed by Fourier methods, a step function is found to include all frequencies, and since no amplifier has an infinite response band, then an amplified pulse will not be a faithful reproduction of an input step.

An input pulse and the effects of distortion are shown in Fig. 11-2.

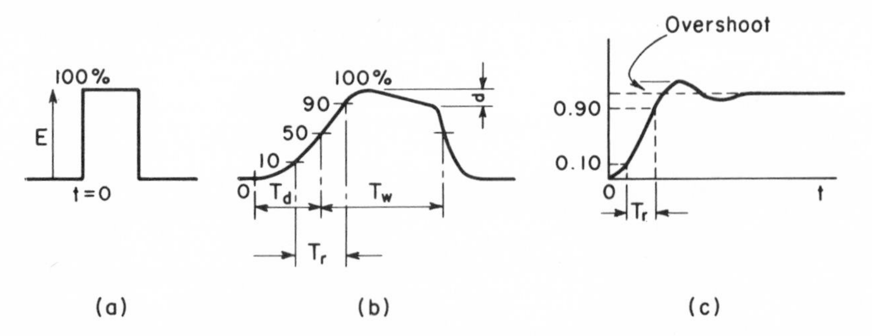

Figure 11-2. Definitions of pulse characteristics: (a) ideal pulse; (b) non-oscillatory response; (c) overshoot present.

Time $t = 0$ is that time at which the unit step function is applied. The output does not immediately start to rise, due to *delay* in the amplifier circuits. Measurement of time delay T_d is difficult because it involves subjective determination of the time at which the amplitude departs from the base line on an oscilloscope trace. Thus the delay is arbitrarily defined as the time for the pulse to rise to 50 per cent of final value, starting from $t = 0$.

As shown in the figure, the *rise time* of practical amplifiers is always finite, because of amplifier frequency distortion. Again because of measurement difficulties, the *rise time* T_r is defined as the time of rise of the leading edge of the output between 10 per cent and 90 per cent of the final amplitude.

The output may *overshoot*, as shown in Fig. 11-2(c). The magnitude of overshoot is designated γ and stated in per cent of the pulse amplitude. Overshoot is usually due to inclusion of inductive circuit elements, producing a second-order response.

While the width or duration of the input pulse is definite, the duration of the output pulse T_w is usually defined as the time between 50 per cent points on the rising and falling edges of the output pulse.

In a step function, the output amplitude remains indefinitely at unity, and no ac coupled amplifier can provide this output. Thus the output pulse top *tilts* or *sags* with time, and the amount of tilt or sag is stated in per cent of the maximum amplitude.

11-3 BANDWIDTH REQUIREMENTS FOR PULSE AMPLIFICATION

A general pulse train is shown in Fig. 11-3. The period of the wave is $2\pi/\omega_R$ or $T_R = 1/f_R$, where f_R is called the *pulse repetition frequency*. Such a pulse train may be interpreted on a frequency basis by use of a Fourier series. Amplifier bandwidth can then be specified so as to provide gain for all frequencies of appreciable amplitude.

A Fourier representation can be stated as

$$f(t) = A_o + \sum_{k=1}^{\infty} a_k \cos k\omega_R t + \sum_{k=1}^{\infty} b_k \sin k\omega_R t \tag{11-1}$$

By the usual Fourier methods the constant or average term is

$$A_o = \frac{\omega_R}{2\pi} \int_{-\pi/\omega_R}^{\pi/\omega_R} f(t)\, dt \tag{11-2}$$

and the coefficients are

$$a_k = \frac{\omega_R}{2\pi} \int_{-\pi/\omega_R}^{\pi/\omega_R} f(t) \cos k\omega_R t\, dt \tag{11-3}$$

$$b_k = \frac{\omega_R}{2\pi} \int_{-\pi/\omega_R}^{\pi/\omega_R} f(t) \sin k\omega_R t\, dt \tag{11-4}$$

If $f(t)$ is an odd function, then the a_k coefficients are zero, and the Fourier series for an odd function includes only sine terms. Conversely, if $f(t)$ is even, then $b_k = 0$, and the series will contain only cosine terms. The constant term may be present with either result.

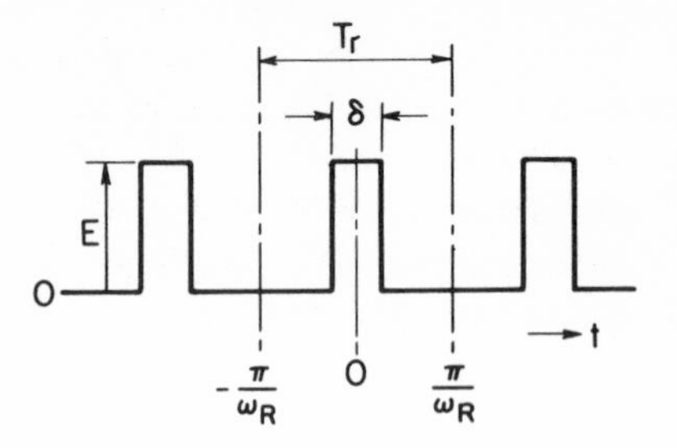

Figure 11-3. Recurrent rectangular pulses.

This may be demonstrated for the wave of Fig. 11-3, with the origin located to yield even symmetry, namely $f(x) = f(-x)$. Noting that $f(t) = E$ inside the pulse interval given by $t = \pm\delta/2$, and $f(t) = 0$ elsewhere in the period, then

$$A_o = \frac{\omega_R}{2\pi}\int_{-\delta/2}^{\delta/2} E\,dt = \frac{\delta}{T_R}E = f_R\,\delta E \tag{11-5}$$

The a_k coefficients are

$$a_k = 2f_R\int_{-\delta/2}^{\delta/2} E\cos 2\pi f_R kt\,dt = 2Ef_R\delta\frac{\sin(k\pi f_R\delta)}{k\pi f_R\delta} \tag{11-6}$$

Thus we have determined the amplitude of the kth harmonic for a series of pulses of duration δ and repetition frequency f_R.

The harmonic amplitudes are functions of $(\sin x)/x$, where $x = k\pi f_R\delta$. This relation appears in Fig. 11-4, as the envelope of the harmonic amplitudes.

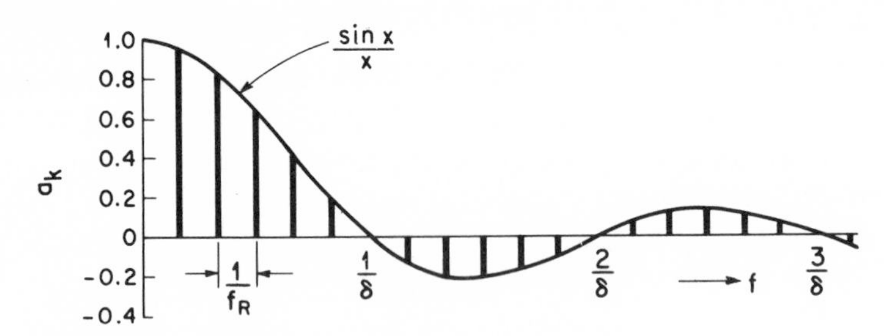

Figure 11-4. Distribution of harmonic amplitudes of a pulse chain.

By setting Eq. 11-6 to zero, the crossover frequencies can be found as

$$kf_R = \frac{n}{\delta} \tag{11-7}$$

Thus the loop crossovers occur at frequencies $1/\delta$, $2/\delta$, $3/\delta$, . . . and the frequency width of a loop increases as the pulse narrows.

The number of harmonic frequencies in each loop is $k = 1/f_R\delta$. For a pulse of 1 μs duration, the first zero is at 1 megahertz, the second at 2 megahertz, etc. If the repetition frequency is 1000 per second, there will be 1000 frequencies in each loop of the envelope.

As the repetition rate is reduced, more harmonic frequencies appear in each loop of the envelope, and in the limit where only a single pulse is transmitted, or $f_R = 0$, a continuous spectrum of frequencies appears. This spectrum is predicted by the Fourier integral, and for the even pulses of trainthe under discussion, the spectrum is given by

$$S(\omega) = \frac{E\delta}{2\pi}\left[\frac{\sin(\omega\delta/2)}{\omega\delta/2}\right] \tag{11-8}$$

Since this is also a function of $(\sin x)/x$, the spectrum of harmonic amplitudes for the single pulse is bounded by the envelope curve of Fig. 11-4.

The fidelity of pulse reproduction will be dependent on the number of harmonics accepted in the pass band of the amplifier. It is found that if the pass band includes the fourth zero of the signal envelope, the pulse reproduction may be considered excellent. Reasonably accurate reproduction of the pulse can be obtained with a pass band including only the first zero at a frequency of $1/\delta$.

11-4 THE POLE-ZERO DIAGRAM

It should now be apparent that the gain functions of amplifiers can be expressed as the ratio of two polynomials in $j\omega$, multiplied by a scale factor. As an example

$$A(\omega) = K_1 \frac{(j\omega + \omega_1)(j\omega + \omega_3) + \cdots}{(j\omega + \omega_2)(j\omega + \omega_4) + \cdots} \tag{11-9}$$

where $\omega_1 \ldots \omega_n$ are the cutoff frequencies or corner points previously utilized in the study of frequency response of amplifiers.

Further information about amplifier performance may be developed if we substitute the variable s for $j\omega$, where $s = \sigma + j\omega$. The response then obtained is that of a circuit excited by an exponential input $i = I\epsilon^{st}$. The portion of the response due to $\epsilon^{\sigma t}$ provides information on transient output, decaying with time for negative σ, while that due to $\epsilon^{j\omega t} = \cos\omega t + j\sin\omega t$ provides the desired knowledge of steady-state sinusoidal frequency response. The use of s may be extended to the writing of the circuit differential equation, transforming that from a time function into an algebraic expression in the complex variable s.

For example, consider the simple RLC circuit of Fig. 11-5. The impedance is

$$\frac{E}{I} = \frac{1}{C}\,\frac{j\omega + (R/L)}{(j\omega)^2 + j\omega(R/L) + 1/LC} \tag{11-10}$$

Substitution of $s = j\omega$ and factoring of the denominator leads to the transfer function in the general form illustrated by Eq. 11-9:

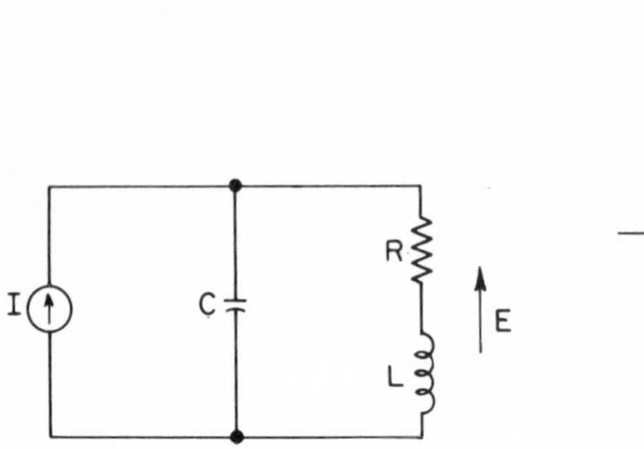

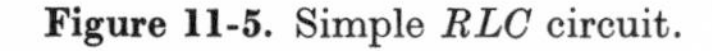

Figure 11-5. Simple *RLC* circuit.

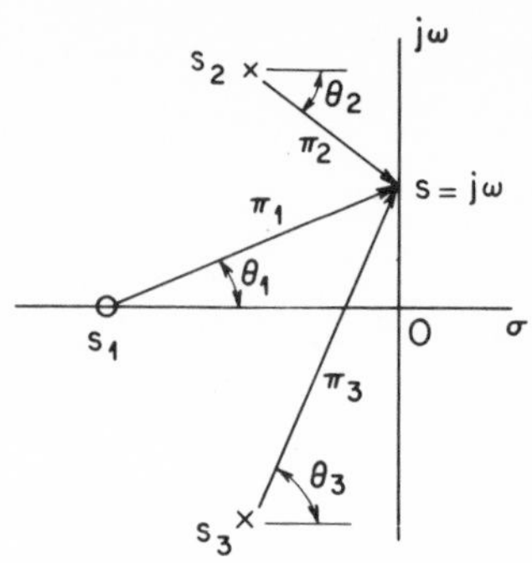

Figure 11-6. Pole-zero diagram for Eq. 11-11.

$$H(s) = \frac{1}{C}\frac{s - s_1}{(s - s_2)(s - s_3)} \tag{11-11}$$

where

$$s_1 = \frac{-R}{L}, \qquad s_2, s_3 = -\frac{R}{2L} \pm \sqrt{\left(\frac{R}{2L}\right)^2 - \frac{1}{LC}}$$

The response then has a numerator zero at $s = s_1$, and when $s = s_2$ or $s = s_3$ the denominator has zeros and the transfer function has poles. Complex poles and zeros must always occur in conjugate pairs, as for s_2 and s_3 above, since the expression for $H(s)$ is real.

A geometric interpretation for Eq. 11-11 follows if the poles and zeros are plotted on the complex plane as in Fig. 11-6, with the zero at $s_1 = -R/L$ indicated by a circle on the negative σ axis, and the poles indicated by crosses. Rewriting Eq. 11-11 as

$$H(\omega) = K_2\frac{j\omega - s_1}{(j\omega - s_2)(j\omega - s_3)} \tag{11-12}$$

it can be seen that the function is represented by a scale factor K_2 and a quotient of vector distances on the complex plane. The magnitude $|j\omega - s_1|$ is measured from s_1 to the value of ω on the j axis, the angle of the vector being indicated as θ_1 in Fig. 11-6. That is,

$$|H(\omega)| = K_3\left(\frac{\pi_1}{\pi_2\pi_3}\right) \tag{11-13}$$

where $\pi_1 = |j\omega - s_1|$, $\pi_2 = |j\omega - s_2|$, $\pi_3 = |j\omega - s_3|$ as measured on the diagram. Likewise

$$\theta(\omega) = \theta_1 - \theta_2 - \theta_3 \tag{11-14}$$

completing the determination of $H(\omega)$. The point ω traverses all points on the $j\omega$ axis, and the diagram permits visualization of the changes of $H(\omega)$ with frequency.

As an example employing a familiar relation, consider a general *RC*-

coupled amplifier gain expression, which can be formed from past results as

$$A_v(\omega) = K_4\left(\frac{1}{j\omega + \omega_2}\right)\left(\frac{j\omega}{j\omega + \omega_1}\right) = K_4\left(\frac{1}{s + s_2}\right)\left(\frac{s}{s + s_1}\right) \quad (11\text{-}15)$$

where $s_2 = \omega_2$, and $s_1 = \omega_1$, as the cutoff angular frequencies. It is here assumed that the low-frequency cutoff is determined by only one RC time constant, for reasons of simplicity.

The function thus has a numerator zero at the origin, and poles due to the denominator terms at $s_2 = -\omega_2$ and $s_1 = -\omega_1$ on the negative real axis, and this arrangement appears in Fig. 11-7.

Figure 11-7. Pole-zero diagram for the low-frequency range of an amplifier.

The ω point moves from the origin at zero frequency, up the positive $j\omega$ axis, reaching infinite frequency. The vectors $\pi_0\underline{/\theta_0}$, $\pi_1\underline{/\theta_1}$, $\pi_2\underline{/\theta_2}$ then vary as ω changes, and permit development of the circuit response function. For usual low-pass amplifiers, the high-frequency cutoff, f_2, may be very much greater than the low-frequency cutoff, f_1, possibly by a ratio of 100 to 1000 times. Thus the actual geometric picture must be extended off the page to the left if the true location of s_2 is to be shown; this extension must be left to the imagination, but it is important in the following discussion. That is, the amplification function changes more rapidly for a pole or zero close to the $j\omega$ axis than for a pole or zero far to the left. With a combination of a pole very close to the $j\omega$ axis and one far from the axis, the remote vector may be considered as constant over a broad range of frequencies, while the near pole produces a rapid variation of the function.

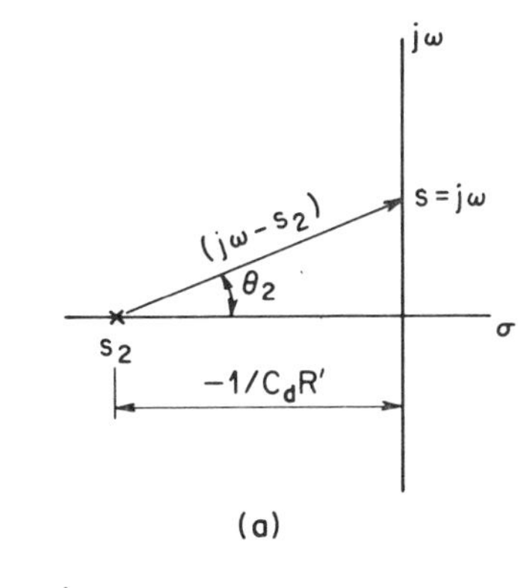

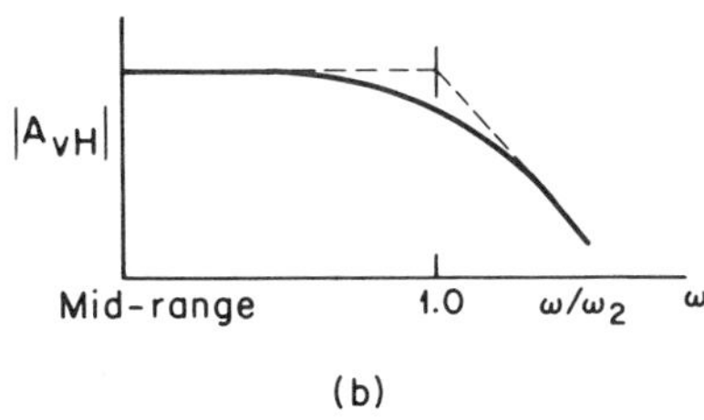

Figure 11-8. (a) Plot for high-frequency transistor gain, with $R' = R_s r_{ie}/(R_s + r_{ie})$. (b) Gain-frequency plot.

From the pole-zero diagram, the gain and phase angle may be written as

$$A_v = K\frac{\pi_0}{\pi_1\pi_2}, \qquad \theta_v = 90^\circ - \theta_1 - \theta_2 \quad (11\text{-}16)$$

For low frequencies, with the point ω near the origin, π_2 can be considered constant. The numerator π_0 increases linearly with frequency and π_1 more slowly, so that the gain increases with frequency. At $\omega = \omega_1$, the relation $\pi_0/\pi_1 = 1/\sqrt{2}$, and the low-frequency cutoff is passed. The phase angle is $\theta_v = 90° - 45° - 0° = +45°$.

At higher frequency, it can be seen that $\pi_0 \cong \pi_1$, and these magnitudes substantially cancel. Magnitude π_2 is still relatively constant, so that mid-range frequencies have been reached. Above mid-range, π_0 and π_1 will continue to cancel, and the value of the gain becomes a function of π_2. When $\pi_2 = \sqrt{2}$, the gain is $1/\sqrt{2}$ and the high-frequency band limit has been reached. The phase angle is then $\theta_2 = 90° - 90° - 45° = -45°$. In this region, with π_0 and π_1 considered as cancelling each other, the variation due to π_2 is illustrated in Fig. 11-8, as well as the resultant high-frequency response curve.

In this case, the pole-zero plot confirms the reasonableness of the previous assumptions which divided the frequency range of an amplifier into three regions. Other properties of pole-zero plots will be developed.

11-5 RISE TIME OF THE RC AMPLIFIER

Most wide-band pulse amplifiers use the common-emitter or common-cathode circuit, because of the high gain available, and the study of pulse performance will begin with the *RC*-coupled form. If desired, the results from the C-E circuit can be easily extended to the pentode amplifier, because of the similarity of the equivalent circuits.

The high-frequency interstage network for the C-E amplifier in Fig. 11-9 shows the first transistor as Q_1 and the second as Q_2. The voltage gain is most convenient, and for the high-frequency range can be written from Eq. 11-16 as

$$A_{vH} = K_5\left(\frac{1}{s + \omega_2}\right)$$

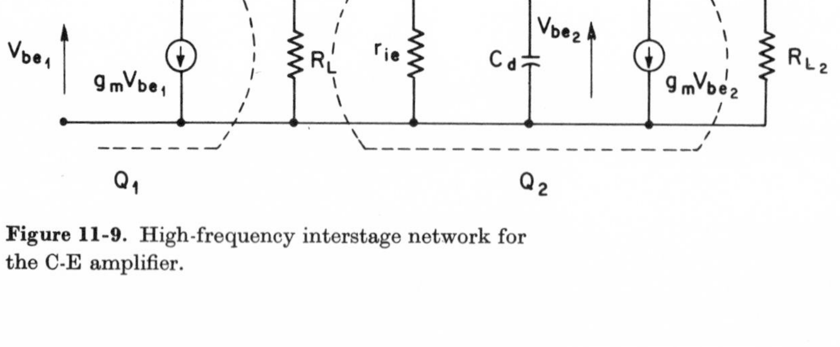

Figure 11-9. High-frequency interstage network for the C-E amplifier.

The leading edge of a pulse can be represented by a step function and the response of an amplifier with such an instantaneous rise of voltage is measured by the output pulse *rise time*. This criterion may be determined for the RC common-emitter amplifier by applying a step input voltage, as $v_i(s) = 1/s$. The output voltage is then expressible as

$$v_o(s) = \frac{-g_m}{sC_d}\left[\frac{1}{s + (1/R_T C_d)}\right] v_i \tag{11-17}$$

where $R_T = R_L r_{ie}/(R_L + r_{ie})$.

Taking the inverse transform, the result is given in the time domain by

$$v_o(t) = -g_m R_T v_i (1 - \epsilon^{-t/R_T C_d}) \tag{11-18}$$

which indicates an exponential rise of voltage with time, as the result of a step input.

The ultimate 100 per cent level is $v_o = -g_m R_T v_i$, or the value of the mid-range gain times the input. The output will reach the 10 per cent level at $t_{10} = 0.106 R_T C_d$, and the 90 per cent level at $t_{90} = 2.303 R_T C_d$. Therefore, the defined rise time of the amplifier is

$$T_r = 2.2 R_T C_d \tag{11-19}$$

Both gain and rise time are proportional to R_T, or the transistor load; to design an amplifier for high gain requires large R_T, but this lengthens the rise time. This dependence is shown by taking the ratio of gain and rise time as

$$\frac{\text{gain}}{\text{rise time}} = \frac{A}{T_r} = \frac{g_m}{2.2 C_d} \tag{11-20}$$

and this expression is a constant for a given transistor or pentode. It is sometimes referred to as a *gain-rise time figure of merit*.

Considering the RC amplifier as a low-pass device, the bandwidth may be taken from zero to a frequency given by $\omega_2/2\pi$. Since $\omega_2 = 1/R_T C_d$, then the *bandwidth-rise time product* is a constant, as

$$BWT_r = \frac{2.2 R_T C_d}{2\pi R_T C_d} = 0.35 \tag{11-21}$$

While defined for RC amplifiers, this ratio is approximately correct for many amplifier types even with a small overshoot permitted.

For cascaded amplifiers of little overshoot, the overall rise time for n stages has been found to approximate

$$T_{rn} \cong \sqrt{T_{r1}^2 + T_{r2}^2 + T_{r3}^2 + \cdots} \tag{11-22}$$

If the overshoot exceeds the five per cent level, the total rise time will be somewhat less than predicted.

11-6 TILT OR SAG OF THE PULSE TOP

If the duration δ of the pulse is long with respect to the rise time, it is possible to analyze for the rise time and then separately to consider the tilt or sag of the pulse top. This can be done because the rise time is a function of the shunt capacitances and the high-frequency characteristics, whereas the sag is dependent on the series capacitances and their capability in maintaining a charge during the pulse duration. This capability is related to the low-frequency response of the amplifier.

Neglecting rise time, Fig. 11-10 shows the pulse form during the duration δ. For constant output $v_o = -r_{ie}i_{b2}$ in Fig. 11-11, the current must be constant and C must be large.

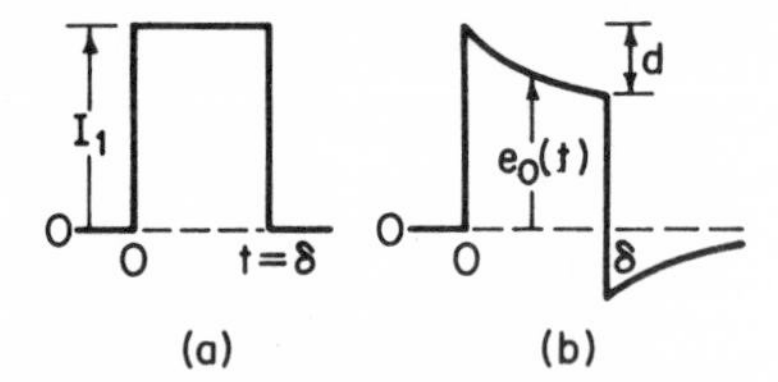

Figure 11-10. Pulse forms.

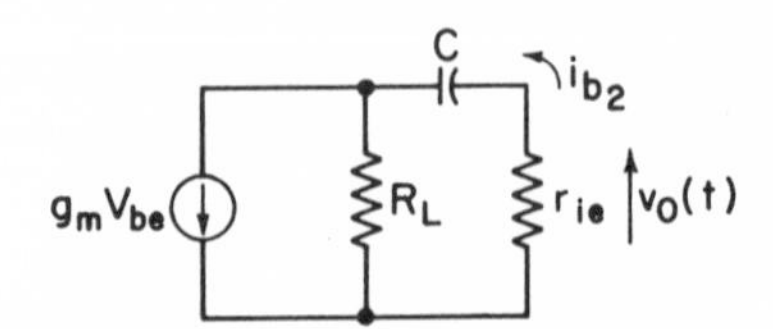

Figure 11-11. Transistor output circuit for study of sag.

The output v_o can be written as the product of gain and the input current i_{b1}, so that

$$v_o(\omega) = \frac{-g_m R_L i_{b1}}{R_L + r_{ie}}\left(\frac{j\omega}{j\omega + \omega_1}\right) \tag{11-23}$$

where $\omega_1 = 1/C(R_L + r_{ie}) = 1/\tau$. Replacing $j\omega$ with s, and introducing a current step input as $i_{b1} = I_1/s$, the output voltage is

$$v_o(s) = \frac{g_m R_L}{R_L + r_{ie}}\left(\frac{I_1}{s}\right)\left[\frac{s}{s + (1/\tau)}\right]$$

Then

$$v_o(t) = \mathscr{L}^{-1}v_o(s) = \frac{g_m R_L}{R_L + r_{ie}} I_1 \epsilon^{-t/\tau}, \qquad 0 < t < \delta \tag{11-24}$$

where I_1 is the input pulse amplitude. Thus the sag of the pulse top is exponential, as indicated in Fig. 11-10. The maximum amplitude of the pulse is given by $[g_m R_L/(R_L + r_{ie})]I_1$, which is $A_{iM}I_1$.

The output wave should have negligible sag and zero output for $t > \delta$. Small sag is obtained by making the time constant τ large, or δ/τ small; pulses obtained for several ratios of δ/τ appear in Fig. 11-12. Small δ/τ also provides small output after $t = \delta$ as well.

At the time $t = \delta-$, the amplitude has fallen to

$$v_o(\delta) = A_{iM} I_1 \epsilon^{-\delta/\tau} \tag{11-25}$$

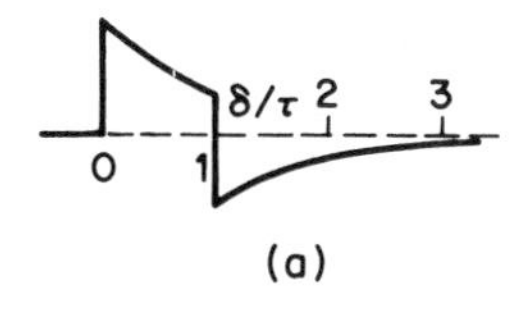

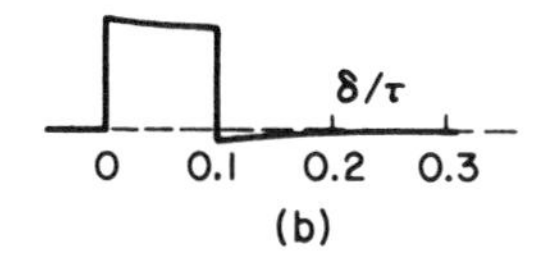

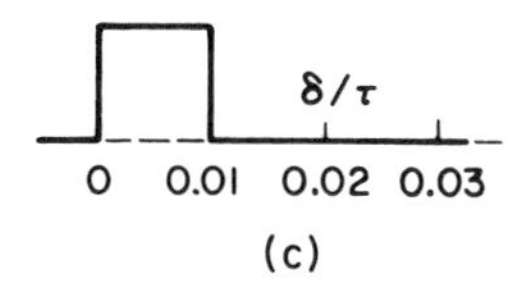

Figure 11-12. Effect of δ/τ ratio: (a) $\delta/\tau = 1$; (b) $\delta/\tau = 0.1$; (c) $\delta/\tau = 0.01$.

The sag magnitude may be approximated by replacing the exponential with its series as

$$\frac{v_o(\delta)}{A_{iM}} = I_1\left[1 - \frac{\delta}{\tau} + \frac{1}{2}\left(\frac{\delta}{\tau}\right)^2 - \ldots\right] \tag{11-26}$$

and for small values of δ/τ the terms beyond the second may be neglected. The unity term gives the output without sag, and the second term is the sag magnitude in time δ. The ratio of these terms is

$$\begin{aligned} \text{per cent sag} &\cong \frac{\delta}{\tau} \times 100 \text{ per cent} \\ &= \frac{\delta}{C(R_L + r_{ie})} \times 100 \text{ per cent} \end{aligned} \tag{11-27}$$

By previous definition, the cutoff frequency due to C was $\omega_C = 1/C(R_L + r_{ie})$, and so the per cent sag of the pulse response can be related to the steady-state frequency response of the amplifier as

$$\text{per cent sag} = \delta\omega_C \times 100 \text{ per cent} \tag{11-28}$$

As an example, if $\omega_C = 20$ hertz in a given amplifier, and if the sag is to be held to two per cent, then the longest pulse which can be amplified is

$$\delta = \frac{0.02}{2\pi \times 20} = 1590\ \mu\text{s}$$

as limited by the coupling capacitor C.

A second factor in tilt or sag determination is contributed by the emitter or cathode bias circuit, shown in Fig. 11-13. Analysis of the circuit leads to

$$v_o(s) = -g_m R_L v_i \frac{s + (1/R_E C_E)}{s + (1 + g_m R_E)/R_E C_E} \tag{11-29}$$

The output for $0 < t < \delta$ is given by

$$v_o(t) = \frac{-g_m R_L v_i}{1 + g_m R_E}(1 + g_m R_E \epsilon^{-t/\tau'}) \tag{11-30}$$

where

$$\tau' = R_E C_E/(1 + g_m R_E)$$

Again replacing the exponential with its series at the end of the pulse:

$$v_o(\delta) \cong -g_m R_L v_i\left(1 - \frac{\delta g_m}{C_E}\right) \tag{11-31}$$

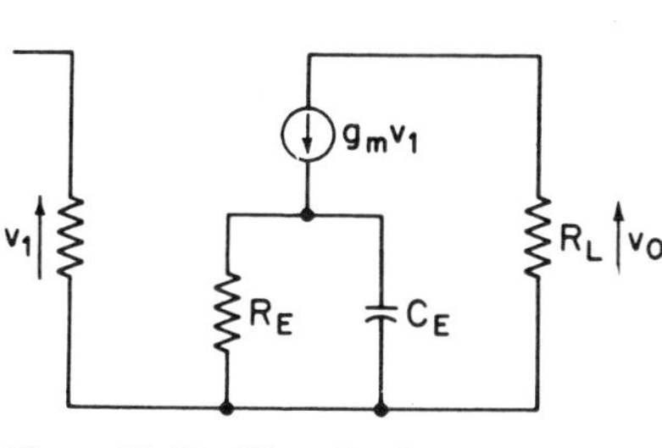

Figure 11-13. Bias circuit.

after dropping higher-order terms, thus restricting the result to small values of δ/τ'.

The per cent sag can be found as before, by taking the ratio of the variable term to the initial term, giving

$$\text{per cent sag} \cong \frac{\delta g_m}{C_E} \times 100 \text{ per cent} \tag{11-32}$$

While this result appears to be independent of R_E, this is true only for the approximation made, as R_E would enter through the higher-order terms.

The order of inaccuracy introduced by use of the shortened series appears above $\delta/\tau' = 0.20$ in Fig. 11-14. The straight line is plotted for Eq. 11-32, and the curve follows the accurate result of Eq. 11-30. Both are for a circuit with $R_E = 200$ ohms, $C_E = 25\ \mu$F, $R_L = 10^4$ ohms, and a transistor having $g_m = 7.5 \times 10^{-3}$ mho.

The sags of successive stages are additive in multistage amplifiers, and so the requirements on a single stage of such an amplifier may be severe. As was suggested for the steady-state solution, transient requirements are often best satisfied by elimination of the emitter or cathode bypass capacitor. A small, possibly 100 pF, bypass capacitor is then found to improve the rise time, without being a source of additional sag. This usage is known as emitter peaking, and gives a spike at the leading edge of the wave which raises the wave in a manner favorable to reduction of the rise time.

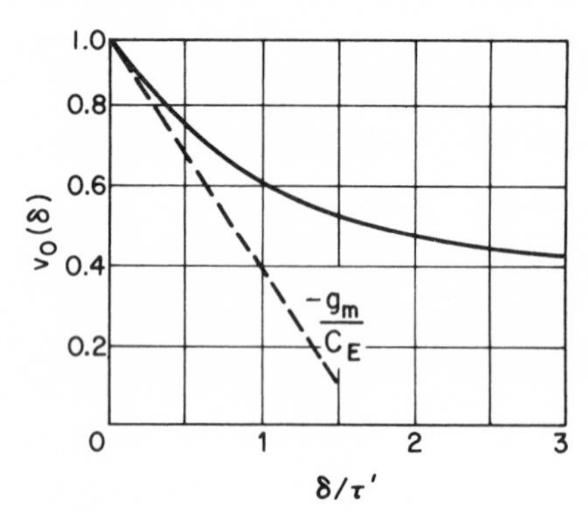

Figure 11-14. Pulse voltage at end of the sag vs. δ/τ'.

11-7 *INCREASE OF VIDEO AMPLIFIER BANDWIDTH BY SHUNT PEAKING*

The most obvious way to increase amplifier bandwidth with a given transistor or pentode is to decrease the load resistance, thus trading gain for frequency band, in accordance with the reasoning which accompanied the derivation of the gain–bandwidth figure. Circuit-wise, by a reduction of load, the reactance of the shunt capacitances becomes commensurate with R_L at a higher frequency or f_2 is raised, but the gain is decreased. With a transistor, reduction of R_L is also felt in C_d, since C_d is modified by a change in load through the Miller effect. In fact, f_2 may increase from 0.8 MHz to 8 MHz, and the gain may fall by only a factor of 5, as the load on a transistor amplifier is reduced from 10^4 to 10^2 ohms.

Such small loads may produce less than needed gains, however, and there are methods for extending f_2 values with higher loads. These employ load circuits with additional circuit elements, so placed as to counter the drop in high-frequency gain, or to raise the load impedance as the reactance of the shunting capacitances falls with frequency. Such circuits also improve the rise time of the amplifier, as should occur with higher f_2 values.

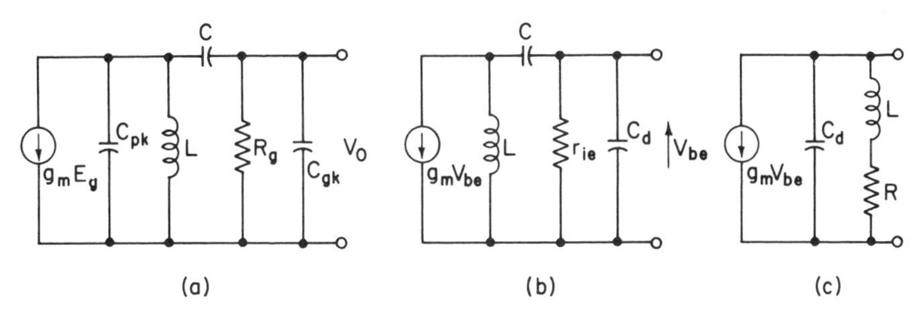

Figure 11-15. Shunt compensation: (a) pentode equivalent circuit; (b) transistor C-E circuit; (c) general equivalent circuit.

One such compensated circuit places an inductor in shunt with the internal capacitances, as in Fig. 11-15, and is referred to as a *shunt-peaked circuit*. Therein it is convenient to reflect r_{ie} into a series connection with L, as shown at (c); the value shown as R is the equivalent input resistance plus the resistance of the inductor. Additional resistance may be added as well, for increased mid-range gain. Capacitor C is dropped in the high-frequency range, and the circuit of Fig. 11-15(c) can serve for either transistor or pentode.

The equivalent current generator drives a load as

$$Z(s) = \frac{1}{1/(R_L + s) + sC_d} \tag{11-33}$$

and the gain is given by

$$A_{vH}(s) = \frac{-g_m}{C_d}\left[\frac{s + R/L}{s^2 + (sR/L) + (1/LC_d)}\right] \tag{11-34}$$

Of particular interest is the behavior of the gain function as L is varied, since that is the element added for frequency compensation. Note that

$$\omega_2 = \frac{1}{RC_d} \tag{11-35}$$

is the defined upper half-power point for the amplifier without compensation, for $L = 0$ and load $= R$. Thus ω_2 serves as a base for performance comparison with the compensated circuit. Other parameters can be defined as

$$a = \frac{R}{2L}$$

$$\omega_0^2 = \frac{1}{LC_d} = \frac{R}{L}\frac{1}{RC_d} = 2a\omega_2$$

and the gain function can be wrritten as

$$A_{vH}(s) = \frac{-g_m}{C_d}\frac{s + 2a}{s^2 + 2sa + 2a\omega_2} \tag{11-36}$$

This expression shows two poles due to denominator zeros at

$$s_1, s_2 = -a \pm \sqrt{a^2 - 2a\omega_2} \tag{11-37}$$

and with a numerator zero at

$$s_0 = -2a \tag{11-38}$$

For $L = 0$ and a very large, the poles exist at $s_1 \cong -\infty$ and $s_2 = -\omega_2 = -1/RC_d$, with the zero at $-\infty$, as indicated in Fig. 11-16(a). Since the effects of the pole and zero at $-\infty$ will cancel, the remaining pole, s_2, will represent the arrangement which gave the high-frequency range of the uncompensated amplifier, of Fig. 11-8. As L increases and a decreases, the poles move toward each other, and s_1 separates from s_0.

When $a = 2\omega_2 = 2/RC_d$, the radical in Eq. 11-37 vanishes, and there is a double pole at $-2\omega_2 = -2/RC_d$. The value of inductance L is then

$$L_{\text{crit}} = \frac{R^2C_d}{4} \tag{11-39}$$

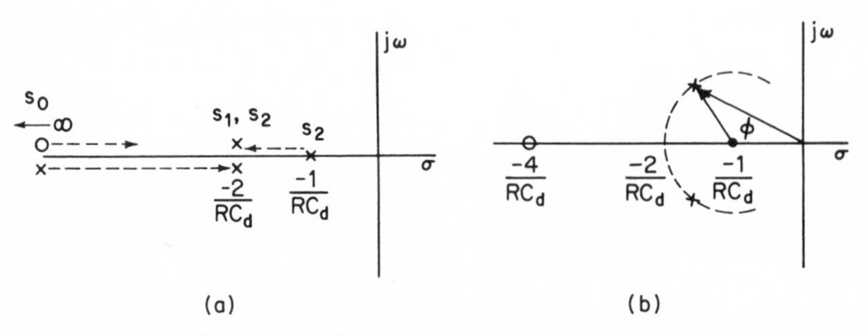

Figure 11-16. (a) Migration of the poles for $L < L_{\text{crit}}$; (b) separation of the complex poles.

If L is further increased the poles become complex conjugates because

$$s_1, s_2 = -a \pm j\sqrt{2a\omega_2 - a^2} \tag{11-40}$$

The variations in pole and zero position, or the *root-locus plot* of Fig. 11-16 can be obtained from

a	s_0	s_1	s_2	L/L_{crit}
∞	$-\infty$	$-\infty$	$\dfrac{-1}{RC_d}$	0
$\dfrac{4}{RC_d}$	$\dfrac{-8}{RC_d}$	$\dfrac{-6.8}{RC_d}$	$\dfrac{-1.2}{RC_d}$	0.5
$\dfrac{2}{RC_d}$	$\dfrac{-4}{RC_d}$	$\dfrac{-2}{RC_d}$	$\dfrac{-2}{RC_d}$	1
$\dfrac{1}{RC_d}$	$\dfrac{-2}{RC_d}$	$\dfrac{(-1 - j1)}{RC_d}$	$\dfrac{(-1 + j1)}{RC_d}$	2

From Eq. 11-40 the coordinates of the poles give

$$x = -a, \qquad y^2 = 2a\omega_2 - a^2$$

from which

$$y^2 + 2\omega_2 x + x^2 + \omega_2^2 = \omega_2^2$$

after adding ω_2^2 to each side of the equation to complete the square. Then

$$y^2 + (x + \omega_2)^2 = \omega_2^2 \tag{11-41}$$

shows that the conjugate poles move on the indicated locus circle of Fig. 11-16(b), as L is further increased. The circle has a radius $r = 1/RC_d = \omega_2$, and a center at $x = -1/RC_d = -\omega_2$.

With $L < L_{crit}$ and one pole partially cancelled by the nearby zero, the gain function obtained is similar to that of the high-frequency response of the RC amplifier, but with the response extended by the presence of the second pole and zero.

For $L > L_{crit}$ the situation changes as the complex poles separate. As L increases from L_{crit}, the effect of pole separation raises the high-frequency portion of the gain curve, but when the poles pass the position making $\varphi =$ 39° in Fig. 11-16(b), a peak develops in the gain curve. Because of the oscillatory nature of the circuit with complex poles, the pulse response tends to overshoot.

The above reasoning may be quantified by consideration of the steady-state response. Adding another parameter

$$m = \frac{\omega_2 L}{R} = \frac{L}{R^2 C_d} = \frac{L}{4L_{crit}} \tag{11-42}$$

Eq. 11-34 for the gain can be written

$$\mathbf{A}_{vH}(\omega) = -g_m R\left[\frac{1 + jm(\omega/\omega_2)}{1 - m(\omega/\omega_2)^2 + j\omega/\omega_2}\right] \tag{11-43}$$

$$\mathbf{A}_{vH}(\omega) = -g_m R\left[\frac{1 + m^2(\omega/\omega_2)^2}{1 + (1 - 2m)(\omega/\omega_2)^2 + m^2(\omega/\omega_2)^4}\right]^{1/2} \tag{11-44}$$

It is desired to find that combination of poles and the zero which will yield the flattest gain-frequency curve. A function of the form

$$H(\omega) = \frac{1 + a_1\omega^2 + a_2\omega^4 + \ldots}{1 + b_1\omega^2 + b_2\omega^4 + \ldots}$$

can be made to have the flattest monotonic response by equating coefficients of equivalent powers of the variable; that is, $a_1 = b_1$, $a_2 = b_2$, ...

Equating the coefficients of $(\omega/\omega_2)^2$ in Eq. 11-44 yields

$$m = -1 \pm \sqrt{2} \tag{11-45}$$

and choosing the positive sign, since the negative sign has no physical meaning, we find for maximal flatness of gain

$$m = 0.414$$

As a result, the value of inductance should be

$$L = 1.656L_{\text{crit}} = 0.414R^2C_d \tag{11-46}$$

The pole positions can be determined by writing the tangent of the angle φ, indicated in Fig. 11-16(b), as

$$\tan \varphi = \frac{\sqrt{2a\omega_2 - a^2}}{a} = \sqrt{4m - 1} \tag{11-47}$$

and for $m = 0.414$, the angle $\varphi = 39°$ will determine the pole positions on the locus circle. Since

$$a = \frac{R}{2L} = \frac{1}{2mRC_d} \tag{11-48}$$

and the zero is always located at $-2a$, then the zero location for the desired condition of flat response is $-1/mRC_d = -2.42/RC_d$.

The conditions for maximum bandwidth without a rise above the gain value at mid-frequency have been obtained. The results are confirmed by the gain magnitude curves, plotted as a function of m in Fig. 11-17. The curve for $m = 0$ is for the uncompensated amplifier, or for $L = 0$. The value of $m = 0.414$ gives a uniformly falling characteristic, larger m values produce a peak, as shown for $m = 0.5$ in the figure.

With $m = 0.414$, the bandwidth of the compensated amplifier is given by $f_2' = 1.72f_2$, and there is little further improvement with higher m values.

The phase angle of the shunt-peaked circuit is

$$\theta = -\tan^{-1}\frac{\omega}{\omega_2}\left[1 - m + m^2\left(\frac{\omega}{\omega_2}\right)^2\right] \tag{11-49}$$

Taking the derivative of θ with respect to ω, it is found that constant time

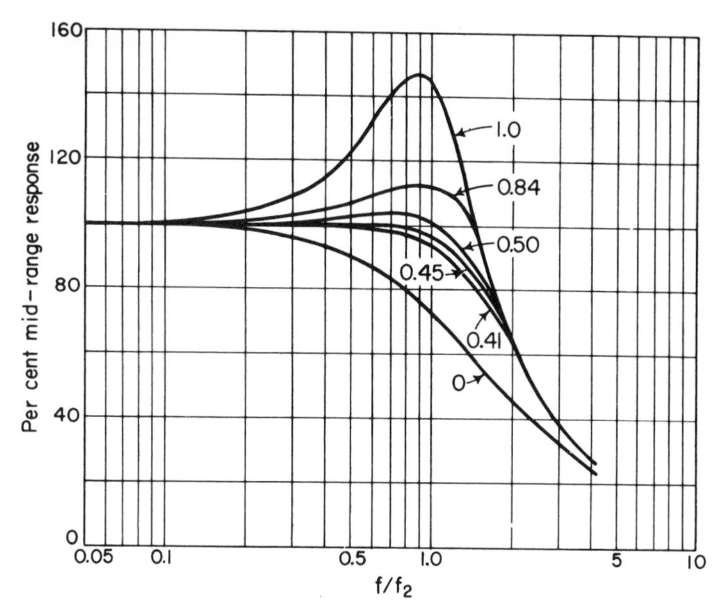

Figure 11-17. High-frequency response of the shunt-compensated amplifier, with m as a parameter.

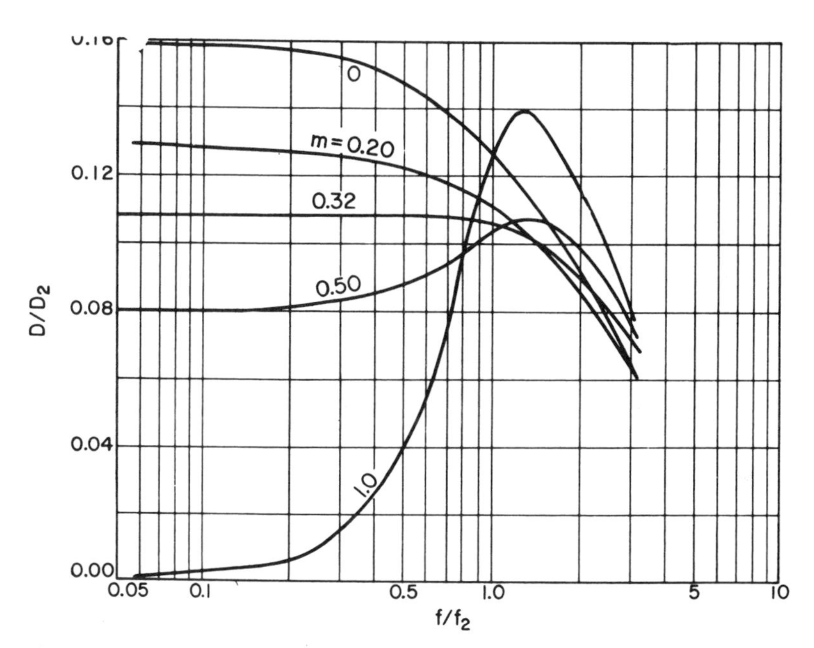

Figure 11-18. Time-delay curves for the shunt-compensated amplifier, with m as a parameter.

delay is achieved with $m = 0.32$. The time delay at ω vs. the delay at ω_2 is plotted in Fig. 11-18, for various values of m.

Equation 11-47 places the poles for $m = 0.32$ at an angle $\varphi = 28°$ from the real axis, and the zero at $-3.12/RC_d$. These are the loci for constant delay.

With $m = 0.41$ for flat gain and $m = 0.32$ for constant delay a compromise is necessary; a value of $m = 0.35$ is often considered suitable. The response to pulses, considered in the next section, will show that $m = 0.25$ may be desirable when minimum rise time without overshoot is important. With $m = 0.25$ the inductance has the critical value, with a double pole on the real axis.

11-8 RISE TIME OF THE SHUNT-PEAKED AMPLIFIER

Consideration must also be given to the effect of shunt-peaking on the amplifier rise time, since the response can become oscillatory with complex conjugate poles. That is, the response will then be of the form $K\epsilon^{(\sigma+j\omega)t}$. For a unit step input $V_i(s) = 1/s$ and the output voltage can be written from the gain expression of Eq. 11-36 for the shunt-peaked amplifier as

$$v_o(s) = \left(\frac{-g_m}{C_d}\right)\left\{\frac{s + 2a}{s[s + a(1 + \sqrt{1 - 4m})][s + a(1 - \sqrt{1 - 4m})]}\right\} \tag{11-50}$$

The above may be solved by reference to a table of transform pairs, and for the case in which $m < 0.25$, the output voltage becomes

$$e_o(t) = -g_m R\left[1 - \epsilon^{-t/2m\tau}\left(\frac{1 - 2m}{\sqrt{1 - 4m}} \sinh \frac{\sqrt{1 - 4m}}{2m}\frac{t}{\tau} + \cosh \frac{\sqrt{1 - 4m}}{2m}\frac{t}{\tau}\right)\right] \tag{11-51}$$

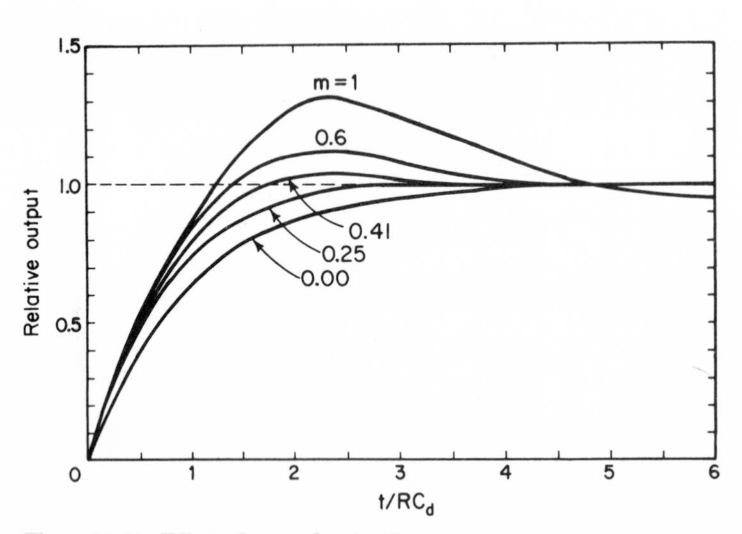

Figure 11-19. Effect of m on the rise time.

and if $m > 0.25$

$$e_o(t) = -g_m R\left[1 - \epsilon^{-t/2m\tau}\left(\frac{1-2m}{\sqrt{4m-1}} \sin \frac{\sqrt{4m-1}}{2m} \frac{t}{\tau} + \cos \frac{\sqrt{4m-1}}{2m} \frac{t}{\tau}\right)\right] \tag{11-52}$$

where $\tau = RC_d$. The pulse height will be given by $g_m R$.

In Fig. 11-19, the output rise is plotted as a function of $t/\tau = t/RC_d$, for m as a parameter. Here again, $m = 0$ is the case of the uncompensated amplifier, so that the reduction in rise time produced by peaking is readily seen. However, overshoot appears for $m > 0.25$, when the poles become complex conjugates.

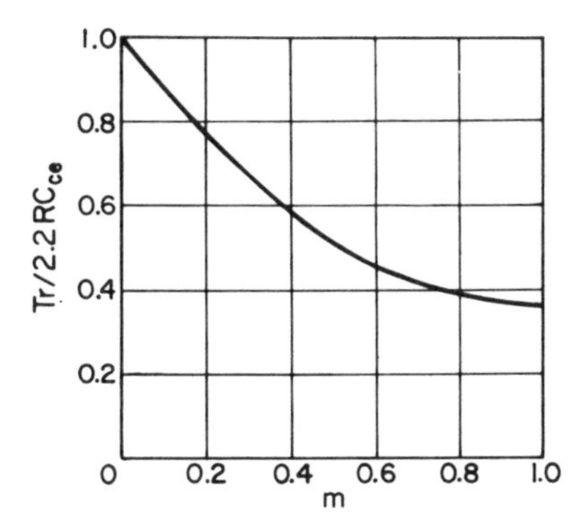

Figure 11-20. Rise time compared to that of the uncompensated amplifier.

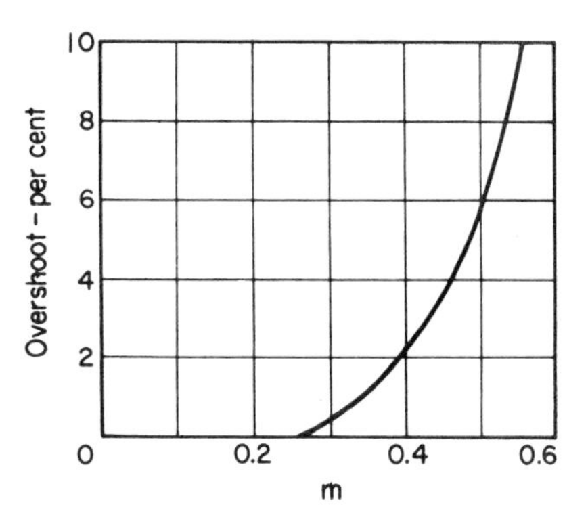

Figure 11-21. Variation of overshoot with m.

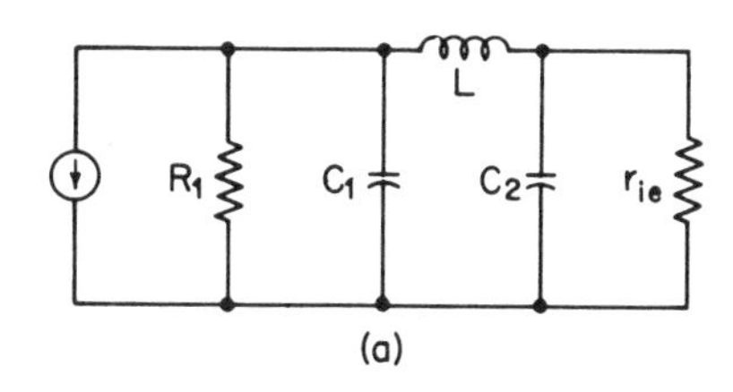

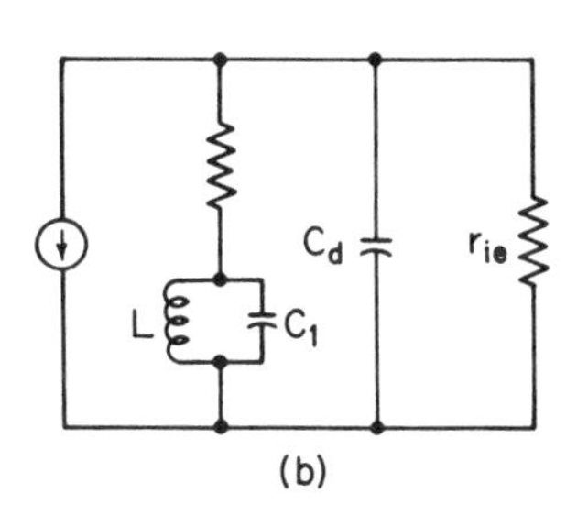

Figure 11-22. (a) Series-peaked circuit; (b) modified shunt-peaked circuit.

Comparison of the rise time to that of the uncompensated amplifier is given in Fig. 11-20. Thus at $m = 0.41$ the rise time is about 60 per cent of that of the uncompensated amplifier. However, an overshoot is encountered at the top of the leading edge of the pulse, and the amount of this overshoot, versus m, is indicated in Fig. 11-21. In multistage amplifiers a small overshoot in each stage produces a significantly larger overall overshoot, and the compromise value $m = 0.35$ is often used to get optimum bandwidth, phase delay, and rise time.

More sophisticated circuits are available, employing additional circuit elements, and one occasionally employed is the *series-peaking circuit* of Fig. 11-22(a). The design requires a specific ratio $C_2/(C_1 + C_2)$ and the bandwidth is then $2\omega_2$. However, if the natural capacitance division of the transistors or tubes does not meet this requirement, much of the theoretical advantage is not obtained. The rise time of the shunt-peaked circuit is less than that of the series-peaked circuit, and the former circuit, with its smooth drop off of high-frequency gain and its reduced complexity, is rather generally used in video-frequency service.

11-9 OPTIMUM NUMBER OF VIDEO AMPLIFIER STAGES

To make the results of this section general, we will consider $C = C_d$ for transistors, and $C = C_g$ for pentodes, with R being the effective output load for each. With G as the overall gain of an n-stage amplifier, then

$$G = (g_m R)^n$$

Equation 11-22 gives the rise time for an amplifier of n identical stages

$$\text{overall } T_r = T_{r1}\sqrt{n} = 2.2\ RC\sqrt{n}$$

so that

$$\text{overall } T_r = 2.2\frac{CG^{1/n}\sqrt{n}}{g_m} \tag{11-53}$$

Since the rise time is a nonlinear function of n, it is desirable to predict the optimum number of stages for a specified overall gain G, with a minimum rise time. Taking dT_r/dn and setting it to zero gives

$$\frac{2.2CG^{1/n}}{g_m\sqrt{n}}\left(\frac{1}{2} - \frac{1}{n}\ln G\right) = 0 \tag{11-54}$$

which leads to

$$\text{optimum } n = 2 \ln G \tag{11-55}$$

From this result $G = \epsilon^{n/2}$ and since the gain per stage is $G^{1/n}$, then

$$\text{optimum gain per stage} = \sqrt{\epsilon} = 1.65 \tag{11-56}$$

Using this result, the minimum overall rise time follows as

$$\text{minimum } T_r = \frac{2.2C\sqrt{2\epsilon \ln G}}{g_m} = \frac{5.1\ C\sqrt{\ln G}}{g_m} \tag{11-57}$$

Therefore an amplifier with an overall gain of 500, $g_m = 20{,}000\ \mu$mho, $C =$ 200 pF, optimally should be built with 13 stages, and would have an overall rise time approximating 0.125 μs. The load resistance would be 80 ohms and the bandwidth 6.4 megahertz. Actually there is a broad optimum for the number of stages for short rise time, and these results merely indicate that it is better to distribute a specified overall gain among a number of stages than to achieve the gain in a few stages.

11-10 BAND-PASS AMPLIFIERS

In the *radio-frequency* (R-F) *amplifier*, the input signal spectrum consists of a center frequency f_c around which may be grouped other frequency components, as indicated in Fig. 11-23(a). The center frequency may range from one to many megahertz and the side frequencies may extend from 5 to 10 kilohertz for amplitude modulated voice or music transmission, to several hundred kilohertz for other purposes. When the desired band is only a fraction of a per cent of the center frequency the response is known as *narrow band*.

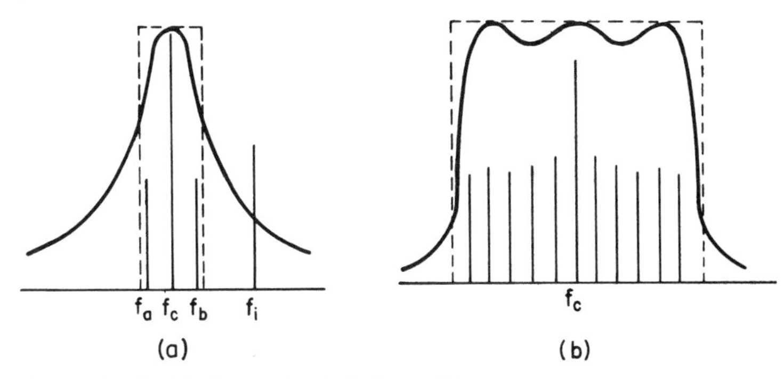

Figure 11-23. (a) Narrow-band R-F amplifier response; (b) response of an I-F amplifier.

Other combinations of resonant circuits operating at high radio frequencies may employ bandwidths of 10 to 30 megahertz at center frequencies of 100 megahertz or more. The *intermediate-frequency* (I-F) *amplifier* is a form designed for a bandwidth of a few kilohertz at a center frequency of a few hundred kilohertz for voice modulated signals, up to bandwidths of 6 megahertz at a center frequency of 40 megahertz for television reception. These systems, in which the pass band ranges from a few per cent to as much as 30 per cent of the center frequency, are considered as *broad band*.

Radio-frequency amplifiers are also expected to provide *selectivity* for frequencies in a desired spectrum, and an ideal response is indicated by the dashed rectangle of (a), Fig. 11-23. All frequencies in the ideal response band would then be amplified equally, and all frequencies outside the desired band, such as the interfering signal at f_1, would produce zero response.

Another basic requirement of an R-F amplifier is usually *tunability*, or that it be possible to shift or *tune* the response rectangle over a considerable range of frequencies at the will of the operator. Circuit simplicity is then important, since cascaded stages must be simultaneously tuned. The tunability requirement is most easily met through use of parallel-resonant circuits, with variable capacitance often, and variable inductance more rarely, used for changing the resonant response frequency.

Such circuits provide only an approximation to the desired rectangular frequency response, and the discrimination against adjacent frequencies is often inadequate. Because of the shape of the response of the resonant circuit, the so-called "skirt" portions of the curve will yield some response for signals at adjacent frequencies.

A measure of relative selectivity is sometimes taken as the ratio of the frequency width at 60 dB down from the center frequency response, to the bandwidth at which a signal is only 6 dB down. For a single parallel-resonant circuit this *shape factor* may approximate 600:1, or with a 6 dB bandwidth of 10 kilohertz, the 60 dB bandwidth may be found to be 5 or 6 megahertz. Several parallel-resonant circuits may be cascaded to reduce the 60 dB bandwidth, but other circuit means are necessary to achieve a closer coincidence between actual response and the ideal selectivity rectangle.

The needed additional selectivity is usually provided in the communications system by the intermediate-frequency amplifiers, operating with responses as indicated in Fig. 11-23(b). Because these circuits are usually tuned to fixed frequencies, greater circuit complexity can be tolerated. Each stage of these I-F amplifiers may employ a resonant circuit, but with the group tuned to staggered frequencies, to achieve fairly uniform response across a wide pass band. In other circuit combinations steeper sides or narrower skirts can be obtained, thus more closely approaching the desired rectangular response shape.

11-11 THE PARALLEL-RESONANT CIRCUIT

Because of the wide use of the parallel *RLC* circuit for tuning and loading of narrow-band radio-frequency amplifiers, we will develop a few properties of the circuit, shown in Fig. 11-24(a). The resistance R_s is that of the inductor, since at radio frequencies the capacitors ordinarily employ air as the dielectric and have very low losses. The reactances are energy storage ele-

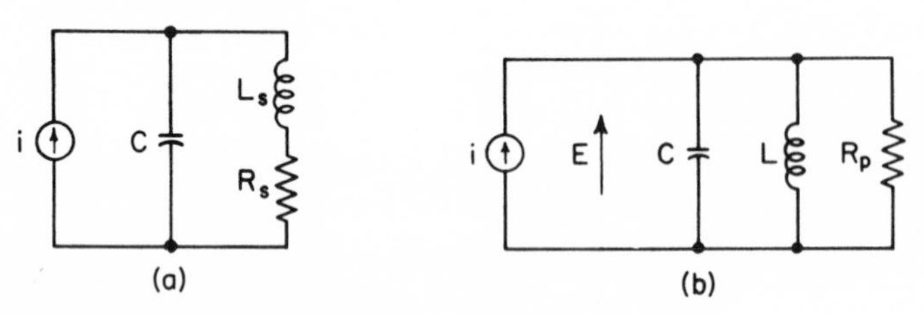

Figure 11-24. (a) Series resistance form; (b) parallel resistance form.

ments, and a merit factor Q has been defined to indicate the relative efficiency; thus

$$Q = \frac{2\pi \times \text{maximum energy stored per cycle}}{\text{energy dissipated per cycle}}$$

If I_m is the peak current in the inductor, then

$$Q = \frac{2\pi L_s I_m^2}{I_m^2 R_s / f_0} = \frac{\omega_0 L_s}{R_s} \tag{11-58}$$

with $f_0 = \omega_0/2\pi$ being the frequency of circuit resonance. Since this expression is the ratio of the reactance of the inductance to its resistance, Q is a quality rating for a given inductor at the stated frequency.

It is well known that the antiresonant frequency of the circuit of (a) is

$$f_r = \frac{1}{2\pi}\sqrt{\frac{1}{L_s C} - \frac{R_s^2}{L_s^2}} = \frac{1}{2\pi}\sqrt{\frac{1}{L_s C}}\sqrt{\frac{Q^2}{1 + Q^2}} \tag{11-59}$$

The value of Q at radio frequencies usually exceeds 10 and may reach values of several hundred; for the high-Q case the resonant frequency reduces to the undamped natural frequency of the circuit

$$f_0 = \frac{\omega_0}{2\pi} = \frac{1}{2\pi}\sqrt{\frac{1}{L_s C}} \tag{11-60}$$

The impedance of the circuit of (a) can be written as

$$\mathbf{Z} = \frac{(R_s + j\omega L_s)(1/j\omega C)}{R_s + j[\omega L_s - (1/\omega C)]} = \frac{L_s/C[1 - (jR_s/\omega L_s)]}{R_s\{1 + j[(\omega L_s/R_s) - (1/\omega C R_s)]\}} \tag{11-61}$$

The reactive term in the denominator involves the difference of two numbers which are nearly equal when close to the resonant frequency. For greater accuracy in computation, the expression can be put in a different form by defining a parameter δ as the frequency variation from resonance, where

$$\delta = \frac{\omega - \omega_0}{\omega_0} \tag{11-62}$$

Then

$$\frac{\omega L_s}{R_s} = (1 + \delta)Q_0, \qquad Q_0 = \frac{1}{R}\sqrt{\frac{L}{C}} \tag{11-63}$$

and Eq. 11-61 can be written as

$$\mathbf{Z} = \frac{(L_s/R_s C)\{1 - j/[Q_0(1 + \delta)]\}}{1 + jQ_0[1 + \delta - 1/(1 + \delta)]} \tag{11-64}$$

Assuming that $Q^2 \gg 1$ for the high-Q case, then

$$\mathbf{Z} = \frac{R_s Q_0^2}{1 + j[Q_0 \delta(2 + \delta)/(1 + \delta)]} \tag{11-65}$$

which is a form better suited to computation.

At resonance, $\delta = 0$, and the impedance of the circuit becomes

$$\mathbf{Z}_0 = R_s Q_0^2 = \frac{L_s}{CR_s} \tag{11-66}$$

the latter relation indicating that the resonant circuit impedance is a function of the L/C ratio chosen.

The circuit impedance may be normalized against the impedance at resonance, and then Eq. 11-65 becomes

$$\frac{\mathbf{Z}}{R_s Q_0^2} = \frac{1}{1 + j[Q_0\delta(2+\delta)/(1+\delta)]} \tag{11-67}$$

and this result is plotted in Fig. 11-25.

An important factor in parallel-circuit performance is the rapid change in phase at f_0, from inductive below to capacitive above resonance. When the circuit is employed as a load for transistor or tube, the change in sign of the load reactance at resonance may affect the input circuit through the variable elements present there by the Miller effect.

In application it is often more convenient to use the parallel resistance form of Fig. 11-24(b); R_p may then be readily combined with other circuit

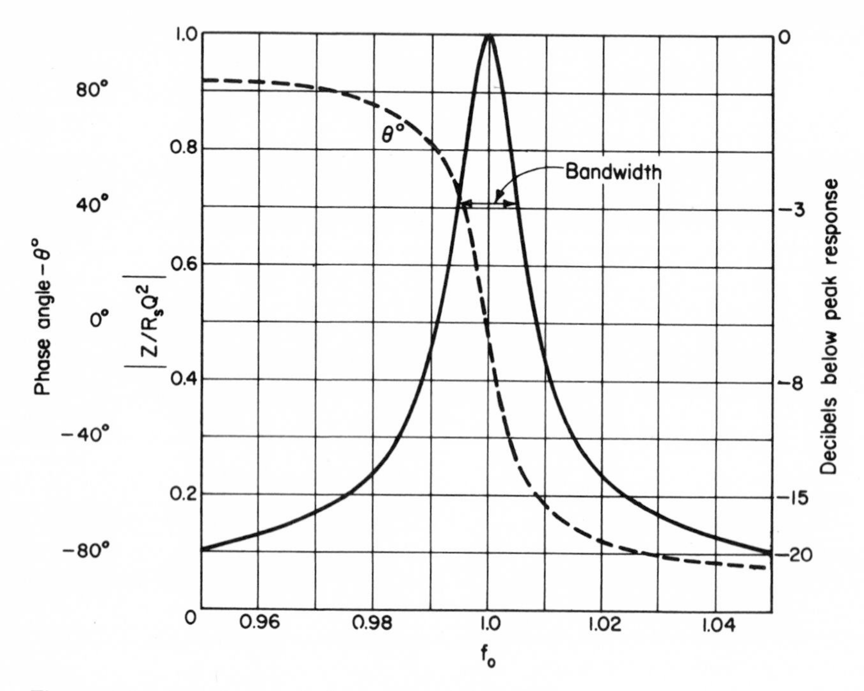

Figure 11-25. Impedance and phase angle of the parallel-resonant circuit ($Q = 100$).

resistances. The L and R_p branches in (b) are equivalent to the L_s, R_s series arm of (a), and the admittance can be written as

$$\mathbf{Y} = \frac{1}{R_s + j\omega L_s} = \frac{1}{R_s}\left(\frac{1}{1 + \omega^2 L_s^2/R_s^2}\right) - \frac{j}{\omega L_s}\left(\frac{1}{1 + R_s^2/\omega^2 L_s^2}\right) \quad (11\text{-}68)$$

For the high-Q case, or with $Q > 10$, the parallel resistance and inductance of (b) are

$$R_p = \frac{\omega_0^2 L_s^2}{R_s} = Q_0 \omega L_s, \qquad L \cong L_s \quad (11\text{-}69)$$

where Q_0 is that of the coil itself, at the resonant frequency.

By use of the Q definition, and with E_m as the peak voltage across the circuit in Fig. 11-24(b), one can write

$$Q_0 = \frac{2\pi C E_m^2}{E_m^2/R_p f_0} = \omega_0 C R_p \quad (11\text{-}70)$$

This expression is the Q of the resonant circuit. With R_p equivalent to R_s in effect, the Q value should be identical with that calculated by Eq. 11-58.

11-12 BANDWIDTH OF THE PARALLEL-RESONANT CIRCUIT

The definition of resonant circuit bandwidth is consistent with previous definitions as the width in cycles of the resonant-response curve (of Fig. 11-25) between the frequencies at which the power is one-half of the resonant power.

Equation 11-67 reduces to

$$\frac{\mathbf{Z}_0}{R_s Q_0^2} = \frac{1}{1 + jQ_0[1 + \delta - 1/(1 + \delta)]} \quad (11\text{-}71)$$

At the band limits the reactive term must equal unity. From the definition of δ

$$\frac{\omega_0}{\omega_1} = 1 + \delta_1, \qquad \frac{\omega_2}{\omega_0} = 1 + \delta_2$$

and so at the band limits the reactive term yields

$$Q_0\left(\frac{\omega_2}{\omega_0} - \frac{\omega_0}{\omega_2}\right) = Q_0\left(\frac{\omega_0}{\omega_1} - \frac{\omega_1}{\omega_0}\right) = 1 \quad (11\text{-}72)$$

from which

$$\omega_0 = \sqrt{\omega_1 \omega_2} \quad (11\text{-}73)$$

and the curve of response exhibits geometric symmetry about f_0. Substitution in Eq. 11-72 yields

$$\text{BW} = f_2 - f_1 = \frac{f_0}{Q_0} \quad (11\text{-}74)$$

and this provides a second definition of Q of the circuit. In fact, it is often most convenient to determine the in-place Q of a circuit by measuring the bandwidth of the resonant response.

Since $Q_0 = (1/R)\sqrt{L/C}$, the bandwidth is influenced by the selection of the ratio of L to C for the circuit. High C gives greater selectivity or narrower bandwidth, but it also reduces the circuit impedance at resonance.

11-13 SINGLY-TUNED AMPLIFIERS—ROOT LOCUS

The parallel-resonant circuit is employed as a coupling means between cascaded radio-frequency amplifiers, where some frequency selectivity is desired. The resultant circuit for a narrow-band amplifier is shown in Fig. 11-26. The circuit will have a gain of $-g_m R$ at resonance, and small gain at frequencies far from resonance.

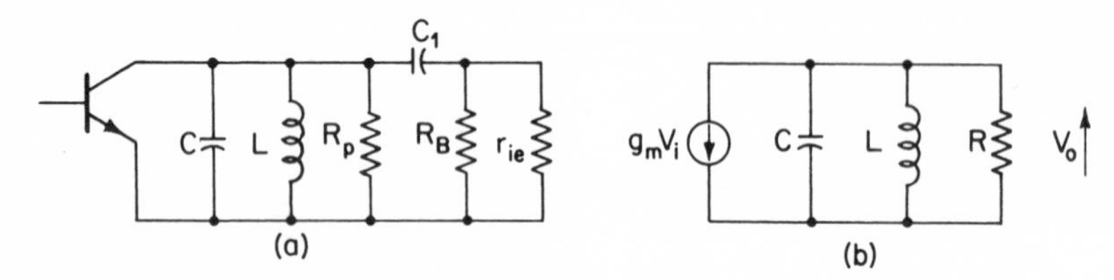

Figure 11-26. (a) Narrow-band interstage coupling; (b) equivalent circuit.

The internal device and circuit stray capacitances are absorbed into the tuning capacitance C. Resistor R_s represents the tuned-circuit losses, including losses added by eddy currents in metal shields surrounding the inductor. When combined in parallel with R_B and r_{ie} of the transistor following-stage input, the total parallel value becomes R of Fig. 11-26(b). In the case of a pentode amplifier, R is the effective combination of R_s, R_g, and R_{in} of the following amplifier. In either case the overall circuit Q is

$$Q_e = \omega_0 CR$$

The impedance of the load circuit of 11-26(b) is

$$\mathbf{Z}(\omega) = \frac{1}{(1/R) + (1/j\omega L) + j\omega C}$$

from which the gain can be written as

$$A_v(s) = -\frac{g_m}{C}\frac{s}{s^2 + s/RC + 1/LC} \tag{11-75}$$

As before, we define several parameters:

$$\omega_0^2 = \frac{1}{LC}, \qquad b = \frac{1}{2RC}$$

and so

$$A_v(s) = \frac{-g_m}{C}\left(\frac{s}{s^2 + 2sb + \omega_0^2}\right) \tag{11-76}$$

The circuit has a zero at $s = 0$, and two poles

$$s_1, s_2 = -b \pm \sqrt{b^2 - \omega_0^2}$$

At $b = -\omega_0$ on the negative real axis these become a double pole. For narrow-band response, we are interested in a system with large values of parallel R, or with pole locations near to the $j\omega$ axis, and the poles are then complex conjugates at

$$s_1, s_2 = -b \pm j\sqrt{\omega_0^2 - b^2} \tag{11-77}$$

We see that the poles are located on a locus having

$$x = -b, \qquad y^2 = \omega_0^2 - x^2$$

or

$$x^2 + y^2 = \omega_0^2 \tag{11-78}$$

which describes the root-locus as a circle of radius ω_0, with center at the origin, as shown in Fig. 11-27(a).

From Eq. 11-70, $Q_0 = \omega_0 CR$ for the circuit of Fig. 11-26(b), and so $b = \omega_0/2Q_0$. With Q_0 values ranging up to several hundred for high-Q systems, the distance of the poles from the $j\omega$ axis is only a fraction of a per cent of the distance ω_0. This is also confirmed by

$$\tan\varphi = \frac{y}{x} \cong \frac{\omega_0}{b} = 2Q_0$$

indicating that φ is an angle in excess of 89°, in most cases. To accurately portray such a pole location would extend Fig. 11-27(b) off the page, and so the figure is only an indication of the effect to be expected when frequency passes the ω_0 point. With the gain magnitude given by

$$|A| = \frac{-g_m}{C}\left(\frac{\pi_0}{\pi_1\pi_2}\right) \tag{11-79}$$

it can be seen that near the resonant frequency location on the $j\omega$ axis, π_0 and π_1 are large and $\pi_1 \cong 2\pi_0$. These vectors change very slowly with ω, and thus introduce only the factor $\frac{1}{2}$ in the result. However, π_2 changes

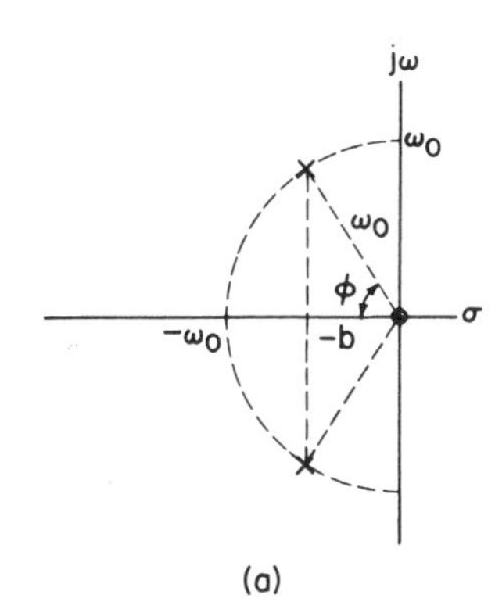

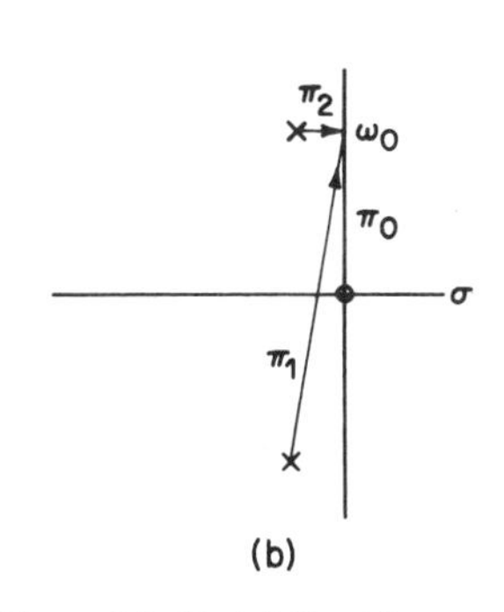

Figure 11-27. (a) Root-locus of narrow-band amplifier; (b) variation of factors.

very rapidly in magnitude as ω_0 is passed, and the phase angle goes rapidly from inductive to capacitive.

Following this reasoning it is possible to predict the resonant-frequency gain of the amplifier as

$$A_v(\omega) = \frac{-g_m}{C}\left(\frac{1}{2\pi_2}\right) = \frac{-g_m}{2Cb} = -g_m R \tag{11-80}$$

Since the bandwidth is measured by the frequencies at which the phase angle is $\pm 45°$, and since this angle is determined by vector π_2 in the narrow-band case, it follows that the magnitude of π_2 is $\sqrt{2}\, b$, as shown in Fig. 11-28. From the two triangles formed by π_2 in going to ω_1 and ω_2, the bandwidth must be

$$\frac{\text{BW}}{\omega_0} = \frac{2b}{\omega_0} = \frac{1}{\omega_0 RC} = \frac{1}{Q_0} \tag{11-81}$$

as a per cent of the resonant frequency; this is the same result previously reached by algebraic methods for Eq. 11-74.

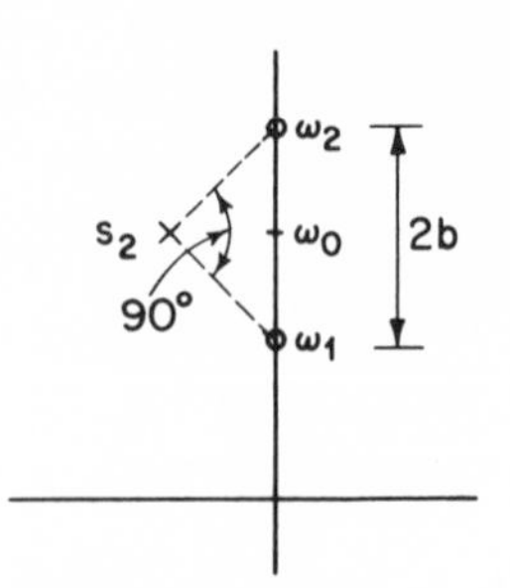

Figure 11-28. Vectors at the band limits.

11-14 TRANSIENT RESPONSE OF THE RLC-TUNED AMPLIFIER

Applying a step input $v_i(p)$ to the circuit having a response of the form of Eq. 11-76 leads to a solution

$$v_o(t) = -g_m R(1 - \epsilon^{-t/2RC}) \sin \omega_0 t \tag{11-82}$$

This result appears as the leading edge of the wave envelope in Fig. 11-29. For an input pulse, the output envelope would decay similarly, as

$$v_o(t + \delta) = -g_m R\epsilon^{-t/2RC} \sin \omega_0 t \tag{11-83}$$

and the trailing portion of the wave is obtained.

A tuned amplifier will not instantaneously follow a pulsed input, but will

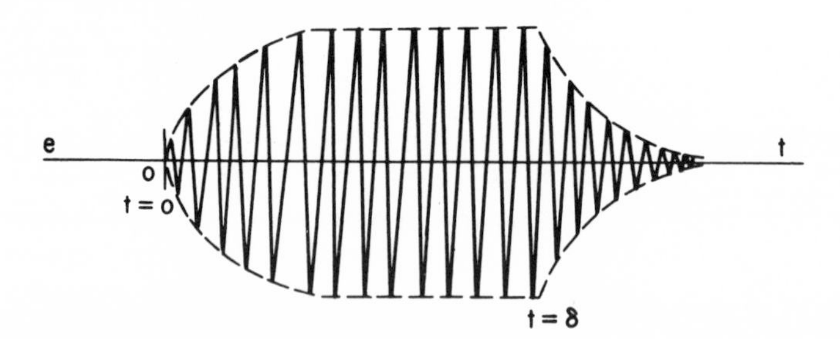

Figure 11-29. Rise and decay of a pulse in a tuned amplifier.

have an envelope rise and fall time dependent on the time constant $2RC$. Thus for fast rise and fall times, the tuning capacitance used should be small. The circuit is said to "ring" as a bell, when it continues in oscillation after $t = \delta$ at the completion of the input pulse, and this effect will have later application.

11-15 PENTODE SINGLY-TUNED AMPLIFIERS

The circuit discussed in the previous sections is used as a frequency-selective load for an amplifier with impedance matching of input and output achieved by a transformer. The resonant circuit is placed in the circuit of higher impedance which, for a pentode, is the grid circuit of the following tube, as in Fig. 11-30, where (b) is derived from Chapter 5.

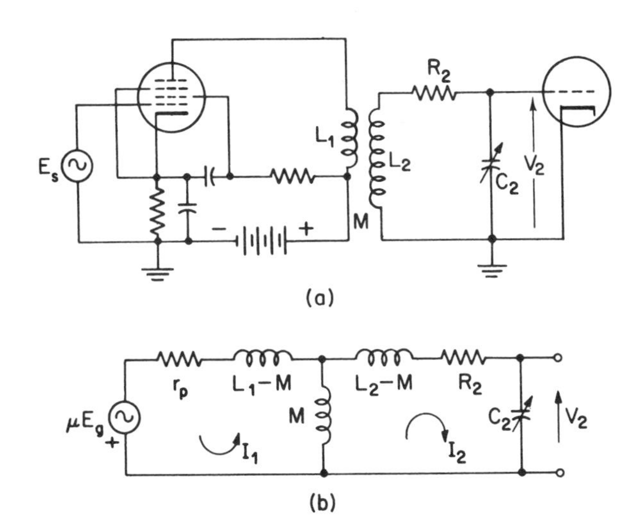

Figure 11-30. The pentode inductively coupled amplifier.

Since we wish a voltage gain expression, it is convenient to use an equivalent voltage generator for the pentode, and the mesh equations are

$$\begin{bmatrix} \mu \mathbf{E}_g \\ 0 \end{bmatrix} = \begin{bmatrix} r_p + j\omega L_1 & -j\omega M \\ -j\omega M & R_2 + j(\omega L_2 - 1/j\omega C_2) \end{bmatrix} \begin{bmatrix} \mathbf{I}_1 \\ \mathbf{I}_2 \end{bmatrix} \tag{11-84}$$

Using the transfer impedance from Chapter 5, $\mathbf{I}_2 = \mathbf{E}_g/\mathbf{Z}_T$ and $\mathbf{V}_2 = -j\mathbf{I}_2/\omega C_2$, so that

$$\mathbf{A}_v(\omega) = \frac{\mathbf{V}_2}{\mathbf{E}_g} = \frac{-j\mu}{\omega C_2 Z_T}$$

$$= \frac{-j\mu}{\omega C_2}\left\{\frac{-j\omega M}{-\omega^2 M^2 - (r_p + j\omega L_1)[R_2 + j(\omega L_2 - 1/\omega C_2)]}\right\} \tag{11-85}$$

after neglecting R_1 of the transformer, since $R_1 \ll r_p$. For the pentode $\omega L_1 \ll r_p$, also, and so

$$\mathbf{A}_v(\omega) = \frac{-g_m \omega M Q_2(1 + \delta)}{\left(1 + \dfrac{\omega^2 M^2}{r_p R_2}\right)\left[1 + j\dfrac{Q_2\delta(2 + \delta)/(1 + \delta)}{1 + \omega^2 M^2/r_p R_2}\right]} \tag{11-86}$$

The second denominator term follows from the process that developed Eq. 11-65 from Eq. 11-61. The primary circuit resistance is reflected into the secondary circuit by ωM, and the effective circuit Q is

$$Q_e = \frac{Q_2}{1 + \omega^2 M^2/r_p R_2} \tag{11-87}$$

Thus

$$\mathbf{A}_v(\omega) = \frac{-g_m \omega M Q_e(1 + \delta)}{1 + jQ_e\delta(2 + \delta)/(1 + \delta)} \tag{11-88}$$

for frequencies near resonance, and Q_e determines the band width.

At resonance $\delta = 0$, $\omega = \omega_0$, $\omega_0 L_2 = 1/\omega_0 C_2$, and the maximum output voltage and gain are

$$\mathbf{A}_v(\omega) = -g_m \omega_0 M Q_e \tag{11-89}$$

The optimum value of M is $\omega_0 M = \sqrt{r_p R_2}$, but this usually cannot be attained because of the large r_p, and M is often adjusted for a value of Q_e which will give a desired band width.

Since $M = k\sqrt{L_1 L_2}$, where k is the *coefficient of coupling*, the gain is also a function of secondary inductance. For some transformers using powdered-iron cores, the values of k will approximate unity. While high L_2 is desirable, it often cannot be used, because C_2 must be large enough to swamp out variations in device capacitances due to temperature or to the Miller effect.

For instance, with $f_0 = 1$ megahertz, then $\omega_0 = 6.28 \times 10^6$. For a resonant impedance of 100,000 ohms $= R_p$, and with $Q_0 = 100$ for a bandwidth of 10,000 hertz, then from Eq. 11-70

$$\frac{1}{\omega_0 C} = \frac{R_p}{Q_0} = 1000 \qquad \text{ohms}$$

from which

$$C = 159 \qquad \text{pF}$$

and this would be considered a satisfactory value. The inductor would have $L = 159\ \mu$H.

11-16 TRANSISTOR SINGLY-TUNED AMPLIFIERS

For transistor interstage coupling, the resonant circuit is placed in the collector circuit, that being of the higher impedance. The transformer then reflects the high value of resonant circuit impedance as a lower value of impedance to the transistor input shown at 1 in Fig. 11-31.

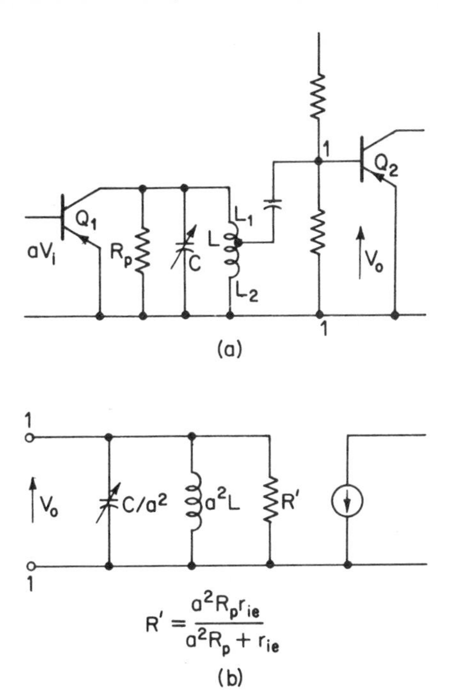

Figure 11-31. Ferrite core coupling transformer for transistor amplifier.

Most such transformers employ ferrite or powdered-iron cores, and the coefficient of coupling k will approximate unity. Using the usual definitions for such circuits, $a = E_1/E_2 = I_2/I_1 = N_1/N_2$. Then

$$a^2 = \frac{E_1 I_2}{E_2 I_1} \tag{11-90}$$

but $Z_1 = E_1/I_1$, $Z_2 = E_2/I_2$, and so the secondary current and voltage are due to a reflected primary impedance

$$Z_2 = Z_1/a^2 \tag{11-91}$$

That is, primary impedances seen from the secondary side appear as divided by a^2. Likewise L_1 appears as L_1/a^2 on the secondary side, along with a^2C_1, and R_1/a^2.

Resistance r_{oe} of the transistor is combined with R of the tuned circuit as R_p. The bias resistances R_B may be neglected as large and so the reflected resistance R_p/a^2 appears in parallel with r_{ie} of the transistor, and

$$R' = \frac{R_p r_{ie}/a^2}{R_p/a^2 + r_{ie}} \tag{11-92}$$

The gain expression of Eq. 11-75 was written for just such a circuit as now appears in Fig. 11-31(b). Therefore

$$A_v(s) = \frac{V_o}{aV_i} = -\frac{g_m}{aC}\frac{s}{s^2 + s/a^2R'C + 1/LC} \tag{11-93}$$

The transformation has not altered the resonant frequency, $\omega_0 = 1/\sqrt{LC}$, but it has changed Q, the effective bandwidth, and the resonant gain. The value of Q_e is

$$Q_e = a^2\omega_0 R'C \tag{11-94}$$

and the bandwidth is now determined by this value of Q_e. The resonant gain is given by

$$A_v(\omega) = -g_m R'/a \tag{11-95}$$

The condition for maximum power transfer from transistor Q_1 to transistor Q_2 requires that

$$r_{ie} = R_p/a^2 \tag{11-96}$$

or

$$a = \sqrt{R_p/r_{ie}} \tag{11-97}$$

However, such a value of a may not lead to desired values of bandwidth. It is possible to free the design from this limitation by connecting the collector to a tap on the coil as shown in Fig. 11-32(b).

It would also be possible to provide the desired impedance transformation by use of an air-core transformer as in Fig. 11-32(a), or by use of the capacitive voltage-divider of (c).

When the resonant circuit of an amplifier load is tuned on the capacitive

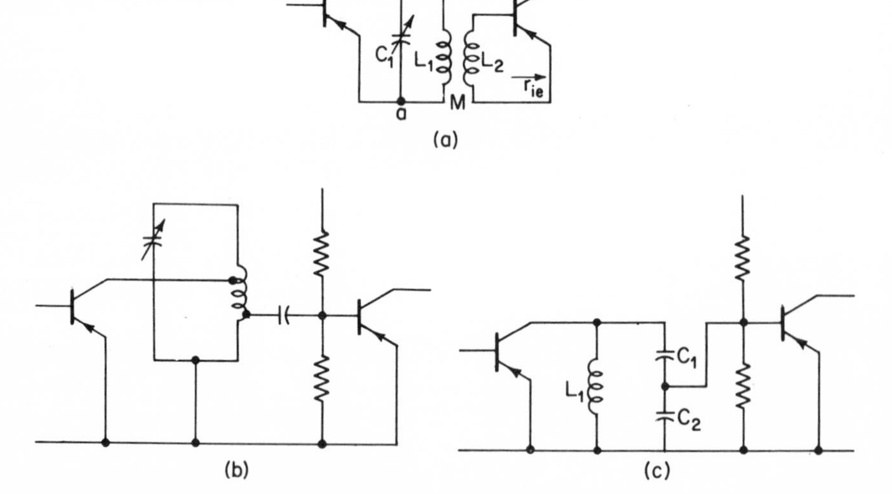

Figure 11-32. Impedance matching for transistor amplifiers.

side of resonance, the Miller effect introduces a positive resistance and a capacitance in shunt with r_{ie} in the input circuit. When tuned on the inductive side of resonance, a negative resistance appears with the capacitance. The effect of these varying elements must often be minimized by reduction of the load and gain from the optimum.

11-17 STAGGER-TUNED AMPLIFIERS

Use of two resonant-load amplifiers in cascade will provide a sharpening of the response pattern. However, in many fixed-frequency amplifiers such as those employed in the intermediate-frequency amplifiers of radio receivers, a wider and flatter pass-band response is desired, combined with a reduction in skirt response, to give a closer approach to the ideal response rectangle. For fixed frequencies, where tuning operations can be performed in manufacture, the response band may be broadened by staggering the resonant frequencies of several narrow-band circuits, as indicated in Fig. 11-33(a). The flatness of the top of the response, as well as the magnitude of any dip at the top, is dependent on the spacing of the separate resonant frequencies, or the placement of the mathematical poles of the analysis, as fixed in the design.

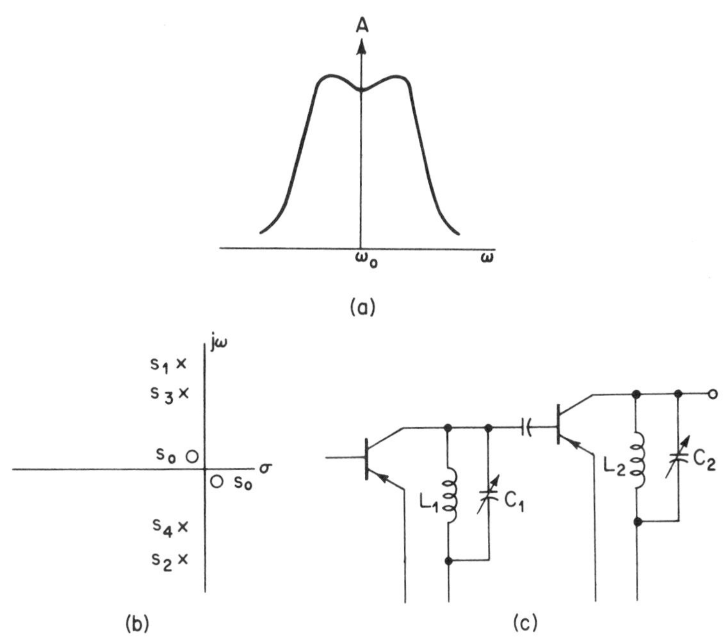

Figure 11-33. (a) Response of a stagger-tuned pair of resonant circuits; (b) pole-zero diagram; (c) stagger-tuned pair.

The response of the stagger-tuned pair of Fig. 11-33(c) will be that of two narrow-band amplifiers, since the active device between will provide circuit isolation. The response then involves a fourth-degree polynomial and there will be four poles and two zeros for the root-locus plot, as at (b), Fig. 11-33. Poles s_1 and s_2 are the conjugate pair from one stage, poles s_3 and s_4 are contributed by the second stage; the zeros of both will lie at the origin, as was the case for the single resonant stage of Section 11-13.

Again, we are interested in the high shunt R case, that of high Q_e and narrow-band response from each circuit. As was reasoned before, the conjugate poles will be far from the origin and very close to the $j\omega$ axis. Pole s_2 and one zero will contribute almost fixed vectors for the variation of ω near the poles s_1 and s_3. The ratio of vector magnitudes π_0 to π_2 will then be approximately $\frac{1}{2}$, and the angles will be equal and cancel. Similarly, pole s_4 and the other zero will contribute a second factor of $\frac{1}{2}$, and their angles will cancel. The gain, for ω near the poles at s_1 and s_3, is then a variable function of those poles only and can be written as

$$|A_v(\omega)| = \frac{g_m^2}{C_1C_2}\left(\frac{1}{4\pi_1\pi_3}\right) \tag{11-98}$$

where C_1 and C_2 are the respective tuning capacitances.

In general, the frequency response characteristic of such an amplifier is due to the poles concentrated near the $j\omega$ axis and remote from the origin on the positive $j\omega$ axis; the zeros at the origin and the poles on the negative $j\omega$ axis then participate only as a constant factor.

Figure 11-34(a) expands the region around the poles, s_1 and s_3, which are effective in producing the gain and phase variation. As before, the pole coordinate on the negative real axis is given by

$$b = \frac{-1}{2R_1C_1} = \frac{-1}{2R_2C_2}$$

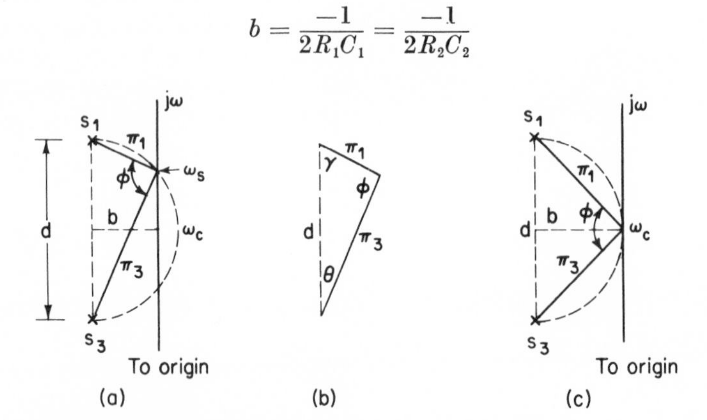

Figure 11-34. (a) Geometry around s_1 and s_3; (b) the vector triangle; (c) the limiting case for peak gain.

The value $\omega = \omega_c$ is the center frequency of the two poles, with their frequencies separated a distance d. The gain varies as $1/\pi_1\pi_3$, and π_1 and π_3 vary as ω moves along the $j\omega$ axis. We can achieve a better understanding of the manner in which the response varies with frequency if the variations of π_1 and π_3 are reduced to the change of a single variable.

The triangle formed by the distance d and the vectors π_1 and π_3 is presented separately in Fig. 11-34(b). From the law of sines it is known that

$$\frac{\pi_1}{\sin\theta} = \frac{\pi_3}{\sin\gamma} = \frac{d}{\sin\varphi} \tag{11-99}$$

However

$$\sin\theta = \frac{b}{\pi_3}, \qquad \sin\gamma = \frac{b}{\pi_1}$$

and so

$$\pi_1\pi_3 = \frac{bd}{\sin\varphi} \tag{11-100}$$

thus reducing the gain variation to a function of one variable, $\sin\varphi$.

Equation 11-98 can then be written

$$|A_v(\omega)| = \frac{g_m^2}{4C_1C_2}\frac{\sin\varphi}{bd}$$

By rearrangement, this becomes

$$|A_v(\omega)| = g_m^2 R_1 R_2 \frac{b}{d}\sin\varphi \tag{11-101}$$

Each amplifier stage has a peak gain given by $g_m R$.

The gain is maximum when $\varphi = 90°$ and varies as the ω point travels on the $j\omega$ axis. As ω moves from ω_p toward ω_c in (a), the angle φ increases above 90°, and the gain falls. As ω moves above ω_p, the angle φ decreases below 90° and the gain also falls. At a frequency remote from ω_c, the angle φ approaches zero and so does the gain.

The angle $\varphi = 90°$ requires that π_1 and π_3 intersect on the dashed circle of diameter d. However, the intersection of π_1 and π_3 must also occur on the $j\omega$ axis as a locus, so that there are two frequencies, $\pm\omega_p$, of peak gain for the case of (a), Fig. 11-34, where the so-called *peaking circle* intersects the $j\omega$ axis. That is,

$$\omega_p = \omega_c \pm \omega_\delta \tag{11-102}$$

and with $d/2$ being the radius of the peaking circle, it follows from the triangle that

$$\omega_\delta^2 + b^2 = \left(\frac{d}{2}\right)^2$$

$$\omega_\delta = \sqrt{\frac{d}{2}^2 - \frac{1}{4R^2C^2}} \tag{11-103}$$

When the pole separation is reduced to $d = 2b$, or

$$d = \frac{1}{RC}$$

the peaking circle has the location shown in Fig. 11-34(c), just tangent to the $j\omega$ axis. The value of φ at $\omega = \omega_c$ is 90°, and the two response peaks have collapsed to one at ω_c. This is called the condition of *maximally flat response* for such a double resonant circuit amplifier. For values of $d < 1/RC$ the peaking circle does not intersect the $j\omega$ axis, angle φ is maximum but less than 90° at $\omega = \omega_c$, and so there is a single-peak response with a lower value of gain.

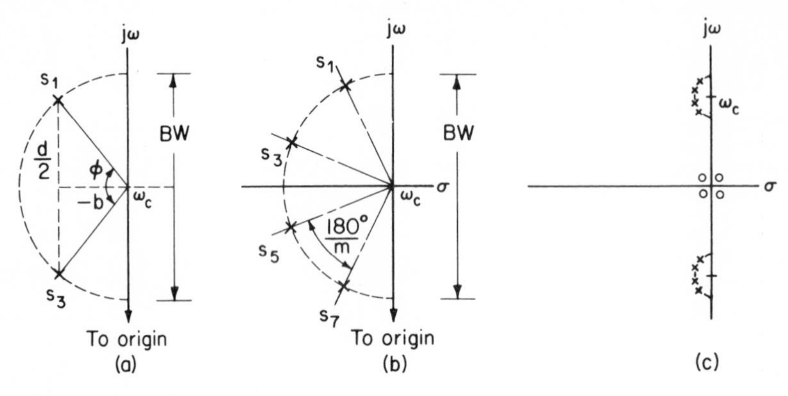

Figure 11-35. (a) Bandwidth circle for maximally flat case; (b) bandwidth circle for quadrupole; (c) pole-zero pattern for quadrupole.

The bandwidth may be determined by drawing another circle from the origin through the poles, as in Fig. 11-35. The radius is $d/\sqrt{2}$ and the bandwidth is $\text{BW} = 2d/\sqrt{2} = \sqrt{2}\,d$. For the maximally flat case the poles should be at $\pm 45°$ from the real axis, so that the gain is at maximum, at the single peak. For the overstaggered case, the ratio of peak gain to center frequency gain is

$$\left|\frac{A_p}{A_c}\right| = \frac{1}{\sin \varphi_c} = \lambda$$

and from the figure

$$\tan \frac{\varphi_c}{2} = \frac{d}{2b} \tag{11-104}$$

The parameter λ is the peak-to-dip ratio or the amplitude ripple in the pass band, and is usually specified. It thus fixes the pole angles.

This discussion indicates the method to be followed for design of amplifiers employing additional resonant circuits. A semi-circular locus, with a radius BW/2, is drawn with a center at ω_c of the pass band. The poles are then symmetrically disposed on this locus, as shown in Fig. 11-35(b). From the pole coordinates, the individual stage resonant frequencies ω_0 and bandwidth, $BW/\omega_0 = 1/Q_e = 2b/\omega_0$, can be computed, resulting in Table 11-1.

Table 11-1

DATA FOR n STAGGER-TUNED CIRCUITS†
OVERALL BANDWIDTH$=BW_0$

n	Number of stages tuned	Tuned to	Stage bandwidth
2	2	$f_0 \pm 0.35\ BW_0$	$0.70\ BW_0$
3	1	f_0	BW_0
	2	$f_0 \pm 0.43\ BW_0$	$0.50\ BW_0$
4	2	$f_0 \pm 0.46\ BW_0$	$0.50\ BW_0$
	2	$f_0 \pm 0.19\ BW_0$	$0.92\ BW_0$

†From Valley and Wallman, *Vacuum Tube Amplifiers*. McGraw-Hill Book Company, New York, 1948.

The total pole and zero picture for a four-stage amplifier would have the appearance of Fig. 11-35(c). Four zeros are at the origin for four resonant loads, and four poles are in each group concentrated around ω_c and $-\omega_c$. The reasoning which applies to the pole and zero vectors can then be repeated leading to the consideration of the poles surrounding ω_c. The more stages cascaded, the flatter will be the pass-band characteristic and the steeper the skirt cutoff, as indicated in Fig. 11-36(c). The aforementioned distribution of poles is known as a *Butterworth* arrangement.

Higher numbers of stages are not usually practical, since the poles close to the $j\omega$ axis may require values of Q_e higher than can be secured with practical components.

To make such amplifiers possible under the conditions of heavier loading

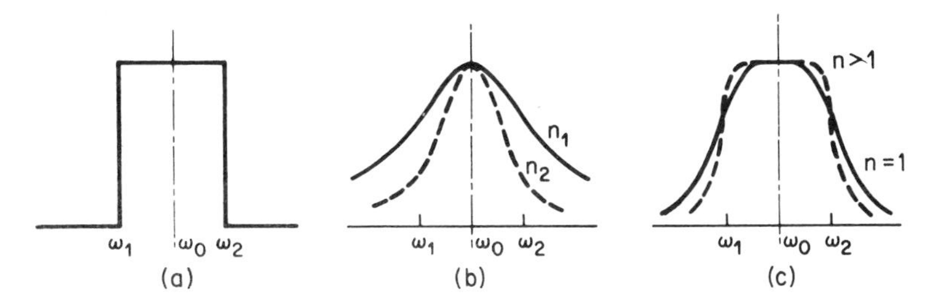

Figure 11-36. (a) Ideal selectivity; (b) narrow-band tuned circuit response; (c) maximally flat or Butterworth response.

imposed by transistors, the circuits will usually employ coil taps for connection to the collector circuits. The circuits can be individually aligned to frequency, and the order of resonant frequency distribution among the successive stages is unimportant.

11-18 DOUBLY TUNED AMPLIFIERS

The doubly tuned transformer is widely used for intermediate-frequency amplification at fixed frequencies, below several megahertz, as shown in Fig. 11-37. Its analysis leads to the same pole-zero pattern discussed for the stagger-tuned amplifier with $n = 2$. The mutual-inductance coupling becomes the means for positioning the associated resonant frequencies. If M is very small, the two resonant response frequencies are merged, but separate as M is made larger. The circuit furnishes an analytic demonstration paralleling the pole-zero methods of the previous section.

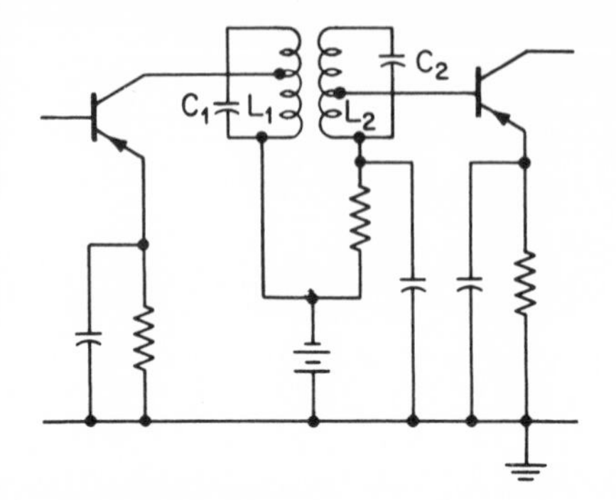

Figure 11-37. A doubly tuned transistor amplifier.

The gain function can be arrived at from Fig. 11-38(b). With device internal resistance large:

$$\mathbf{I}_2 = \frac{jg_m\mathbf{E}_g}{\omega C_1\mathbf{Z}_{T12}} = \frac{jg_m\mathbf{E}_g}{\omega C_1[\mathbf{z}_r - \mathbf{z}_i(\mathbf{z}_o + \mathbf{Z}_L)/\mathbf{z}_f]} \tag{11-105}$$

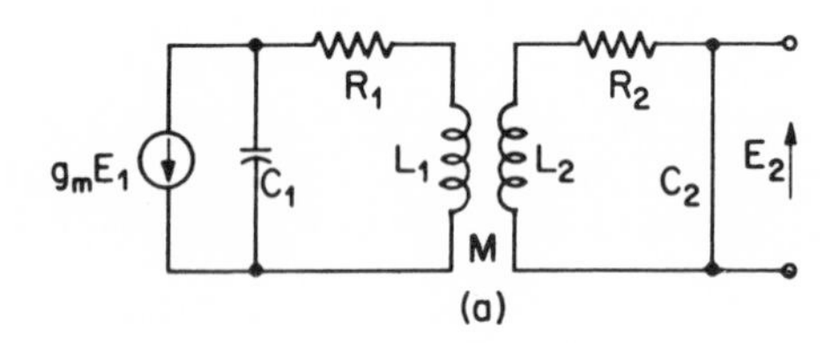

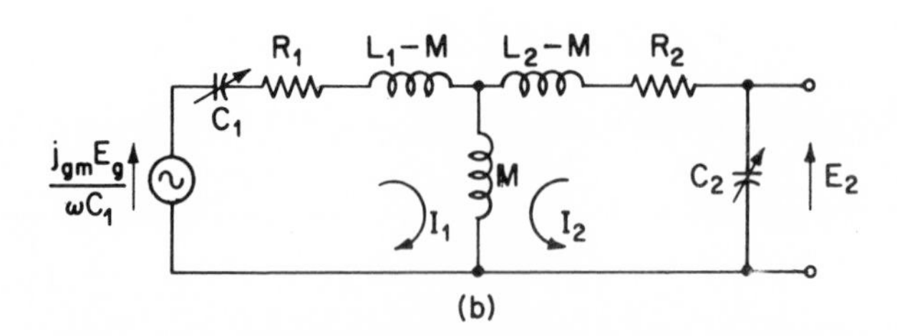

Figure 11-38. Doubly tuned transformer and equivalent circuit.

The transfer impedance may be assembled, starting with

$$\mathbf{z}_i = R_1 + j\left(\omega L_1 - \frac{1}{\omega C_1}\right) \cong \frac{\omega_0 L_1}{Q_1}(1 + j2Q_1\delta) \tag{11-106}$$

after noting that $1/(1 + \delta) \cong 1 - \delta$ for δ small. We also define

$$\omega_0^2 = \frac{1}{L_1C_1} = \frac{1}{L_2C_2}$$

The second mesh impedance can be written

$$\mathbf{z}_o + \mathbf{Z}_L \cong \frac{\omega_0 L_2}{Q_2}(1 + j2Q_2\delta) \tag{11-107}$$

The mutual impedance is

$$\mathbf{z}_r = \mathbf{z}_f = j\omega M = j\omega k\sqrt{L_1L_2} \tag{11-108}$$

by reason of the definition of the coefficient of coupling.

The transfer impedance then may be written, with δ small, as

$$\mathbf{Z}_T \cong \frac{j\omega_0\sqrt{L_1L_2}}{kQ_1Q_2}[k^2 + (1 + j2Q_1\delta)(1 + j2Q_2\delta)] \tag{11-109}$$

The secondary voltage then is

$$\mathbf{E}_2 = \frac{j\mathbf{I}_2}{\omega C_2} = \frac{-g_m\omega_0^2L_1L_2\mathbf{E}_g}{(1 + \delta)^2 Z_T}$$

after use of the resonance relations. The gain follows as

$$\mathbf{A}_v(\omega) = \frac{\mathbf{E}_2}{\mathbf{E}_g} = \frac{jg_m\omega_0k\sqrt{L_1L_2}Q_1Q_2}{k^2Q_1Q_2 + (1 + j2Q_1\delta)(1 + j2Q_2\delta)} \tag{11-110}$$

These circuits are frequently designed with $Q_1 = Q_2 = Q$. Then

$$\mathbf{A}_v(\omega) = \frac{g_m\omega_0k\sqrt{L_1L_2}Q^2}{4Q\delta - j(1 + k^2Q^2 - 4Q^2\delta^2)} \tag{11-111}$$

It is possible to find the value of the frequency variation δ at which the gain is maximum by taking the derivative of the absolute gain magnitude, and setting it equal to zero. This yields

$$\delta(4Q^2\delta^2 + 1 - Q^2k^2) = 0$$

and

$$\delta_{1,2} = \pm\frac{1}{2}\sqrt{k^2 - \frac{1}{Q^2}} \tag{11-112}$$

These are the values of δ at the maxima, and $\delta = 0$ produces a minimum between the two peaks.

For the special case in which

$$k = k_c = \frac{1}{Q}$$

the maximum gain occurs only at the value of $\delta = 0$. This value, $k_c = 1/Q$, is called *critical coupling*, and corresponds to the conditions which made $\varphi = 90°$ at $\omega = \omega_c$ in the pole-zero solution. For values of $k < 1/Q$ the peak will occur at $\omega = \omega_c$, but the peak gain will be less than the maximum. The condition of $k < 1/Q$ is called *insufficient coupling*.

The gain in terms of $k/k_c = kQ$ is

$$|A_v(\omega)| = g_m Q \omega_0 \sqrt{L_1 L_2} \left[\frac{kQ}{\sqrt{(1 + k^2Q^2 - 4Q^2\delta^2)^2 + 16Q^2\delta^2}} \right] \quad (11\text{-}113)$$

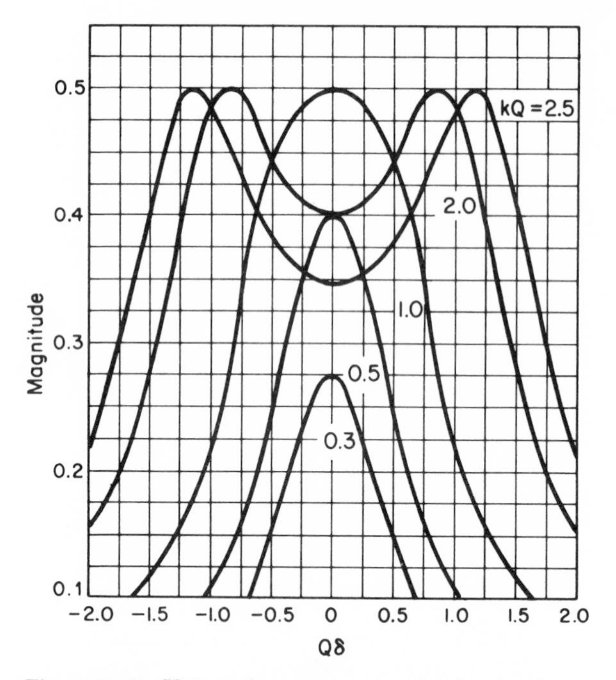

Figure 11-39. Universal response curves for the circuit of Figure 11-38.

The magnitude of the term in brackets is drawn in Fig. 11-39, with kQ as a parameter. These curves illustrate the identity of performance of the doubly-tuned circuit and the stagger-tuned amplifier with $n = 2$. The widening of the single peak for the maximally-flat case with $kQ = 1$ can be seen.

The gain is a function of Q, L_1 and L_2, with the choice of k fixing the frequency response characteristic.

11-19 THE OVERCOUPLED CIRCUIT

For values of $kQ > 1$, the circuit is *overcoupled*, and the situation is that in which the peaking-circle locus has two intersections with the $j\omega$ axis. The frequencies of the response peaks in Fig. 11-40 can be found as

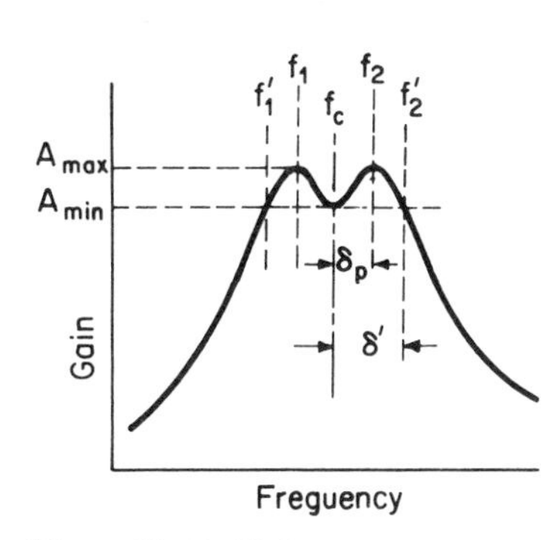

Figure 11-40. Reference frequencies for the overcoupled transformer.

$$f_1 = f_0\left(1 - \frac{1}{2Q}\sqrt{k^2Q^2 - 1}\right) \quad (11\text{-}114)$$

$$f_2 = f_0\left(1 + \frac{1}{2Q}\sqrt{k^2Q^2 - 1}\right) \quad (11\text{-}115)$$

The gain magnitude at either peak is

$$|A_p| = \frac{g_m Q \omega_0 \sqrt{L_1 L_2}}{2} \quad (11\text{-}116)$$

The gain at the dip at center frequency can be written for $\delta = 0$ as

$$|A_c| = g_m Q \omega_0 \sqrt{L_1 L_2}\left(\frac{kQ}{1 + k^2Q^2}\right) \quad (11\text{-}117)$$

Two additional frequencies of calculable response can be identified as f_1' and f_2' on the curve shoulders, at which the gain has fallen to a level equal to $|A_c|$. This result provides the bandwidth in which the variation in the gain is not greater than $|A_p/A_c| \leqq \lambda$. Equating $|A_c|$ to Eq. 11-113 gives

$$\frac{kQ}{1 + k^2Q^2} = \frac{kQ}{\sqrt{(1 + k^2Q^2 - 4Q^2\delta'^2) + 16Q^2\delta'^2}}$$

where δ' is the frequency deviation at f_1' and f_2'. Solving for δ' gives

$$\delta' = \pm\frac{1}{\sqrt{2}\,Q}\sqrt{k^2Q^2 - 1} \quad (11\text{-}118)$$

However, f_1 and f_2 are separated as

$$f_2 - f_1 = 2\delta_p = \frac{1}{Q}\sqrt{k^2Q^2 - 1} \quad (11\text{-}119)$$

so it appears that the bandwidth between the δ' frequencies is

$$2\delta' = 2\sqrt{2}\,\delta_p = \sqrt{2}\,(f_2 - f_1) \quad (11\text{-}120)$$

If the ratio $|A_p/A_c| = \lambda$ is made 3 dB then the normal 3 dB bandwidth is expressed by Eq. 11-120; it is $\sqrt{2}$ times the peak separation. For a circuit with critical coupling and only one peak, the double-tuned circuit with the maximally flat response will have a bandwidth $\sqrt{2}$ times as great as a singly tuned circuit. This was also predicted by the pole-zero analysis.

For the overcoupled case a means of control of the dip magnitude or the response ripple may be developed. The ratio λ is

$$\lambda = \left|\frac{A_p}{A_c}\right| = \frac{1 + k^2Q^2}{2kQ} \quad (11\text{-}121)$$

Solving for the value of kQ which will give a specified value of λ,

$$kQ = \lambda + \sqrt{\lambda^2 - 1} \quad (11\text{-}122)$$

We choose the positive sign, since kQ is always greater than unity for the overcoupled case. The designer may then find the necessary kQ value for a specified λ ratio.

The phase response is most linear for $k = 0.577/Q$, while the maximally-flat gain case requires $k = 1/Q$, and so a compromise in design is necessary.

The frequency band in which the gain deviation remains less than λ is given by

$$\Delta f_\lambda = \frac{\sqrt{2} f_0}{Q} \sqrt{k^2Q^2 - 1} \tag{11-123}$$

If the 3 dB bandwidth is desired, the value of $\lambda = \sqrt{2}$. Then kQ should be 2.032 and

$$\text{BW} = \frac{2.5 f_0}{Q} \tag{11-124}$$

Thus by choosing the Q value and determining k from Eq. 11-122, any desired value of bandwidth and ripple magnitude λ may be obtained. However, if the top of the response curve is to be nearly flat, or the ripple is to be small and BW large, then the gain will be reduced.

11-20 THE CHEBYSHEV RESPONSE

The overcoupled doubly-tuned amplifier is actually one of a general class of filter-loaded amplifiers: those with equal-ripple or *Chebyshev response* in the pass band. The gain function employs Chebyshev polynomials, a functional form developed in the search for a linkage for steam locomotive use that would give a uniform straight-line motion within a specified tolerance.

The response of singly-tuned and multiply-tuned superimposed resonant frequency amplifiers is shown in Fig. 11-36(b), with a resultant narrowing of the pass band with increasing n, the number of tuned circuits. The gain function can be generalized as

$$|A(\delta)| = (g_m Z_L)^n \frac{1}{[1 + (2Q\delta)^2]^{n/2}} \tag{11-125}$$

where the bandwidth is $(\sqrt{2^{1/n} - 1})^{-1}$ times the width of a single stage, or is less than the single-stage band width.

If Eq. 11-125 is modified to

$$|A(\delta)| = (g_m Z_L)^n \frac{1}{[1 + (2Q\delta)^{2n}]^{1/2}} \tag{11-126}$$

the maximally flat response function is obtained as shown in Fig. 11-36(c), resulting from appropriate placement of response poles on a circular locus. As n increases, the shape of the response nears the ideal rectangle and the bandwidth remains constant; i.e., the number of poles is increased but the diameter of the pole-locus circle remains constant. The band limit will still occur at $2Q\delta = \pm 1$, or where $|A(\delta)| = 1/\sqrt{2}\, g_m Z_L$ as defined.

It has been found that if the gain function be further modified to

$$A(\delta) = (g_m Z_L)^n \frac{1}{[1 + \epsilon C_n^2(2Q\delta)]^{1/2}} \tag{11-127}$$

where $C_n(2Q\delta)$ is a Chebyshev polynomial, the response will be one of equal ripple as in Fig. 11-41. The Chebyshev polynomials, which for the first few orders are

$$C_1(x) = x, \qquad C_3(x) = 4x^3 - 3x$$

$$C_2(x) = 2x^2 - 1, \qquad C_4(x) = 8x^4 - 8x^2 + 1$$

give values of $C_n^2(x)$ which oscillate between 0 and 1 for the band $-1 \leq x \leq 1$. The magnitude $|A|$ then oscillates between 1 and $1/\sqrt{1 + \epsilon}$, and therefore ϵ is the ripple parameter, defined as

$$\epsilon = \left[\left(\frac{A_p}{A_c}\right)^2 - 1\right]^{1/2} \tag{11-128}$$

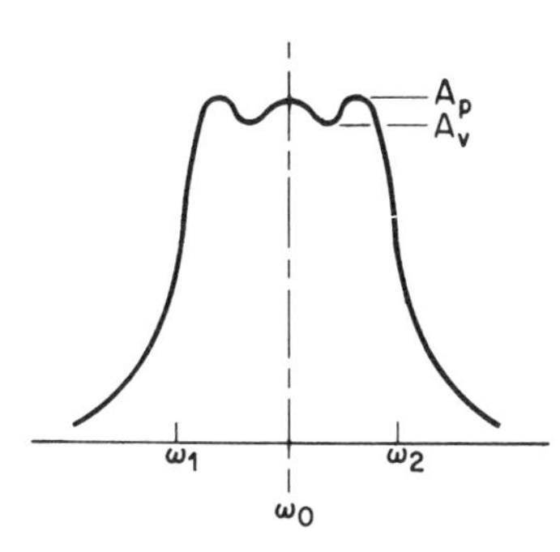

Figure 11-41. Chebyshev response.

This value of ripple can be selected by the designer, as was λ for the overcoupled circuit. Outside the pass band, $C_n^2(x)$ varies as x^{2n}, and the gain falls off as it does for the maximally flat response, that response being a special case of equal ripple response with ripple of zero. The Chebyshev response provides somewhat more gain for the same pass bandwidth, and a closer approach to constant gain over the band than is possible with the maximally flat approximation.

Expansion of these gain functions with $j\omega = s$, will yield an expression of the form

$$A(s) = \frac{H}{(s - s_1)(s - s_2)(s - s_3) \ldots} \tag{11-129}$$

The form of the denominator which results for various n values, and either the maximally flat or the equal-ripple basic equations, is given in tables, of which the first few orders are reproduced in Table 11-2.

Equations of the form of 11-129 will, in turn, locate the poles for the tuned circuits. In the s plane the equal-ripple response is a result of pole locations spaced along an ellipse as a locus. The major axis will equal the bandwidth on the $j\omega$ axis, and this is constant for varying n. Thus the maximally flat response, with a circular locus, is a special case of an elliptical locus.

It is possible through these methods to develop tuned multistage amplifiers with overall bandwidths as high as 30 per cent of the center frequency. However, the development of the complete theory is beyond the scope undertaken by this text. Further information may be found in the references.

Table 11-2(a)

BUTTERWORTH RESPONSE

n	
1	$s + 1$
2	$s^2 + \sqrt{2}\,s + 1$
3	$(s + 1)(s^2 + 1.000s + 1)$
4	$(s^2 + 0.765s + 1)(s^2 + 1.848s + 1)$

Table 11-2(b)

CHEBYSHEV RESPONSE
($\frac{1}{2}$ dB ripple; $\epsilon = 0.349$)

n	
1	$s + 2.863$
2	$s^2 + 1.426s + 1.516$
3	$(s + 0.626)(s^2 + 0.626s + 1.142)$
4	$(s^2 + 0.351s + 1.064)(s^2 + 0.845s + 0.356)$

11-21 NEUTRALIZATION

The Miller effect in both transistor and triode is a variable which creates design and performance problems. With inductive load always possible with resonant circuits, the appearance of negative resistance in the input circuit leads to unstable gain and possibly to oscillation. The introduction of a varying capacitance also adversely affects the tuning of resonant circuits in the input.

Since the feedback is due to either C_{gp} or C_{cb}, these capacitances should be made small. The pentode avoids the problem because C_{gp} is made small by design. If the feedback capacitances cannot be made small, it is possible to *neutralize* that effect by circuit alteration. When triodes are used in transmitters at high frequencies, they are neutralized by the circuit of Fig. 11-42(a). For transistors, the gain may

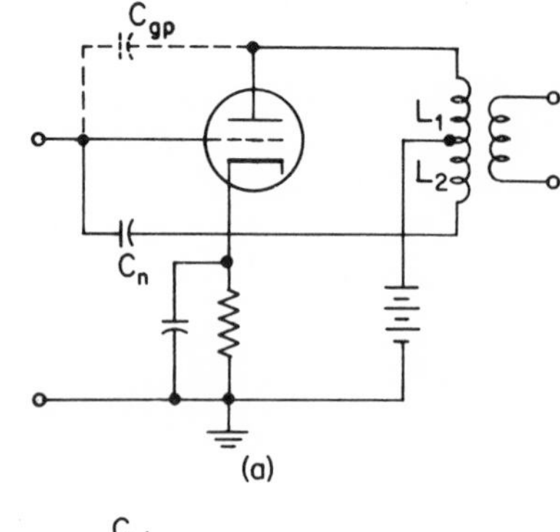

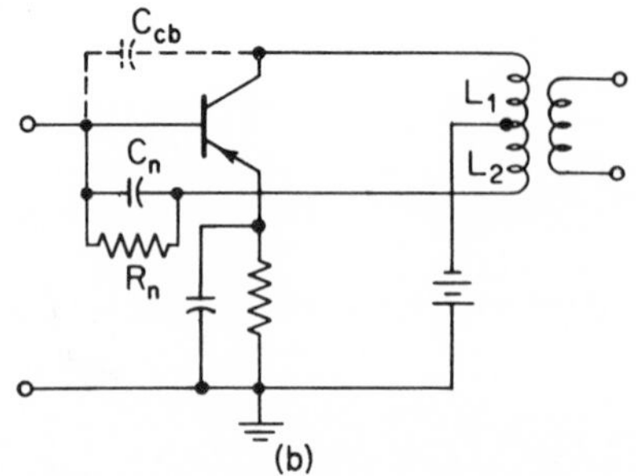

Figure 11-42. Neutralizing circuits.

be reduced to decrease the effect on the input circuit, or transistors may also be neutralized as in Fig. 11-42(b).

It can be seen from the circuits that the effect of the neutralizing circuit is to place the base-emitter input across the points of a bridge composed of C_{cb}, C_n, L_1, L_2 in the transistor case. If the bridge is balanced, or $C_n = C_{cb}$ for $L_1 = L_2$, then the C_n path feeds back to the input an equal and opposed current to that transmitted through C_{cb}. For the transistor, C_{cb} is not always purely capacitive, and a small resistive component of current may have to be introduced by R_n.

Neutralization is somewhat a function of frequency, and is affected by change of the active device. As a result, it is often more desirable to design the circuit with a lower load and gain, and avoid the neutralization problem.

PROBLEMS

11-1. A rectangular pulse of duration 1.57μs is repeated 3000 times per second. Determine the amplifier bandwidth if good reproduction requires the inclusion of the second zero of the harmonic amplitude envelope. What is the order of the highest transmitted harmonic?

11-2. Tabulate the amplitude of all harmonics resulting from a 100 hertz square wave, out to the ninth harmonic. Plot these against frequency, along with the enclosing spectrum envelope. Explain the absence of even harmonics.

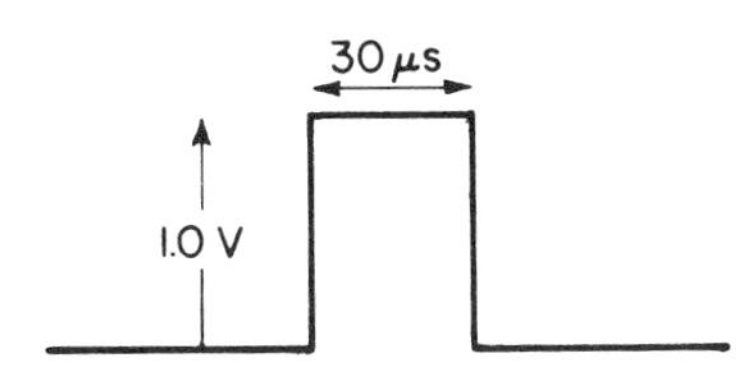

Figure 11-43.

11-3. A rectangular pulse of the form of Fig. 11-43 is applied to an *RC*-coupled amplifier. If the pentode has $g_m = 4500$ μmho, r_p and R_g are large, and $C_g =$ 22 pF, find:

(a) The value of R_L for a rise time of 0.25 μs.

(b) The value of the coupling capacitor to make the sag less than 10 per cent, if $R_g = 0.5$ megohm.

(c) The output peak voltage.

11-4. The circuit of Fig. 11-15(a) is used with a pentode having $g_m = 5000$ μmho and high r_p; $R_g = 1$ megohm, and the total of C_{pk} and C_{gk} is 50 pF. If the mid-frequency gain is to be 15, find the value of R and L needed to shunt-compensate the circuit to as high a frequency as possible, with negligible rise of gain over A_m. Plot a gain-frequency curve over the critical region.

11-5. Using the coefficient-equating method of Section 11-7, determine an expression for R in Fig. 11-44 which will give the flattest low-frequency response, and lower f_1. The expression is to be in terms of R, L, C, and the transistor parameters.

11-6. (a) Find the output voltage for an input signal of 0.01 v_i in the circuit of Fig. 11-26(c). The transistor has $g_m = 7000$ μmho, and $1/h_{oe}$ over 1 megohm. The inductor has $L = 70$ μH, $Q = 200$; $R_B = 2$ megohms; and the reactance of C is negligible. The resonant frequency is 450 kHz.

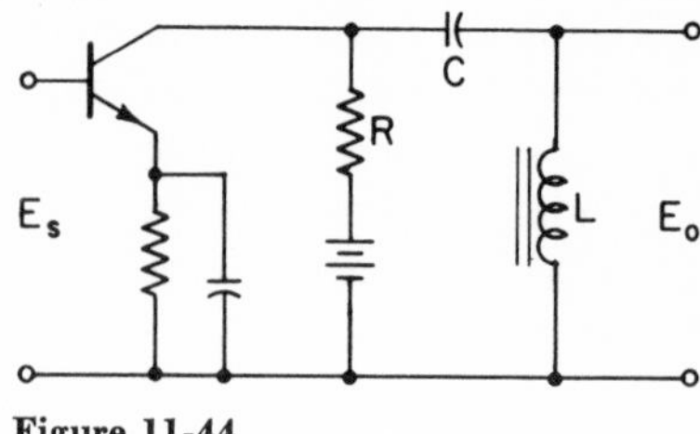

Figure 11-44.

(b) Find the output voltage for a simultaneously received equal signal at 500 kHz.

11-7. The circuit of Fig. 11-30 is used with a pentode having $g_m = 5000$ μmho, and $r_p = 0.5$ megohm. The transformer has $L_1 = 30$ μH, $L_2 = 200$ μH, $M = 10$ μH, $R_1 = 5$ ohms, $R_2 = 20$ ohms. The input is 0.1 V, 10^6 Hz, and the secondary circuit is resonant at that frequency. Find E_2.

11-8. The doubly tuned transformer of Fig. 11-38 has $L_1 = L_2 = 100$ μH, $Q_1 = Q_2 = 100$, with coupling at 150 per cent of critical. The source has $g_m = 4500$ μmho and high internal resistance.
(a) Find the values for C_1 and C_2 for resonance at 1 megahertz.
(b) Find the maximum gain available.
(c) Compute the bandwidth and the value of λ achieved.

11-9. A doubly tuned transformer of the form in Fig. 11-38 is to have $f_c = 450$ kilohertz, with bandwidth between response peaks of 100 kilohertz, the ratio $A_p/A_c = \sqrt{2}$, and $A_p = 155$, $g_m = 10^{-3}$ mho. Determine the required values for L, C, and k and sketch the response characteristic to scale.

11-10. (a) Obtain the transfer function E_2/I_1 for the circuit of Fig. 11-45(a).
(b) If $R_1 = 20{,}000$ ohms, $R_2 = 2000$ ohms, $C = 0.1$ μF, sketch the pole-zero pattern to scale.
(c) Using scale and protractor, determine graphically the gain and phase characteristics up to mid-frequency and plot these curves.

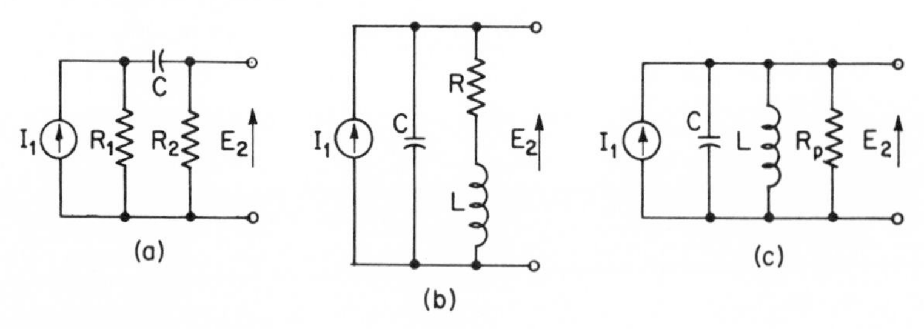

Figure 11-45.

11-11. (a) The circuit of Fig. 11-45(b) has $L = 0.01$ H, $C = 1$ μF, $R = 3$ ohms. Determine ω_0, the bandwidth, and the resonant Q.
(b) Reassemble the circuit, using the parallel R form of Fig. 11-45(c), and determine the needed element values.

11-12. The circuit of Fig. 11-45(b) has a transfer function

$$\frac{E_2}{I_1}(s) = \frac{1}{C}\frac{s + (R/L)}{s^2 + (R/L)s + 1/LC}$$

(a) Draw and dimension the pole-zero diagram and pole loci as R is varied from zero.

(b) Repeat for C increasing from zero.

11-13. The generator of Fig. 11-45(c) has a g_m of 10 mA/V. The center frequency is to be 20 megahertz, bandwidth 500 kilohertz. Select values for R_p and L if $C = 16$ pF, and calculate the voltage gain.

11-14. A two-stage wide-band amplifier is to be designed for maximally-flat response. One stage will be shunt-compensated, with a pole-zero pattern of Fig. 11-16; the second stage will be uncompensated, and contributes a pole located at the zero of the compensated stage. The poles of the compensated stage will be so located that angle $\varphi = 45°$ to the origin, capacitance shunted across each stage is 10 pF, and $g_m = 10{,}000$ μmho.

(a) Find the values of R and L as needed for loads.

(b) What is A_m for a frequency of 3 megahertz as f_2 of the uncompensated stage?

11-15. A double-tuned circuit couples a transistor of $R_o = 40{,}000$ ohms output to the input of a transistor having $R_{ie} = 1000$ ohms. The center frequency is to be 450 kilohertz, and the bandwidth is to be 15 kilohertz. If the circuit has $k/k_c = 1.5$ find the circuit Q, and the needed values of L and C and M. The secondary will be tapped to match $R_{ie} = 1000$ ohms.

REFERENCES

1. Bedford, A. V., and G. L., Fredendall, "Transient Response of Multi-Stage Video-Frequency Amplifiers," *Proc. IRE*, Vol. 27, p. 277 (1939).

2. Wheeler, H. A., "Wideband Amplifiers for Television," *Proc. IRE*, Vol. 27, p. 429 (1939).

3. Landon, V. D., "Cascade Amplifiers with Maximal Flatness," *RCA Rev.*, Vol. 5, pp. 347, 481 (1941).

4. Darlington, S., "The Potential Analog Method of Network Synthesis," *Bell System Tech. Jour.*, Vol. 30, p. 315 (1951).

5. Pettit, J. M., and M. M. McWhorter, *Electronic Amplifier Circuits*. McGraw-Hill Book Company, New York, 1961.

6. Shea, R. F., *Transistor Applications*. John Wiley & Sons, Inc., New York, 1964.

7. Angelo, E. J., Jr., *Electronic Circuits*. McGraw-Hill Book Company, New York, 1964.

8. Glasford, G. M., *Linear Analysis of Electronic Circuits*. Addison-Wesley Pub. Co., Reading, Mass., 1965.

12

FEEDBACK

Negative or inverse feedback in amplifiers implies a comparison between output and input wave forms. Any difference between the wave forms generates an error signal which is then used to control the amplifier so as to reduce the error toward zero. The operations of comparison and feedback of the error signal are of broad application in amplification and control, and through feedback it is possible to stabilize amplifier gain and to reduce the distortion created in amplifiers.

12-1 FEEDBACK IN AMPLIFIERS

For an amplifier the gain has been defined as

$$\mathbf{A} = \frac{\mathbf{V}_o}{\mathbf{V}_s}$$

An amplifier with feedback is diagrammed in Fig. 12-1, with a fraction of the output voltage fed back into the input circuit. The overall gain with feedback present is identified as $\mathbf{A}'$, and

$$\mathbf{A}' = \frac{\mathbf{V}'_o}{\mathbf{V}_s} \tag{12-1}$$

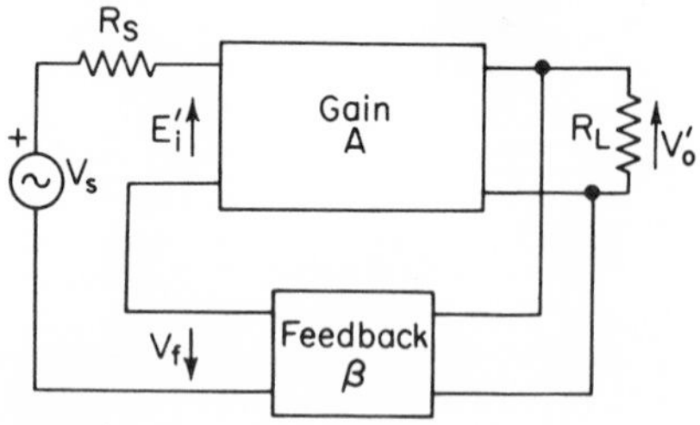

Figure 12-1. Series input, voltage feedback.

The voltage fed back to the input circuit is

$$\mathbf{V}_f = \beta \mathbf{V}'_o \tag{12-2}$$

where $\boldsymbol{\beta}$ is the reverse transfer ratio† of the feedback network, complex in general. The input voltage with feedback is

$$\mathbf{E}_i' = \mathbf{V}_s + \mathbf{V}_f \tag{12-3}$$

This is the controlling error signal. So

$$\mathbf{V}_o' = \mathbf{A}(\mathbf{V}_s + \boldsymbol{\beta}\mathbf{V}_o') \tag{12-4}$$

$$\mathbf{A}' = \frac{\mathbf{V}_o'}{\mathbf{V}_s} = \frac{\mathbf{A}}{1 - \mathbf{A}\boldsymbol{\beta}} \tag{12-5}$$

which becomes a defining relation for feedback amplifiers.

Examination of Eq. 12-5 shows that if

$$|1 - \mathbf{A}\boldsymbol{\beta}| < 1$$

then gain $|A'|$ is greater than $|A|$ before feedback was added. The feedback is then said to be *positive* and the circuit is *regenerative*. In general, positive feedback causes an increase in gain, decreased gain stability, and higher distortion. It is avoided for most applications.

If

$$|1 - \mathbf{A}\boldsymbol{\beta}| > 1$$

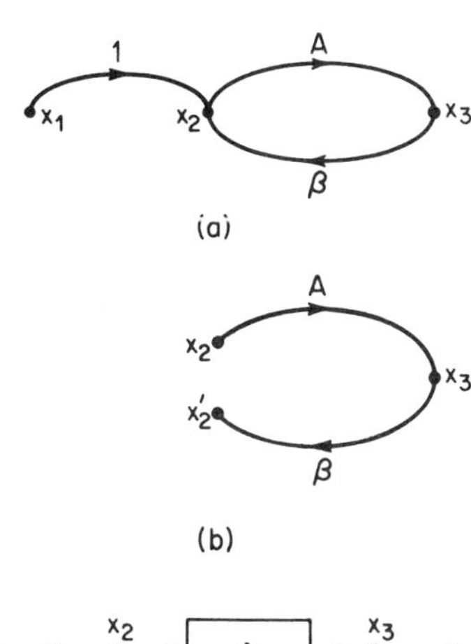

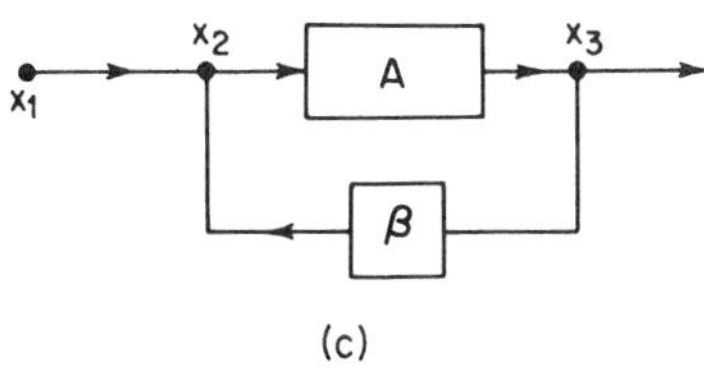

Figure 12-2. Signal-flow graphs of the feedback amplifier.

then gain $|A'|$ is less than $|A|$. The feedback is said to be *negative* or *inverse* and the circuit is *degenerative*. Negative feedback improves gain stability, and reduces distortion generated in an amplifier.

When

$$\mathbf{A}\boldsymbol{\beta} = 1 + j0$$

the gain becomes infinite and the amplifier becomes an oscillator, giving an output voltage which is independent of any external input voltage.

While the term $\mathbf{A}\boldsymbol{\beta}$ is complex in general, it is usually desirable to maintain its components as real numbers, either positive or negative, by circuit design. If $|1 - \mathbf{A}\boldsymbol{\beta}| > 1$, then $\mathbf{A}\boldsymbol{\beta}$ must carry an angle of 180° for negative feedback. That is, the gain must be $-\mathbf{A}$ over the frequency range or $\boldsymbol{\beta}$ must introduce an out-of-phase voltage. Equation 12-3 defines the

†The notation $\boldsymbol{\beta}$ is a long-standing convention, and to avoid confusion with the symbol for the current amplification factor for the common-emitter circuit, the latter will be referred to as h_{fe} in this chapter.

comparison process by convention, and the actual comparison of input and feedback signals creates a difference signal for negative feedback.

Figure 12-2 illustrates the signal-flow diagram for a feedback amplifier. In (b) the feedback circuit is removed from the input node and a signal $\mathbf{V}_i$ applied at x_2. The voltage appearing at the open $\boldsymbol{\beta}$-circuit terminal at x_2' is $\mathbf{A}\boldsymbol{\beta}\mathbf{V}_i$. The ratio of voltages at x_2' and x_2 is then

$$\mathbf{A}_o = \frac{\mathbf{A}\boldsymbol{\beta}\mathbf{V}_i}{\mathbf{V}_i} = \mathbf{A}\boldsymbol{\beta} \tag{12-6}$$

and this result is known as the *open-loop gain.*

If a unit signal is applies at x_2, the output at x_2' is $\mathbf{A}\boldsymbol{\beta}$, so that the difference between the input signal and that returned by the open feedback loop is $1 - \mathbf{A}\boldsymbol{\beta}$. This term is therefore called the *return difference.*

The overall gain, with the feedback loop closed as in (a), is $\mathbf{A}'$, as a function of $\mathbf{A}$ and $\boldsymbol{\beta}$ and their phase angles. Since this gain is measured with the feedback loop in place, A' is known as the *closed-loop gain.*

The amount of feedback employed is sometimes stated in decibels, as the ratio of the power output without feedback to that with feedback, for constant input. Thus:

$$\text{dB feedback} = 10 \log \frac{V_o^2/R}{V_o'^2/R} = 20 \log \frac{V_o}{V_o'}$$

$$= 20 \log \frac{V_o/V_s}{V_o'/V_s} = 20 \log \frac{A}{A'} \tag{12-7}$$

and the dB of feedback is equal to the return difference, expressed in dB. If an open-loop gain curve be plotted in dB, with a second plot for the closed-loop gain, the return difference is identifiable as the dB difference between the curves, since subtracting on the logarithmic gain scale is equivalent to division.

As the overall gain is reduced by negative feedback, the input signal $\mathbf{V}_s$ must be increased if the same output is to be obtained. That is,

$$\frac{\mathbf{A}'}{\mathbf{A}} = \frac{1}{1 - \mathbf{A}\boldsymbol{\beta}}$$

and for equal output voltages

$$\mathbf{A}\mathbf{V}_s = \mathbf{A}'\mathbf{V}_s'$$

from which

$$\mathbf{V}_s' = (1 - \mathbf{A}\boldsymbol{\beta})\mathbf{V}_s \tag{12-8}$$

Example: An amplifier has a gain $A = 70 \underline{/180^\circ}$, and a normal input signal of 1.0 V rms. Feedback with $\beta = 0.1$ is added, and with $A\beta = -7.0$ the overall or closed-loop gain is

$$A' = \frac{A}{1 - A\beta} = \frac{-70}{1 + 7.0} = -8.75$$

An input voltage of

$$V_s' = (1 - A\beta)V_s = 1.0(1 + 7) = 8.0 \quad \text{V}$$

will then be needed.

The output voltage without feedback was

$$V_o = AV_s = 70 \times 1.0 = 70.0 \quad \text{V}$$

and with feedback and the increased input

$$V_o' = A'V_s' = 8.75 \times 8.0 = 70.0 \quad \text{V}$$

With feedback the actual input voltage is

$$V_i' = V_s' + \beta V_o' = 8.0 - 0.10 \times 70 = 1.0 \quad \text{V}$$

The dB of feedback introduced can be found as

$$\text{dB} = 20 \log \frac{70}{1.0 \times 8.75} = 18$$

With negative feedback V_s' becomes 8.0 V for the 70 V output, but −7.0 V is returned through the feedback loop for comparison with the input, and V_i' is reduced to the original 1.0 V level. Thus the external signal requirements are raised in the presence of negative feedback, but inside of the feedback loop the amplifier is operating under conditions identical with those existing without feedback.

12-2 NEGATIVE FEEDBACK AND GAIN STABILITY

With negative feedback and with $|A\beta| \gg 1$, Eq. 12-5 becomes

$$A' \cong -\frac{1}{\beta} \tag{12-9}$$

showing that with very large internal gain, the closed-loop gain can be made essentially independent of device parameters and supply voltage variations.

If β is made a function of frequency, the overall closed-loop gain will be an inverse of that frequency function. If the circuit determining β is designed to reject some particular frequency range, the amplifier gain will be high in that range, and vice versa.

Even when the open-loop gain $A\beta$ is not large, considerable stability of gain can be achieved with negative feedback. For an incremental change dA in the internal gain, Eq. 12-5 leads to

$$dA' = \frac{(1 - A\beta)dA + A\beta\, dA}{(1 - A\beta)^2}$$

$$\frac{dA'}{A'} = \frac{1}{1 - A\beta}\frac{dA}{A} \tag{12-10}$$

The *gain sensitivity* may be defined as

$$S_A = \frac{dA'/A'}{dA/A} = \frac{1}{1 - A\beta} \tag{12-11}$$

Thus the per cent change in feedback amplifier gain as related to internal gain change depends in inverse fashion on the return difference.

12-3 EFFECT OF FEEDBACK ON OUTPUT AND INPUT RESISTANCES

The choice of series or shunt connection for the source of feedback voltage, and the choice of series or shunt introduction into the input circuit, leads to four basic feedback circuit forms, as in Fig. 12-3. If the feedback source is the load voltage or is derived in shunt with the load, the result is *voltage feedback*, as in (a) and (b). If the feedback source is derived from the load current or is in series with the load, the result is *current feedback*, as in (c) and (d).

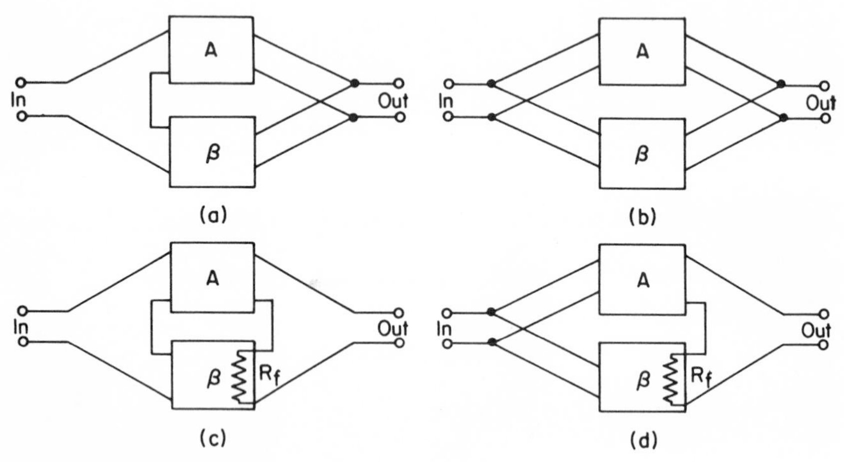

Figure 12-3. Voltage or shunt feedback: (a) series input; (b) shunt input. Current or series feedback: (c) series input; (d) shunt input.

For constant input, the voltage feedback circuit tends to maintain constant load voltage, independent of load resistance. With $\mathbf{V}_s$ short-circuited, a test voltage $\mathbf{V}_t$ applied to the amplifier output as in Fig. 12-4 will measure the output resistance as

$$R_o' = \frac{\mathbf{V}_t}{\mathbf{I}_t} \tag{12-12}$$

assuming the $\boldsymbol{\beta}$ circuit takes negligible current. With $\mathbf{V}_i = \boldsymbol{\beta}\mathbf{V}_t$, the amplifier output will be $\mathbf{V}_o = \mathbf{A}\boldsymbol{\beta}\mathbf{V}_t$, and in the output mesh

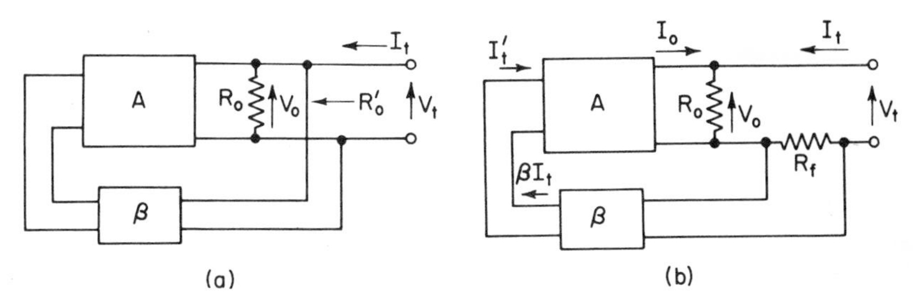

Figure 12-4. (a) Voltage feedback; (b) current feedback.

$$\mathbf{I}_t = \frac{\mathbf{V}_t - \mathbf{V}_o}{R_o} = \frac{\mathbf{V}_t(1 - \mathbf{A}\boldsymbol{\beta})}{R_o} \tag{12-13}$$

where $\mathbf{A}\boldsymbol{\beta}$ will normally be negative and real by design. The output resistance with voltage feedback then is

$$R'_o = \frac{\mathbf{V}_t}{\mathbf{I}_t} = \frac{R_o}{1 - A\beta} \tag{12-14}$$

This result indicates *a reduction of the output resistance* of the amplifier by the factor $1/(1 - A\beta)$ because of the voltage feedback.

The current feedback circuit with constant input tends to keep a constant voltage across R_f and therefore maintains a constant load current. In Fig. 12-4(b) we apply a test voltage $\mathbf{V}_t$ at the output and short-circuit the normal input source voltage. Then $\mathbf{I}_i = -\boldsymbol{\beta}\mathbf{I}_t$ as connected, and $\mathbf{I}_o = -\mathbf{A}\boldsymbol{\beta}\mathbf{I}_t$. However,

$$\mathbf{V}_o = (\mathbf{I}_o + \mathbf{I}_t)R_o = \mathbf{I}_t R_o(1 - \mathbf{A}\boldsymbol{\beta}) \tag{12-15}$$

and if $R_f \ll R_o$, then $\mathbf{V}_o = \mathbf{V}_t$ and so

$$R'_o = \frac{\mathbf{V}_t}{\mathbf{I}_t} = R_o(1 - \mathbf{A}\boldsymbol{\beta}) \tag{12-16}$$

which indicates that the *output resistance increases* with current feedback.

Series or shunt input of the feedback voltage or current alters the input resistance of the feedback amplifier. Without feedback, or with $\mathbf{V}_f = 0$, the input resistance is

$$R_{\text{in}} = \frac{\mathbf{V}_s}{\mathbf{I}_i} \tag{12-17}$$

With feedback introduced in series with the signal,

$$\mathbf{V}'_s = (1 - \mathbf{A}\boldsymbol{\beta})\mathbf{V}_s$$

from Eq. 12-8, and

$$\mathbf{I}_i = \frac{\mathbf{V}'_s}{\mathbf{R}'_{\text{in}}} = \frac{(1 - \mathbf{A}\boldsymbol{\beta})\mathbf{V}_s}{R'_{\text{in}}}$$

$$R'_{\text{in}} = R_{\text{in}}(1 - \mathbf{A}\boldsymbol{\beta}) \tag{12-18}$$

Series introduction of the feedback voltage *raises the input resistance* of the amplifier with negative feedback, as in Fig. 12-3(a) or (c).

A similar analysis can be carried out for the shunt input circuit of Fig. 12-3(b) or (d), leading to

$$R'_{\text{in}} = \frac{R_{\text{in}}}{1 - \mathbf{A}\boldsymbol{\beta}} \tag{12-19}$$

The *input resistance is reduced* by the factor $1/(1 - \mathbf{A}\boldsymbol{\beta})$ through shunt input of the feedback voltage.

Thus the circuit designer can control the output resistance of an amplifier by choice of the source of feedback voltage, and can alter the amplifier input resistance by the manner in which the feedback is introduced into the input circuit. As a result, voltage feedback is often used to reduce the output resistance of vacuum-tube amplifiers, and series introduction of the feedback voltage is often applied to raise the input resistance of transistor amplifiers, as by use of an unbypassed emitter resistor. The cathode follower and emitter follower serve as examples of voltage feedback, with $\beta = 1$, and with series input.

12-4 EFFECT OF NEGATIVE FEEDBACK ON NONLINEAR DISTORTION

Negative feedback may be applied to large-signal amplifiers to reduce the generation of nonlinear distortion components. In Fig. 12-5, $\mathbf{V}_h$ is the distortion output voltage, not present in the input signal. Without feedback $\mathbf{E}'_i = \mathbf{V}_s$ and the output is

$$\mathbf{V}_o = \mathbf{A}\mathbf{V}_s + \mathbf{V}_h \tag{12-20}$$

To add feedback and keep the same output level, $\mathbf{V}'_o = \mathbf{V}_o$, will require an increased input signal given by Eq. 12-8 as

$$\mathbf{V}'_s = (1 - \mathbf{A}\boldsymbol{\beta})\mathbf{V}_s$$

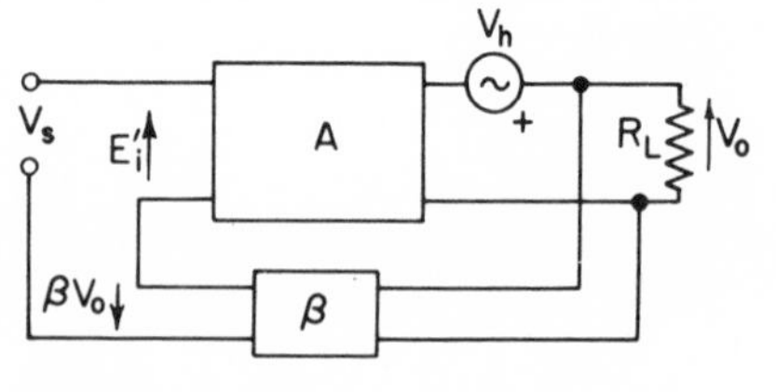

Figure 12-5. Effect of feedback on distortion, $\mathbf{V}_h$.

Then, with feedback

$$\begin{aligned}\mathbf{V}'_o = \mathbf{V}_o &= \mathbf{A}(\mathbf{V}'_s + \boldsymbol{\beta}\mathbf{V}'_o) + \mathbf{V}_h \\ &= \mathbf{A}(1 - \mathbf{A}\boldsymbol{\beta})\mathbf{V}_s + \mathbf{A}\boldsymbol{\beta}\mathbf{V}'_o + \mathbf{V}_h \end{aligned} \tag{12-21}$$

from which

$$\mathbf{V}'_o = \mathbf{V}_o = \mathbf{A}\mathbf{V}_s + \frac{\mathbf{V}_h}{1 - \mathbf{A}\boldsymbol{\beta}} \tag{12-22}$$

Comparison of Eqs. 12-20 and 12-22 indicates that for equal outputs the harmonic distortion is reduced with negative feedback by the factor $1/(1 - \mathbf{A}\boldsymbol{\beta})$, the reciprocal of the return difference. However, the distortion is reduced by the same factor as the gain, and this seems like a useless sacrifice, but such may not be the case. Nonlinear distortion is usually a result of signals traversing a large portion of the dynamic characteristic of the amplifying device, and such swings occur only in the output stage. Reduction of distortion there by feedback, and recovery of the lost gain in small signal stages earlier in the amplifier, are useful practices. It is almost always necessary to use negative feedback when distortion of less than five per cent is desired.

Generator $\mathbf{V}_h$ can also be considered to represent noise generated in the stage, and this component will be reduced, provided that the input signal level can be raised in accordance with Eq. 12-8, without introducing additional noise in the input.

12-5 IMPROVEMENT OF FREQUENCY RESPONSE WITH NEGATIVE FEEDBACK

In Section 10-6 it was shown that the high-frequency current response of an *RC*-coupled common-emitter amplifier is

$$\mathbf{A}_{iH} = \frac{\mathbf{A}_{iM}}{1 + j\omega/\omega_2} \tag{12-23}$$

If a resistive network is used for feedback, it will result in β real for all frequencies. Then use of Eq. 12-23 in the basic feedback gain expression of Eq. 12-5 leads to

$$\mathbf{A}'_{iH} = \frac{\mathbf{A}_{iH}}{1 - \mathbf{A}_{iH}\boldsymbol{\beta}} = \frac{\dfrac{\mathbf{A}_{iM}}{1 + j\omega/\omega_2}}{1 - \dfrac{\mathbf{A}_{iM}\boldsymbol{\beta}}{1 + j\omega/\omega_2}} = \frac{\mathbf{A}'_{iM}}{1 + j\omega/\omega'_2} \tag{12-24}$$

where

$$\mathbf{A}'_{iM} = \frac{\mathbf{A}_{iM}}{1 - \mathbf{A}_{iM}\boldsymbol{\beta}} \tag{12-25}$$

and

$$\omega'_2 = \omega_2(1 - \mathbf{A}_{iM}\boldsymbol{\beta}) \tag{12-26}$$

With negative feedback $\omega'_2 > \omega_2$, or the high-frequency range of the amplifier has been extended.

A similar analysis permits development of the low-frequency band limit expression, for a single time-constant circuit, as

$$\omega'_1 = \frac{\omega_1}{1 - A_{iM}\beta} \tag{12-27}$$

and negative feedback also extends the low-frequency range of the amplifier. A resultant frequency response curve is drawn in Fig. 12-6.

Often, increased bandwidth is a secondary result of the use of feedback for reduction of distortion, as discussed in the preceding sections. Since a

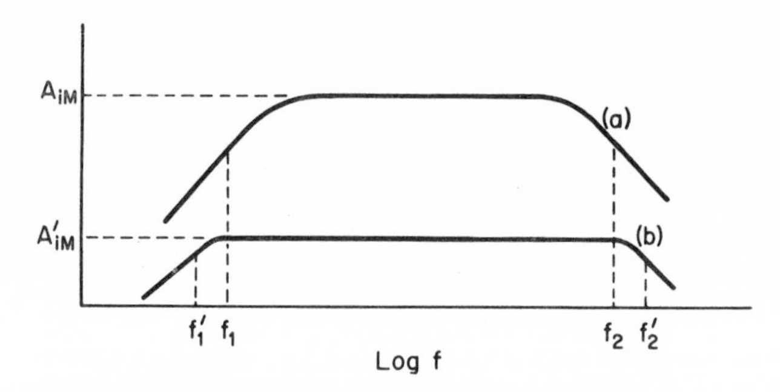

Figure 12-6. Frequency response: (a) no feedback; (b) with negative feedback.

high value of $A\beta$ is needed in such cases, it should not be surprising that quite wide bandwidths are incidentally obtained, often much greater than are required by the frequency spectrum of the signals being amplified. For instance, audio amplifiers are found with bandwidths from 6 to 100,000 hertz, as a result of distortion reduction by feedback, even though the audio spectrum only requires a response from 20 to 20,000 hertz.

Feedback networks often must utilize reactive elements to block dc potentials. In such cases the ideal resistive nature of the feedback network cannot be maintained at all frequencies, β becomes complex, and the feedback at frequency extremes may become positive. For reasons of stability, it is found necessary to control the phase angles of the β network and of the gain function to the wider bandwidths resulting from use of negative feedback for distortion reduction; this is often a difficult design problem.

12-6 TYPICAL FEEDBACK CIRCUITS

Current feedback is readily added to a common-emitter amplifier by use of an unbypassed emitter resistor R_E, this also giving series feedback input, as illustrated in Fig. 12-7. Resistor R_E is usually small with respect to R_L, and so has little direct effect on the load current. From the circuit:

$$\mathbf{V}_1 = \mathbf{I}_b r_{ie} + (\mathbf{I}_b + \mathbf{I}_c)R_E = [r_{ie} + (1 + h_{fe})R_E]\mathbf{I}_b \qquad (12\text{-}28)$$

$$\mathbf{V}_2 = -\mathbf{I}_c R_L = -h_{fe} R_L \mathbf{I}_b \qquad (12\text{-}29)$$

The voltage gain then is

$$\mathbf{A}_{ve} = \frac{\mathbf{V}_2}{\mathbf{V}_1} = \frac{-h_{fe}R_L}{r_{ie} + (1 + h_{fe})R_E} \qquad (12\text{-}30)$$

By definition the feedback factor is

$$\beta = \frac{(1 + h_{fe})R_E}{h_{fe}R_L} \cong \frac{R_E}{R_L} \qquad (12\text{-}31)$$

the approximation following if $h_{fe} \gg 1$.

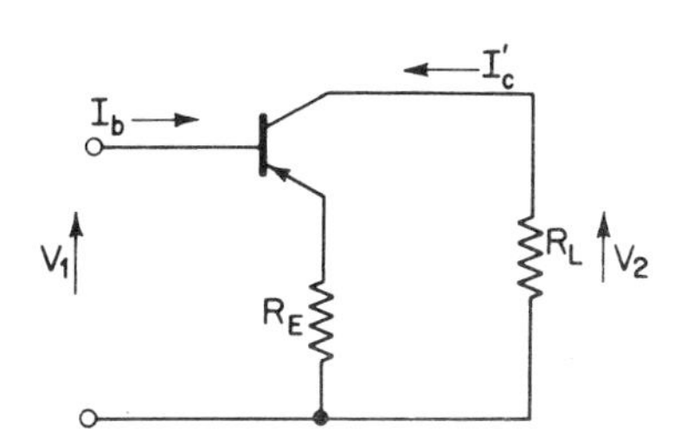

Figure 12-7. Current feedback, series input.

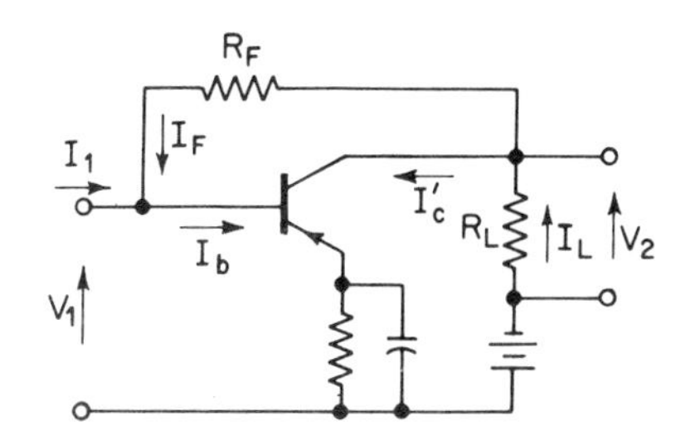

Figure 12-8. Voltage feedback, shunt input.

The addition of the term $(1 + h_{fe})R_E$ in the denominator of Eq. 12-30 reduces the voltage gain, and it also stabilizes the gain against changes in r_{ie} with temperature or as between different transistors. If $(1 + h_{fe})R_E$ is made large with respect to r_{ie} by design, then Eq. 12-30 reduces to

$$\mathbf{A}_{ve} \cong -\frac{1}{\beta} \tag{12-32}$$

as the ultimate in gain stability.

The input resistance with series feedback can be written directly from Eq. 12-28 as

$$R'_{\text{in}} = \frac{\mathbf{V}_1}{\mathbf{I}_b} = r_{ie} + (1 + h_{fe})R_E \tag{12-33}$$

Thus we find an increased input resistance, as predicted when series input of the feedback voltage is employed. The gain of the preceding stage may then be increased, due to improved load matching.

A second simple feedback application, giving voltage feedback with shunt input, is shown in Fig. 12-8. Resistor R_F is usually large with respect to R_L, so that the load current is not appreciably altered. By use of the load voltage, $-I_L R_L = -I'_c R_L$.

$$\mathbf{I}_F = \frac{R_L}{r_{ie} + R_F}\mathbf{I}'_c \cong \frac{R_L}{R_F}\mathbf{I}'_c \tag{12-34}$$

since $r_{ie} \ll R_F$.

Summing the input currents,

$$\mathbf{I}_1 + \mathbf{I}_F - \mathbf{I}_b = 0 \tag{12-35}$$

Since $\mathbf{I}_b = \mathbf{I}'_c/\mathbf{A}_i$, then by use of Eq. 12-34

$$\frac{\mathbf{I}'_c}{\mathbf{I}_1} = \mathbf{A}'_{iM} = \frac{\mathbf{A}_i}{1 - \mathbf{A}_i(R_L/R_F)} \tag{12-36}$$

and it is immediately recognizable that

$$\beta = \frac{R_L}{R_F} \tag{12-37}$$

Since $\mathbf{A}_i$ is negative for the usual C-E amplifier, the feedback is negative and the current gain is reduced.

The input resistance is obtainable by writing Eq. 12-35 as

$$\mathbf{I}_1 = \mathbf{I}_b - \frac{R_L}{R_F}\mathbf{I}'_c = \mathbf{I}_b(1 - \mathbf{A}_i\boldsymbol{\beta})$$

from which

$$R'_{\text{in}} = \frac{\mathbf{V}_1}{\mathbf{I}_1} = \frac{\mathbf{V}_1}{\mathbf{I}_b(1 - \mathbf{A}_i\boldsymbol{\beta})} = \frac{R_{\text{in}}}{1 - \mathbf{A}_i\boldsymbol{\beta}} \tag{12-38}$$

as a decrease from the nonfeedback value R_{in}, as predicted for shunt input of the feedback quantity.

These basic principles may then be applied in multistage amplifiers as shown in Fig. 12-9. For circuits of the C-E type, there is an inherent phase reversal in each stage. The β voltage is of such phase in odd-stage amplifiers that it may be introduced into the base or grid circuit, and for amplifiers having an even number of stages the feedback voltage is conveniently introduced into the emitter or cathode circuit. In Figs. 12-9(a) and (c) some current feedback is also present due to the current of the first stage passing through the unbypassed emitter or cathode resistor, along with the output

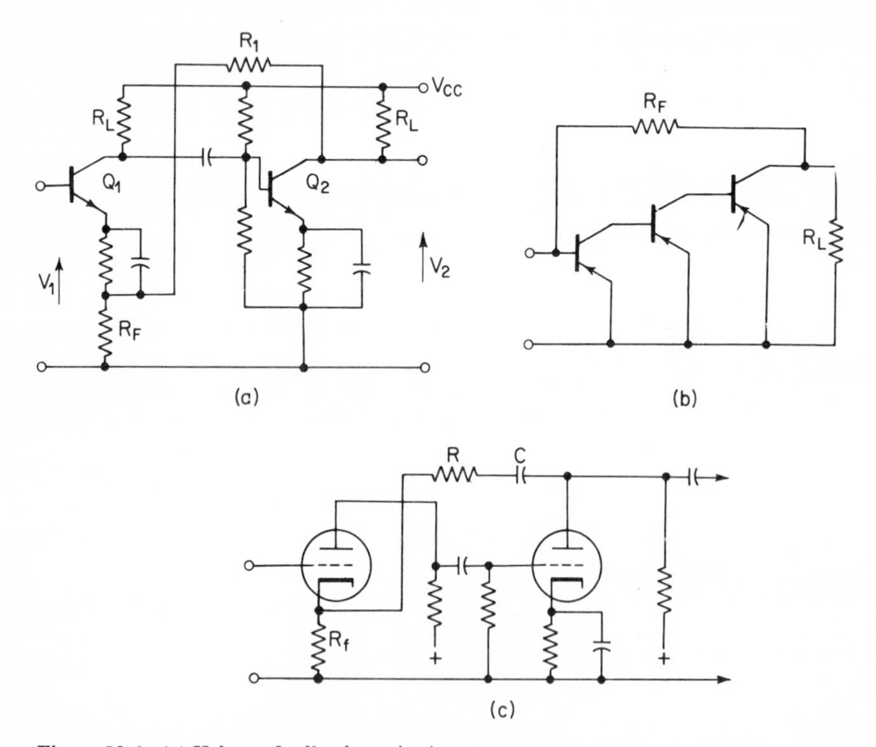

Figure 12-9. (a) Voltage feedback, series input, two stages; (b) voltage feedback, shunt input, three stages; (c) voltage feedback, series input—triodes.

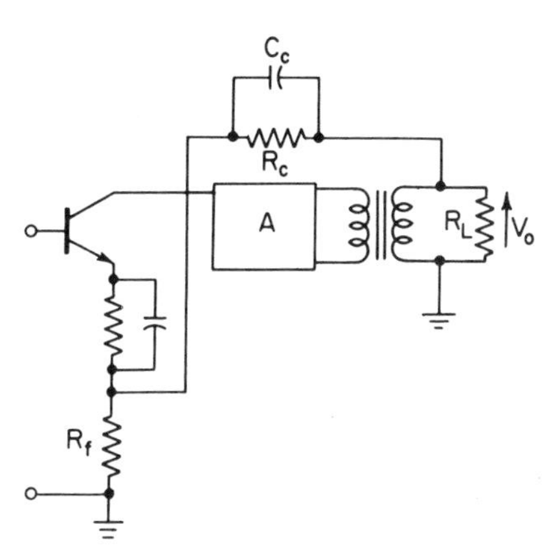

Figure 12-10. Feedback from the secondary of the output transformer.

stage current. Calculation of the amount of feedback is difficult, and it is most easily brought to the desired value by experimental adjustment of R_F.

Output resistance properties of these amplifiers are functions of feedback voltage source just as in the simple circuit forms. Similarly, variation of input resistance is dependent on the manner of input circuit introduction of the feedback voltage.

Feedback from the secondary of an output transformer may also be employed as in Fig. 12-10, the polarity of the windings being such as to assign the proper sign to $A\beta$. This arrangement has the advantage that frequency distortion introduced by the transformer may be reduced by the feedback. Because of the wide variation in impedance phase angle to be expected in a transformer, due to the magnetizing inductance at low frequencies and to resonance between leakage inductance and winding capacitance at high frequencies, this use of feedback often leads to problems of circuit stability at the frequency extremes. Corrective networks to alter the β angle at various frequencies are used for improvement, as indicated by R_c and C_c of the figure.

12-7 STABILITY OF NEGATIVE FEEDBACK CIRCUITS

Since A and β are complex and frequency functions, it cannot be expected that because $|1 - A\beta| > 1$ at some frequency, the relation will also be true at other frequencies. An amplifier may be stable over a certain frequency range, but in some other range $|1 - A\beta|$ may be less than unity, or $A\beta$ may even be equal to $1 + j0$. Of course, for the first case the amplifier will be unstable with positive feedback, and will oscillate under the second condition. The requirements for stability must then be further investigated.

Consider a plot of $A\beta \underline{/\theta^\circ} = a + jb$ (θ is the total phase angle) in the complex plane for all frequencies, both positive and negative, from 0 to ∞. Nyquist has shown such a plot is a closed curve. A critical condition exists where

$$|1 - A\beta| = 1$$

or

$$(1 - a)^2 + b^2 = 1 \tag{12-39}$$

This equation defines a circle of unit radius, with center at the point 1, $j0$. If the $A\beta$ vector terminates inside this circle, then obviously

$$|1 - A\beta| < 1$$

and the amplifier is unstable, or the feedback is positive. If the $A\beta$ vector terminates outside this circle

$$|1 - A\beta| > 1$$

and the amplifier is stable, and the feedback is negative.

If the $A\beta$ vector terminates at $1, j0$, previous discussion shows that the amplifier will oscillate. Nyquist has shown that *if the $A\beta$ curve encloses the $1, j0$ point*, the amplifier will oscillate.

An $A\beta$ plot for a resistance-coupled common-emitter amplifier may be constructed from the data of Section 10-7, with an assumption of mid-range gain of -60 and $\beta = 0.1$. The following table results and is plotted in Fig. 12-11(a). The polar plot for such an amplifier is a circle, with frequency progressing clockwise from the origin. The mid-frequencies appear at the 180° point, where the gain is constant at 6.0. The circle is the locus of $\mathbf{A\beta}$ for all frequencies.

A phasor for $1 \underline{/0°}$ is drawn to the point $1, j0$ so that the term $1 - \mathbf{A\beta}$ may be determined graphically. The unit circle with center at $1, j0$ is also shown. The only way for the $1 - \mathbf{A\beta}$ phasor to have a magnitude less than unity is for the polar $\mathbf{A\beta}$ locus to pass

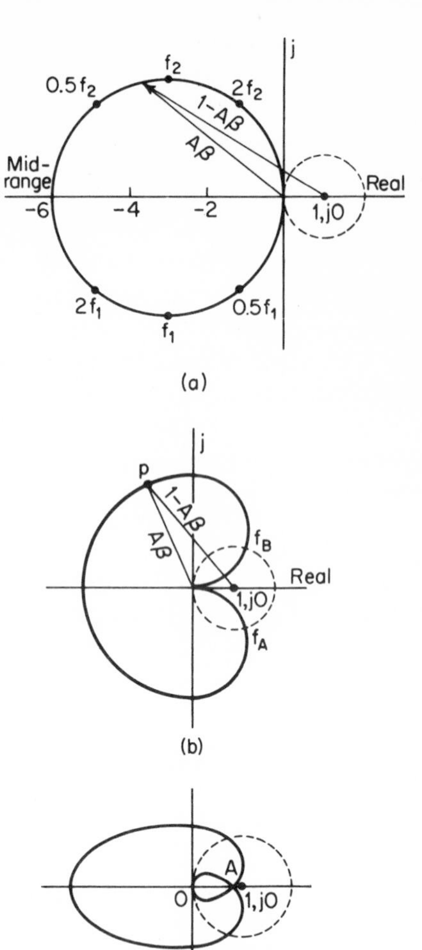

Figure 12-11. Nyquist plots: (a) *RC* amplifier, one stage; (b) showing region of instability; (c) conditionally stable.

Frequency	$\mathbf{A\beta}$	Frequency	$\mathbf{A\beta}$
$0.1f_1$	$0.72 \underline{/264°}$	$0.5f_2$	$5.40 \underline{/154°}$
$0.5f_1$	$2.70 \underline{/243°}$	f_2	$4.20 \underline{/135°}$
f_1	$4.20 \underline{/225°}$	$2f_2$	$2.70 \underline{/117°}$
$2f_1$	$5.40 \underline{/207°}$	$8f_2$	$0.72 \underline{/96°}$
Mid-range	$6.00 \underline{/180°}$		

inside the unit circle. In Fig. 12-11(b) the $A\beta$ locus, as traced by point P, will lie inside the unit circle for frequencies below f_A and above f_B. The magnitude of $1 - A\beta$ is then less than unity and so the area of the unit circle corresponds to a region of positive feedback and increasing gain. In an amplifier gain vs. frequency plot, peaks of gain would exist in the gain curve in the frequency regions below f_A and above f_B, although the amplifier would be stable.

In Fig. 12-11(c) a similar gain-frequency curve is obtained. However, should the gain at 0° increase, then the amplifier may oscillate. This amplifier is conditionally stable.

With its inherent 90° phase shift, a one-stage feedback RC amplifier is absolutely stable, but the gain in a single stage may be so low that much of the feedback advantage is not obtained. Higher gain amplifiers of two stages, with phase angles near 180°, may be unstable, and those with additional stages require careful choice of components and compensation networks in the feedback loop at extreme frequencies.

In general, instability will not occur if the phase angle of the voltage fed back never differs as much as 180° from its mid-range value, or if the absolute value of $A\beta$ becomes less than unity when the phase angle approaches zero degrees. It is usually possible to design an amplifier so that at least the latter condition may be met.

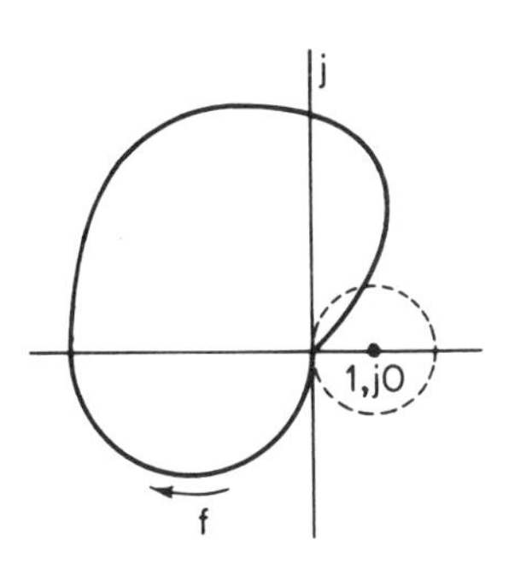

Figure 12-12. Nyquist plot for a transformer output stage.

Figure 12-12 represents a transformer-output amplifier in which the feedback is obtained from the secondary of the transformer, as in Fig. 12-10. Although the $1, j0$ point is not encircled, the $A\beta$ vector terminates inside the unit circle at high frequencies, and a gain peak with positive feedback will be expected. Reduction of this peak by a load shunting capacitor or use of R_c, C_c in the feedback loop would be desirable.

12-8 ANALYTIC CONSIDERATIONS IN STABILITY

Stability may also be determined directly by study of the gain function. Again using the operator s as a complex frequency

$$\epsilon^{st} = \epsilon^{\sigma t}\epsilon^{j\omega t} \tag{12-40}$$

where the expression describes a growing or decaying sinusoid. An unstable circuit results from a growing sinusoid with a positive σ; a positive σ in the value for s locates a pole in the right half-plane. A decaying sinusoid results from a negative σ, with the pole located in the negative half-plane of the s plot.

Consider a feedback amplifier with a gain expression

$$G(s) = \frac{A(s)}{1 - A(s)\beta} \tag{12-41}$$

Since $A(s)$ may be expressed in polynomial form, the gain expression will have poles and zeros due to denominator and numerator roots. Therefore, the gain may grow, or the amplifier will be unstable, if any pole contributed by a denominator zero has a positive σ, or if such a pole lies in the right half-plane of the s plot. Thus, *for stability there should be no zeros of the denominator* $[1 - A(s)\beta]$ *in the right half-plane.*

This result may be illustrated by consideration of a resistance-capacitance coupled amplifier of two stages, with overall feedback of magnitude β and zero phase shift in the feedback circuit. The high-frequency gain of one stage is

$$A(s) = \frac{-g_m R}{1 + sRC} = -\frac{g_m}{C}\left(\frac{1}{s + 1/RC}\right) \tag{12-42}$$

The denominator has a zero, or the function has a pole, at $s = -1/RC$, and the amplifier of one stage is evidently stable. The term $-g_m/C$ is the scale factor or mid-frequency gain.

Cascading two identical stages gives

$$A(s) = \left(\frac{g_m}{C}\right)^2 \frac{1}{(s + 1/RC)^2}$$

Using this result in Eq. 12-41 to find the effect of feedback around this amplifier gives

$$G(s) = \frac{\left(\dfrac{g_m/C}{s + 1/RC}\right)^2}{1 - \left(\dfrac{g_m/C}{s + 1/RC}\right)^2 \beta}$$

$$= \left(\frac{g_m}{C}\right)^2 \frac{1}{s^2 + 2s/RC + (1/RC)^2 - (g_m/C)^2\beta} \tag{12-43}$$

The stability of the amplifier can be determined from a study of the zeros of the denominator. These zeros are

$$s_1 = -\frac{1}{RC} + \frac{g_m}{C}\sqrt{\beta}, \qquad s_2 = -\frac{1}{RC} - \frac{g_m}{C}\sqrt{\beta}$$

The zero due to s_1 will lie in the right half-plane if

$$g_m R > \frac{1}{\sqrt{\beta}} \tag{12-44}$$

Since the mid-frequency gain of the two-stage amplifier without feedback is equal to $(g_m R)^2$, it is apparent that for stability the gain must be less than $1/\beta$, inside the feedback loop.

12-9 THE ROUTH-HURWITZ CRITERION

It is apparent that one problem of feedback amplifier stability revolves around the characteristic equation or denominator of the feedback gain function, and the presence or absence of right half-plane zeros in the denominator term. A stable denominator will have no right half-plane zeros and no zeros on the j axis in the s plane; the Routh-Hurwitz criterion provides a means for determining if the characteristic equation describes a stable feedback system. Application of the method may be easier than factoring a polynomial of high order.

Consider a denominator polynomial such as

$$a_0 s^n + a_1 s^{n-1} + a_2 s^{n-2} + \cdots + a_n \tag{12-45}$$

All coefficients $a_0, a_1, \ldots, a_{n-1}, a_n$ must be real and of the same algebraic sign. Otherwise roots in the right half-plane are indicated. By arranging the coefficients in a specified pattern, it is possible to indicate the presence or absence of positive real roots, or zeros located in the right half-plane.

The following array of terms is constructed:

$$\left.\begin{array}{lll} a_0 & a_2 & a_4 \\ a_1 & a_3 & a_5 \\ b_1 & b_3 & \\ c_1 & c_3 & \\ d_1 & & \\ e_1 & & \end{array}\right\} \tag{12-46}$$

where the coefficients b_1, b_3, c_1, c_3, d_1, e_1 are computed from

$$b_1 = \frac{a_1 a_2 - a_0 a_3}{a_1}$$

$$b_3 = \frac{a_1 a_4 - a_0 a_5}{a_1}$$

$$c_1 = \frac{b_1 a_3 - a_1 b_3}{b_1}$$

$$c_3 = \frac{b_1 a_5 - a_1 \times 0}{b_1} = a_5$$

$$d_1 = \frac{c_1 b_3 - b_1 c_3}{c_1}$$

$$e_1 = \frac{d_1 c_3 - c_1 \times 0}{d_1} = c_3$$

The array terminates when the remaining terms in the left column of Eq. 12-46 are found to be zero.

The signs of the terms in the left column of Eq. 12-46 are then inspected. If all terms are positive, the characteristic equation has all roots in the left half-plane. If negative elements exist, the number of right half-plane roots will be equal to the number of changes of sign existing in the column.

As an example, consider

$$s^4 + 3s^3 + 6s^2 + 3s + 1 \tag{12-47}$$

from which an array is formed:

$$\begin{array}{ccc} 1 & 6 & 1 \\ 3 & 3 & 0 \\ 15 & 1 & \\ \frac{42}{15} & 0 & \\ 1 & & \\ 0 & & \end{array}$$

The amplifier, for which Eq. 12-47 is the characteristic equation, will be stable.

A row may have all elements equal to zero; this indicates conjugate roots on the $j\omega$ axis. Such a characteristic equation indicates incipient instability.

12-10 GAIN AND PHASE MARGIN

To determine the stability of an amplifier, it is necessary to show that $|\mathbf{A}\boldsymbol{\beta}|$ is less than unity when the overall or loop phase angle approaches 0°, and plots of $\mathbf{A}\boldsymbol{\beta}$ magnitude, usually in decibels, and of phase angle, may be used directly. Such curves also allow determination of the closeness of approach of actual performance to the stability limits.

The *gain margin* is the decibel value of $\mathbf{A}\boldsymbol{\beta}$ at the frequency at which the phase angle of $\mathbf{A}\boldsymbol{\beta}$ is 0°. If negative, this indicates the decibel rise in gain which is theoretically permissible without oscillation. If positive, the amplifier is unstable.

The *phase margin* is the angle of $\mathbf{A}\boldsymbol{\beta}$ at the frequency at which $\mathbf{A}\boldsymbol{\beta}$ has a value of zero decibels, or unity magnitude.

The desirable magnitude of gain and phase margins varies with the application, but it is usual to require 10 dB of gain margin and 30° of phase margin to take care of unavoidable shifts due to component variation. Figure 12-13 illustrates these criteria, showing a high-range plot for an amplifier with a gain margin of 8.2 dB and a phase margin of 36°, the amplifier being stable. Various combinations of *RC* circuits may be employed in the feedback loop or in the coupling networks to reduce the high-frequency response if further gain margin is considered desirable.

The frequency at which the gain and phase margins are measured is far outside the region of flat response of the amplifier, and beyond the normally used range. This again shows the necessity for control of phase and gain well

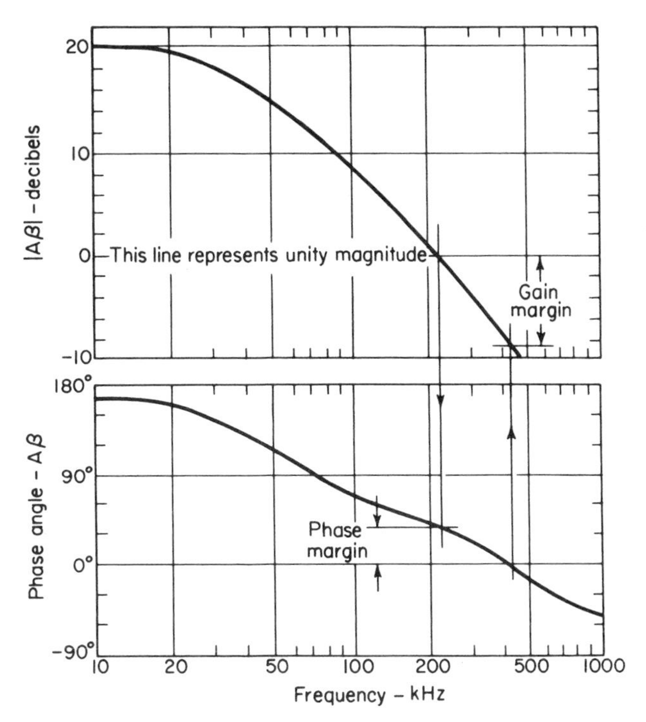

Figure 12-13. High frequency values of $A\beta$, showing gain and phase margin.

beyond the operating band in feedback amplifiers, and indicates that reduced distortion and increased stability have been purchased at the expense of problems associated with greater bandwidth.

PROBLEMS

12-1. With negative feedback an amplifier gives an output of 10.5 V with an input of 1.12 V. When feedback is removed it requires 0.15 V input for the same output. Find the value of β, and of the gain without feedback. Input and output are in phase, and β is real.

12-2. An amplifier has a gain of 30 dB and an output of 100 V with 10 per cent distortion. If feedback is to be used to reduce the distortion to 1.5 per cent, what value of β should be used, and what will be the new gain if A and β are real?

12-3. (a) Neglecting the reactance of C, develop an expression for gain in terms of tube and circuit constants, for the circuit of Fig. 12-14.

(b) Show that β is equal to the sum of the current and voltage feedback values of β.

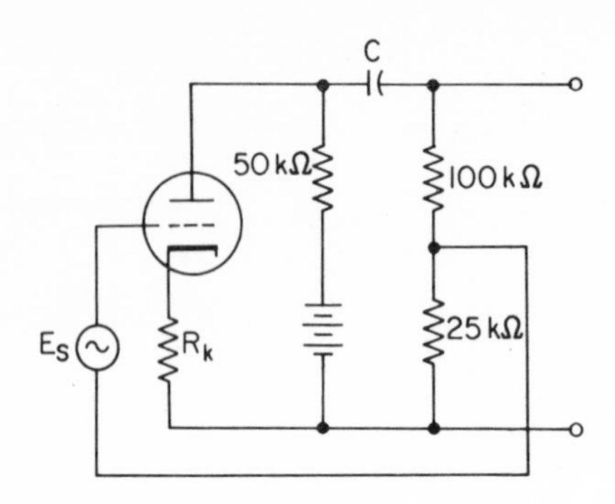

Figure 12-14.

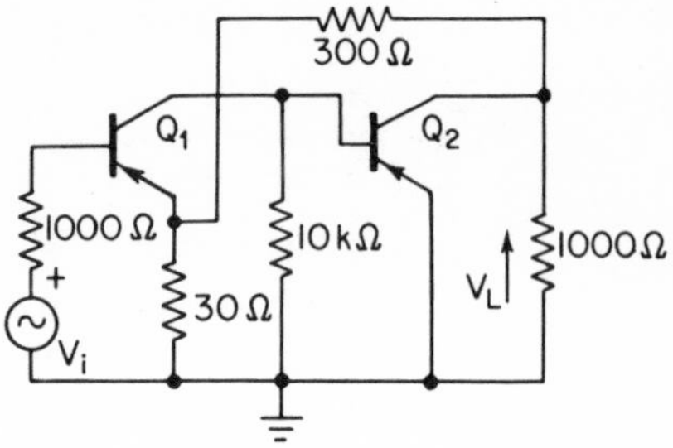

Figure 12-15.

12-4. A precision amplifier is to employ feedback, and must have an overall gain of 60. If the guaranteed voltage gain accuracy is to be 1 per cent, and if internal gain changes of 12 per cent are found probable without feedback, determine the value of β required for stability, and the gain before feedback is added.

12-5. A three identical stage negative feedback amplifier has a closed-loop voltage gain of 50. Determine the individual stage gain and β if the overall gain changes 0.5 per cent when the individual stages each change by 5 per cent in gain, due to supply voltage variation.

12-6. Find the closed-loop voltage gain and R_{in} of the feedback amplifier of Fig. 12-15. Q_1 and Q_2 have $r_{ie} = 2000$ ohms and $h_{fe} = 40$.

12-7. The circuit of Fig. 12-14 is used with a triode having $\mu = 20$, $r_p = 7700$ ohms, and with $R_k = 0$. With $C = 0.05\ \mu\text{F}$, determine the low-frequency band limit.

12-8. In the circuit of Fig. 12-7, the load resistor is $R_L = 10{,}000$ ohms, and the signal source has a resistance of 100 ohms. The transistor has $h_{ie} = 1700$ ohms, $h_{re} = 5 \times 10^{-4}$, $h_{fe} = 44$, $h_{oe} = 22\ \mu$mhos.

(a) Find the voltage gain without feedback or $R_E = 0$.

(b) Using $R_E = 1000$ ohms, find the voltage gain with feedback.

12-9. In Fig. 12-16, the triode has $\mu = 35$, $r_p = 22{,}000$ ohms. Plot a Nyquist diagram for the circuit over the low-frequency range; check this against a gain-frequency plot and explain the rise in gain.

12-10. An RC amplifier of three identical stages has $f_1 = 48$ hertz, $f_2 = 140$ kilohertz for each stage. If

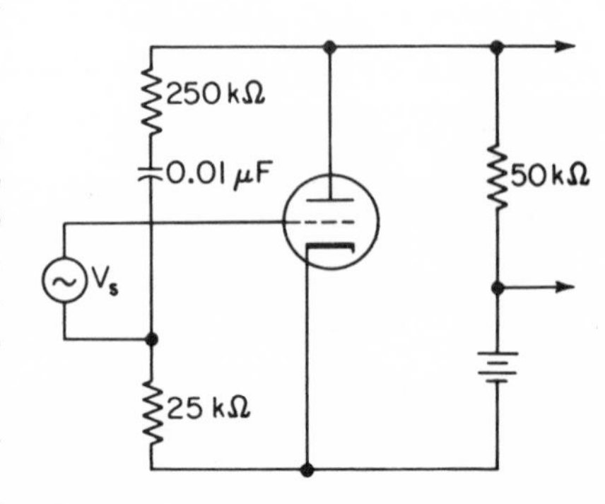

Figure 12-16.

the overall gain inside the feedback loop is -450, and $\beta = 0.1$ is applied, determine f_1' and f_2' for the amplifier with feedback.

12-11. The transistor of Problem 12-8 is used in the circuit of Fig. 12-8 with source resistance of 200 ohms and $R_L = 45{,}000$ ohms. Specify R_F to provide 16 dB of feedback.

12-12. Apply the Routh-Hurwitz criterion and determine whether the amplifiers having the following characteristic equations are stable:

(a) $4s^4 + 3s^3 + 7s^2 + 3s + 4 = 0.$

(b) $s^4 + s^3 + s^2 + 12s + 6 = 0.$

(c) $s^3 + 11s^2 + 11s + 10 = 0.$

(d) $4s^4 + 5s^3 + 9s^2 + 5s + 4 = 0.$

12-13. A given feedback amplifier has a feedback factor given by

$$A\beta = \frac{-25(1 + 0.05s)}{1 + 0.20s + 0.02s^2}$$

Determine whether the amplifier is stable, and state limiting conditions.

12-14. A feedback amplifier has

$$A\beta = \frac{45}{s^2 + 3s + 1}$$

Is the amplifier stable?

REFERENCES

1. Black, H. S., "Stabilized Feedback Amplifiers," *Elec. Eng.*, Vol. 53, p. 114 (1934).
2. Learned, V., "Corrective Networks for Feedback Circuits," *Proc. IRE*, Vol. 32, p. 403 (1944).
3. Bode, H. W., *Network Analysis and Feedback Amplifier Design*. D. Van Nostrand Co., Inc., Princeton, N. J., 1945.
4. Guillemin, E. A., *The Mathematics of Circuit Analysis*. John Wiley & Sons, Inc., New York, 1949.
5. Shea, R. F., *et al.*, *Principles of Transistor Circuits*. John Wiley & Sons, Inc., New York, 1953.
6. Truxal, J. G., *Automatic Feedback Control System Synthesis*. McGraw-Hill Book Company, New York, 1955.
7. Linvill, J. G., and J. F. Gibbons, *Transistors and Active Circuits*. McGraw-Hill Book Company, New York, 1961.
8. Thornton, R. D., *et al.*, *Multistage Transistor Circuits* (SEEC, Vol. 5). John Wiley & Sons, Inc., New York, 1965.
9. Hakim, S. S., *Feedback Circuit Analysis*. John Wiley & Sons, Inc., New York, 1966.

13

DIRECT-COUPLED AMPLIFIERS

Slowly varying signals are encountered in analog computation, in regulation of dc power supplies for electronic equipment, and in measurement. For amplification of such signals in which a zero-frequency component may appear and must be preserved, it is necessary to eliminate the coupling capacitance of the *RC* amplifier, resulting in the *direct-coupled* or *dc amplifier*.

It is apparent that much previous analysis has been concerned with problems created by the coupling and bypass capacitors, and their elimination might seem a solution to these other problems. As is so often true in engineering application, however, an apparently simple solution introduces new problems and that of operating point stability is paramount in the dc amplifier. The transistor differential dc amplifier is introduced here as the best present solution for dc amplifier stability. It is also of importance in integrated circuit form.

13-1 PROBLEMS OF DIRECT-COUPLED AMPLIFIERS

When the coupling capacitor of a low-frequency amplifier is eliminated, the input element of a second stage is placed at the potential of the anode or collector of the preceding stage. To restore proper bias levels once required a number of voltage sources, as in Fig. 13-1, and two sources are still needed in many circuits.

However, a more basic problem of the dc amplifier is that such circuits are unable to discriminate between a 1 mV change in input signal and a 1 mV change in bias potential. Such changes in supply voltages are common, as are equivalent changes in device parameters and currents with temperature or time.

It is usually desired that a dc amplifier give zero output for zero input voltage, and drift of output away from zero as a function of time is a limiting

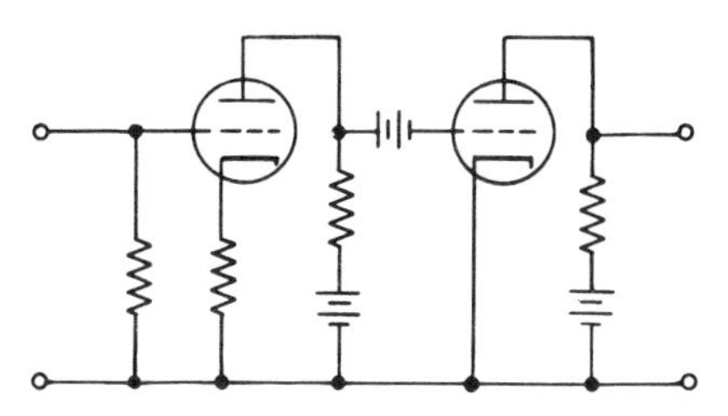

Figure 13-1. dc amplifier with separate voltage supplies.

performance factor. Drift may be considered as equivalent to a change of dc bias on the first stage, and is frequently defined as a voltage, which, if placed on the input of an equivalent ideal amplifier, will produce the same change in output as given by the actual drift. A *drift figure* is then stated as *the equivalent input voltage change*, usually per hour of operation.

In some applications drift can be compensated by occasional manual adjustment, as for the very common balanced dc voltmeter of Fig. 13-2. With a power supply having negative and positive voltages to ground, the vacuum tube circuit of Fig. 13-3 once had use as an input circuit for dc amplifiers. The output at A can be brought to zero with zero input by adjustment of R_k, thus compensating for drift.

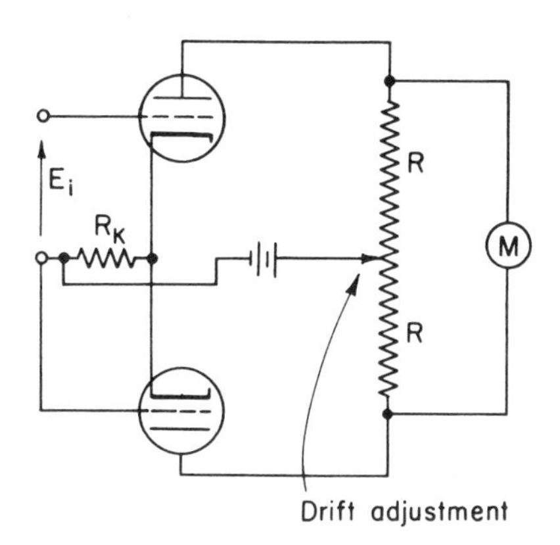

Figure 13-2. dc voltmeter with manual adjustment of zero.

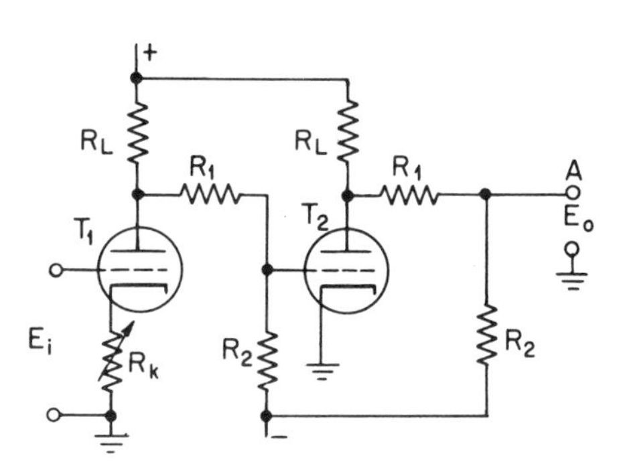

Figure 13-3. dc amplifier with resistance coupling.

13-2 CHOPPER AMPLIFIERS

One solution to the drift problem in dc amplifiers is to chop the slowly varying signal into a form of ac, amplify this with an ac amplifier, and rectify the output back to a slowly varying signal. Such a system appears in Fig. 13-4.

Many samples must be taken per cycle of input to give good reproduction of the input variation, or the sampling frequency of Fig. 13-5 must be large with respect to the highest frequency present in the signal. This places a limit of about 6 hertz on signal frequencies if 60 hertz power is used to supply a chopper magnet. For higher signal frequencies transistors may be used as choppers.

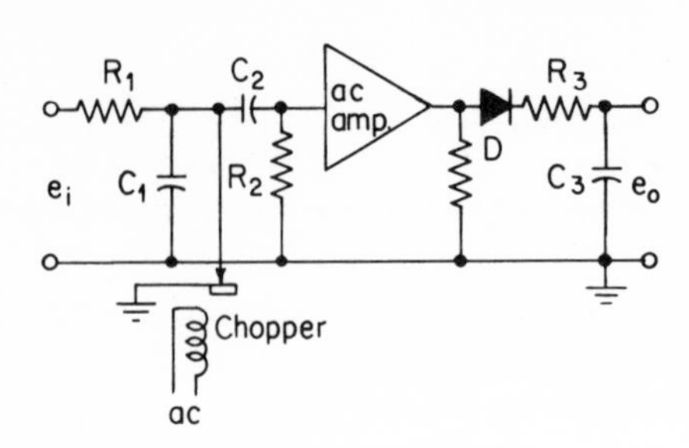

Figure 13-4. Chopper-modulated amplifier.

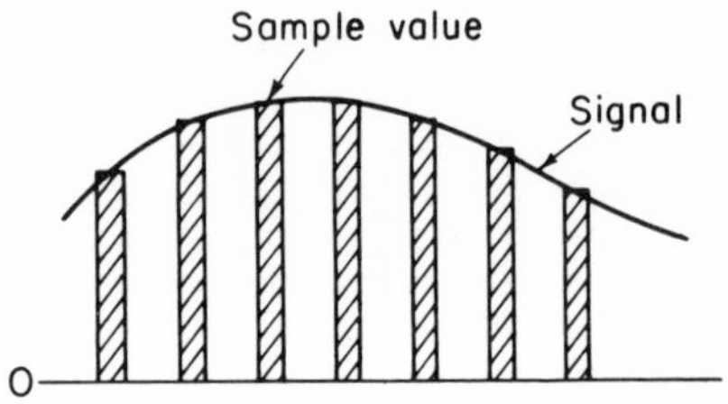

Figure 13-5. Method of sampling a slowly varying wave.

The amplitude of the amplifier input will equal the average of the input voltage during the period that the chopper is open. To provide this average, the input is filtered by R_1, C_1, which removes variation frequencies above one-tenth of the chopper frequency. The amplifier output is rectified by diode D, and harmonics of the chopping frequency removed by the R_3, C_3 output filter.

Greater stability against drift is possible in the *chopper-stabilized dc amplifier* of Fig. 13-6. It uses a chopper ac amplifier of gain A_2, inside a feed-

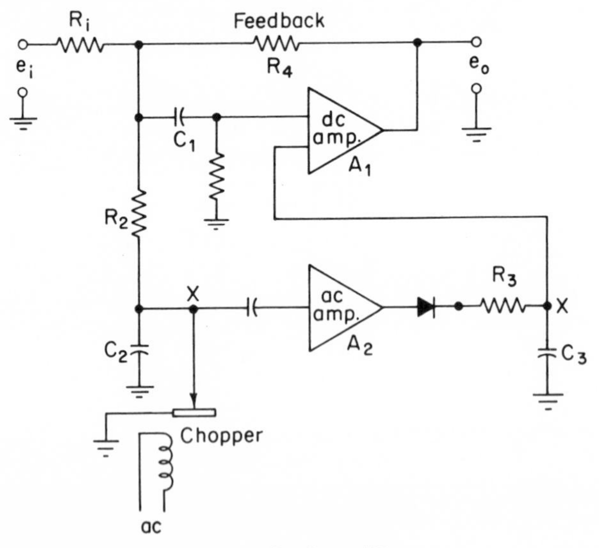

Figure 13-6. Chopper-stabilized amplifier.

back loop, at the same time the higher-frequency signal components are directly amplified in a dc amplifier of gain A_1. These higher frequencies do not go through the chopper, and the bandwidth limitation imposed by the sampling frequency does not apply.

Any drift in output voltage is fed back through R_4, and charges C_2 through R_2, the RC value being chosen for a long time constant. The voltage on C_2 is sampled, amplified in the ac amplifier, rectified, filtered, and applied to the balancing input of the dc amplifier. There it counteracts the drift voltage of the amplifier and restores the output voltage toward zero.

Because of C_1, medium and higher frequencies pass through the dc amplifier with a gain A_1. The drift and very low frequency signals pass through both amplifiers in series, with a gain A_1A_2. The open-loop gain of the two paths differs, but with large voltage feedback through R_4, the closed loop gain is unaffected so long as the internal gain is maintained large at all operating frequencies, in accordance with Eq. 12-9.

The driftless ac amplifier reduces the drift by the factor $1/A_2$. It is possible to obtain drift figures of only a few microvolts per hour.

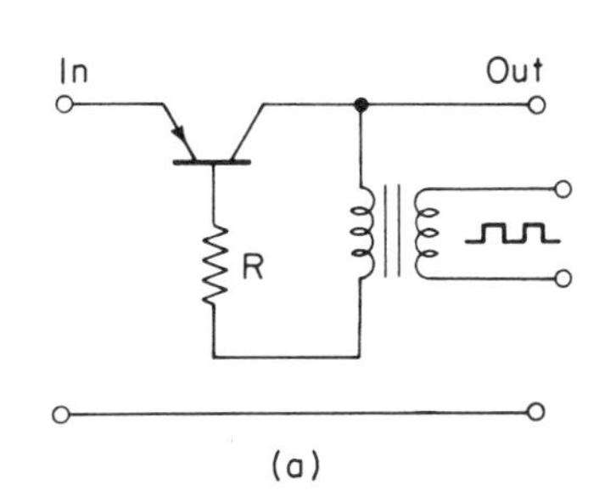

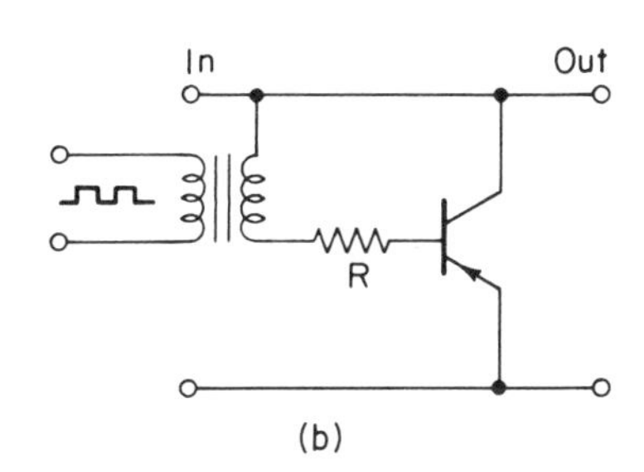

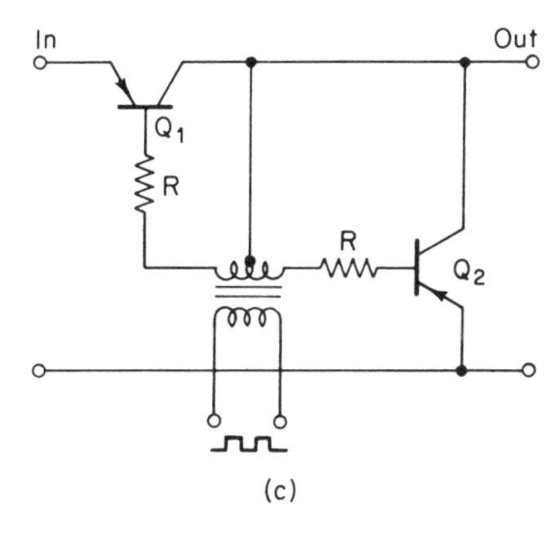

Figure 13-7. Choppers: (a) series; (b) shunt; (c) series-shunt.

13-3 TRANSISTOR CHOPPERS

The transistor with its low saturation resistance for the "on" case, and its high cutoff resistance for "off," can be used as a signal chopper at rates up to 100 kilohertz. Figure 13-7 illustrates a series switch for use with high-resistance sources, and a shunt switch for use with low-resistance sources. A square wave switches the transistor alternately on and off. The output sample will differ from the input by an inherent offset voltage, even at zero current, due to the presence of the active junction.

The series-shunt circuit at Fig. 13-7(c) is used to eliminate the offset voltage. With Q_1 on and Q_2 off or vice versa, there is a junction always in circuit. A switching transformer must be used to isolate the base voltage from the square-wave source.

The field-effect transistor inherently provides isolation for the square-wave source, by reason of its insulated gate as in Fig. 13-8. There is no offset voltage, since the path from source to drain is simply a modulated resistance, varying from a low on value to a very high resistance in the off condition. The series resistance present in the on state exceeds the saturation resistance of a junction transistor, however. The FET may also be used in a series-shunt circuit, as in Fig. 13-8(b).

Phototransistors are employed in similar circuits for isolation of the switching source, the light from a flashing gas-discharge lamp being used for switching the transistor.

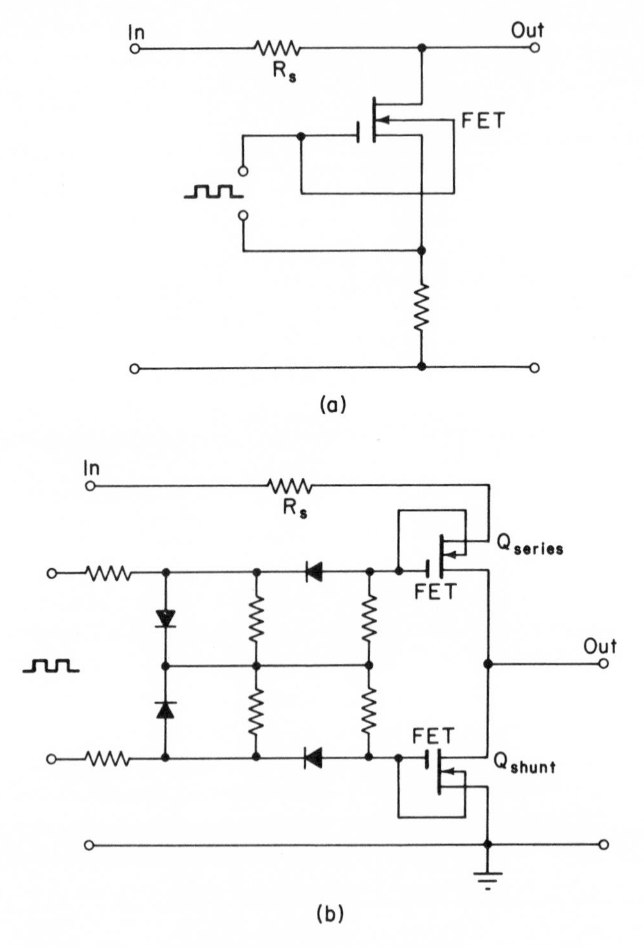

Figure 13-8 (a) FET shunt chopper; (b) FET series-shunt chopper.

13-4 DIFFERENTIAL AMPLIFIERS

Balancing of changes in one active device against those in a second and matched device in a bridge circuit was illustrated in Fig. 13-2. As applied with transistor elements in the *differential amplifier*, the principle has permitted solution of many of the drift problems previously encountered in the dc amplifier. A basic version of the amplifier appears in Fig. 13-9.

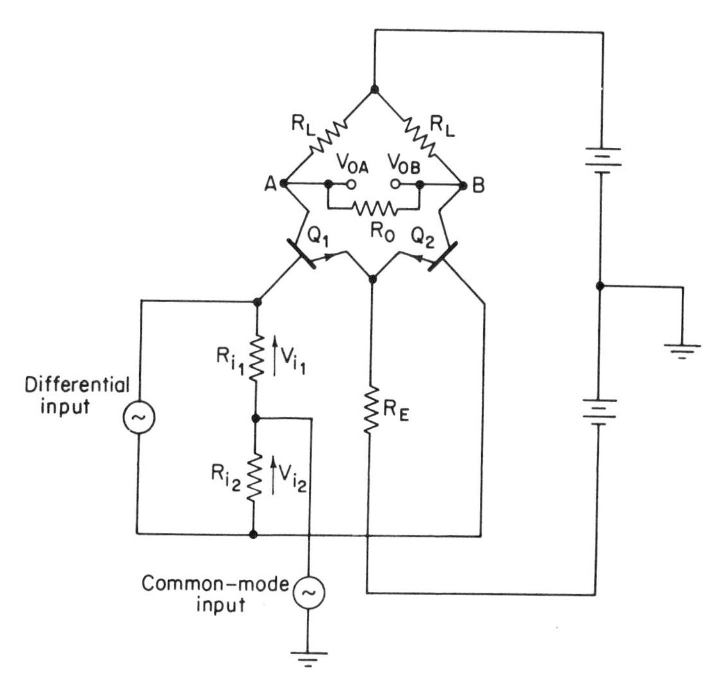

Figure 13-9. Differential amplifier with differential and common-mode inputs.

The two transistors and their load resistors form the bridge, which is balanced with identical transistors and resistors. With zero input the output voltage is zero. If a *differential input signal* is now applied by the generator, the input voltages V_{i1} and V_{i2} will be equal in amplitude and opposite in phase, assuming that $R_{i1} = R_{i2}$. The collector current of Q_1 might increase, in which case an identical decrease will occur in the current of Q_2. The result will be a difference voltage $V_{oA} - V_{oB}$ between the two output terminals of the bridge.

If a *common-mode signal* is introduced, as indicated in Fig. 13-9, such a signal being caused by hum pickup or other interference, the input signal to both transistors is equal in amplitude and in phase. The change in current in the transistors is identically up or down, the bridge remains balanced, and the output voltage between terminals A and B is zero.

If the common-mode signal causes the transistor currents to rise, the voltage drop across R_E also rises. This represents negative feedback and reduces the gain for the common-mode signal. Differential-mode signals are not affected by the emitter resistance since the current through one transistor rises while that in the other transistor falls by an equal amount; the current in R_E remains constant for differential signals. It is usually desired that the difference-mode signal be amplified and the common-mode signal be suppressed, so a high value of R_E is desirable. High R_E is also desirable if there happens to be a mismatch of transistor gains, since there will be signal feedback of such phase as to degenerate the transistor with the larger current, forcing a better balance.

It is to the balancing of equal changes that the circuit owes its freedom from drift due to supply voltage and temperature changes. By forming the two transistors side-by-side on a single small silicon chip, the characteristics of v_{BE} and h_{FE} can be matched over an extended temperature range. Since the devices are mounted in the same envelope, they will be closely maintained at the same temperature, and net v_{BE} temperature coefficients as low as 3 μV per °C can be obtained.

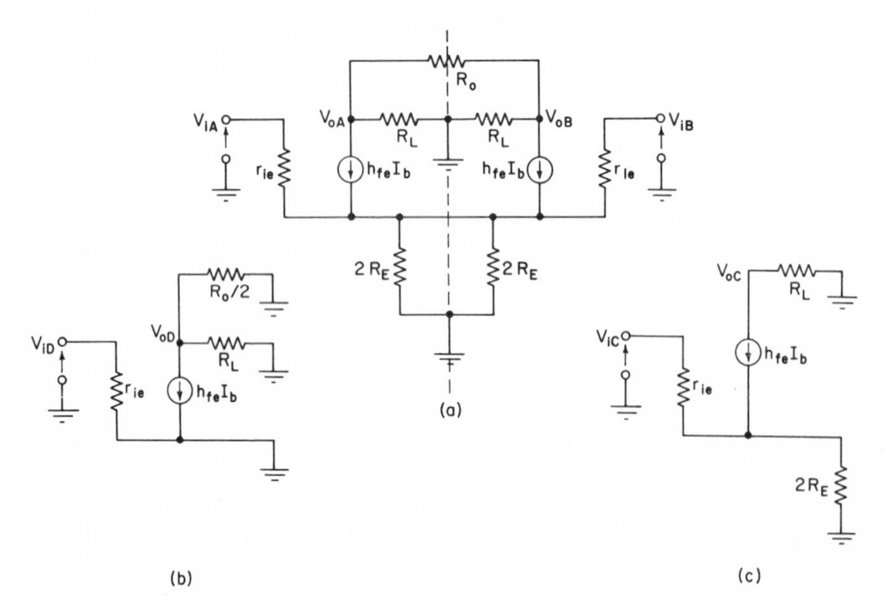

Figure 13-10. (a) Equivalent circuit for a differential amplifier; (b) differential-mode equivalent half-circuit; (c) common-mode equivalent half-circuit.

The differential-mode and the common-mode outputs can be written as

$$V_{oD} = A_{DD} V_{iD} \tag{13-1}$$

$$V_{oC} = A_{CC} V_{iC} \tag{13-2}$$

where A_{DD} and A_{CC} are the respective differential- and common-mode gains in the equivalent circuit of Fig. 13-10(a). Because of the symmetry of the circuit, its analysis can be carried out by bisection of the circuit into halves, along the line of signal ground potential. In the full circuit in Fig. 13-10(a) resistor R_o represents any load introduced by a following amplifier stage. The differential-mode half circuit at (b) contains $R_o/2$, but does not contain R_E, since no differential-mode voltage appears across that resistor. The common-mode half circuit at (c) contains $2R_E$, the factor 2 being necessitated by the circuit bisection, and R_o does not appear since zero common-mode voltage appears across A and B.

It is possible to write the differential output-differential input half gain from (b), Fig. 13-10, as

$$A_{DD} = \frac{V_{oD}}{V_{iD}} = -\frac{h_{fe} R_p}{h_{ie}} = -g_m R_p \tag{13-3}$$

where

$$R_p = \frac{R_L R_o/2}{R_L + R_o/2}$$

or R_p is the equivalent of R_L and $R_o/2$ in parallel. The total output voltage is

$$V_{oB} - (-V_{oA}) = -2 g_m R_p V_{iA}$$

If all components and transistors are matched, and with $V_{iA} = V_{iB}$ for a common-mode signal, the two outputs are equal and zero common-mode output will be obtained across R_o.

It is often desirable to have one side of the output at ground potential, and so the output is taken between one output terminal and ground. The common-mode signal will then produce some output, although there will be a substantial reduction of the gain for the common-mode signal as compared to the gain for the differential-mode signal, because of the emitter degeneration. Figure 13-11 illustrates the several alternatives in choice of input and output terminals.

A measure of the discrimination obtained against common-mode signals is given by the *common-mode rejection ratio*, defined as

$$\text{CMRR} = \frac{\text{common-mode input voltage}}{\text{differential-mode input voltage}} \tag{13-4}$$

for the same differential output voltage. This might also be worded as the ratio of the gain with differential input to the gain with both inputs connected together.

With differential input and single-ended output, as in Fig. 13-11(b),

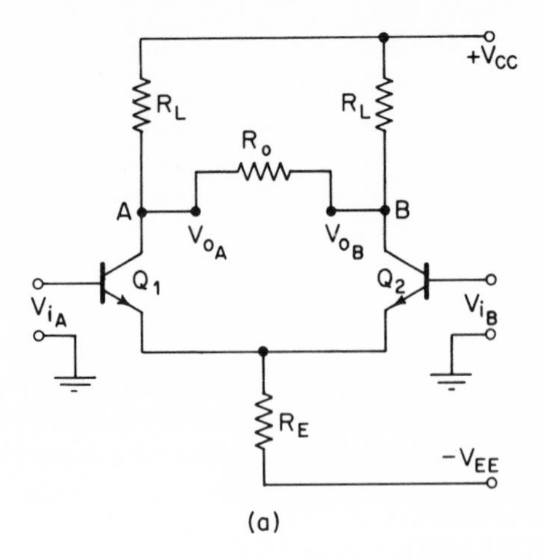

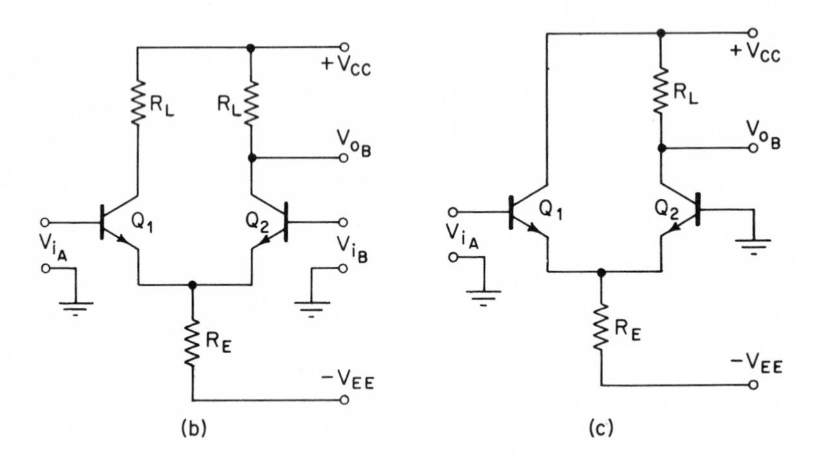

Figure 13-11. Forms of differential amplifiers.

the output with $V_{iA} = -V_{iB}$ is given by $-h_{fe}R_L/h_{ie}$ (assuming no loading by R_o). For the common mode, with $V_{iA} = V_{iB}$, the circuit at Fig. 13-11(c) gives

$$A_{CC} = \frac{V_{oC}}{V_{iC}} = \frac{-h_{fe}R_L}{h_{ie} + 2R_E(1 + h_{fe})} \tag{13-5}$$

The CMRR value for the differential-input, single-ended output amplifier then is

$$\text{CMRR} = \frac{h_{ie} + 2R_E(1 + h_{fe})}{h_{ie}} \simeq \frac{2h_{fe}R_E}{h_{ie}} \tag{13-6}$$

Common-mode rejections may be as high as 100 dB in some amplifiers.

The value of R_E should be large to create large CMRR values, and this may require use of a separate V_{EE} supply. Another method of obtaining an effective high emitter resistance is shown in Fig. 13-12(b), combined with a Darlington input stage for high input resistance. The voltage at the emitters of Q_3 and Q_4 is constant for a differential signal, but varies with a common-mode signal. Transistor Q_5 and the diode form a constant-current source, but any common-mode drop across the $3K$ resistor is amplified and alters the constant-current value, giving negative feedback to the emitters of Q_1 and Q_2 and a reduction of the common-mode output signal.

The diode supplies temperature compensation for Q_5. Tendency of the transistor current to increase with temperature is counterbalanced by reduction of diode forward resistance and reduction of the forward base bias of the transistor. This tends to decrease the emitter current, and excellent compensation is obtained by matching the diode resistance characteristics to those of the base-emitter diode of Q_5.

The differential amplifier is now the basic unit in the design of dc amplifiers. The circuit is virtually insensitive to temperature variation, especially when manufactured in integrated circuit form, because all components on a chip are processed simultaneously and the elements are so closely related that appreciable temperature differentials cannot occur. The circuit is stable and versatile and can be used with floating inputs and outputs, or where grounded inputs and outputs are required.

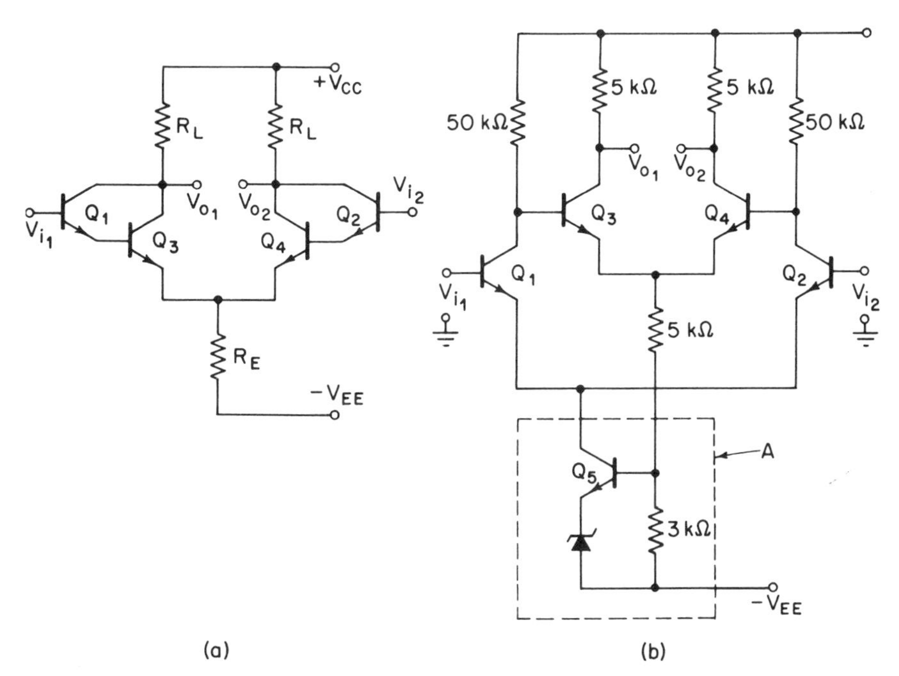

Figure 13-12. (a) Darlington input for a differential amplifier; (b) with common-mode feedback.

13-5 THE OPERATIONAL AMPLIFIER

A common usage of the dc amplifier is as an *operational amplifier*, capable of performing the basic mathematical operations of addition, subtraction, differentiation, and integration in analog computation. This performance is achieved by a feedback amplifier, as in Fig. 13-13, wherein any stable high-gain dc amplifier may be used.

The amplifier has an input element Z_i and a feedback element Z_f, which are usual R, L, or C elements or combinations, where

$$v_R = iR, \qquad v_L = L\frac{di}{dt} = Lsi, \qquad v_C = \frac{1}{C}\int i\,dt = \frac{i}{Cs}$$

the elements then being referred to as $Z_i(s)$, $Z_f(s)$ or operational impedances.

If the current into the amplifier is made negligible, then $i_1 = i_2$, and

$$\frac{v_1 - v_i}{Z_i(s)} = \frac{v_i - v_2}{Z_f(s)} \tag{13-7}$$

The amplifier has an internal gain A given by

$$A = -\frac{v_2}{v_i}$$

assuming an odd number of phase reversals. Eliminating v_i from Eq. 13-7 gives

$$v_2 = -\left[\frac{Z_f(s)}{Z_i(s)}\right]\frac{v_1}{1 + \dfrac{1}{A}\left[\dfrac{Z_f(s)}{Z_i(s)}\right]} \tag{13-8}$$

If the gain A is made large so

$$|A| \gg \left|1 + \frac{Z_f(s)}{Z_i(s)}\right|$$

then

$$v_2 = -\frac{Z_f(s)}{Z_i(s)}v_1 \tag{13-9}$$

This equation indicates that the input and output voltages are related as the negative of $Z_f(s)/Z_i(s)$, the ratio of the selected impedances.

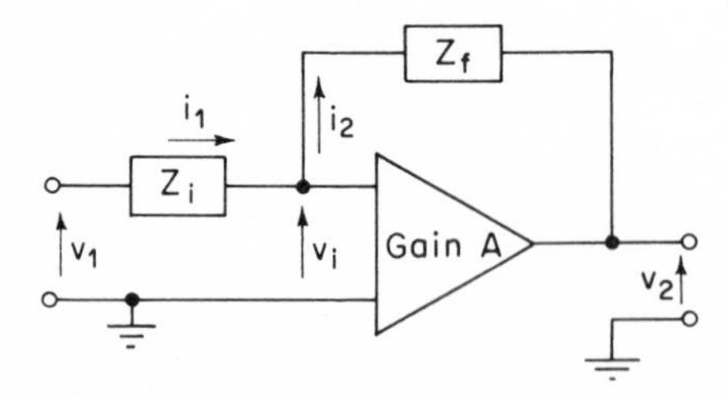

Figure 13-13. The operational amplifier.

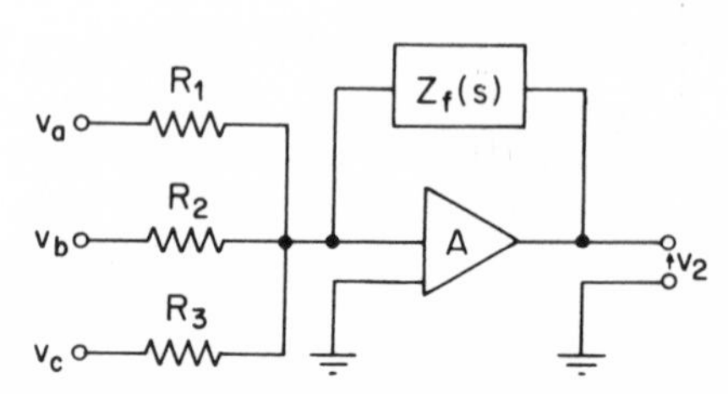

Figure 13-14. Summing amplifier.

If, for example, $Z_f(s) = R_f$, $Z_i(s) = R_i$, then

$$v_2 = -\frac{R_f}{R_i}v_1 \tag{13-10}$$

and v_2 represents v_1 multiplied by a constant R_f/R_i, with the sign changed.

If $Z_f(s)$ is made a resistor R_f and $Z_i(s)$ a capacitor C_i, then

$$v_2 = -R_f C_i s v_1 \tag{13-11}$$

and the output of the amplifier is the derivative of the input voltage, multiplied by a constant and with the sign changed. If $R_f = 1$ megohm, $C_i = 1\ \mu$F, then the $R_f C_i$ factor becomes unity.

Further, if $Z_f(s)$ is a capacitor C_f, and $Z_i(s)$ a resistor R_i, then

$$v_2 = -\frac{1}{R_i C_f}\frac{v}{s} \tag{13-12}$$

and the output is equal to the negative of the integral of the input, multiplied by a constant $1/R_i C_f$.

Several input voltages may be simultaneously operated upon in the circuit of Fig. 13-14. If A is again very large

$$-v_2 = \frac{Z_f(s)}{R_1}v_a + \frac{Z_f(s)}{R_2}v_b + \frac{Z_f(s)}{R_3}v_c$$

If $R_1 = R_2 = R_3$, then

$$-v_2 = \frac{Z_f(s)}{R_1}(v_a + v_b + v_c) \tag{13-13}$$

and the output represents a summation of the inputs.

Various combinations of R and C may be employed as $Z_i(s)$ and $Z_f(s)$, leading to a considerable variety in the mathematical operations which may be performed.

Because of the multiplication by ω which occurs in the differentiation process, the output grows in proportion to frequency, and high-frequency random noise in the input is increased. Because of this difficulty the process of differentiation is usually avoided by algebraic manipulation of the equations which are to be solved.

13-6 THE DIFFERENTIAL AMPLIFIER AS AN OPERATIONAL AMPLIFIER

The flexibility of input arrangement makes the differential amplifier of considerable value as a basic circuit for operational amplifier use. Figure 13-15 represents the development of the differential amplifier at (a) into a practical operational amplifier at (b). The output is single-ended, by non-use of terminal 4 of (a) at signal frequency. Internal gain A must be large, and because of the large gain and the feedback, certain interesting results appear.

For instance, with voltage feedback the output resistance falls, and would become zero for infinite gain. With shunt feedback input, the input resistance rises, and would become infinite for infinite gain. With feedback, the negative output fed back from 3 to 1 opposes the positive input on the external terminal. With infinite gain, the output can increase until the voltage at 1 becomes negligibly small. Therefore $I_i = 0$ and $V_i = 0$ for the ideal operational amplifier. Since the internal gains can be made very large, these results can be applied in the analysis of several simple uses of precision amplifiers.

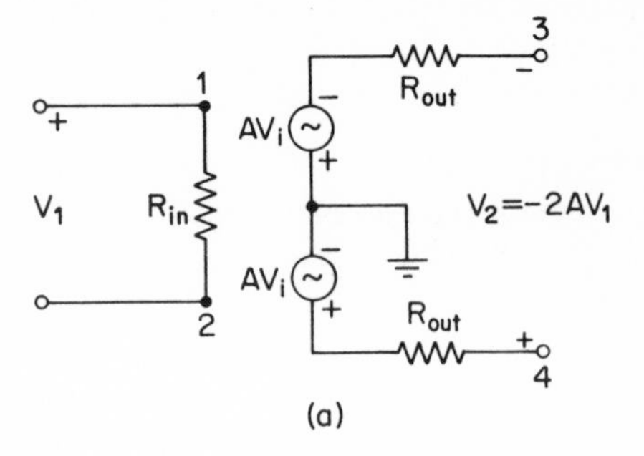

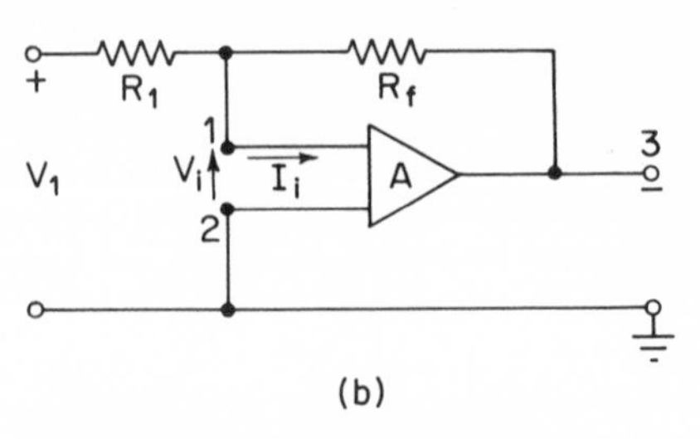

Figure 13-15. (a) Differential input, differential output amplifier; (b) single-ended output with feedback.

An isolating buffer or voltage following amplifier of unity voltage gain is possible with the circuit of Fig. 13-16(a). By definition

$$\mathbf{V}_2 = -\mathbf{A}\mathbf{V}_i$$

and from the circuit

$$\mathbf{V}_1 + \mathbf{V}_i = \mathbf{V}_2$$

Then

$$\mathbf{V}_1 - \frac{\mathbf{V}_2}{\mathbf{A}} = \mathbf{V}_2$$

but with $\mathbf{A} \longrightarrow \infty$, this becomes

$$\mathbf{V}_1 = \mathbf{V}_2 \tag{13-14}$$

and so the output voltage follows the input voltage. Since the output resistance is theoretically zero, any desired output current can be obtained.

Figure 13-16(b) illustrates the inverting operational amplifier with resistive operational impedances. Equal currents are present in R_1 and R_f, thus the input resistance of the amplifier is infinite. So

$$\frac{\mathbf{V}_1 - \mathbf{V}_i}{R_1} - \frac{\mathbf{V}_i - \mathbf{V}_2}{R_f} = 0$$

With very large internal gain, we know that $\mathbf{V}_i \longrightarrow 0$, and so

$$\mathbf{V}_2 = -\frac{R_f}{R_1}\mathbf{V}_1 \tag{13-15}$$

which is the result of Eq. 13-10. An amplifier so connected becomes a precision voltage inverter, with multiplication by a constant. The input resistance is now R_1, since input 1 is at zero or virtual ground potential.

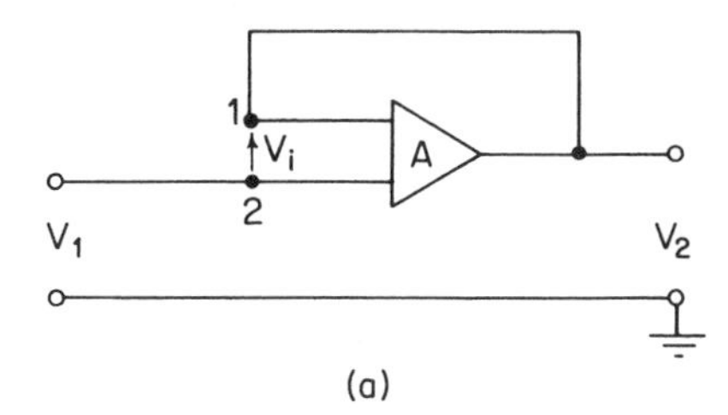

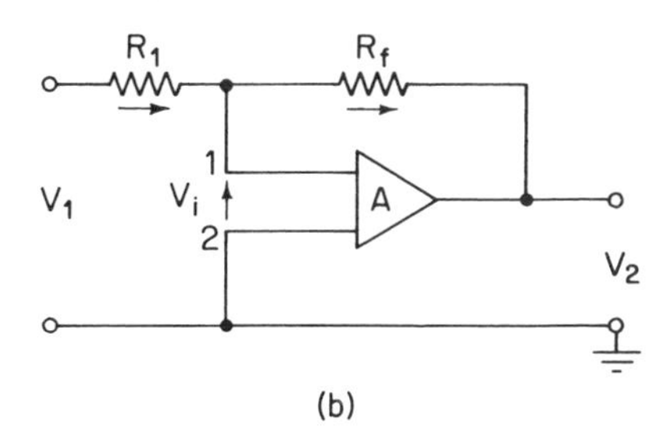

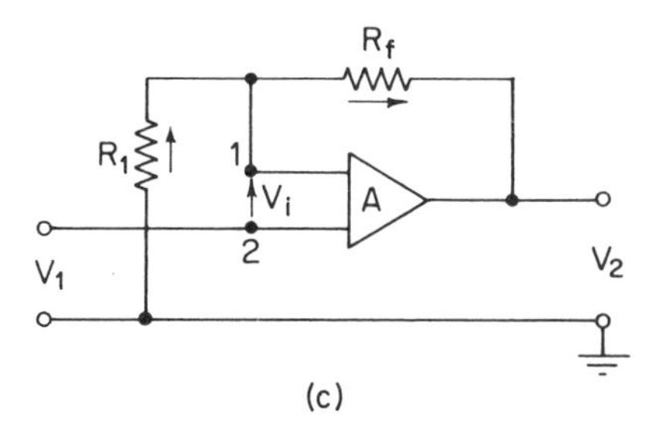

Figure 13-16. (a) Unity-gain amplifier; (b) inverting amplifier; (c) non-inverting amplifier.

Now consider the non-inverting circuit at Fig. 13-16(c). Resistors R_f and R_1 represent a voltage divider and

$$V_1 + V_i = \frac{R_1}{R_f + R_1} V_2 \quad (13\text{-}16)$$

However, V_i is negligibly small with A very large, and so

$$V_2 = \frac{R_f + R_1}{R_1} V_1 \quad (13\text{-}17)$$

and this provides multiplication of the input voltage by a constant, without voltage inversion. From Eq. 12-18, the input resistance to terminal 2 is equal to the internal input resistance multiplied by the return difference, and this can be very large. For inverting applications the input resistance will be that of R_1. The output resistance will be given by Eq. 12-14 as the internal output resistance divided by the return difference and can be very small.

The ac internal frequency response is an important design element, and will be limited in frequency range by the internal capacitances, as indicated in Fig. 13-17. The internal fall in gain at high frequencies may be too great for feedback amplifier stability, because of the relation between rate of gain rolloff and phase angle. For instance, if the gain falls as fast as 40 dB per frequency decade, the phase angle will reach 180°; with a 60 dB fall per decade, the phase angle reaches 270°. These conditions in a feedback amplifier loop will lead to instability if the gain is above unity. It is then necessary to reduce the gain so that it falls below unity (0 dB) before the phase angle exceeds 180° and this shaping of the gain and phase characteristics may be done with external *RC* network compensation, giving -20 dB per frequency decade rolloff in the gain curve, and with an accompanying phase shift approaching $-90°$.

The rolloff from the compensating network must begin at a frequency below that at which the amplifier introduces additional phase shift. The resultant curves illustrate the constant gain–bandwidth product obtainable, and show that often the main function of a large internal bandwidth is to permit phase compensation at the higher frequencies. The internal band-

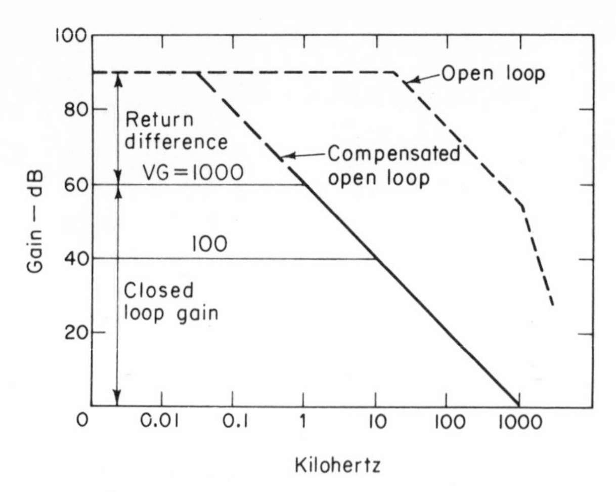

Figure 13-17. Bode plot for an operational amplifier with unity-gain bandwidth of 1 MHz.

width may have to be more than one megahertz to provide a closed-loop gain of 60 dB at 1000 hertz.

If R_f is replaced by a capacitor C, the circuit becomes an integrator as previously discussed for the basic operational amplifier; that is, Eq. 13-15 leads to

$$V_2 = -\frac{1}{sR_1C}V_1 = -\frac{1}{R_1C}\int V_1\,dt \tag{13-18}$$

For the integrator, unity gain (0 dB) is reached at the frequency at which $X_C = R_1$. Below that frequency the amplifier gain should vary as $-1/\omega$, or it should change at the rate of -20 dB per decade, as the integral of $E \sin \omega t = -(E/\omega) \cos \omega t$. For a perfect integrator the gain should go to infinity at 0 Hz, but practical limited internal gains will cut off the curve as in Fig. 13-17. This cutoff of gain should occur at a frequency below the lowest expected input frequency.

For summing, each input is applied to terminal 1 through an appropriate resistor R_a, R_b, ... Current in R_f will be equal to the sum of the input currents, or

$$V_2 = -R_f\left(\frac{V_a}{R_a} + \frac{V_b}{R_b} + \dots\right) \tag{13-19}$$

so that each input is multiplied by a scaling factor $-R_f/R_x$ before the addition is performed.

A summing amplifier should have a flat frequency response and the roll-off should occur above the highest expected operating frequency. To obtain sufficient bandwidth may require operation with limited gain.

More involved circuit forms for Z_i and Z_f can be employed to yield useful results in control systems. Examples are introduced by Problem 13-6, wherein Fig. 13-23(c) functions as a *lag network*, and Fig. 13-23(b) would become a *lead network* if C_2 were short-circuited. A lag network acts as a low-pass filter and a lead network as a high-pass form. A transfer function simulating both high- and low-pass, or band-pass action, may be achieved with a parallel RC network at Z_f and a series RC branch at Z_i. Other transfer functions may also be simulated, and these operational amplifier forms are particularly useful as active filters in integrated circuits, since only R and C circuit elements need be employed.

13-7 VOLTAGE REGULATORS USING dc AMPLIFIERS

The dc amplifier is employed in circuits which stabilize the output voltages of rectifier circuits supplying dc voltages. Basic shunt and series forms of these circuits appear in Fig. 13-18. These circuits will include a means of comparing the load voltage V_o with a standard, usually the voltage of a Zener diode. The error or difference voltage is passed through a dc amplifier and its output signal adjusts the magnitude of a resistive loss element.

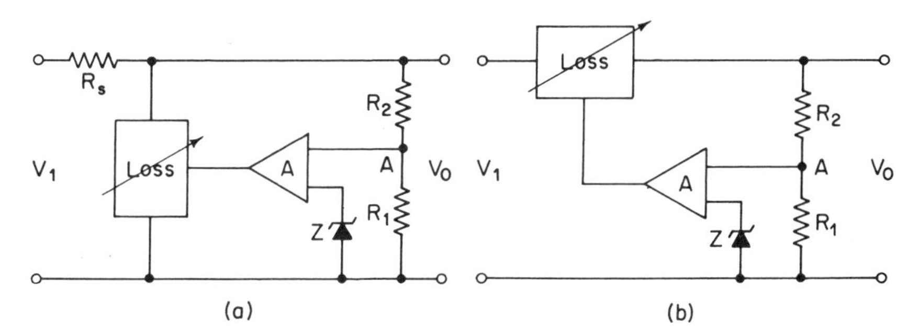

Figure 13-18. (a) Shunt loss regulator; (b) series loss regulator.

In (a) the voltage at point A is proportional to V_o, and this is compared to the fixed standard voltage Z. An increase in potential at A causes the loss element to take a greater shunt current, increasing the drop in R_s and restoring V_o toward its desired level.

At (b), an increase in potential at A causes the series loss element to increase in resistance and to create a greater series voltage drop, restoring V_o toward its desired value.

Simplified operating circuits appear in Fig. 13-19. In the shunt form at (a) any change in V_o is transmitted to R_1, appearing as input to transistor Q_1. Its output controls the effective shunt resistance of Q_2, and causes the voltage drop in R_s to change so as to restore V_o toward standard. The tran-

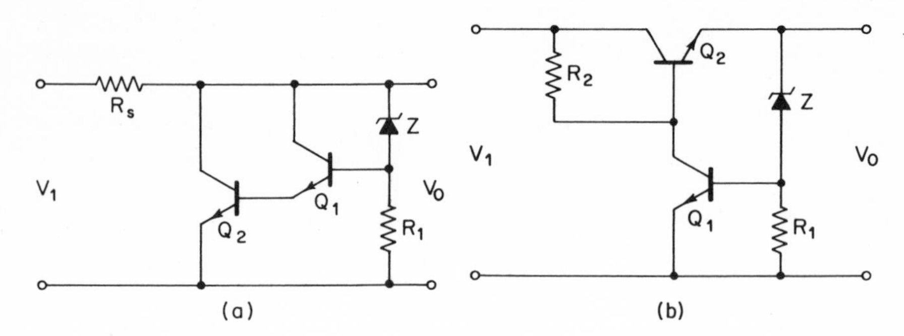

Figure 13-19. (a) Shunt regulator; (b) series regulator.

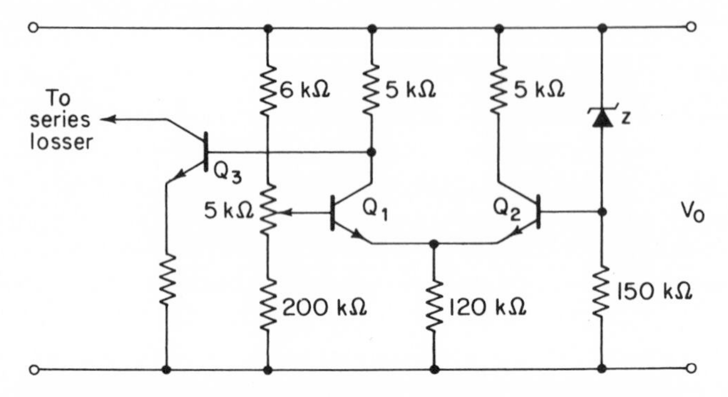

Figure 13-20. Differential amplifier input for a regulator.

sistor carries only a shunt current, and the circuit has the advantage of being short-circuit proof, since an output short circuit simply unloads the transistor.

In (b), a change in load voltage V_o is compared to Z and the difference controls the dc amplifier Q_1. Its output controls the series loss element Q_2, which may consist of several transistors in parallel to handle a large output current. A load short circuit can destroy the series element, Q_2.

These circuits are feedback devices, in which the differential amplifier is often employed because of its high stability, as in Fig. 13-20. Improvement in sensitivity is possible by inclusion of additional amplification inside the feedback loop. The regulator also discriminates against ripple as a form of output voltage variation, so that a regulator serves as a partial filter.

The circuit of Fig. 13-21 can be used for determination of the effect of series regulation on the output resistance of a power supply. Then for small changes

$$\Delta I_{b1} = \frac{\Delta V_o}{R_D + h_{ie}} \tag{13-20}$$

$$\Delta I_o = h_{fe2}\Delta I_{b2} + \Delta I_{b2} = (1 + h_{fe2})\Delta I_{b2} \tag{13-21}$$

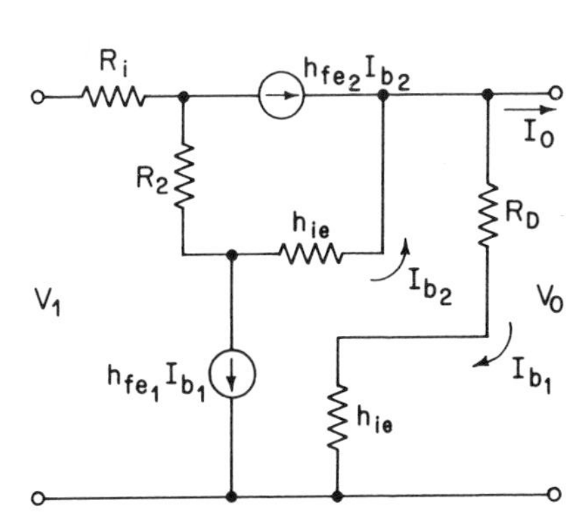

Figure 13-21. Equivalent circuit for analysis of Fig. 13-19(b).

The output resistance is determined by small changes in load voltage and current

$$R_o = \frac{\Delta V_o}{\Delta I_o} = \frac{R_D + h_{ie}}{1 + h_{fe2}} \frac{\Delta I_{b1}}{\Delta I_{b2}} \quad (13\text{-}22)$$

However, I_{b2} is the output current of Q_1, if R_2 is large with respect to h'_{ie}. Then

$$\frac{\Delta I_{b1}}{\Delta I_{b2}} = \frac{1}{h_{fe1}}$$

and

$$R_o = \frac{R_D + h_{ie}}{h_{fe1}(1 + h_{fe2})} \cong \frac{R_D + h_{ie}}{h_{fe1}h_{fe2}} \quad (13\text{-}23)$$

For good regulation of the output voltage the output resistance of the power supply should be low; large current gains are therefore desirable in both transistors. In fact, transistor Q_1 may be replaced with a more elaborate dc amplifier to achieve higher h_{fe1}. Very low values of output resistance and voltage regulations of a fraction of a per cent are possible.

Transistor Q_2 must be designed for and mounted so as to dissipate the power lost as a series dropping resistor. Silicon units are almost universal, because of the low I_{CBO} values obtainable at the usual high ambient temperatures in power supply service.

PROBLEMS

13-1. In Fig. 13-22, the identical tubes have $\mu = 20$, $r_p = 8000$ ohms, and the current through meter M is zero for $e_i = 0$. Find the current through the meter per volt of input.

13-2. The transistors in the circuit of Fig. 13-11(b) have $h_{ie} = 5000$ ohms, $h_{fe} = 50$, and $R_L = 50\text{k}\Omega$, $R_E = 5\text{k}\Omega$. Transistor Q_1 has a temperature coefficient of $v_{BE} = -2.1$ mV/°C, transistor Q_2 has a coefficient of $v_{BE} = -3.6$ mV/°C. Determine the effective overall coefficient to output if h_{oe} is small.

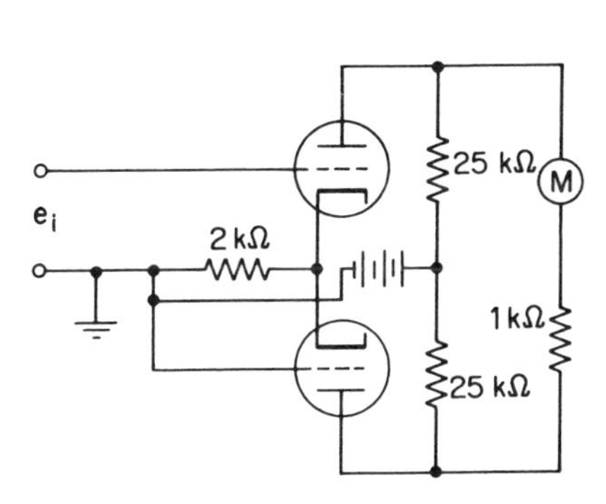

Figure 13-22.

13-3. For the circuit of Fig. 13-13, $Z_i = 1$ megohm, $Z_f = 0.2$ μF. For an input of $v_i = 0.3t$, determine the output voltage as a time function, for $|A|$ very large.

13-4. Repeat Problem 13-3, if Z_f is composed of 1 μF in parallel with 1.2 megohm resistance.

13-5. In Fig. 13-14, $R_1 = 0.2$ megohm, $R_2 = 1.3$ megohms, $R_3 = 0.75$ megohm, and Z_f is a 0.5 μF capacitor. The inputs are $v_a = 0.8 \sin 477t$, $v_b = 60t$, $v_c = 1.2$ V. Plot the output wave over $\frac{1}{60}$ second.

13-6. Determine the relation of v_2 to v_1 for each of the operational circuits of Fig. 13-23. Gain $|A|$ is very large.

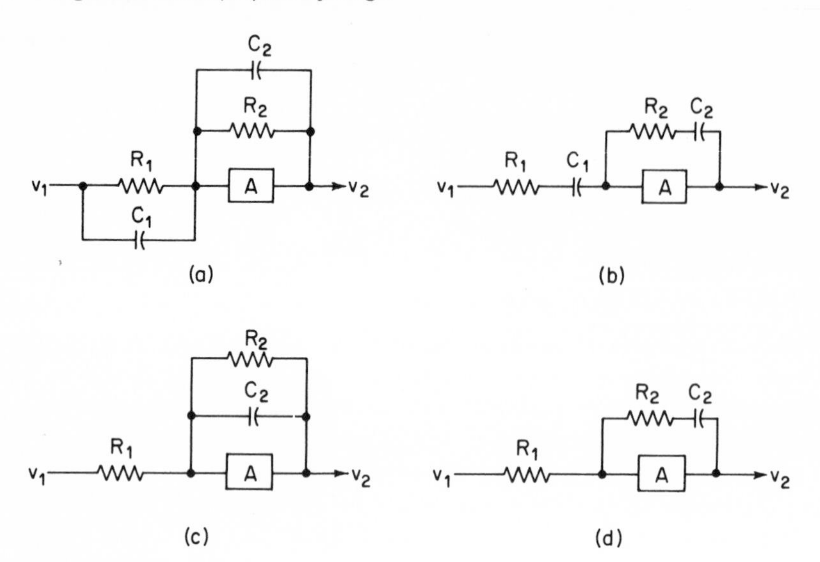

Figure 13-23.

REFERENCES

1. Regazzini, J. R., R. H. Randall, and F. A. Russell, "Analysis of Problems in Dynamics by Electronic Circuits," *Proc. IRE*, Vol. 35, p. 444 (1947).

2. Chaplin, G. B., and A. R. Owens, "Some Transistor Input Stages for High-Gain D-C Amplifiers," *Proc. IRE*, p. 105 (May 1958).

3. Warfield, J. W., *Electronic Analog Computers*. Prentice-Hall, Inc., Englewood Cliffs, N. J., 1959.

4. Okada, R. H., "Stable Transistor Wide-Band D-C Amplifiers," *Communications and Electronics, AIEE* (March 1960).

5. Middlebrook, R. D., *Differential Amplifiers*. John Wiley & Sons, Inc., New York, 1963.

6. *Transistor Manual*. General Electric Co., Semiconductor Products Dept., Syracuse, N. Y., 1964.

7. Thornton, R. D., *et al.*, *Multistage Transistor Circuits* (SEEC, Vol. 5). John Wiley & Sons, Inc., New York, 1965.

8. *RCA Linear Integrated Circuit Fundamentals*. Radio Corporation of America, Harrison, N. J., 1966.

14

POWER AMPLIFIERS WITH LARGE SIGNALS

Large input signals to transistor or vacuum tube are required if appreciable power output is to be obtained. Dynamic characteristics are nonlinear for such wide swings of voltage, and current conduction may not be continuous, so that equivalent circuit methods of analysis cannot be used. Graphical analysis is therefore applied here to large-signal audio- and radio-frequency amplifiers.

14-1 OPERATING CLASSIFICATIONS

A universal system of classification of active device operating conditions is based on the operating point location on the dynamic curve of the device.

Referring to Fig. 14-1, placement of the operating point near the middle of a linear region of the dynamic curve results in continuous output current and *Class A operation*. Distortion is low because of the linear operating region, and voltage amplification is high, but the efficiency of conversion of dc power to ac power cannot exceed a theoretical maximum of 50 per cent.

Selection of operating point location at cutoff places the forward half-cycle of input signal on the linear characteristic, with current then existing for 180° of the input cycle; this is *Class B operation*. The reverse half-cycles of current are missing, and distortion is high, but the maximum theoretical power conversion efficiency can reach 78 per cent.

If the operating point is set at a bias considerably beyond cutoff, output current is present for less than 180° of the input cycle; this is *Class C operation*. Distortion is very high, but large input signals may be used with high power conversion to ac; the maximum theoretical conversion efficiency can be 100 per cent.

In vacuum tube operation, when the drive is restricted so that the grid

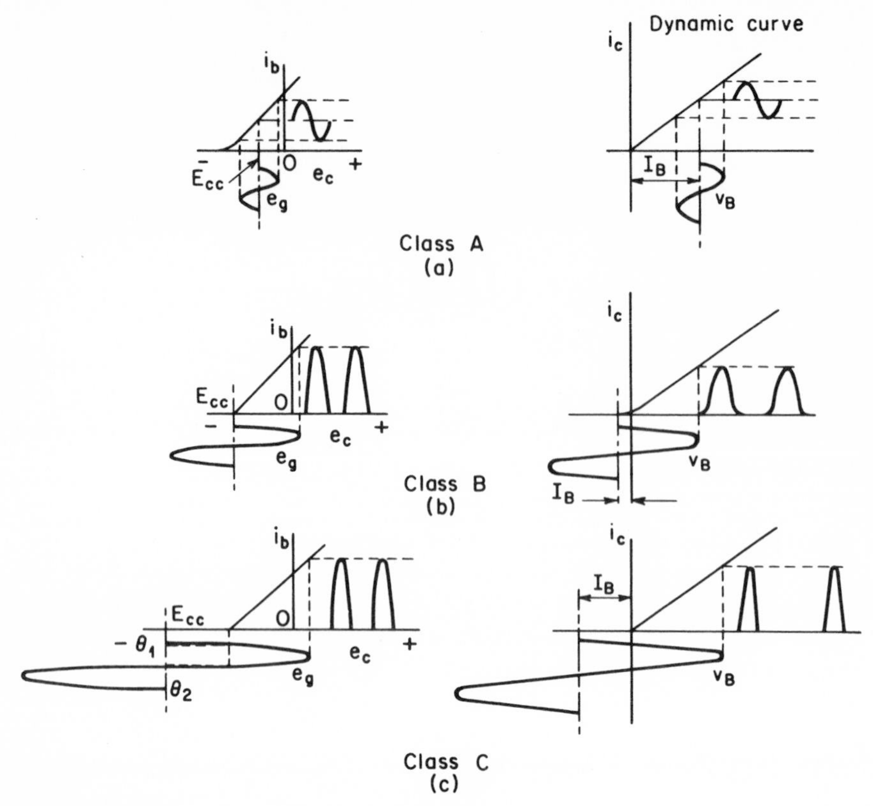

Figure 14-1. Class A, B, C operation. Left—tube; right—transistor.

is not driven positive or grid current does not exist, a subscript 1 may be added to the class designation, as A_1; when drive is increased and grid current is present, the condition is designated by a subscript 2, as A_2.

14-2 OUTPUT CIRCUITS

In the amplifier of Fig. 14-2(a) the direct current component produces a power loss in R_L which serves no useful purpose, and must be dissipated. It is also not to be expected that the available load R_L will necessarily be of a magnitude suited as a load for the transistor or tube. The circuit at (b) eliminates most of the dc power loss, and provides a means of adjusting the apparent load R'_L to the value desired for tube or transistor operation.

Ideally the output transformer will have small losses and the voltages and currents are related by the turns ratio $a = N_1/N_2$, as

$$a = \frac{V_1}{V_2} = \frac{I_2}{I_1} \tag{14-1}$$

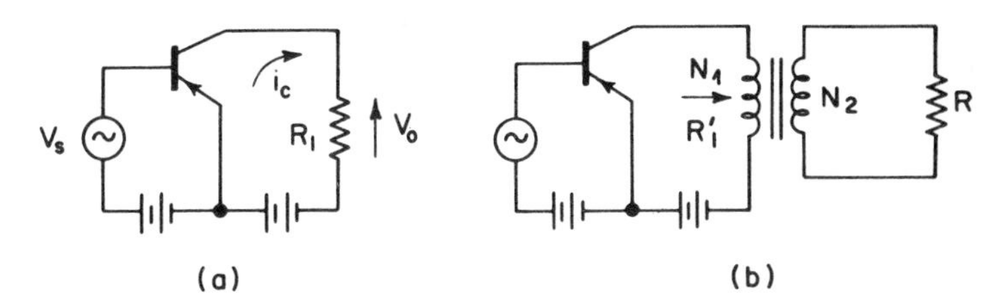

Figure 14-2. Output circuits.

The impedance seen from the primary side is

$$Z_1 = \frac{V_1}{I_1} = \frac{V_2 I_2}{I_1^2} = \frac{a^2 V_2}{I_2} = a^2 Z_2 \tag{14-2}$$

since the secondary load is $Z_2 = V_2/I_2$. For the circuit under discussion the apparent primary load is then

$$R'_L = a^2 R$$

and a can be selected to make a load R appear as a desired value R'_L.

The above analysis applies to an *ideal* or perfect transformer. However, the primary of the transformer of Fig. 14-3 will usually have a small dc resistance R_p. A well-designed transformer will also have low iron losses. Direct current in the primary tends to saturate the iron, and creates wave form distortion. Push-pull circuits are used to cancel the dc magnetomotive force on the core, or the magnetic circuit may be built with an air gap. The leakage inductance L'_s should be made small by interleaving primary and secondary windings, so that the response peak, created by series resonance with the winding capacitance C, will be moved to a high frequency outside the useful range. Such transformers will be found to have power efficiencies ranging from 50 to 85–90 per cent.

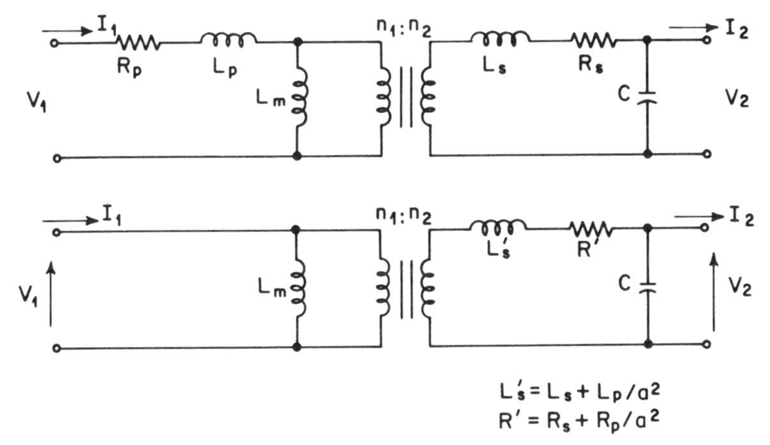

Figure 14-3. Top: transformer equivalent circuit; bottom: series quantities referred to the secondary.

14-3 POWER RELATIONS

A power amplifier is a frequency converter, changing dc power into ac power. The power input is supplied by the dc source, and the average power in the circuit will be distributed according to

$$\text{power input} = \text{ac power output} + \text{losses}$$

$$\left.\begin{aligned} V_{CC}I_C &= I_c^2R_L' + \text{losses} \\ E_{bb}I_b &= I_p^2R_L' + \text{losses} \end{aligned}\right\} \tag{14-3}$$

The losses are those in the dc resistances of the load and the bias resistor, and those internal to the transistor or tube. The internal loss in Class A operation is found from the device input, less the signal output, or

$$\left.\begin{aligned} \text{transistor internal loss} &= V_{CE}I_{Cs} - I_c^2R_L' \\ \text{tube internal loss} &= E_bI_{bs} - I_p^2R_L' \end{aligned}\right\} \tag{14-4}$$

where I_{Cs} and I_{bs} are average currents with signal present. The internal loss is sometimes called the *power dissipation* of the transistor or tube.

If there is no input signal and consequently zero ac output, the entire input must be dissipated by the transistor or tube. Designs must be based on the worst case expected, and so in Class A service the tube or transistor must be capable of dissipating the power loss expected with zero applied signal as a maximum. As the ac signal output increases, the internal loss decreases and the device runs cooler.

While the maximum permissible collector loss is stated as a rating of a transistor, it is also a function of ambient temperature and the cooling properties of the mounting or *heat sink*. The power limit is established by the maximum permissible germanium or silicon temperature of about 85°C and 200°C, respectively. The difference between the collector temperature and the surrounding ambient temperature is relatively small, and radiation will not be effective. Conduction must be relied upon for heat removal from the collector. The latter is always mounted in good thermal contact with the case, and the case must be placed in good thermal contact with the heat sink, often with fins provided for additional convection heat transfer.

A maximum value of anode dissipation is also stated for a vacuum tube intended for power output service, to insure that the anode temperature does not become excessive. Temperatures are high and effective heat transfer is possible by radiation or conduction.

Thus such devices are limited by their losses, not by maximum output, as for other types of electrical equipment—this is a fundamental difference.

The efficiency of conversion from dc power to ac power is

$$\eta = \frac{\text{ac power output}}{\text{dc power input}} \times 100\% \tag{14-5}$$

where average power is ordinarily understood.

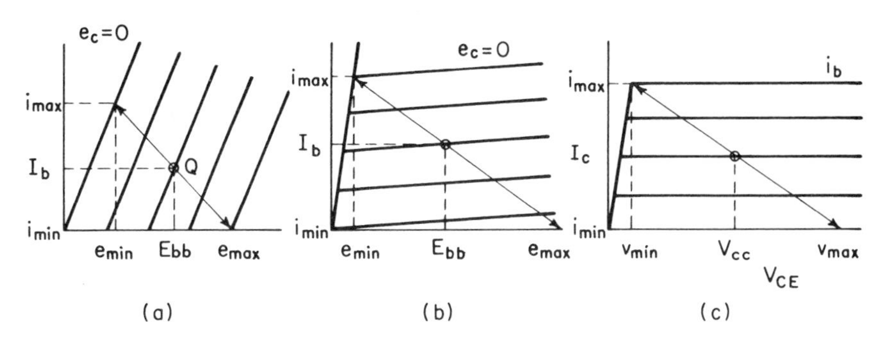

Figure 14-4. Idealized characteristics: (a) triode; (b) pentode; (c) transistor.

Choice of operating conditions for tube or transistor places a limit on operating efficiency. For the assumed Class A case, linearized characteristics are presented in Fig. 14-4. The power efficiency can be written as

$$\eta = \frac{I_c^2 R_L'}{V_{CC} I_C} \times 100\% = \frac{(i_{\max} - i_{\min})^2 R_L'}{8 V_{CC} I_C} \times 100\% \tag{14-6}$$

where R_L' is the ac resistance reflected from the load and appearing in the collector circuit. The saturation voltage of a transistor is very low, and in a limit situation we can say that $i_{\min} = 0$, after which $I_C = i_{\max}/2$. Then

$$\eta = 50 \frac{i_{\max} R_L'}{2 V_{CC}} \% \tag{14-7}$$

but in the limit $i_{\max} R_L' = 2 V_{CC}$, so that the *theoretical maximum value for the conversion efficiency* in a transformer-coupled Class A amplifier is 50 per cent, and actual efficiencies in transistor amplifiers approach that figure. Because of the distortion due to the curved low voltage characteristics of vacuum tubes, tube efficiencies rarely exceed 25 per cent. Higher minimum anode voltages make triodes less efficient than pentodes, as indicated by Fig. 14-4(a) and (b).

14-4 THE CONSTANT-LOSS HYPERBOLA; THE ac LOAD LINE

Prediction of transistor or tube performance under large-signal Class A conditions is usually graphical, because of the nonlinearity of the dynamic transfer curve in general; this method was initially discussed in Section 7-3. However, a new concept has been added here; with transformer coupling the direct and alternating current loads are not necessarily identical in impedance, and an additional step must be added to the graphical procedure.

Consider the circuit of Fig. 14-5, to be used in Class A operation with the transistor whose characteristics appear in Fig. 14-6. As a first step, the allowable collector loss must be established. The designer must determine the expected operating temperature at the expected loss value; this will be discussed later as a thermal problem. At the moment let us assume that the transistor and its mounting can dissipate 30 watts at an allowable temperature. A hyperbola can then be drawn as in Fig. 14-6 for

$$\text{loss} = V_{CC}I_C = \text{constant} \qquad (14\text{-}8)$$

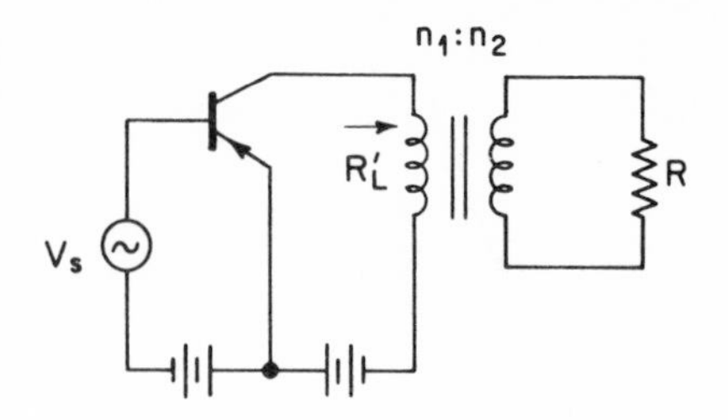

Figure 14-5. Transformer-output amplifier.

It is customary to use transformer-coupled loads to achieve high power efficiency, and therefore $V_{CE} = V_{CC}$. Any allowable operating point must lie on or below this *constant-loss hyperbola,* to avoid overheating of the active device. With a transistor rated for maximum $V_{CE} = 60$ V, it is reasonable to choose one-half of this for collector supply, or $V_{CC} = 30$ V. The *dc load line* may then be drawn vertically from $v_{CE} = 30$ V, assuming the transformer dc resistance is negligible. The

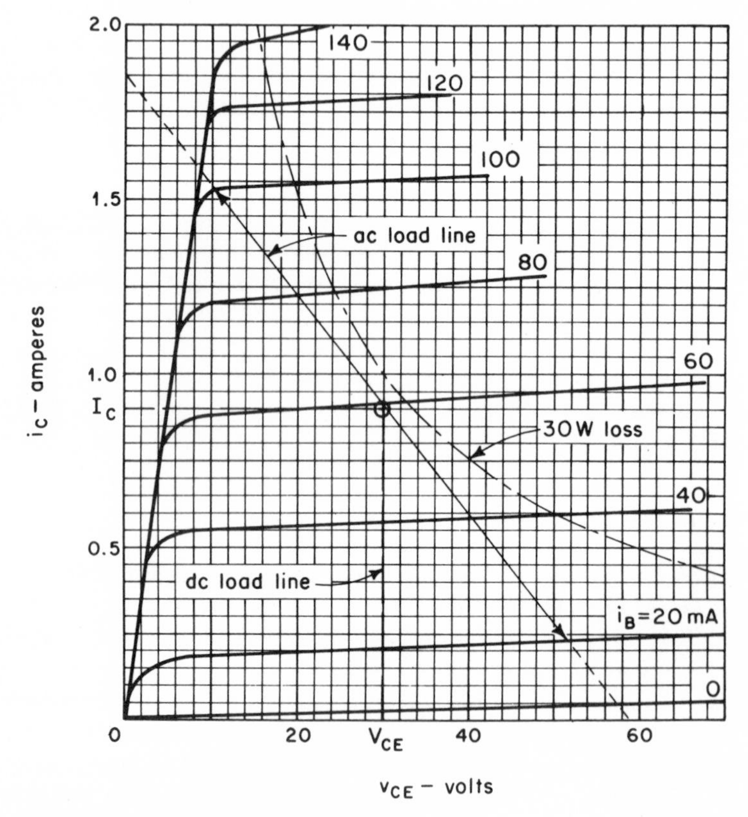

Figure 14-6. ac load line construction.

operating point Q must be located on this line and will be placed below the constant-loss curve, as at 0.9 A, for a slight factor of safety. The Q point is in the middle of a linear operating region, as desired for Class A use.

Point Q represents zero ac voltage on an *ac load line*, needed to define the path of operation with ac present in the output circuit. A suitable ac load may be specified from a transformer secondary load $R = 4$ ohms, with a transformer ratio $a = \sqrt{8}$, or $R_L' = 4a^2 = 32$ ohms. A positive swing of 30 V/32 ohms $= 0.94$ A above the Q point and this fixes the y intercept of the ac load line at $i_C = 1.84$ A, $v_{CE} = 0$ V. The line can then be drawn as in the figure. The resultant slope is $-1/R_L'$.

For minimum distortion the positive-going signal cannot carry past the knee of the curves, or beyond $i_B = 100$ mA, and the symmetrical negative-going swing then is limited to the $i_B = 20$ mA curve, as shown by the solid portion of the ac load line.

With the input signal at 40 mA peak, the output will be

$$V_o = \frac{51 - 10}{2\sqrt{2}} = 14.5 \qquad \text{V rms}$$

and the power output is

$$P_o = \frac{14.5^2}{32} = 6.6 \qquad \text{W}$$

The power efficiency is

$$\eta = \frac{6.6 \times 100}{0.9 \times 30} = 26.9\ \%$$

A transfer or current-gain curve resulting from the 32-ohm load line can be plotted from points chosen on the load line, and appears in Fig. 14-7. It is linear for a considerable region on each side of the operating point but drive with a large signal would introduce odd harmonics. This current transfer curve will be used in predicting transistor output performance.

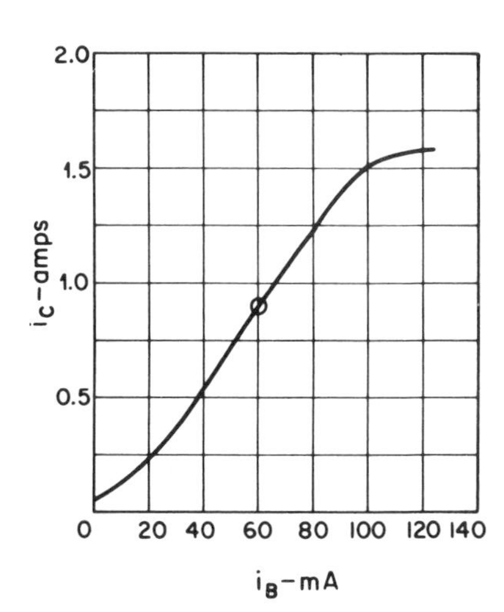

Figure 14-7. Transfer curve for the load line of Fig. 14-6.

14-5 DETERMINATION OF NONLINEAR DISTORTION

A dynamic transfer curve relating output and input quantities can be used to predict output wave forms resulting from arbitrary driving functions; Fig. 7-3 shows this as a graphical process. With sinusoidal input, the departure of the output wave form from sinusoidal form indicates nonlinear distortion and the presence of harmonics created by the nonlinear transfer curve. An analytical process allows the harmonic amplitudes to be calculated by writing an equation for the transfer curve of the device, and using the ac load line to provide voltage and current values for determination of the coefficients.

That is, the dynamic curve can be expressed by a Taylor's series expanded about the Q point as

$$i_C = a_o + a_1(e - e_Q) + a_2(e - e_Q)^2 + a_3(e - e_Q)^3 + a_4(e - e_Q)^4 + \ldots \tag{14-9}$$

Considering the input signal as a cosine function, where

$$E_m \cos \omega t = e - e_Q$$

the output current can be predicted as

$$i_C = a_o + a_1 E_m \cos \omega t + a_2 E_m^2 \cos^2 \omega t + a_3 E_m^3 \cos^3 \omega t + a_4 E_m^4 \cos^4 \omega t + \ldots \tag{14-10}$$

For most transistors and tubes the terms above the fourth power may be dropped as small; this does not invalidate the principle of the analysis. By use of trigonometric identities and by combination of the resultant frequency terms, the above becomes

$$i_C = a_o + A_0 + A_1 \cos \omega t + A_2 \cos 2\omega t + A_3 \cos 3\omega t + A_4 \cos 4\omega t \tag{14-11}$$

In Eq. 14-10, if the input signal is zero, operation is at the Q point, $i_C = I_C$, and therefore $a_o = I_C$. The two dc terms appearing in Eq. 14-11 represent the previously-defined average current with signal present, or

$$a_o + A_0 = I_C + A_0 = I_{Cs}$$

The amplitudes of the distortion components can be determined by evaluation of the A coefficients. The load line of Fig. 14-8 can be used to determine the current at five instants in time, spaced over the positive and negative half-cycles. These instants are chosen at $\omega t = 0$ with current designated $i_{\max}$ when the input voltage is at positive maximum; $\omega t = \pi/3$ with a current designated i_x when the input is at one-half the maximum; $\omega t = \pi/2$ with current I_C at zero voltage input at the Q point; $\omega t = 2\pi/3$ with current i_y at input one-half of the negative maximum; and $\omega t = \pi$ with current $i_{\min}$ when the input is at negative maximum.

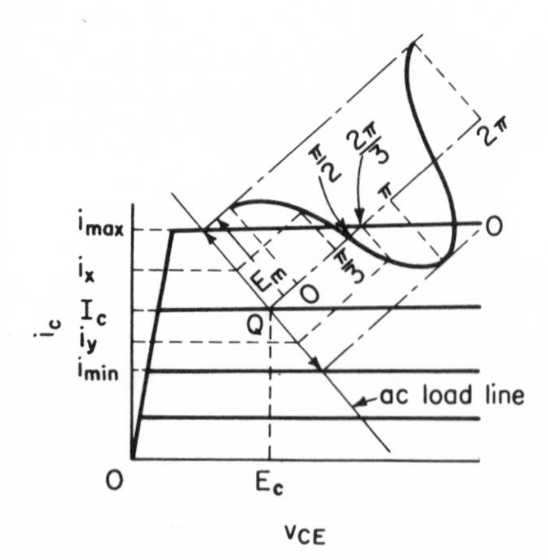

Figure 14-8. Calculation of nonlinear distortion from the transistor load line.

Substitution of these values of time and current into Eq. 14-11 yields five equations:

$$\left.\begin{aligned}
\omega t &= 0^\circ & i_{\max} &= I_C + A_0 + A_1 + A_2 + A_3 + A_4 \\
\omega t &= 60^\circ & i_x &= I_C + A_0 + \frac{A_1}{2} - \frac{A_2}{2} - A_3 - \frac{A_4}{2} \\
\omega t &= 90^\circ & I_C &= I_C + A_0 \qquad\quad - A_2 \qquad\quad + A_4 \\
\omega t &= 120^\circ & i_y &= I_C + A_0 - \frac{A_1}{2} - \frac{A_2}{2} + A_3 - \frac{A_4}{2} \\
\omega t &= 180^\circ & i_{\min} &= I_C + A_0 - A_1 + A_2 - A_3 + A_4
\end{aligned}\right\} \tag{14-12}$$

Elimination among these five equations results in the following expressions for the amplitudes of the various components:

$$A_0 = \tfrac{1}{6}(i_{\max} + i_{\min}) + \tfrac{1}{3}(i_x + i_y) - I_C \tag{14-13}$$

$$A_1 = \tfrac{1}{3}(i_{\max} - i_{\min}) + \tfrac{1}{3}(i_x - i_y) \tag{14-14}$$

$$A_2 = \tfrac{1}{4}(i_{\max} + i_{\min}) - \tfrac{1}{2}I_C \tag{14-15}$$

$$A_3 = \tfrac{1}{6}(i_{\max} - i_{\min}) - \tfrac{1}{3}(i_x - i_y) \tag{14-16}$$

$$A_4 = \tfrac{1}{12}(i_{\max} + i_{\min}) - \tfrac{1}{3}(i_x + i_y) + \tfrac{1}{2}I_C \tag{14-17}$$

It is possible to define percentage distortion due to a particular harmonic as the ratio of the harmonic amplitude to that of the fundamental times 100 per cent, or

$$D_2 = \frac{A_2}{A_1} \times 100\,\%, \qquad D_3 = \frac{A_3}{A_1} \times 100\,\%, \qquad D_4 = \frac{A_4}{A_1} \times 100\,\%$$

The total harmonic distortion is defined as the ratio of the effective value of all harmonics to that of the fundamental, or

$$D = \frac{\sqrt{A_2^2 + A_3^2 + A_4^2 + \cdots}}{A_1} \times 100\,\% \tag{14-18}$$

The above may be simplified if all harmonics above the second are considered negligible, in which case Eq. 14-11 may be written as

$$i_C = I_C + C_0 + C_1 \cos \omega t + C_2 \cos 2\omega t$$

By use of a three-point voltage-current-time table, the values of the amplitude coefficients can then be obtained as

$$C_0 = \tfrac{1}{4}(i_{\max} + i_{\min}) - \tfrac{1}{2}I_C \tag{14-19}$$

$$C_1 = \tfrac{1}{2}(i_{\max} - i_{\min}) \tag{14-20}$$

$$C_2 = \tfrac{1}{4}(i_{\max} + i_{\min}) - \tfrac{1}{2}I_C \tag{14-21}$$

These equations lead to some saving in time, if only the second harmonic is to be considered.

The value of I_{Cs} defines a new Q point, and the ac load line will shift to this new position. New values of current can then be found and the A coefficients reevaluated, as a second-order approximation. The shift in Q point may not be large and such accuracy is usually not necessary.

14-6 INTERMODULATION DISTORTION

If there is curvature of the dynamic characteristic, and two or more signal frequencies are simultaneously applied, an additional form of distortion known as *intermodulation* results. This type of distortion adds frequencies in the output which are not harmonically related to the input frequencies.

If an input of two nonharmonically related frequencies, ω_1 and ω_2, is introduced, as

$$e_i = E_1 \sin \omega_1 t + E_2 \sin \omega_2 t$$

then Eq. 14-10 becomes

$$i_c = I_C + a_1(E_1 \sin \omega_1 t + E_2 \sin \omega_2 t) + a_2(E_1 \sin \omega_1 t + E_2 \sin \omega_2 t)^2 + \dots \tag{14-22}$$

After expansion and introduction of identities including

$$\sin \omega_1 t \sin \omega_2 t = \tfrac{1}{2}[\cos(\omega_1 - \omega_2)t - \cos(\omega_1 + \omega_2)t]$$

the relation for current becomes

$$\left.\begin{aligned} i_c = I_C &+ \frac{a_2}{2}(E_1^2 + E_2^2) + a_1E_1 \sin \omega_1 t + a_1E_2 \sin \omega_2 t \\ &- \frac{a_2E_1^2}{2}\cos 2\omega_1 t - \frac{a_2E_2^2}{2}\cos 2\omega_2 t \\ &- a_2E_1E_2 \cos(\omega_1 + \omega_2)t + a_2E_1E_2\cos(\omega_1 - \omega_2)t \end{aligned}\right\} \tag{14-23}$$

The first line includes a modified steady-state term, and the amplified input signal; the second line represents second-order distortion, and the third line represents additional distortion as the sum-and-difference frequencies of the input. If other frequencies are present, then additional sum-and-difference terms appear, so that analysis on the basis of two input frequencies is only indicative of the general situation.

These intermodulation frequencies produce background noise which is objectionable in audio service. The elimination of this distortion provides much incentive for the improvement of linearity with negative feedback.

14-7 TRANSISTOR CLASS A POWER AMPLIFIER: INPUT CONSIDERATIONS

Power output and distortion of Class A audio-frequency amplifiers are functions of input signal level, source resistance and circuit bias, and of the output load selected. A given choice of input and output circuit constants may give satisfactorily large power output but contribute more than allowable distortion, due to nonlinearities in the input-output dynamic current transfer curve.

The limit on allowable distortion is dependent on subjective factors. For audio-frequency use, a distortion content of 1 or 2 per cent is considered maximum for high-fidelity equipment, and a maximum of 5 per cent is sometimes set for radio receivers and public address service, but some equipment will exceed these figures.

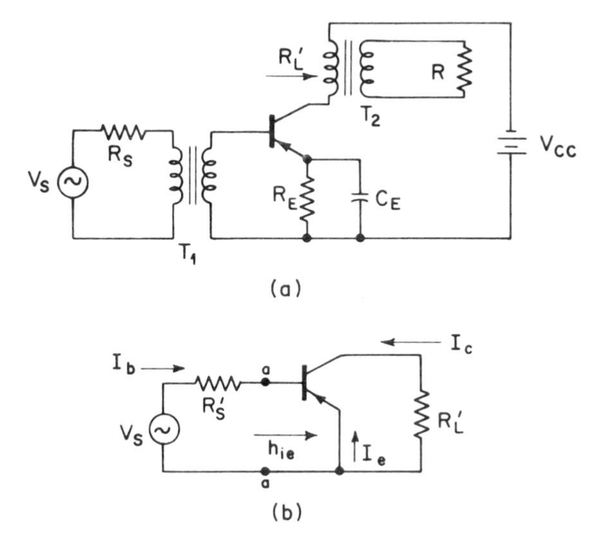

Figure 14-9. Class A audio amplifier.

The circuit of a transistor Class A audio power amplifier includes the elements of Fig. 14-9. The transistor input circuit has a typically nonlinear relation between base current and base-emitter voltage, and ideally should be supplied by a current source. However, practical amplifiers supplying the input current and voltage are not ideal, and it has been found possible to utilize nonlinearity in the input relation to partially compensate for nonlinearity in the output characteristic.

From the circuit, the power to the input is

$$P_s = I_b^2(R'_s + h_{ie})$$

The power output is

$$P_o = I_c^2 R'_L$$

and the power gain of the amplifier is

$$\text{stage P.G.} = \frac{P_o}{P_s} = \frac{I_c^2 R'_L}{I_b^2(R'_s + h_{ie})} = \frac{h_{fe}^2 R'_L}{R'_s + h_{ie}} \tag{14-24}$$

It is interesting to note that for the transistor alone, looking into the a, a terminals in Fig. 14-9(b), the power gain is

$$\text{transistor P.G.} = \frac{h_{fe}^2 R'_L}{h_{ie}} = h_{fe} g_m R'_L \tag{14-25}$$

This equation demonstrates the dependence of power gain on transistor parameters.

Maximum power transfer to the base-emitter circuit will occur when $R'_s = h_{ie}$. Then the stage power gain will be

$$\text{matched input P.G.} = \frac{h_{fe}^2 R'_L}{2h_{ie}} = h_{fe}\frac{g_m R'_L}{2} \tag{14-26}$$

Now consider the input characteristic of a power transistor in Fig. 14-10. With an ideal voltage source and $R'_s = 0$, a bias of 1.35 V will result in an I_B value of 0.15 A. A signal of 0.45 V peak will cause base current variation from $I_{\max} = 0.3$ A to $I_{\min} = 0.05$ A. These values indicate a distorted current wave, and using Eqs. 14-20 and 14-21, the second harmonic is found to be 28 per cent.

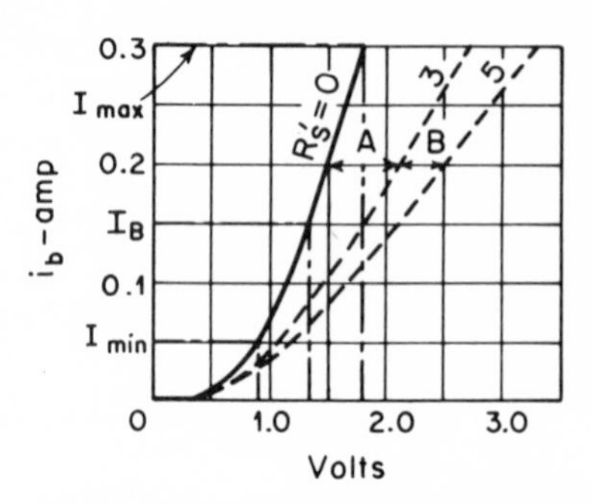

Figure 14-10. Input characteristics and effect of source resistance (2N1530).

Input curves may be drawn for $R'_s = 3$ ohms and 5 ohms, where the abscissa value A is found by adding $3i_b$ to the curve value, and $5i_b$ for $A + B$. Using the same I_B operating point, the input voltage is increased progressively to swings of ± 0.9 V and ± 1.2 V, to use the major portion of the input curve in each instance. Calculations then indicate

R'_s ohms	0	3	5
D_2	28%	10%	7.7%

By inference, a current source with $R'_s = \infty$ would have zero input distortion but would be wasteful of input power.

It can be concluded that input distortion is going to be present with practical driving sources. The linear upper portion of the $i_b - v_{be}$ curve can be found to have a slope representative of 3 ohms. Since the input distortion is indicated as not too severe with that resistance, and since a power-matched source has the advantage of optimum use of driver power, it is customary to match the source. The intent is then to compensate for some of the output characteristic nonlinearity with input circuit distortion components of opposite phase.

The input characteristic is a temperature function, with v_{be} changing as much as 20 per cent for a 100°C change, and the source resistance should be large enough to maintain input stability. The matched value seems sufficient for this purpose also.

14-8 TRANSISTOR CLASS A POWER AMPLIFIER: OUTPUT AND DISTORTION

The pnp transistor whose input characteristics were used in the previous section has the output characteristics of Fig. 14-11, with a maximum v_{CE} rating of 45 V, and maximum i_C value of 5 amperes. The thermal considerations of the next section will establish that 40 watts dissipation at a case temperature of 70°C is allowable for the conditions under which the transistor is to operate. The 40 watt constant-loss locus is then drawn. For eco-

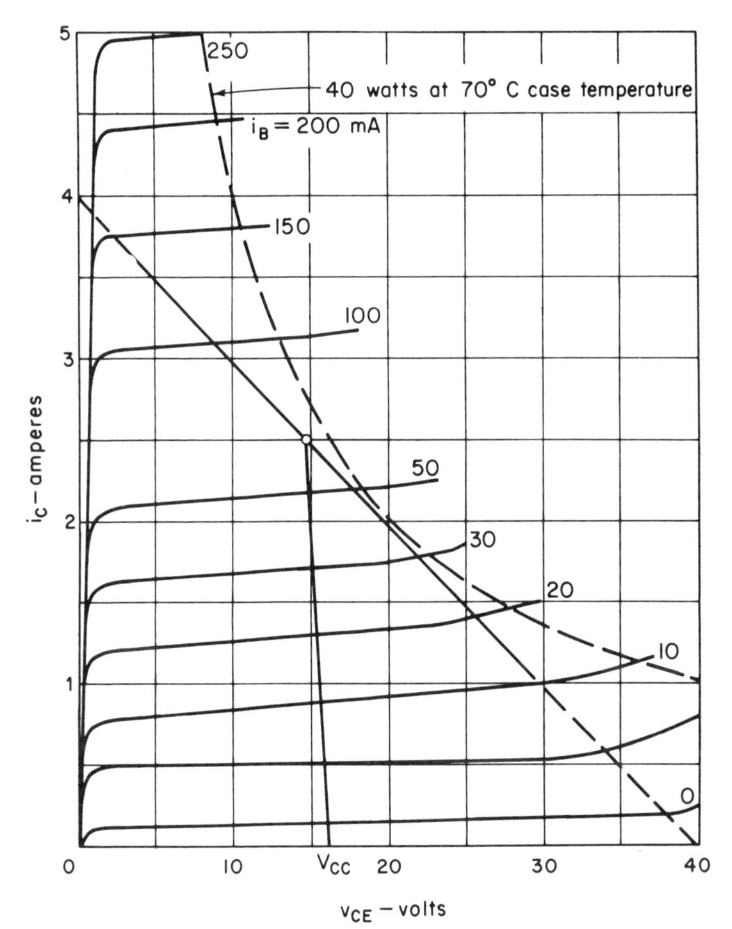

Figure 14-11. Output characteristics (2N1530).

nomical use of the transistor the largest possible power output is desired, and thus a load line should be sketched which utilizes the highest voltage and highest current values, within ratings, and is tangent to or below the constant-loss hyperbola. This will be a trial load, and its actual feasibility will be determined by further graphical analysis. The load line of Fig. 14-11 is so drawn and found to represent a resistance $R'_L = 10$ ohms, with the circuit of Fig. 14-9.

We now transfer the input characteristic for the matched condition, $R'_s = 3$ ohms, from Fig. 14-10 to the lower portion of Fig. 14-12. In the upper part of that figure is drawn the current-transfer curve using i_B, i_C data from the load-line intersections on Fig. 14-11.

Choice of the operating point is arbitrary, and it has been so placed at $I_B = 63$ mA that the dynamic current-transfer curve and the input characteristic are both essentially linear for base currents above the I_B value. Thus nonlinear distortion is confined to the lower part of the curves, where it is expected there will be compensatory action. The Q-point bias is determined as

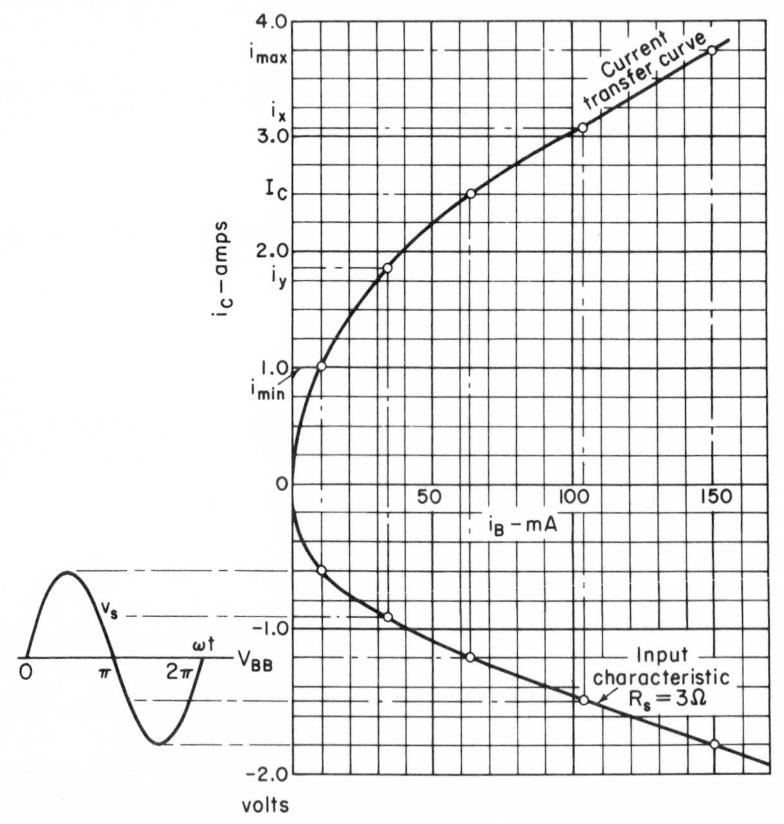

Figure 14-12. Transistor distortion and output calculations (2N1530).

-1.2 V for the pnp transistor. Because of the extreme curvature of the characterististics, it is evident that operation should not carry the base current below 10 mA, and this restricts the input voltage to 0.6 V peak, or $V_s = 0.42$ V rms.

The remaining steps consist of locating the negative input peak at -1.8 V, and 150 mA base current, locating the input 30° or half-voltage points, and carrying all these points to the current transfer curve and to the i_C axis. Thus i_{max}, i_{min}, i_x, i_y, and I_C are determined.

Using these data and the methods of Section 14-5, the performance of the transistor with the selected load can be found as:

$R'_L = 10$ ohms	$i_{max} = 3.75$ A
$R'_s = 3$ ohms	$i_x = 3.05$ A
$V_{BB} = -1.2$ V	$I_C = 2.50$ A
$V_{CC} = 14.7\text{ V} + I_C R_E = 16.0$ V	$i_y = 1.85$ A
$V_s = 0.42$ V rms	$i_{min} = 1.00$ A
$A_0 = 2.39$ A	$I_{C_s} = 2.30$ A
$A_1 = 1.31$ A	$I_c = 0.93$ A rms
$A_2 = -0.06$ A	$D_2 = -4.6$ per cent
$A_3 = 0.06$ A	$D_3 = 4.6$ per cent
$A_4 =$ negligible	$D_4 =$ negligible

total $D = 6.5$ per cent
$P_{out} = 0.93^2 \times 10 = 8.65$ W
$P_{dc} = 2.39 \times 14.7 = 35.1$ W
power efficiency $= 24.7$ per cent
stage power input $= 0.42^2/6 = 30$ mW
power gain $= 288 = 24.6$ dB

In view of the previously computed distortion level due to the matched input source, it is apparent that considerable compensation has taken place between output and input characteristics. This will vary with transistor type.

With $V_{CE} = 14.7$ V from the curves, and $R_E = 1.2/2.5 \cong 0.5$ ohm, the dc load line can be drawn, calling for $V_{CC} = 16$ V.

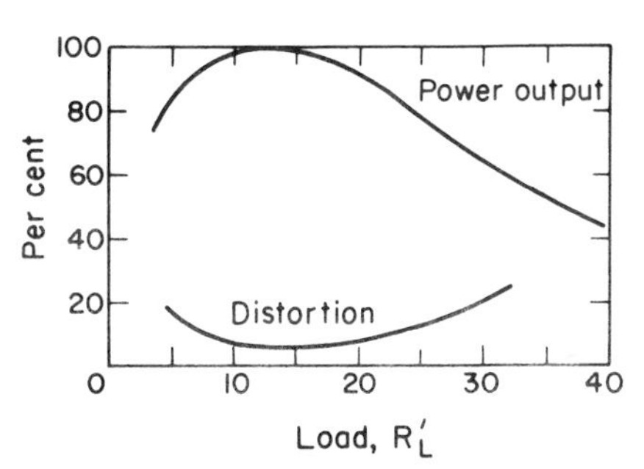

Figure 14-13. Transistor amplifier performance vs. load.

Other load lines of differing slopes may be experimentally drawn and calculations carried through to determine if improved conditions can be found; the work may also be done experimentally in the laboratory, using the conditions of this example as a starting point. In general, such experiments will show that power output and distortion vary with load somewhat as in Fig. 14-13. While the value of 6.5 per cent total distortion obtained in the example is reasonable for a single-ended amplifier, the figure could be reduced to less than one per cent by use of negative feedback.

Without feedback the frequency response of power transistors may be limited to 10 kilohertz or less, due to the internal capacitances which result from the large electrode areas needed to handle the heavy currents. Negative feedback, although applied to reduce nonlinear distortion, will expand the frequency-response range as well.

In theory, the output transformer reflects to the primary circuit a resistive load of value $R'_L = a^2R$. However, no transformer is actually ideal and at both low and high frequencies it may introduce a reactive component in addition to R'_L. With a phase angle between current and voltage, the ac load line opens to become an ellipse, with a circle as the limit for a pure reactance load. Such a reactive load line may carry the operation into a region of nonlinear characteristics, with ensuing distortion. Thus reactive loads are avoided whenever possible.

14-9 DETERMINATION OF TRANSISTOR OPERATING TEMPERATURE

For small-signal transistors, heat conduction along the leads or direct convection from the case to the surrounding air is sufficient to remove the small losses. With transistors giving outputs above the milliwatt level, provision must be made for conduction of the internal dissipation from the case.

The rate of conduction of heat is proportional to the difference in temperature of source and surroundings, in this case the junction and the ambient, and so

$$T_J - T_A = \theta_{JA} P_d \qquad (14\text{-}27)$$

where θ is a *thermal resistance*, expressed in units of °C per watt.

The thermal circuit, analogous to the transistor and its mounting, is represented in Fig. 14-14, where θ_{JC}, θ_{CS}, θ_{SA} are shown as thermal resistances in the paths: junction-to-case, case-to-sink, and sink-to-ambient, and

$$\theta_{JA} = \theta_{JC} + \theta_{CS} + \theta_{SA} \qquad (14\text{-}28)$$

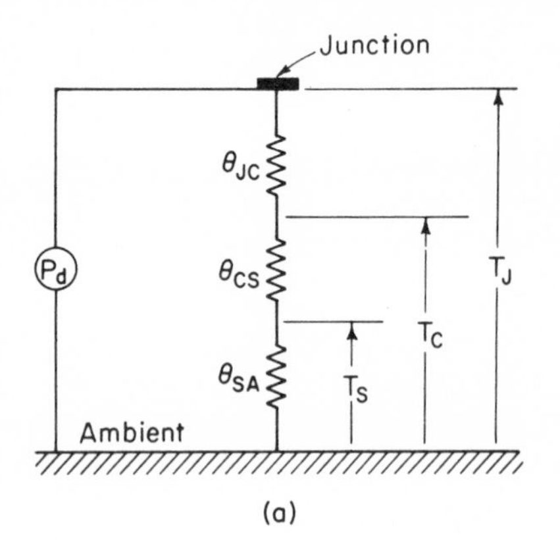

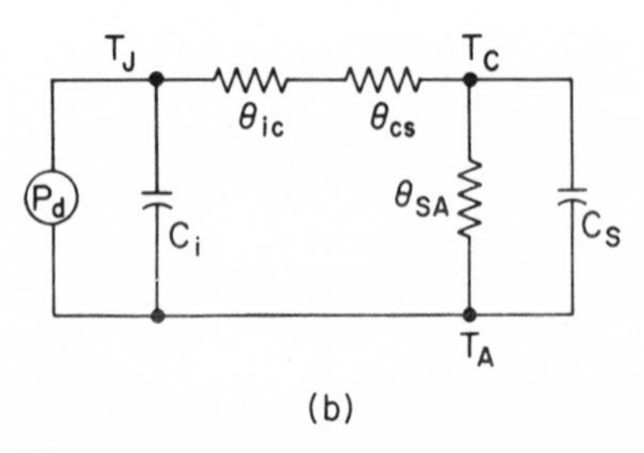

Figure 14-14. Thermal–electrical analogy.

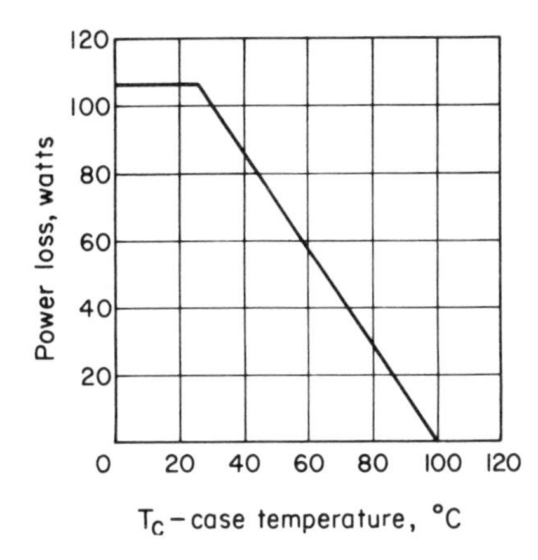

Figure 14-15. Power vs. temperature derating curve (2N1530).

Resistance θ_{CS} is often small. The value of θ_{JC} is specified for power transistors and indicated by a *derating curve*, as in Fig. 14-15, for the transistor employed in the previous section. The reciprocal slope of the curve gives

$$\theta_{JC} = \frac{T_{J\,\max} - T_C}{P_d} \tag{14-29}$$

where T_J is the junction operating temperature at rated dissipation, T_C is the case temperature, and 25°C is assumed as the normal ambient temperature.

The transistor of the previous section may be used in an example to illustrate static design procedure. That transistor is rated at $P_d = 110$ W, $T_{J\,\max} = 100$°C, $\theta_{JC} = 0.8$°C/W, and $I_{CBO\,\max} = 20$ mA. The contribution to the collector current is SI_{CBO}, where S is the instability factor and

$$S = \frac{R_E + R_B}{R_E + R_B/(1 + h_{FE})} \tag{14-30}$$

With $R_E = 0.5$ ohm, $R_B = 3$ ohms, $h_{FE} = 40$ maximum, the value of S is 6.1. The operating point power input is

$$\begin{aligned} P_d &= V_{CC}(I_C + SI_{CBO\,\max}) - R_E(I_C + SI_{CBO\,\max})^2 \\ &= 16(2.5 + 0.122) - 0.5(2.5 + 0.122)^2 = 38.6 \qquad \text{W} \end{aligned}$$

at $T_J = 100$°C.

Expanding Eq. 14-27

$$\theta_{SA} = \frac{T_J - T_A}{P_d} - \theta_{JC} \tag{14-31}$$

The amount of cooling that must be furnished by the heat sink is a function of the ambient temperature T_A. An ambient of $T_A = 35$°C may be assumed, and with $\theta_{JC} = 0.8$°C/W

$$\theta_{SA} = \frac{100 - 35}{38.6} - 0.8 = 0.88 \qquad \text{°C/W}$$

This value of thermal resistance can be obtained by use of forced-air cooling with a standard heat sink, whose heat-dissipating ability is indicated in Fig. 14-16.

The heat-sink temperature will be

$$T_S = \theta_{SA}P_d + T_A = 0.88 \times 38.6 + 35 = 69 \qquad \text{°C}$$

Thus it is now possible to support the assumption which was used to locate the constant-loss hyperbola for 40 watts at 70° ambient in Section 14-8.

For transient thermal conditions, as encountered in pulse and switching applications, allowance may be made for the *thermal capacitance*, or storage of heat, present in the device and mounting. While a true representation involves distributed thermal constants, the analog circuit of Fig. 14-14(b) may be used for pulse times long with respect to $\theta_i C_i$, the internal transistor thermal time constant.

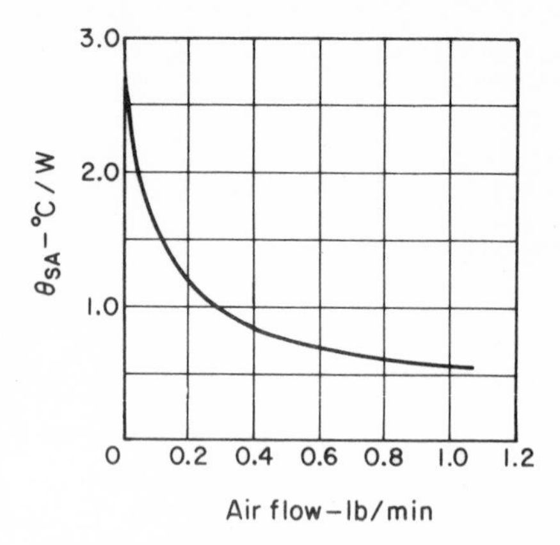

Figure 14-16. Heat sink with forced-air cooling.

14-10 THERMAL RUNAWAY

It was shown that

$$T_J = T_A + \theta_{JA} P_d \tag{14-32}$$

This equation indicates that, for constant power loss in the transistor, the temperature T_J of the junction will increase 5° for every 5° increase in the ambient temperature T_A. That is, the junction temperature seems to float at some constant differential above the ambient temperature. This would be true if it were not for an added loss resulting from the increase of I_{CBO} with T_J. With each increment in T_J the value of I_{CBO} will rise by an amount $\Delta I_C = SI_{CBO}$, where S is the circuit instability factor. Thus P_d will not remain constant with an increase in T_A. Equation 14-32 may be written to show this as

$$T_J = T_A + \theta_{JA}(P_c + P_v) \tag{14-33}$$

where P_c is the constant power loss due to I_C and V_{CE}, and P_v is the additional loss that arises due to ΔI_C.

At an ambient temperature $T_A = T_0 = 25$°C, the value of the varying term may be taken as zero, and it will remain negligible in effect with respect to P_c to some higher value of T_A. In this temperature range any change in T_A will be reflected in an almost equal change in T_J, or $dT_J/dT_A = 1$. As T_A rises and carries T_J upward, the resultant increase in ΔI_C and P_v becomes appreciable, T_J rises further and dT_J/dT_A becomes greater than unity. At some critical temperature T_R, the rate of increase in P_v per unit change in T_A becomes so large that $dT_J/dT_A \to \infty$, due to regenerative heating of the junction created by the rise of ΔI_C with T_J. The transistor then self-heats to ultimate destruction, a phenomenon known as *thermal runaway*.

Collector current can also increase with temperature due to changes in h_{FE}, or due to the temperature coefficient of v_{BE}, but the major cause of

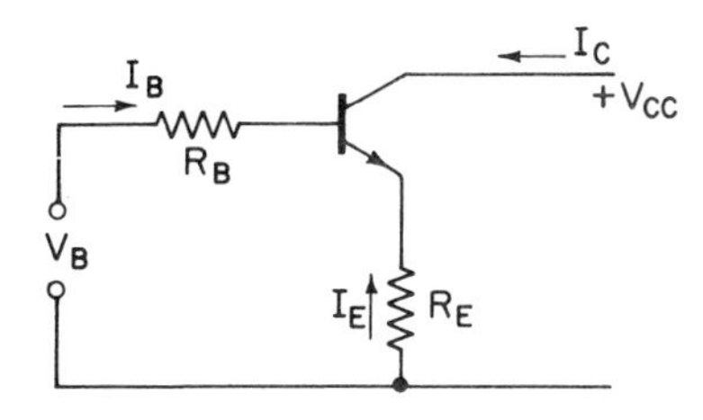

Figure 14-17. Power amplifier basic circuit.

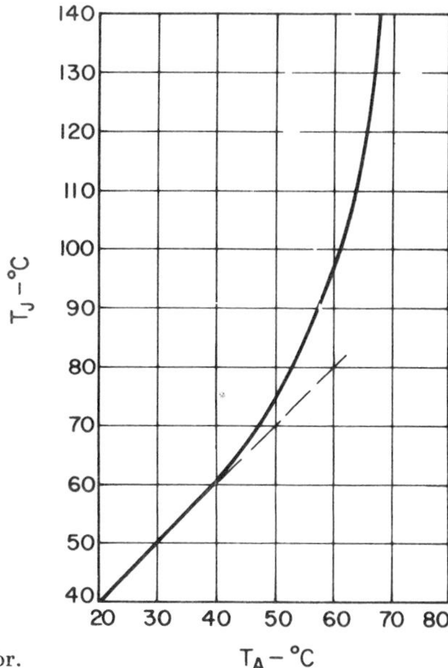

Figure 14-18. Thermal runaway in a transistor.

temperature-induced changes in P_d in power amplifier service is that of change in I_{CBO}.

The basic dc circuit of a transistor power amplifier may appear as in Fig. 14-17. For "worst-case" study, we may assume that $v_{BE} = 0$, h_{FE} is large, and there is zero dc load resistance with transformer output coupling. Since R_E will be small for good dc power efficiency, there will be little resistance in the collector-emitter circuit and I_C will be determined largely by the transistor. Power amplifiers are often close to thermal instability, therefore, in contrast to small-signal amplifiers in which I_C is limited to a safe value by circuit resistances.

The collector dissipation of a Class A power amplifier may be written at the Q point as

$$P_d = [V_{CC} - (I_C + \Delta I_C)(R_E + R_L)](I_C + \Delta I_C) \tag{14-34}$$

where ΔI_C represents the change in collector current as a result of junction temperature rise. Resistance R_L will be negligible with transformer output coupling, as would be normal for good power efficiency at large power output. From Eq. 14-27, the junction temperature rise above ambient is then

$$T_J - T_A = \theta_{JA} P_d = \theta_{JA}(V_{CC} I_C - R_E I_C^2) + \theta_{JA} \Delta I_C (V_{CC} - 2R_E I_C) = \theta_{JA}(P_c + P_v)$$

The result follows by neglect of the second order term in ΔI_C. The last expression is obtained from Eq. 14-33; $\theta_{JA} P_v$ is the regenerative term.

It is generally assumed that I_{CBO} doubles for every 11°C in germanium and every 18°C in silicon, so that

$$I_{CBO} = I'_{CBO} \epsilon^{a(T_J - T_0)} \tag{14-35}$$

with I'_{CBO} the rated value at $T_0 = 25°\text{C}$, and $a = 0.063$ and 0.038 for germanium and silicon units, respectively. Then

$$T_A = T_J - \theta_{JA}(V_{CC}I_C - R_E I_C^2) - \theta_{JA} S I'_{CBO} \epsilon^{a(T_J - T_0)}(V_{CC} - 2R_E I_C) \quad (14\text{-}36)$$

This relation between T_J and T_A is plotted in Fig. 14-18, for an unstable situation. As runaway is approached, the increase in T_J with respect to change in T_A can be noted.

Thermal runaway in a power amplifier can be predicted by determination of the conditions causing dT_J/dT_A to become large. Taking the derivative dT_A/dT_J of Eq. 14-36 and inverting, the result is

$$\frac{dT_J}{dT_A} = \frac{1}{1 - \theta_{JA} a S I'_{CBO} \epsilon^{a(T_J - T_0)}(V_{CC} - 2R_E I_C)} \quad (14\text{-}37)$$

The denominator goes to zero for thermal runaway and it is possible to predict that the transistor will be thermally stable if

$$\frac{1}{\theta_{JA} a S I_{CBO\,\max}} < (V_{CC} - 2R_E I_C) \quad (14\text{-}38)$$

where $I_{CBO\,\max}$ is the reverse saturation current at $T_{J\,\max}$, obtained by solution of Eq. 14-36.

The junction temperature at runaway can be found from the denominator of Eq. 14-37 as

$$T_{JR} = T_0 + \frac{1}{a} \ln \frac{1}{\theta_{JA} a S I'_{CBO}(V_{CC} - 2R_E I_C)}$$

The maximum allowable thermal resistance for stability is

$$\theta_{JA\,\max} = \frac{T_{J\,\max} - T_{A\,\max}}{V_{CC}I_C - R_E I_C^2 + S I'_{CBO} \epsilon^{a(T_J - T_0)}(V_{CC} - 2R_E I_C)} \quad (14\text{-}39)$$

Thermal runaway will not occur if the thermal resistance of the transistor case and mounting is kept below this value.

14-11 VOLTAGE LIMITATIONS ON THE TRANSISTOR

Power output from a transistor amplifier increases with applied voltage, but the peak transistor voltage has two physical limits.

In *punch-through*, as the voltage is increased across the collector-base diode the depletion layer becomes wider, narrowing the base region. This change in width is greater in high resistivity base materials. At some high collector voltage the collector depletion layer may extend completely through the base, causing a short circuit from collector to emitter. Punch-through of the base does not damage the transistor if the current is limited, but the transistor is inoperative above the limiting voltage.

First breakdown of a transistor occurs with the avalanching of collector current at high collector voltages, an occurrence without serious effects on

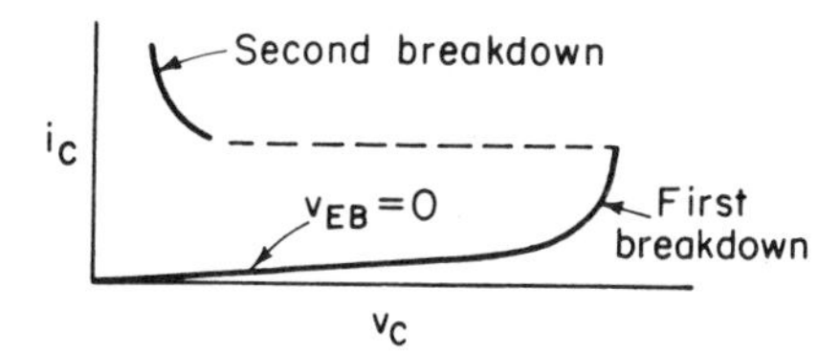

Figure 14-19. Volt–ampere curve showing second breakdown.

the transistor. *Second breakdown*, which may follow avalanching, can lead to transistor destruction. A negative resistance region develops and the current can rise to high values if not circuit-limited, as indicated in Fig. 14-19. The damage appears to follow when the high current channels into a small area and the resultant concentrated temperature rise melts a small hole through the base layer. Safe operating voltage and current limits are usually shown by charts for each power transistor type.

14-12 MAXIMUM POWER OUTPUT—THE TETRODE OR PENTODE

The tetrode or beam tube will be used to illustrate the method of power output determination and distortion analysis for the vacuum tube family. As for the transistor, the first step is the construction of the constant-loss

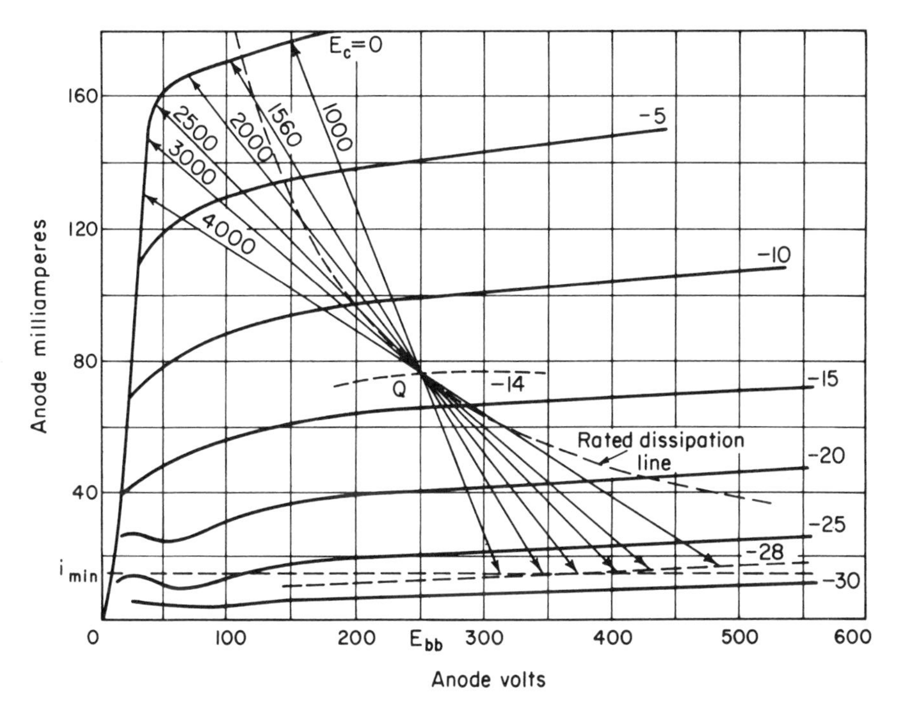

Figure 14-20. Graphical determination of power output, beam tetrode: $P_d = 19$ W; $E_{bb\,\max} = 250$ V; $g_m = 6000$ μmho; $r_p = 22{,}000$ ohms.

hyperbola for $E_b I_b = P_d$, where P_d is the rated anode dissipation. Transformer coupling can be assumed and $E_b = E_{bb}$. For maximum output, E_{bb} will be chosen as the highest rated supply voltage and the Q point will be fixed on the hyperbola as in Fig. 14-20.

The first experimental load line should pass through the $e_c = 0$ line near the knee of the curve. Then calculate the load resistance represented and draw a set of load lines for values ranging from one-half to twice the first load. As the next step, choose a value of $i_{\min}$, so that operation will not carry into the nonlinear characteristics at low anode current. A line at $i_{\min}$ will serve as an operation bound, and for Class A operation the other end of operation will carry to an equal positive input voltage above the Q point.

It is then possible to determine the required currents and to compute

Table 14-1
DETERMINATION OF OPTIMUM LOAD—DATA FOR FIG. 14-21

R_L ohms	$i_{\max}$ mA	$i_{\min}$ mA	i_x mA	i_y mA	A_1	I_p mA	Power watts	D_2 %	D_3 %	D_4 %	Total Harm. %
1000	177	15	122	38	82	58	3.4	11.6	1.2	1.8	11.8
1560	171	15	121	39	79	56	4.9	10.1	1.6	1.2	10.3
2000	166	15	120	39	77	54	5.9	8.8	2.3	0.8	9.1
2500	158	16	119	39	74	52	6.8	6.8	3.9	—	7.8
3000	146	16	117	40	69	49	7.2	2.9	5.8	—	6.5
4000	130	17	113	40	62	44	7.8	−2.9	8.7	—	9.2
5000	120	18	109	41	57	40	8.0	−6.5	10.7	—	12.5

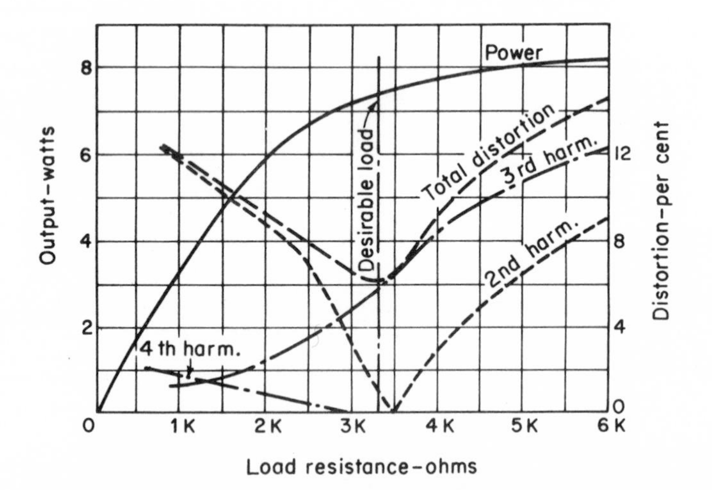

Figure 14-21. Power output and distortion of a tetrode.

the power output and distortion content for each load, giving the results of Table 14-1 and the curves of Fig. 14-21. A suitable load might be chosen as 3300 ohms. The total harmonic minimum near that load is due to the fact that the second harmonic goes to zero when

$$I_b = \frac{i_{\max} + i_{\min}}{2}$$

this result following from Eq. 14-15. This occurs for a load line drawn such that the Q point falls halfway between $i_{\max}$ and $i_{\min}$.

The choice of load is made to yield the highest power output with minimum practical distortion. A higher load would increase the power slightly but at considerably higher distortion.

14-13 THE CLASS A PUSH-PULL AMPLIFIER

By moving the operating point toward cutoff, a larger input signal can be accommodated and higher power output achieved in a power amplifier, but at the cost of increased even-order distortion. By use of two transistors or two tubes in the push-pull connection of Fig. 14-22, this even-order distortion can be canceled and greater output obtained.

The characteristics of the circuit can be demonstrated by a general analysis. Assuming identical device characteristics, the output current of one may be written as

$$i = I_o + a_1 e_s + a_2 e_s^2 + a_3 e_s^3 + \dots \tag{14-40}$$

As connected, the input secondary voltages will be

$$e_1 = E_2 \sin \omega t$$

$$e_2 = E_2 \sin (\omega t + \pi)$$

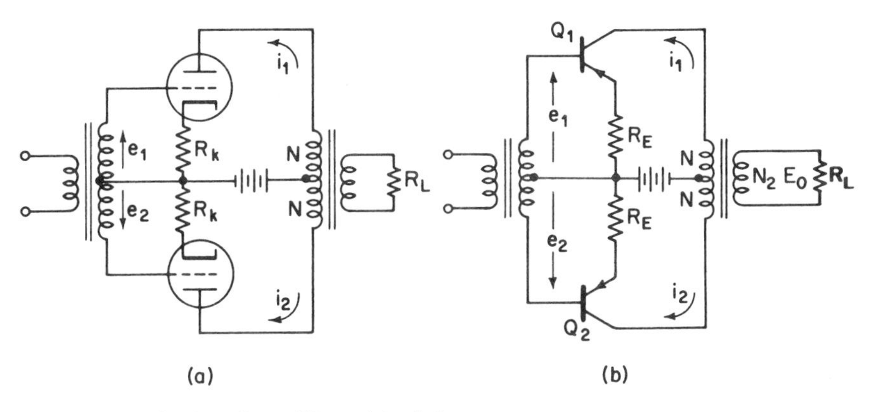

Figure 14-22. Push-pull amplifiers: (a) triode; (b) transistor.

The output current of Q_1 is then

$$i_1 = I_{01} + a_1E_2 \sin \omega t + a_2E_2^2 \sin^2 \omega t + a_3E_2^3 \sin^3 \omega t + \ldots$$

and that of device Q_2 is,

$$i_2 = I_{02} + a_1E_2 \sin (\omega t + \pi) + a_2E_2^2 \sin^2 (\omega t + \pi) + a_3E_2^3 \sin^3 (\omega t + \pi) + \ldots$$

By use of trigonometric identities in terms of multiple angles,

$$\left.\begin{aligned} i_1 &= I_{01} + B_0 + B_1 \sin \omega t - B_2 \cos 2\omega t + B_3 \sin 3\omega t \\ &\quad - B_4 \cos 4\omega t + \ldots \\ i_2 &= I_{02} + B_0 + B_1 \sin (\omega t + \pi) - B_2 \cos 2(\omega t + \pi) \\ &\quad + B_3 \sin 3(\omega t + \pi) - B_4 \cos 4(\omega t + \pi) + \ldots \end{aligned}\right\} \tag{14-41}$$

From trigonometry,

$$\left.\begin{aligned} \sin (\omega t + \pi) &= -\sin \omega t \\ \cos 2(\omega t + \pi) &= \cos 2\omega t \end{aligned}\right\} \tag{14-42}$$

and similarly for all even and odd harmonics, so that i_2 can be written as

$$i_2 = I_{02} + B_0 - B_1 \sin \omega t - B_2 \cos 2\omega t - B_3 \sin 3\omega t - B_4 \cos 4\omega t - \ldots \tag{14-43}$$

Since the positive currents are assumed as shown, the ampere turns acting on the transformer core are $N(i_1 - i_2)$. The secondary voltage E_o will be proportional to the ampere turns, or

$$E_o = 2K(B_1 \sin \omega t + B_3 \sin 3\omega t + B_5 \sin 5\omega t + \ldots) \tag{14-44}$$

All harmonics will have signs determined by Eq. 14-42 and so the push-pull connection with matched tubes or transistors will eliminate all even-order distortion frequencies from the output. Power supply ripple will also cancel. Figure 14-23 illustrates the results graphically.

Dynamic Class A push-pull operation may be better understood by reference to the output characteristics of Fig. 14-24. To indicate the subtractive arrangement, the curves for Q_2 are plotted inverted and aligned at the value of V_{CC}. Individual transistor (or pentode) load lines may be drawn through operating points 1 and 2 to i_m and $2V_{CC}$ as extremes of operation. The composite load line for push-pull operation is the algebraic sum of the individual load lines, and is shown as a solid line.

The slope of the composite line gives $R_L = V_{CC}/i_m$, and the load should be matched to this value for maximum power output. Thus

$$R_L = \left(\frac{2N}{N_2}\right)^2 R \tag{14-45}$$

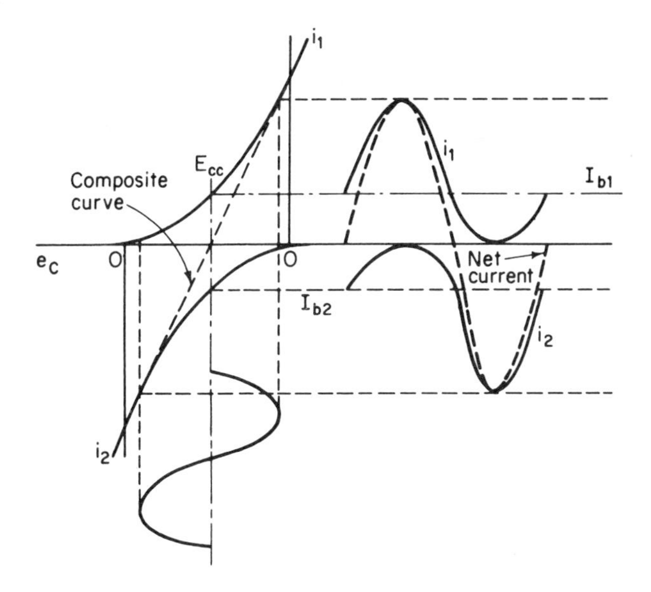

Figure 14-23. Use of transfer curves to explain push-pull action.

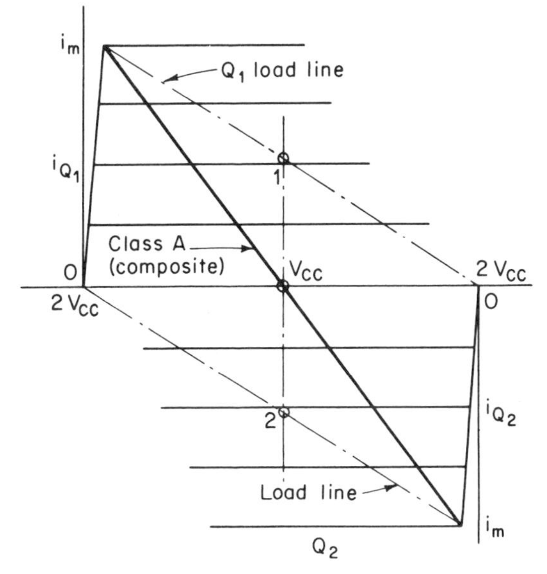

Figure 14-24. Composite output characteristics.

where N is the number of turns in one-half the primary winding. The individual transistors are operating with loads $R_A = V_{CC}/(i_m/2) = 2V_{CC}/i_m$. The "per transistor" load reflected by the impedance ratio of the transformer is $R_L/4$, so that $R_A = R_L/2$ and each transistor is operating into a load twice its matching value. Since the two transistors operate simultaneously and in parallel, the effect is that of a matched load for the circuit.

14-14 ELIMINATION OF THE OUTPUT TRANSFORMER

The output load of most audio power amplifiers is a loudspeaker, and these are low impedance devices, usually in the range from 3 to 16 ohms. To match such loads to tube amplifiers requires the use of an output transformer, a large and costly item for high fidelity operation. However, because of the low output impedance of transistors, loudspeakers may be directly employed as transistor loads; thus the output transformer, with its problems of size, cost, and inherent frequency distortion, may be eliminated in appropriate push-pull circuits.

The usual circuits stem directly from the properties of the bridge of Fig. 14-25(a), formed from superposition of two conventional push-pull amplifiers, composed of Q_1 and Q_2, and Q_3 and Q_4, respectively, with a common load. Transistors Q_1 and Q_3 are driven upward in current and Q_2 and Q_4 simultaneously driven downward by a common transformer; an alternating signal appears in the load.

The circuit at (b) uses a tapped power supply, and only two transistors in a half bridge. The ac potential at A in (b) and (c) is the same, and the circuit at (c) follows. However, it requires a large capacitor C, with reactance small in comparison to the 3 to 16 ohm load at the lowest frequency of interest.

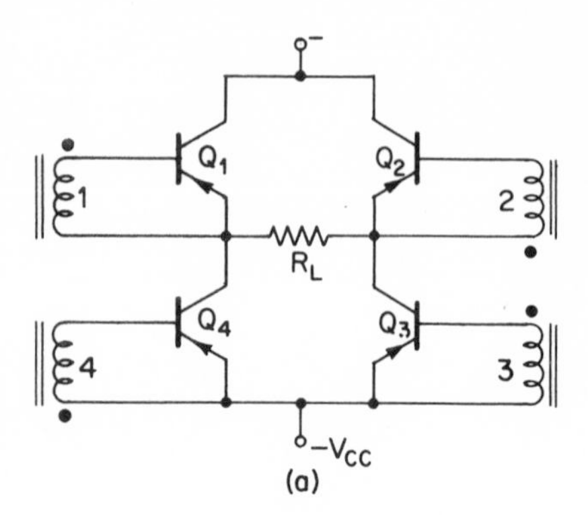

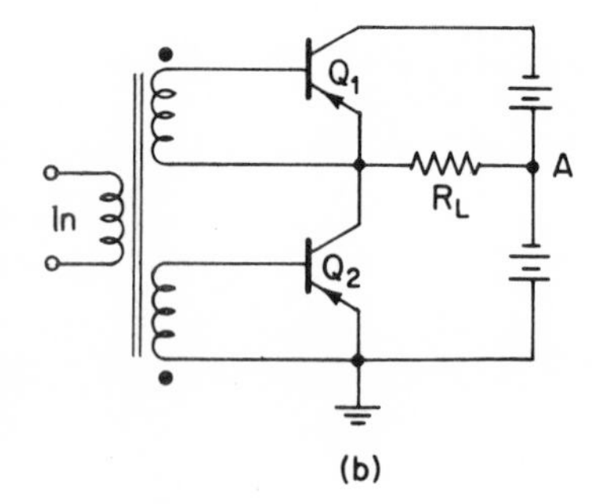

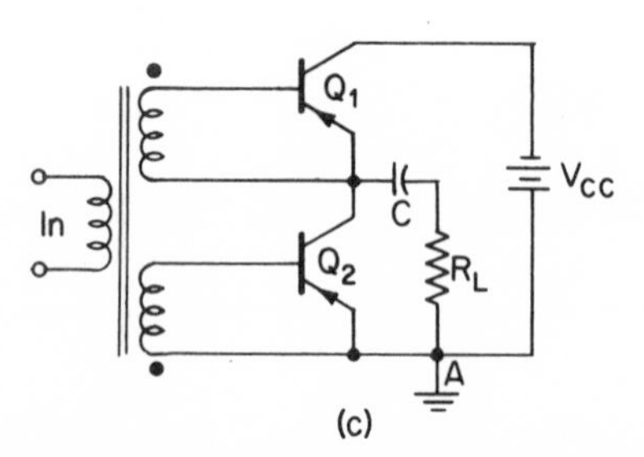

Figure 14-25. Development of circuits without the output transformer.

14-15 THE CLASS B PUSH-PULL AMPLIFIER

Because of the increased power efficiency, and the possibility of larger power output from small devices, operation under Class B conditions is often desirable. While the reverse half-cycle is then cut off, the even harmonic distortion so generated can be canceled if the push-pull connection is utilized.

A dynamic transfer characteristic is plotted in Fig. 14-26(a). The characteristic of Q_2 is plotted upside down to show the subtractive nature of the push-pull relation. Transistor or tube Q_1 conducts on the first input half-cycle and

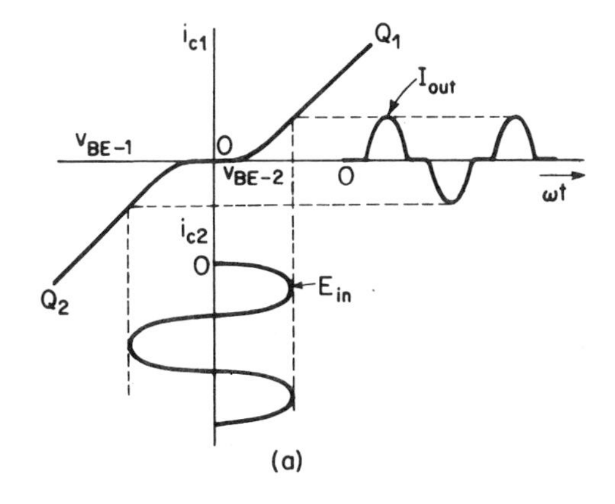

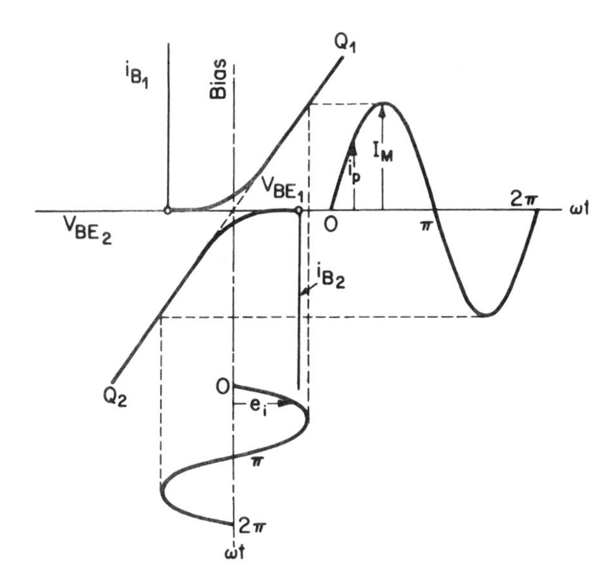

Figure 14-26. (a) Crossover distortion; (b) effect of forward bias.

Q_2 conducts on the second half-cycle. As shown, transistors and tubes do not cut off sharply; for small input voltages there will be no output, and larger inputs will produce the stepped output current of the figure. The result is known as *crossover distortion*.

Crossover distortion can be eliminated by use of a forward bias, as in Fig. 14-26(b). This has the effect of sliding the dynamic curve for Q_2 to the right in (a). Since some current will be present at all times, the operation is not strictly Class B and properly should be called Class AB. So-called "zero bias" triode tubes also are operated under a bias less than cutoff at zero grid volts; in tubes this is known as *extended-cutoff operation*.

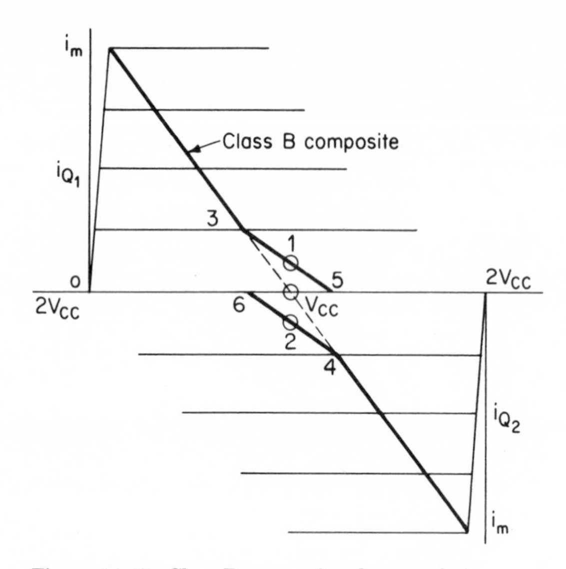

Figure 14-27. Class B composite characteristic.

An output characteristic is plotted in Fig. 14-27, the composite line including the dashed portion. The slope of this line will establish the output load as

$$R_o = \frac{R_L}{4} = \frac{V_{CC}}{i_m}$$

where R_L is the collector-to-collector or anode-to-anode load. One of the individual device load lines follows the composite curve down to 3, switches to Class A operation through the actual operating point at 1, and cuts off at 5. In the 3–5 region the slope is that of a Class A load, since both devices operate in parallel in that region.

Analysis of Class B performance on an ideal basis can be undertaken from the composite line of Fig. 14-27. The total dc input current is twice the average of one-half sine wave and so

$$I_{dc} = \frac{2I_M}{\pi} \quad \text{and} \quad P_{dc} = \frac{2I_M V_{CC}}{\pi}$$

The ac output current and power are

$$I_{rms} = \frac{I_M}{\sqrt{2}} \quad \text{and} \quad P_o = \frac{I_M^2}{2} R_o$$

For a transistor with low saturation voltage the maximum peak output voltage will approximate V_{CC}. Defining the output level as a fraction of this possible maximum leads to $V_o = KV_{CC} = KI_M R_o$, where K takes on values from 0 to 1. The power output and dc input may then be written as

$$P_o = \frac{K^2 V_{CC}^2}{2R_o}, \qquad P_{dc} = \frac{2KV_{CC}^2}{\pi R_o} \tag{14-46}$$

The power efficiency of a linear Class B amplifier is

$$\eta_p = \frac{\pi K}{4} \times 100\% \tag{14-47}$$

Maximum efficiency obviously occurs for maximum output signal of V_{CC} peak value, or for $K = 1$. Then

$$\text{maximum } \eta_p = \frac{\pi}{4} \times 100\% = 78.5\% \tag{14-48}$$

This is a considerable improvement over Class A conditions, and is responsible for the widespread use of Class B amplifiers.

The power output may be expressed in terms of the dc current input as

$$P_o = \frac{\pi^2 I_{dc}^2 R_o}{8} \tag{14-49}$$

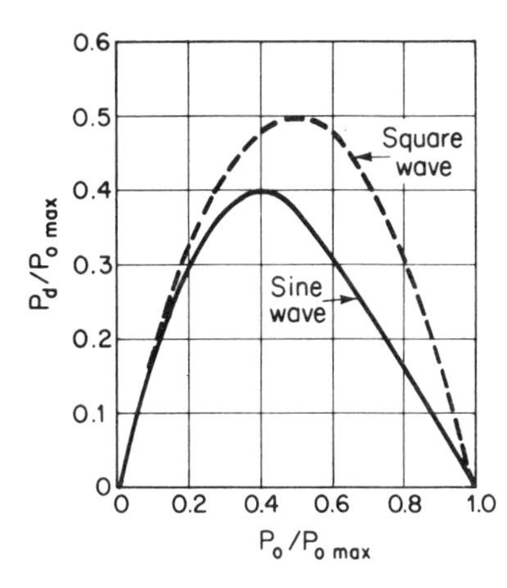

Figure 14-28. Power loss as a function of power output, Class B.

Therefore, the power input is large only for large signals, and the conditions leading to peak loss must be determined. By use of Eqs. 14-46, the dissipation is

$$P_d = \frac{2V_{CC}^2}{\pi R_o}\left(K - \frac{K^2\pi}{4}\right) \tag{14-50}$$

Taking the derivative with respect to K, and setting the result equal to zero shows that maximum dissipation occurs when $K = 2/\pi$. The maximum dissipation for both transistors is then

$$\text{maximum } P_d = \frac{2V_{CC}^2}{\pi^2 R_o} = 0.20\frac{V_{CC}^2}{R_o} \quad \text{W} \tag{14-51}$$

at the condition of $2/\pi$ or 40 per cent of the

maximum possible power output. The efficiency is then 50 per cent. The variation of loss with power output is shown in Fig. 14-28.

Linearity of output for large signals is dependent on the linearity of the large-signal transfer curves; there is no compensation of one device by the other as in Class A operation. Saturation effects at large currents can introduce distortion, and there may be some crossover distortion. Thus negative feedback is necessary to provide freedom from these sources of distortion.

All the above relations have been derived on the basis of sinusoidal inputs. Speech, not requiring continuous peak output, will be less demanding; conditions with a square wave or switched input will be more severe.

A typical transistor Class B amplifier, driven by an emitter follower and a common-emitter stage with complementary devices, is shown in Fig. 14-29. This provides a low source resistance to the Class B base circuits, and permits direct coupling.

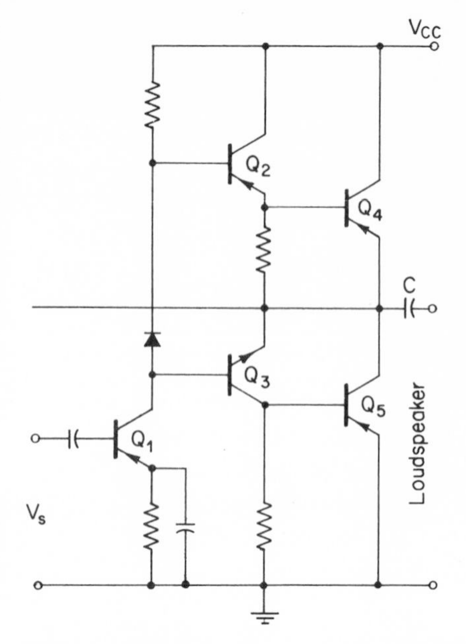

Figure 14-29. Class B push-pull amplifier and driver circuit.

14-16 PHASE INVERTERS FOR PUSH-PULL INPUT

Two equal voltages in phase opposition are needed for push-pull amplifier input. Center-tapped impedance matching transformers are used with transistors, but are avoided in tube service because of cost and poor performance at the high impedance levels needed.

Circuits which provide the necessary voltages and phase relation are called *phase inverters*; examples appear in Fig. 14-30. The phase-splitting forms at (a) and (b) are cathode or emitter followers with anode or collector loads added. Changes in device characteristics affect both outputs equally, and voltage balance is maintained.

The signals at terminals 1 and 2 are at 180°, as required. The gain to the cathode or emitter load is less than unity, and with $R_1 = R_2$, a similar gain is obtained to output 2. However, the output impedance at terminal 2 is that of a common-emitter stage, while terminal 1 provides that of a follower. An additional transistor may be added as in (c), to match both voltages and source impedances.

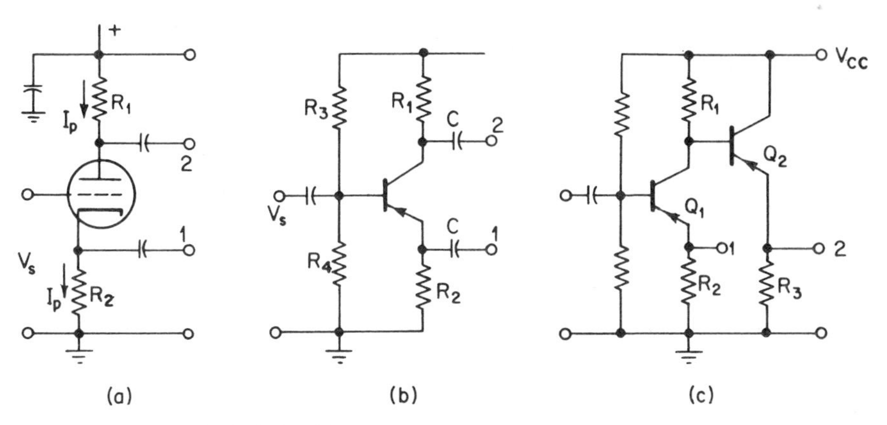

Figure 14-30. (a), (b) Split-load inverters; (c) two-stage inverter.

A differential amplifier may also be used as a phase inverter for greater gain.

14-17 TRANSISTOR PHASE-INVERSION WITH COMPLEMENTARY TRANSISTOR TYPES

The push-pull amplifier input requirement for equal and phase-opposed voltages can be eliminated by use of a pair of npn and pnp *complementary transistors*. As in Fig. 14-31, a signal will simultaneously drive the base of one unit into the operating region and the base of the second unit beyond cutoff. Thus a common input will give Class B performance. Figure 14-29 employs such a connection for driver transistors Q_2 and Q_3. Matched complementary pairs are available.

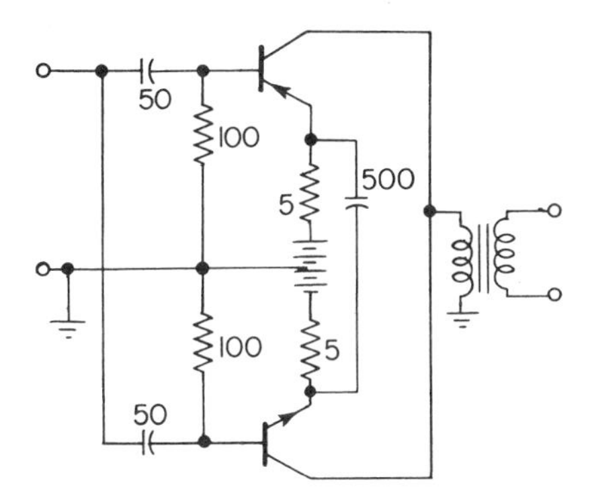

Figure 14-31. Push-pull by complementary symmetry.

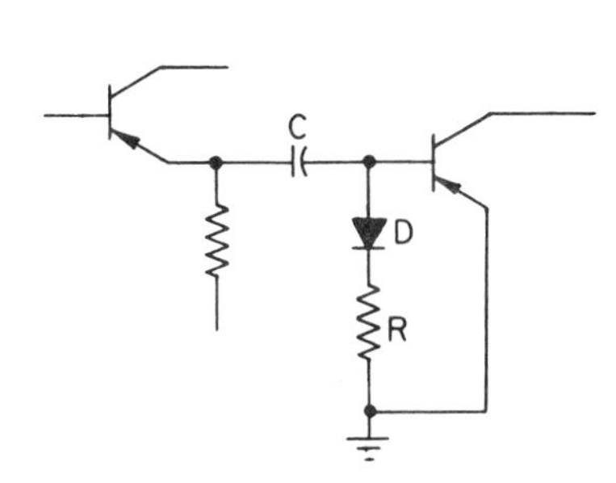

Figure 14-32. Class B driver circuit.

Figure 14-32, as an input to a Class B stage, corrects the asymmetrical base current which causes a dc voltage to develop across C, upsetting the base bias. The diode and R provide a circuit that gives a load on the driver during the half-cycle when the base is driven to cutoff.

14-18 CLASS B AND CLASS C RESONANT-LOAD AMPLIFIERS

Power in a narrow band of radio frequencies is generated by Class B and Class C amplifiers with a resonant circuit or *tank* as a load. The tubes or transistors operate as synchronous switches, supplying dc power in pulses to the resonant circuit, in synchronism with the voltage across the load. After supplying a pulse of energy, the switch disconnects the energy source from the load, and the energy in the load continues in oscillation at the circuit resonant frequency. The action is comparable to that of a pendulum, driven by short energy pulses, and swinging freely at its own rate for most of the cycle. Because of the Class B or Class C operation, the power efficiency is high, reaching levels of 65 per cent in Class B and 85 per cent in Class C amplifiers in practice.

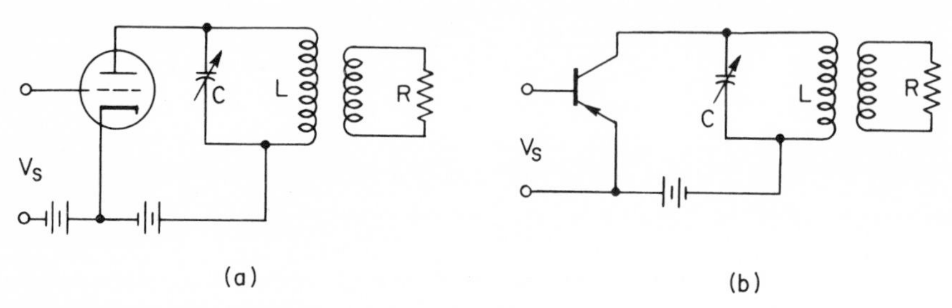

Figure 14-33. Power amplifiers with resonant loads.

Basic circuits are as shown in Fig. 14-33. The load R is inductively coupled into the tank circuit which is assumed resonant at the frequency of V_s. The resonant circuit acts as a filter to reject the harmonics generated in Class B or C operation. High radio-frequency power is more often generated with vacuum tubes and the analysis will be given for those devices, but the methods are suited to the transistor.

The amplifier bias is set at cutoff for Class B and at two or more times cutoff for Class C service. Anode conduction is discontinuous, as indicated for Class C operation in Fig. 14-34, where the switch action connects the source to the load for the conduction angle $2\theta_1$, less than 180°. In this interval, sufficient energy must be delivered to the tank to supply the losses and power output of the resonant circuit. The anode voltage appears as e_b, oscillating

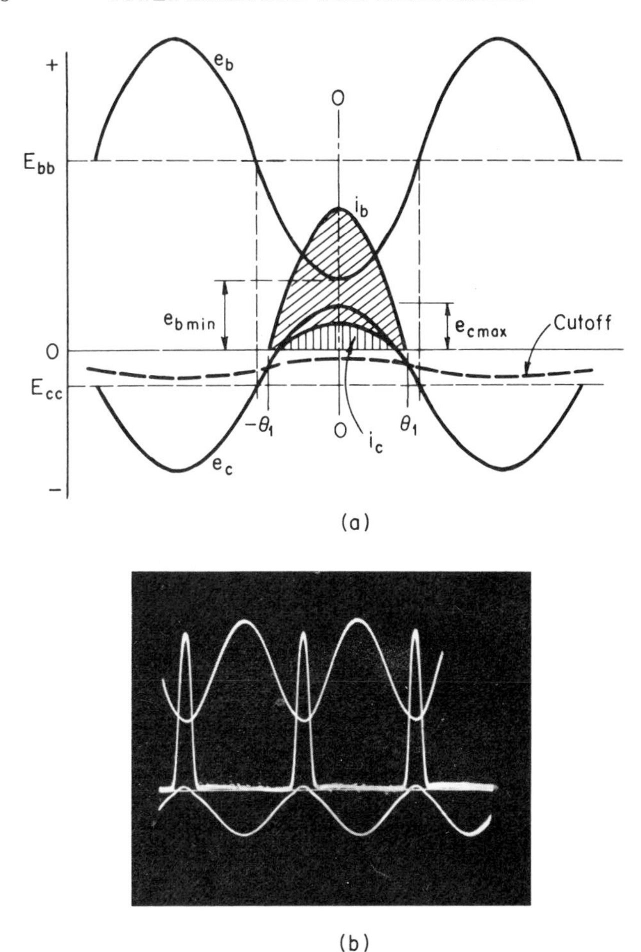

Figure 14-34. (a) The synchronous switch; (b) current and voltage in a Class C amplifier.

from $E_{bb} + \hat{E}_o$ to $E_{bb} - \hat{E}_o$, where $\hat{E}_o$ is the peak of the load voltage. As the grid input rises above the cutoff line at $-\theta_1$, conduction starts to the anode. When the grid goes positive at $-\theta_2$, a current i_c begins to the grid.

For the vacuum tube the power loss internal to the tube is $e_b i_b$, on an instantaneous basis. To reduce the loss or raise the efficiency, it is desirable to reduce e_b during the conduction interval. This may be done by increasing the load impedance or by increasing the bias and driving voltage. However, if the minimum anode voltage is driven below the maximum grid voltage, the grid becomes the most positive element and the grid current rises abruptly. The grid driving power increases but, with the anode current at or near saturation, the positive grid robs the anode of current and the power output fails to increase proportionally. In Class C it is usually found that an input which causes the maximum value of grid voltage to equal, but not to exceed, the minimum value of anode voltage during the cycle, leads to the highest efficiency.

To avoid nonlinearity near saturation, it is desirable to use somewhat less input drive for the linear Class B amplifier, and the maximum drive is usually restricted to $e_{b\,\min} = 2e_{c\,\max}$.

14-19 GRAPHICAL CLASS C ANALYSIS

In Fig. 14-34 it can be seen that anode current is cut off until the grid voltage

$$e_c = E_{cc} + \hat{E}_g \cos\theta \tag{14-52}$$

equals cutoff. Here $\hat{E}_g$ represents the peak of the ac grid voltage wave. The anode voltage is

$$e_b = E_{bb} - \hat{I}_1 R_L \cos\theta \tag{14-53}$$

where R_L is the resonant impedance of the tuned load, and $\hat{I}_1$ is the peak of the fundamental component of anode current.

The above equations may be used to obtain

$$e_b = E_{bb} + \frac{E_{cc}\hat{E}_o}{\hat{E}_g} - \frac{\hat{E}_o}{\hat{E}_g} e_c \tag{14-54}$$

This is the equation for a straight line of slope equal to $-\hat{E}_o/\hat{E}_g$ on the e_b, e_c family of *constant-current curves* of Fig. 14-35.

A Q point may be selected at twice cutoff, -370 V bias, and 2000 anode volts, on the dashed two-times-cutoff locus. A second point will determine the operation line and may be selected at $e_{b\,\min} = e_{c\,\max} = +200$ V at point A, using the criterion for optimum driving voltage. The operation line may now be drawn from A to Q. It conforms to Eq. 14-54.

The selection of A has fixed $\hat{E}_g = 570$ V, $\hat{E}_o = 2000 - 200 = 1800$ V. Point C is marked at zero anode current, where conduction starts when the anode voltage swings down to 1150 V. The angle of conduction is

$$2\theta_1 = 2\cos^{-1}\frac{(E_{bb} - E_{co})}{\hat{E}_o} = 2\cos^{-1}\frac{(2000 - 1150)}{1800} = 124^\circ$$

Grid conduction starts at B when the grid is at zero volts and going positive. Calling this angle of conduction $2\theta_2$, then

$$2\theta_2 = 2\cos^{-1}\frac{E_{cc}}{\hat{E}_g} = 2\cos^{-1}\frac{370}{570} = 99^\circ$$

Values of current for any angle may be read by marking points on the operating line corresponding to increments of θ, intervals of 10° being used for the data of Table 14-2.

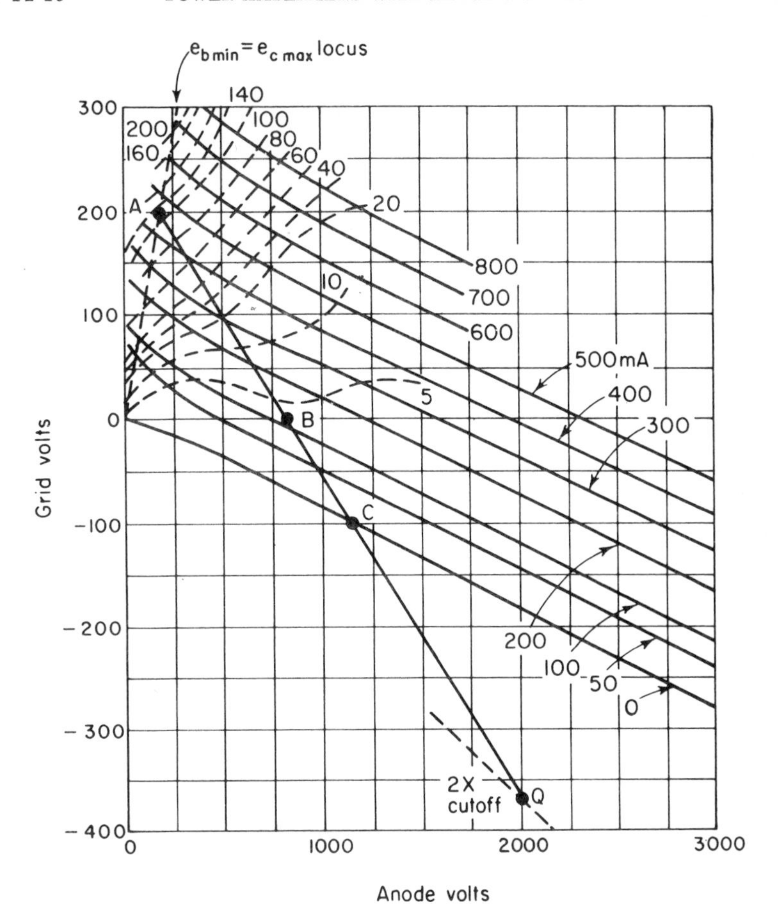

Figure 14-35. Triode constant-current curves. Solid lines: i_b; dashed lines: i_c.

Table 14-2

DATA FROM OPERATION LINE OF FIG. 14-35

$E_{bb} = 2000$ V $\quad \mu = 11 \quad e_{b\,min} = 200$ V

$E_{cc} = -370$ V $\quad \hat{E}_g = 570$ V $\quad e_{c\,max} = 200$ V

θ	0°	10°	20°	30°	40°	50°	60°	63°
$\cos\theta$	1.00	0.985	0.940	0.866	0.766	0.643	0.500	0.342
$\hat{E}_o \cos\theta$	1800	1770	1690	1560	1380	1160	900	610
$E_{bb} - \hat{E}_o \cos\theta$	200	230	310	440	620	840	1100	1190
i_b, mA	460	450	415	350	230	120	10	0
i_c, mA	110	95	75	35	10	0	0	0

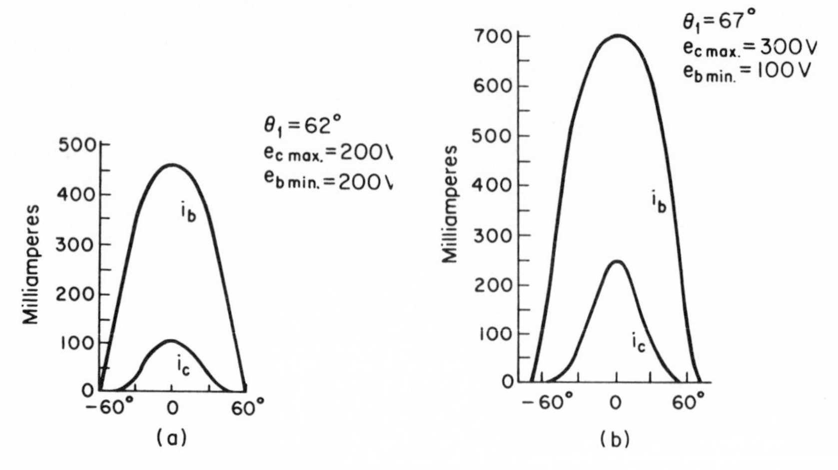

Figure 14-36. Wave forms of grid and anode current.

Because of the nonsinusoidal wave forms of Fig. 14-36(a) plotted from the above data, graphical integration must be used to arrive at the average current values. That is, taking 10° or $\pi/18$ radian intervals over half of the cosine wave, with i_o the current ordinate at $\theta = 0°$, and $i_n = 0$, it is possible to derive

$$I_b = 2 \times \frac{1}{2\pi}\int_0^{\pi} i_b\, d\theta = \frac{1}{\pi} \times \frac{\pi}{18}\left(\frac{i_o}{2} + i_1 + i_2 + \ldots + i_{n-1}\right) \tag{14-55}$$

the coefficient 2 being present because of the symmetry of the cosine wave. A similar method yields the average grid current, I_c.

A Fourier expansion will give the peak value of the fundamental current component $\hat{I}_1$, and graphical integration must again be employed as

$$I_1 = \frac{2}{\pi}\int_0^{\pi} i_b \cos\theta\, d\theta = \frac{2}{\pi} \times \frac{\pi}{18}\left[\frac{i_o \cos 0°}{2} + i_1 \cos 10° + i_2 \cos 20° + \ldots + i_{n-1} \cos(10n - 10°)\right] \tag{14-56}$$

A similar process yields $\hat{I}_{g1}$.

14-20 POWER CONSIDERATIONS IN CLASS B OR CLASS C AMPLIFIERS

The anode dissipation is

$$P_d = P_{dc} - P_o = E_{bb}I_b - \frac{\hat{I}_1^2 R_L}{2} \tag{14-57}$$

assuming a reasonable Q figure of merit for the tuned load of resonant resistive impedance

$$R_L = \frac{\hat{E}_o}{\hat{I}_1} \tag{14-58}$$

which is 1800/0.180 = 10,000 ohms for the operating line of Fig. 14-35.

Part of the input power is used to energize the bias source, the remainder being dissipated on the grid. The total is

$$P_g = \frac{2}{2\pi}\int_0^{\pi} i_c \hat{E}_g \cos\theta \, d\theta = \frac{\hat{E}_g}{\pi}\int_0^{\pi} i_c \cos\theta \, d\theta \tag{14-59}$$

The final integral can be recognized as one-half that which gives the value of $\hat{I}_g$, so that

$$P_g = \frac{\hat{E}_g \hat{I}_g}{2} \tag{14-60}$$

The value of grid input power allows determination of expected power gain, an important criterion of performance, since input power is expensive in terms of equipment needed for its generation. Tetrodes and pentodes excel over triodes in this respect.

The results are tabulated in Table 14-3 for the operating line of the figure, and a second line on which $e_{c\,max} > e_{b\,min}$. While Class C conditions have been used in this example, the method is equally suitable for Class B analysis where $2\theta_1 = 180°$.

Constant-current characteristics for transistors can be drawn from the usual curve families, and are found to be quite linear up to saturation. The graphical design of transistor Class B or C power amplifiers then follows the methods just discussed for the vacuum tube.

Table 14-3

CALCULATED VALUES FROM FIG. 14-35

	(a)	(b)		(a)	(b)
E_{bb}	2000 V	2000 V	$\hat{I}_1$	180 mA	303 mA
E_{cc}	−370 V	−370 V	$\hat{I}_{g1}$	28.5 mA	65 mA
$e_{c\,max}$	+200 V	+300 V	R_L	10,000 ohms	5940 ohms
$e_{b\,min}$	+200 V	+200 V	Power input	200 W	348 W
θ_1	62°	67°	Power output	162 W	272 W
θ_2	49.5°	56.5°	Anode loss	38 W	76 W
$\hat{E}_g$	570 V	670 V	η_p	81%	78.5 W
$\hat{E}_o$	1800 V	1800 V	Grid driving power	8.1 W	21.8 W
I_b	100 mA	174 mA	Grid loss	3.0 W	9.9 W
I_c	15 mA	33 mA	Power gain	20	12.5

14-21 THE CLASS B LINEAR AMPLIFIER

In Class C amplifiers the output E_o is an involved function of E_g, since as E_g changes, the angle θ_1 also changes. However, there is a need for a frequency selective power amplifier with a linear input-output relation to amplify, without distortion, the input signals of varying amplitude as are encountered with amplitude-modulated radio-frequency signals. Such a linear relationship can be developed in the Class B resonant-load amplifier if the tuned circuit has a reasonable Q (usually 12 or more), since the current pulses are then half-sinusoids due to the fixed conduction angle.

An analytical method due to Everitt is based on the functional relation between anode current and the electrode voltages in a vacuum tube. If linearity is assumed

$$i_b = f(\mu e_c + e_b) = g_p(\mu e_c + e_b) + C$$

If the origin is translated to cutoff, then $C = 0$, and the above becomes

$$i_b = g_m\left(e_c + \frac{e_b}{\mu}\right), \qquad i_b \geqq 0 \tag{14-61}$$

This relation predicts the linear transfer curve of Fig. 14-37. An actual characteristic may be curved near zero, but this has negligible effect on the current pulse form. By driving the input only to $e_{b\,\min} = 2e_{c\,\max}$, saturation effects and nonlinearity are avoided at the upper end of the curve.

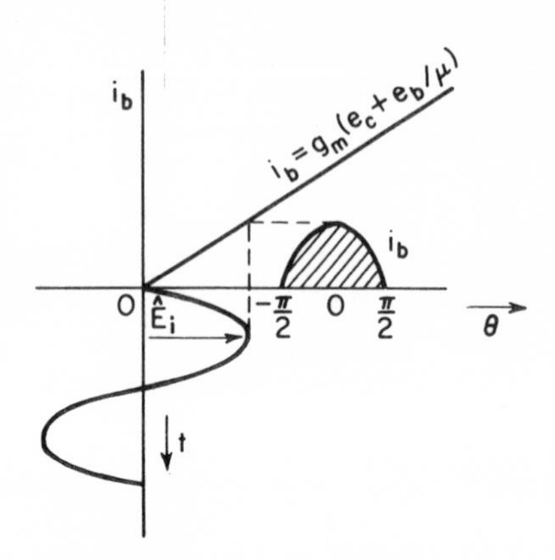

Figure 14-37. Relation between i_b and $e_c + e_b/\mu$ in the linear amplifier.

Applied to the linear characteristic is a composite voltage of peak value

$$\hat{E}_i = E_{cc} + \hat{E}_g + \frac{E_{bb} - \hat{I}_1 R_L}{\mu}$$

but for Class B operation the bias will be $E_{cc} = -E_{bb}/\mu$, and so

$$\hat{E}_i = \hat{E}_g - \frac{\hat{I}_1 R_L}{\mu} \tag{14-62}$$

As a cosine function the current is given by

$$i_b = g_m \hat{E}_i \cos\theta$$

leading to the output pulse of the figure, where the conduction angle is $2\theta_1 = 180°$.

As before, the coefficient of the fundamental frequency current component may be found as a Fourier coefficient

$$\hat{I}_1 = \frac{2}{\pi}\int_0^{\pi/2} i_b \cos\theta\, d\theta = \frac{g_m \hat{E}_g}{2}\left(\frac{1}{1 + R_L/2r_p}\right) \tag{14-63}$$

the final expression resulting from use of Eq. 14-62. The output voltage, in rms value, is

$$E_o = \frac{g_m \hat{E}_g R_L}{2\sqrt{2}}\left(\frac{1}{1 + R_L/2r_p}\right) \tag{14-64}$$

The output voltage, and therefore the output power, will maximize at $R_L = 2r_p$. Load value is found critical if saturation is to be avoided, and if the desired linear transfer curve is to be obtained.

Equation 14-64 shows that the output voltage E_o is a linear function of the input voltage E_g, since the tube parameters can be maintained near constants over the range of operation, if saturation is avoided. Thus we justify the use of the Class B linear amplifier for the linear amplification of amplitude-modulated radio-frequency signals.

The dc anode current follows by averaging the half-sine pulses over the full cycle, giving

$$I_b = \frac{g_m \hat{E}_g}{\pi}\left(\frac{1}{1 + R_L/2r_p}\right) \tag{14-65}$$

For reasonable Q values, the impedance of the tuned circuit will increase rapidly near resonance, and Eq. 14-65 shows that when the load is in resonance with resultant high R_L, the direct anode current will have a dip as in Fig. 14-38. Final adjustment of the load for resonance is usually done by tuning L or C for the minimum in the reading of the dc anode ammeter. The same variation of I_b near resonance is observed in Class C amplifiers as well.

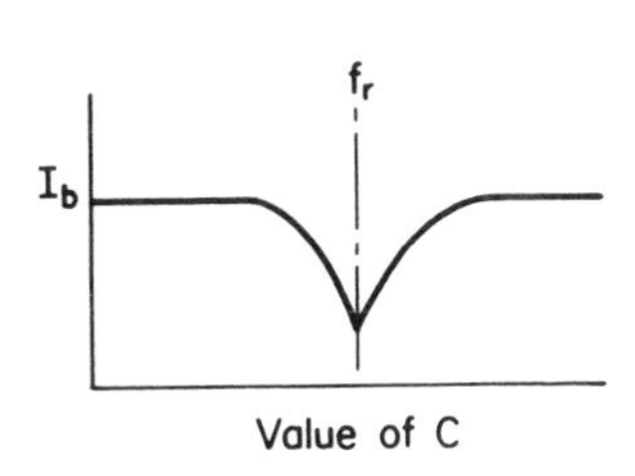

Figure 14-38. Variation of dc anode current at resonance.

From the above, a complete performance analysis can be made. Power output and dissipation follow the usual Class B derivations.

14-22 *RESONANT LOAD REQUIREMENTS*

In Class B or C power amplifiers the resonant circuit load need only be selective enough to discriminate against the harmonic frequencies; channel selectivity as in small-signal amplifiers is not required. Thus resonant circuit Q values under output load are usually in the range from 12 to 20. These values will also satisfy the requirement for a minimum number of circulating volt-amperes in the resonant circuit, to maintain the pendulum effect and a sinusoidal voltage across the pulse-excited circuit.

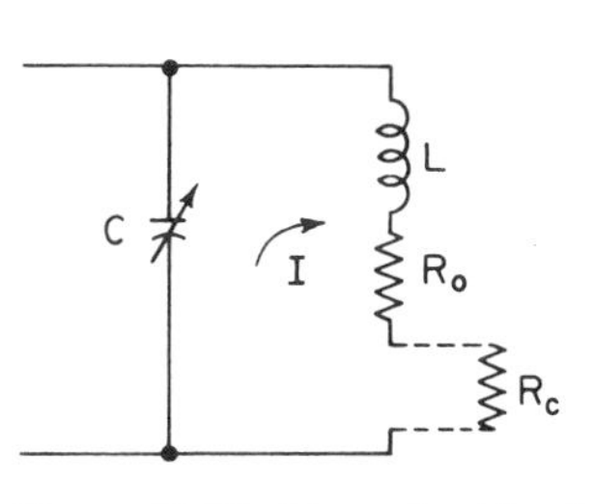

Figure 14-39. The output load circuit.

For high overall power efficiency, the coupled-in load R_c should be large in comparison to the internal circuit losses represented by R_o of the inductor in Fig. 14-39. Then

$$R_o = \frac{\omega_0 L}{Q_o}, \qquad R_c = \frac{\omega_0 L}{Q}$$

and the circuit Q_o and the loaded Q are

defined. The efficiency is then

$$\eta = \frac{I^2 R_c}{I^2(R_c + R_o)} = \frac{Q_o - Q}{Q_o} \times 100\% \tag{14-66}$$

Therefore the inductor should have a high Q_o, and then the coupled-in load can be adjusted to establish the loaded Q in the range of 12 to 20 for good power efficiency to the output.

Since the resonant impedance is given by $R_L = E_o^2/P_o$, and $R_L \cong L/CR_o$, the tuned circuit parameters can be

$$X_C = \frac{E_o^2}{QP_o}, \qquad X_L = X_C\left(\frac{Q^2}{Q^2 + 1}\right) \tag{14-67}$$

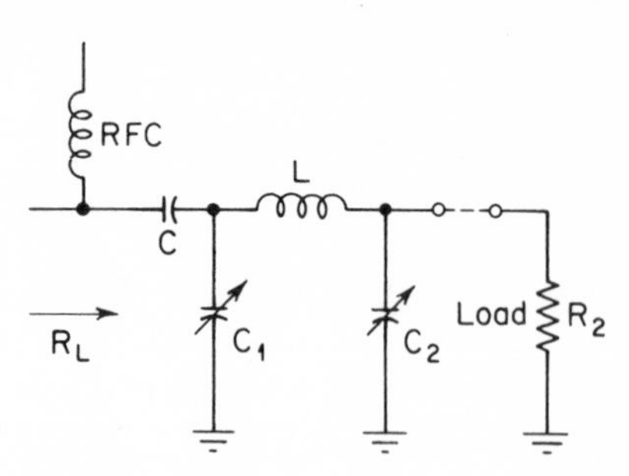

Figure 14-40. Pi-network output circuit.

A common modification of the resonant tank circuit is the π network of Fig. 14-40, which provides increased harmonic suppression. The output power is normally supplied to an antenna load, of 50 to 75 ohms, and the load presented to the generator will approximate several thousand ohms. With Q greater than 12, the elements should be

$$\left.\begin{aligned} X_{C1} = \frac{R_L}{Q}, \qquad X_{C2} = \sqrt{\frac{R_L R_2}{Q^2 + 1 - R_L/R_2}} \\ X_L = \frac{R_L}{Q^2 + 1}\left(\frac{R_L}{X_{C1}} + \frac{R_2}{X_{C2}}\right) \end{aligned}\right\} \tag{14-68}$$

14-23 THE GROUNDED-GRID POWER AMPLIFIER

The reasons of shielding between input and output and consequent stability that make the grounded-grid amplifier suitable for small-signal use also make it useful as a power amplifier in Class B or C. General power operation will be as studied for that particular class of service, but the input is somewhat different, as

$$P_{\text{in}} = \frac{\hat{E}_g \hat{I}_g}{2} + \frac{\hat{E}_g \hat{I}_1}{2} \tag{14-69}$$

by observation of Fig. 14-41. The ac power available in the output circuit is

$$P_o = \frac{\hat{E}_g \hat{I}_1}{2} + \frac{\hat{E}_o \hat{I}_1}{2} \tag{14-70}$$

The term $\hat{E}_g \hat{I}_1/2$, while required as additional input, is not lost but is added to the power made available in the load.

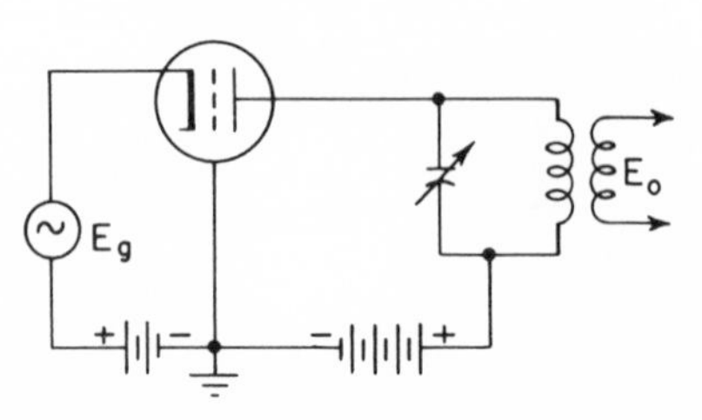

Figure 14-41. Grounded-grid Class C amplifier.

14-24 AMPLIFIER NEUTRALIZATION

In the generation of power at high radio frequencies triodes are often preferred in the grounded-cathode circuit, and for stability it is necessary to neutralize the feedback of energy from output to input through the grid-anode capacity C_{gp}. The problem was discussed in Section 11-21 for small-signal circuits, and that discussion applies.

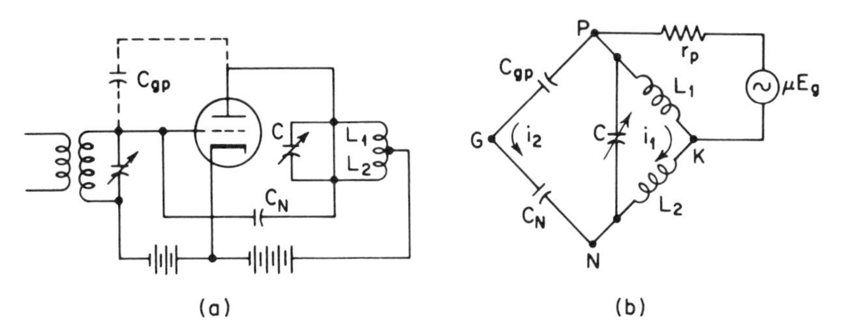

Figure 14-42. Hazeltine neutralizing circuit and equivalent.

One form of neutralizing circuit for Class B or C power amplifiers is shown in Fig. 14-42. For neutralization of feedback the output generator μE_g should produce no voltage between G and K. The source, while connected across the P-K arm, is coupled into the K-N arm by capacitance C and the mutual inductance between L_1 and L_2. In effect, μE_g appears across the P-N points in a normal bridge manner. For neutralization, balance requires

$$\frac{C_N}{C_{gp}} = \frac{L_1}{L_2} \tag{14-71}$$

Adjustment is carried out by balancing for energy feed in reverse—from grid to anode. Grid excitation is supplied but the anode voltage is removed. If the circuits are unbalanced, energy will feed to the anode tank through C_N or C_{gp}. By searching for energy in the tank circuit it is possible to find the adjustment of C_N which reduces energy transfer to a negligible amount. The circuit is then balanced for energy transfer or feedback in the reverse direction.

Transistor neutralization to reduce internal feedback may be carried out in similar circuits.

PROBLEMS

14-1. The transistor whose characteristics appear in Figs. 14-10 and 14-11 is operated from a source of 5 ohms impedance at a Q point of $v_{CE} = 20$ V, $i_B = 20$ mA. The collector voltage is limited to 40 V. With input signal of 1 V peak, determine:

(a) Possible power output.
(b) Second and third harmonic distortion.
(c) Input power and the power gain.

14-2. A 6L6 beam tetrode, with characteristics in the Appendix, is transformer-coupled to a 10 ohm load. With $E_{bb} = E_{c2} = 250$ V, $I_b = 65$ mA, and $i_{\min} = 10$ mA, plot curves of power output and total distortion against reflected load value.

14-3. Using the transistor of Fig. 14-6, with $v_{CE} = 30$ V, $i_B = 60$ mA, and a transformer-connected load reflecting 20 ohms, calculate:
(a) The Class A power output with a signal of 40 mA peak from a current source.
(b) The total distortion.
(c) The power dissipated and the power efficiency.

14-4. Two transistors are operated at $V_{CC} = 15$ V. Each transistor is mounted on a heat sink capable of dissipating 5 W. Calculate the current i_m at maximum loss, the push-pull collector-to-collector load, and the maximum power output of the two transistors in Class B.

14-5. A power transistor has $\theta_{JC} = 0.8$°C/W, $\theta_{CS} = 0.8$°C/W, and the heat sink of Fig. 14-16 has an air flow of 0.2 lb/min. With T_J not to exceed 90°C, how much power can be dissipated at 35°C $= T_A$?

14-6. A germanium transistor is operated at $I_C = 0.5$ A, $V_{CE} = 10$ V, and is limited to $T_J = 100$°C and $T_{A\max} = 50$°C. Because of distortion, ΔI_C cannot exceed 0.15 A from a 25°C ambient temperature. With $I_{CBO} = 0.0002$ A at 25°C, find the maximum S and θ_{JA} to prevent thermal runaway.

14-7. Using a germanium transistor with $\theta_{JA} = 250$°C/W and $I_{CBO} = 16$ μA at 25°C, with $R_E = 100$ ohms, $R_B = 1000$ ohms, $V_{CC} = 9$ V, $I_C = 0.002$ A, $h_{FE} = 40$, determine if thermal runaway can happen. If so, at what temperature?

14-8. A push-pull circuit is operating Class B with an anode-to-anode load of 5000 ohms. The dc meter in the common anode circuit reads 212 mA for a sinusoidal signal. What is the ac power output? If $E_{BB} = 1560$ V, what are the anode circuit efficiency and the anode dissipation? What is the peak dissipation of the amplifier?

14-9. Characteristics for a triode are shown in Fig. 14-43. One tube is operated with $E_{bb} = 2000$ V, $E_{cc} = -200$ V, and $e_{b\min} = e_{c\max} = 300$ V. Compute:
(a) I_b and I_c.
(b) Output power.
(c) Power gain.
(d) Power efficiency.
(e) Anode dissipation.
(f) The required load resistance.

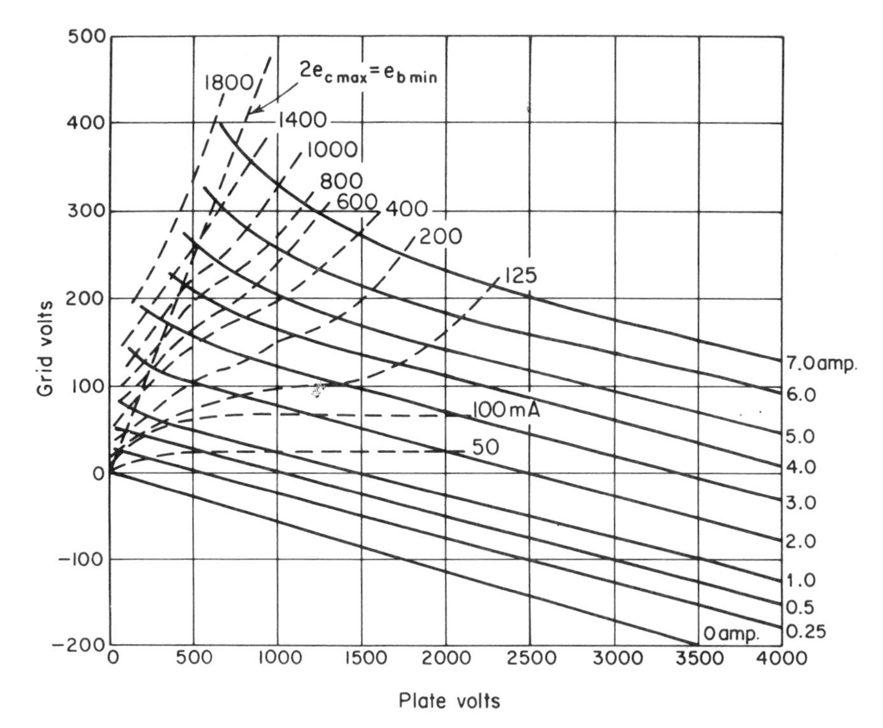

Figure 14-43.

14-10. Repeat Problem 14-9 with $e_{b\,\min} = 200$ V and $e_{c\,\max} = 100$ V. Compare results with Problem 14-9.

14-11. A parallel-resonant circuit of $Q = 20$ has $C = 27$ pF to tune the circuit to 10 megahertz. It is excited by a current

$$i = 0.41 + 1.62 \sin 2\pi \times 10^7 t + 0.45 \sin 4\pi \times 10^7 t$$

in amperes. Find the ratio of the fundamental to second harmonic voltages across the circuit.

14-12. The tank circuit of Fig. 14-39 operates at 10 megahertz with $C = 50$ pF; coil $Q_o = 300$. If the efficiency of the circuit is to exceed 90 per cent, what is the minimum value of R_C?

14-13. A certain triode tube may be considered as having linear characteristics expressed by

$$i_b = 0.003(5e_c + e_b) \quad \text{A}$$

(a) For Class A operation with a transformer-coupled load R'_L of 500 ohms with $R_{\text{dc}} = 0$, $E_b = 300$ V, $E_c = -30$ V, find the power output, efficiency, and dc power input for a sinusoidal signal of 30 V peak.

(b) Find the value of the load which will give the maximum power output under the conditions of (a).

14-14. A triode with $\mu = 35$ and $r_p = 3000$ ohms is operated as a Class B linear amplifier with $E_b = 2500$ V, $\hat{E}_g = 90$ V, $E_{cc} = -70$ V, $\theta_1 = 90°$, $e_{b\,\min} = 300$ V, and with a tuned load of 4000 ohms at resonance. Find I_b, I_1, the power output, the anode input, and the anode-circuit efficiency.

REFERENCES

1. Everitt, W. L., "Optimum Operating Conditions for Class C Amplifiers," *Proc. IRE*, Vol. 22, p. 152 (1934).
2. McProud, C. G., and R. T. Wildermuth, "Phase Inverter Circuits," *Electronics*, Vol. 13, p. 47 (December 1940).
3. Ebers, J. J., and J. L. Moll, "Large Signal Behavior of Junction Transistors," *Proc. IRE*, Vol. 42, p. 1761 (1954).
4. *Motorola Power Transistor Handbook*. Motorola Semiconductor Products Div., Inc., Phoenix, Ariz., 1960.
5. Glasford, G. M., *Linear Analysis of Electronic Circuits*. Addison-Wesley Pub. Co., Inc., Reading, Mass., 1965.
6. Inbar, G. F., "Thermal and Power Considerations in Class B Transistorized Amplifiers," *Trans. on Audio, IEEE*, AU-13 No. 4 (July-August 1965).
7. Stover, W. A., *et al.*, *Audio and AM/FM Circuit Design Handbook*. Texas Insts., Inc., Dallas, Texas, 1966.
8. Thornton, R. D., *et al.*, *Characteristics and Limitations of Transistors* (SEEC, Vol. 4). John Wiley & Sons, Inc., New York, 1966.
9. *Proceedings of the IEEE*, special issue on large power transistors, Vol. 55, p. 1247 *et seq.* (1967).

15

OSCILLATOR PRINCIPLES

Oscillators are the generating sources for sinusoidal frequencies, and stability of frequency is the most rigid requirement on oscillator performance. Almost any oscillator will be stable if operated in isolation, at constant temperature, with unchanging components and tubes or transistors, supplied by constant voltages, and if no power be taken from the circuit. The real problem arises in the attempt to obtain reasonable power output from oscillators in normal environments when built with practical components.

We leave to a later chapter the discussion of relaxation forms of oscillators.

15-1 FEEDBACK REQUIREMENTS FOR OSCILLATION

An oscillator can be studied as a feedback amplifier, operating in the unstable mode. In the circuit of Fig. 15-1 the voltage fed back supplies the entire amplifier input and

$$\mathbf{V}_f = \mathbf{V}_i = \boldsymbol{\beta}\mathbf{V}_o = \mathbf{A}\boldsymbol{\beta}\mathbf{V}_i$$

from which

$$(1 - \mathbf{A}\boldsymbol{\beta})\mathbf{V}_i = 0 \tag{15-1}$$

If an output is to be obtained, then $\mathbf{V}_i \neq 0$; therefore $(1 - \mathbf{A}\boldsymbol{\beta}) = 0$, and

$$\mathbf{A}\boldsymbol{\beta} = 1 + j0 \tag{15-2}$$

which was the situation leading to oscillation in the feedback amplifier.

The expression presents two basic requirements for oscillation:

1. that the open-loop gain $|A\beta| = 1$;
2. that the net phase shift around the loop $= 2n\pi$, where $n = 0, 1, 2, 3, \ldots$.

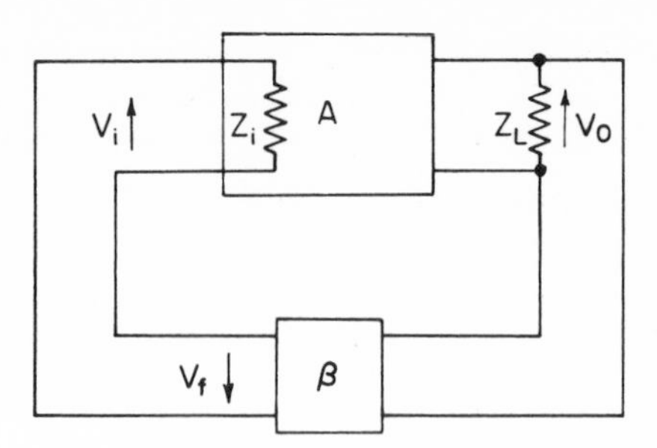

Figure 15-1. Basic feedback oscillator.

An oscillating amplifier self-adjusts to meet both of these criteria. An initial switching surge may be assumed to start the operation. A voltage resulting from this disturbance propagates back to the input, and appears again as an amplified output. The process is repeated at greater amplitude, and as the amplifier works into saturation and cutoff regions, the average gain falls to the level necessitated by Eq. 15-2, and a steady amplitude is reached.

Limitation of amplitude by saturation and cutoff implies distortion, which must be filtered out by a high Q resonant load circuit. When high purity of wave form is desired, the gain may be limited by an auxiliary means which varies β.

The circuit adjusts in frequency so as to bring the phase shift around the loop to $2n\pi$. For good frequency stability $d\theta/d\omega$ should be large, so that a large change in phase angle is obtained for a small shift in frequency. This also makes a high Q resonant circuit desirable.

15-2 CIRCUIT REQUIREMENTS FOR OSCILLATION

Figure 15-2 illustrates a basic feedback oscillator circuit, utilizing a reactance network to provide feedback between the collector and base terminals through Z_3. A similar circuit is used with triodes and pentodes. The resonant network

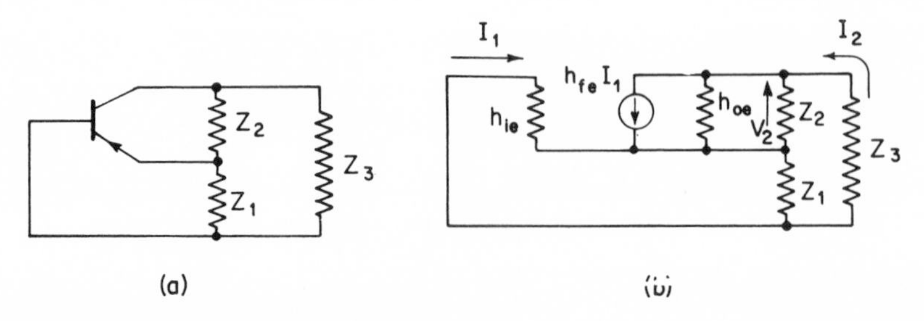

Figure 15-2. Resonant-load oscillator.

supplies 180° of phase shift between the output collector voltage and the voltage fed back to the base, and the characteristics of the common-emitter amplifier provide an additional 180° shift between base and collector voltages; thus the phase requirement for oscillation is met.

Design requirements can be found by analysis, assuming linear operation with the equivalent circuit, the method being equally applicable to the vacuum tube. Oscillator operation is usually in Class C, but use of the equivalent circuit is justified to provide qualitative understanding. The transistor may be assumed to have negligible h_{re}; then using a current summation in Fig. 15-2(b):

$$\mathbf{V}_2 = (\mathbf{I}_2 - h_{fe}\mathbf{I}_1)\left(\frac{\mathbf{Z}_2}{1 + h_{oe}\mathbf{Z}_2}\right) \tag{15-3}$$

Also

$$\mathbf{V}_2 + \mathbf{I}_2\left(\mathbf{Z}_3 + \frac{\mathbf{Z}_1 h_{ie}}{\mathbf{Z}_1 + h_{ie}}\right) = 0 \tag{15-4}$$

It is then possible to write the gain as

$$\mathbf{A} = \frac{\mathbf{I}_2}{\mathbf{I}_1} = \frac{h_{fe}\mathbf{Z}_2}{(h_{oe}\mathbf{Z}_2 + 1)\left(\mathbf{Z}_3 + \dfrac{\mathbf{Z}_1 h_{ie}}{\mathbf{Z}_1 + h_{ie}} + \dfrac{\mathbf{Z}_2}{h_{oe}\mathbf{Z}_2 + 1}\right)} \tag{15-5}$$

Using current division

$$\mathbf{I}_1 = \frac{-\mathbf{Z}_1}{\mathbf{Z}_1 + h_{ie}}\mathbf{I}_2$$

and so the feedback is

$$\boldsymbol{\beta} = \frac{\mathbf{I}_1}{\mathbf{I}_2} = \frac{-\mathbf{Z}_1}{\mathbf{Z}_1 + h_{ie}} \tag{15-6}$$

We are then ready to apply the oscillation condition of Eq. 15-2 to obtain

$$\mathbf{A}\boldsymbol{\beta} = \frac{-\mathbf{Z}_1}{\mathbf{Z}_1 + h_{ie}} \frac{h_{fe}\mathbf{Z}_2}{\mathbf{Z}_2 + (h_{oe}\mathbf{Z}_2 + 1)\left(\mathbf{Z}_3 + \dfrac{\mathbf{Z}_1 h_{ie}}{\mathbf{Z}_1 + h_{ie}}\right)} = 1 \tag{15-7}$$

from which

$$\begin{aligned} -h_{fe}\mathbf{Z}_1\mathbf{Z}_2 &= \mathbf{Z}_1\mathbf{Z}_2 + \mathbf{Z}_1\mathbf{Z}_3 + h_{ie}(\mathbf{Z}_1 + \mathbf{Z}_2 + \mathbf{Z}_3) \\ &\quad + h_{ie}h_{oe}(\mathbf{Z}_1\mathbf{Z}_2 + \mathbf{Z}_2\mathbf{Z}_3) + h_{oe}\mathbf{Z}_1\mathbf{Z}_2\mathbf{Z}_3 \end{aligned}$$

If the impedances are assumed to be purely reactive, or

$$\mathbf{Z}_1 = j\mathbf{X}_1, \qquad \mathbf{Z}_2 = j\mathbf{X}_2, \qquad \mathbf{Z}_3 = j\mathbf{X}_3$$

then the above becomes

$$\begin{aligned} h_{fe}\mathbf{X}_1\mathbf{X}_2 &= -\mathbf{X}_1\mathbf{X}_2 - \mathbf{X}_1\mathbf{X}_3 + jh_{ie}(\mathbf{X}_1 + \mathbf{X}_2 + \mathbf{X}_3) \\ &\quad - h_{ie}h_{oe}(\mathbf{X}_1\mathbf{X}_2 + \mathbf{X}_2\mathbf{X}_3) - jh_{oe}\mathbf{X}_1\mathbf{X}_2\mathbf{X}_3 \end{aligned}$$

Equating the real terms leads to

$$h_{fe}\mathbf{X}_1\mathbf{X}_2 = -[\mathbf{X}_1\mathbf{X}_2 + \mathbf{X}_1\mathbf{X}_3 + h_{ie}h_{oe}(\mathbf{X}_1\mathbf{X}_2 + \mathbf{X}_2\mathbf{X}_3)]$$

This relation can be valid only if one of the three reactances is made of opposite sign to the other two. That is, the three circuit reactances must include capacitive and inductive forms if the condition for oscillation in Eq. 15-7 is to be satisfied.

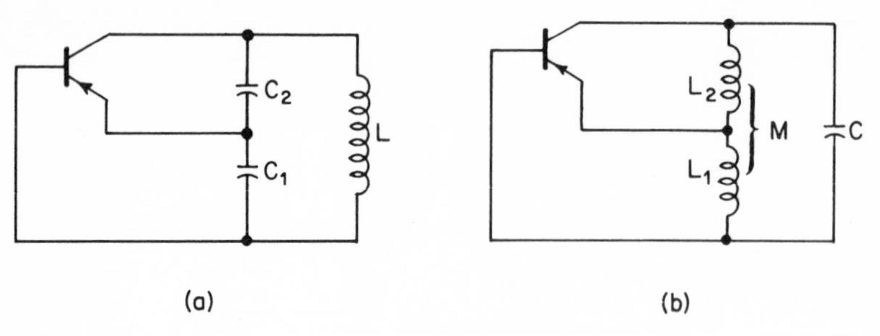

Figure 15-3. (a) Colpitts oscillator; (b) Hartley oscillator.

The requirement on the reactances logically leads to two standard forms of oscillator circuit, as shown in Fig. 15-3. Choosing the *Colpitts circuit* of (a), we have

$$\mathbf{Z}_1 = \frac{-j}{\omega C_1}, \qquad \mathbf{Z}_2 = \frac{-j}{\omega C_2}, \qquad \mathbf{Z}_3 = j\omega L$$

Substitution in Eq. 15-7 yields

$$\frac{h_{fe}}{\omega^2 L(C_2 + C_1 h_{ie} h_{oe}) - h_{ie} h_{oe} - 1 + j\omega[\omega^2 L C_1 C_2 h_{ie} - h_{ie}(C_1 + C_2) - L h_{oe}]} = 1 \tag{15-8}$$

For oscillation to occur, the angular frequency must adjust to a value ω_0 at which the phase shift in the loop is $2n\pi$ and the reactive term in the denominator becomes zero. So

$$\omega_0^2 L C_1 C_2 - C_1 - C_2 - \frac{h_{oe}}{h_{ie}} L = 0$$

$$\omega_0 = \sqrt{\frac{1}{L}\left(\frac{1}{C_1} + \frac{1}{C_2}\right) + \frac{h_{oe}}{h_{ie}} \frac{1}{C_1 C_2}} \tag{15-9}$$

Since the ratio h_{oe}/h_{ie} will be very small, the resonant angular frequency is closely that of the Colpitts tuned circuit of L and capacitors C_1 and C_2 in series.

At the angular frequency ω_0 the real part of Eq. 15-8 must be equal to unity, and the gain condition for oscillation can be found by reducing the real part to

$$h_{fe} \geq \frac{C_2}{C_1} + h_{ie} h_{oe} \frac{C_1}{C_2} + \frac{h_{oe}}{h_{ie}} \frac{L}{C_1} \tag{15-10}$$

after neglecting second-order terms. As a matter of convenience the ratio

C_2/C_1 is often made unity in the Colpitts circuit, so that the above condition is easily satisfied.

A similar analysis can be carried out for the *Hartley circuit* of Fig. 15-3(b), and the oscillation conditions found to be

$$\omega_0 = \sqrt{\frac{1}{C\left[L_1 + L_2 + 2M - \dfrac{h_{oe}}{h_{ie}}(L_1L_2 - M^2)\right]}} \tag{15-11}$$

$$h_{fe} \geqq \frac{L_1 + M}{L_2 + M}\left\{\frac{1 - (h_{oe}/h_{ie})[L_1(L_1L_2 - M^2)/(L_1 + M)^2]}{1 - (h_{oe}/h_{ie})[M(L_1L_2 - M^2)/(L_1 + M)(L_2 + M)]}\right\} \tag{15-12}$$

where again the correction terms containing h_{oe}/h_{ie} are small.

15-3 *OTHER BASIC OSCILLATOR CIRCUITS*

The previous section introduced the basic Hartley and Colpitts forms of oscillator circuits. The Colpitts circuit is repeated at (a), Fig. 15-4, in a practical form, including a choke *RFC* to prevent the power supply from short-

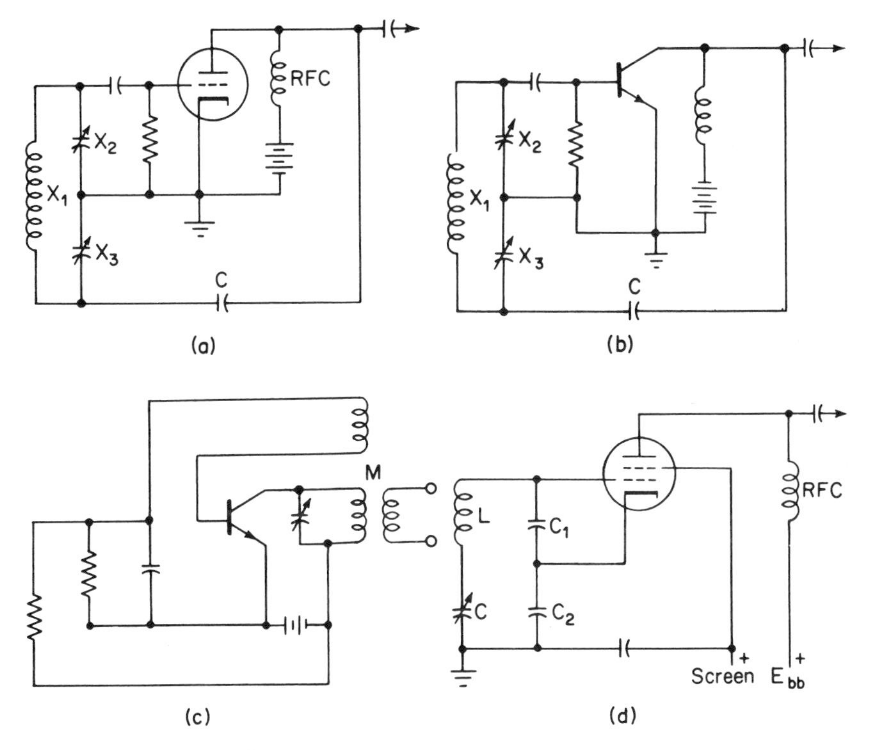

Figure 15-4. (a), (b) Colpitts circuits; (c) tuned-collector circuit; (d) Clapp version of the electron-coupled oscillator.

circuiting part of the tuned circuit. A further adaptation of the Hartley circuit appears at (c) as a tuned collector circuit, where feedback to the base is by mutual coupling.

Changes in output loading which may cause changes in collector or anode voltage and thus affect the frequency-determining circuit through C_{gp} may be eliminated by use of an isolating amplifier or *buffer* between the oscillator circuit and the output load. An oscillator with a built-in buffer is the *electron-coupled Colpitts circuit* of (d). This oscillator uses the cathode, control grid, and screen of a pentode as a triode, with the resonant circuit shielded from the output load by the grounded screen. The anode or output is coupled to the resonant circuit only through the electron stream.

15-4 PIEZOELECTRIC CRYSTALS FOR CONTROL OF FREQUENCY

When frequencies must be maintained within a few hertz per megahertz, piezoelectric quartz crystals are often used to replace the frequency-determining resonant circuit of an oscillator. In such crystals an alternating electrical potential placed across one pair of faces causes mechanical vibration on another axis; at mechanical resonance these vibrations can have large amplitude. The converse effect of mechanical excitation and electrical charge variation is also present—a crystal is a form of electro-mechanical coupling. The frequency of vibration is dependent on crystal dimensions, and is quite stable because of the stability of quartz.

With axes designated as in Fig. 15-5, x- and y-cut crystals are obtained by slicing crystal sections as shown, with thickness perpendicular to the axis determining the resonant frequency. The resonant frequency is a temperature function, but by appropriate orientation of the angle at which the crystal is cut, a zero temperature coefficient of frequency can be achieved over usual temperature ranges.

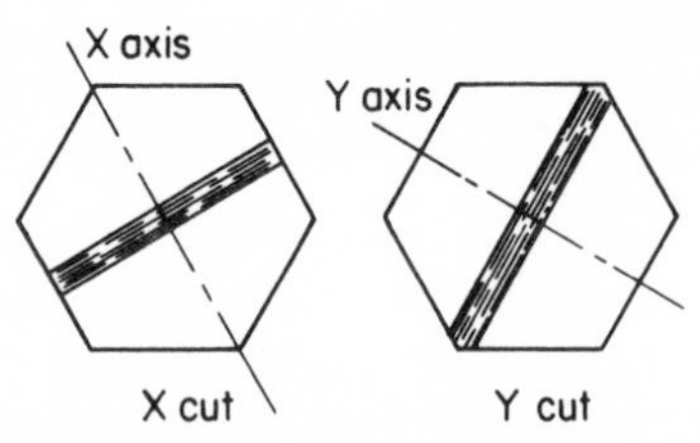

Figure 15-5. Quartz crystal cuts; Z-axis normal to the page.

A crystal appears as an analog of the electrical circuit of Fig. 15-6(b), where C_h is the shunting capacitance due to the crystal electrodes, usually electroplated in position. The valuable properties of a crystal are its very high ratio of mass to damping (analogous to Q) and its high ratio of mass to elastance (analogous to high L/C ratio). The Q may be in the range from 10,000 up to 500,000 for some special designs.

The crystal has series and parallel resonances, shown in Fig. 15-6(c) and expressed by

$$X = \frac{-(\omega^2 LC - 1)}{\omega(C + C_h)[\omega^2 LCC_h/(C + C_h) - 1]} \tag{15-13}$$

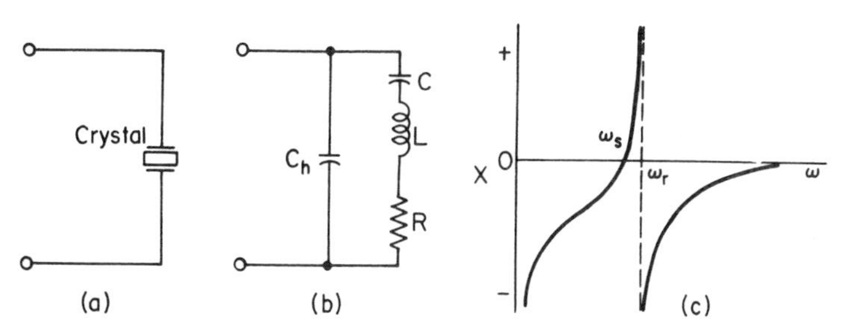

Figure 15-6. (a) Crystal symbol; (b) equivalent electrical circuit; (c) reactance curve for the crystal.

The ratio of C_h to C may be several hundred or more so that the series resonance or the zero due to the numerator is very close to the antiresonance or pole due to the denominator. Resonant frequencies from about 10 kilohertz to 10 megahertz can be conveniently obtained; higher frequencies are possible by shaping and mounting the crystals so as to accentuate harmonic modes of vibration.

Several circuits employing crystals appear in Fig. 15-7. The Pierce oscillator of (b) utilizes the crystal at parallel resonance as a feedback element, but is basically a Hartley oscillator, with the center point of the resonant circuit at the base. The circuit at (a) has a tuned output-tuned input form, with the grid circuit controlling frequency as the circuit of highest Q. Oscillation can occur only when the Miller effect resistance in the input circuit is negative, and this requires that the anode circuit be tuned on the inductive side of resonance. Feedback is through the C_{gp} capacitance of the tube.

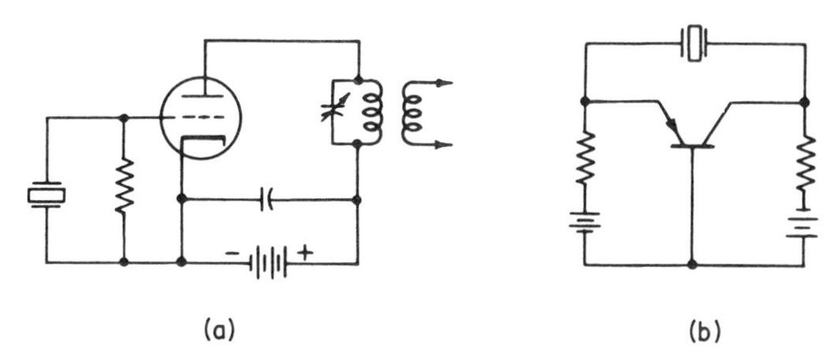

Figure 15-7. (a) Tuned-plate oscillator; (b) Pierce oscillator.

15-5 FREQUENCY STABILITY OF OSCILLATORS

The analyses of Section 15-2 indicate frequency effects due to the shunting parameters of transistor or tube. While the frequency of oscillation is largely controlled by the resonant circuit, the ratios h_{oe}/h_{ie} for transistors, or R_L/r_p

for triodes, should be small if changes in device parameters and supply voltages are not to shift frequency.

Circuit loading will also shift frequency, and its primary cause is the change in Miller effect input capacitance C_{in} with load impedance and its phase angle. The effect of input capacitance change can be reduced in the Hartley or Colpitts circuits by making the tank capacitors large and L small; this reduces C_{in} as a per cent of total circuit C. The result is a high Q circuit, which also is desirable because of a large value of $d\theta/d\omega$, needed for frequency stability.

The crystal oscillator is much less affected by change in C_{in}. That capacitance appears in parallel with C_h, and from the circuit of Fig. 15-6 the effective capacitance of the crystal is then

$$C_e = \frac{C(C_h + C_{in})}{C + C_h + C_{in}} \tag{15-14}$$

For a given crystal resonant at 474 kilohertz, the value of $C = 0.0316$ pF, $C_h = 5.7$ pF, $C_{in} = 30$ pF. Since C is so small, a change in C_{in} cannot have much effect on C_e. In fact, with a change of 10 per cent in C_{in}, the change in frequency of the crystal is only 0.0033 per cent, or 15.6 hertz at crystal resonance.

For improvement of frequency stability, an oscillator may be followed by a buffer Class A amplifier, to isolate the load from the resonant circuit. This is the feature of the electron-coupled circuit. Temperature effects may be reduced or eliminated by use of negative temperature coefficient capacitors, as inductors have positive temperature coefficients.

15-6 RESISTANCE-CAPACITANCE OSCILLATORS

Resistance-capacitance networks may be employed at low frequencies, as in Fig. 15-8, to provide feedback and phase shift for oscillation. The circuit oscillates at a frequency at which the total RC network gives 180° phase shift, the tube or transistor giving another 180° shift. Since there is no resonant circuit for filtering, the amplitude must be kept low if a good sinusoid is to be obtained. Because of the complexity of the network, the circuit is not suited to variable frequency use.

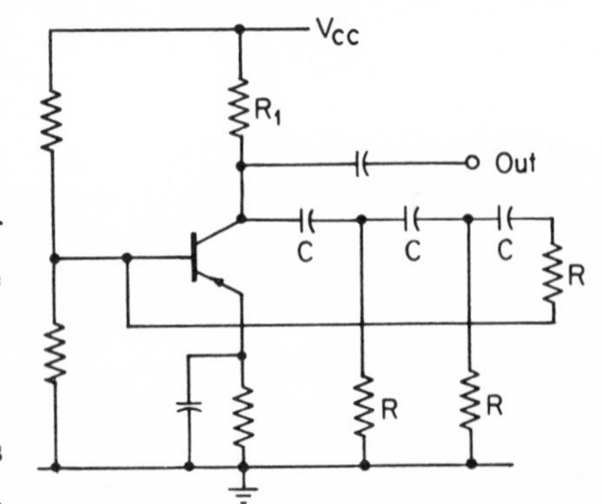

Figure 15-8. Phase-shift oscillator.

A tunable circuit is that at Fig. 15-9. The two-stage circuit employs positive feedback for oscillation, and negative feedback for amplitude limi-

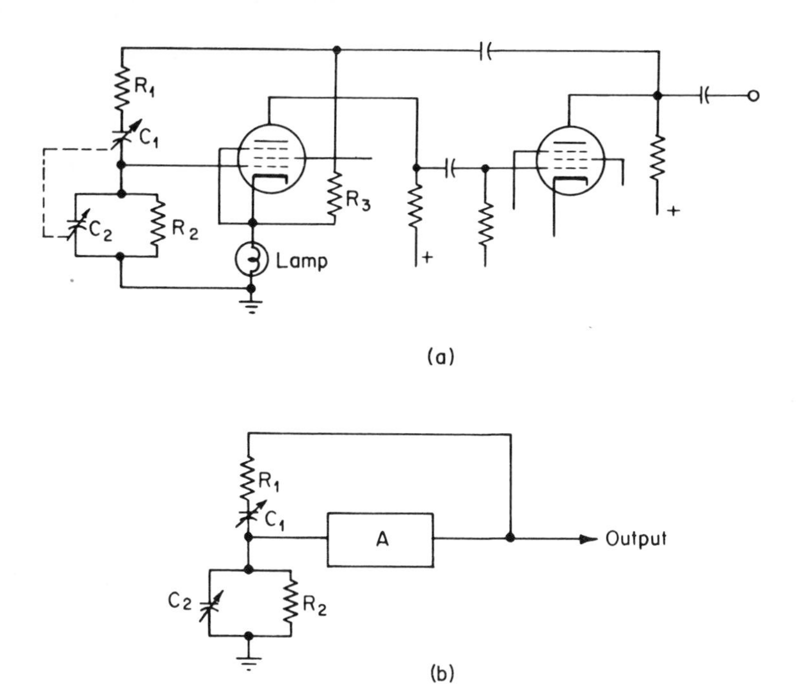

Figure 15-9. Resistance–capacity oscillator.

tation and wave form improvement. Ordinarily C_1 and C_2 are simultaneously varied to change frequency. Negative feedback is provided by R_3 and the tungsten lamp cathode resistor. At low output the current fed back to the lamp is small and its resistance is low. As the output tends to increase, the lamp current and resistance increase, with greater negative feedback. As a result, the amplitude and gain are stabilized, and good wave form is maintained.

For analysis this RC oscillator may be reduced to the circuit at (b); this leaves the positive feedback necessary for oscillation, but removes the negative feedback gain-stabilization elements. As before, we may write $\boldsymbol{\beta}$ as

$$\boldsymbol{\beta} = \frac{\mathbf{Z}_{\text{parallel } R_2C_2}}{\mathbf{Z}_{\text{parallel } R_2C_2} + \mathbf{Z}_{\text{series } R_1C_1}}$$

It is usual to make $R_1 = R_2 = R$ and $C_1 = C_2 = C$, so that with internal gain $\mathbf{A}$

$$\mathbf{A}\boldsymbol{\beta} = \frac{\mathbf{A}}{RC}\left\{\frac{1}{(3/RC) + j\omega[1 - (1/\omega^2R^2C^2)]}\right\} = 1 \tag{15-15}$$

Equating the reactive term to zero leads to

$$\omega = \sqrt{\frac{1}{R_1R_2C_1C_2}} = \frac{1}{RC} \tag{15-16}$$

and setting the real term to unity,

$$A = 3 \tag{15-17}$$

The negative feedback provided by the cathode lamp resistor will maintain this value of gain. Excellent output wave form is possible in such laboratory oscillators.

15-7 TUNNEL DIODE NEGATIVE-RESISTANCE OSCILLATOR

For the circuit of Fig. 15-10(a) it is possible to write

$$(s^2LC + sL/R + 1)i_L = 0$$

with a solution

$$i_L = K\epsilon^{[-(1/2RC)+j\omega]t} \tag{15-18}$$

This is the equation of a decaying oscillation at frequency ω. To maintain the oscillation the real term of the exponent must be zero or positive; this is possible if negative resistance is added to the circuit as at (b).

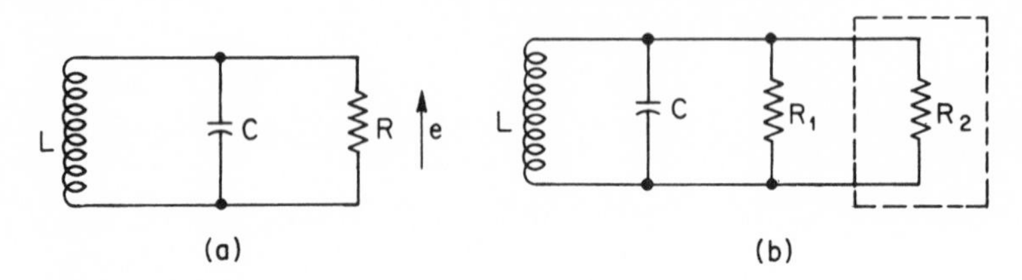

Figure 15-10. Negative-resistance oscillator.

For oscillations to build up in the tuned circuit of feedback oscillators, negative resistance must also be introduced and this is done by energy fed back to supply the circuit losses. For the previous analyses it seemed more desirable to stress the feedback conditions rather than the concept of negative resistance.

Now however, we introduce the tunnel diode as a negative-resistance device, operating without feedback. It has the characteristics of Fig. 15-11(a), in which the *A*-*B* region has negative slope and therefore can furnish a negative resistance to maintain oscillation in a resonant circuit. A dc load line, characteristic of the series resistance R_B, is so chosen along with V_s, as to place operation at C in the middle of the negative resistance region. Since the slope of the characteristic is $-g_d$, then $R_B < |1/g_d|$ for an intersection to occur. The frequency of oscillation will be

$$\omega_0 = \sqrt{\frac{1}{L(C_1 + C_2)} - \frac{g_d^2}{C_2(C_1 + C_2)}} \tag{15-19}$$

At C the negative diode conductance is greater than the positive circuit conductance and oscillations will start when V_s is applied, the current point

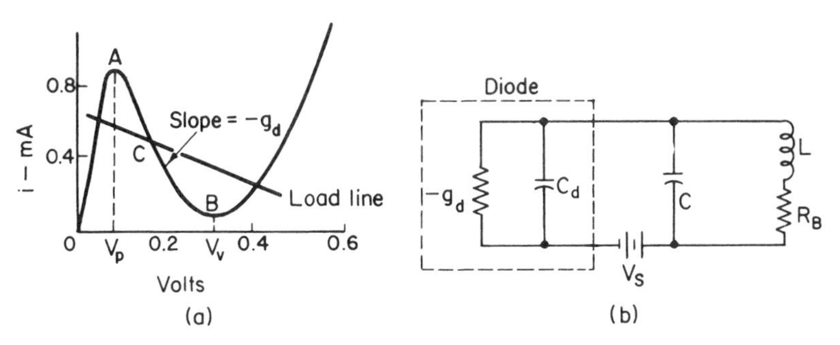

Figure 15-11. (a) Tunnel diode volt–ampere curve; (b) oscillator circuit.

moving along the device characteristics from C to beyond A and B. As the amplitude becomes greater, the average slope of the diode characteristic decreases, and oscillation stabilizes at the amplitude limits at which $|g_d| = R_B$ and the net circuit resistance is zero.

PROBLEMS

15-1. For the Hartley oscillator of Fig. 15-3(b), and neglecting mutual inductance M, derive the resonant frequency and gain conditions.

15-2. For the collector-tuned oscillator of Fig. 15-4(c), derive the condition for resonant frequency and determine the gain needed for oscillation, with load R in series with L.

15-3. A triode Colpitts oscillator has $L = 37\ \mu$H, $C_1 = C_2 = 300$ pF, $r_p =$ 15,000 ohms, and Q of the resonant circuit $= 15$. Find the value of μ needed for oscillation and the frequency of oscillation. Include the effect of r_p.

15-4. In Fig. 15-8, $R = R_1 = 30{,}000$ ohms, $C = 0.002\ \mu$F. Find the frequency of oscillation and the value of gain required, neglecting I_b.

15-5. In Fig. 15-12, find the gain A required if the circuit is to oscillate, and the frequency of oscillation. $C = 0.1\ \mu$F, $L = 0.15$ H, $R = 20{,}000$ ohms.

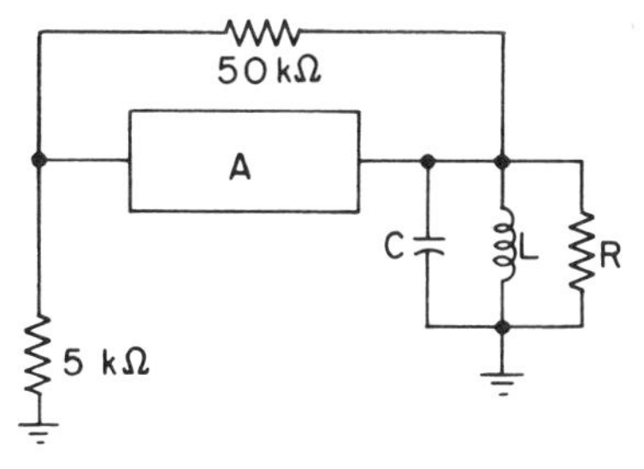

Figure 15-12.

15-6. A crystal has analogous electrical values as: $L = 250$ H, $C = 0.04$ pF, $R = 1800$ ohms, and $C_h = 8$ pF. When used in the grid circuit of a triode with $\mu = 30$, $r_p = 10{,}000$ ohms, $C_{gk} = 5$ pF, $C_{gp} = 3$ pF, and a resistive load of 27,000 ohms, find the frequency of oscillation.

15-7. Develop Eq. 15-15.

15-8. A quartz crystal has $L = 3.66$ H, $C = 0.032$ pF, $C_h = 6$ pF, $R = 4500$ ohms.

(a) Find series ω_0 and parallel ω_0.

(b) What is the Q?

REFERENCES

1. Terman, F. E., *et al.*, "Some Applications of Negative Feedback with Particular Reference to Laboratory Equipment," *Proc. IRE*, Vol. 27, p. 649 (1939).
2. Mason, W. P., "Low Temperature Coefficient Quartz Crystals," *Bell Syst. Tech. Jour.*, Vol. 19, p. 74 (1940).
3. Edson, W. A., *Vacuum Tube Oscillators*. John Wiley & Sons, Inc., New York, 1953.
4. Lo, A. W., *et al.*, *Transistor Electronics*. Prentice-Hall, Inc., Englewood Cliffs, N. J., 1955.
5. Ko, Wen Hsiung, "Designing Tunnel Diode Oscillators," *Electronics*, Vol. 34, No. 6, p. 68 (February 10, 1961).
6. Cote, A. J., and J. B. Oakes, *Linear Vacuum Tube and Transistor Circuits*. McGraw-Hill Book Company, New York, 1961.
7. Shea, R. F., *Transistor Applications*. John Wiley & Sons, Inc., New York, 1965.

16

AM AND FM MODULATION PROCESSES

In order to transmit information by a sinusoidal wave form, it is necessary to vary the characteristics of the wave in some manner; this variation of some property of the signal is called *modulation*. One of the most common uses of the process of modulation is in radio transmission, where the frequency of the usual audio informational signal is not suited to direct radiation. By modulation, the information is translated to a suitable frequency band, and subsequently radiated. In other systems the signal is transformed by a suitable code before modulation. At the receiving end of the circuit, a reverse process of translation, *demodulation* or decoding, is employed to strip the information from the transmitted wave forms. This latter process will be studied in Chapter 17.

16-1 FUNDAMENTALS OF MODULATION

Whenever an alternating voltage is written

$$e = A \cos (\omega t + \theta_o)$$

the angular position of the rotating phasor is defined in terms of an angle θ with respect to a reference, and as a function of time, since the phasor is assumed in rotation. Angular velocity is $\omega = d\theta/dt$ and so

$$\theta = \int \omega \, dt \times \theta_o$$

showing that the phasor position is a function of all preceding values of angular velocity.

An alternating voltage may then be defined as

$$e = A \cos \theta = A \cos \left(\int \omega \, dt + \theta_o \right) \tag{16-1}$$

There appear two parameters of an ac wave which may be varied or modulated. These are the amplitude A and the angle, resulting in two basic types of modulation:

I. *Variation of amplitude with time, giving amplitude modulation* (AM).

II. *Variation of the angle with time, resulting in angle modulation.* There are two subtypes:

(a) Variation of $\int \omega \, dt$ with time, resulting in *frequency modulation* (FM).

(b) Variation of the phase angle θ_o with time, giving *phase modulation* (PM).

For the case of *amplitude modulation*, the angular velocity ω is a constant with value ω_c, and the *carrier frequency* is $f_c = \omega_c/2\pi$. Use of Eq. 16-1 then gives

$$e_c = A \cos (\omega_c t + \theta_o) \tag{16-2}$$

If the signal carrying the information is

$$e_m = E_m \cos \omega_m t$$

the amplitude A should vary as

$$A = E_c + k_a E_m \cos \omega_m t = E_c\left(1 + \frac{k_a E_m}{E_c} \cos \omega_m t\right) \tag{16-3}$$

The complete amplitude-modulated voltage is then written as

$$e = E_c(1 + m_a \cos \omega_m t) \cos \omega_c t \tag{16-4}$$

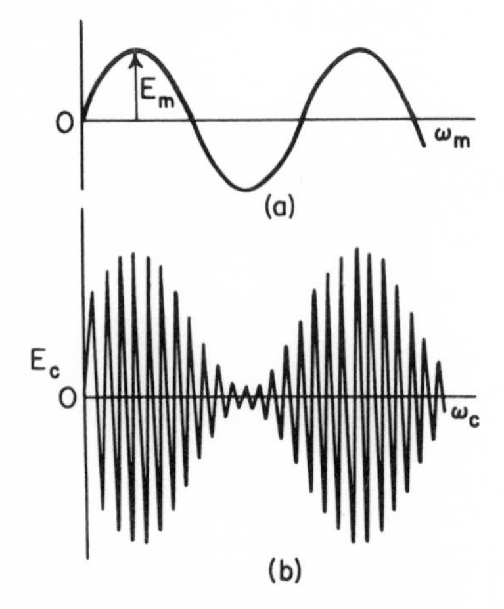

Figure 16-1. (a) The modulating wave; (b) amplitude-modulated wave.

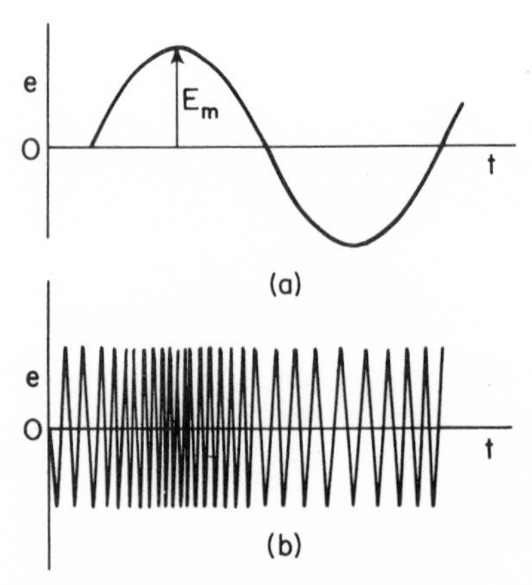

Figure 16-2. (a) Modulating wave; (b) frequency-modulated wave.

where the constant phase angle θ_o has been dropped. The *modulation factor* is defined as

$$m_a = \frac{k_a E_m}{E_c} \tag{16-5}$$

When stated in per cent, this becomes the *modulation percentage*. In usual amplitude modulation systems it is not desirable for this to exceed 100 per cent.

For case II(a), that of *frequency modulation*, the angular velocity ω is made to vary in accordance with the amplitude of the modulating signal, as

$$\omega = \omega_o + k_f E_m \cos \omega_m t \tag{16-6}$$

where $f_o = \omega_o/2\pi$ is called the *center frequency*, ω_m is the modulation angular frequency, and $k_f E_m$ is the degree of frequency variation proportional to the modulating amplitude. Then

$$e = E_c \cos \int \omega \, dt + \theta_o = E_c \cos \left(\omega_o t + \frac{k_f E_m}{E_c} \sin \omega_m t \right) \tag{16-7}$$

the constant θ_o being dropped.

The maximum value of $k_f E_m/2\pi$ is called the *frequency deviation* and designated Δf. This is the deviation in hertz from the center frequency f_o.

The ratio of Δf to the maximum modulating frequency is called the *deviation ratio* of the system, as

$$\delta = \frac{\Delta f_{\max}}{f_{m \max}} \tag{16-8}$$

Finally, the frequency-modulated wave can be written

$$e = E_c \cos (\omega_o t + \delta \sin \omega_m t) \tag{16-9}$$

The frequency deviates from the center frequency in proportion to the amplitude of the modulating signal. The amount of deviation is independent of the modulating frequency, and the modulated wave is of constant amplitude.

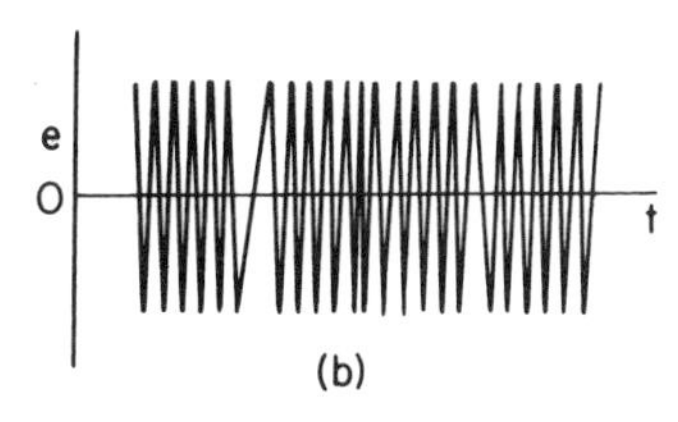

Figure 16-3. (a) Modulating wave; (b) phase-modulated wave.

For case II(b), that of *phase modulation*, ω is constant at ω_o, and the result of Eq. 16-1 is

$$e = E_c \cos (\omega_o t + \varphi)$$

Then the angle φ may be given a time variation as

$$\varphi = \varphi_o + k_p E_m \cos \omega_m t \tag{16-10}$$

where k_p is the phase angle variation per unit signal amplitude. Then

$$e = E_c \cos (\omega_o t + k_p E_m \cos \omega_m t)$$

Defining the *phase deviation* as

$$\varphi_d = k_p E_m \tag{16-11}$$

the general expression for a phase-modulated wave is

$$e = E_c \cos (\omega_o t + \varphi_d \cos \omega_m t) \tag{16-12}$$

Since both frequency and phase modulation are types of angle modulation, the similarity between Eq. 16-9 for frequency modulation and Eq. 16-12 for phase modulation is not unexpected. The difference lies in a 90° shift in the modulation phase angle and in the coefficients:

$$\text{FM:} \quad \delta = \frac{k_f E_m}{\omega_m}, \qquad \text{PM:} \quad \varphi_d = k_p E_m$$

Therefore δ is an inverse function of modulating frequency, whereas φ_d is independent of modulating frequency. Because of the similarity of the two modulation forms, it is relatively easy to convert from one form to the other, and this is the principle of the Armstrong type of frequency modulation.

16-2 FREQUENCY AND POWER SPECTRA IN AMPLITUDE MODULATION

An expression for an amplitude-modulated wave has been written as

$$e = E_c(1 + m_a \cos \omega_m t) \cos \omega_c t \tag{16-13}$$

By trigonometric identity

$$\cos (\omega_m t) \cos (\omega_c t) = \tfrac{1}{2}[\cos (\omega_c + \omega_m)t + \cos (\omega_c - \omega_m)t]$$

so that an amplitude-modulated wave becomes

$$e = E_c \cos \omega_c t + \frac{m_a E_c}{2} \cos (\omega_c + \omega_m)t + \frac{m_a E_c}{2} \cos (\omega_c - \omega_m)t \tag{16-14}$$

A wave that is amplitude-modulated by a single frequency actually consists of three frequencies. One is the original carrier frequency, ω_c, and the other two represent the sum and difference of the carrier and modulation frequencies. The frequency represented by the sum of carrier and modulation frequencies is called the *upper side frequency;* that resulting from the difference is the *lower side frequency*. For modulation by the complex wave forms of speech or music, many such side frequency pairs will exist, one pair for each frequency component in the signal; these groups of side frequencies are called *sidebands*.

The bandwidth required for transmission of an amplitude-modulated radio-frequency signal is equal to twice the highest modulating frequency. Figure 16-4 shows the frequency spectrum resulting from simultaneous modulation of a carrier with 1000 cycles and 5000 cycles. The carrier envelope

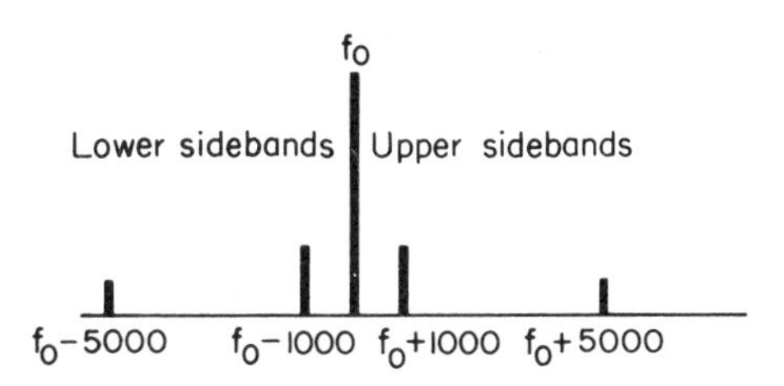

Figure 16-4. Carrier amplitude modulated at 1000 and 5000 hertz.

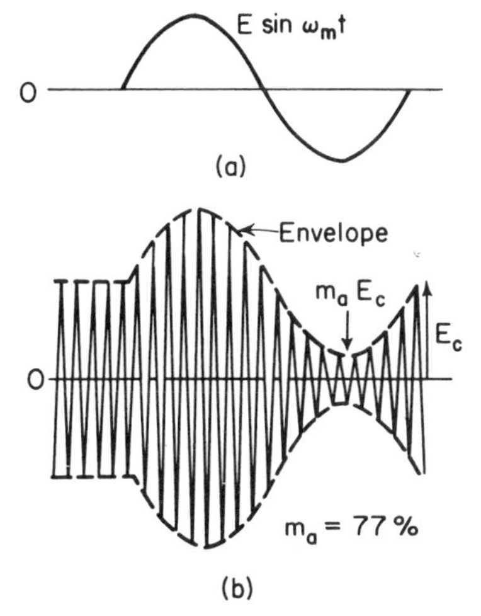

Figure 16-5. Amplitude modulated wave with $m = 77$ per cent.

produced by amplitude modulation appears in Fig. 16-5. Such a wave is not the result of addition of carrier and modulation frequencies, but occurs because of the *product* of two trigonometric functions appearing in Eq. 16-13.

From Eq. 16-14 it is apparent that the amplitude of either sideband is $m_a/2$ times the carrier amplitude. Power is proportional to the square of voltage, so that

$$\text{carrier power} = \frac{KE_c^2}{2}$$

The total sideband power is $m_a^2/2$ times the carrier power so that the total power in an amplitude-modulated wave is

$$P_{AM} = \frac{KE_c^2}{2}\left(1 + \frac{m_a^2}{2}\right) \qquad (16\text{-}15)$$

At 100 per cent modulation with a sinusoidal wave form the sideband power is 50 per cent and the total power is 150 per cent of average carrier power.

On positive modulation peaks the amplitude of the modulated wave is double the carrier level, so that instantaneous peak power is four times the carrier level.

16-3 AMPLITUDE MODULATION PRINCIPLES

Amplitude modulation may be accomplished by any process whereby an output frequency is a result of the *product* of two input frequencies. Two general methods are available for achieving this product: one involves a nonlinear relation between voltage and current in a circuit device; the second permits use of a circuit having a linear relation between input and output power.

A transistor or vacuum tube can be assumed as having a nonlinear volt-ampere relation between input and output as

$$i = a_o + a_1e_s + a_2e_s^2 + \dots \tag{16-16}$$

and if a two-frequency input be applied as

$$e_s = E_c \cos \omega_c t + E_m \cos \omega_m t$$

an expression of the following form results:

$$\begin{aligned} i = a_oI_o &+ \frac{a_2(E_c^2 + E_m^2)}{2} + a_1E_m \cos \omega_m t + \frac{a_2E_m^2}{2} \cos 2\omega_m t \\ &+ a_1E_c \cos \omega_c t + a_2E_cE_m \cos (\omega_c - \omega_m)t + a_2E_cE_m \cos (\omega_c + \omega_m)t \\ &+ \frac{a_2E_c^2}{2} \cos 2\omega_c t + \dots \end{aligned} \tag{16-17}$$

The first line of the expression represents dc and the modulation frequency and its second harmonic; the second line appears as a set of frequencies grouped about f_c and predicted by Eq. 16-14 as the carrier and upper and lower sidebands of an amplitude-modulated wave, and the third line represents second-order distortion. If higher-order terms had been included in Eq. 16-16, then higher harmonics of f_m would appear in the first line, and higher harmonic frequencies and sidebands would also be added to the third line.

The frequencies other than the carrier and first-order sidebands are usually unwanted and considered spurious products of the modulation process. It is usual to make f_c large with respect to f_m, and the wanted frequencies can normally be separated from the spurious products by tuned-circuit filters. Should the tuned circuits have too high a Q value, the impedance may fall at some of the further side frequencies and this will lower the effective modulation percentage for these frequencies; this result is known as *sideband clipping*.

One linear method of amplitude modulation is by power conversion. If a Class C device amplifying one frequency has its input power varied at a second frequency, the output will be modulated at the second frequency. This process includes the systems of anode or collector modulation, and grid or screen forms of efficiency modulation, in which the modulating power is made available by changing the operating efficiency of the circuit.

Another linear form of modulation causes a current of one frequency to pass through an impedance whose magnitude is varied at a second frequency. That is,

$$E = I_m \sin \omega t \times Z \sin \beta t$$

The product term appears and may be expanded to show the side frequencies. This method is used in some systems where ac current is applied to an inductor, and the reactance of the inductor varied at a second frequency.

16-4 MODULATED CLASS C AMPLIFIER

Assume a Class C amplifier with the input circuit driven by a source at carrier frequency f_c. In Class C operation the driving voltage can be increased until

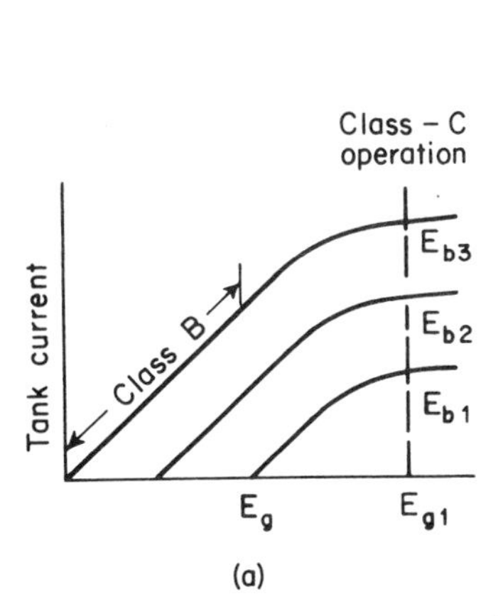

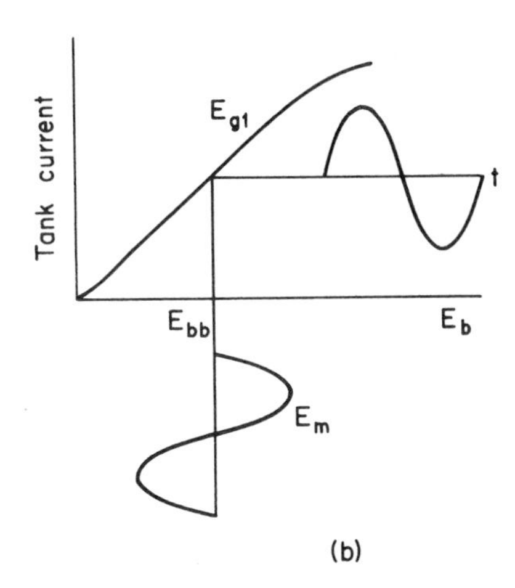

Figure 16-6. (a) Tank current vs. E_g for Class C amplifier; (b) tank current vs. E_b for a saturated Class C amplifier.

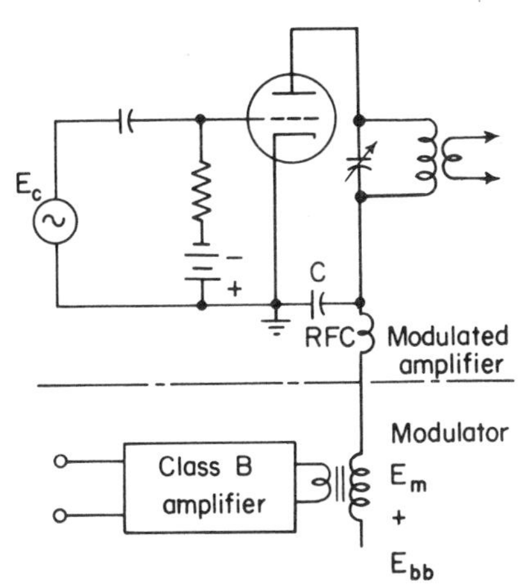

Figure 16-7. Anode modulation. C is a radio-frequency bypass capacitor.

saturation conditions are reached, as shown by the dashed line at E_{g1} in Fig. 16-6(a). Replotting of data taken for E_{g1} gives the curve of (b). This curve indicates that a linear relation can be expected between output tank current and supply voltage in a Class C amplifier driven into saturation.

If the supply voltage is then varied at a rate slow with respect to a cycle of the carrier frequency, the alternating tank current will vary linearly with E_b, and the envelope of variation will be that of the modulating signal. A sufficiently long region of linearity exists on the curve.

A circuit for varying E_b at the modulation rate, where $f_m \ll f_c$, is shown in Fig. 16-7. The *modulated amplifier* is that portion of the circuit supplied by carrier E_c, and the *modulator* is a Class B audio amplifier supplying the modulating audio signal in series with E_{bb}. The total voltage supplied to the modulated amplifier is

$$E_b = E_{bb} + E_m \cos \omega_m t = E_{bb}(1 + m \cos \omega_m t)$$

Should m exceed 1.0, or the modulation percentage exceed 100 per cent, then E_b would become negative at intervals. A time of zero output would appear, since no current can exist with E_b negative. This would cause transient damping currents in the tank circuit, creating spurious frequencies over a band near the carrier frequency and interference in adjacent radio channels.

The current, when averaged over a cycle of the carrier frequency, has two components

$$I_b' = I_b + mI_b \cos \omega_m t \tag{16-18}$$

If the modulation is linear, the cosine term adds nothing to a time average over a cycle of f_m, and a dc ammeter reads I_b' independent of the value of m. If nonlinear effects enter, the average of I_b' is different from I_b and the resultant fluctuation in dc input current is taken as an indication of improper modulation. The power input to the modulated amplifier is

$$P_{\text{in}} = \frac{1}{2\pi} \int_0^{2\pi} E_{bb}(1 + m \cos \omega_m t) I_b(1 + m \cos \omega_m t)\, d\omega_m t$$

$$= E_{bb} I_b + \frac{m^2}{2} E_{bb} I_b \tag{16-19}$$

The dc source can supply only the steady component of power, and the variation term of power input is obtained from the modulator, or

$$P_{\text{out modulator}} = \frac{m^2}{2\eta} E_{bb} I_b \tag{16-20}$$

where η is the efficiency of the modulator output transformer. Equation 16-20 represents sideband power input to the modulated amplifier.

The Class C amplifier power efficiency η_p is approximately constant over the modulation cycle and so the power output to the tank circuit is

$$P_{\text{out}} = \eta_p E_{bb} I_b \left(1 + \frac{m^2}{2}\right)$$

The loss in the modulated amplifier is then

$$P_d = P_{\text{in}} - P_{\text{out}} = E_{bb} I_b \left(1 + \frac{m^2}{2}\right)(1 - \eta_p) \tag{16-21}$$

indicating that the loss increases with m. In the unmodulated condition the loss should be only 67 per cent of dissipation rating. However, 100 per cent modulation is not continuous in speech and music, and the allowable dissipation may be somewhat increased for that service.

The impedance into which the modulator transformer works is that of the Class C stage. It can be found by

$$R_b = \frac{\text{modulation component of secondary voltage}}{\text{modulation component of secondary current}}$$

$$= \frac{mE_{bb} \cos \omega_m t}{mI_b \cos \omega_m t} = \frac{E_{bb}}{I_b} \tag{16-22}$$

Knowing the load desired for the Class B modulator it is then possible to determine the turns ratio of the modulator transformer.

16-5 EFFICIENCY MODULATION

With a sufficiently high load, a Class C amplifier driven at carrier frequency f_c will develop a linear relation between output current and bias voltage. If the bias then is made to vary at frequency f_m, as would happen in the circuit of Fig. 16-8, amplitude modulation will be obtained. The circuit operates through changing the current conduction angle at the f_m rate.

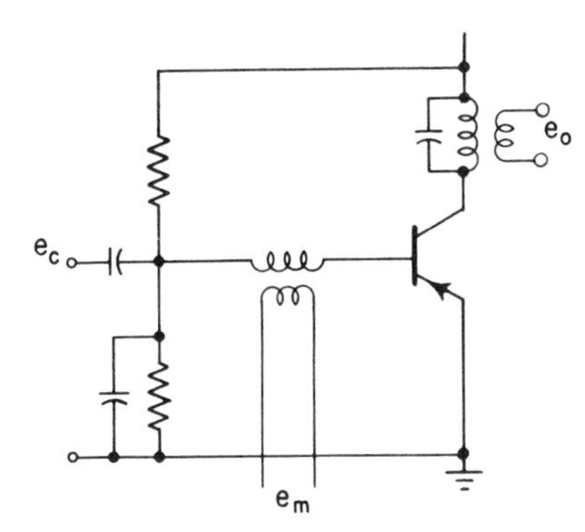

Figure 16-8. One form of efficiency or bias modulation.

The output current will be

$$i_C = Kg_mE_c \cos \omega_c t$$

but g_m will depend on the bias as

$$g_m = kV_{BB} + k_mE_m \cos \omega_m t$$

so that

$$i_C = KE_c(kV_{BB} + k_mE_m \cos \omega_m t) \cos \omega_c t \tag{16-23}$$

which is of the basic product form, indicating amplitude modulation.

Under carrier only, or zero modulation conditions, the power output will be one-fourth the peak value and the voltage peak across the load will be one-half the supply voltage. The minimum collector or anode voltage will be high and the efficiency of the order of 30 per cent. On positive peaks, as in Fig. 16-9, the output voltage doubles or approaches the supply voltage and the efficiency doubles to 60 per cent. The power supplied for the sidebands of the modulated wave is obtained by changes in circuit efficiency, shifting

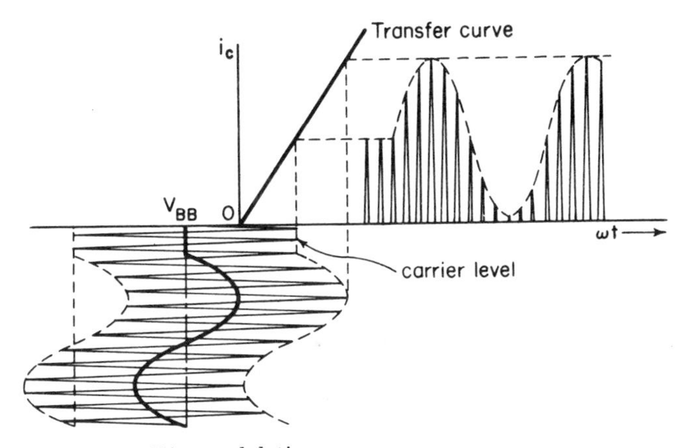

Figure 16-9. Bias modulation.

power between the losses and the load. For this reason the system is sometimes called *efficiency modulation*. Because of the poor average efficiency the method is only employed in simple systems at low power level.

Screen or suppressor modulation of tetrodes or pentodes is also a form of efficiency modulation. Suppressor modulation is fairly easy to adjust because the bias and signal voltages are applied in separate circuits, reducing interaction.

16-6 SSB SYSTEMS

The carrier in amplitude modulation conveys no information and thus might be considered an inefficient use of power. It is also apparent that the second sideband conveys redundant information (overlooking frequency-selective fading in an ionospheric transmission path which may distort one sideband at times). The band occupied by an AM wave in the frequency spectrum can be halved if only one sideband is transmitted without the carrier. The result is a *single-sideband suppressed-carrier* signal, usually referred to as *single-sideband* (SSB) *transmission*.

The carrier must be reintroduced at the receiver for such systems, and must be synchronized closely with the original carrier oscillation to avoid signal distortion. Lack of exact synchronism produces a shift in each frequency component which is additive or subtractive, so that harmonic components are no longer harmonically related. It is found that the reintroduced carrier must be within a few tens of hertz of the original carrier to preserve intelligibility.

Two major methods are used at present to separate the sidebands. Both systems employ a *balanced modulator* to remove the carrier from the signal. The *phasing method* of SSB generation develops the two sideband sets without a carrier, but with the sidebands out of phase. Addition or subtraction of these sidebands from the original set then removes the upper or lower sideband as desired.

The *filter method* generates the sidebands by removal of the carrier, and then discards one sideband by use of a narrow-band filter. A filter problem arises because the lowest frequency in the unwanted sideband is separated from its counterpart frequency in the wanted sideband by only twice the lowest modulating frequency. The separation of these frequencies is a severe filter requirement, since it may be desired that the unwanted sideband be suppressed with respect to the wanted sideband by a ratio as high as 50 dB.

To solve the problem many systems cut off the speech pass band at 200 hertz, so that a 400 hertz difference exists between the nearest frequencies. Suitable filters can then be designed to separate the sidebands when in the range from 20 kilohertz to 5 megahertz. The desired final frequency of the sideband to be transmitted is obtained by translation or heterodyning, a process to be discussed later.

Single-sideband signals have other communications advantages beyond the reduction of transmission bandwidth. In double-sideband amplitude modulation it is the average carrier power that is limiting. The total sideband power is 50 per cent of the carrier power at 100 per cent modulation and the total DSB AM average power is then 150 per cent of carrier power.

For SSB the transmitter does not need to generate carrier power, and ratings are in terms of *peak envelope power* (PEP), the power capability at the peak of the modulating signal with linearity of the amplifier maintained. For equal sideband power, a DSB AM system will require 1.5 times the average carrier power at 100 per cent modulation, whereas the SSB system will require only 0.5 of the carrier power of the DSB system. Thus the total average power of the DSB system must be three times the average power of the SSB system, assuming equal modulation powers.

At the receiver, each sideband of the DSB signal carries 25 per cent of the AM carrier power, and provides an output voltage proportional to $\sqrt{0.25}$ or 0.5. The two sideband output voltages are additive and so the receiver output can be taken as 1.0. The SSB signal has a power equal to 50 per cent of the AM carrier, and provides an output voltage proportional to $\sqrt{0.5} = 0.707$. It appears that AM has an advantage of 3 dB (20 log 1.0/0.707) in received signal. However, this advantage is lost when practicalities enter and signal-to-noise ratio is introduced. Since the DSB AM signal requires a received bandwidth twice that of the SSB signal and

$$\text{for DSB: } \frac{S}{N} = \frac{1.0}{n\sqrt{2B}} \qquad \text{for SSB: } \frac{S}{N} = \frac{0.707}{n\sqrt{B}}$$

the two systems will produce equal received signal-to-noise ratios. If the average input to the SSB system is made equal to the AM carrier power input, the SSB signal has a 3 dB advantage.

However, speech is not a continuous sine wave, and its average power is rather low with respect to its peak requirements. A peak-to-average power ratio of 10:1 is often indicated for speech. Under that condition, a DSB signal would require an average input of 1.05 units compared to an SSB input of 0.05 units.

Another limiting factor on transmitter performance is the voltage breakdown of output circuit components. At 100 per cent modulation the DSB powers are respectively 1.0, 0.25, 0.25 for carrier and sidebands. The voltages are additive at the peak and the circuit voltage for which insulation must be provided is proportional to the square root of the powers as $1.0 + \sqrt{0.25} + \sqrt{0.25} = 2.0$. For an SSB signal the relative power is 0.5 and the circuit voltage is proportional to 0.707. The relative system voltages are in the ratio of $2.0/0.707 = 2.83$. Thus components designed for lower working voltages can be used in SSB equipment, reducing both size and cost.

The received carrier represents an unneeded signal in reception, and two adjacent carriers, differing by an audio frequency, can produce an unwanted audio tone in reception. The freedom from such "beats" or whistle frequencies is a major subjective factor giving SSB another advantage over AM transmission with carrier.

16-7 THE BALANCED MODULATOR

A *balanced modulator* circuit in the form of Fig. 16-10 is employed to generate an AM signal without carrier, and provides the first step in SSB generation. Operation occurs over a device region having a transfer characteristic expressible by

$$i = a_1 e_i + a_2 e_i^2 + \ldots \qquad (16\text{-}24)$$

The modulation frequency and carrier frequency voltages introduced to Q_1 are

$$e_1 = E_m \sin \omega_m t + E_c \sin \omega_c t$$

and to Q_2

$$e_2 = -E_m \sin \omega_m t + E_c \sin \omega_c t$$

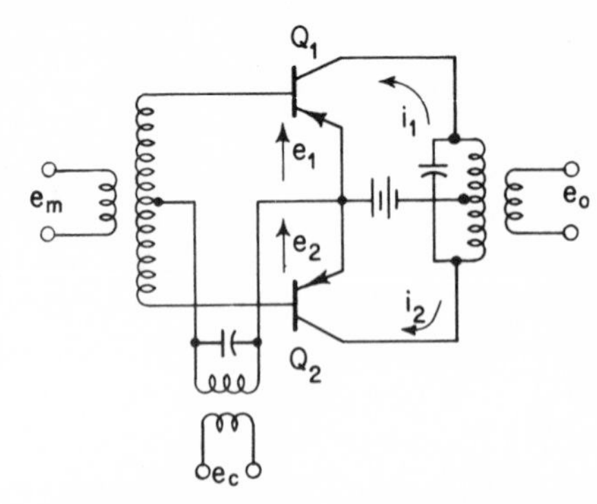

Figure 16-10. The balanced modulator.

The ac output currents then are

$$i_1 = a_1 E_m \sin \omega_m t + a_1 E_c \sin \omega_c t + a_2 E_m^2 \sin^2 \omega_m t + 2a_2 E_m E_c \sin \omega_m t \sin \omega_c t + a_2 E_c^2 \sin^2 \omega_c t + \ldots \qquad (16\text{-}25)$$

$$i_2 = -a_1 E_m \sin \omega_m t + a_1 E_c \sin \omega_c t + a_2 E_m^2 \sin^2 \omega_m t - 2a_2 E_m E_c \sin \omega_m t \sin \omega_c t + a_2 E_c^2 \sin^2 \omega_c t + \ldots \qquad (16\text{-}26)$$

The output from the transformer is proportional to the difference of these two currents, and is given by

$$e_o = 2k(a_1 E_m \sin \omega_m t + 2a_2 E_m E_c \sin \omega_m t \sin \omega_c t + \ldots) \qquad (16\text{-}27)$$

A tuned circuit at the output, resonant to the carrier frequency, will reject the modulation frequency term ω_m, provided that $\omega_c \gg \omega_m$. It will also reduce the effect of any components at multiples of ω_c, due to higher even-order terms in the transfer characteristic. Odd-power terms are balanced out by the nature of the circuit.

The remaining frequency term may then be considered as

$$e_o = 2ka_2 E_m E_c[\cos (\omega_c + \omega_m)t - \cos (\omega_c - \omega_m)t] \qquad (16\text{-}28)$$

which represents a signal containing only the sidebands of an AM wave.

The active devices of a balanced modulator are often diodes, since it is easier to match their characteristics and to approximate the desired square-law characteristic. Equivalence of the diodes in terms of power series coefficients is assumed, and it is also necessary to have a balanced circuit in terms of couplings, resistances, and capacitances to avoid a feed-through of the unwanted carrier. Manual balance adjustments are usually required.

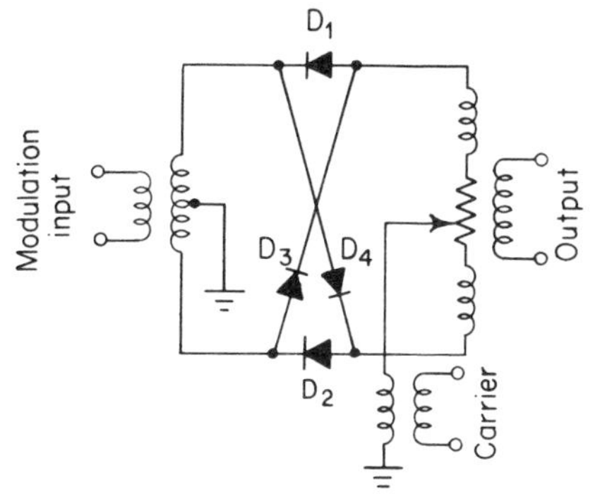

Figure 16-11. Diode-balanced modulator.

Carrier voltage should be large with respect to the peak amplitude of the modulating signal, so that the diodes are quickly switched from open to closed condition.

Figure 16-11 illustrates a *ring modulator* commonly used in telephony for carrier elimination. It makes unnecessary an exact balance of transformers, if the carrier balance control is properly set.

16-8 SSB BY THE PHASING METHOD

Generation of an SSB signal by the phasing method requires two balanced modulators and two phase-shift networks, one giving a 90° shift over the modulation band, the other a 90° shift for the carrier frequency. Because of the necessity for phase shifting over a frequency band, the shift of modulation frequencies is best accomplished with +45° and −45° shifters in the respective branches, as shown in the block diagram of Fig. 16-12.

It may again be assumed that the modulator devices have square-law characteristics with equations of the form

$$i = a_1 e_s + a_2 e_s^2 \tag{16-29}$$

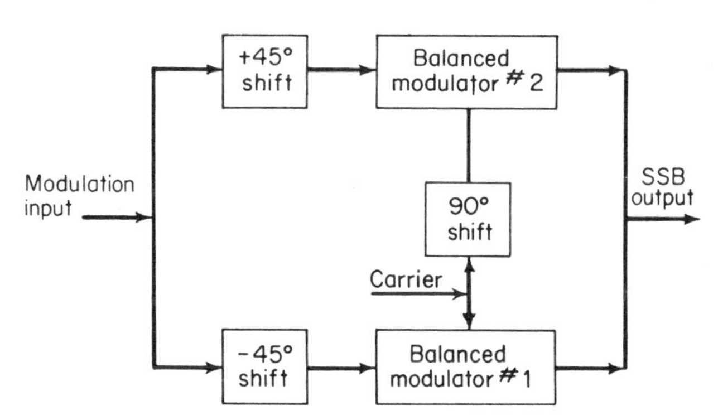

Figure 16-12. Block diagram of phasing method of SSB generation.

For modulator (1) the two inputs are

$$e_{s1} = E_m \sin \omega_m t + E_c \sin \omega_c t$$

$$e_{s2} = -E_m \sin \omega_m t + E_c \sin \omega_c t$$

Using these inputs and with an output circuit resonant to ω_c, the output is

$$E_{o1} = 2ka_2 E_m E_c[\cos (\omega_c + \omega_m)t - \cos (\omega_c - \omega_m)t] \qquad (16\text{-}30)$$

For modulator (2) the two inputs are

$$e'_{s1} = E_m \cos \omega_m t + E_c \cos \omega_c t$$

$$e'_{s2} = -E_m \cos \omega_m t + E_c \cos \omega_c t$$

and again using Eq. 16-29, the output of the second modulator can be written as

$$E_{o2} = 2ka_2 E_m E_c[\cos (\omega_c + \omega_m)t + \cos (\omega_c - \omega_m)t] \qquad (16\text{-}31)$$

If the outputs of the two modulators are added

$$E_{o1} + E_{o2} = Ka_2 E_m E_c \cos (\omega_c + \omega_m)t \qquad (16\text{-}32)$$

and the upper sideband is obtained. Subtraction of the outputs yields the lower sideband as

$$E_{o1} - E_{o2} = Ka_2 E_m E_c \cos (\omega_c - \omega_m)t \qquad (16\text{-}33)$$

16-9 FILTER METHOD OF SSB GENERATION

When the sidebands of the balanced modulator output are separated by a filter, that network must have a cutoff frequency and attenuation slope which will separate signals differing only by twice the lowest modulation

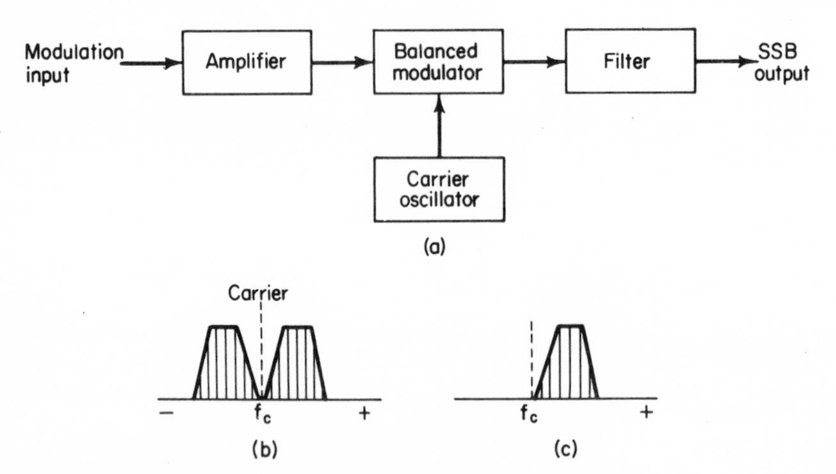

Figure 16-13. (a) Filter method of SSB generation; (b) balanced modulator output; (c) output after filter.

frequency. There are available quartz crystal or mechanical resonator filters which will reduce the unwanted sideband as much as 50 dB.

A block diagram appears in Fig. 16-13. Without infinitely steep cutoff slope, it is customary to place the filter cutoff so that the carrier frequency would be about 20 dB below the wanted sideband level, as is demonstrated in Fig. 16-14. The response curve is for a mechanical resonator filter. Operating at 455 kHz, it employs elastances (capacitance) or spring elements, and masses (inductance) in an assembly analogous to an electrical filter, with the Q of mechanical elements very high. Using well-designed mechanical or quartz crystal filters, a typical response for SSB is 2.6 kHz wide at 6 dB below pass band level, and only 4.2 kHz wide at 60 dB below that level.

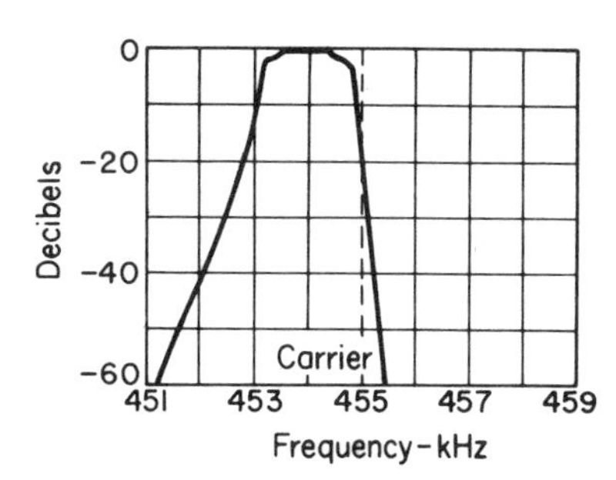

Figure 16-14. Response of mechanical filter at 455 kHz.

16-10 FREQUENCY SPECTRUM IN FREQUENCY MODULATION

It was shown in Section 16-1 that a frequency-modulated wave could be written as

$$e = E_c \sin(\omega_o t + \delta \sin \omega_m t)$$

This expression is of the form $\sin(a + b)$, and can be expanded as

$$\sin(\omega_o t + \delta \sin \omega_m t) = \sin(\omega_o t)\cos(\delta \sin \omega_m t) + \cos(\omega_o t)\sin(\delta \sin \omega_m t) \qquad (16\text{-}34)$$

These functions can be evaluated by means of Bessel coefficients, where

$$\cos(\delta \sin \omega_m t) = J_o(\delta) + 2J_2(\delta)\cos 2\omega_m t + 2J_4(\delta)\cos 4\omega_m t + \dots \qquad (16\text{-}35)$$

$$\sin(\delta \sin \omega_m t) = 2J_1(\delta)\sin \omega_m t + 2J_3(\delta)\sin 3\omega_m t + 2J_5(\delta)\sin 5\omega_m t + \dots \qquad (16\text{-}36)$$

The J_n's are Bessel functions of the first kind and order n, and are defined by an infinite series given by

$$J_n(\delta) = \sum_{k=0}^{\infty} \frac{(-1)^k}{k!(k+n)!}\left(\frac{\delta}{2}\right)^{n+2k} \qquad (16\text{-}37)$$

Values of $J_n(\delta)$ are tabulated in various mathematical tables and a brief table is given in the Appendix.

By use of

$$\sin a \cos b = \tfrac{1}{2}[\sin(a + b) + \sin(a - b)]$$

$$\cos a \sin b = \tfrac{1}{2}[\sin(a + b) - \sin(a - b)]$$

the frequency-modulated wave becomes

$$e = E_c\{J_o(\delta) \sin \omega_o t + J_1(\delta)[\sin (\omega_o + \omega_m)t - \sin (\omega_o - \omega_m)t]$$
$$+ J_2(\delta)[\sin (\omega_o + 2\omega_m)t + \sin (\omega_o - 2\omega_m)t]$$
$$+ J_3(\delta)[\sin (\omega_o + 3\omega_m)t - \sin (\omega_o - 3\omega_m)t]$$
$$+ J_4(\delta)[\sin (\omega_o + 4\omega_m)t + \sin (\omega_o - 4\omega_m)t] + \ldots\} \tag{16-38}$$

The frequency-modulated wave is composed of a center frequency $\omega_o/2\pi$, and an infinite set of side frequencies, each pair spaced by an amount equal to the modulating frequency f_m. Fortunately, for $\delta \gg 5$, and $n > \delta$, the coefficients of $J_n(\delta)$ decrease quite rapidly and the useful FM bandwidth can then be taken as

$$BW \cong 2\delta f_{m\ \max} = 2\Delta f_{\max} \tag{16-39}$$

Observation of Fig. 16-15 confirms this. More exact analysis, including all side frequencies having amplitudes over 0.01, shows the following required bandwidths:

δ	BW	δ	BW
1.0	$3.0 \times 2f_m$	8.0	$11.2 \times 2f_m$
2.0	$4.3 \times 2f_m$	10.0	$13.5 \times 2f_m$
4.0	$6.8 \times 2f_m$	15.0	$20 \times 2f_m$
		20	$24 \times 2f_m$

Thus Eq. 16-39 is a reasonable approximation at $\delta = 20$, and above.

For small deviation ratios with $\delta <$ 0.6, the spectrum reduces to

$$e = E_c J_o(\delta) \cos \omega_c t + J_1(\delta)[\sin (\omega_o + \omega_m)t - \sin (\omega_o - \omega_m)t]$$

which is similar to double-sideband AM, but with phase-shifted carrier. The bandwidth then is $BW = 2f_{m\ \max}$, and the system is termed *narrow-band FM*. For $\delta >$ 0.6, the system is considered *wide-band*.

For FM broadcasting the Δf range is usually ± 75 kHz, and receiver bandwidths are designed for ± 100 kHz. The maximum modulating frequency is considered 15 kHz, so that $\delta \cong 5$. The signal band would then be near 200 kHz, and

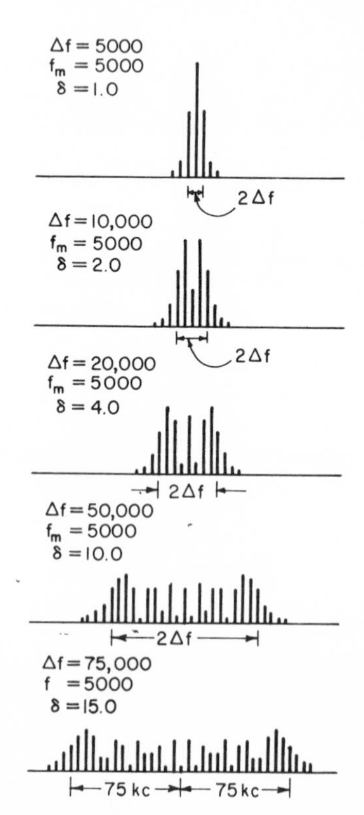

Figure 16-15. Spectra of FM waves, showing effect of changing δ.

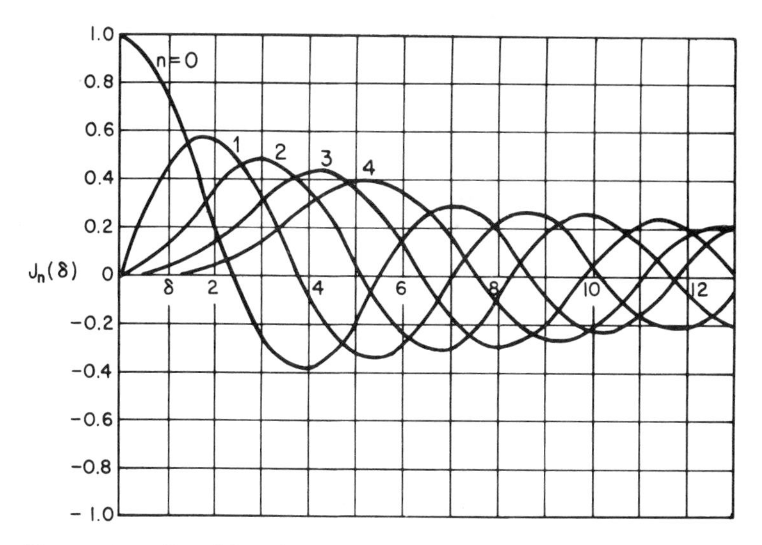

Figure 16-16. Bessel functions.

would fit within the receiver design band.

The function J_0 becomes zero or has roots at δ equal to 2.40, 5.52, 8.65, 11.79, 14.93, . . . , as shown in Fig. 16-16. At those values the center frequency does not exist. In Fig. 16-15 for $\delta = 15$, the center frequency term is small, since $\delta = 15$ is close to the root at 14.93.

16-11 FM GENERATION BY REACTANCE VARIATION

Variation of the reactance in the tuned circuit of an oscillator at the modulation frequency will frequency modulate the oscillator frequency. It is possible to employ a tube or transistor as such a variable reactance, one form appearing in Fig. 16-17. The variable reactance acts across terminals

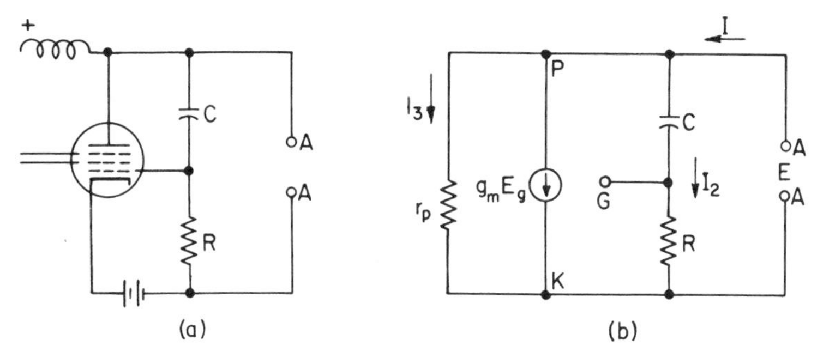

Figure 16-17. (a) Reactance-tube circuit; (b) equivalent circuit.

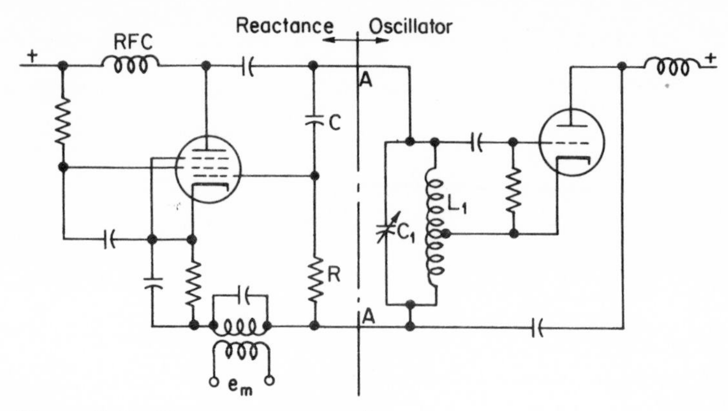

Figure 16-18. Use of a reactance tube.

A, A when these terminals are shunted across the oscillator tank as in Fig. 16-18.

The reactance of C must be large with respect to R, so that the current $\mathbf{I}_2$ will be in quadrature to $\mathbf{E}$. Neglecting r_p of the pentode, the current $\mathbf{I}$ is

$$\mathbf{I} = j\omega C\mathbf{E} + g_m\mathbf{E}_g$$

The voltage at the grid is $\mathbf{E}_g \cong j\omega C\mathbf{E}R$, so that

$$\mathbf{I} \cong j\omega C(1 + g_m R)\mathbf{E} \tag{16-40}$$

Therefore the circuit represents an effective capacitance

$$C_e = C(1 + g_m R) \cong g_m RC \tag{16-41}$$

The capacitance is controllable by g_m, which is a function of the grid bias, and g_m can be varied by the modulating signal, as in Fig. 16-18.

Linearity of frequency deviation vs. bias voltage can be secured by making the bias change small. The value of g_m under modulation is

$$g_m = g_e(1 + k_f E_m \cos \omega_m t)$$

and the oscillator frequency will be

$$\omega = \sqrt{\frac{1}{L_1[C_1 + RCg_e(1 + k_f E_m \cos \omega_m t)]}} \tag{16-42}$$

The center frequency with zero modulation will be

$$\omega_0 = \sqrt{\frac{1}{L_1(C_1 + RCg_e)}}$$

The shift due to modulation then is

$$\frac{\omega}{\omega_0} = \sqrt{\frac{1}{1 + \dfrac{k_f E_m \cos \omega_m t}{1 + C_1/RCg_e}}} \tag{16-43}$$

The frequency shift is small, and the above may be expanded by the binomial theorem. The first two terms give

$$f \cong f_o\left[1 - \frac{k_f}{2(1 + C_1/RCg_e)} E_m \cos \omega_m t\right] \tag{16-44}$$

which is a frequency-modulated wave with

$$\delta = \frac{k_f E_m}{2(1 + C_1/RCg_e)} \tag{16-45}$$

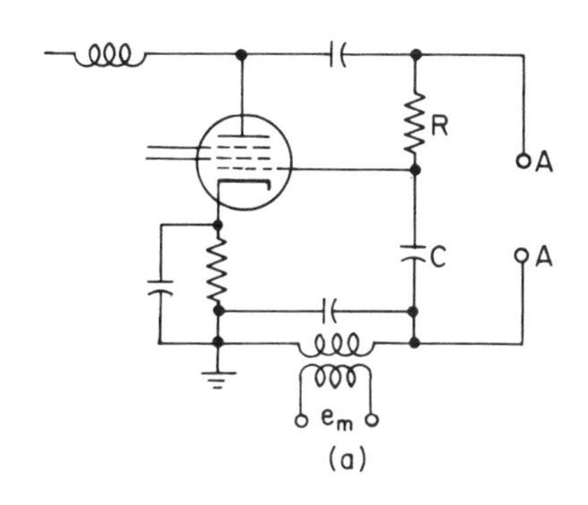

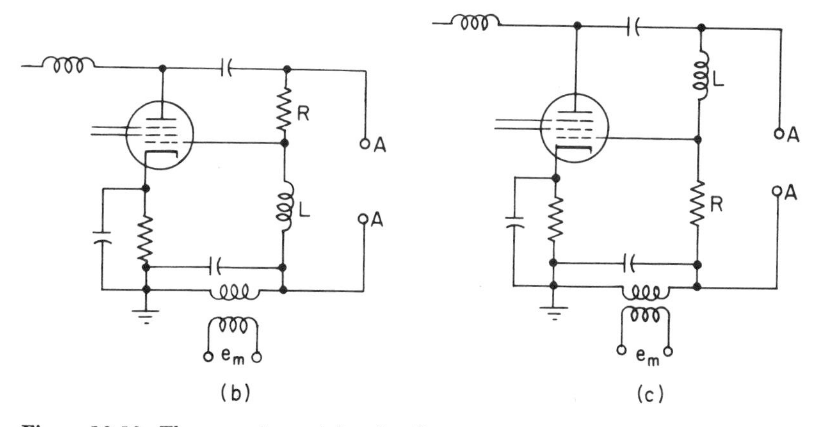

Figure 16-19. Three reactance-tube circuits.

Three other basic reactance-tube circuits are shown in Fig. 16-19. Those at (a) and (c) produce variable inductance, that at (b) is capacitive in effect.

16-12 DIODE FM GENERATION

Since semiconductor diode capacitance is a function of reverse voltage, the diode forms a suitable means of frequency modulation of an oscillator, as shown in Fig. 16-20. A polarizing voltage V_R maintains reverse bias in series with e_m. The choke RFC keeps the polarizing circuit from affecting the higher frequency oscillator currents. The capacitance of the junction is proportional to $V_B^{-1/2}$, and the oscillator frequency then is

$$\omega = \frac{1}{L(C + KV_B^{-1/2})} \qquad (16\text{-}46)$$

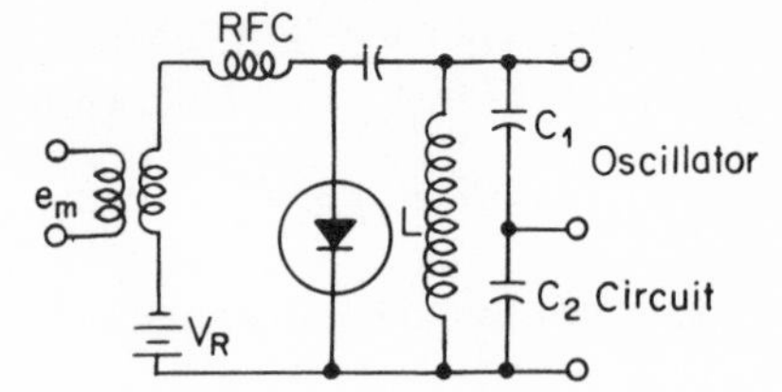

Figure 16-20. Diode FM modulator.

In order to achieve linearity of frequency deviation with V_B, it is necessary to keep the effect of $KV_B^{-1/2}$ small.

16-13 FM SYSTEMS

A block diagram of a complete FM transmitter, using a reactance-tube circuit is shown in Fig. 16-21. The stability of the center frequency is dependent on the stability of oscillator and reactance tube, and does not usually meet channel tolerances which allow ± 2 kilohertz at 100 megahertz. It is customary to employ a frequency-stabilizing system, in which a portion of the output is sampled and applied to a circuit called a *frequency discriminator*, which gives a dc output voltage proportional in magnitude and polarity to the instantaneous shift of the modulated wave from the center frequency. This dc voltage is averaged and applied as additional bias to the reactance tube, so as to shift the average frequency in a direction to correct the drift from true center.

The above method employs a tuned circuit in the discriminator as the standard against which the frequency is compared. In other equipment, the output frequency is compared against a crystal oscillator, with the difference signal being rectified and averaged for use as a correction bias for the reactance tube modulator.

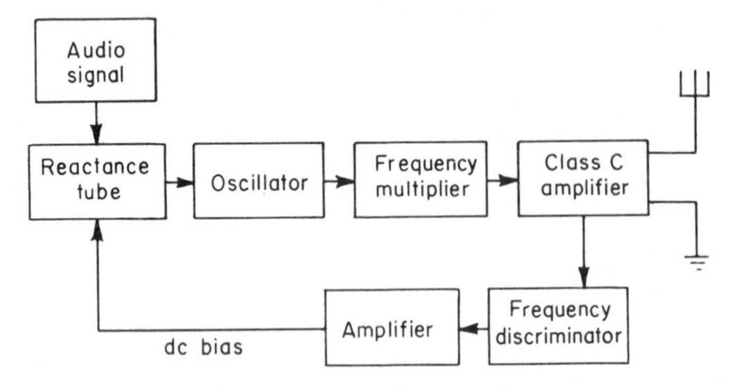

Figure 16-21. Block diagram of an FM transmitter.

The *Armstrong system* employs a crystal oscillator, with stable center frequency. The system operates on the audio signal with a weighting network, so that the modulating signal amplitude becomes inversely proportional to frequency. By phase modulation with this signal, it is possible to generate a synthetic FM wave.

It may be seen from Fig. 16-16 that the Bessel coefficients for $\delta < 0.5$ are significant only for $n \leqq 2$. Replacing δ with φ_d, and for $\varphi_d < 0.5$, $n \leqq 2$, Eq. 16-38 can be written

$$e = E_c\{J_o(\varphi_d) \cos \omega_o t + J_1(\varphi_d)[\sin (\omega_o + \omega_m)t - \sin (\omega_o - \omega_m)t] \\ + J_2(\varphi_d)[\sin (\omega_o + 2\omega_m)t + \sin (\omega_o - 2\omega_m)t)\} \qquad (16\text{-}47)$$

If φ_d is small, then J_2 will be small, and this PM wave will seem to be the AM wave of Eq. 16-14, with the carrier shifted 90° with respect to the sidebands. Thus an amplitude-modulated wave may be produced, the carrier removed in a balanced modulator, and the sidebands recombined with a 90° shifted carrier, the result being a PM wave. If, then, the modulating frequency is given an amplitude weighting which varies inversely with frequency, or φ_d replaced with δ, the wave becomes a frequency-modulated one.

Figure 16-22 illustrates the result. At (a) the two side frequency phasors rotate in unison but in opposite directions, and the sum of the carrier and

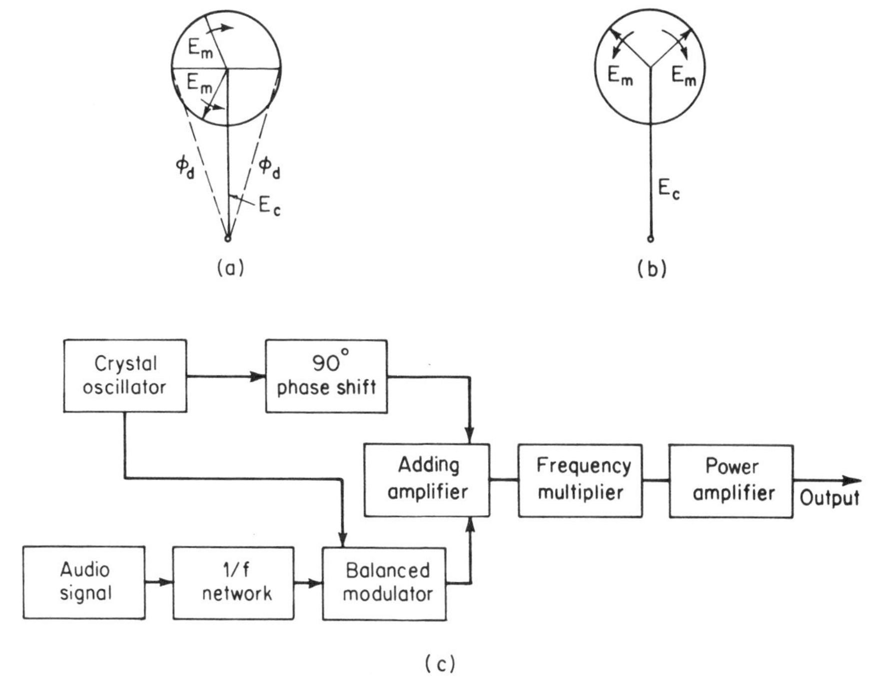

Figure 16-22. (a) PM; (b) AM; (c) the Armstrong system.

sidebands causes the carrier to change phase by the angle φ_d. It will be noted that the phase relation of carrier and side frequencies is 90° with respect to the AM situation at (b); in the latter case only the amplitude of the carrier is varied, and not the phase.

In the block diagram at (c), the audio signal is frequency-weighted and applied to the balanced modulator. The output E_B then consists only of side frequencies, and these are recombined with the shifted carrier in the adding amplifier. The whole process is performed at low frequency, and frequency multiplication is used to produce the desired frequency deviations of thousands of hertz.

16-14 AM VERSUS FM WITH RESPECT TO INTERFERENCE

Noise signals caused by atmospheric static or man-made electrical discharges create reception problems for radio signals. If a receiver is designed for reception of FM signals with wide frequency bandwidth and is made insensitive to amplitude variations of the signal, such random interference can be largely eliminated. Thus FM has a major advantage over AM for noisy channel use, particularly in the mobile service.

Interference between two simultaneously received signals on the same or adjacent channels is a major difficulty in radio reception. In this respect FM also has advantages over AM. In Fig. 16-23(a), one signal as represented by phasor A rotates about point O. A second signal of amplitude B is received simultaneously, so that the sum voltage applied to the receiver input is R. As phasor B rotates through its cycle, the amplitude of R will change greatly; thus the input to the receiver is varied in amplitude by signal B, and in an AM receiver the B signal may cause appreciable amplitude interference with A even though B is quite small.

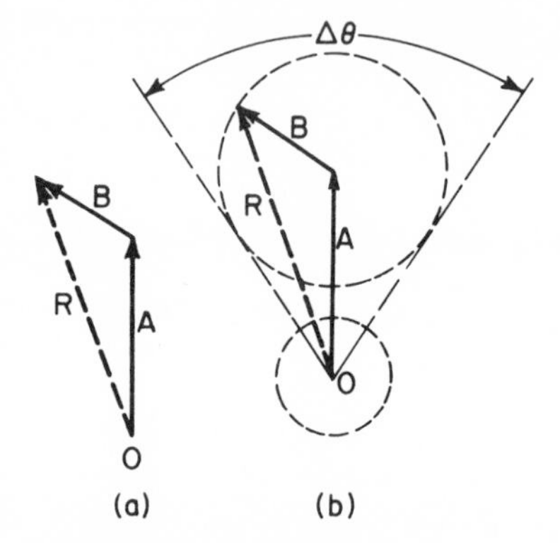

Figure 16-23. Effect of interference between two FM signals.

If signal A is frequency-modulated, as in Fig. 16-23(b), and an interfering FM signal B is added, the resultant input to the receiver will be R. Phasor B will rotate about the end of A, and the resultant R will vary in amplitude and also in relative phase angle. For the amplitudes shown, the variation $\Delta\theta$ approximates 1 radian, as indicated by the circle locus. Even if the two signals are equal in amplitude, the phase-angle variation cannot be much greater, although the amplitude variation would then be so severe that AM reception would be impossible.

An FM receiver can be made insensitive to amplitude variations by clipping all signals to a common low level; this effect is indicated by reducing the signals to an amplitude fixed by the small circle with center at O. Amplitude variations, due either to amplitude modulation on the signal or to amplitude variation of R caused by the interfering FM signal, can be removed. The voltage effective on the receiver would then be fixed by this dashed circle, but varying about $\pm\frac{1}{2}$ radian in phase with respect to phasor A. If the original frequency modulation causes its phase angle to vary through $2\pi\ \Delta f/f_m$ radians per hertz, or 10π radians for $\Delta f = 75{,}000$ hertz and $f_m =$ 15,000 hertz, it can be seen that the phase variation of $\pm\frac{1}{2}$ radian due to interference is quite small. If $f_m = 150$ hertz for the same value of Δf, then the desired signal phase variation would be 500π radians, still further reducing the effect of the interference.

The effect of the interfering signal can be reduced by widening the frequency band of the signal or increasing Δf, sufficiently strong signals being assumed. E. H. Armstrong is credited with demonstrating that increased bandwidth of signal does reduce interference when the signal is definitely above the noise level.

16-15 AMPLIFICATION OF MODULATED SIGNALS

It would appear that amplitude modulation at a low signal level, followed by amplification to raise the power level, would provide a saving in power and cost over the high-level AM methods discussed. If such amplification is undertaken with a Class C amplifier, operating into saturation, variations in input voltage will have only small effect on the output, or the modulation would be stripped off the carrier.

However, the output voltage of the Class B radio-frequency amplifier is a direct function of input voltage, so that such a circuit will properly amplify an AM signal. Such linear Class B circuits are widely used, particularly to raise the power level of SSB signals originally generated at low level.

There is no such problem in amplification of FM signals, since the amplitude is constant. Class C amplifiers, with their high efficiency, are used to raise the power level of FM signals.

16-16 INFORMATION CONTENT AND CHANNEL CAPACITY

It may be assumed that the cost of a given system for transmission of information is proportional to the bandwidth required. A system requiring a given bandwidth may be justified, however, if it can be shown that the information transmittable per unit time is greater than for some other narrower band system. A measure of the information content of signals and their relation to bandwidth or channel requirements is an economic necessity in the selection of optimum modulation and transmission methods.

There are three signal characteristics which must be included in such a measure: the frequency band required, the time needed for transmission, and the system requirements for recognition of levels of signal intensity.

Speech is our most usual method of information exchange. As a continuous function its bandwidth is equal to the difference between maximum and minimum frequency components present. Its speed of transmission is limited largely by the muscles of the mouth, and its range of intensity variation may be 1000:1 or 30 dB. Thus speeeh becomes a system of limited bandwidth but very large intensity range.

Signals coded into pulses may be found to have much different requirements. The bandwidth for perfect pulse reproduction is theoretically infinite, but the number of recognizable signal levels may be reduced to two. Thus pulse transmission is a system requiring wide bandwidth, but with limited requirements for recognition of signal intensity levels.

Approximations may be made in these methods, which reduce the system requirements. It is found that in speech the ear can recognize threshold changes in intensity which approximate twice the power; these steps of just-audible intensity change may be taken as approximately 2.5 dB. For a signal-intensity ratio of 30 dB, it is then sufficient that we recognize about 12 levels. Naturalness or fidelity may be lost, but intelligibility will not be decreased.

Sensitivity of the several systems to noise or random signals must also be considered, since this will control the rate of information accurately received, and determine the power level required for transmission. Such sensitivity to noise is much greater for speech, where we are attempting to recognize a large number of intensity levels, than it is for pulses where we note only their presence or absence, i.e., only two levels.

It is apparent that measures of information content of a signal, rate of information transmittal, and efficiency of use of channel bandwidth would be very helpful in system design. Such measures are available in a portion of our science known as *information theory*.

The measure of information quantity is defined by the number of possible values, states, or levels which a signal may have in a basic interval of the signal. This number of states or levels is defined as the equivalent number of ideal on-off signals, or binary digits, which would be required if it were to be so transmitted. That is, the *quantity of information* in a basic signal interval is

$$I_o = \log_2 L \tag{16-48}$$

where the unit of I_o is the *bit* (*bi*nary digi*t*) and L is the number of states or levels distinguished by the receiver.

As an example, consider speech which has been sampled at 2.5 dB levels, as discussed above. There would be 12 such volume levels in the 30 dB dynamic range covered by the voice and recognized by the receiver. The information content of speech would then be *defined* as

$$I_o = \log_2 12 = 3.58 \qquad \text{bits}$$

For a teletypewriter code in which five digit positions are represented by holes punched in the transmission tape, there are $2^5 = 32$ possible states of information per basic interval, or per code. This provides for the 26 letters of the alphabet plus start, stop, shift, and other control signals. The quantity of information per basic interval is then

$$H = \log_2 32 = 5 \qquad \text{bits}$$

The quantity of information transmitted per unit time is the *information rate* in bits per second. For speech, it has been demonstrated that signals should be sampled at a rate equal to twice the highest frequency component present (thus providing at least one sample for each half-cycle), and this provides a measure of the number of speech intervals present. Although speech contains frequencies up to 6000 hertz, it is found that intelligibility is maintained even when the frequency band is limited to 250–3000 hertz, as is done in telephony. With 12 distinguishable levels and $2750 \times 2 = 5500$ basic intervals per second, the rate of transmission of speech information is,

$$R = 5500 \log_2 12 = 19{,}700 \qquad \text{bits/s}$$

For the teletypewriter code, the usual transmission speed is 60 five-letter words per minute. Allowing an additional space code for each word gives a rate of six codes per second. Then

$$R = 6 \log_2 32 = 30 \qquad \text{bits/s}$$

It is of interest to analyze a television signal in a similar manner. It may be assumed that the eye can distinguish 10 gradations of light intensity. The quantity of information contained in a given picture element or dot is

$$H = \log_2 10 = 3.32 \qquad \text{bits/picture element}$$

There are 525 lines per frame, and it may be assumed that resolution and circuit bandwidth are such that each line will have about 500 identifiable dots. There are 30 frames per second, so that the number of picture elements or dots per second is

$$n = 525 \times 500 \times 30 = 7.875 \times 10^6$$

and the information rate then is

$$R = 7.875 \times 10^6 \log_2 10 = 2.62 \times 10^7 \qquad \text{bits/s}$$

We may reason that teletypewriter service transmits information slowly, but requires only a narrow frequency band. Speech is of intermediate speed with moderate band requirements, and television transmits information at a high rate, but requires a wide frequency channel.

It is found that the capacity of a channel to transmit information is a function not only of bandwidth, but also of *signal-to-noise ratio*, expressed as S/N. Noise power is ordinarily assumed to be uniformly distributed over the received band, or is said to be "white" in character. The problem is to estimate the probability of being able to distinguish a given number of levels or states in the presence of noise.

Assume that in a stated bandwidth there is a noise power of N watts, and a signal power of S watts. The received power is then $S + N$ watts, or proportional to $\sqrt{S + N}$ volts, and we are attempting to receive this in the presence of $\sqrt{N}$ volts of noise. A signal less than the noise will not usually be distinguishable, whereas a signal $\sqrt{S}$ greater than $\sqrt{N}$ will probably be separable. The number of distinct states or levels which can be distinguished, *on the average*, may be taken as proportional to the ratio of the voltages, or

$$L = \frac{\sqrt{S + N}}{\sqrt{N}} = \sqrt{1 + \frac{S}{N}} \tag{16-49}$$

The maximum information available in a basic interval of the signal is then

$$H = \log_2 \sqrt{1 + \frac{S}{N}} = \frac{1}{2} \log_2 \left(1 + \frac{S}{N}\right) \tag{16-50}$$

We have already assumed that the number of basic information intervals per second is twice the frequency bandwidth or $2B$. The total information which can be transmitted in time T over a noisy channel of bandwidth B is then

$$C = BT \log_2 \left(1 + \frac{S}{N}\right) \tag{16-51}$$

This result, which relates the information transmitted to the bandwidth of the channel, the time available for transmission, and the signal-to-noise ratio, is called the *Hartley-Shannon law.* As originally suggested by Hartley in 1928 it had the form BT, but this was modified to the above form many years later by Shannon, to include the effect of signal-to-noise ratio.

It may be seen that the total information can be increased by widening the band, allowing greater time for transmission, or increasing the signal-to-noise ratio. The latter increase may be very expensive in terms of equipment complexity or power, because of the exponential increase rate.

Equation 16-51 shows that it is possible to receive a signal and information even though S is less than N, or the signal is apparently buried in the noise. This is the subject of major efforts in the communication art, but will not be gone into here except to indicate that such reception may be possible by slowing down the rate of transmission, or increasing T or the time of transmission, or by so coding the information as to require a much greater bandwidth for transmission. These techniques are employed in space communications.

Speech, when transmitted by double-sideband AM, occupies a band twice its own base frequency range, and requires a large S/N for successful reception. At 100 per cent modulation, double-sideband AM employs three times as much power as SSB. Single-sideband SSB requires a channel only equal to its base frequency range. Another advantage of SSB is in the avoidance of heterodyne interference from the carrier, a factor not related to channel capacity. Inefficient use of channel and power by AM is compensated for by its simplified circuitry and apparatus.

When speech is transmitted by FM the bandwidth may be enlarged indefinitely by change of deviation ratio. It has been shown that such enlargement reduces the sensitivity of the system to noise interference, or permits lower S/N ratios.

It must also be remembered that the frequency spectrum is limited, and system designers must always fit their systems into finite frequency channels, or face trade-offs in performance against the limitation of the frequency spectrum. Thus ideal design results can rarely be achieved in practice.

PROBLEMS

16-1. If a voltage of form

$$e = E_a(1 + m \sin \omega_m t) \sin \omega_c t$$

is applied to a resistor R, find the power dissipated. If $m = 0.35$, find the power disspated by each of the frequency components present.

16-2. A modulated wave is expressed by the equation

$$e = 25(1 + 0.27 \cos 1250t + 0.18 \cos 3000t) \cos 10^7 t$$

State all frequencies present, and give the amplitudes of m for each frequency.

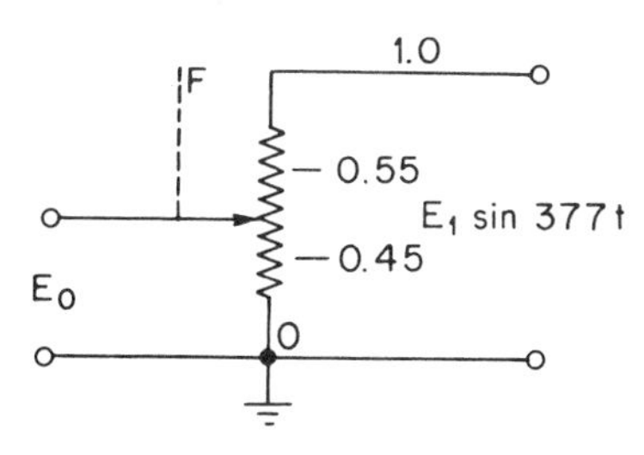

Figure 16-24.

16-3. In Fig. 16-24 the force F moves the slider between the 45 and 55 per cent points on the resistor at a rate of six times per second. Write the expression for the voltage E_o, and state all frequencies present.

16-4. An amplitude-modulated carrier of 2.5 MHz is 60 per cent modulated at 10,000 hertz. Write the expression for the voltages appearing across a resonant circuit, tuned to 2.5 MHz, and having Q of 5, through which this current flows. What is the value of m_a for the voltage across the circuit? Repeat for a tuned circuit of $Q = 150$. $L = 4$ microhenrys in both cases.

16-5. An amplitude-modulated Class C amplifier has a plate efficiency of 76 per cent and a carrier power input of 1000 W.

(a) What modulator power output will be required for 90 per cent modulation?
(b) If the modulated amplifier operates at 2250 plate volts, what ratio will be required for a transformer to couple a Class B modulator requiring 7500 ohms plate-to-plate load into the Class C modulated amplifier?
(c) If the transformer is 78 per cent efficient, find the power output of the modulator tubes for 90 per cent modulation.

16-6. A plate-modulated amplifier operating at 2000 plate volts has a rated plate dissipation of 100 W. When it is operated at 100 per cent amplitude modulation and full allowable input, the radio-frequency output to the tank circuit is 467 W.
(a) Find the unmodulated carrier power output.
(b) Determine the value of plate current taken by the modulated amplifier.
(c) A Class B modulator has an output impedance of 7500 ohms. What transformer ratio is required to couple this to the above modulated amplifier if the transformer is 92 per cent efficient?
(d) What power output must the Class B tubes supply for 100 per cent modulation?

16-7. A wave with peak amplitude of 100 V and frequency of 100 MHz is frequency-modulated at 10,000 hertz with a frequency deviation of 75 kHz. Find the amplitude of the center frequency and all side frequencies up to the sixth. Plot the spectrum to scale.

16-8. A frequency-modulated broadcasting station is assigned a frequency channel of 92.1 to 92.3 megahertz.
(a) What is the maximum permissible value of δ for a modulating frequency of 5000 hertz?
(b) For this modulating frequency, how many sideband frequencies could be permitted to exist on each side of the center frequency?
(c) Plot a line diagram of amplitudes of all sidebands which will fill the channel.

16-9. A 5000 Hz modulating frequency provides $\delta = 2$ in an FM wave.
(a) What bandwidth is required for the transmitter resonant circuits to pass this wave?
(b) What bandwidth is required if f_m is changed to 200 Hz?

16-10. Find expressions for the resistance and inductance produced at the A, A terminals by the circuit of Fig. 16-19(a). Neglect r_p.

16-11. Find expressions for the resistance and capacity produced at the A, A terminals by the circuit of Fig. 16-19(b). Neglect r_p.

16-12. Find expressions for the resistance and inductance produced at the A, A terminals by the circuit of Fig. 16-19(c). Neglect r_p.

16-13. In the reactance-tube circuit of Fig. 16-18, the g_m of the reactance tube can be varied over the range of 800 to 2000 μmho. With $f_c = 500$ kHz, and C chosen as 2 pF, what value should R have if it is desired to shift the frequency over a range of ± 7.5 kHz? The capacity C_1 is 25 pF.

16-14. The relation between modulating voltage and frequency deviation produced may be found by varying the voltage and noting the values of voltage at which the center-frequency amplitude becomes zero. Explain why this may be done.

16-15. The reactance-tube circuit of Fig. 16-19(b) is connected to a Hartley oscillator operating at 10 megahertz (before the modulator is connected). With $R = 10{,}000$ ohms and $L = 10 \times 10^{-6}$ H, what frequency deviation will be produced if the g_m of the reactance tube is changed from 300 to 1200 μmho and oscillator $C = 70$ pF?

16-16. A wirephoto picture of size 5×7 inches is scanned at a rate of 100 lines per inch, with equal horizontal resolution; we assume that we can see 12 density gradations. Determine the S/N ratio in dB required for the channel, of bandwidth 1200 hertz, if the picture is transmitted in 60 seconds.

16-17. If an FM station with $f_{m\,\max} = 15$ kHz and $\delta = 4$ operates at 1 MHz in the broadcast band, how many 10 kHz broadcast channels would it cover? What would be the 3 dB bandwidth of a tuned circuit to pass this FM station spectrum?

REFERENCES

1. Roder, H., "Amplitude, Phase, and Frequency Modulation," *Proc. IRE*, Vol. 19, p. 2145 (1931).
2. Armstrong, E. H., "A Method of Reducing Disturbances in Radio Signaling by a System of Frequency Modulation," *Proc. IRE*, Vol. 24, p. 689 (1936).
3. Crosby, M. G., "Reactance-Tube Frequency Modulators," *RCA Rev.*, Vol. 5, p. 89 (1940).
4. Jahnke, E., and F. Emde, *Tables of Functions*. Dover Publications, Inc., New York, 1945.
5. Shannon, C. E., "A Mathematical Theory of Communication," *Bell System Tech. Jour.*, p. 27 (July and October 1948).
6. Goldman, S., *Information Theory*. Prentice-Hall, Inc., Englewood Cliffs, N. J., 1953.
7. Black, H. S., *Modulation Theory*. D. Van Nostrand Co., Inc., New York, 1953.
8. *Proceedings of the IRE*, special issue on single sideband, Vol. 44, pp. 1661–1914 (1956).
9. Schwarz, M., *Information, Transmission, Modulation, and Noise*. McGraw-Hill Book Company, New York, 1959.
10. Hancock, J. C., *An Introduction to the Principles of Communication Theory*. McGraw-Hill Book Company, New York, 1961.

11. Russell, G. M., *Modulation and Coding in Information Systems*. Prentice-Hall, Inc., Englewood Cliffs, N. J., 1962.

12. Raisbeck, G., *Information Theory*. The M. I. T. Press, Cambridge, Mass., 1963.

13. Pappenfus, E. W., W. B. Breune, and E. O. Schoenike, *Single Sideband Principles and Circuits*. McGraw-Hill Book Company, New York, 1964.

17

AM AND FM DEMODULATION

It is now possible to discuss the process of *demodulation* or *detection*, whereby a modulated AM or FM wave is translated to recover the original information. The diode is the usual device involved.

Demodulators for AM are large-signal or *linear envelope detectors*, or of small-signal *square-law* form. The latter has a level of distortion which makes its use undesirable; the linear envelope detector will be considered here.

Demodulators for FM utilize circuits which convert frequency excursions about a center frequency into amplitude variations, and are designed to be insensitive to amplitude variations in the signal, thus reducing noise and interference.

17-1 LINEAR ENVELOPE DEMODULATION

A diode in series with a large load resistor will have a linear volt-ampere characteristic as in Fig. 17-1(a). If an amplitude-modulated voltage is applied, rectification will occur and the original modulation can be recovered as a variation of the average value of the output, when averaged over a time long with respect to a carrier cycle and short with respect to a modulation cycle. The average current over a carrier cycle is

$$I_{\text{av}} = \frac{1}{2\pi}\int_{-\pi/2}^{\pi/2} \frac{E_c}{r_d + R}(1 + m_a \cos \omega_m t) \cos \omega_c t \, d\omega_c t$$

where r_d is the diode internal resistance. The current is zero over the negative portion of the carrier cycle. Then with load R,

$$e_d = \frac{E_c R}{\pi(r_d + R)} + \frac{k_a E_m R \cos \omega_m t}{\pi(r_d + R)} \tag{17-1}$$

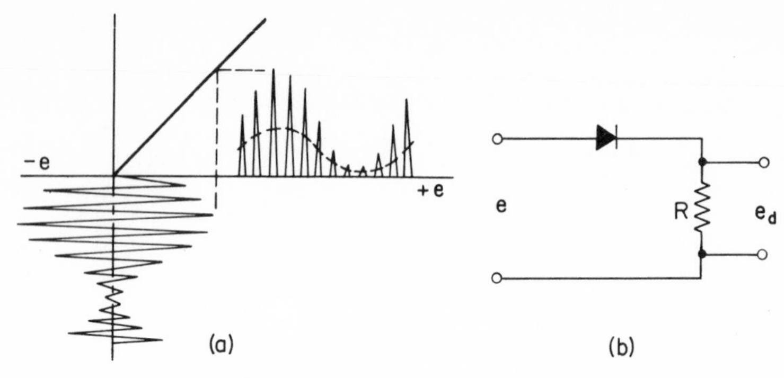

Figure 17-1. Linear diode detection.

The process has recovered a direct component proportional to E_c and a modulation component.

The simple circuit of Fig. 17-1(b) is inefficient because the $1/\pi$ coefficient limits the output to about one-third of the input. To improve the efficiency a filter capacitor can be added across the load, as in Fig. 17-2(a). Action is similar to an RC filter with a rectifier, the capacitor C charging to the peak of each carrier cycle and discharging through the resistance R on the negative half-cycle. The voltage across C and the load tends to follow the *envelope* of the modulated wave.

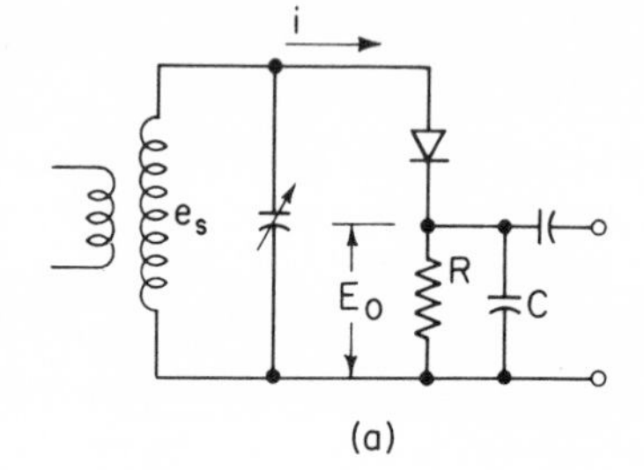

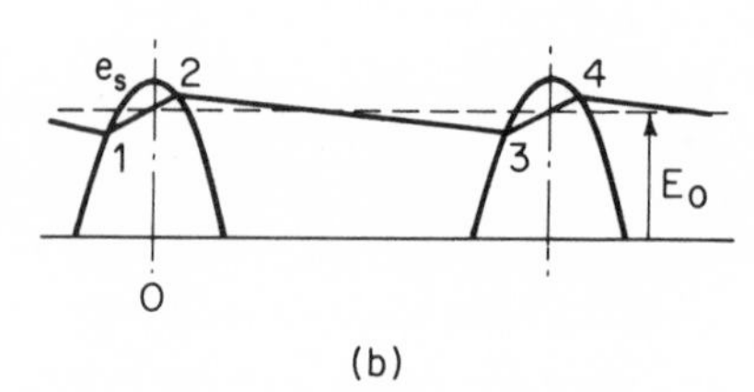

Figure 17-2. Addition of C to the diode load.

The variation of voltage over the interval 1, 2, 3, 4 is at carrier frequency. The variation is small and it can be assumed that the voltage across C is a constant E_o, over a few cycles of carrier, but a variable over a cycle of modulation frequency. That is, $f_m \ll f_c$.

An amplitude-modulated wave

$$e_s = E_c(1 + m_a \cos \omega_m t) \cos \omega_c t$$

is applied. The diode voltage will be biased by E_o, as indicated in Fig. 17-3, and the current is discontinuous, but methods of analysis similar to those used for the Class B linear amplifier may be applied. The voltage across the diode can be stated as

$$e_d = E_c(1 + m_a \cos \omega_m t) \cos \omega_c t - E_o$$

which may be more compactly written for one carrier cycle, as

$$e_d = E' \cos \theta - E_o \tag{17-2}$$

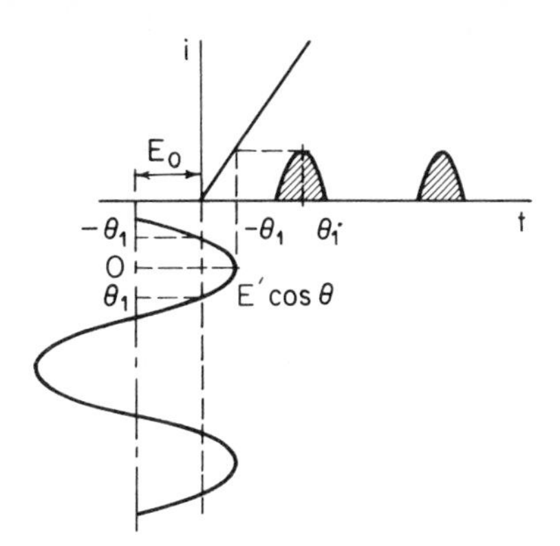

Figure 17-3. Operation of the linear diode.

Since the term E' is practically constant over a few cycles of f_c, the current can be written as $i_d = e_d/r_d$ and averaged as

$$I_o = \frac{1}{\pi r_d}(E' \sin \theta_1 - E_o\theta_1) \qquad (17\text{-}3)$$

noting that the conduction limits are $-\theta_1$ to θ_1.

From the figure it is apparent that $\cos \theta_1 = E_o/E'$, and so

$$E_o = RI_o = \frac{E'R}{\pi r_d}(\sin \theta_1 - \theta_1 \cos \theta_1) \qquad (17\text{-}4)$$

that is,

$$E_0 = \frac{RE_c}{\pi r_d}[(\sin \theta_1 - \theta_1 \cos \theta_1) + m_a(\sin \theta_1 + \theta_1 \cos \theta_1) \cos \omega_m t] \qquad (17\text{-}5)$$

in view of the definition of E'. The first term on the right is a direct voltage with magnitude dependent on E_c. The second term has an amplitude proportional to $m_aE_c = E_m$, and varies at modulation frequency. This represents recovery of the original modulation, without distortion.

Equation 17-4 may be transformed to

$$\frac{R}{r_d} = \frac{\pi}{\tan \theta_1 - \theta_1} \qquad (17\text{-}6)$$

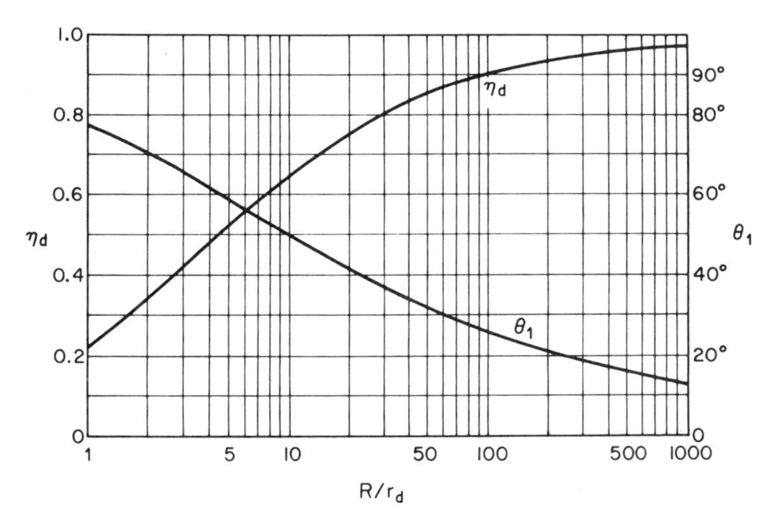

Figure 17-4. Efficiency and conduction angle of the linear diode.

This ratio is a function of a design choice, and θ_1 may be determined from Fig. 17-4. Since R/r_d is a constant for a given circuit, then θ_1 is also a constant.

If the load R were infinite, then $E_o = E'$ and this output could not be exceeded. The ratio of E_o actually obtained to the value E' which could be obtained in theory is defined as the *detection efficiency*, η_d. That is,

$$\eta_d = \frac{E_o}{E'} = \frac{R}{\pi r_d}(\sin\theta_1 - \theta_1\cos\theta_1) \tag{17-7}$$

and this result is plotted in Fig. 17-5 as a function of R/r_d. The detection efficiency of a linear diode detector can be found from Eq. 17-1, and the result is plotted for $\omega RC = 0$, for comparison with the detection efficiency achieved by envelope detection with the bypass capacitor.

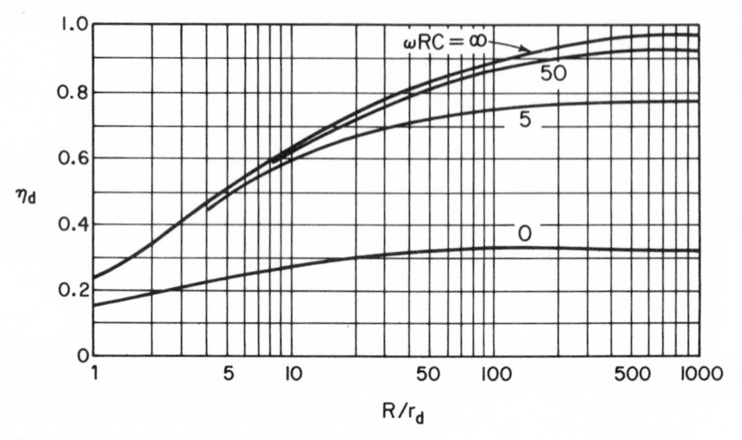

Figure 17-5. Variation of efficiency with C.

17-2 INPUT IMPEDANCE OF THE ENVELOPE DETECTOR

The usual detector input is from a tuned circuit, and the shunting effect of the diode and load reduce the circuit Q. It is thus desirable to determine the magnitude of the loading effect. The average input power to the diode and load is

$$P_{\text{in}} = \frac{1}{2\pi}\int_{-\pi}^{\pi} ei\,dt = \frac{(E')^2}{2\pi r_d}(\theta_1 - \sin\theta_1\cos\theta_1) \tag{17-8}$$

The input resistance of the diode and load is equal to the resistance which would dissipate P_{in} with the same voltage applied. Since E' is a peak value

$$R_i = \frac{(E')^2}{2P_{\text{in}}} = \frac{\pi r_d}{\theta_1 - \sin\theta_1\cos\theta_1} \tag{17-9}$$

By use of Eq. 17-6 it is possible to obtain

$$\frac{R_i}{R} = \frac{\tan\theta_1 - \theta_1}{\theta_1 - \sin\theta_1 \cos\theta_1} \tag{17-10}$$

This function is plotted in Fig. 17-6. With usual circuits employing R/r_d values of 50 or more, the equivalent load resistance of the diode tends toward $0.5R$ in the envelope detector. In the linear diode detector this load appears as equal to $2R$.

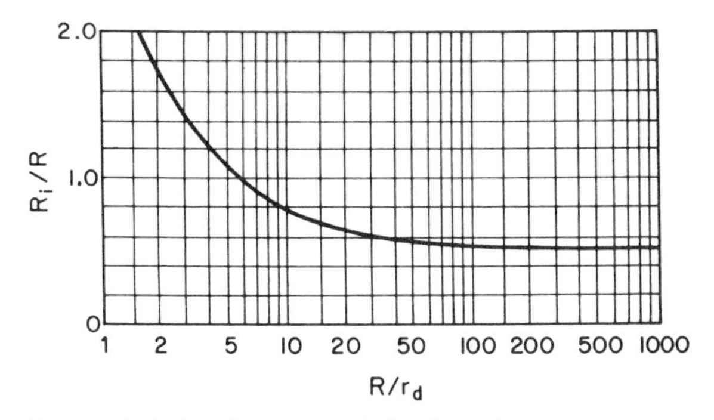

Figure 17-6. Loading effect of the diode detector.

17-3 DISTORTION IN ENVELOPE DETECTORS

The RC time constant of the diode load determines the rapidity with which the load voltage can change. If the voltage cannot change fast enough to follow the modulation envelope, amplitude distortion will result. This occurs only as the modulation envelope is decreasing, since C must discharge through R and the time constant is large. On the increasing portion of the envelope the capacitor is being charged through r_d, much smaller than R, and the time constant is not limiting.

The rate of change of the modulation envelope is

$$\frac{de}{dt} = \frac{d}{dt}[E_c(1 + m_a \cos\omega_m t)] = -E_c\omega_m m_a \sin\omega_m t \tag{17-11}$$

The capacitor C discharges according to $e_c = E_c\epsilon^{-t/RC}$, so that

$$\frac{de_c}{dt} = \frac{-e_c}{RC} = -\frac{E_c(1 + m_a \cos\omega_m t)}{RC} \tag{17-12}$$

In order that the voltage across C may follow variations in the modulation envelope, the rate of fall of the voltage across C must be potentially greater then the rate of fall of the signal envelope. The critical condition which will just permit following changes in the envelope voltage can be found by setting Eq. 17-11 equal to Eq. 17-12; this leads to the condition that

$$RC \leq \frac{1}{\omega_m}\left(\frac{1 + m_a \cos\omega_m t}{m_a \sin\omega_m t}\right) \tag{17-13}$$

Maximizing the right side with respect to t gives

$$\cos \omega_m t = -m_a, \qquad \sin \omega_m t = \sqrt{1 - m_a^2}$$

Consequently the value of the load time constant is limited as

$$RC \leq \frac{1}{\omega_m m_a} \sqrt{1 - m_a^2} \tag{17-14}$$

The choice of circuit parameters is fixed by the maximum modulation frequency and the expected modulation percentage. However, distortion is not excessive if RC does not exceed $1/\omega_m m_a$, even though RC should approach zero for 100 per cent modulation.

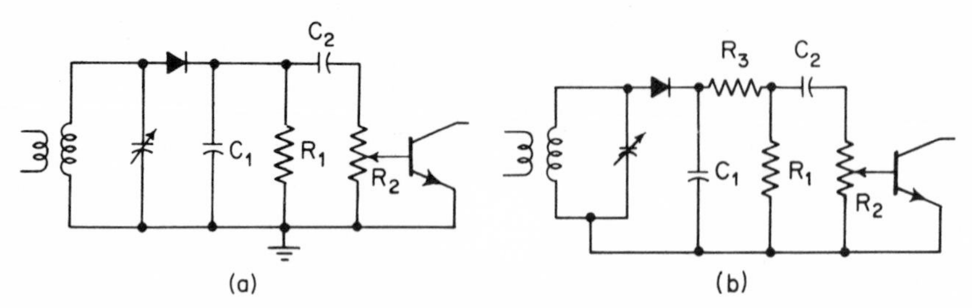

Figure 17-7. Distortion reduction in the envelope detector.

A second form of distortion, peak clipping, may occur when the complete circuit of Fig. 17-7(a) is used. Here C_2 blocks the dc component, and R_2 serves as a gain control. The dc and ac load impedances are different, the ac resistance being due to R_1 and R_2 in parallel, the reactance of C being small. The ac resistance is smaller than R_{dc}, and the ac current might exceed the dc component. However, negative currents are impossible in the diode and so the negative peak would be clipped at zero, and this represents distortion.

From Eq. 17-5 it is possible to write

$$\frac{\text{peak } E_{ac}}{E_{dc}} = \frac{(\text{peak } I_{ac})R_{ac}}{R_{dc}I_{dc}} = m_a \tag{17-15}$$

If clipping is to be avoided, the maximum value of peak I_{ac}/I_{dc} is unity and so the maximum value of m_a follows as

$$\text{maximum } m_a = \frac{R_{ac}}{R_{dc}} \tag{17-16}$$

In the circuit of Fig. 17-7(a) this becomes $R_2/(R_1 + R_2)$ which makes it desirable to lower R_1 to permit $m_a = 1$, but this lowers the detection efficiency. In Fig. 17-7(b) an attempt is made to equalize the dc and ac load values by introducing a resistor R_3 in series with the ac load. This expedient sacrifices some output voltage, but additional gain can be supplied elsewhere.

With a large input signal and a large R/r_d to linearize the diode curve, it is possible to hold distortion to one or two per cent in envelope detectors.

17-4 AUTOMATIC GAIN CONTROL (AGC)

In radio reception the strength of the received signal varies, and to provide nominally constant output power, it is desirable to vary the gain of the receiver, in inverse proportion to the received signal strength. This is done with *automatic gain control* (AGC) or automatic volume control (AVC) circuits.

A measure of the strength of the received carrier is available in the dc voltage output of the detector diode, written in Eq. 17-5 as

$$e_{dc} = \frac{E_c R}{\pi r_d}(\sin \theta_1 - \theta_1 \cos \theta_1)$$

A low-pass *RC* filter, with time constant long with respect to the lowest modulation cycle, as R_3C_3 in Fig. 17-8, will separate this dc voltage from the modulation. The dc voltage may then be applied as a control bias to vary the gain of amplifiers ahead of the detector.

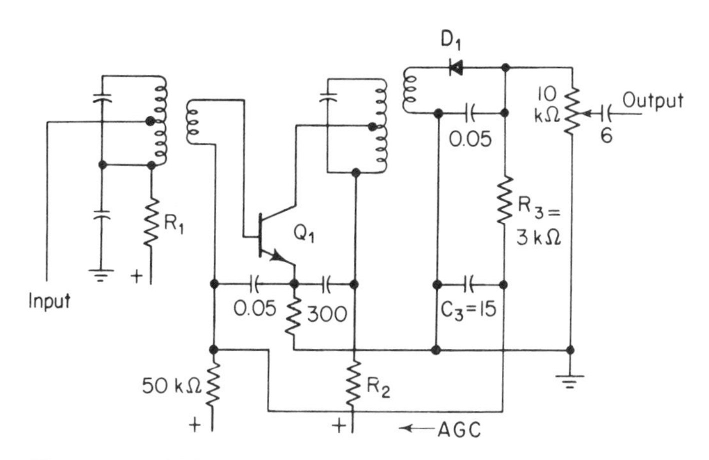

Figure 17-8. AGC system (capacities in μF).

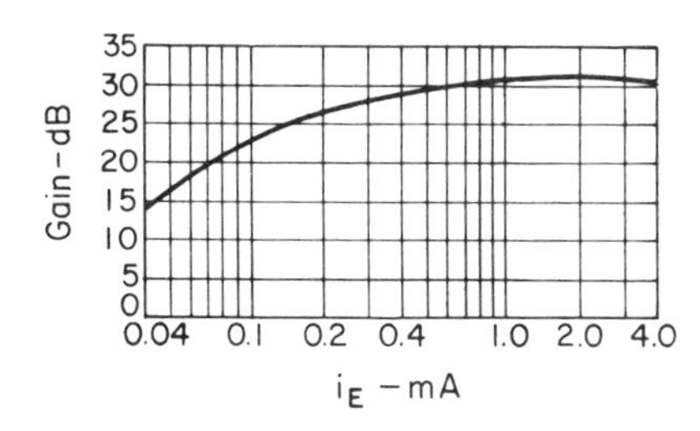

Figure 17-9. Gain vs. I_E for the 2N168 transistor.

In transistor amplifiers several currents might be varied to change the gain, but the most desirable seems to be I_E. For a typical npn transistor, Fig. 17-9 shows the range of gain change available. In Fig. 17-8 the negative voltage from D_1 changes the base bias of Q_1 and reduces the emitter current with increasing signal strength. As the emitter current is reduced, h_{fe} falls and the gain is reduced; h_{ie} and

h_{oe} also change and impedance mismatches result which further reduce the overall gain.

In vacuum-tube circuits, the negative AGC voltage from the detector is applied as grid bias to several variable-gain amplifier stages, reducing the g_m and dropping the overall gain.

Since the control must allow a slight change to obtain a working bias voltage, it is not possible to produce absolute constancy of detector output. The system will reduce the strength of all signals, even the extremely weak ones. To remedy this, a separate diode may be used for AGC action; by a threshold bias, this second diode may be held from conduction until a certain signal level is reached. Thus full gain is available for weak signals, as shown in Fig. 17-10, and the output is more constant above the threshold level.

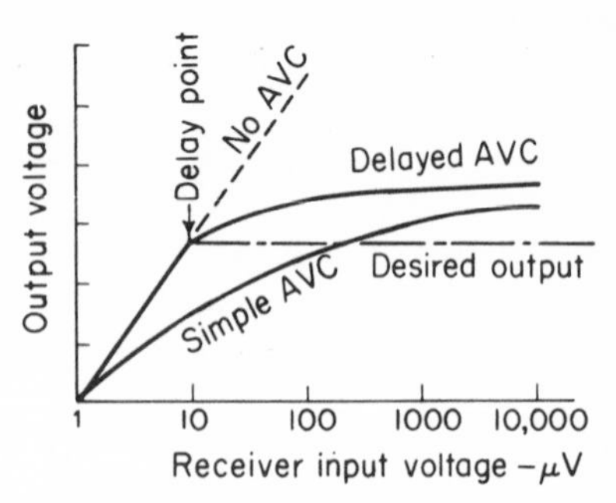

Figure 17-10. Operation of a receiver with AGC.

17-5 DYNAMIC RECTIFICATION CHARACTERISTICS

It is sometimes convenient to determine envelope detector performance from *rectification characteristics*, as in Fig. 17-11 for a 6H6 vacuum diode.

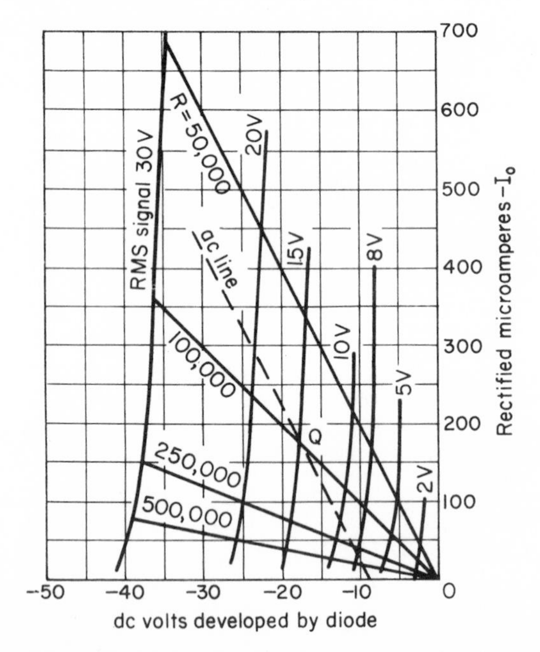

Figure 17-11. Rectification characteristics for the 6H6 diode.

Load lines may be superimposed to show the relation between E_o and I_o for any load R and rms carrier voltage. From the value of rms unmodulated carrier voltage, a Q point is determined. If the ac load is different from the dc load R, then operation is along an ac line drawn with appropriate slope through the Q point.

Such a line is drawn dashed through a Q point fixed by a 100,000 ohm dc load and an rms carrier voltage of 15 V. The ac load line has a slope determined for a 50,000 ohm load. The ac load line reaches zero current before the applied ac voltage reaches zero; that is, the current reaches zero when the ac rms voltage reaches 6 V. The ac load line indicates that the anode current would be zero for all values of radio-frequency voltage less than 6 V, showing the effects of negative peak clipping. For lower percentages of modulation the voltage will not fall to such low values, and distortion will not occur. This result corresponds to that of the previous analytical discussion on peak clipping.

17-6 THE AMPLITUDE LIMITER

The frequency discriminator used for one form of frequency demodulation is sensitive to amplitude variation. If the noise reduction properties of the FM system are to be exploited, any amplitude variation due to noise or interference must be removed by reducing the signal to a constant amplitude. Then the only varying property of the signal is frequency, and the frequency variations can be converted to the amplitude variations of the original speech or music.

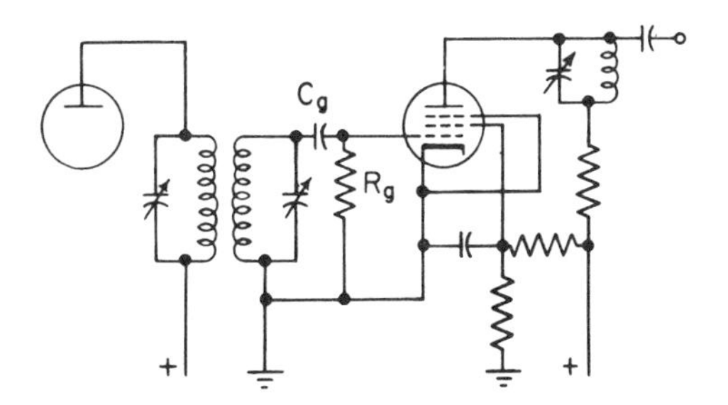

Figure 17-12. The amplitude limiter.

In the pentode *limiter* of Fig. 17-12, as the input swings positive the grid current through R_g creates a negative bias which shifts the zero axis of the wave downward, holding the positive peaks of the signal near zero, and maintaining the bias by small pulses of grid current. Cutoff occurs at a small negative voltage, and the output peaks are limited by zero volts and cutoff. Performance is indicated in Fig. 17-13, above 10 V input.

The emitter-coupled limiter of Fig. 17-14(a) is biased so that both transistors are in conduction. For any input v_i more negative than some level

$-V_x$, Q_1 is cut off. With Q_2 conducting the output voltage is V_{sat}. As v_i rises, Q_1 begins to conduct and raises the emitter voltage, reducing the current in Q_2. At some positive input level $v_i = +V_x$, transistor Q_2 cuts off and the output voltage is V_{CC}. Thus for an input with peak-to-peak value exceeding $2V_x$, the output is limited at a peak-to-peak value $V_{CC} - V_{sat}$. This action is illustrated in Fig. 7-14(b).

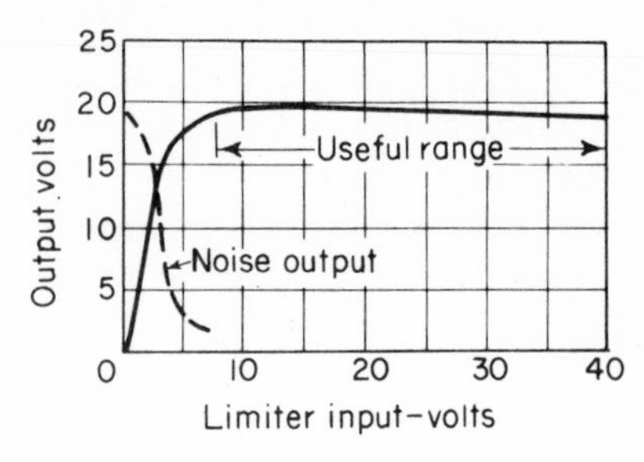

Figure 17-13. Limiter performance.

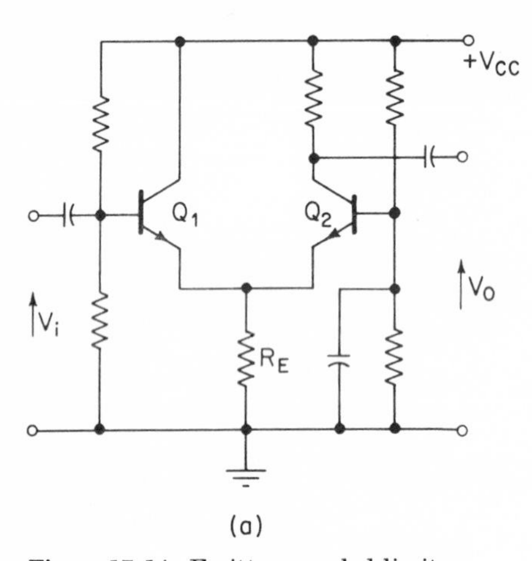

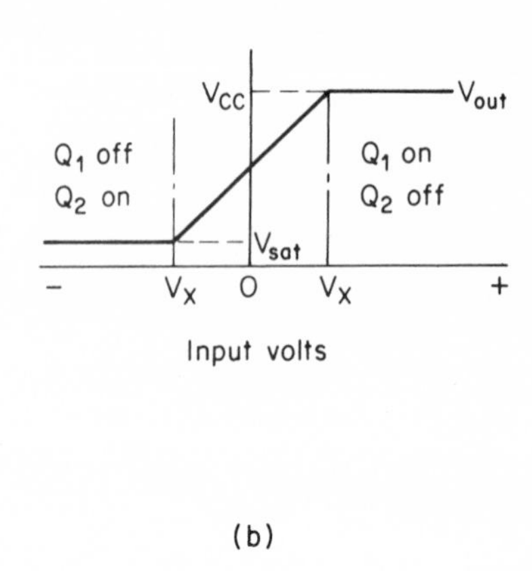

Figure 17-14. Emitter-coupled limiter.

17-7 THE FREQUENCY DISCRIMINATOR

One means of conversion of the frequency variations in an FM signal to amplitude variations is a *discriminator* circuit, which employs a double-tuned transformer, with equivalent in Fig. 17-15. The primary and secondary are resonant, or $\omega_o L_1 = 1/\omega_o C_1$, and $\omega_o L_2 = 1/\omega_o C_2$. The internal resistance of the preceding amplifier is neglected, and resistance R_s includes that of detector diodes to be connected. Using previously derived methods and the frequency variation from resonance, δ, it is possible to obtain the primary voltage $\mathbf{E}_1$ at terminals a, a as

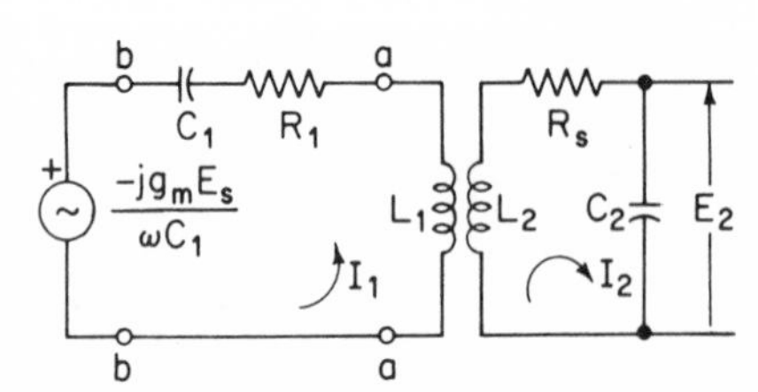

Figure 17-15. Circuit for the doubly tuned amplifier.

$$\mathbf{E}_1 = \frac{-jg_m\mathbf{E}_s}{\omega_o(1+\delta)C_1}\frac{\mathbf{Z}_{aa}}{\mathbf{Z}_{bb}} = g_mQ\omega_oL_1\mathbf{E}_s\left[\frac{1+j2Q\delta-jk^2Q}{(1+j2Q\delta)^2+k^2Q^2}\right]$$

The term jk^2Q is usually negligible and so

$$\mathbf{E}_1 = g_mQ\omega_oL_1\mathbf{E}_s\left[\frac{1+j2Q\delta}{(1+j2Q\delta)^2+k^2Q^2}\right] \tag{17-17}$$

The secondary voltage $\mathbf{E}_2$ of the doubly tuned circuit was previously obtained as

$$\mathbf{E}_2 = \frac{jg_mkQ^2\omega_o\sqrt{L_1L_2}\mathbf{E}_s}{(1+j2Q\delta)^2+k^2Q^2} \tag{17-18}$$

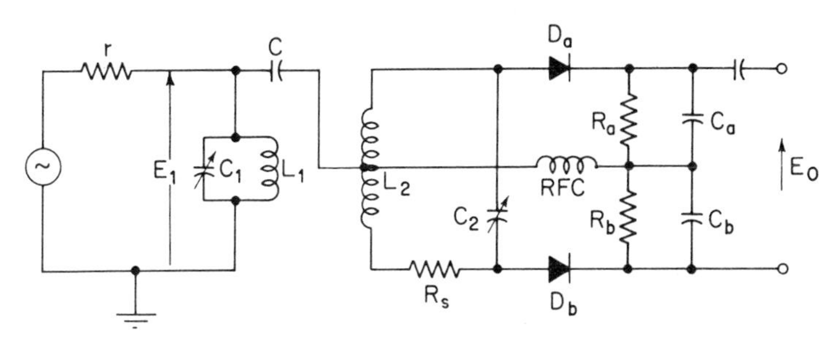

Figure 17-16. The Foster–Seeley frequency discriminator.

In the frequency discriminator the secondary circuit operates into two diodes, connected as in Fig. 17-16, and with capacitor C connected between the primary and the center point of the secondary, making the voltages additive.

Diodes D_a and D_b have applied voltages to ground which can be called $\mathbf{E}_a$ and $\mathbf{E}_b$, written as

$$\mathbf{E}_a = \mathbf{E}_1 + \frac{\mathbf{E}_2}{2}, \qquad \mathbf{E}_b = \mathbf{E}_1 - \frac{\mathbf{E}_2}{2}$$

At f_o, where $\delta = 0$, the primary and secondary voltages are

$$\mathbf{E}_1 = g_mQ\omega_oL_1\mathbf{E}_s\left(\frac{1}{1+k^2Q^2}\right) \tag{17-19}$$

$$\mathbf{E}_2 = jg_mkQ^2\omega_o\sqrt{L_1L_2}\mathbf{E}_s\left(\frac{1}{1+k^2Q^2}\right) \tag{17-20}$$

and $\mathbf{E}_2$ is in quadrature with $\mathbf{E}_1$. These expressions lead to the resonant frequency phasor diagram of (a), Fig. 17-17, showing $\mathbf{E}_a$ and $\mathbf{E}_b$. At frequencies above resonance the phasor diagram becomes that at (b), and below resonance that at (c).

The diodes are connected to provide an output E_o proportional to the difference of $|E_a|$ and $|E_b|$; at resonance this difference is zero. Frequency

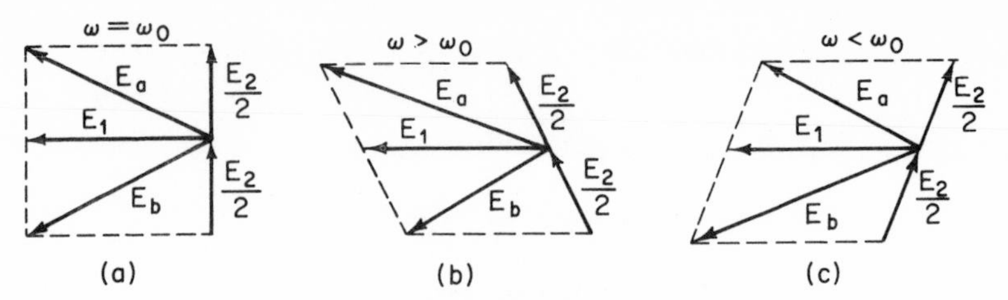

Figure 17-17. Phasor diagrams of discriminator performance.

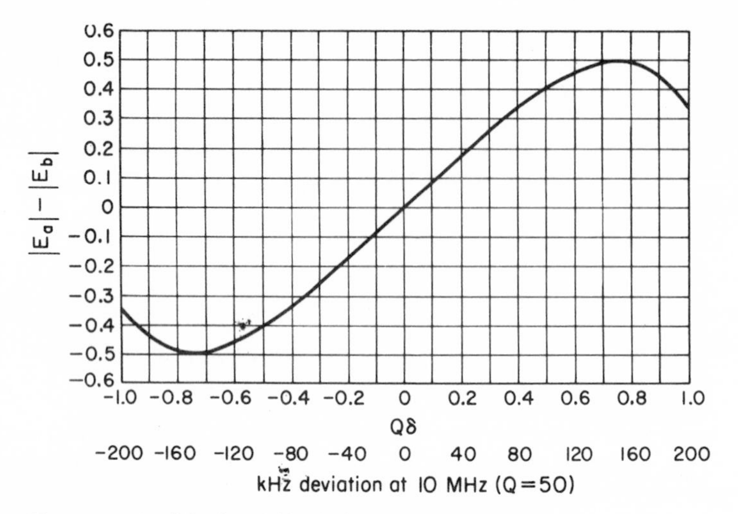

Figure 17-18. Calculated discriminator performance.

deviations from resonance are translated into amplitude and polarity information which leads to the curve of Fig. 17-18; over an appreciable frequency range the FM demodulator provides an output E_o linear with frequency.

The diode voltages may be written as

$$\mathbf{E}_a = \frac{g_m Q\omega_o L_1 \mathbf{E}_s[1 + j2Q\delta + j(kQ/2)\sqrt{(L_2/L_1)}]}{(1 + j2Q\delta)^2 + k^2Q^2} \tag{17-21}$$

$$\mathbf{E}_b = \frac{g_m Q\omega_o L_1 \mathbf{E}_s[1 + j2Q\delta - j(kQ/2)\sqrt{(L_2/L_1)}]}{(1 + j2Q\delta)^2 + k^2Q^2} \tag{17-22}$$

If two new parameters are introduced as

$$X = Q\delta, \qquad K = kQ = \frac{k}{k_c}$$

then the voltage magnitude difference appears as

$$|E_a| - |E_b| = \frac{g_m Q\omega_r L_1 |E_s|}{\sqrt{(1 + K^2)^2 + 8X^2(1 - K^2 + 2X^2)}}$$
$$\times \left[\sqrt{1 + \left(2X + \frac{K}{2}\sqrt{\frac{L_2}{L_1}}\right)^2} - \sqrt{1 + \left(2X - \frac{K}{2}\sqrt{\frac{L_2}{L_1}}\right)^2}\right] \tag{17-23}$$

The ratio of the magnitudes of E_2 to E_1 at resonance is

$$\left|\frac{E_2}{E_1}\right| = kQ\sqrt{\frac{L_2}{L_1}} \tag{17-24}$$

and it is customary to design the transformer to give a value of 2 for this ratio. For the usual case kQ is chosen equal to 1.5, so that $L_2/L_1 = 1.77$.

The value of Q may be determined from the relation

$$Q = \frac{f_r}{2\,\Delta f}$$

where $2\,\Delta f$ is the desired range of linear frequency deviation. For speech and music this is usually chosen as 200 kilohertz, and at an operating frequency of 10 megahertz the value of Q will be 50.

Since $kQ = k/k_c$ has been assumed as 1.5, then $k = 0.03$. One of the terms of Eq. 17-23 may be assembled as

$$\frac{K}{2}\sqrt{\frac{L_2}{L_1}} = \frac{1.5}{2}\sqrt{1.77} = 1.0$$

and the reason for the choice of parameters is apparent. The size of L_2 is limited by the need of a reasonable value for C_2; this may be taken as 50 pF. The transformer design may be summarized as

$$L_2 = \frac{1}{\omega_r^2 C_2} = 5.08\ \mu\text{H}, \qquad L_1 = \frac{L_2}{1.77} = 2.87\ \mu\text{H},$$

$$C_1 = \frac{1}{\omega_r^2 L_1} = 88.3\ \text{pF}, \qquad Q_1 = Q_2 = Q = 50$$

$$kQ = 1.5, \text{ therefore } k = 0.03, \qquad M = k\sqrt{L_1 L_2} = 0.12\ \mu\text{H}$$

The design has been used to draw the output relation of Fig. 17-18. The curve has a linear region running over values of $Q\delta = \pm 0.5$ or $\delta = \pm 0.010$, i.e., 100 kilohertz each side of a 10 megahertz resonant frequency.

Equation 17-23 shows the output proportional to E_s; thus the output would be affected by amplitude variations or noise if the circuit were not preceded by a limiter.

17-8 THE RATIO DETECTOR FOR FM

For a particular signal, undeviated in frequency, a dc voltage of 10 V may appear across resistors R_a and R_b of Fig. 17-16(a), the polarities being opposite and E_o equal to zero. If the input frequency then deviates so that the voltage applied to D_a increases and that to D_b decreases, the voltage across R_a may increase to +15 V and that across R_b may decrease to −5 V. The output voltage is +10 V, or the difference between the two, with ground as reference.

A stronger signal may produce 15 V when undeviated. When it is deviated the same amount as before, the voltage across R_a becomes +22.5 V and that across R_b becomes -7.5 V, with $+15$ V output. In both cases the same frequency deviation gave rise to differing outputs, and the circuit responded to amplitude variations. However, in both cases the *ratio* of the voltages across R_a and R_b was the same, namely 3:1. If a detector was made responsive to this ratio and not to respective magnitudes, then it would be insensitive to amplitude variations and a limiter would not be needed.

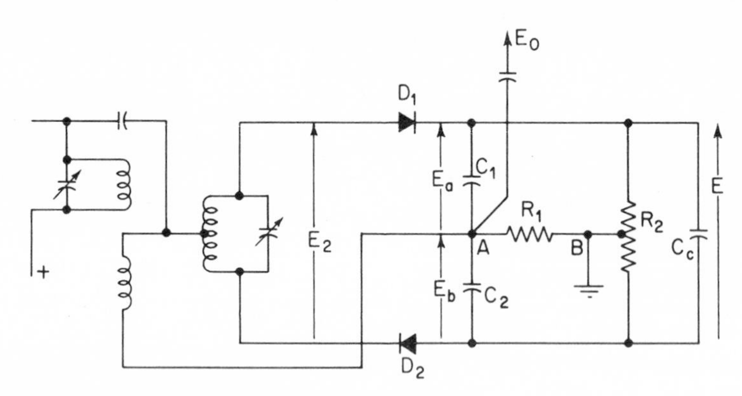

Figure 17-19. The ratio detector.

A circuit for obtaining the desired ratio is shown in Fig. 17-19, the main change from the discriminator being made by reversal of one diode so that D_1 and D_2 are in series. The full secondary voltage is applied and results in a dc output voltage E which is constant, almost irrespective of amplitude modulation because of the large time constant in R_2 and C_c.

For an undeviated signal the voltages on C_1 and C_2 are equal, and no voltage appears across R_1. When frequency deviation takes place the unbalanced diode voltages charge C_1 and C_2 unequally, and point A shifts in potential. Point B is fixed in potential by the R_2C_2 combination, and consequently a voltage appears across R_1. Since E is relatively fixed for a given signal, the sum of E_1 and E_2 must also be fixed, and it is then the ratio of E_1 to E_2 which changes at the frequency of the modulating signal. Thus the circuit is a *ratio detector*. Since it is slightly affected by amplitude, its performance can be improved if it is preceded by a limiter.

17-9 FREQUENCY CONVERSION

The function of the *frequency converter* is to translate or shift a band of frequencies centering on f_1 so as to create a similar band of frequencies centering on frequency f_2. The principle of product frequencies is employed, through use of *product modulators* or *product detector* circuits.

For the pentagrid converter of Fig. 17-20, the cathode and first and second grids constitute a triode oscillator at frequency f_x, causing the electron

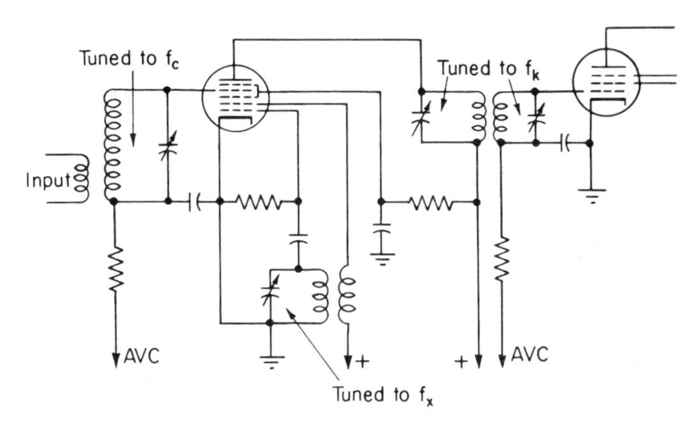

Figure 17-20. Converter circuit, with pentagrid tube.

stream to vary with considerable amplitude. The band of frequencies to be translated, centering on f_c, is applied to the second control grid and further modulates the electron stream, producing a product frequency for each frequency present. Isolation of the several signals is accomplished by the shield grids.

The transistor converter of Fig. 17-21 constitutes both an oscillator at frequency f_x and a converter. The oscillator has a tuned emitter-base circuit, and the signal frequencies to be converted are also supplied to the base. The output product frequencies appear in the collector circuit. Isolation of the several circuits is not as good as with the pentagrid tube, and some shift of oscillator frequency with input changes in the base may be encountered.

If the signal to be translated includes a carrier f_c, and sidebands $f_c - f_m$ and $f_c + f_m$, then from the previously developed modulation theory, it is

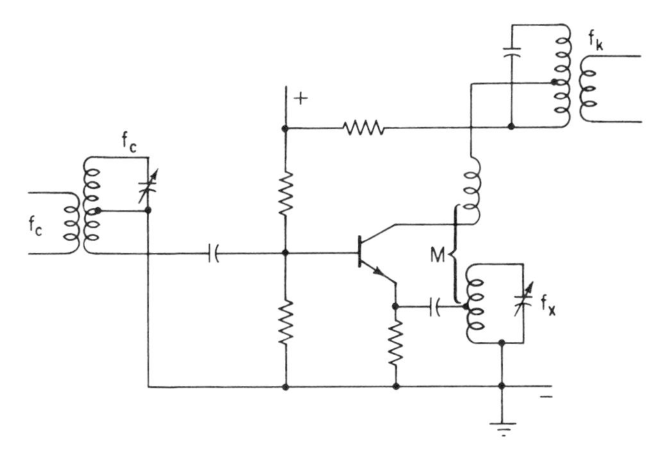

Figure 17-21. Transistor frequency converter.

known that in the converter output will be frequencies

$$
\begin{array}{ll}
f_c & \\
f_x & \\
f_c + f_x = f_q & f_c - f_x = f_k \\
f_c + f_x - f_m = f_q - f_m & f_c - f_x - f_m = f_k - f_m \\
f_c + f_x + f_m = f_q + f_m & f_c - f_x + f_m = f_k + f_m
\end{array}
$$

The sum and difference terms will have amplitudes proportional to the products of the individual component amplitudes.

The frequencies centered about f_q or f_k constitute two translated frequency bands, each containing the frequencies of the original modulating signal. Thus signals received with a carrier in one frequency range may be translated and amplified at more favorable frequencies before detection and extraction of the information content. This principle receives extensive application.

There may also be present unwanted higher-order modulation product frequencies discussed in Section 16-3. These spurious frequencies can usually be reduced to small amplitude by circuit design and filtering, but can cause interfering signals in some systems.

Various pentagrid or transistor converter circuits are available, with internal oscillator action or with a separate oscillator circuit to supply f_x. In any case, the output at the new frequency is given by a *conversion transconductance* g_c, where

$$g_c = \frac{\Delta I_p(f_k)}{\Delta E_c} \tag{17-25}$$

as a measure of the efficiency with which the input frequency is converted to the new frequency. Conversion gain then is

$$A_c = g_c Z_L$$

where Z_L is the load impedance at the output carrier frequency, f_q or f_k.

The production of higher-order modulation products can be greatly reduced by use of a field-effect transistor as the converter device. The FET was described as a "square-law" device in which terms higher than the second order in the input-output characteristic have negligible coefficients. The use of the FET decreases the amplitude of the unwanted conversion frequencies and simplifies the filter problems associated with frequency conversion. The FET may be used in circuits similar to those indicated for the pentode as a converter.

17-10 AM RADIO RECEIVING SYSTEMS

Based on the frequency translation principle, the early *heterodyne system* of radio reception employed a local oscillator frequency f_x, which was adjusted to equality with f_c of the input signal. Then, for AM inputs of frequency

$f_c \pm f_m$, the output was $f_c - f_x \pm f_m$ or f_m only, the original modulation. This simple system lacked selectivity and instability of the local oscillator made the receiver hard to adjust. It was replaced by the *superheterodyne*, a more elaborate adaptation of the principle. In the superheterodyne, frequency f_x is generated so that $f_c - f_x = f_k$, where f_k is a constant *intermediate frequency* (I.F.), intermediate in value between f_c and f_m.

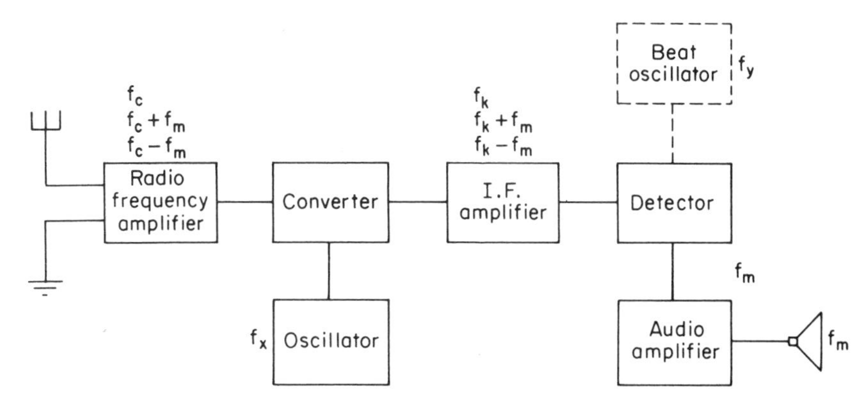

Figure 17-22. Block diagram of the AM superheterodyne.

A block diagram of such a receiving system appears in Fig. 17-22. After amplification at signal frequency the signal is translated in the converter to a new frequency group f_k, and $f_k \pm f_m$. Amplification at the fixed I.F. frequency follows. Since the intermediate frequency is usually lower than the original signal frequency the tuned circuits can be more selective (in terms of width in cycles) than could be obtained at signal frequency. The circuits are coupled with doubly-tuned transformers, or stagger tuning is employed to give a flat top and steep skirts to the overall frequency response curve.

After I.F. amplification the signal is applied to a demodulator, usually an envelope or product detector, and the modulation frequency is derived and given further amplification at an audio frequency before it reaches the final output device, usually a loudspeaker.

It is equally possible to translate up in frequency, by use of the sum-frequency band centering on f_q, but this is not often done because stable gain and desired selectivity characteristics are easier to obtain at a lower frequency.

Because of unintentional positive feedback, there will be a limit to the amount of stable gain achievable on any one frequency. By shifting the frequency band of the signal, it is possible to approach this limiting stable gain in each of several frequency bands in cascade. Thus high gain and controllable selectivity through fixed-frequency transformer design are advantages of the superheterodyne system. The avoidance of image interference, mentioned below, justifies additional frequency translation, often resulting in double- and triple-conversion systems.

Because of the requirement that f_k be constant for all received frequencies in a given band, the oscillator frequency must continuously differ from f_c by the amount of the I.F., i.e., f_k. For any oscillator frequency f_x, two signals differing by twice the intermediate frequency can give output at the intermediate frequency. If the desired signal is at 545 kHz, the oscillator will be tuned to 1000 kHz with a 455 kHz intermediate frequency. A second signal at 1455 kHz could also be received and would be called the *image signal.* By introducing additional selectivity at the signal frequency to lower the strength of the undesired 1455 kHz image, and by the use of higher intermediate frequencies to separate the image further from the desired signal, it is possible to suppress these image responses.

A continuous wave or CW telegraph signal, broken into dots and dashes, carries no amplitude information and will not give an output from a linear detector, other than the dc from the carrier. It is necessary to create an audio tone from the signals, and a separate *beat-frequency oscillator* is used, as shown by the dashed lines in Fig. 17-22. This oscillator has a fixed frequency output f_y differing from f_k by one or two thousand hertz. Modulation of f_k by f_y creates an audible difference frequency tone for each dot or dash.

17-11 FM RECEIVING SYSTEMS

Receiving systems for FM employ the superheterodyne circuit, shown in block form in Fig. 17-23. Most frequency-modulated signals use center frequencies above 80 MHz, and the intermediate frequency must be high enough that bandwidth is available to pass bands of possibly 200 kHz. An I.F. of 10.7 MHz is often employed.

Except for these frequency differences, the FM receiver follows the design of the AM type until the limiter is reached. There the varying amplitude signals are reduced to a common level and the signal passed to the discrimi-

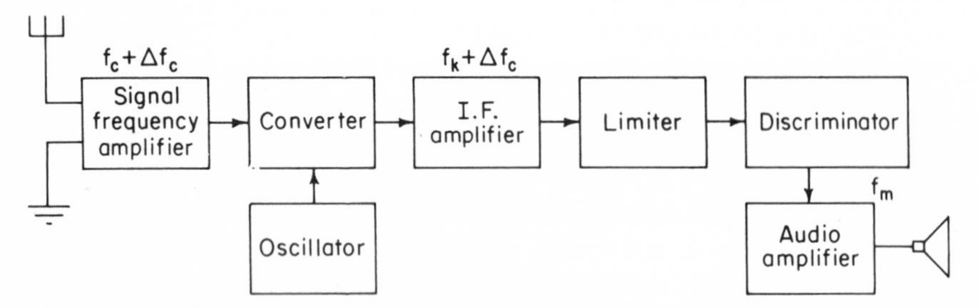

Figure 17-23. Block diagram of a limiter–discriminator FM receiver.

nator. The output of the discriminator is the original signal, varying in amplitude at f_m, which is further amplified and applied to the output device, usually a loudspeaker.

17-12 PHASE SYNCHRONIZATION

There are needs for specialized detectors in which it is desirable that a local oscillator be synchronized and locked in phase to a received sinusoidal signal. This operation is possible by employing the basic phase-comparison circuit of Fig. 17-24, with its variational output connected to a variable reactance phase-correction circuit in the local oscillator.

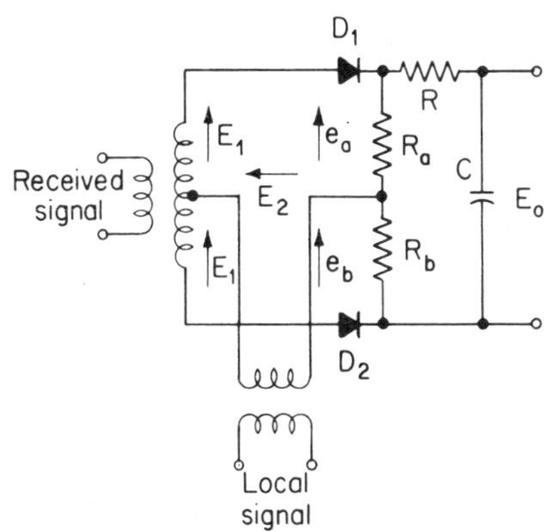

Figure 17-24. Phase comparison circuit.

The inputs to the diodes are

$$e_a = E_{2m} \sin \omega t + E_{1m} \sin (\omega t + \varphi)$$

$$e_b = E_{2m} \sin \omega t - E_{1m} \sin (\omega t + \varphi)$$

After rectification, assuming a linear dynamic characteristic, the voltages across R_a and R_b are subtracted and applied to the RC low-pass filter circuit. The resultant voltage term is

$$E_o = \frac{2E_{1m}}{\pi} \cos \varphi$$

This voltage will be positive for a leading phase angle and negative for a lagging angle, and will vary smoothly with φ.

This varying voltage is applied to a voltage-variable capacitance diode, across the oscillator resonant circuit, as for FM modulation. This diode will cause the oscillator resonant frequency to vary so as to maintain a constant phase relation to the input signal.

17-13 FREQUENCY SYNTHESIS TECHNIQUES

The generation of accurate, preselected frequencies over a wide range is a systems problem—that of combining various circuits and techniques into a useful device. In equipment of this nature the frequency is not selected by a continuous tuning operation, but by push button or other digital means to yield the desired frequency to a sufficient degree of precision.

Most directly, the desired frequency is the result of frequency division, mixing, and filtering as indicated in Fig. 17-25. A standard fixed frequency, usually a crystal oscillator operating at 1 MHz, is used with harmonic generators to generate the first nine harmonics which are made available on a

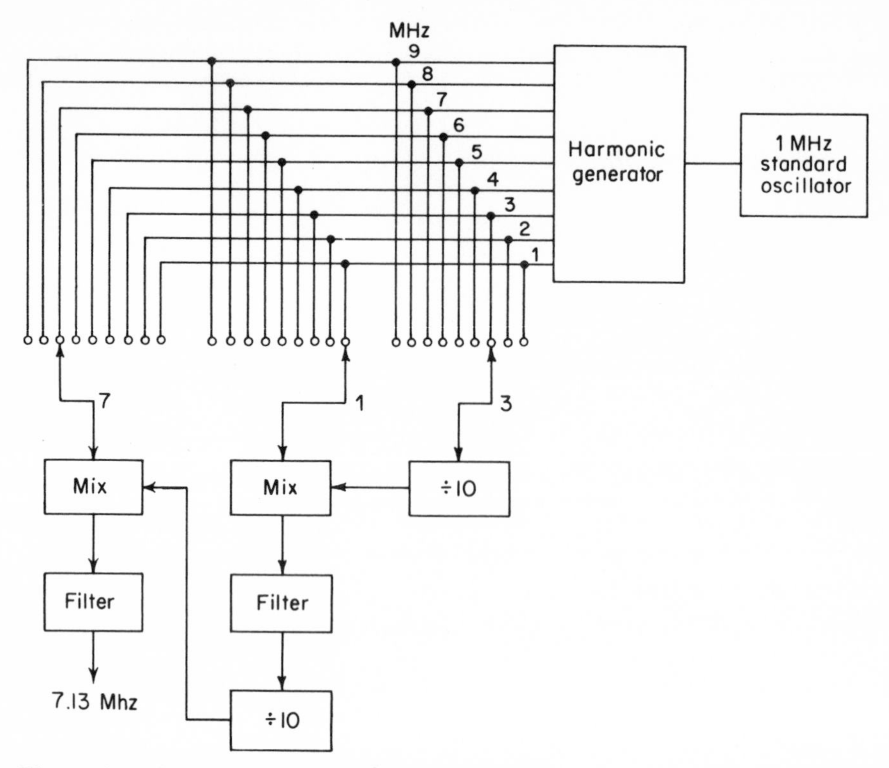

Figure 17-25. Frequency synthesis by direct methods.

set of switches. To generate a frequency of 7.13 MHz, the harmonic at 3 MHz is selected and divided by 10, in circuits described in Chapter 18. The result is added to the 1 MHz frequency chosen by the second switch, and the result divided by 10. Finally this result is added to the 7 MHz harmonic to yield 7.13 MHz as the output. Filters are used to eliminate spurious modulation products.

A second method of synthesis is diagrammed in Fig. 17-26, where the frequency generated by a voltage-controlled oscillator is compared to the output of a low-frequency synthesizer. The error signal from the frequency comparator or discriminator is used to change the frequency of the controlled oscillator with a variable reactance circuit. This system is suited to the generation of higher frequencies than are possible with the system of Fig. 17-25, since the latter is limited to the range in which stable decade frequency dividers will operate.

Resettability of tuned-circuit oscillators has always been a problem, and the frequency synthesizer offers a solution. By use of sufficient dividers and mixers, the degree of frequency resolution can be made as high as desired.

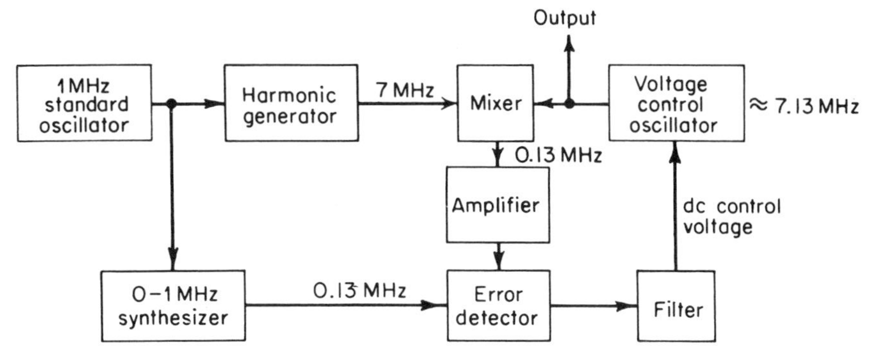

Figure 17-26. Indirect frequency synthesis.

The absolute accuracy of frequency is equal to that of the reference oscillator, and this can be made high because it operates at a fixed frequency.

17-14 PARAMETRIC CONVERTERS

Frequency converters at microwave frequencies must be low in internal noise if the weak signals encountered in radar or space communications are to be received. The output of such circuits includes the signal, thermal noise power $kT\,\Delta f$ from the antenna, and thermal and shot noise contributed by the amplifying device.

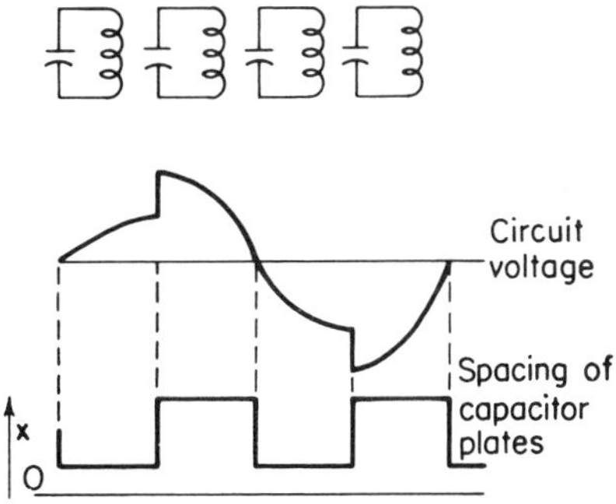

Research to develop converters with lower noise output has led to several devices, among which is the *parametric amplifier*. This device is an energy converter which operates with nonlinear reactances or parameters, the nonlinear solid-state capacitor being considered here.

As an analogy, consider a resonant circuit, including a capacitor whose plates may be mechanically moved, thus changing its capacitance. As indicated in Fig. 17-27, the capacitor plates are pulled apart at the voltage maximum, decreasing C and raising E because $E = q/C$ and the charge is unchanged. When the voltage E goes through zero the plates are returned to their original close spacing. Whenever the capacitance is decreased the voltage increases and voltage amplification is obtained.

Circuit voltage

Spacing of capacitor plates

x

0

Figure 17-27. Principle of the parametric amplifier.

The energy required to pull the plates apart against the attraction of the electric field is "pumped" into the circuit by an auxiliary source, and is stored in the electric field. Since the plates are pushed together at a time of zero charge or voltage, the energy is not returned to the pump but is ultimately transferred to the resonant circuit.

It can be seen that the frequency of the energy pump f_p must be exactly twice, and in phase with, the signal frequency f_s. This is a situation difficult to achieve and is called the degenerate case. In practical situations, f_s may be somewhat different from $f_p/2$, and the signal and pump drift into and out of periods of favorable interaction, giving beats and somewhat less than the full theoretical gain.

Any nonlinear reactance which can be time varied may be employed to transfer the energy, but the device which made such systems practical was the *varactor*, or solid-state diode. Under reverse bias, the thickness of the depletion layer and its capacitance are nonlinear functions of the applied voltage.

In general, parametric amplifiers are frequency converters; they are amplifiers when the output is at the same frequency as the input. In the circuit of Fig. 17-28(a) the pump frequency f_p is twice the signal frequency f_s. Since E_p is made large with respect to E_s, the voltage-variable capacitance is changed at pump frequency, and when the two oscillations are in proper phase relation, the signal builds up. Since exact equality of frequency of $f_p = 2f_s$ and the appropriate phase cannot be met in actuality, it is convenient to allow $f_p \cong 2f_s$, and a difference frequency $f_i = f_p - f_s$ will appear at the output.

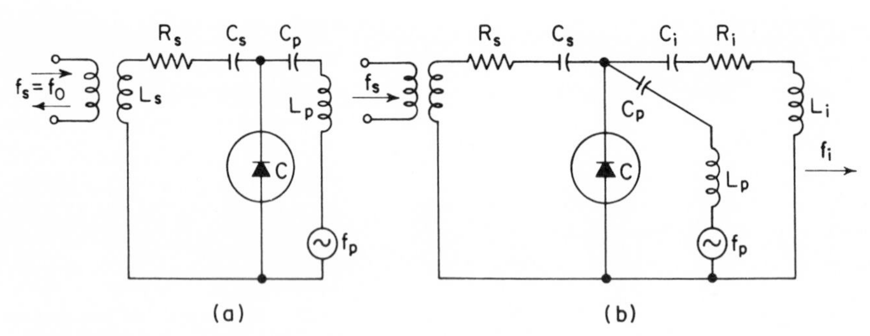

Figure 17-28. Parametric amplifiers: (a) synchronous; (b) general form.

The phase and frequency limitations can be relaxed if we add an *idler* circuit $L_iC_iR_i$, as at Fig. 17-28(b), resonant at the difference frequency $f_i = f_p \pm f_s$. This idler circuit serves as an additional energy reservoir, accepting energy from the pump or f_s circuit and storing it until needed, and releasing it at the proper time and phase to provide gain in the signal circuit. In fact, with f_p differing greatly from f_s, gain can be obtained only if the idler circuit is present.

The variable capacitance may be assumed to change as

$$\frac{1}{C} = \frac{1}{C_d} \sin \omega_p t$$

in step with the pump frequency, and the current in the signal circuit is

$$i_s = I_s \sin \omega_s t \tag{17-26}$$

The voltage across C, due to this signal current, is

$$v_C = \frac{1}{C} \int i_s \, dt = \frac{1}{C_d} \sin \omega_p t \int I_s \sin \omega_s t \, dt$$

This circuit is a modulator of the variable impedance form, and the result shows two sidebands, as

$$v_C = \frac{I_s}{2\omega_s C_d} [\sin (\omega_p t + \omega_s t) + \sin (\omega_p t - \omega_s t)] \tag{17-27}$$

after neglecting constant phase angles.

If we now require that

$$\omega_p = \omega_s + \omega_i$$

where $\omega_s = 1/\sqrt{L_s C_s}$ and $\omega_i = 1/\sqrt{L_i C_i}$, then

$$v_C = \frac{I_s}{2\omega_s C_d} [\sin \omega_i t + \sin (2\omega_p - \omega_i) t] \tag{17-28}$$

The current in the idler circuit is

$$i_i = \frac{v_C}{R_i} = \frac{I_s}{2\omega_s C_d R_i} \sin \omega_i t \tag{17-29}$$

the other sideband being far from idler resonance.

The current i_i in the idler circuit will create an additional voltage across C as

$$\Delta v_C = \frac{1}{C} \int i_i \, dt = \frac{I_s}{4\omega_s \omega_i R_i C_d^2} [\sin (\omega_p + \omega_i) t + \sin \omega_s t] \tag{17-30}$$

But Δv_C will then introduce an additional current component $\Delta i_s = \Delta v_C / R_s$ in the signal circuit. Neglecting the first frequency term as far from signal frequency resonance, this additional current component is

$$\Delta i_s = \frac{I_s}{4\omega_s \omega_i R_i R_s C_d^2} \sin \omega_s t \tag{17-31}$$

and this demonstrates the ability of the idler circuit to automatically produce the proper phase relations. This result is additive to the signal current, and gain has been obtained.

If the output frequency is chosen as $f_o > f_s$, the amplifier is known as an *up-converter*, and when $f_o < f_s$, it is known as a *down-converter*. Frequency relations leading to useful gain are shown in Fig. 17-29.

For Fig. 17-29(b) and (c), with f_o higher than f_p and output taken from the idler tank, the up-converter is stable and the gain is the ratio f_o/f_s. In

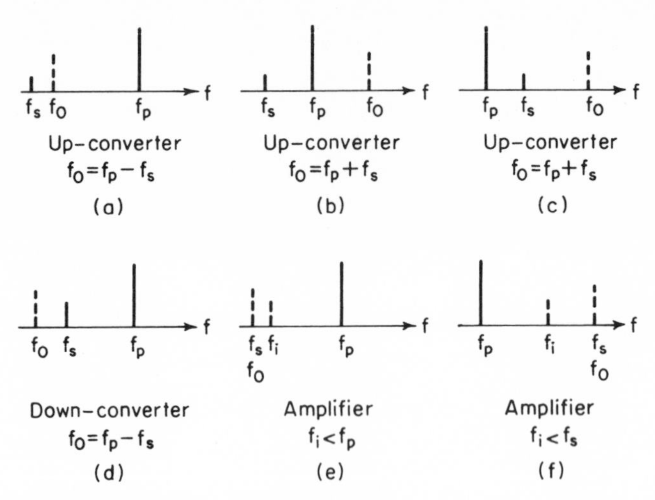

Figure 17-29. Parametric amplifier forms.

(a), the gain is $-f_o/f_s$, and the negative sign indicates that the feedback is positive, with gain obtained at the cost of stability. The theoretical gain of the down-converter at (d) is also f_o/f_s, but less than unity in this case. The circuit is regenerative and the gain may be unstable.

In general, ΔC values of diodes will range up to 10 pF, and will usually be found larger in high reverse-voltage units. Silicon appears better than germanium. Usual pump voltages will drive the diode up to about one-half of the reverse voltage rating.

The noise introduced by vacuum tubes and transistors is due to the inherent particle nature of current or charge flow. In a voltage-variable capacitor there is theoretically no noise because there is no current, only a variation of the width of the space-charge region. The diode is not perfect, and a small amount of thermal noise will be introduced by the bulk resistance of the diode material. If the complication is warranted, this internal noise can be reduced by refrigeration of the diode to liquid nitrogen temperatures.

PROBLEMS

17-1. A diode and load have a characteristic

$$i = 10^{-5}e_s + 3.5 \times 10^{-6}e_s^2$$

for small positive inputs. If an amplitude-modulated wave, having a carrier peak amplitude (unmodulated) of 2.5 V is modulated 80 per cent and applied to the diode, find all frequency components and their amplitudes across a load of 60,000 ohms. Also compute the per cent second harmonic distortion introduced into the output signal.

17-2. A sinusoidal modulated signal is applied to an envelope detector with $R = 100{,}000$ ohms. If the per cent of modulation is varied for a carrier of 10 V, plot the variation of output against per cent modulation, using the chart for the diode of Fig. 17-11.

17-3. A carrier voltage of 10 V rms is applied to an envelope detector. The load R is 250,000 ohms and the current I_o is 30 μA.
(a) Find the efficiency of detection.
(b) Find the load the detector presents to the tuned circuit.
(c) What voltage is available for AVC use?

17-4. An amplitude-modulated wave of carrier peak value of 12 V, 10^6 Hz, with 65 per cent modulation, is applied to an envelope detector with $R = 200{,}000$ ohms, $r_d = 27{,}000$, and $C = 150$ pF. Find the modulation-output voltage, the voltage available for AVC action, the detection efficiency, and the load which the circuit imposes on the tuned input circuit.

If the Q of the unloaded resonant circuit is 150, and the inductance is 40 μH, find the bandwidth of the circuit when it is loaded with the diode detector.

17-5. In an envelope detector circuit, with $R = 250{,}000$ ohms and $C = 250$ pF, plot as a function of modulation frequency the highest value of m that can be employed without introduction of distortion due to inability to follow the modulation envelope.

17-6. A 6H6 diode is used in the circuit of Fig. 17-7(a). If $R_1 = 250{,}000$ ohms, $R_2 = 500{,}000$ ohms (tap at top), $C_1 = 100$ pF, and $C_2 = 0.05$ μF, with a carrier amplitude of 10 V at 1 megahertz frequency and modulation of 12,000 hertz, find:
(a) The limiting modulation percentage above which envelope distortion increases rapidly.
(b) The circuit changes that should be made to reduce this difficulty.

17-7. In the circuit of Fig. 17-7(a), $R_1 = 200{,}000$ ohms, $R_2 = 250{,}000$ ohms, $C_1 = 100$ pF, and $C_2 = 0.001$ μF. For the tap on R_2 at maximum find the highest permissible value of m if peak clipping is to be avoided.

17-8. (a) In the circuit of Fig. 17-7(a), $R_1 = 200{,}000$ ohms, $C_1 = 100$ pF, and $C_2 = 0.1$ μF. Plot a curve of the value of R_2 against the corresponding value of m which may be employed without peak clipping being present.
(b) If the circuit of Fig. 17-7(b) is used and $R_1 = 50{,}000$ ohms, again plot a curve of R_2 against the value of m that may be used without clipping distortion; $R_3 = 100{,}000$ ohms.

17-9. A resonant circuit of $Q = 50$ and $L = 10$ μH is used as a detector of FM waves. If the center frequency is 10 MHz and the circuit is tuned so that the center frequency is placed below the resonant frequency where the power input is one-half the maximum, find the change in voltage across the resonant circuit for a frequency deviation of 50 kHz above and below the center frequency if a constant $I = 50$ μA is applied.

17-10. A frequency discriminator operates at 10.7 MHz. If the primary is tuned with a 50 pF capacitor, $Q_1 = Q_2 = 80$, find the output to the diodes in terms of volts per kilohertz deviation from the center frequency, using usual design parameters and $g_m = 0.0035$.

17-11. (a) Draw a block diagram of a superheterodyne circuit for AM reception with one radio-frequency amplifier, two intermediate-frequency stages, and one audio amplifier. If the received signal has a carrier of 950 kHz, modulated at 2000 Hz, and the intermediate-frequency amplifier is tuned to 455 kHz, designate the frequencies present in each amplifier stage.

(b) If the oscillator frequency is above the signal frequency, what image frequency may cause interference?

17-12. Capacitor C_2 in Fig. 17-16 is chosen to be 50 pF. If the detection efficiency is 0.90, and $R_a + R_b = 250K$, $Q_1 = Q_2$, determine an appropriate design for the input transformer for the discriminator to receive a band of ± 200 kHz at a center frequency of 22 MHz. By plotting a response curve, show whether the design will yield satisfactory linearity.

17-13. A single-sideband wave is applied to a frequency translator along with a single frequency $E_q \sin \omega_q t$. Show that the SSB signal may be raised or lowered in frequency by this method.

REFERENCES

1. Foster, D. E., and S. W. Seeley, "Automatic Tuning, Simplified Circuits, and Design Practice," *Proc. IRE*, Vol. 25, p. 289 (1937).
2. Roder, H., "Theory of the Discriminator Circuit for AFC," *Proc. IRE*, Vol. 26, p. 590 (1938).
3. Seeley, S. W., "The Ratio Detector," *RCA Rev.*, Vol. 8, p. 201 (1947).
4. Black, H. S., *Modulation Theory*. D. Van Nostrand Co., Inc., Princeton, N. J., 1953.
5. Manley, J. M., and H. E. Rowe, "Some General Properties of Nonlinear Elements," *Proc. IRE*, Vol. 44, p. 904 (1956).
6. *Proceedings of the IRE*, special issue on single sideband modulation, Vol. 44 (December 1956).
7. Uhlir, A., Jr., "The Potential of Semiconductor Diodes in High-Frequency Communication," *Proc. IRE*, Vol. 46, p. 1099 (1948).
8. Schwarz, M., *Information, Transmission, Modulation, and Noise*. McGraw-Hill Book Company, New York, 1959.
9. Blackwell, L. A., and K. L. Kotzebue, *Semiconductor-Diode Parametric Amplifiers*. Prentice-Hall, Inc., Englewood Cliffs, N. J., 1961.
10. Hakim, S. S., and R. Barrett, *Transistor Circuits in Electronics*. Hayden Book Co., New York, 1964.
11. Chang, K. K. N., *Parameteric and Tunnel Diodes*. Prentice-Hall, Inc., Englewood Cliffs, N. J., 1964.

18

WAVE-SHAPING AND SWITCHING

Another important area of electronic study is that concerned with the generation of nonsinusoidal voltage or current wave forms, in which the ability of the transistor or tube to serve as a fast and almost resistance-less switch is emphasized. Applications for circuits of this nature occur in television, computers, and pulse communication.

18-1 EXPONENTIAL CIRCUIT RESPONSE

A series combination of R and C appears in Fig. 18-1. If v_o is the initial potential across C, then

$$\left(\frac{1}{sC} + R\right)i + v_o = E$$

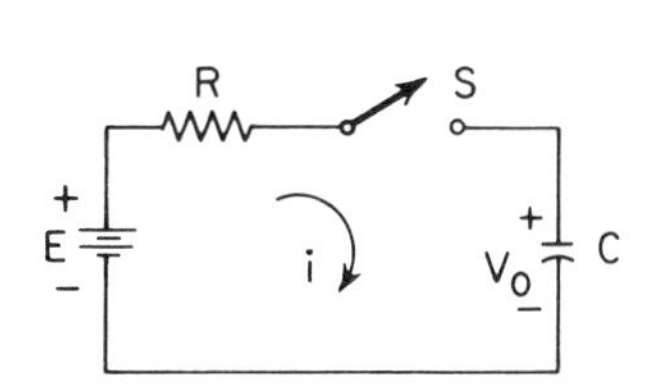

Figure 18-1. Simple RC circuit.

and the well-known solution is

$$i = \frac{E - v_o}{R}\epsilon^{-t/RC} \tag{18-1}$$

The potential across the capacitor is

$$v_C = E - Ri$$
$$= v_o + (E - v_o)(1 - \epsilon^{-t/RC}) \tag{18-2}$$

The currents and voltages in such circuits respond in accordance with $\epsilon^{-t/RC}$ and $1 - \epsilon^{-t/RC}$, involving the time constant RC.

At $t = 0$ the current is $i_o = (E - v_o)/R$ and at $t = \infty$ the current is zero. At $t = 0$ the capacitor potential is v_o and at $t = \infty$ it is E. Between these limits the changes follow an exponential path with values given by Fig. 18-2. For estimation purposes a change of voltage or current in an RC circuit is 63 per cent completed in RC seconds or one time constant, 88 per cent com-

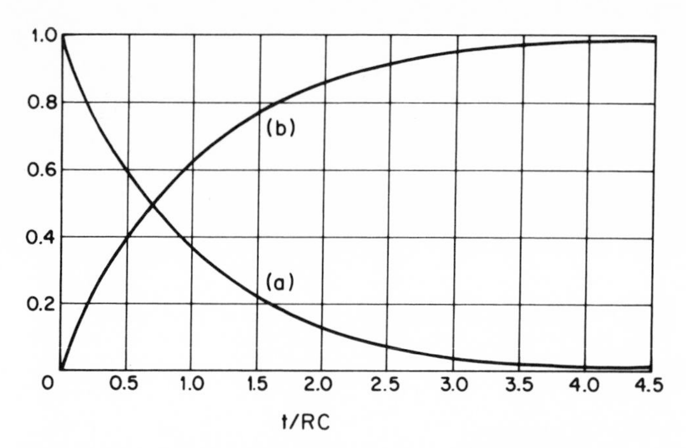

Figure 18-2. (a) Value of $\epsilon^{-t/RC}$; (b) value of $1 - \epsilon^{-t/RC}$.

pleted in $2RC$ seconds, and may be considered fully completed in $4RC$ seconds when the variable has reached 98 per cent of its infinite time value.

Similar changes occur in the R, L circuit when measured in terms of the time constant L/R.

18-2 THE DIFFERENTIATOR OR HIGH-PASS RC FILTER

Summing the currents at A, Fig. 18-3, leads to

$$e_o = \frac{RCs}{1 + RCs} e_i \tag{18-3}$$

The effect of the circuit on an input rectangular pulse is illustrated in Fig. 18-4. After an initial transient, the input step changes the output abruptly by an amount E, since the voltage across C cannot change instantaneously.

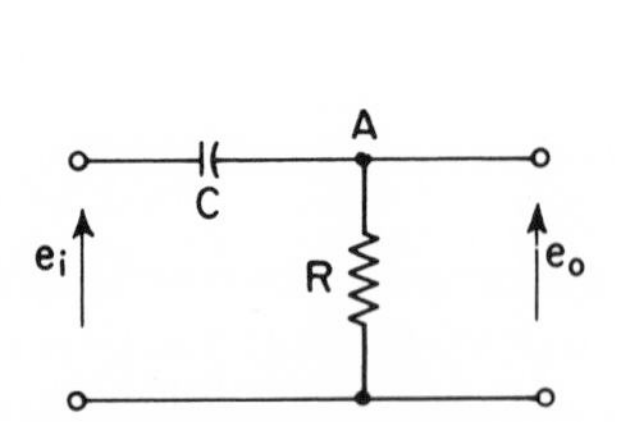

Figure 18-3. An RC differentiating network.

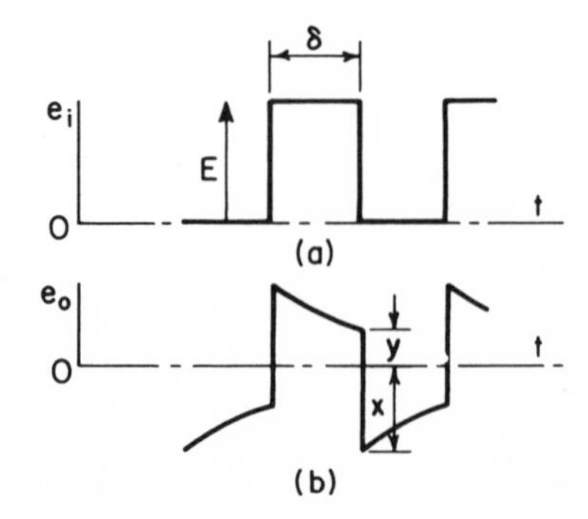

Figure 18-4. Square wave applied to the circuit of Fig. 18-3.

As the capacitor charges, i and e_o fall exponentially. When e_o has fallen to value y at $t = \delta$, the input voltage drops E volts, and carries e_o with it in an amount $y + x = E$. Capacitor C now charges exponentially with reverse polarity. After the transient the zero axis must adjust so that the positive and negative areas are equal, since they represent charge moved, and the net must be zero.

It is apparent that

$$y = x\epsilon^{-\delta/RC}$$

so that

$$x = \frac{E}{1 + \epsilon^{-\delta/RC}} \tag{18-4}$$

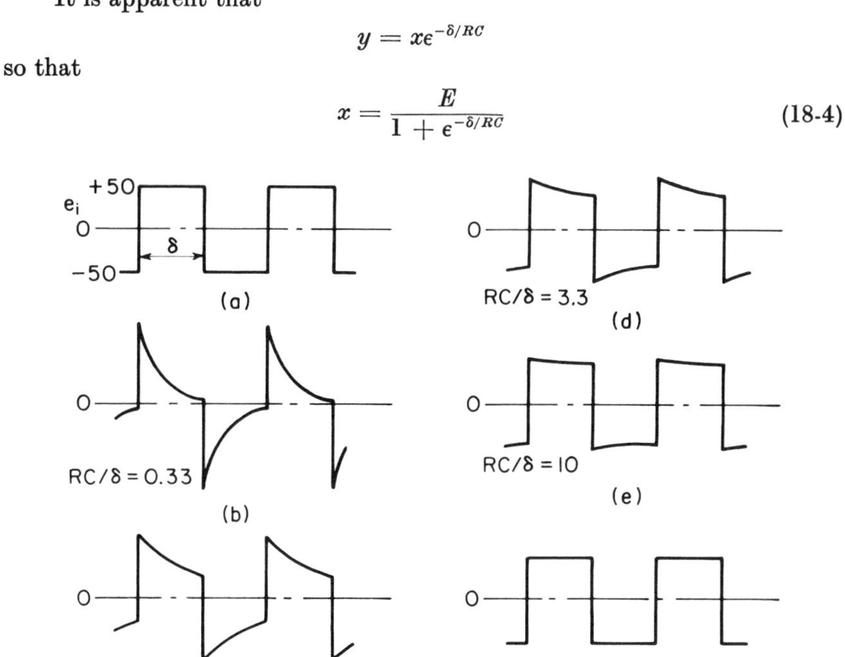

Figure 18-5. Response of an *RC* circuit to a square wave.

Figure 18-5 shows the effect of the ratio RC/δ on the output voltage from the circuit. For good fidelity of output, $RC \gg \delta$ and this is the requirement when the circuit is used as an *RC* amplifier coupling. For $RC \ll \delta$, Eq. 18-3 reduces to

$$e_o \cong RCse_i \tag{18-5}$$

and the output voltage approaches the form of a spike, to be expected from the derivative of a rectangular wave. With $RC \ll \delta$ the circuit is a *differentiator*.

In a steady-state sinusoidal form, Eq. 18-3 can be written

$$E_o = \frac{E_i}{\sqrt{1 + \omega_1^2/\omega^2}} \underline{/\tan^{-1}(\omega_1/\omega)} \tag{18-6}$$

where $\omega_1 = 1/RC$. This is the equation of the usual RC coupling and of a high-pass RC filter, with the voltage ratio $1/\sqrt{2}$ and the angle 45° at the low-frequency band limit $f_1 = \omega_1/2\pi$, where $R = 1/\omega C$.

18-3 THE INTEGRATOR OR LOW-PASS RC FILTER

The *RC integrator* or low-pass RC filter is shown in Fig. 18-6. Summing the currents at A leads to

$$e_o = \frac{e_i}{1 + RCs} \cong \frac{e_i}{RCs} \tag{18-7}$$

Figure 18-7 illustrates the effect on a rectangular wave applied as input e_i. While the output amplitude is small, for $RC \gg \delta$ as at (b), the output is an integral of the input, and this requirement that RC be large with respect to δ leads to the approximation in Eq. 18-7 above. The wave is then a close approach to a sawtooth.

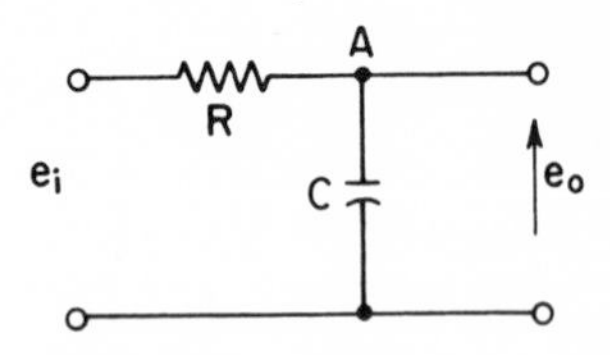

Figure 18-6. An RC integrating network.

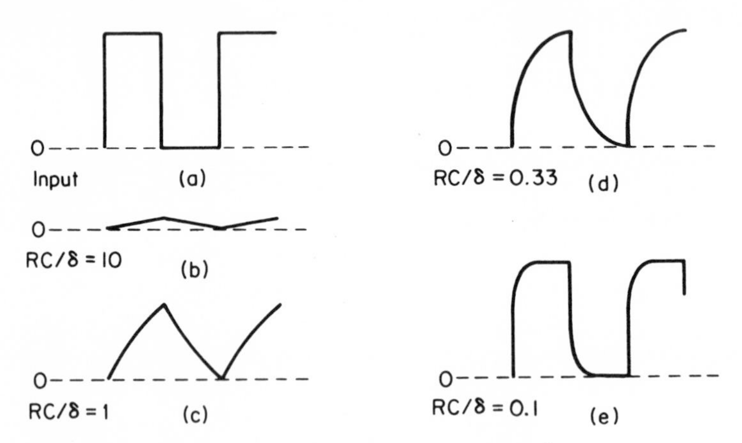

Figure 18-7. Effect of an RC integrator on a rectangular wave.

For good fidelity of output wave form it is necessary that $RC \ll \delta$; a rule of thumb calls for $RC/\delta \ll 0.16$.

With sinusoidal input

$$E_o = \frac{E_i}{\sqrt{1 + \omega^2/\omega_2^2}} \underline{/\tan^{-1}(\omega/\omega_2)} \tag{18-8}$$

where $\omega_2 = 1/RC$ and is the upper half-power or band limit frequency for the circuit when viewed as a low-pass filter. At ω_2, $R = 1/\omega C$.

18-4 THE DIODE AS A SWITCH

The diode serves as a polarity-controlled switch. While having the desirable property of simplicity, it is not quite perfect electrically. In the forward direction it has a small resistance, with a voltage drop in the range of 0.2 to 1.0 V. In the reverse direction its resistance may vary from several hundred thousand ohms to several megohms.

Charge storage effects serve to limit the speed of switching, or put "tails" on signals when the applied potential reverses. Under forward bias, the density of the minority charge on each side of the junction is very high, as indicated in Fig. 18-8(a). This situation was previously discussed in Chapter 2. When the potential reverses as at t_o in (b), this minority charge must be swept out across the junction, until the charge densities at the junction become essentially zero. This reverses the current as shown in (c). The current is assumed to be determined by the diode load R_L, and if the reverse voltage magnitude equals the previous forward voltage magnitude, then the current $|-i_F| = |i_F|$ and the current will be constant until the stored charge is removed at t_1. Time $t_1 - t_o$ is called the *diode storage time.*

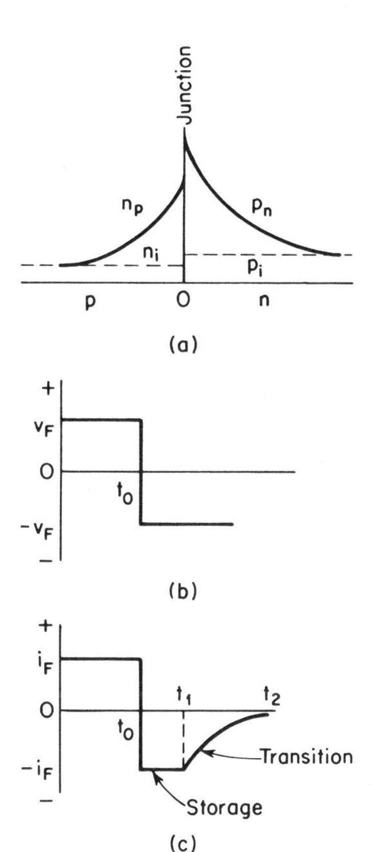

Figure 18-8. (a) Stored charge under forward bias; (b) voltage applied to the diode; (c) diode current after potential reversal.

At $t = t_1$, there remain only a few stored charges still to be swept out of remote regions and the transition capacitance across the reverse-biased junction begins to charge. The reverse current ceases when the potential across the capacitance reaches $-v_F$ at some time t_2. Since this is indefinite, the defined *reverse recovery time* ends when the diode back resistance has risen to a value specified by the manufacturer, and switching diodes with recovery times as short as a fraction of a nanosecond are available.

During forward conduction a current i crosses the junction and provides minority charges for the storage region. At the same time, the minority charges in the region are diffusing from the junction and recombining with majority charges at more remote locations. The rate of change of charge in the storage region is then expressible by

$$\frac{dQ}{dt} = -\frac{Q}{\tau} + i$$

where Q is the stored charge and τ is the minority carrier lifetime, as before. This is the charge continuity equation.

The diode current I far from the junction is made up of majority charges moving to recombination. At steady state, with $dQ/dt = 0$, it is necessary that $i = I$ and so

$$I = \frac{Q}{\tau}$$

That is, the diode current supplies carriers as needed for the process of recombination, or recombination occurs at a rate as called for by a diode current I of majority carriers.

Taking the derivative with respect to v and letting $dQ/dv = C_T$, then

$$\frac{dI}{dv} = \frac{C_T}{\tau}$$

Recognizing τ as a time constant, it is possible to identify its components as

$$r_d = \frac{0.026}{I}, \qquad C_T = \frac{\tau dI}{dv}$$

Resistance r_d has been previously discussed as the diode forward resistance. Values of C_T are found in the range from 20 pF upwards, and are proportional to the current.

It is possible to vary the impurity concentration in such a way that the minority carriers are restricted to a region very close to the junction. If the transition capacitance is also small, such diodes are suitable for use with fast rising wave forms and are known as *snap-off diodes*.

An even faster form of diode switch, operating with a recovery time of perhaps 50 picoseconds, is the *hot-carrier diode*. This diode may consist of a junction of n semiconductor and a metal, with the metal positive in the forward direction. Electrons then pass readily from the n semiconductor and join the free electrons of the metal as a majority-carrier current. There is no storage of minority carriers, and when the potential is reversed the major effect is that of charging the junction capacitance.

In the forward direction the semiconductor electrons enter the metal with energies higher than the mean electron energy there, and are referred to as *hot electrons*. Hot-hole diodes can also be made, but the electron form is preferred because of the increased speed of operation with the higher electron mobility.

18-5 WAVE CLIPPERS

A simple circuit for shaping a voltage wave by *limiting* or *clipping* is shown in Fig. 18-9. A diode in series with an emf is shunted across the circuit. The source should have a resistance R, large with respect to the forward resistance

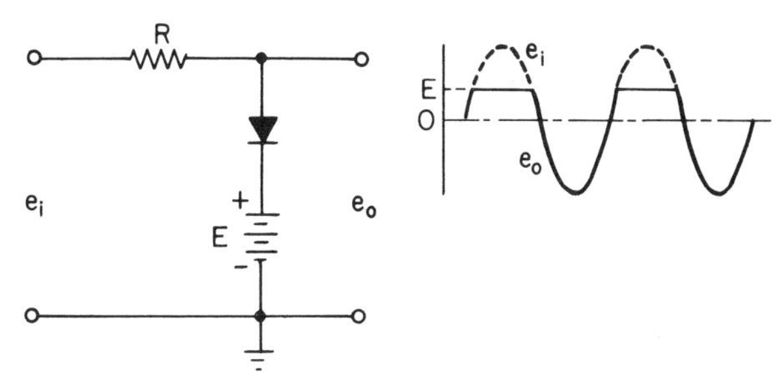

Figure 18-9. Clipping circuit, operating at a level other than zero.

of the diode, r_d. For all values of e_i above E the diode appears as a short circuit, and the output is at E volts, the excess of e_i over E being iR drop. For values of e_i less than E, the diode has a reversed potential and is open, and the full input voltage is passed to the output.

The dual clipper of Fig. 18-10 is used to limit a signal to constant amplitude by clipping both positive and negative excursions; erratic pulses or noise may be reduced in this way. Triode or transistor limiting by driving the device into saturation and beyond cutoff, as shown in Fig. 18-11, has been previously discussed for amplitude limiting in FM reception.

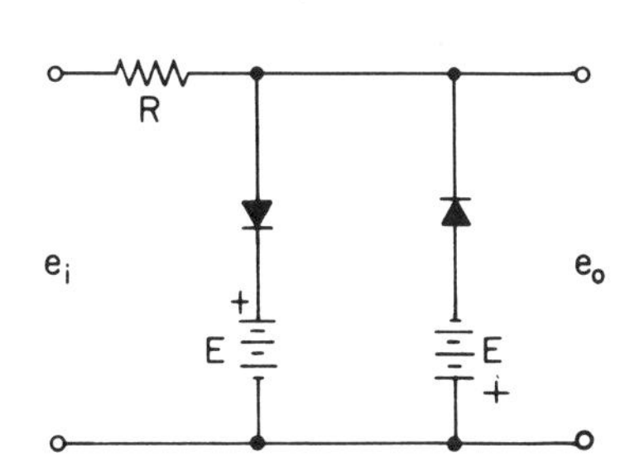

Figure 18-10. Double-diode clipper for removing amplitude variation.

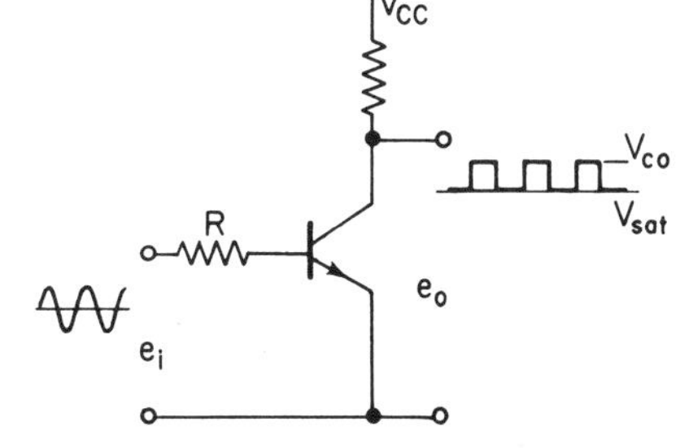

Figure 18-11. Overdriven transistor for saturation and cutoff limiting.

18-6 CLAMPING CIRCUITS

The *diode clamp* or *dc restorer* is used to insert a dc component into a wave after ac amplification. A simple circuit is shown in Fig. 18-12, as an *RC* coupling circuit of long time constant and a shunt diode. The source and diode should be of low resistance.

When the wave of (b) goes positive, the diode conducts and connects point A to the ground potential. The full voltage e_i then appears across C and the output is clamped at zero, as in (c). When the input falls, the right side of C must instantaneously fall by the same amount, taking point A to

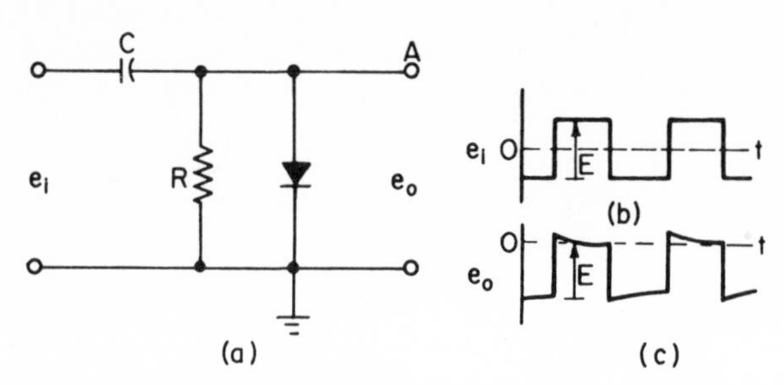

Figure 18-12. Diode clamping.

the value of negative E. If the RC time constant is very long with respect to the wave period, then C cannot appreciably discharge, so that A remains essentially at $-E$ during the half period. When the input rises, the output again goes up to zero. Any charge which C loses during the negative interval, through the large RC time constant, is immediately resupplied through the low time constant source circuit. Therefore the small positive pips represent the recharging actions.

The output may be clamped at any desired level by inserting a potential of appropriate sign and magnitude in series with the diode.

18-7 THE TRANSISTOR AS A SWITCH

As a switch, a transistor is normally operated in the *cutoff region* as an open circuit, and in the *saturation region* as a closed circuit. The active or linear region, of importance in amplifiers, is passed through abruptly along the load line in switching from cutoff to saturation, and is not of importance in switching applications.

The cutoff region finds both junctions reverse-biased, and only reverse saturation currents are present, or $I_E = 0$ and $I_C = I_{CO}$ in the common-base circuit. In the common-emitter circuit $I_C = I_{CO}/(1 - \alpha)$. In germanium, α is appreciable at small currents and a small reverse base bias may be needed to bring the transistor to cutoff. For silicon, α is close to zero at currents of the order of I_{CO}, and with $I_B = 0$ the transistor is cut off.

For ideal switching the value of $v_{CE\,\mathrm{sat}}$ should be small. The saturation resistance $R_{CE\,\mathrm{sat}} = v_{CE\,\mathrm{sat}}/i_{\mathrm{sat}}$ should also be small, and is usually specified by the transistor manufacturer. This resistance is represented by the reciprocal of the slope of the straight i/v characteristic to the left of the knee. For more precision, some manufacturers plot this region in detail as a *saturation characteristic*. Saturation voltages range from a few tenths of a volt to several volts, depending on the material and manufacturing method of the transistor. In switching from saturation to cutoff, large-signal operation of the transistor is implied, and the approximation of the Ebers-Moll circuit of Section 6-11 is useful in quantitative performance prediction.

Voltage limitations and thermal dissipation restrictions, as previously discussed for large signal performance, also apply; because operation is at either cutoff or saturation with low $v_{CE\ \mathrm{sat}}$, the thermal considerations are not usually limiting.

The transient effects in turn-on and turn-off of a transistor are illustrated in Fig. 18-13. The input base-emitter voltage is indicated by the dashed pulse, with the output collector current shown by the solid curve. Time T_d is the *delay time*, during which the emitter transition capacitance $C_{b'e}$ and the collector-base capacitance $C_{b'c}$ are being charged from below cutoff to a level at which the current is ten per cent of its final value. In addition to the time of charge of these capacitances, there is a finite time needed for the charges to propagate across the base width and begin to affect the collector current.

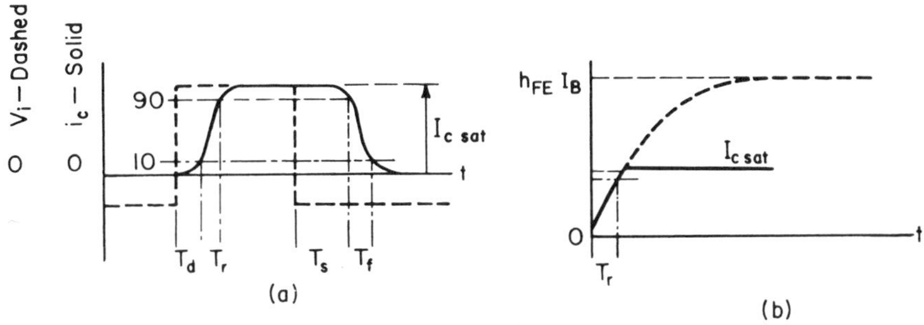

Figure 18-13. (a) Input and output wave forms for a transistor switch; (b) effect of overdrive on rise time.

The effective charging time constant is that due to the capacitance $C_{b'e} + C_{b'c} = C_d$ and the input resistance $R_s + r_{bb'}$. If the transistor has not been previously driven far into cutoff, then the amount of charge needed, and the delay time, are reduced.

The *rise time* T_r between the 10 per cent and 90 per cent points of the rising wave, is determined by the time constant due to the collector load resistance R_L and the collector-base transition capacitance as $\tau = h_{FE}C_{b'c}R_L$. The charging curve is illustrated in Fig. 18-13(b), and shows the effect of providing overdrive in the base current. If $h_{FE}I_B$ is made large with respect to the base current needed to drive to saturation, or with respect to the saturation collector current, then the turn-on time is reduced.

The *storage time* T_s is related to the excess minority charge in the base plus an additional base charge resulting from the base being overdriven into saturation. The base current reverses at the beginning of T_s, but the collector current remains constant until the excess charge is swept out of the base region.

The *fall time* T_f is related to the same factors that caused T_r, the rise time of the pulse.

All these times serve to delay transistor response to pulses; the reduction of switching time is the subject of much device and circuit research.

18-8 CATHODE-RAY SWEEP VOLTAGE GENERATION

The linear time-base or sweep wave form for electrostatic deflection of a cathode-ray tube beam must move the spot across the screen as a linear function of time, and return the spot rapidly to the starting point at the end of each sweep. This action requires a wave of sawtooth form, with a linear ramp rise and an abrupt fall to zero.

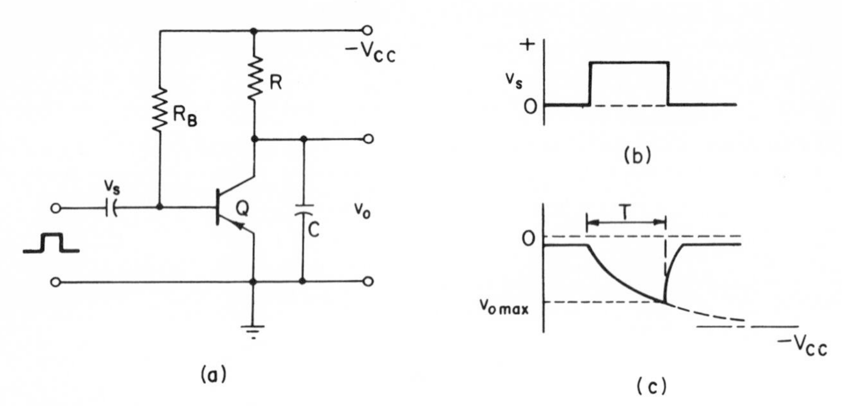

Figure 18-14. Triggered *RC* sweep and wave forms.

A simple sweep wave form can be generated by the rise of potential across a capacitor when C is charged from a constant potential through a resistor R, as in Fig. 18-14(a). Resistor R_B biases the transistor into saturation and the collector output voltage is close to zero. Capacitor C is connected across the saturated transistor and is discharged. A positive-going pulse of duration T is now applied to the input, driving the transistor into cutoff. The capacitor charges through R from $-V_{CC}$ and its potential changes as in (c), according to

$$v_o = -V_{CC}(1 - \epsilon^{-t/RC})$$

At the end of the input pulse or *gate* signal the transistor goes back to saturation, discharges C, and v_o rises rapidly to the saturation level. The output voltage, being an exponential, is a compromise with linearity; reasonable linearity can be approached if $T \ll RC$, or if $v_{o\,\max}/V_{CC}$ is small, where

$$T = RC \ln \frac{1}{(1 - v_o/V_{CC})} \tag{18-9}$$

A repetitive sawtooth can be obtained if the input is a repetitive rectangular wave.

18-9 THE MILLER-INTEGRATOR SWEEP

The *Miller integrator* of Fig. 18-15 provides a more linear sawtooth wave. Capacitor C charges with a nearly linear rate of potential rise, due to the integration process of the operational amplifier, to which the circuit is related.

With Q_2 in saturation and Q_1 at cutoff, point A is at V_{CC} and capacity C is charged to V_{CC}. When a square wave is applied to cut off Q_2, then Q_1 is turned on. The collector potential starts to fall, but it cannot fall rapidly because if it did so the potential change across C would drive the base negative, in turn causing the collector current to decrease and the collector potential to rise. In effect, the collector potential is forced to decrease at almost the same rate as does the capacitor potential, resulting in C being discharged at a nearly constant potential difference. This creates a quasi-constant discharge current and an almost linear change in capacitor potential.

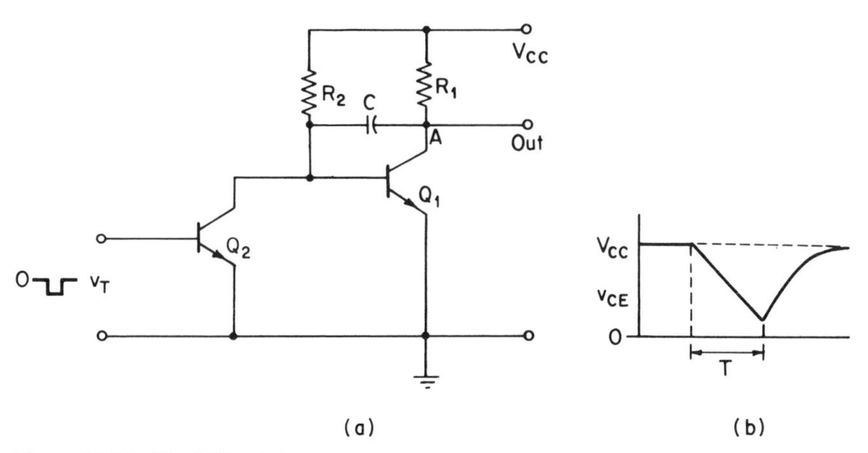

Figure 18-15. The Miller integrator sweep.

The capacitor discharges from an initial potential of $V_{CC} - v_{BE}$ with a discharge current

$$i = \frac{V_{CC} - v_{BE}}{R_2} \tag{18-10}$$

The potential across C is

$$v_{CE} - v_{BE} = \frac{1}{C}\int i\,dt = \frac{i}{sC}$$

Using Eq. 18-10 it is possible to write

$$v_{CE}\left(1 - \frac{1}{A} + \frac{1}{sACR_2}\right) = \frac{V_{CC}}{sCR_2} \tag{18-11}$$

After dropping $1/A$, since A must be large,

$$v_{CE} = \frac{AV_{CC}}{1 + sACR_2} \tag{18-12}$$

Subject to the condition that $v_{CE} = V_{CC}$ at $t = 0$, the solution of Eq. 18-12 is

$$v_{CE} = V_{CC}\left[1 - \frac{1}{CR_2}(1 - \epsilon^{-t/ACR_2})\right] \tag{18-13}$$

The effective sweep voltage is

$$v_{CE} = \frac{V_{CC}}{CR_2}(1 - \epsilon^{-t/ACR_2}) \tag{18-14}$$

In effect the circuit increases the time constant by the voltage gain A, and makes efficient use of V_{CC}.

The sweep continues until the end of the applied gating pulse, or until $v_{CE} = v_{\text{sat}}$. The latter condition is to be avoided, since it represents bottoming of the wave form. The return transient can be much shorter, since $R_1 \ll R_2$.

A pentode version of the circuit is similar, except that the gating or turn-on signal can be most easily introduced by applying a positive gating voltage to the suppressor grid.

18-10 THE BOOTSTRAP SWEEP

Feedback may be employed in another way to linearize the rise of potential across the capacitor of an RC sweep circuit. A capacitor voltage v_o rises with time as

$$v_o = \frac{1}{C}\int i\,dt$$

and if it is charged from a potential which rises similarly

$$v_i = E_{bb} + \frac{1}{C}\int i\,dt$$

then, in effect, the capacitor is being charged from a constant potential difference E_{bb}, and the capacitor voltage will rise linearly with time.

In the circuit of Fig. 18-16, T_1 is normally conducting but at the beginning of the sweep it is cut off by a negative gating signal; this leaves R_1, C_1, and C free with C uncharged. The cathode follower T_2 is caused to feed back a voltage equal to the voltage on

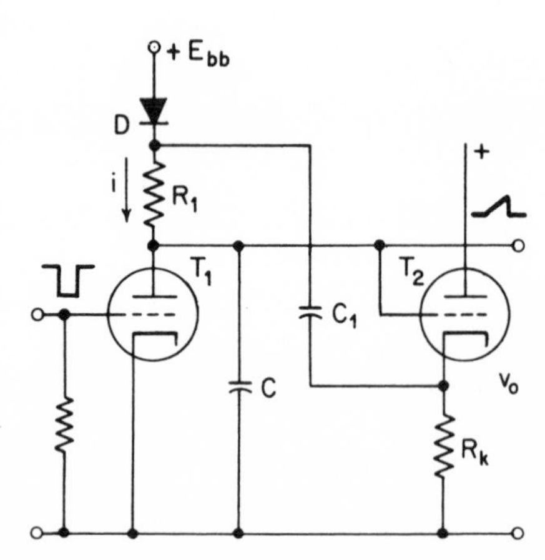

Figure 18-16. A form of bootstrap sweep circuit.

C, to the circuit charging C. Writing an equation through capacitors C and C_1:

$$iR_1 + v_o = v_{C2}(0) + Av_o$$

where A is the voltage gain of the cathode follower. However, $v_{C2}(0) = E_{bb}$ since initially $v_o = 0$. Then

$$iR_1 = E_{bb} - (1 - A)v_o \tag{18-15}$$

Since A is near unity for a cathode follower,

$$i \cong \frac{E_{bb}}{R_1} \tag{18-16}$$

indicating that the current charging C is nearly constant and v_o will rise almost linearly with time; the actual linearity is dependent on how close the gain A is to unity in Eq. 18-15.

It is assumed that C_2/C is large so that the potential of C_2 remains at E_{bb} during the sweep interval. The circuit feeds its output back to the input; thus it raises itself by its own bootstraps.

18-11 DEVIATION FROM LINEARITY IN RC SWEEPS

Severe accuracy requirements are often imposed on sweep wave forms in applications where time or distance is being measured. The inaccuracy or departure from linearity of the wave form may be defined in several ways, as shown in Fig. 18-17. A useful definition is that at (b), where the actual curve is compared to a straight line chosen to pass through the origin and the maximum output voltage v_1. This is realistic, since v_1 is often specified. The equation of such a line is

$$v = \left(\frac{v_1}{t_1/RC}\right)\frac{t}{RC} \tag{18-17}$$

making it convenient to measure t in time-constant units.

For the simple exponential sweep wave form of Section 18-8

$$v_o = (1 - \epsilon^{-t/RC}) \tag{18-18}$$

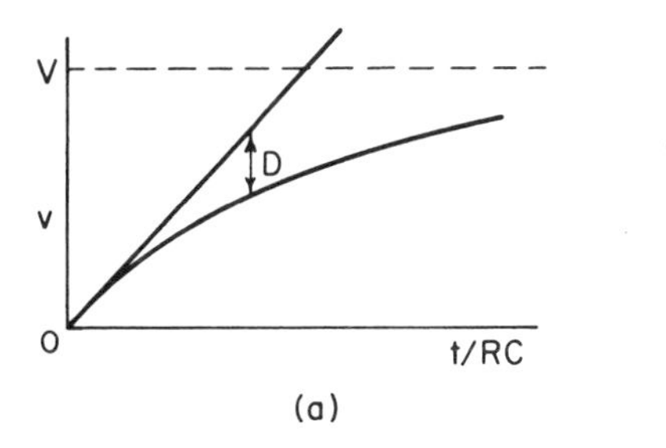

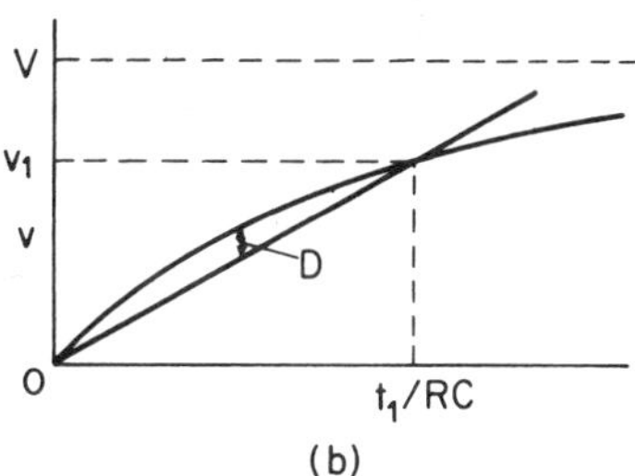

Figure 18-17. (a) Transmission error; (b) displacement error.

The point of maximum departure from linearity may be assumed to occur at the middle of the sweep, where $t = t_1/2$. The first two terms of the series expansion for the exponential are

$$v = V(1 - \epsilon^{-t/RC}) = V\left(\frac{t}{RC} - \frac{t^2}{2R^2C^2}\right) \tag{18-19}$$

The straight line gives for the voltage at $t = t_1/2$:

$$v = \frac{V[1 - (t_1/2RC)]t_1}{2RC}$$

and the exponential curve gives

$$v_{\text{mid}} = \frac{[V1 - (t_1/4RC)]t_1}{2RC}$$

The maximum deviation from linearity can then be written as

$$D = \frac{v_{\text{mid}} - v}{v_{\text{mid}}} \times 100\% \cong \frac{t_1/4RC}{1 - (t_1/4RC)} \times 100\% \tag{18-20}$$

This percentage departure from linearity is plotted in Fig. 18-18, as a function of t_1/RC. The curve shows that the exponential sweep must use small values for t_1/RC, and then linearity better than one per cent is obtainable only with very small output voltages.

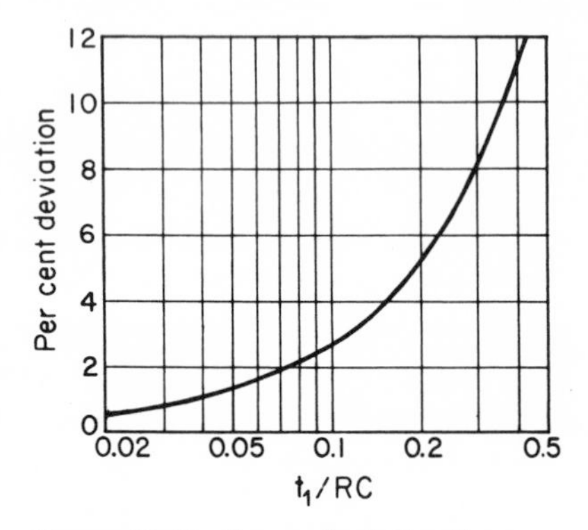

Figure 18-18. Sweep deviation from linearity.

The Miller integrator replaces t/RC with t/ARC, or multiplies the time constant by the gain. To obtain the actual value of t_1/RC which will give a certain error, the abscissa values may be multiplied by A. The bootstrap sweep improves the accuracy by the ratio $1/(1 - A)$ where A is the cathode-follower gain, near unity; the abscissa value may then be divided by $1 - A$. These two circuits are equivalent in results and a gain of 33 in the Miller circuit is analogous to a cathode-follower gain of 0.97 in the bootstrap circuit.

18-12 A CURRENT SWEEP

In television and radar service the cathode-ray beam is deflected by a magnetic field produced by a set of coils, known as a yoke, surrounding the neck of the cathode-ray tube. In air-core coils, or in iron-core coils below saturation, the value of B is proportional to i and a linear time-varying current is required for sweep deflection. If a voltage V is applied to an inductance L at $t = 0$, the current will increase as $i_L = Vt/L$. This suggests the simple current sweep generator of Fig. 18-19, analogous to the RC voltage sweep circuit.

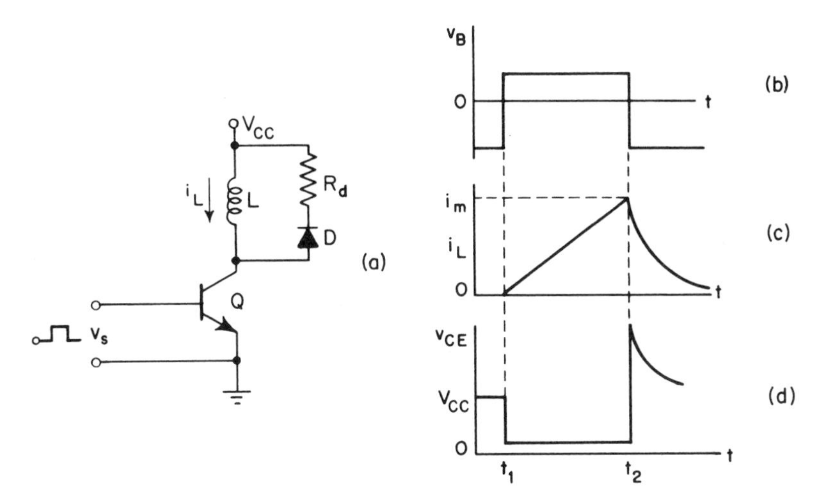

Figure 18-19. (a) Current sweep circuit; (b) input wave form; (c) inductor current; (d) collector voltage.

The transistor Q serves as a switch and is normally cut off, but is turned on by a positive signal as at t_1 in (b). If the ratio L/R of the coil is high, then the current i_L increases linearly with time to t_2. At this instant the transistor is cut off, but the current continues through R_d and diode D, the $L\ di/dt$ voltage having reversed and biased the diode in the forward direction. This current serves to dissipate the stored magnetic energy and decays to zero with a time constant due to L and the sum of R_d and the diode forward resistance. The value of R_d is chosen to limit the v_{CE} voltage spike at t_2, as shown in (d).

While the L/R ratio will be made high, some resistance is inevitable in the yoke, and the current actually rises along a portion of an exponential curve:

$$i_L = \frac{V_{CC}}{R_L}(1 - \epsilon^{-R_L t/L}) \tag{18-21}$$

Thus the simple current sweep has linearity problems which are similar to those discussed for voltage sweeps. Methods are available for improving the linearity of the sweep current.

18-13 RELAXATION CIRCUIT FOR PULSE GENERATION

If i_o is the current in L at t_o, the parallel RLC circuit of Fig. 18-20 has the following current equation:

$$\left(s^2 + \frac{s}{RC} + \frac{1}{LC}\right)v + i_o = 0 \tag{18-22}$$

At $t = 0$, $v = 0$, and $dv/dt = -i_o/C$, a solution is obtained for the underdamped case as

$$v = -\frac{i_o}{C}\epsilon^{-t/2RC}\sin \omega t \qquad (18\text{-}23)$$

where

$$\omega = \sqrt{\frac{1}{LC} - \frac{1}{4R^2C^2}} \qquad (18\text{-}24)$$

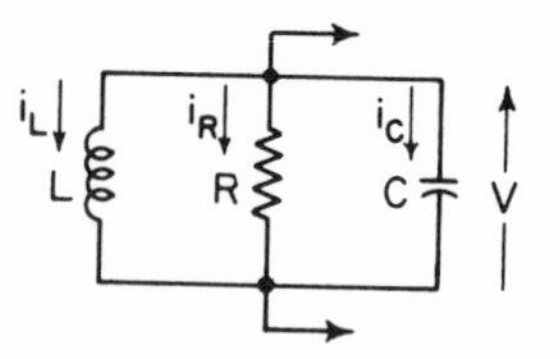

Figure 18-20. Currents in the *RLC* circuit.

The circuit voltage is damped by the factor $\epsilon^{-t/2RC}$. The losses are small for the underdamped case, and the peak energy stored in the inductor by i_o is transferred substantially without loss to the capacitor during the first cycle. Thus

$$\frac{Li_o^2}{2} = \frac{Cv_{\max}^2}{2}$$

and the peak voltage of the first cycle will be

$$v_{\max} = i_o\sqrt{\frac{L}{C}} \qquad (18\text{-}25)$$

The oscillatory circuit can be used as a load for a tube or transistor. The input should be a square wave, and during the "on" half-cycle, the current stabilizes at i_o, and due to damping from the saturation resistance, the *RLC* circuit cannot oscillate. At the instant the input wave cuts the device off, the *RLC* circuit is isolated and left free to oscillate at its natural frequency, Eq. 18-24; the result appears in Fig. 18-21.

Pulses of short duration can be obtained by clipping the output at a level equal to the peak of the second cycle.

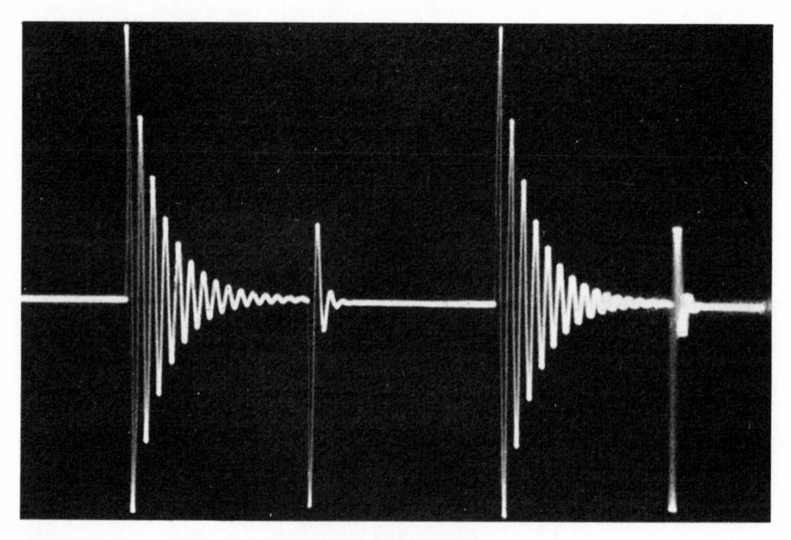

Figure 18-21. Output voltage of a ringing circuit.

18-14 THE BLOCKING OSCILLATOR

Another relaxation circuit used to generate short pulses is the *blocking oscillator*. Suited to either transistor or tube, the basic form is that of the base- or grid-tuned oscillator, with large mutual coupling for feedback, as in Fig. 18-22. The time constant R_eC_e must be long with respect to the time of a cycle of the resonant frequency between L_1 and its internal winding capacity.

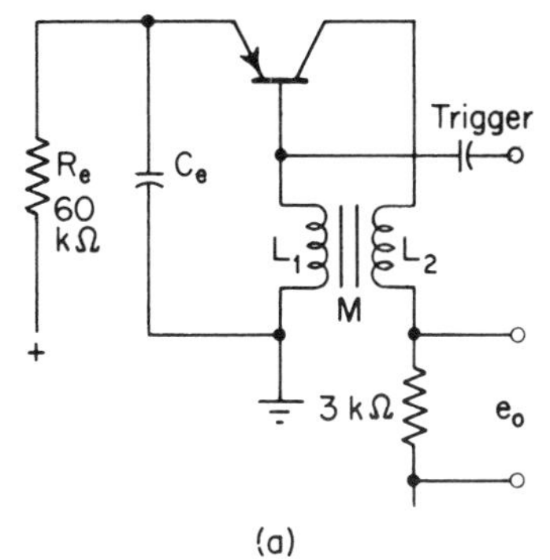

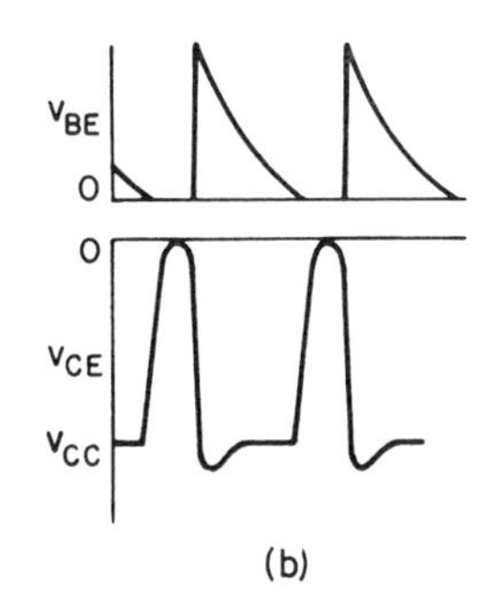

Figure 18-22. Transistor blocking oscillator.

Initially C_e is charged negative to the emitter and the transistor is cut off. The capacitor discharges through R_e, and when the emitter voltage reaches zero, the emitter current starts. This current through the closely coupled transformer induces a voltage which drives the base negative, further increasing the emitter current.

The emitter current charges C_e negatively, and as its voltage rises the collector current starts to fall. As the collector current falls out of saturation it induces a positive base voltage through the transformer, driving the transistor back to cutoff. Capacitor C_e is left negatively charged, ready to repeat the cycle after its discharge through R_e.

The repetition rate for short pulses is approximately

$$f_r = \frac{nV_{EE}}{V_{CC}R_eC_e} \tag{18-26}$$

where n is the collector-to-base transformer ratio, and it is assumed that the transistor swings to saturation.

Intermittent oscillations of this nature may develop in any tuned-circuit oscillator, and are an indication that the RC time constant of the bias circuit is excessive. The phenomenon is occasionally referred to as *squegging*.

18-15 THE ASTABLE MULTIVIBRATOR

A number of regenerative circuits capable of fast switching from one state to another, with lock-up possibilities, are known as *multivibrators*. Three basic types are classified according to the stability of their operating states. They are:

1. The *astable multivibrator*, which continuously oscillates back and forth between two semi-stable states, providing a useful source of square waves.
2. The *monostable multivibrator*, with one semi-stable and one stable state. It can be switched to the semi-stable state and returns to the stable state after a fixed delay period.
3. The *bistable multivibrator*, with two stable states, between which it can be switched.

Multivibrators employ a pair of active devices, and the stable or semi-stable states between which switching occurs are found to be (1) device A conducting and B nonconducting, (2) device A nonconducting and B conducting. Each of the astable multivibrators of Fig. 18-23 is a two-stage RC amplifier, with output coupled back to input in a symmetrical fashion which creates positive feedback and regeneration. The rate of switching between the semi-stable states will depend on the several RC time constants in the circuit.

Consider the transistor circuit of (a), and assume that transistor Q_1 has just switched to conducting, as at t_1 in Fig. 18-24. The collector voltage

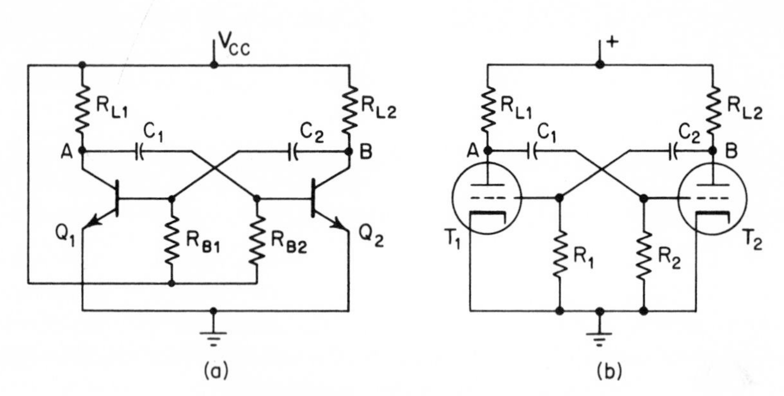

Figure 18-23. Astable multivibrators.

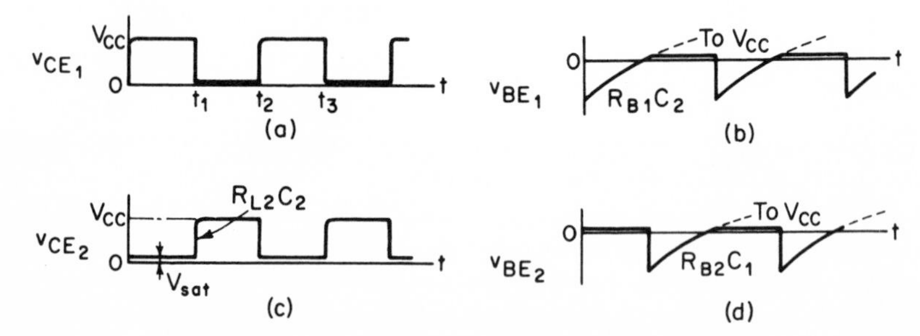

Figure 18-24. Astable multivibrator wave forms.

of Q_1 has just gone from V_{CC} to V_{sat} and this negative voltage step has been transmitted through C_1 to the base of Q_2, cutting off Q_2.

Capacitor C_1 has $V_{\text{sat}} \cong 0$ at A, with $-V_{CC}$ at the base of Q_2 as in (a), Fig. 18-25. The base potential will rise from $-V_{CC}$ as the capacitor charges through R_{B2} toward $+V_{CC}$, or with a total charging potential of $2V_{CC}$. When v_{BE2} reaches zero volts at t_2, transistor Q_2 turns on. The potential at B then drops from V_{CC} to V_{sat}, and a negative pulse is transmitted to the base of Q_1, turning it off as at t_2, Fig. 18-24(b). The potentials are then as illustrated in (b), Fig. 18-25.

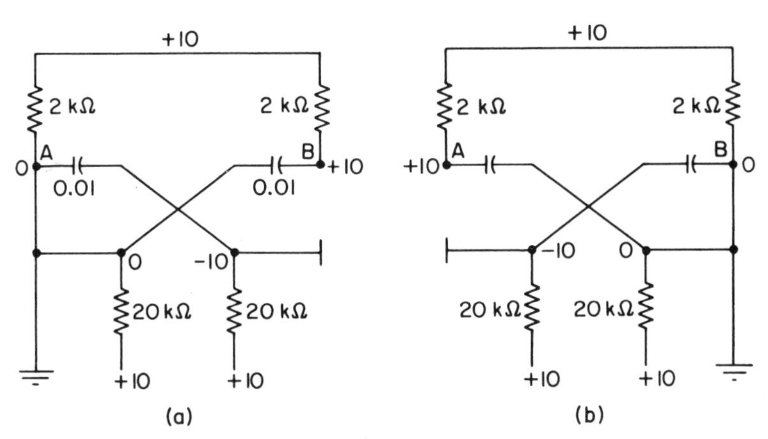

Figure 18-25. Astable multivibrator equivalent circuits: (a) at t_1; (b) at t_2.

Capacitor C_2 then charges toward $+V_{CC}$ with a time constant C_2R_{B1} and at t_3 the potential of v_{BE1} reaches zero, Q_1 turns on and Q_2 turns off, and one cycle has been completed.

Consider $V_{CC} = 10$ V, and $C_1R_{B2} = C_2R_{B1} = 200\ \mu$s. Since v_{BE} is rising toward $2V_{CC}$, the time needed for the base potential to rise 10 V can be found as

$$10 = 2V_{CC}\epsilon^{-(t_2-t_1)/C_1R_{B2}} = 20\epsilon^{-(t_2-t_1)/200}$$

or

$$\frac{T}{2} = t_2 - t_1 = C_1R_{B2} \ln 2 = 0.693C_1R_{B2} \qquad (18\text{-}27)$$

$$= 200 \ln 2 = 139 \qquad \mu\text{s}$$

The period of the output wave is twice this interval and the frequency is $\frac{1}{278} \times 10^{-6} = 3.6$ kilohertz.

If the base resistors are returned to a variable source instead of V_{CC}, it is possible to change the delay time, as illustrated in Fig. 18-26. There is almost a linear relation between the charging voltage and time delay, and this furnishes an effective means of controlling the frequency or the length of the pulse by a voltage.

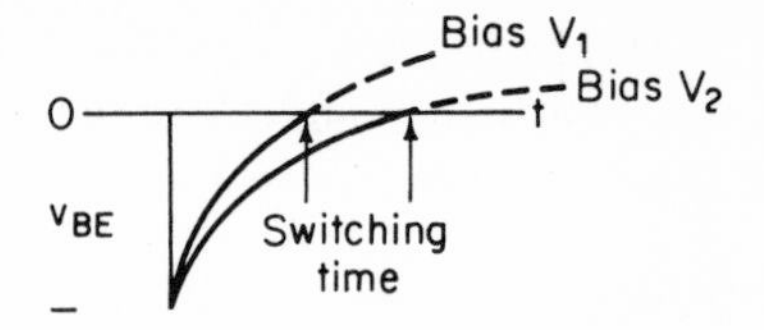

Figure 18-26. Effect of changing base bias.

Unsymmetrical waves may be obtained by making the C_1R_{B2} and the C_2R_{B1} time constants unequal. The output may be taken from A or B; after clipping the output is an excellent square wave.

The vacuum-tube version is similar, except that the grid resistors may return to ground or to a varying voltage.

18-16 SYNCHRONIZATION OF THE MULTIVIBRATOR

It is possible to synchronize a multivibrator with a driving signal having a frequency approximately n times as great as the free-running multivibrator frequency. The synchronizing signal is preferably a pulse, as shown in Fig. 18-27(a), although a sine signal may be used. The pulse is inserted into the base or grid circuits of the multivibrator, adding to the rising exponential voltage there.

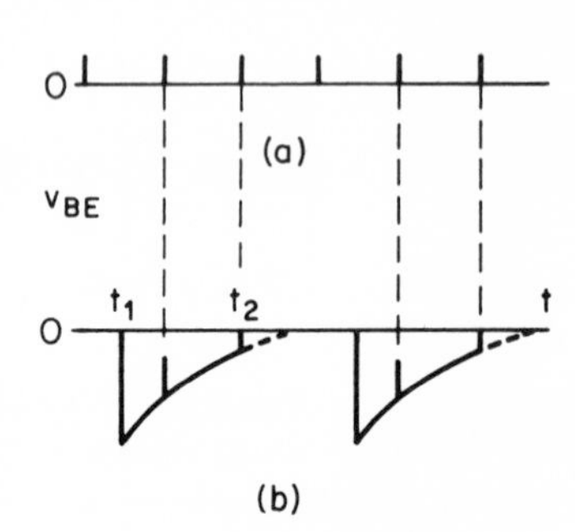

Figure 18-27. (a) Synchronizing pulses; (b) base voltage in a synchronized multivibrator.

Stable operation results when the exponential voltage reaches the triggering point a little late, the added pulse then forcing triggering at an exact earlier instant. For the case shown, the *count-down factor* or the ratio between pulse rate and output frequency is three.

18-17 THE MONOSTABLE MULTIVIBRATOR

If one of the switching capacitors is changed to a resistor, the astable circuit becomes a *monostable multivibrator*. One of the operating points becomes stable, the second remaining semi-stable.

A transistor version appears in Fig. 18-28, and the triode circuit is similar. Because of the negative bias applied through R_{B1} to the base of Q_1, the stable state exists with Q_1 off, Q_2 on. A negative trigger pulse is applied at A, and to the base of Q_2 through C_1. Transistor Q_2 cuts off and the positive rise of voltage at B, transmitted through R_2 to the base of Q_1, turns on that transistor. The circuit is then in a semi-stable state.

After time Δt the capacitor C_1 has discharged through R_{B2} and raised the base voltage of Q_2 back to zero, when the circuit returns to its initial

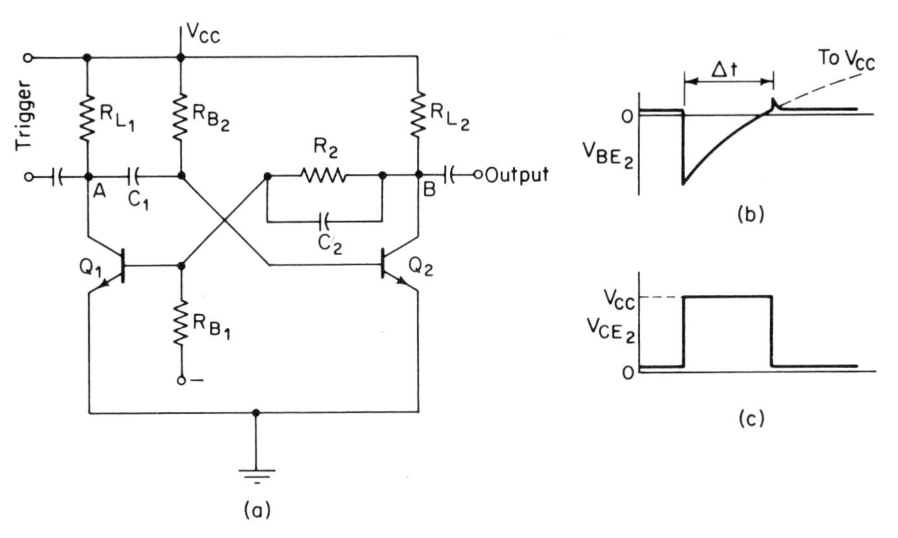

Figure 18-28. Transistor monostable circuit.

stable state of Q_1 off, Q_2 on. The period is calculated as before,

$$T = \Delta t = C_1 R_{B2} \ln 2 = 0.693 C_1 R_{B2} \tag{18-28}$$

When the voltage pulse is transmitted to the base of Q_1, the input capacitance of Q_1 must be charged through R_2 and this slows the change in base voltage. A *speed-up* capacitor C_2, of the order of 50 pF, charges the base input capacitance rapidly; it also applies the full value of the change in voltage to the base of Q_1.

It is important to select components so that Q_2 is definitely on and Q_1 is off in the stable state. The circuit to be considered is that of Fig. 18-29 in which the base current of Q_2 is

$$I_{B2} = \frac{V_{CC}}{R_{B2}} = \frac{10}{80{,}000} = 125 \quad \mu\text{A}$$

The transistor should saturate with less than this value. The base voltage of Q_1 is determined by $V_{BB} = -6$ V, and R_{B1} and R_2, since the potential at B is zero. Then

$$v_{BE1} = \frac{R_{B1}}{R_2 + R_{B1}} V_{BB}$$

$$= \frac{40}{120} \times (-6) = -2.0 \quad \text{V}$$

With its base at -2.0 V, Q_1 is definitely cut off. Therefore the stable state is established.

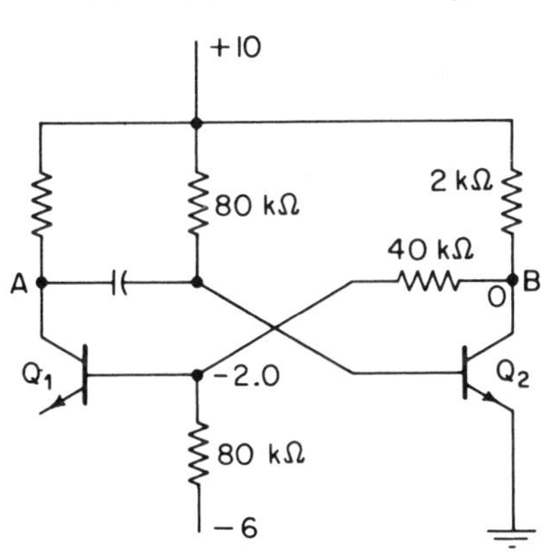

Figure 18-29. Potentials in the stable state.

By taking an output from point B, an accurate square pulse may be obtained as in Fig. 18-28(c). The output is useful as an adjustable *gating* voltage, suited to switch some other circuit for a predetermined interval after an initial trigger pulse.

18-18 THE BISTABLE MULTIVIBRATOR

The third of the general class of regenerative multivibrators is the *bistable* circuit or *flip-flop*, in which there are two stable operating states separated by an unstable region. The circuit is derived from the astable form by replacing both switching capacitors with resistors R_1 and R_2, as in Fig. 18-30 or Fig. 18-31. Capacitors are added to furnish a speed-up function, as for the monostable circuit.

Assume that Q_1 is conducting and Q_2 is cut off, as in Fig. 18-32. With point A at zero volts, the base voltage of Q_2 is

$$v_{BE2} = \frac{R_1}{R_1 + R_{B2}}(-6) = \frac{40}{120} \times (-6) = -2.0 \qquad \text{V}$$

quite sufficient to cut off the npn transistor. The current to the base of Q_1 is $I_{B1} = I_1 - I_2$, and

$$I_1 = \frac{V_{BB}}{R_{B1}} = \frac{6}{80{,}000} = 75 \qquad \mu\text{A}$$

$$I_2 = \frac{V_{CC}}{R_L + R_2} = \frac{10}{42{,}000} = 238 \qquad \mu\text{A}$$

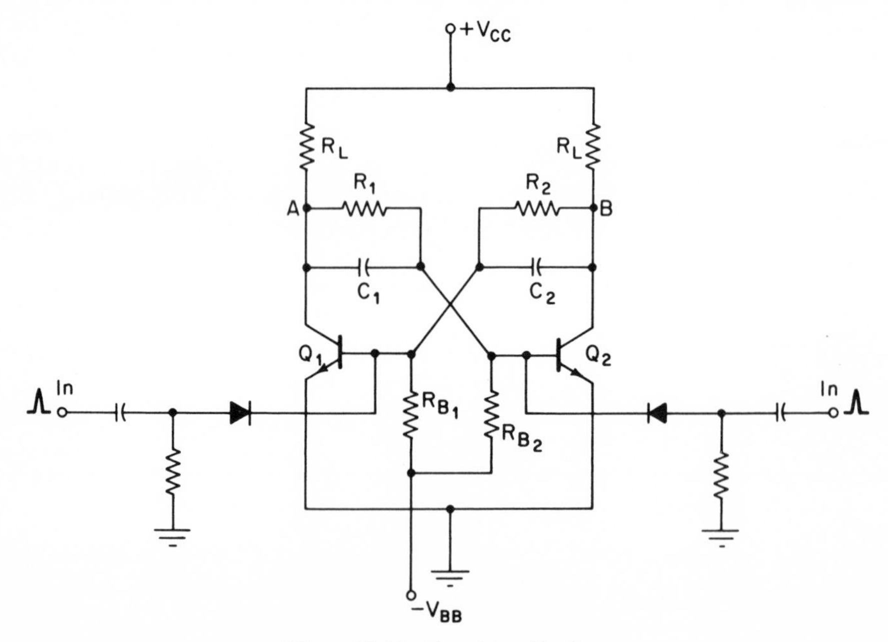

Figure 18-30. Transistor flip-flop.

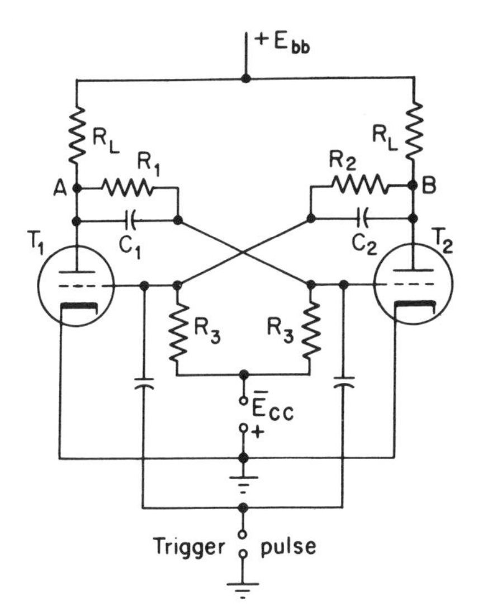

Figure 18-31. Triode bistable circuit.

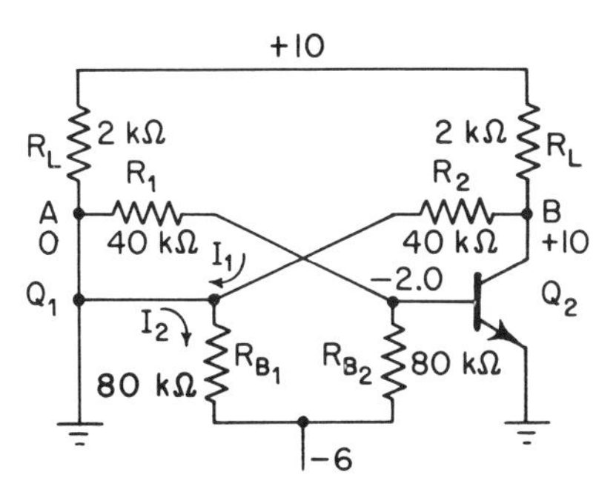

Figure 18-32. Potentials in the stable state.

Then $I_{B1} = -163\ \mu$A. This value may be compared with the transistor characteristics to insure that it is in excess of the saturation base current. Also

$$v_{CE2} = \frac{R_2}{R_L + R_2} V_{CC}$$

$$= \frac{40{,}000}{42{,}000} \times 10 = 9.5 \quad \text{V}$$

which compares with the 10 V assumption in the circuit diagram. Thus the circuit is in a stable state.

A positive trigger pulse to Q_2 will turn on Q_2. This action generates a negative voltage step at B which is transmitted to the base of Q_1, turning it off. The circuit voltages and currents are now those computed, but with reversed subscripts, and the circuit is in its second stable state.

Most sensitive triggering will occur with positive pulses applied to the base, but switching may also be accomplished with larger pulses directed to the collector.

The time required to transfer conduction from one state to the other is known as the *resolving time*, that is, the time needed to separate two input pulses. The time constants associated with the speed-up capacitors should be short with respect to the expected time between pulses for good pulse resolution.

18-19 SATURATION IN FLIP-FLOPS

In the study of transistors it was mentioned that charges are stored in the base region while a transistor is in saturation, and when a transistor is switched from saturation to cutoff, these charges come out of storage slowly and so delay the switch operation.

In an attempt to improve the switching speed a technique using the *nonsaturating clamp* circuit of Fig. 18-33 has evolved. Diode D_1 is germanium and D_2 is silicon. As transistor T_1 turns on and the voltage at A falls below the voltage at C, diode D_1 shunts some of the base driving current into the collector. Since the silicon diode D_2 has a forward voltage drop of about 0.7 V over a consi'derable current range, it acts as a voltage regulator, holding V_{CE} at about 0.7 V, above saturation level and above V_{BE}. Thus, saturation is approached but not reached, and the time delay produced by base storage in saturated circuits is reduced. Conversely, the saturation voltage V_{sat} and the saturation v_{BE} are only about 0.5 V and this permits the assumption that all saturation voltages are negligible with respect to supply levels. This makes saturated circuit design simple, because it is unnecessary to check actual operating potentials. Saturated operation also reduces the transistor dissipation during the conducting interval; however, it does require large trigger power to reverse a saturated transistor.

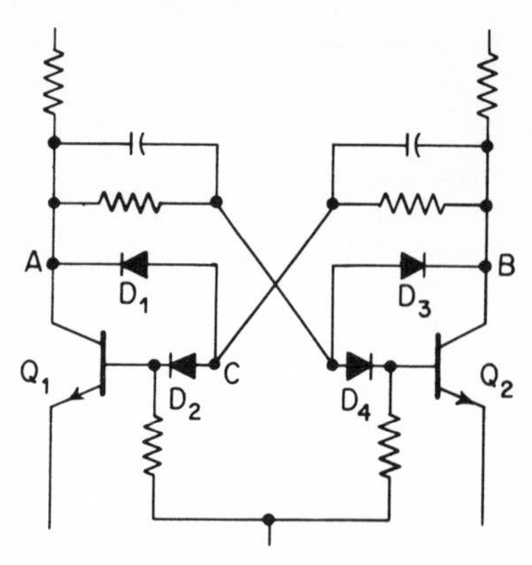

Figure 18-33. Nonsaturating clamp diodes in a bistable circuit.

A circuit which takes advantage of the stability of saturated design is the *direct-coupled flip-flop* of Fig. 18-34. The voltages indicated are for germanium transistors. In saturation the voltage on the base of the conducting transistor Q_1 is larger than its v_{CE}, and v_{CE} is less than the base voltage required to switch transistor Q_2 on; thus the base bias of the off transistor is so low that negligible current exists. The large base current in the saturated transistor helps to reduce charge storage and improves the switching speed.

Since the switching operation involves only the commutation of base current and saturation collector current and since these currents may be nearly equal, the output voltage will be low. For the circuit shown, this output will swing only from 0.02 V to 0.50 V.

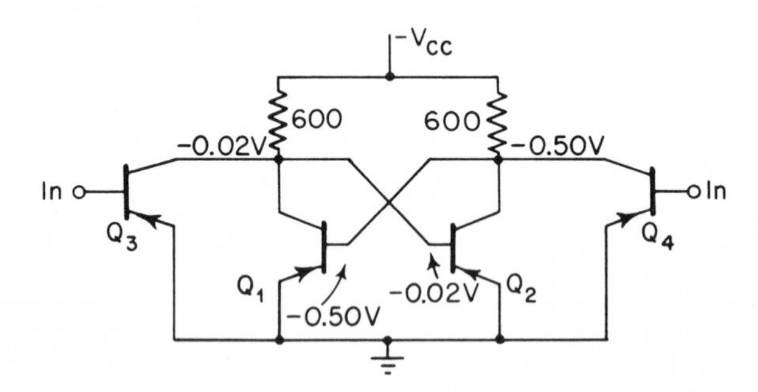

Figure 18-34. DCT flip-flop.

18-20 THE SCHMITT TRIGGER

Figure 18-35 shows a useful variation of the bistable multivibrator which triggers on change in input voltage level. This *Schmitt trigger circuit* is derived by removing one of the cross-coupling networks and adding emitter or cathode bias.

Initially the input signal v_i may be zero; Q_1 is cutoff by the voltage V_E across the emitter resistor R_E, transistor Q_2 being conducting. Transistor Q_1 remains cutoff until v_1 exceeds V_E; v_{BE} is neglected by assumption. Then as Q_1 turns on, a negative-going voltage appears at A and is transmitted to the base of Q_2, reducing its collector current. Voltage V_E falls, further increasing I_{C1} and reducing the voltage at A; the effect is regenerative and Q_2 is carried into cutoff and Q_1 into active conduction.

Reduction of v_i will now cause a reverse switching action, but hysteresis may appear in the level of switching voltage. That is, with a low value of R_{L1} and consequently low loop gain, after v_i reaches V_E the output voltage rises along the dashed curve of (d) from a to x. With increased R_{L1} and unity

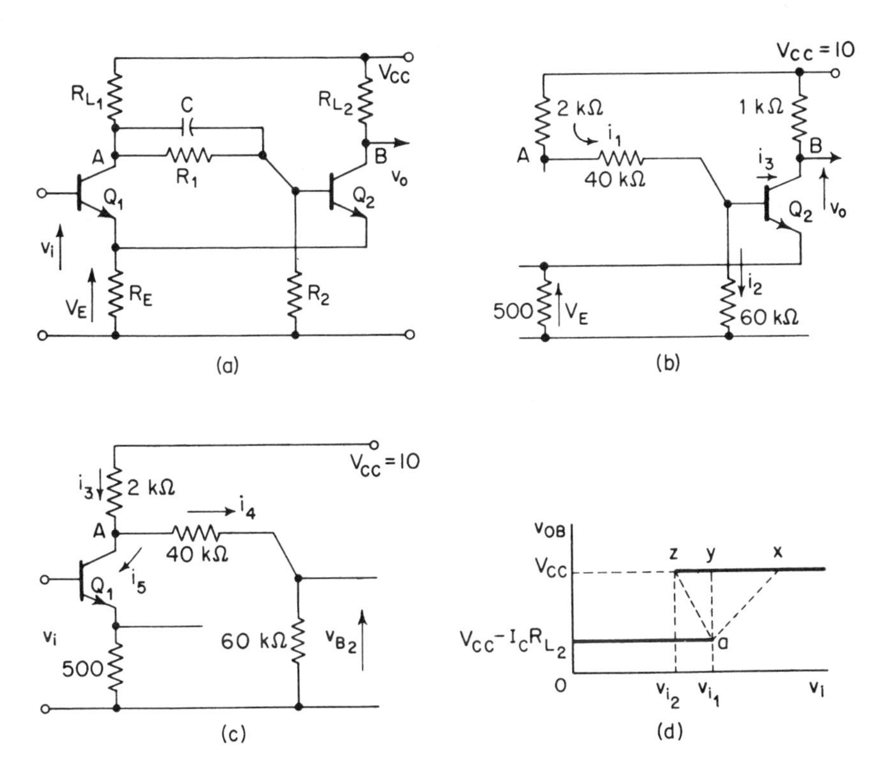

Figure 18-35. The Schmitt trigger.

gain, regenerative action begins and the output rises along the curve from a to y. With gain greater than unity the output follows the curve from a to z, the hysteresis effect being evident.

The amount of hysteresis can be determined. Neglecting v_{BE} of the transistors, the currents defined in Fig. 18-35(b) lead to

$$i_1 = i_2 + i_3$$

where

$$i_1 = \frac{V_{CC} - V_E}{R_{L1} + R_1}, \qquad i_2 = \frac{V_E}{R_2}, \qquad i_3 = \frac{V_E}{R_E(1 + h_{FE})}$$

Using these current relations, V_E can be obtained. At the change of state in an upward direction $v_i = V_E$, and so

$$v_{i1} = V_{E1} = \frac{V_{CC}}{1 + \dfrac{R_{L1} + R_1}{R_2} + \dfrac{R_{L1} + R_1}{R_E(1 + h_{FE})}} \tag{18-29}$$

For the parameters of the figure and with a transistor having $h_{FE} = 50$,

$$v_{i1} = \frac{10}{1 + \dfrac{42{,}000}{60{,}000} + \dfrac{42{,}000}{25{,}500}} = 3.0 \qquad \text{V}$$

With $V_E = 3.0$ V the original assumption that Q_1 is cutoff is supported.

Reverse switching starts with Q_2 cut off and Q_1 conducting. The applicable currents from Fig. 18-35(c) are

$$i_3 = i_4 + i_5$$

Then

$$i_3 = \frac{V_{CC} - v_A}{R_{L1}}, \qquad i_4 = \frac{v_A}{R_1 + R_2}, \qquad i_5 \cong \frac{V_E}{R_E}$$

the last expression being justified since h_{FE} will be large with respect to unity. Also

$$v_A = v_{B2}\left(\frac{R_1 + R_2}{R_2}\right) \tag{18-30}$$

and the current summation then leads to

$$\frac{V_{CC}}{R_{L1}} = \frac{v_{B2}}{R_2}\left(1 + \frac{R_1 + R_2}{R_{L1}}\right) + \frac{V_E}{R_E} \tag{18-31}$$

Again neglecting v_{BE}, transistor Q_2 will begin to turn on when $v_{B2} \cong V_E$. Thus $v_{i2} = V_{E2} \cong v_{B2}$, and so Eq. 18-31 leads to

$$v_{i2} = \frac{V_{CC}}{1 + \dfrac{R_{L1} + R_1}{R_2} + \dfrac{R_{L1}}{R_E}} \tag{18-32}$$

Comparison of Eqs. 18-29 and 18-32 shows that $v_{i1} > v_{i2}$ and thus hysteresis exists. The amount of hysteresis in our example may be found from

$$v_{i2} = \frac{10}{1 + \frac{42{,}000}{60{,}000} + \frac{2000}{500}} = 1.76 \qquad \text{V}$$

as compared to 3.0 V for v_{i1}.

Since $i_{C2} = h_{FE}i_E/(1 + h_{FE})$, then

$$i_{C2} = \frac{h_{FE}}{1 + h_{FE}} \frac{V_E}{R_E} = 5.9 \qquad \text{mA}$$

With Q_2 conducting, the output voltage at B is

$$v_{oB} = V_{CC} - i_{C2}R_{L2} = 10 - 0.0059 \times 1000 = 4.1 \qquad \text{V}$$

With Q_2 cut off the output at B is 10 V, and so the output voltage jump at switching is 5.9 V.

From Eqs. 18-29 and 18-32 it follows that the input switching levels would be equal if

$$h_{FE}R_{L1} = R_1$$

This is an effective means of fixing the gain. It is possible to reduce the hysteresis by introducing a resistor R_x in series with the emitter of Q_1. This does not affect the value of v_{i1} but will raise v_{i2}. The voltage V_E at which Q_2 begins conducting will be as before, because that voltage is determined by the unchanged base potential of Q_2. Therefore $V_E = 1.76$ V. However, v_{i2} will be greater than V_E by the amount of the voltage drop in R_x. For $v_{i2} = 3.0$ V

$$\frac{R_x + R_E}{R_E} = \frac{3.0}{1.76}$$

from which $R_x = 352$ ohms for $R_E = 500$ ohms. Then $v_{i1} = v_{i2} = 3.0$ V and the condition is one of zero hysteresis.

Since v_{BE} has been neglected, the actual trigger voltage v_{i1} will be greater than calculated by about 0.2 V for germanium and 0.5 V for silicon transistors; v_{i2} will be reduced by similar amounts. Due to variation in device parameters, the gain cannot be reduced to unity or the hysteresis entirely eliminated, because the triggering action then becomes uncertain.

The Schmitt circuit is of value in clipping distorted pulse trains, giving waves with fast rise and fall times, provided that the input amplitude is considerably greater than the triggering level. The circuit is also useful as a *voltage amplitude comparator*, to give a sharp output pulse at the instant the input voltage reaches a particular reference level.

18-21 FLIP-FLOP TRIGGERING

Several methods for triggering of flip-flop circuits have been indicated. In general, a flip-flop may be triggered with a positive pulse to turn an on base of a pnp transistor to the off condition, or a negative pulse into an off base to turn the transistor to conduction, as in Fig. 18-36(a) and (b). Opposite polarities would be used for npn transistors. Greater speed is obtained with the operating pulse applied to the base of the on transistor. The trigger pulse should have a length approximating the turn-off time of the transistor, and overdrive of base current should be supplied for speed in switching.

Diode D_1 eliminates the negative-going spike generated by the differentiating action of the *RC* coupling circuit at the end of the input pulse in (a), or the positive-going spike in (b).

Referring to (a), capacitor *C* is charged through the low base resistance

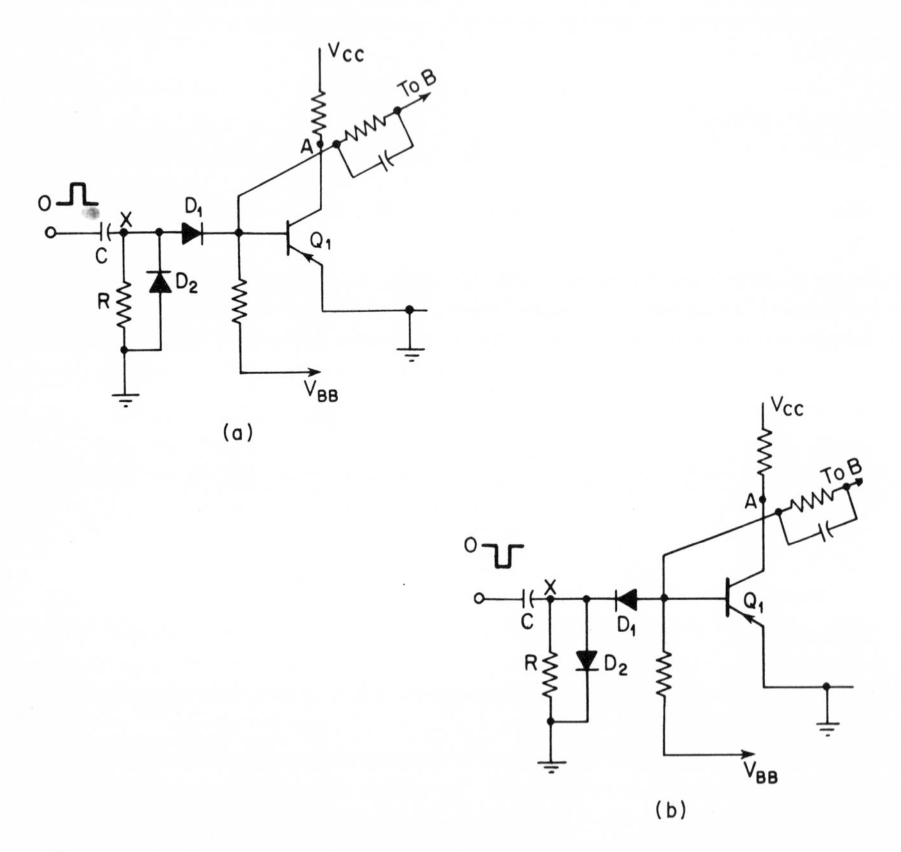

Figure 18-36. Flip-flop triggering: (a) positive pulse turn off; (b) negative pulse turn on.

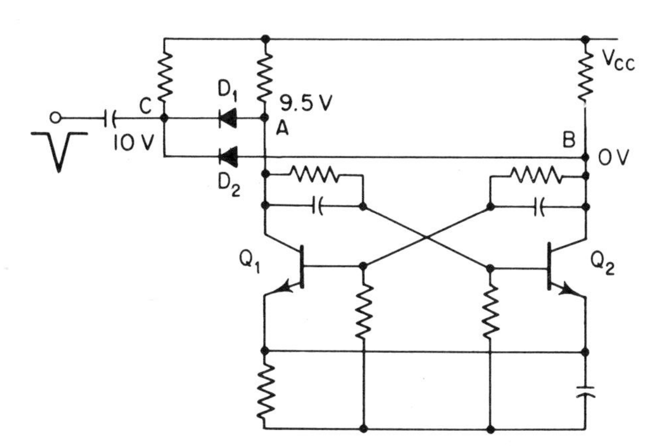

Figure 18-37. Use of steering diodes for guiding switching pulses.

during the pulse rise, and point X is left with a negative charge at the end of the input voltage rise, blocked by the capacitor and the anode of D_1. The time constant for discharge is RC, and this may be longer than the pulse spacing; the following pulse will then have its base clipped by the potential remaining on C, and may not reach sufficient amplitude to trigger. A dc restoring diode is then connected across R as D_2, to discharge C through the low forward diode resistance.

Use of diodes for *steering* of triggering pulses is shown in Fig. 18-37. Therein successive input pulses will switch a bistable circuit between alternate states. With Q_1 off and Q_2 on, a -5 V negative trigger pulse is applied and this takes C to $+5$ V. Diode D_2 remains reverse-biased, but D_1 is forward-biased and the 5 V trigger pulse is sent to point A. It reaches the base of Q_2 and turns it off and Q_1 is turned on in normal fashion. For a second negative trigger pulse diode D_1 is found reverse-biased and the pulse is transmitted by D_2 to B, thus turning off Q_1 through the speed-up capacitor and reversing the circuit. Thus a negative pulse is steered by the diodes to turn off the conducting transistor in each case.

Another *pulse-steering* circuit is shown in Fig. 18-38(a), for base-steering and triggering on the positive-going trailing edge of a negative input pulse. With Q_1 on, the leading edge of the pulse charges C_1 to the pulse amplitude through D_3, diode D_1 being blocked. On the trailing edge of the pulse, the anode of D_1 rises by V volts and C_1 discharges into the base of Q_1, turning that transistor off. At the time of the next pulse C_2 and D_2 act, turning off Q_2 for completion of a cycle.

A circuit for collector triggering is shown in Fig. 18-38(b). The diodes steer the input positive pulse to the off collector and through the feedback network to turn off the base of the conducting transistor.

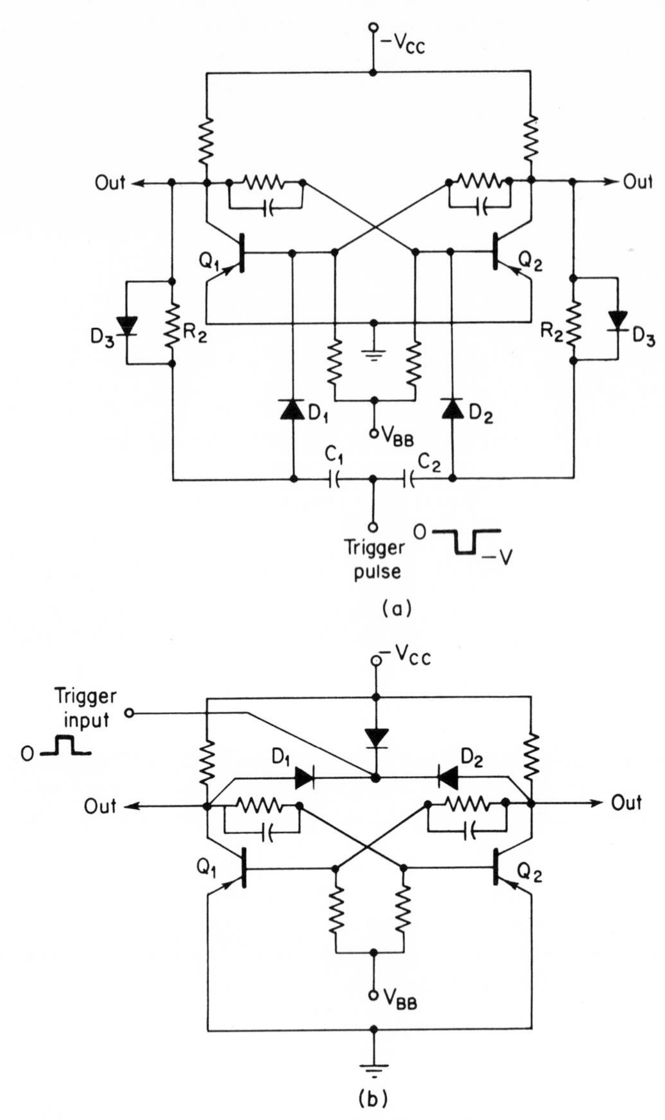

Figure 18-38. (a) Base steering; (b) collector steering.

18-22 SCALERS OR COUNTING CIRCUITS

If Q_2 is initially on, the first applied pulse to a bistable circuit will turn it off and the second pulse will turn it on again. This provides a negative pulse from Q_2 for every two input pulses. A second flip-flop may be cascaded with the first, as in Fig. 18-39 and starting with even-numbered transistors in the on state, a negative pulse will be given from Q_4 for every 2^2 or 4 input pulses. With n bistables in cascade, the scaling factor will be 2^n.

In the figure the odd stages may be provided with indicating lamps which show when that stage is conducting, or the point of connection is at ground. A reset switch is provided so that the even-numbered stages can be initially made conducting.

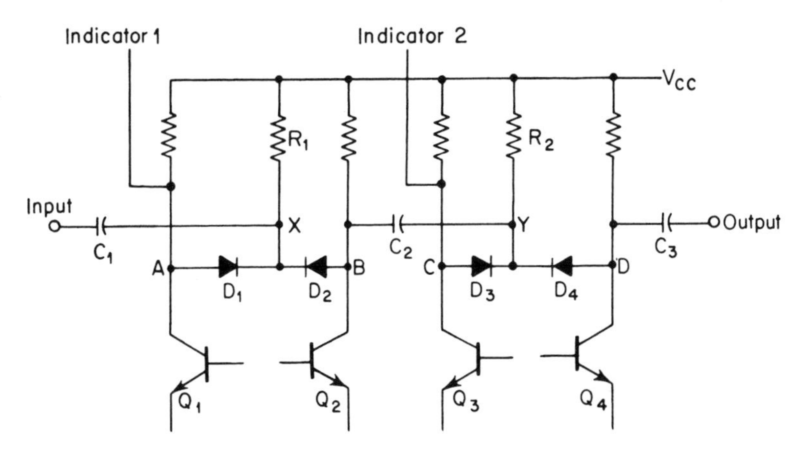

Figure 18-39. Scale-of-four circuit.

The first negative pulse is steered through D_1 and turns on Q_1 and its indicator. The second negative pulse is steered through D_2 and turns on Q_2; a negative pulse is transmitted to switch the second binary, lighting indicator 2 and turning off indicator 1. Successive pulses create the following conditions:

Pulses	T_1	T_2	T_3	T_4	Indicator 1	Indicator 2
Reset	0	1	0	1	0	0
1	1	0	0	1	x	0
2	0	1	1	0	0	x
3	1	0	1	0	x	x
4	0	1	0	1	0	0 + output pulse

If the indicators are given the weights of 1 and 2, their sum indicates the total number of pulses. The output pulse can be used to actuate additional scalers for higher counts, and counters with as many as eight significant digits are available.

Of course, to count in the decimal system would require four binary switches per decade; these would have a potential count of $2^4 = 16$. By use of feedback of pulses it is possible to add m synthetic pulses to the actual input pulse number n so that the scaler will recycle after $n = 10$, or after $10 + m = 16$ synthetic counts.

PROBLEMS

18-1. For the circuit of Fig. 18-40 plot to scale the waves of e_1 and e_2 for a period of 0.5 s after switch s_1 is closed.

18-2. In Fig. 18-40 switch s_1 is closed at $t = 0$, and 0.3 s later switch s_2 is closed. Plot to scale wave forms of e_1 and e_2 for the period from $t = 0$ to $t =$ 0.7 s.

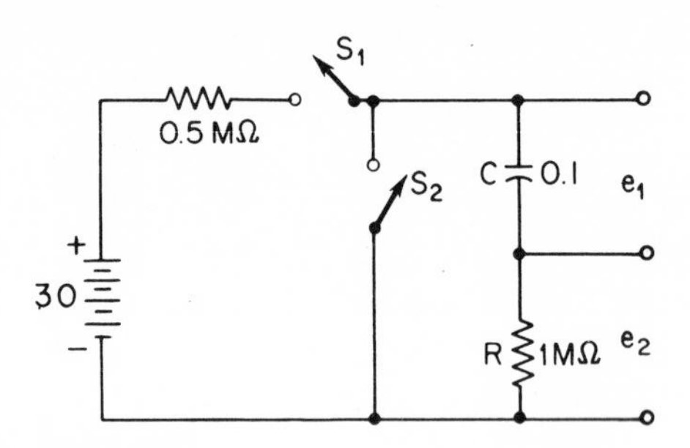

Figure 18-40.

18-3. The circuit of Fig. 18-41(a) is supplied with a rectangular wave input of zero volts for 200 μs and -50 V for 100 μs. The tube has $\mu = 20$, $r_b =$ 7000 ohms. Plot the wave form of $e_o =$ to scale over one cycle of the input.

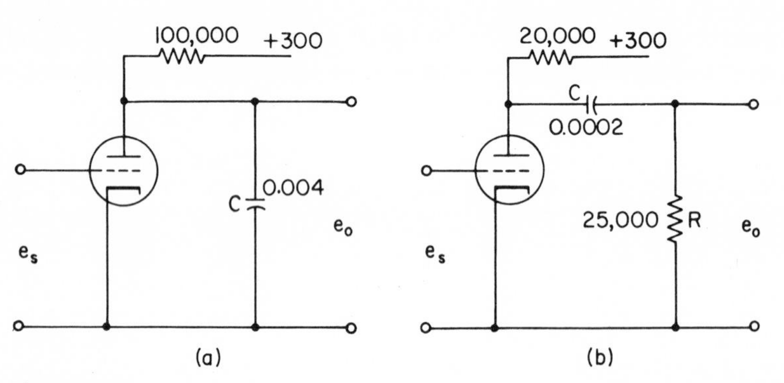

Figure 18-41.

18-4. The circuit of Fig. 18-41(b) has a square-wave input of 200 μs period, swinging from zero to -100 V. With $\mu = 5$, $r_b = 1000$ ohms, plot e_o to scale over one cycle of the input.

18-5. An *RC* integrating circuit of $R = 100{,}000$ ohms, $C = 0.015$ μF has applied a symmetrical triangular wave of $+100$ to -100 V peak-to-peak and period of 0.0001 s. Plot the output wave form.

18-6. (a) A sine wave of 60 V rms is applied to the circuit of Fig. 18-10, with

$R = 10{,}000$ ohms, diode resistance $= 1000$ ohms, and $E = 25$ V, positive grounded. Plot e_o to scale over one cycle.

(b) Repeat (a) with $R = 200{,}000$ ohms.

18-7. Determine the frequency of oscillation of the multivibrator of Fig. 18-23(a) if $C_1 = C_2 = 0.003\ \mu$F, $R_L = 5000$ ohms, $R_{B1} = R_{B2} = 20{,}000$ ohms. Plot a wave form of base voltage on Q_1 during a complete cycle.

18-8. Calculate the frequency of oscillation of Problem 18-7 if $C_1 = 0.015\ \mu$F and $C_2 = 0.02\ \mu$F. Plot the collector-to-collector wave form.

18-9. Design a Miller sweep to give a 500 μs linear output with a linearity error of only 0.1 per cent.

18-10. The circuit of Fig. 18-14(a) utilizes $R = 15{,}000$ ohms, $C = 0.0075\ \mu$F, $V_{CC} = 15$ V. The input positive pulse has a duration of 200 μs. Plot the output wave form and determine the per cent departure from linearity.

18-11. A Miller sweep uses a pentode of $g_m = 2000\ \mu$mho, $R_1 = 50{,}000$ ohms, $R_2 = 1$ megohm, $C = 500$ pF. Determine the per cent departure from linearity for a sweep period of 100 μs.

18-12. Determine the frequency of the output for the circuit of Fig. 18-25(a), where $C = 0.01\ \mu$F.

REFERENCES

1. Eccles, W. H., and F. W. Jordan, "Trigger Relay Utilizing Three-Element Thermionic Vacuum Tubes," *Radio Rev.*, Vol. 1, p. 143 (1919).
2. Soller, J. T., M. A. Starr, and G. E. Valley, Jr., *Cathode Ray Tube Displays.* McGraw-Hill Book Company, New York, 1948.
3. Schultheiss, P. M., and H. J. Reich, "Some Transistor Trigger Circuits," *Proc. IRE*, Vol. 39, p. 627 (1951).
4. McDuffie, G. E., Jr., "Pulse Duration and Repetition Rate of a Transistor Multivibrator," *Proc. IRE*, Vol. 49, p. 1487 (1952).
5. Moll, J. L., "Large Signal Transient Response of Junction Transistors," *Proc. IRE*, Vol. 42, p. 1773 (1954).
6. Moffat, D., "Comparison of RC Sweep and Ideal Sawtooth," *Electronic Ind.*, Vol. 17, p. 64 (September 1958).
7. Pettit, J. M., *Electronic Switching, Timing, and Pulse Circuits.* McGraw-Hill Book Company, New York, 1959.
8. Krakauer, S. M., and S. M. Soshea, "Hot Carrier Diodes Switch in Picoseconds," *Electronics*, Vol. 36, p. 53 (July 19, 1963).
9. Stanton, W. A., *Pulse Technology.* John Wiley & Sons, Inc., New York, 1964.
10. Millman, J., and H. Taub, *Pulse, Digital, and Switching Waveforms.* McGraw-Hill Book Company, New York, 1965.
11. Harris, J. N., P. E. Gray, and C. L. Searle, *Digital Transistor Circuits* (SEEC, Vol. 6). John Wiley & Sons, Inc., New York, 1966.

19

LOGICAL SWITCHING

Boolean algebra, which permits only two values or states for a variable, is well suited for the study of electrical switching circuits. The two permitted values of a variable are usually taken as 0 and 1, which may represent *open* and *closed* conditions of switches or *false* and *true* when applied to logical statements.

We will here consider the Boolean development of logical statements, some of the electronic circuits involving diodes and transistors for implementing these statements, and give several illustrations of the manner in which the simple logical circuits may be combined in complex systems.

19-1 BOOLEAN POSTULATES

As stated above, Boolean algebra allows only two states for the variables. It employs only three operations on these variables: those indicated by the symbol $+$, the symbol $\cdot$, and the operation of complementation indicated by a bar, as $\bar{A}$ is the complement of the variable A. The concepts of addition and multiplication should not be associated automatically with the indicated symbols; meanings consistent with Boolean rules will shortly be discovered.

With only two values permitted for the variables, Boolean theorems are often provable by the process of perfect induction, in which both values of the variables are substituted to find the theorem supported or rejected. This process may be illustrated by a search for the meaning of the symbol $+$.

Using 0 and 1 for the permitted values in our Boolean environment, let us consider the statements:

$$\text{(a}'\text{)}\ 0 + A = \qquad \text{(b}'\text{)}\ 1 + A =$$

Substituting 0 and 1 for the variable A we have the expressions:

$$(a'')\ 0 + 0 = \qquad (b'')\ 1 + 0 =$$

$$(a''')\ 0 + 1 = \qquad (b''')\ 1 + 1 =$$

We seek a meaning for the $+$ operation which will be consistent in this system of equations. It seems logical that operation with two null values (zeros) in (a″) should produce only a null, and so $0 + 0 = 0$. Inverse reasoning would indicate $1 + 1 = 1$, for (b‴), since 2 is not a permitted value in our algebra. These operations suggest the meaning of OR for the $+$ operation; i.e., zero OR zero is zero, one OR one is 1. Employing this meaning in (a‴) and (b″) leads to

$$0 + 1 = 1$$

$$1 + 0 = 1$$

a consistent result, readable as zero OR one are one, and vice versa.

We have then

$$0 + 0 = 0 \qquad 1 + 0 = 1$$

$$0 + 1 = 1 \qquad 1 + 1 = 1$$

or more generally

$$(a)\ 0 + A = A \qquad (b)\ 1 + A = 1 \tag{19-1}$$

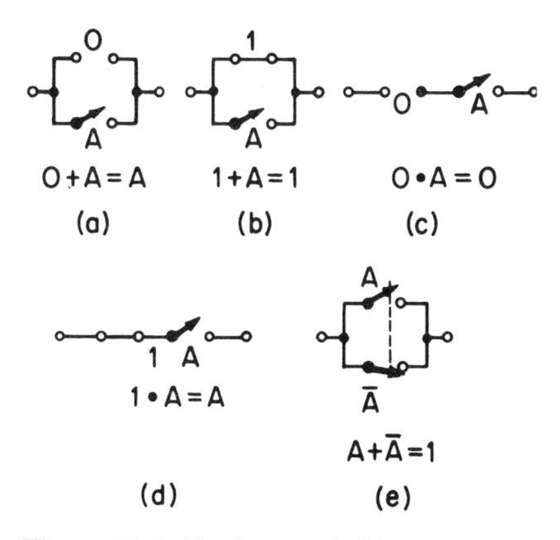

Figure 19-1. Boolean switching functions.

Now let us consider the circuits of Fig. 19-1. In (a) the statement $0 + A$ is represented by an open circuit (0) OR a switch A. If $A = 0$ (open) the result is an open circuit (0); if $A = 1$ (closed) the result is a closed circuit (1). Thus the circuit of (a), Fig. 19-1, operates in accordance with the statement

$$0 + A = A$$

In the circuit at (b), Fig. 19-1, the statement $1 + A$ is represented by a closed circuit (1) OR a switch having open (0) or closed (1) possibilities. The circuit operates in accordance with the statement

$$1 + A = 1$$

The statements of Eq. 19-1 have been proved by induction and given a physical application. These statements are basic Boolean postulates, and *the + sign will be read as* OR.

As a second example, let us find a meaning for the operation indicated by $\cdot$, as in

$$\text{(c}'\text{)}\ 0 \cdot A = \qquad \text{(d}'\text{)}\ 1 \cdot A =$$

Substitution of the permitted values 0 and 1 for the variable A leads to

$$\text{(c}''\text{)}\ 0 \cdot 0 = \qquad \text{(d}''\text{)}\ 1 \cdot 0 =$$

$$\text{(c}'''\text{)}\ 0 \cdot 1 = \qquad \text{(d}'''\text{)}\ 1 \cdot 1 =$$

Trying our usual multiplication processes we have

$$0 \cdot 0 = 0 \qquad 1 \cdot 0 = 0$$

$$0 \cdot 1 = 0 \qquad 1 \cdot 1 = 1$$

and no inconsistencies arise. Therefore, we see by induction that we can consider

$$\text{(c)}\ 0 \cdot A = 0 \qquad \text{(d)}\ 1 \cdot A = A \tag{19-2}$$

as additional basic postulates.

Now consider the switches at Fig. 19-1(c) and (d). Case (c) is represented by an open circuit (0) AND a switch A in series. Case (d) appears as a closed circuit (1) AND a switch in series, and the physical actions support the algebraic statements. It is thus usual *to read the symbol $\cdot$ as* AND.

Complementation is a negative of the statement, that is if $\bar{A} = 1$, then $A = 0$. As a switching function, this can be represented by ganged switches as at (e), such that when A closes, $\bar{A}$ opens. The output potentials of a multivibrator furnish another example of complementation.

The use of OR given above implies the *inclusive* OR meaning. That is, $A + B = 0$ means that A or B are open, or *both* are open. We also employ the *exclusive* OR at times, meaning that A or B are open, but *not both* are open. The latter concept will be indicated when used.

19-2 ADDITIONAL POSTULATES

Other basic relations which can be derived by the use of perfect induction are:

Complementation:

$$A + \bar{A} = 1 \qquad\qquad A \cdot \bar{A} = 0$$

Commutation:

$$A + B = B + A \qquad\qquad AB = BA$$

Associativity:

$$A + (B + C) = (A + B) + C \qquad A(BC) = (AB)C$$

Distributivity:

$$A + BC = (A + B)(A + C) \qquad A(B + C) = AB + AC$$

De Morgan's theorem:

$$\overline{A + B} = \bar{A}\bar{B} \qquad \overline{AB} = \bar{A} + \bar{B}$$

Note that the usual distributive law of addition does not hold in Boolean algebra. This result may be proved as an illustration of the manner in which a *truth table* is set up and used. Such a table is established using all possible combinations of 0 and 1 for the variables involved, and including all terms in the equation. Thus for the statement

$$A + BC = (A + B)(A + C)$$

there will be columns headed for each term as

A	B	C	BC	$A + BC$	$A + B$	$A + C$	$(A + B)(A + C)$
0	0	0	0	0	0	0	0
0	0	1	0	0	0	1	0
0	1	0	0	0	1	0	0
0	1	1	1	1	1	1	1
1	0	0	0	1	1	1	1
1	0	1	0	1	1	1	1
1	1	0	0	1	1	1	1
1	1	1	1	1	1	1	1

The values in the columns for A, B, and C are arbitrarily assumed to include all possible combinations of 1 and 0; all columns to the right are then evaluated using the basic postulates appropriate to the case. It can be seen that the values for the left side of the original equation given in the column headed $A + BC$ exactly correspond to the values for the right side of the equation given in the column headed $(A + B)(A + C)$. The Boolean distributive law is therefore proven.

A few other useful theorems are:

1. $A + AB = A$
2. $A + \bar{A}B = A + B$
3. $A(A + B) = A$
4. $(A + \bar{B})B = AB$
5. $A(B + \bar{B}) = A$
6. $A\bar{B} + \bar{A}B = (A + B)(\overline{AB})$
   $= (A + B)(\bar{A} + \bar{B})$
7. $\overline{AB + \bar{A}\bar{B}} = A\bar{B} + \bar{A}B$

Some of these statements can be proved by algebraic reduction. For example, theorem 6 can be established by including the terms $A\bar{A} = B\bar{B} = 0$ as

$$A\bar{B} + \bar{A}B = A\bar{A} + \bar{A}B + A\bar{B} + B\bar{B} = (A + B)(\bar{A} + \bar{B}) \qquad (19\text{-}3)$$

19-3 MINIMIZATION BY ALGEBRAIC REDUCTION

The algebra of Boolean functions can be employed to derive a simplified form for a switching system, to eliminate redundant operations, and to minimize a system.

Consider the complex switching system presented in Fig. 19-2(a). Questions naturally arise: Must the circuit be this complicated? Are there redundant functions called for? Can the equipment be minimized? Reduction of the circuit form by Boolean algebra is often possible. A truth table may be constructed as:

A	B	C	F
0	0	0	1
1	0	0	0
0	1	0	1
0	0	1	1
1	1	0	0
0	1	1	1
1	0	1	1
1	1	1	1

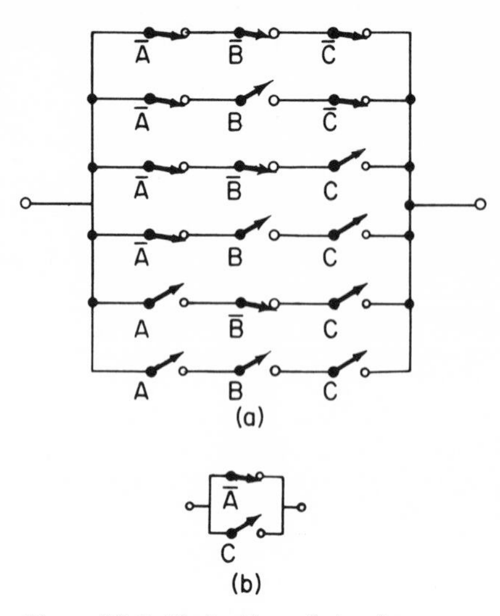

Figure 19-2. Reduction of circuits.

The circuit is closed when $F = 1$, and it is therefore closed for the sum of all such cases. Therefore considering 0's as complements, there results from the six cases of the table in which $F = 1$:

$$\bar{A}\bar{B}\bar{C} + \bar{A}B\bar{C} + \bar{A}\bar{B}C + \bar{A}BC + A\bar{B}C + ABC = 1 \tag{19-4}$$

remembering to read the $+$ signs as OR. Grouping the terms

$$\bar{A}\bar{B}\bar{C} + ABC + \bar{A}(B\bar{C} + \bar{B}C) + C(\bar{A}B + A\bar{B}) = 1$$

The first two terms nullify and the remaining terms reduce by theorem 5 to

$$\bar{A} + C = 1 \tag{19-5}$$

which is represented by the simple circuit in Fig. 19-2(b). The situation chosen was overly complex, but it illustrates the possibility of algebraic minimization of switching circuits.

It would have been simpler to choose the alternative method—to write a set of equations for the situations where $F = 0$, with all included terms complemented, and all indicated operations changed from $+$ to $\cdot$ and vice versa. That is,

$$(\bar{A} + B + C)(\bar{A} + \bar{B} + C) = 1 \tag{19-6}$$

Carrying out the indicated operations leads to

$$\bar{A} + \bar{A}(B + \bar{B}) + \bar{A}C + C(B + 1) = 1$$

$$\bar{A} + C(\bar{A} + 1) = 1$$

$$\bar{A} + C = 1 \qquad (19\text{-}7)$$

which agrees with the previous result.

Thus is illustrated the *duality* of Boolean operations.

19-4 *THE BINARY NUMBER SYSTEM*

Our system of decimal numbers employs ten states or symbols as 0, 1, 2, 3, . . . , 9. To employ electronic devices directly in decimal counting would require that they recognize these ten states. This might require the subdivision of a dynamic curve into ten levels, and this would imply an accuracy and resettability not inherent in electronic devices. However, electronic devices are well suited to counting in a two-state or *binary number system*, using saturation and cutoff or ON and OFF as the recognizable states. The statements of Boolean logic can be implemented with electronic devices operating with two-state or binary signals, and this is the basis for electronic digital computation.

In the decimal system each position indicates an increased power of 10. That is, the decimal number 1206 implies:

$$1206 = 1 \times 10^3 + 2 \times 10^2 + 0 \times 10^1 + 6 \times 10^0 \qquad (19\text{-}8)$$

The binary number system uses only two symbols, usually 0 and 1, and employs the positional concept in powers of the base 2. The base is sometimes referred to as the *radix* of the system. The decimal number 1206 is derived in the binary system as

$$\begin{aligned}\text{decimal } 1206 = \text{binary:}\quad & 1 \times 2^{10} + 0 \times 2^9 + 0 \times 2^8 + 1 \times 2^7 \\ & + 0 \times 2^6 + 1 \times 2^5 + 1 \times 2^4 + 0 \times 2^3 \\ & + 1 \times 2^2 + 1 \times 2^1 + 0 \times 2^0 \\ = 10010110110 \qquad & \qquad (19\text{-}9)\end{aligned}$$

The binary system requires additional digits in trade for the reduced number of states.

We may readily arrive at the properties of addition in the binary number system as

	0	1
0	0	1
1	1	10

since 10 is the binary symbol for decimal $2 = 1 \times 2^1 + 0 \times 2^0$. Thus the sum of binary 1101 (decimal 13) and 1001 (decimal 9) follows as

Binary		*Decimal*
1101		13
1001		9
0100		12
1 1	carry	1
10110		22

For binary multiplication

	0	1
0	0	0
1	0	1

The product of binary 1101 and 1001 can be obtained by addition and left shifting as

Binary		*Decimal*
1101		13
1001		9
1101		117
0000	shift	
0000	shift	
1101		
1110101		

19-5 BINARY CODES

Binary system data are represented as chains of pulses, as in Fig. 19-3. The pulses are operated upon in electronic logic circuits designed to carry out the processes of Boolean algebra. Given the basic idea of a chain of positive and negative, or positive and zero, or zero and negative pulses as representing binary 1's and 0's, there are many possible codes in which the pulses might be transmitted. One of the most common is the *binary-coded-decimal code* (BCD) requiring four pulses or *bits* per decimal digit as

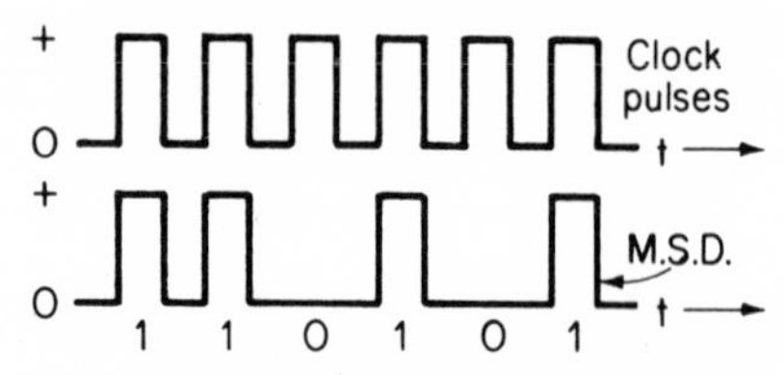

Figure 19-3. Digital-pulse wave forms.

Decimal	8	4	2	1
0	0	0	0	0
1	0	0	0	1
2	0	0	1	0
3	0	0	1	1
4	0	1	0	0

Decimal	8	4	2	1
5	0	1	0	1
6	0	1	1	0
7	0	1	1	1
8	1	0	0	0
9	1	0	0	1

The decimal number 1206 would then be transmitted as four four-bit code groups

0001 0010 0000 0110

retaining the decimal positional system. This is not the binary form of 1206 given in Eq. 19-9, but it is the form in which much data reaches the computer. The human operator works in decimals, but the output of his machine appears in BCD form. It is easily possible for the computer to translate the BCD form to pure binary by arithmetic operations on the BCD input. Decoders, of which Section 19-10 discusses one example, can also be designed to convert BCD or binary output to decimal form.

The above table could be extended to decimal 15 without requiring additional bits; it would then become a *hexadecimal code*, usually employing letters $a, b, \ldots, f$ for the decimal numbers 10 through 15. Another code which is employed in some computer operations is the *octal* or base-8 system. The permitted numbers are then 0, 1, 2, . . . , 7, and the decimal number 24 is written as octal $30(3 \times 8^1 + 0 \times 8^0)$. Binary coding of octal numbers requires only the three least-significant bits of the BCD table above, and binary coding of octal 30 is 011 000.

Since decimal 24 is written 11000 in binary form, and 011 000 in octal coded form, an easy means of conversion from binary numbers to octal numbers is indicated. By setting off the binary number in groups of three bits, each group can be directly translated to an equivalent octal number. For instance, decimal 1206 as written in binary in Eq. 19-9 can be written in groups of three bits as

binary	010	010	110	110
octal	2	2	6	6

and the octal equivalent is 2266.

There are other code arrangements designed to assure greater accuracy, and also including extra checking bits for detecting some errors which may occur in transmission.

There are two basic methods of handling bits in the processing of digital information. In the *series* form the pulses are transmitted and operated upon in sequence; the pulses of Fig. 19-3 may be visualized as sliding off the page to the right in series, with the most significant digit indicated as M.S.D. In *parallel* operation the bits of the figure may be thought of as sliding down the page together in six separate channels; the weight of the pulse is indicated by the path in which the pulse appears. For a signal of 40 bits with series transmission, the time of performing a given operation will be approximately 40 times as long as in parallel operation. However, parallel computation requires 40 sets of equipment to simultaneously process all pulses. Thus speed is purchased at the price of extra equipment.

19-6 BASIC LOGIC BUILDING BLOCKS

While the first major electrical computer used relays as switches, the high speeds of computers today demand that the switching operations be performed electronically in order to reach operating times of a microsecond or less. The circuits which perform the operations of logic are called *logic gates*, and the three operations of OR, AND, and inversion or NOT are carried out in this way.

Digital signals are composed of short voltage pulses, and the level and polarity for a true or false, 1 or 0 signal must be agreed upon, as:

1. Positive logic: the 1 state is more positive than the 0 state.
2. Negative logic: the 1 state is more negative than the 0 state.

In some cases the 0 state may be set at or near zero volts, or the 1 state may be at zero; in other cases the change in levels may be taken as a swing between a positive voltage and a negative voltage.

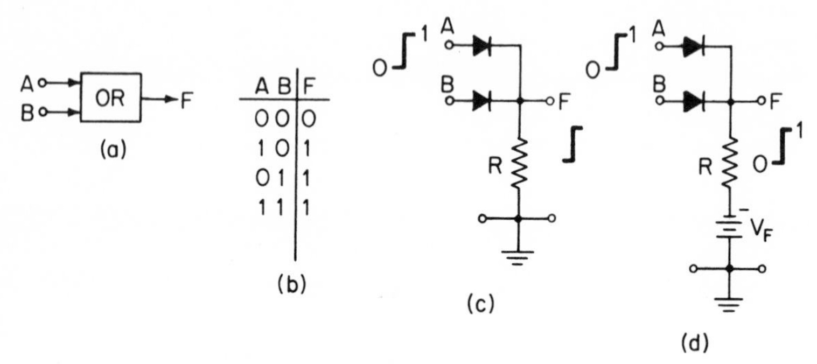

Figure 19-4. (a) OR gate symbol; (b) truth table; (c), (d) DL logic gates.

The OR gate performs the Boolean operation of $A + B$ and *diode logic* (DL) circuit forms are presented in Fig. 19-4. These are shown with two inputs but may have more. With positive logic and +10 V as the logical 1 level, zero volts as the logical 0 level, a logical 1 or +10 V to either input forward biases its diode and places F at a level of a logical 1. Only if both inputs are at 0 level will the output be at logic 0 level.

For faster charging of circuit capacitances source V_F is often added, as in (d); therein V_F is two or more times the logical 1 level. A logical 0 or input at ground level at both A and B gives a logical 0 output, since $-V_F$ turns on both diodes. A logical 1 or +10 V at A or B forward biases the appropriate diode and F goes to the logical 1 level; the B diode is reverse biased and open. Thus the most positive input is capable of controlling the output.

The AND gate performs the Boolean $A \cdot B$ operation, and a diode logic

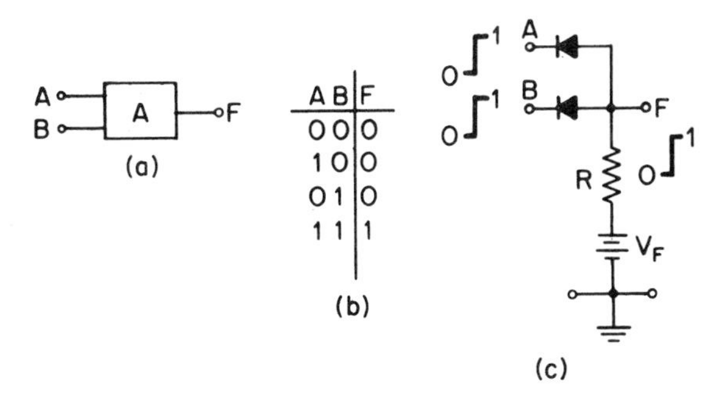

Figure 19-5. (a) AND gate; (b) truth table; (c) diode logic AND gate.

form is shown in Fig. 19-5. With logical 0 or ground potential on both inputs, the diodes conduct due to $+V_F$ and the output is logical 0. Applying a logical 1 only to input A reverse biases and cuts off its diode because F is maintained at ground by the 0 input to diode B. A logical 1 must be applied at A and B to produce a logical 1 output. The most negative input to the diodes is controlling. Resistor R is chosen to provide the desired output voltage level when all inputs are logical 1.

It is of interest to note that if negative logic is employed with the OR and AND circuits, the operating functions reverse and a positive logic OR circuit becomes a negative logic AND circuit, and a positive logic AND circuit becomes a negative logic OR circuit.

Due to diode voltage drops and to varying current demands by the load, the output signals in diode logic will not be exactly at the chosen voltage levels, but the difference between the 1 and 0 levels is great enough for the results to be unambiguous. However, a cascade of several such gates will cause a greater departure from the desired levels and amplifiers with diode clamps may be employed, as shown by diode D in Fig. 19-6. Voltage V_L is the positive logic level and $V_{CC} - I_L R_L > V_L$. Circuits such as the Schmitt trigger may also be inserted for the level-restoring function.

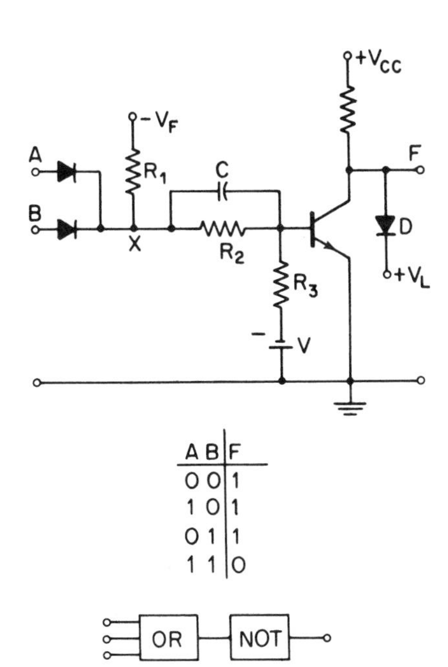

Figure 19-6. NOR circuit, positive logic.

Reverse leakage currents of the

diodes at higher temperatures may limit the number of diode inputs which can be utilized in one gate; the number of input circuits which can be successfully applied is called the *fan-in* property. Likewise, the available output current limits the number of gates that can be driven by the output, and this is known as the *fan-out* capability of the circuit.

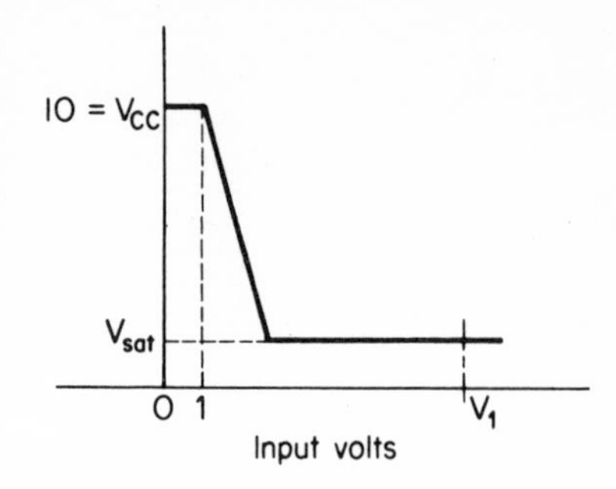

Figure 19-7. Transfer curve for inverter amplifier (positive logic).

The NOT operation or *inversion*, is carried out by adding an amplifier to these circuits, as for NOR in Fig. 19-6. The form of circuit is known as *diode-transistor logic* (DTL). The amplifier has an input-output characteristic as in Fig. 19-7. A 0 level signal less than one volt will leave the transistor cut off by the $-V$ bias and the output will be at V_{CC} or logical 1. A logical 1 signal at about $+10$ V will drive the transistor to saturation with output at V_{sat} or logical 0.

With 0 logic signals at all inputs of the diode OR circuit, both diodes conduct, point X is at ground and the transistor is at cutoff because of $-V$. With one or more inputs at the 1 level, point X is raised to $+10$ V and the transistor is at saturation. Capacitor C is a speed-up unit to increase the rate of charging of the transistor capacitance.

The addition of a NOT circuit to an AND circuit, as in Fig. 19-8, results in the not-and or NAND circuit. The resistors R_1, R_2, R_3 form a network such that when any input is at the 0 level the diode places X at ground, the transistor is cut off by $-V$ and the output rises to V_L. When all inputs are at the 1 level, X is at $+10$ V and the base of the transistor is driven to saturation and the output to the 0 level.

Diode-transistor logic permits fan-in and fan-out figures higher than are possible with DL circuits, because of the current gain provided by the transistor. Potential V_F helps to charge the transistor capacitances and the circuit is usually faster than the DL form.

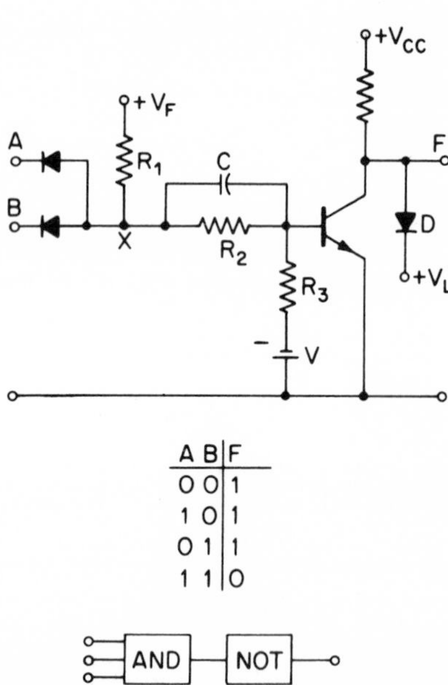

A	B	F
0	0	1
1	0	1
0	1	1
1	1	0

Figure 19-8. NAND circuit, positive logic.

19-7 THE SUMMING RESISTOR

Diode logic gates may be cascaded as in Fig. 19-9, where a positive-logic AND gate supplies one input to an OR gate. Voltage V_F and R tend to operate as a constant-current source, with the current independent of inputs at 1 or 0 levels. The AND gate is inhibited by any one gate at the 0 level or ground. The other inputs at 1 levels are disconnected by their diodes and the current V_F/R passes into the conducting diode. When all inputs are at the 1 level the available input current is $(V_F - V_1 - V_{DF})/R$, where V_1 is the logic 1 level and V_{DF} is the small forward diode drop. The advisability of making V_F several times V_1 is apparent.

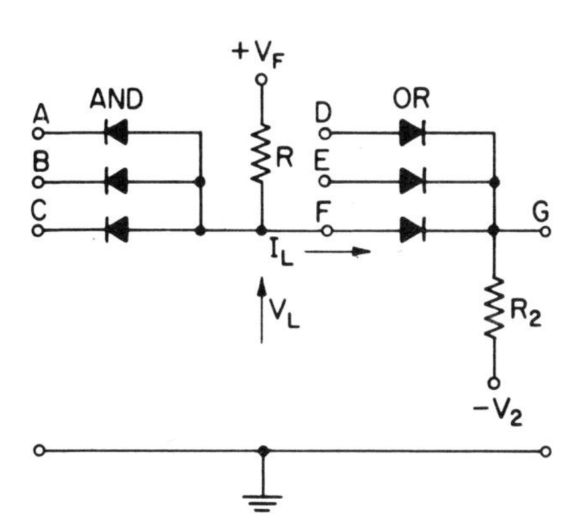

Figure 19-9. Cascaded AND and OR gates.

The summing resistor R places a static limit on the output current from the gate. The current in R must supply I_L and also maintain a small forward current in the diodes, under the condition of 1 level inputs to A, B, C. The load current is $I_L = (V_1 + V_{DF} + V_2)/R_2$. Summing the currents at V_L, and using a minimum forward diode current as I_{DF}, then

$$\frac{V_F - V_1 - V_{DF}}{R} - \frac{V_1 + V_{DF} + V_2}{R_2} - MI_{DF} = 0$$

$$R = \frac{V_F - V_1 - V_{DF}}{(V_1 + V_{DF} - V_2)/R_2 + MI_{DF}} \tag{19-10}$$

where M is the fan-in factor of the AND circuit. For worst-case design it is desirable to use a minimum value of $V_F = 30$ V, the maximum bound on $V_1 = 10$ V, the maximum value for $V_2 = -10$ V, and the minimum value for $R_2 = 35{,}000$ ohms. The current needed to keep a diode in forward conduction, I_{DF}, should be a low temperature value as 0.1 mA, and the fan-in taken as 3. Then

$$R = \frac{30 - 0.5 - 10}{(10 + 0.5 + 10)/35{,}000 + 3 \times 0.1 \times 10^{-3}} = 22{,}000 \qquad \text{ohms}$$

Dynamic considerations may also control the choice of R. Assume that it is desired that the switching operation from the 0 level to the 1 level occur in 1 μs, with a load capacitance of 100 pF. The charge which must be transferred is $\Delta Q = C\Delta E$ and so the average current is

$$i = \frac{C\Delta E}{\Delta t} = \frac{100 \times 10^{-12} \times 10}{1 \times 10^{-6}} = 1 \qquad \text{mA} \quad (19\text{-}11)$$

With $V_F = 30$ V, then

$$R = \frac{30}{0.001} = 30{,}000 \qquad \text{ohms}$$

The static requirement of 22,000 ohms will thus satisfy the dynamic speed requirement as well.

19-8 ADDITIONAL LOGIC-CIRCUIT FORMS

Logic circuits have been developed in other forms known as *resistor-transistor logic* (RTL), *direct-coupled transistor logic* (DCTL), *transistor-transistor logic* (TTL), *current-mode logic* (CML), and others. Examples appear in Figs. 19-10 through 19-13. All are shown for positive logic.

The resistor-transistor logic form in Fig. 19-10 employs resistors for summing and transistors for level restoring and inversion. It does not provide the isolation given by the diodes of the DL circuit, and fan-in capabilities are limited. Transistor parameters are not critical and levels can be held by clamp diodes. Speed is not high because the transistor capacitances must be charged through the isolating resistors.

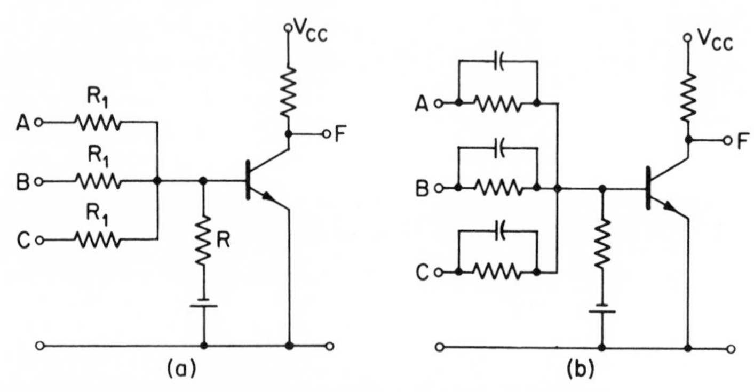

Figure 19-10. NOR circuits: (a) resistor-transistor logic; (b) resistor-capacitor-transistor logic.

A modified form, known as *resistor-capacitor-transistor logic* (RCTL)

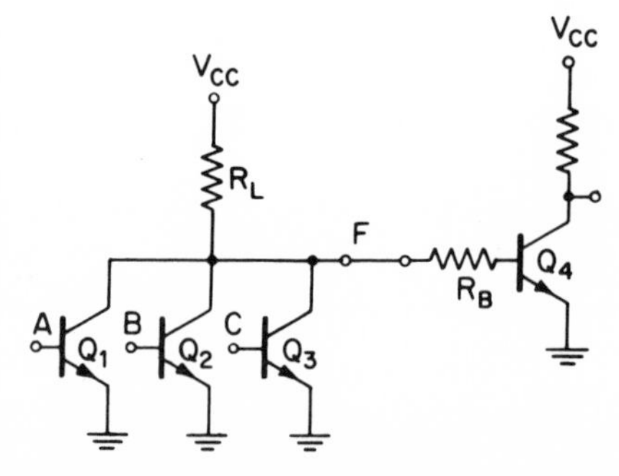

Figure 19-11. NOR gate in DCTL form.

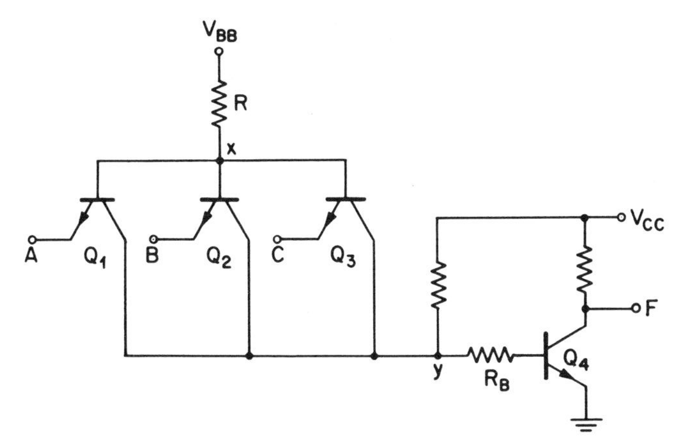

Figure 19-12. NAND gate in TTL form.

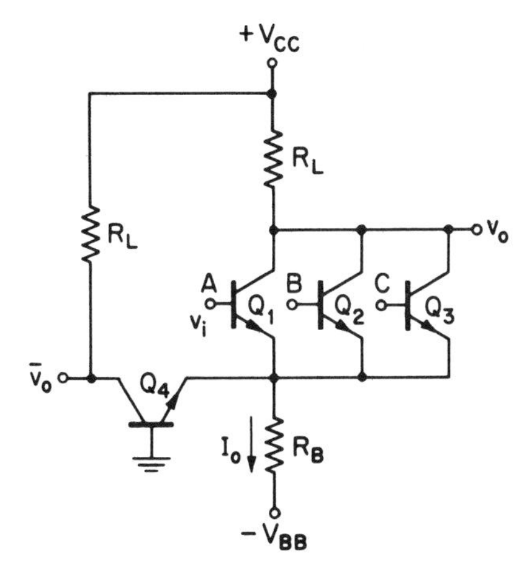

Figure 19-13. NOR circuit in current-mode logic form.

includes speed-up capacitors across the input resistors to help overcome the storage delay of the transistor. Cost is increased but the circuit is satisfactory for medium speed situations.

The DCTL circuit in NOR form appears in Fig. 19-11. A logic 1 input to any of the three gate transistors Q_1, Q_2, Q_3 will turn it on, dropping its output level to V_{sat}; this is the NOR operation. When all inputs drop to logical 0 the output voltage rises to the v_{BE} threshold at which transistor Q_4 turns on. The value of V_{sat} of Q_1, Q_2, or Q_3 must be low enough that Q_4 is off; V_{sat} must be less than v_{BE} at the threshold of conduction. This voltage difference may be small; it is often only a fraction of a volt. Noise or variation in V_{CC} can create spurious switching operations of transistor Q_4. While giving greater speed, the DCTL circuit requires tight specifications on the transistors if it is to be successful.

Transistor-transistor logic employs the common-base circuit, as in Fig. 19-12. If all of the emitters are given a logic 1, where $V_1 < (V_{BB} - v_{BE})$, the connected transistors will be turned off, and a bias through the resistor network turns on Q_4. The output is V_{sat} or a logic 0 in a NAND operation. If a logic 0 is supplied to any input, its transistor will turn on and Q_4 will turn off, giving a logic 1 output at F. This circuit can be faster than any of the circuits so far described.

Because of the relative ease with which transistors can be manufactured in the integrated circuit process, there may be no cost disadvantage in the use of transistors in transistor-transistor logic (TTL) as compared to diodes in DTL circuitry. It is also easier to insure uniformity between transistors manufactured on a single silicon chip than to match discrete diodes or transistors.

Current-mode logic can be made even faster; more components and more power are required, but operation into saturation is avoided, and it is possible to switch in a few nanoseconds.

In the CML circuit of Fig. 19-13 the switching transistors Q_1, Q_2, Q_3 are biased by a voltage across R_B, with the bias current I_o controlled by regulator transistor Q_4. The bias is such that the switching transistors operate midway between logic levels for 0 and 1. A positive logic 1 signal will turn on a switch and drop its collector voltage. The current is fixed by the current through R_B, and is limited to less than the saturation value. In effect, the switch transistor now supplies the bias current previously given by Q_4, and the circuit switches somewhat like a Schmitt trigger.

Logic levels are not preserved, since the output swings between V_{CC} and a lower level given by $V_{CC}(1 - R_L/R_B)$. Speed is improved, since with small voltage swing there is less change in stored charge.

Because the current is not determined by the switching transistors the circuit is not much affected by transistor characteristics. The output may be taken through an emitter follower, whose low impedance provides high fan-out and short rise time for capacitive loads. The circuit provides its own complemented output, since the collector output of Q_4 is complementary to the output from the gates.

The circuit fundamentals described above have led to expansion of each basic logic type into complete compatible circuit families, including inverters and flip-flops. Selections between the various logic forms are made on the basis of switching speed needed for the application, effect of change in device characteristics, noise susceptibility, power output, fan-in and fan-out capabilities, adaptability to integrated circuit manufacture, reliability and certainly cost. Choice of one form over another is not usually clear-cut.

19-9 SERIES TO PARALLEL CONVERSION

Much computer input data, as from radio or wire circuits or magnetic tape, is in serial form but it is possible to use logic circuits to convert it to the parallel form for computer processing. One such logic circuit is indicated in Fig. 19-14.

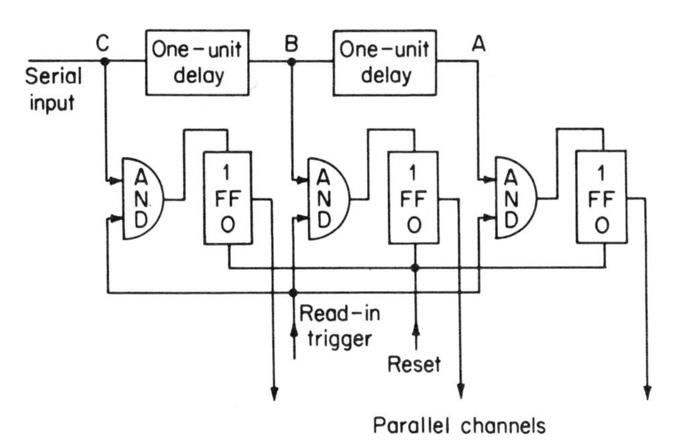

Figure 19-14. Series-to-parallel digital converter for a three-bit signal.

Serial pulses may be considered as fed in a CBA time sequence. Delay units are introduced into the line in which each delay time is equal to the time between bits in the input signal; these delay elements may be sections of simulated transmission line. Thus at a given instant after the serial signal is introduced, the serial signal pulses of CBA will have their pulse centers distributed at the points indicated, as inputs to the AND circuits. At that moment a gating read-in pulse is also applied to the gates, and the corresponding zeros or ones are transferred to the three flip-flops, as inputs to the three parallel processing channels.

19-10 SWITCHING DIODE MATRICES (DECODERS)

A combination of flip-flop counters and diode logic AND networks makes possible a method by which the binary representation of a number may be decoded into signals on decimally-numbered output lines. A translating matrix for BCD code to decimal numbers is drawn in Fig. 19-15.

Inputs $T_1, \ldots, T_5$ are bistable pairs, with conducting states a or b. The diodes, resistor R, and a given vertical bus constitute a diode logic AND gate. With $-V$ input to the matrix, any diode connected to a conducting state will be conducting and put zero volts on the bus; the only bus not at zero will be that bus representing the appropriate decimal number. It may be noted that a 0 in the binary number calls for connection of a diode at a and a 1 calls for the diode at b, so a switching matrix can be readily designed to handle any code and any number of outgoing lines.

Figure 19-16 is the same circuit for a 3-bit number, redrawn to more clearly indicate the AND gate functions.

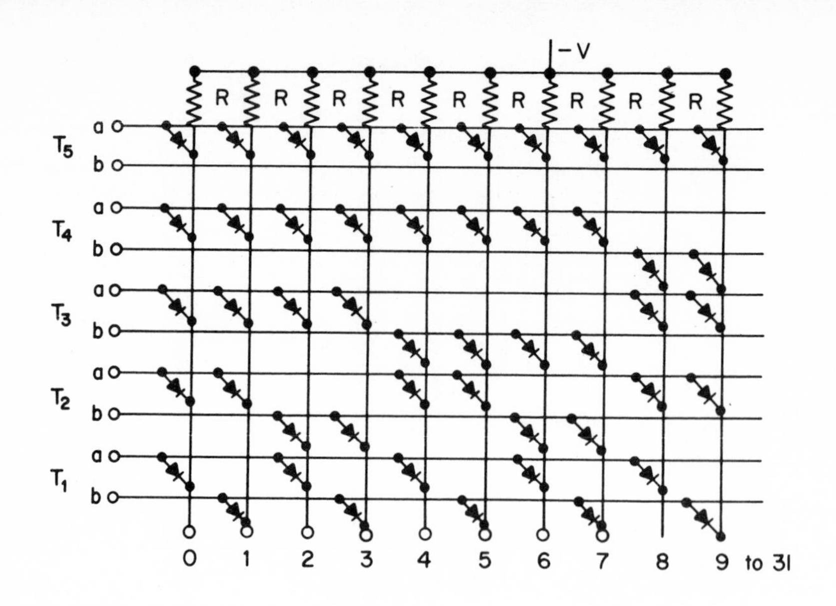

Binary No.	Conducting state T_5	T_4	T_3	T_2	T_1	Output line
0 0 0 0 0	b	b	b	b	b	0
0 0 0 0 1	b	b	b	b	a	1
0 0 0 1 0	b	b	b	a	b	2
0 0 0 1 1	b	b	b	a	a	3
0 0 1 0 0	b	b	a	b	b	4
0 0 1 0 1	b	b	a	b	a	5
0 0 1 1 0	b	b	a	a	b	6
0 0 1 1 1	b	b	a	a	a	7
0 1 0 0 0	b	a	b	b	b	8
0 1 0 0 1	b	a	b	b	a	9

Figure 19-15. BCD decoding matrix.

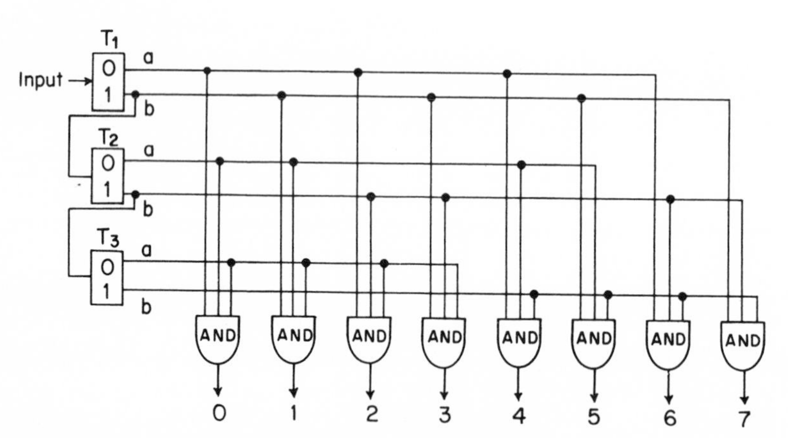

Figure 19-16. Diode logic of Fig. 19-15 combined in AND circuits.

19-11 BINARY COMBINATION ADDERS

The process by which Boolean logic can be applied in the development of a useful circuit for computation can be illustrated by the design of a binary adder. In the table for addition of Section 19-4, the addition of a 0 to a 1 or a 1 to a 0 is simple, but addition of 1 to 1 involves a carry bit, C_2, as well as a sum 0. Also in any logic circuit except the one handling the least significant bit of a binary number, there must be a provision for a carry bit, C_1, coming forward from the addition of the preceding digit.

The results of the addition of two bits, A and B, may be assembled in a Boolean function table, as in Table 19-1.

Table 19-1

OPERATION OF A SINGLE-STAGE BINARY ADDER

Digits		Carry	Sum	Carry
A	B	C_1	S	C_2
0	0	0	0	0
0	0	1	1	0
0	1	0	1	0
0	1	1	0	1
1	0	0	1	0
1	0	1	0	1
1	1	0	0	1
1	1	1	1	1

As before, we may write Boolean functions for all combinations of S or C_2 summing to unity. That is,

$$S = \bar{A}\bar{B}C_1 + \bar{A}B\bar{C}_1 + A\bar{B}\bar{C}_1 + ABC_1 \tag{19-12}$$

and

$$C_2 = \bar{A}BC_1 + A\bar{B}C_1 + AB\bar{C}_1 + ABC_1 \tag{19-13}$$

To this result for C_2 may be added $ABC_1 + ABC_1$, the result not being changed since addition of identical switches in parallel does not change a circuit. However, it is then possible to write

$$\begin{aligned} C_2 &= AB(C_1 + \bar{C}_1) + AC_1(B + \bar{B}) + BC_1(A + \bar{A}) \\ &= AB + AC_1 + BC_1 \end{aligned} \tag{19-14}$$

The complemented terms are available as pulses from the opposite sides of the respective A, B, C flip-flops, so that Eqs. 19-12 and 19-14 may be used directly to build a logical adder, as in Fig. 19-17. The circuit requires seven AND circuits and two OR circuits. Given pulses on inputs A and B as flip-flops, the circuit will give appropriate sum and carry outputs for successive bits

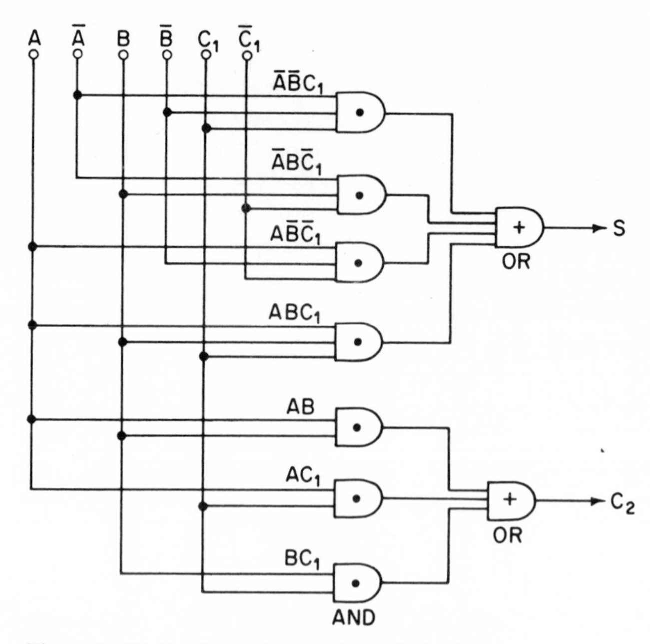

Figure 19-17. Implementation of a logical adder.

in serial operation, or it could serve to add one bit in parallel operation, an identical circuit being required for each bit position of the A and B signals.

Algebraic manipulation of the Boolean function relations obtained from the table can lead to equivalent forms of logic statements, but differing circuits. This can be demonstrated by use of the addition relations of Eqs. 19-12 and 19-13. The expression for S may be written as

$$S = \bar{C}_1(A\bar{B} + \bar{A}B) + C_1(AB + \bar{A}\bar{B})$$

Theorem 7 at the end of Section 19-2 permits this to be changed to

$$S = \bar{C}_1(A\bar{B} + \bar{A}B) + C_1(\overline{A\bar{B} + \bar{A}B}) \tag{19-15}$$

The expression for the carry reduces to

$$C_2 = AB + C_1(A + B)(\overline{AB}) \tag{19-16}$$

These two expressions differ from Eqs. 19-12 and 19-14, but can also be developed into an adder, as shown in Fig. 19-18. The result consists of two identical *half adders* 1, 2, 3, 4, and one additional OR circuit. Figures 19-17 and 19-18 utilize the same number of logic gates, but there may be differences in speed of operation. Boolean algebraic reduction techniques have thus been shown as leading to several alternative solutions for the same problem.

To complete the adder development, there must be a delay added to the carry circuit, so that its output remains available to apply with the next pair

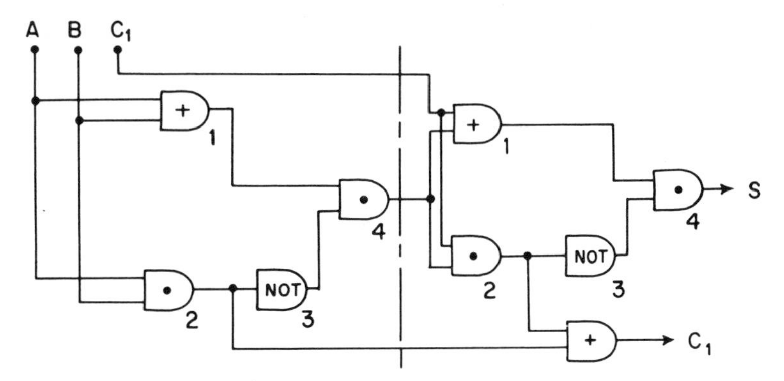

Figure 19-18. Addition with two half adders.

of bits in serial operation, or the sum must be held while the carry bits propagate from one bit circuit to the next in parallel operation.

19-12 BINARY COMPARISON

Much computer operation involves the comparison of two binary numbers to establish identity, as in searching for data filed in a memory or to establish equality in iteration. Comparators can be assembled from the logic building blocks.

The search for equality can be made by feeding the two binary numbers into two flip-flop chains or registers, as for the decoder, and comparing the conditions of the two registers with AND circuits, bit by bit. Such a comparator for three bits is illustrated in Fig. 19-19. Bits from one number are designated A, those from the other as B, and the complementary terms are

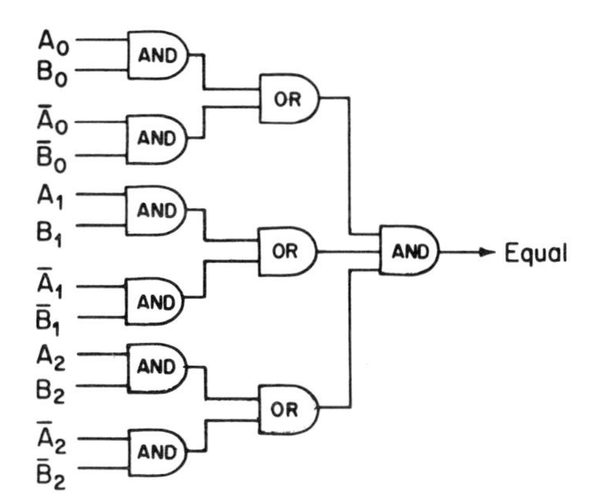

Figure 19-19. Binary number comparison circuit.

derived from the opposite sides of the respective flip-flops. Equality of all three bits, either as ones or zeros, is indicated by output from the final AND circuit, or

$$\text{equality} = (A_0B_0 + \bar{A}_0\bar{B}_0)(A_1B_1 + \bar{A}_1\bar{B}_1)(A_2B_2 + \bar{A}_2\bar{B}_2)$$

PROBLEMS

19-1. Show by use of perfect induction that

$$AB + AC = A(B + C)$$

19-2. Prove that

$$(A + B)(\bar{A} + C)(B + C) = (A + B)(\bar{A} + C)$$

by use of a truth table.

19-3. By algebraic reduction, show that

$$(A + \bar{B})C + A(B + \bar{C}) + \bar{A}B = A + B + \bar{B}C$$

$$(A + C)(\bar{A} + B + C) = (A + C)(B + C)$$

19-4. Simplify:

$$A\bar{B}\bar{C} + A\bar{B}\bar{C}D + \bar{C}A$$

$$\bar{A}\bar{B}C + A\bar{B}C + ABC$$

$$A\bar{B} + \bar{A}B + AB$$

19-5. Decode the following BCD numbers to decimals:

0101 0110 0011
1001 0111 0101
1000 0101 0010
1001 0010 0100

19-6. Write the following decimal numbers in binary:

104, 647, 223, 462

19-7. Octal or base-8 numbers are used in some computing operations. Prepare a table of base-8 equivalents for decimal numbers 0 to 20, and translate the following decimal numbers to base-8 numbers:

123, 1087, 967

19-8. Base-3 numbers have some use. Translate the following decimal numbers to base 3:

101, 97, 245

19-9. Perform the following operations, in which + means addition, × implies multiplication:

$$110110 + 011011 = (a) \qquad 110010 \times 1011 = (c)$$

$$101111 + 100101 = (b) \qquad 101010 \times 1110 = (d)$$

and prove by translating the operations to decimals.

19-10. In Fig. 19-20, write a logic statement and simplify the circuit. Draw the reduced equivalent circuit.

19-11. For Fig. 19-21(a) write the logic statement of the operation.

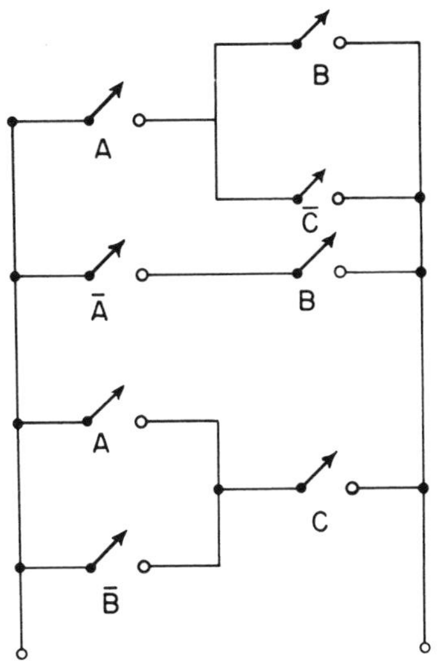

Figure 19-20.

19-12. Write the logic statement for the indicated operation in Fig. 19-21(b).

19-13. The following numbers are in binary code. Write these numbers in octal and decimal forms.

01011101	11001011	11100011
10110010	00101111	01101100
11011011	10010101	10101100

19-14. The *excess-3 code* is sometimes used because no digit is represented by a complete null signal:

0 ⟶ 0011	5 ⟶ 1000
1 ⟶ 0100	6 ⟶ 1001
2 ⟶ 0101	7 ⟶ 1010
3 ⟶ 0110	8 ⟶ 1011
4 ⟶ 0111	9 ⟶ 1100

Design a switching matrix to translate excess-3 coded numbers to their equivalent decimal values.

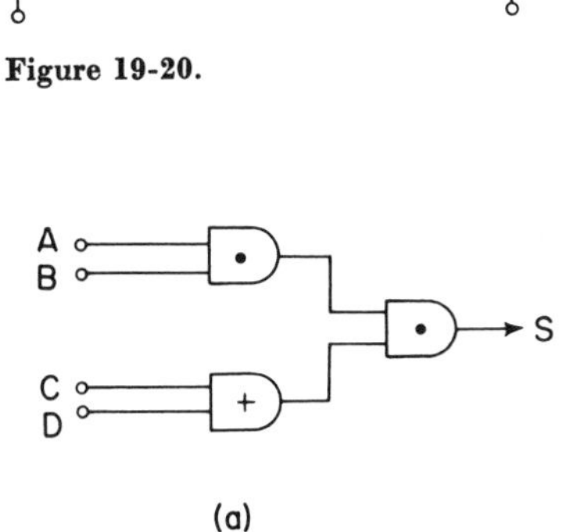

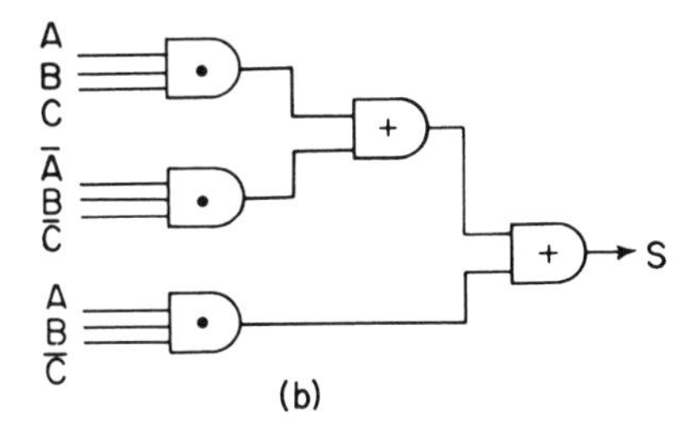

Figure 19-21.

REFERENCES

1. Page, C. H., "Digital Computer Switching Circuits," *Electronics*, Vol. 21, p. 110 (September 1948).
2. Shannon, C. E., "The Synthesis of Two-Terminal Switching Circuits," *Bell System Tech. Jour.*, Vol. 28, p. 59 (1949).
3. Hussey, L. W., "Semiconductor Diode Gates," *Bell System Tech. Jour.*, Vol. 32, p. 1137 (1953).

4. Richards, R. K., *Arithmetic Operations in Digital Computers.* D. Van Nostrand Co., Inc., Princeton, N. J., 1955.

5. Caldwell, S. H., *Switching Circuits and Logical Design.* John Wiley & Sons, Inc., New York, 1958.

6. McCluskey, E. J., Jr., and T. C. Bartes, *A Survey of Switching Circuit Theory.* McGraw-Hill Book Company, New York, 1962.

7. Marcus, M. P., *Switching Circuits for Engineers.* Prentice-Hall, Inc., Englewood Cliffs, N. J., 1962.

8. *Switching Transistor Handbook.* Motorola, Inc., Phoenix, Ariz., 1963.

9. Todd, C. R., "An Annotated Bibliography on NOR and NAND Logic," *IEEE Trans. Electron. Computers,* EC-12, p. 462 (1963).

10. Khambata, A. J., *Integrated Semiconductor Circuits.* John Wiley & Sons, Inc., New York, 1963.

11. Maley, G. A., and J. Earle, *The Logic Design of Transistor Digital Computers.* Prentice-Hall, Inc., Englewood Cliffs, N. J., 1963.

12. Stanton, W. A., *Pulse Technology.* John Wiley & Sons, Inc., New York, 1964.

13. Millman, J., and H. Taub, *Pulse, Digital, and Switching Waveforms.* McGraw-Hill Book Company, New York, 1965.

14. Harris, J. N., P. E. Gray, and C. L. Searle, *Digital Transistor Circuits* (SEEC, Vol. 6). John Wiley & Sons, Inc., New York, 1966.

20

PULSE COMMUNICATION SYSTEMS

Pulse communication systems are extensively employed in telecommunication, radar, and satellite communications, where weak signals must be received in the presence of noise, and in which bandwidth is available to be traded for noise reduction. The study of such systems at this point provides an opportunity to integrate some of the techniques and circuit usages of electron devices previously discussed.

Modulation of a radio-frequency carrier with signal pulses conveys information if some parameter of the pulses is varied, as amplitude, duration, or position in time. An agreed-upon code may also be used as in pulse-code modulation, the earliest form being conventional telegraph on-off modulation; recent usage employs a code related to the amplitude of analog signals. To the present, such pulse-code modulation provides the greatest noise immunity and accuracy of transmission.

20-1 MULTIPLEX USE OF A WIDE FREQUENCY BAND

A given frequency band may be employed to transmit many signals, each of narrow frequency band coverage, or one or a few signals, each having need for wide frequency bands to convey their information.

In *frequency multiplex*, a number of voice signals may be modulated on separate carriers and spaced across the transmission band, as in Fig. 20-1(a). The transmission medium may be a telephone line, or space, as in radio propagation. For instance, a given wire line may have satisfactory transmission characteristics and attenuation over the band from 12 kHz to 60 kHz. Using 3 kHz SSB modulation for voice, it would appear that 16 voice channels would be available. However, the system would require ideally sharp filters to separate the various voice channels at the receiving end of the system.

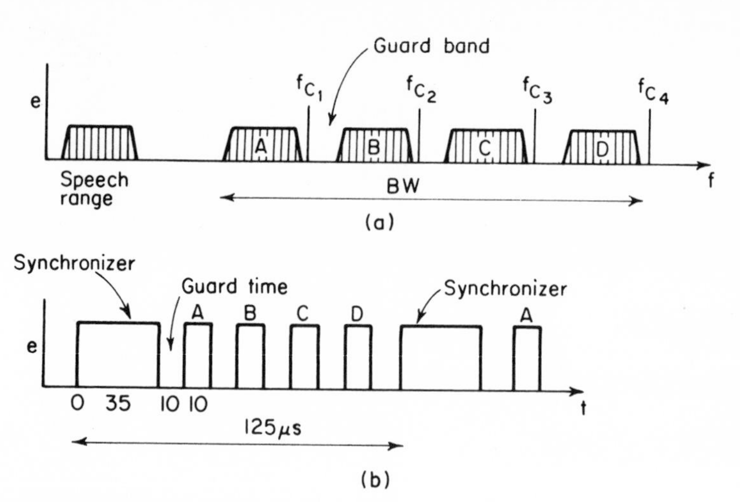

Figure 20-1. (a) Frequency multiplex signal—4 channel; (b) time multiplex—4 channel.

Practical and economic filters might require 1000 hertz as a guard band between signals, and so each channel would need 4 kHz, and 12 signals could be simultaneously transmitted. In microwave radio relay as many as 1200 voice channels are used in one transmission band, requiring at least 4.8 MHz in bandwidth. This is equivalent to one television signal. The basic principles so employed have already been studied, and will not be further discussed here.

In *time multiplex* use of a wide frequency band the information signal is sampled at periodic intervals, to produce sample pulses whose duration is short with respect to the time between pulses. A wide frequency band is then required to accurately transmit these pulses, but it would not be used continuously in time. It is possible to utilize the idle transmission time by insertion of pulses bearing information from other input signals, producing a time-multiplex signal as in Fig. 20-1(b).

Thus frequency multiplex transmits many signals at the same time in different portions of a frequency band, while time multiplex utilizes the entire frequency band to transmit many signals at different times.

Time multiplex systems are based on the transmission of a short pulse of carrier for each signal sample, with some pulse parameter varied in proportion to the signal amplitude at the sampling instant. The rate of sampling, known as the *pulse repetition rate*, is related to $f_{m\,\max}$ of the information signal, and usually must exceed $2f_{m\,\max}$. For voice signals filtered to a maximum frequency of 3.4 kHz, the sampling rate is often made 8000 per second.

Such a repetition rate provides a sample of the analog input signal

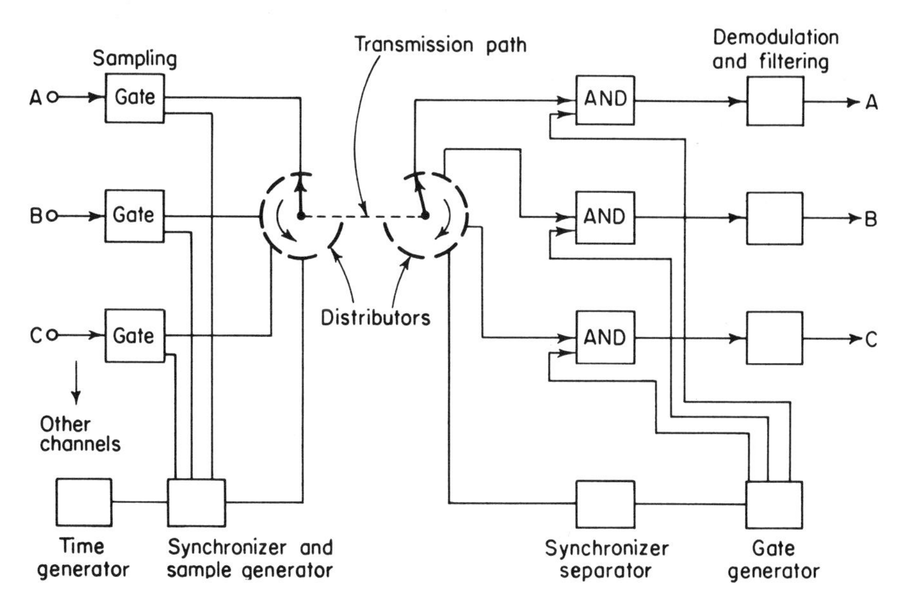

Figure 20-2. Time multiplex system diagram.

every 125 μs. If, for example, the pulse generated by the sample is 10 μs long, then there exists an idle period of 115 μs between pulses. By use of a distributor, schematically indicated in the complete system in Fig. 20-2, it is possible to insert and transmit 10 μs pulses representative of other messages, without interference between signals. To keep the sending and receiving distributors in step, a synchronizing pulse with some distinguishing characteristic such as length or amplitude is added to start a given cycle. *Guard times* are placed between pulses, because sufficient time must be allowed for the transient of a pulse wave to decay and for the necessary gates to open and close. The result is a composite wave form as in Fig. 20-1(b).

An integrating circuit may be used to separate a long synchronizing pulse, or a Schmitt trigger is suitable if amplitude distinguishes the synchronizing pulse. This marker pulse is used to start the switching cycle in a closed set of trigger circuits known as a *ring counter*, in which the on state advances around the ring, making one step and opening the connected signal gate, for each 20 μs.

20-2 SAMPLING

Figure 20-3 illustrates a form of transistor AND gate which will develop amplitude samples of an applied analog signal, v_i. The sampling pulse must be very short compared to the time between samples. When applied to the

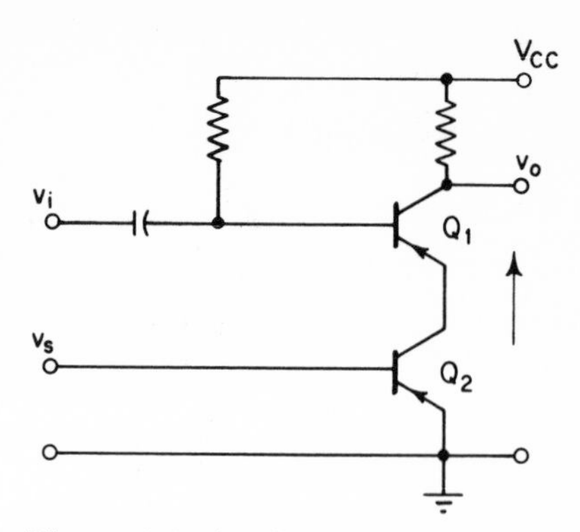

Figure 20-3. Amplitude sampling gate.

gate transistor Q_2, the gate is opened and the output wave form will then be a pulse with amplitude equal to the amplitude of the analog signal at the pulse instant. The rectangular pulse sampling train v_s has a Fourier series expansion

$$v_s = a_0E_s + a_1E_s \cos \omega_s t + a_2E_s \cos 2\omega_s t + \dots$$

The output of the sampling gate is

$$v_o = v_s(1 + kV_i \cos \omega_m t) \quad (20\text{-}1)$$

where ω_m is the information signal frequency. Then

$$v_o = a_0E_s + a_0E_skV_i \cos \omega_m t + a_1E_s \cos \omega_s t + a_1E_skV_i \cos \omega_m t \cos \omega_s t + a_2E_s \cos 2\omega_s t + a_2E_skV_i \cos 2\omega_m t \cos 2\omega_s t + \dots \quad (20\text{-}2)$$

This result indicates that product modulation has taken place, and the output contains the original low-frequency signal plus carrier and sideband terms at the sampling frequency and its harmonics.

The sampling frequency must be at least twice the highest frequency present in the input signal. This statement implies that two samples of each cycle of highest frequency are needed for definition—thus the 8000 samples per second used for a 3.4 kilohertz maximum speech frequency.

The highest frequency present can be estimated from the apparent shortest rise time in the signal by matching to a half-cosine wave, as in Fig. 20-4. If the sampling rate is chosen lower, then the signal must be filtered to remove all components higher than one-half the sampling rate; higher frequency components will alter the wave form of lower frequencies and give inaccurate samples. Because practical filters always have a transition region between pass and stop bands, the actual sampling frequency should be so chosen that

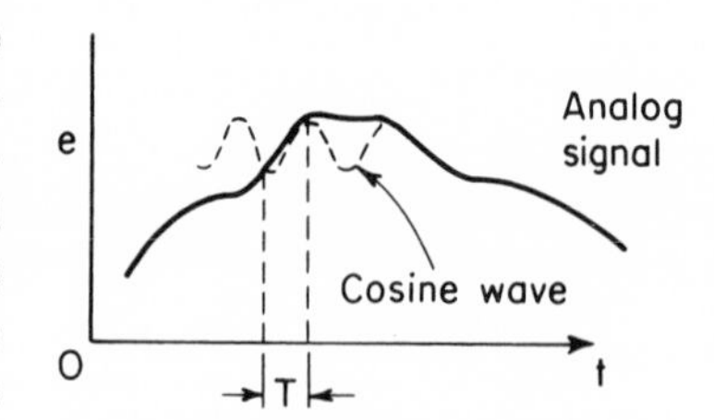

Figure 20-4. Half-cosine fit of fastest rise time: $f_{\max} = 1/2T$.

$$f_s = W_t + 2f_{m\,\max} \quad (20\text{-}3)$$

where W_t is the width of the filter transition band in hertz.

The sample pulses then control some parameter of the pulse train to be transmitted.

20-3 REQUIRED BANDWIDTH FOR TRANSMISSION OF THE PULSES

The bandwidth of a multiplexed pulse signal containing n signal channels must be at least

$$\mathrm{BW} = \frac{1}{(n+1)(\delta + g)} \quad \mathrm{Hz} \tag{20-4}$$

where δ and g are the pulse and guard times, and the synchronizing pulse is assumed to occupy the time of one channel. This bandwidth may be large, but an infinite bandwidth is required for transmission of a perfect pulse. However, the results of Section 10-15 indicated that an infinite channel would be infinitely noisy. To insure minimum noise, signal channels must be frequency limited, and passage of a pulse will then result in pulse degradation, in accordance with the discussion of Section 11-3. The problem then is how much limitation of the bandwidth can be balanced against loss of information content in the signal, as evidenced by deterioration of the pulse shape.

In many systems the time of occurrence of the pulse and its decay time are critical; these parameters are related to the rise time of the pulse. The rise time has been shown as increased by reduction of the effective bandwidth of a channel.

Consider the path of the pulse as including the filtering effect of the usual limiting high-frequency RC integrating time constants, where C represents the circuit and device shunt capacities. We have previously shown that the rise time T_r, defined between the 10 and 90 per cent levels of the pulse, is

$$T_r = 2.2\ RC$$

However, the upper cutoff frequency is $f_2 = 1/(2\pi RC)$, and so

$$T_r = \frac{0.35}{f_2} \tag{20-5}$$

Thus for a pulse to have a 1 μs rise time, the system bandwidth must be at least 350 kilohertz.

A more stringent requirement results if we assume that the cutoff of the filter is abrupt, rather than with the gradual slope of the RC filter circuit. This is equivalent to the assumption of an ideal channel filter, which is closely approximated in many cases. The output voltage then has the form

$$v_o = \frac{1}{\pi} \int_0^x \frac{\sin x}{x}\, dx \tag{20-6}$$

where $x = \omega_c(t - \theta)$, ω_c is the angular cutoff frequency and θ is the filter delay. The integral is known as the *sine-integral* of x abbreviated Si (x) and evaluated in tables. The equation has been normalized on its value at $x = \infty$,

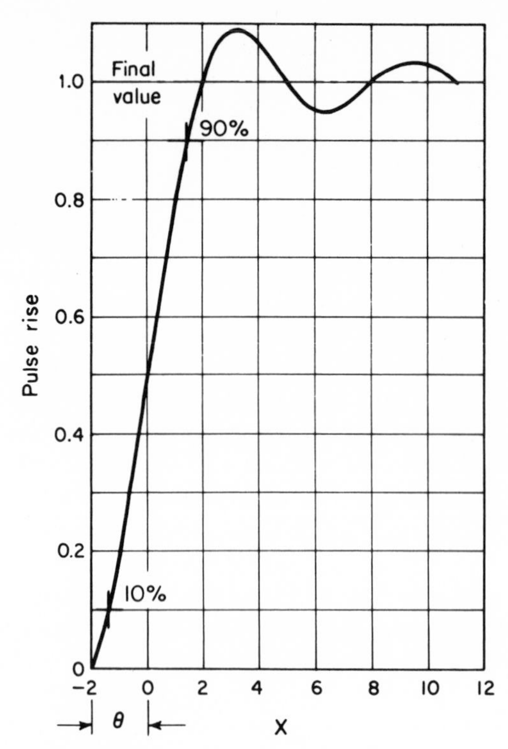

Figure 20-5. Response of an ideal filter to a step input (neglecting response for $t < -\theta$).

and used to plot the response of an ideal filter to a step input in Fig. 20-5.

From the figure it can be seen that the interval on x between the 10 and 90 per cent points is 2.8, so that the defined time of rise is

$$T_r = \frac{2.8}{2\pi f_c} = \frac{0.446}{f_c} \tag{20-7}$$

For a 1 μs rise time, such an ideally filtered system would require a bandwidth of 446 kilohertz.

Thus the pulse wave form will be affected by the bandwidth of the system as well as by the nature of the filter attenuation curve beyond frequency cutoff. Such factors must be considered in choosing among the several methods by which the sampled information may be utilized.

20-4 PULSE-DURATION MODULATION

In *pulse-duration modulation* (PDM), with one-channel wave form shown in Fig. 20-6, the duration of the pulse is varied. It would be desirable to shift both leading and trailing edges as a function of sample amplitude, but in

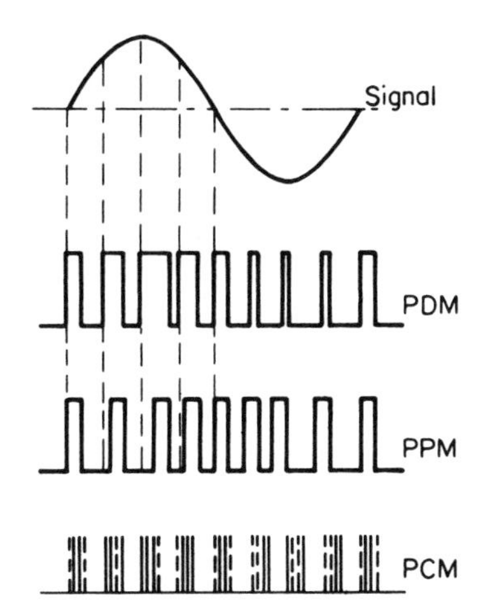

Figure 20-6. Signals in various pulse-modulation systems.

practice it is usually easier to leave the leading edge fixed and vary the time of the trailing edge, as shown in the figure. Since the time of occurrence of both leading and trailing edges must be accurately determined, the rise time must be short.

One method of generating PDM pulses from the PAM samples is that provided by the voltage-variable bistable multivibrator, in which the return voltage for the transistor base or tube grid is varied by the sample voltage. It was previously indicated that the length of pulses so generated was almost linear with the return voltage.

In another system the PAM signal from the sampler is applied to the voltage comparator of Fig. 20-7, where the PAM sample is compared with a sawtooth wave which starts to rise from zero at the beginning of the sample period. At the time the sawtooth wave reaches the amplitude of the sample voltage, transistor Q_2 goes to cutoff, making an end to the output pulse and causing the pulse duration to be proportional to sample amplitude.

Noise effects may be reduced by clipping the received pulses to small constant amplitude. However, some noise effects will remain and appear as "jitter" of the leading or trailing edges. Recovery of the modulation signal can be achieved by passing the detected radio-frequency pulses through a low-pass filter having a cutoff frequency between $f_{m\,\max}$ and the sampling frequency.

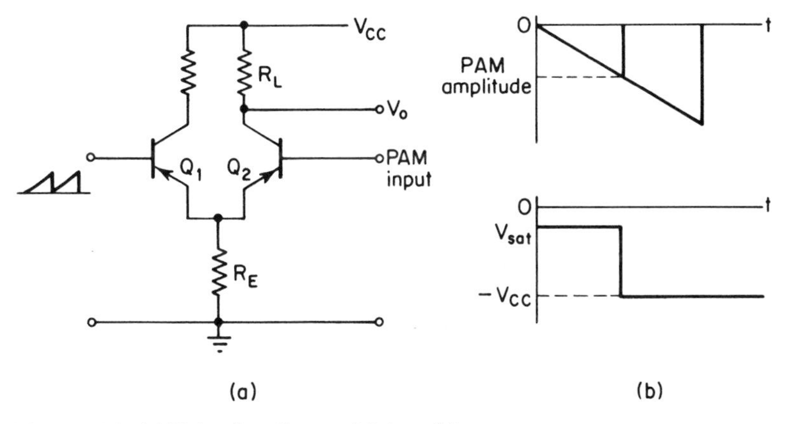

Figure 20-7. (a) Pulse-duration modulator; (b) wave forms.

20-5 PULSE-POSITION MODULATION

Pulse-position modulation (PPM) is a form in which the time position of the pulse is varied around a fixed value and conforms to the signal amplitude at the sampling instant. In effect, the PPM signal can be derived from a PDM signal by differentiating the variable pulse edge to obtain a single pulse of variable timing. PPM is more efficient in the use of power than is PDM, because of the shorter pulse used to convey the same information. However, it requires an accurate indication of pulse occurrence time, and the bandwidth must be sufficient to provide short rise times for the pulses.

Demodulation may be done by a bistable multivibrator, with side A switched on by a pulse from the synchronizer, and side B switched on and side A off by the received pulse. The resultant duration-modulated pulses may be passed through a low-pass filter to separate the signal from the sampling frequency.

20-6 PULSE-CODE MODULATION

Pulse-code modulation (PCM) translates the sampled amplitude data into a code, the code being transmitted as a succession of pulses. The system enjoys great freedom from noise and interfering signals since it is merely necessary to determine that a pulse was or was not transmitted, and it is not necessary to measure any characteristic of the received wave form. If the received signal has sufficient amplitude, it is possible to operate circuits at the receiver which will regenerate the pulse and free it of noise. This permits free use of repeater stations, without the necessity of equalizing for the gain of the transmission medium as a frequency function, as is done for analog signals.

In PCM systems the amplitude of a sample of the signal is measured and rounded off into one of a set of *quantized levels*, which are the only values permitted in the system. Quantization of the sample may be accomplished in a form of cathode-ray tube, as in Fig. 20-8. A horizontal sheet beam of electrons is deflected vertically by the applied signal. Portions of the sheet of electrons pass through the holes in the mask and reach the vertical collecting electrodes behind each column of holes. The signal amplitude is quantized by the position of the holes in the mask. Binary weighting of the sample is used; thus the code for the eight-level mask shown is:

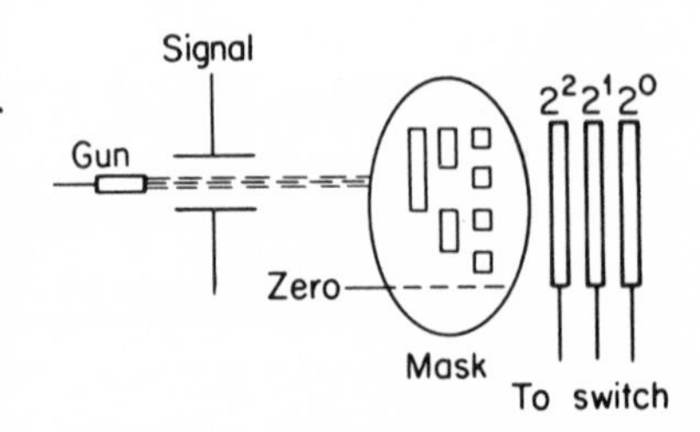

Figure 20-8. Three-pulse, eight-level quantizing tube for PCM.

	Weight 4	2	1		Weight 4	2	1
0	0	0	0	4	1	0	0
1	0	0	1	5	1	0	1
2	0	1	0	6	1	1	0
3	0	1	1	7	1	1	1

By sequential switching of the collecting electrodes, the beam or signal is sampled at the desired frequency and the code pulses assembled into the required serial pulse groups.

An n bit code would provide 2^n levels of sampling, with an increase of n times in the bandwidth required for transmission. The number of levels is determined by the permissible amount of quantizing noise generated by the differences between the stepped output wave and the actual analog signal, as shown in Fig. 20-9. For voice signals it is found that 64 to 128 levels will reduce the noise or erratic output to a satisfactory value. The instantaneous level of the noise has a maximum value of one-half a quantizing level.

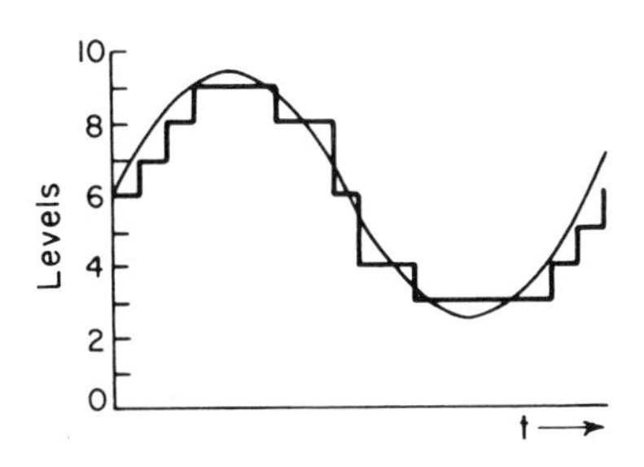

Figure 20-9. Quantizing error.

Figure 20-10 illustrates a wave sampled and quantized into 16 levels, with the resultant code groups at (b). The time at which each code group is transmitted is established by synchronized timers at transmitter and receiver. The required bandwidth will be proportional to the number of quantizing levels per code group, and the number of messages multiplexed on one channel. As previously stated, pulse wave form requirements are not stringent.

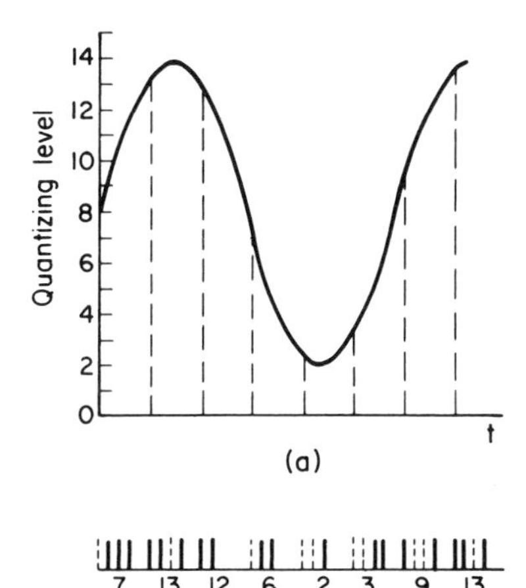

Figure 20-10. Pulse quantizing and coding in PCM.

The PCM system is closest to the ideal in noise tolerance abilities, since it is independent of all pulse characteristics except the presence of the pulse. In PDM the measurement is between the two edges of the pulse, and independent of any marker pulse, but noise can appear to vary the pulse duration. The PPM system is more susceptible to noise than the

PDM system. In PPM it is the time between a marker pulse and the received pulse which is measured, and noise can appear to alter the time of arrival of a pulse. Any jitter in the received marker pulse enters as an added source of noise.

The PCM system seems most capable of meeting the requirements for accurate voice and data communication under conditions encountered in long distance telephony or in space communication.

20-7 *PCM DECODING*

The *Shannon decoder* for PCM is simple in theory, being illustrated in Fig. 20-11. The RC time constant is so chosen that the capacitor charge or voltage decays to one-half its original value in the time interval allowed between successive code pulses. Each 1 in the received code closes the constant-current gate to the RC circuit, and maintains it closed for a short, precise interval. A definite change Q is transferred to C. As the charge decays to $Q/2$ in each pulse period, its value after n pulse periods will be $Q/2^n$.

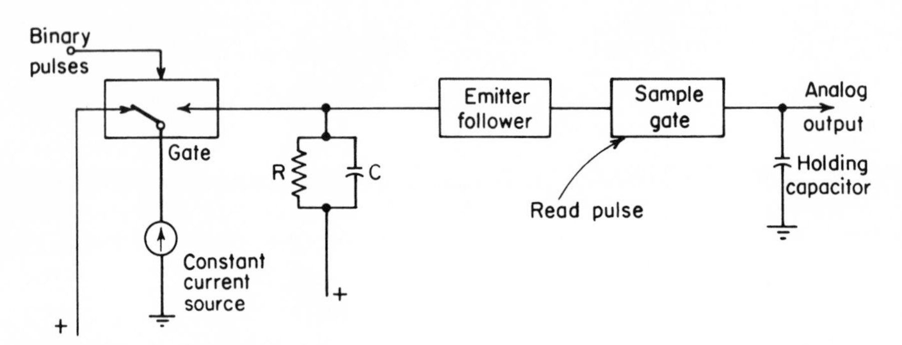

Figure 20-11. Decoder for PCM.

A seven-pulse group might be 1000010, the least significant bit being transmitted first. This first pulse places a charge Q on C, which after seven pulse periods will have decayed to $Q/2^7$. The last 1 will place a charge of value Q on the capacitor which will decay to $Q/2^1$ at the end of the last pulse time. Therefore at the precise time T, at the end of the final pulse time, the output voltage is gated and read as a voltage

$$e_C = \frac{Q_T}{C} = \frac{Q}{2^7C} + \frac{Q}{2C} \tag{20-8}$$

This is equivalent to $(\frac{1}{128} + \frac{1}{2})$, showing that by introducing the proper code symbol, any value from 0 to $\frac{127}{128}$ may be read at the output in terms of e_C.

The holding circuit retains this voltage until the arrival of the weighted sum from the next code group, giving a stepped output voltage having the form of Fig. 20-9. Appropriate filtering will then remove the sampling frequency and quantizing noise.

PROBLEMS

20-1. A wave of $10 + 10 \sin \omega t$ is sampled to four-volt intervals at a rate of $12f$, with a five-level sampling circuit. Write the binary code for the PCM code groups obtained.

20-2. A given amplifier has three stages with f_2 frequencies of 600,000, 450,000, and 325,000 hertz. What is the rise time of a pulse if no overshoot is expected?

20-3. The charge Q in a Shannon PCM decoder raises the voltage of C by 10 V. A binary input of 110110, least significant bit on the right, is applied to the gate. What is the capacitor voltage at the end of the last pulse interval?

20-4. A PDM pulse has a rise time of 1.5 μs. With a duration range of 4 μs from zero to full modulation, determine the per cent error if the pulse duration is measured at 60 per cent amplitude instead of 50 per cent, due to noise jitter.

REFERENCES

1. Sears, R. W., "Electron-Beam Deflection Tube for Pulse-Code Modulation," *Bell System Tech. Jour.*, Vol. 27, p. 44 (1948).
2. Shannon, C. E., and W. Weaver, *The Mathematical Theory of Communication.* University of Illinois Press, Urbana, Ill., 1949.
3. Goldman, S., *Information Theory.* Prentice-Hall, Inc., Englewood Cliffs, N. J., 1953.
4. Black, H. S., *Modulation Theory.* D. Van Nostrand Co., Inc., Princeton, N. J., 1953.
5. Hancock, J. C., *The Principles of Communication Theory.* McGraw-Hill Book Company, New York, 1961.
6. Russell, G. M., *Modulation and Coding in Information Systems.* Prentice-Hall, Inc., Englewood Cliffs, N. J., 1962.

21

POLYPHASE RECTIFICATION AND POWER CONTROL

The first electronic rectifiers capable of handling large currents in industrial applications were gas diodes, employing mercury vapor as the working gas. Heavy-current silicon diodes and four-layer devices have now replaced them because the efficiency of the semiconductor devices is higher and they are more rugged and long-lived.

Single-phase half- and full-wave rectifier circuits were studied in Chapter 4. This chapter will extend that work to polyphase and control rectification, since for large electrical power outputs, the supply is usually polyphase in form.

21-1 SEMICONDUCTOR HIGH-CURRENT DIODES

By increasing the junction area of the alloy type of diode, it is possible to manufacture semiconductor diodes able to handle kilowatts of power. Because of the high industrial ambient temperatures, silicon is almost universally used as the material. Current densities as high as 100 A per sq cm are used, with forward drops of about one volt, as indicated in Fig. 21-1.

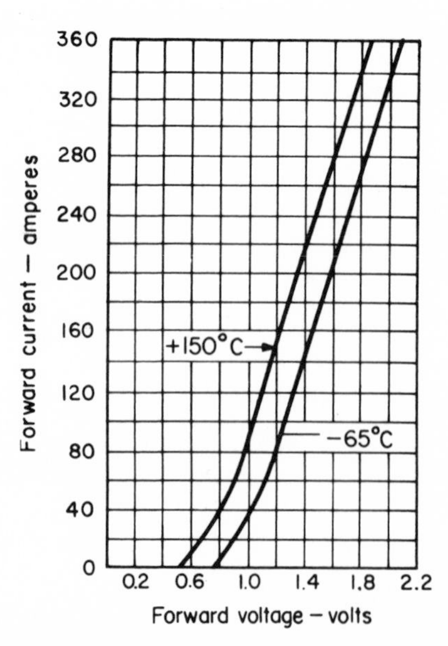

Figure 21-1. Forward volt-ampere curve for a silicon diode.

Internal losses are the sum of forward power dissipation and reverse losses. The forward dissipation is a result of forward current and voltage drop of about one volt, whereas reverse currents are dependent on operating temperatures. For silicon units with forward ratings of several hundred amperes, reverse currents of 50 to 100 mA may be expected. Reverse voltages as high as 600 for germanium and 1000 V for silicon can be allowed. Current surges up to 500 per cent of forward rating may be carried for a second or two, assuming good heat transfer to the heat sink. Operating temperatures of 90°C for germanium and 200°C for silicon are allowable.

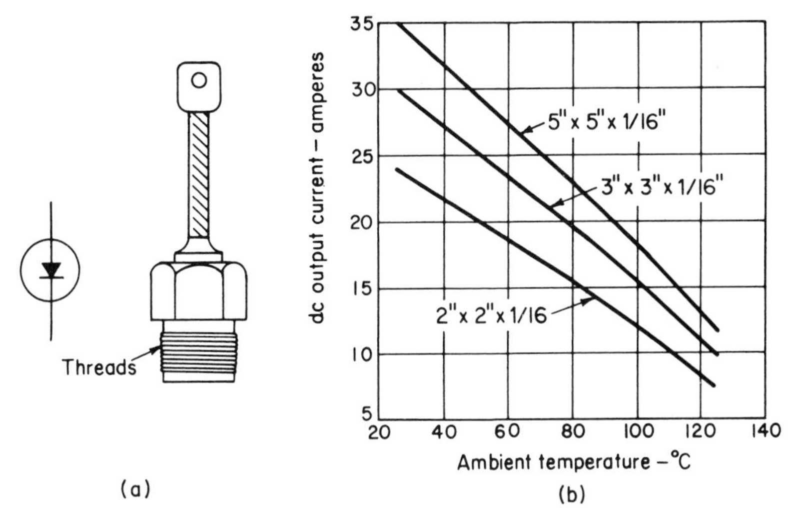

Figure 21-2. Silicon diode and effect of heat sinks in cooling.

The problems of mounting and heat dissipation are the same as those covered in Chapter 14 for power transistors; the effect of aluminum heat sinks is shown in Fig. 21–2. A typical screw-in mounting for a diode is also shown.

In addition to temperature, diodes are rated in terms of average or dc current, peak current, and peak reverse voltage.

21-2 THE THREE-PHASE HALF-WAVE RECTIFIER CIRCUIT

The simplest polyphase rectifier circuit is the three-phase half-wave form of Fig. 21-3, with star-connected transformer secondary. It will be considered in detail, with the many other polyphase circuits given a general treatment.

In the analysis that follows, the effects of transformer resistance and leakage, and internal diode drop are neglected. If D_1 is conducting, then its cathode and the positive end of the load are at essentially the voltage

e_{01}, or e_1 of Fig. 21-4(a). The cathodes of D_2 and D_3 will be positive with respect to their anodes, and cannot conduct. These potential relationships will change at $\pi/6$, $5\pi/6$, $9\pi/6$, . . .; these are points of commutation, as at (b). The current to D_1 is then

$$i_b = \frac{E_m \sin \omega t}{R}, \qquad \frac{\pi}{6} < \omega t < \frac{5\pi}{6} \tag{21-1}$$

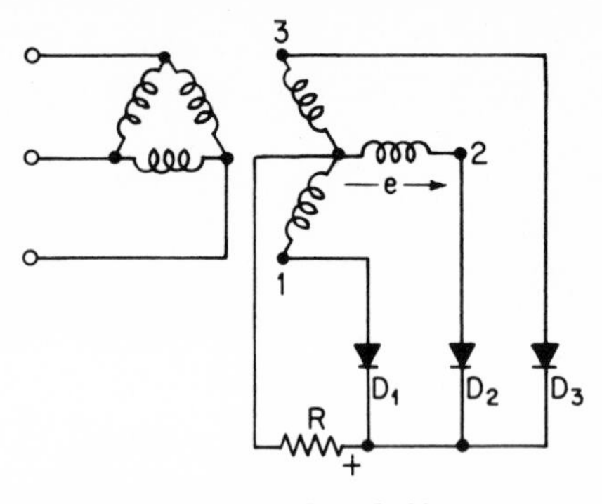

Figure 21-3. Three-phase half-wave rectifier circuit.

and the direct load current is three times the average current of one diode, or

$$\text{total } I_{\text{dc}} = \frac{3}{2\pi} \int_{\pi/6}^{5\pi/6} \frac{E_m \sin \omega t}{R} \, d\omega t = \frac{0.827E_m}{R} \tag{21-2}$$

and

$$E_{\text{dc}} = 0.827E_m$$

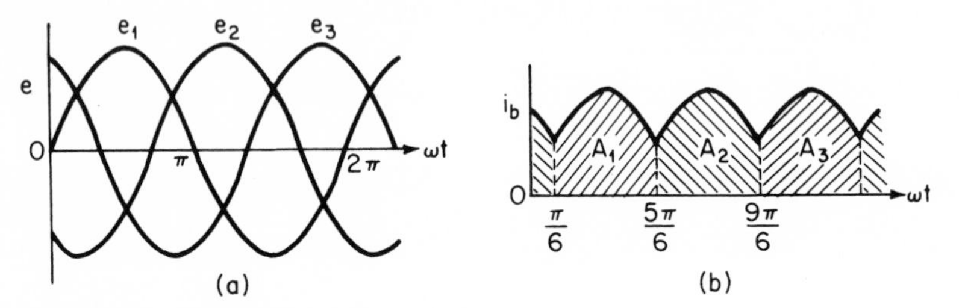

Figure 21-4. Voltages to neutral and anode current, three-phase half-wave circuit.

The power input from the secondary of the transformer is

$$P_{\text{ac}} = \frac{3}{2\pi} \int_{\pi/6}^{5\pi/6} \frac{E_m^2 \sin^2 \omega t}{R} \, d\omega t = \frac{0.706E_m}{R} \tag{21-3}$$

and the efficiency of rectification is readily obtained as

$$\max \eta_R = \frac{(0.827E_m)^2/R}{0.706E_m^2/R} \times 100\% = 96.5\% \tag{21-4}$$

The peak reverse voltage between anode A_1 and the cathode K occurs when $\omega t = 4\pi/3$, and is

$$\text{P.I.V.} = \sqrt{3}\, E_m = 2.09E_{\text{dc}} \tag{21-5}$$

The ripple will be due to three pulses per hertz, so that its lowest frequency is 180 hertz for 60-hertz input. We may find I_{rms} of the load current:

$$I_{\text{rms}} = \sqrt{\frac{3}{2\pi}\int_{\pi/6}^{5\pi/6}\frac{E_m^2 \sin^2 \omega t}{R^2}d\,\omega t} = \frac{0.838E_m}{R} \tag{21-6}$$

and the ripple is then 17 per cent, as computed by

$$r = \sqrt{\left(\frac{I_{\text{rms}}}{I_{\text{dc}}}\right)^2 - 1} = \sqrt{(1.014)^2 - 1} = 0.17 \tag{21-7}$$

The direct current of each anode appears in the secondary phase windings and may cause transformer saturation, with its attendant large primary current. The circuit may be modified to the *zigzag* form, of Fig. 21-5 to avoid this. On each core are two coils which carry dc components I_o in opposite directions, thereby neutralizing the dc magnetomotive force on the core. The anode voltage e is then the sum of two coil voltages at 60°, or each coil must supply $0.575e$.

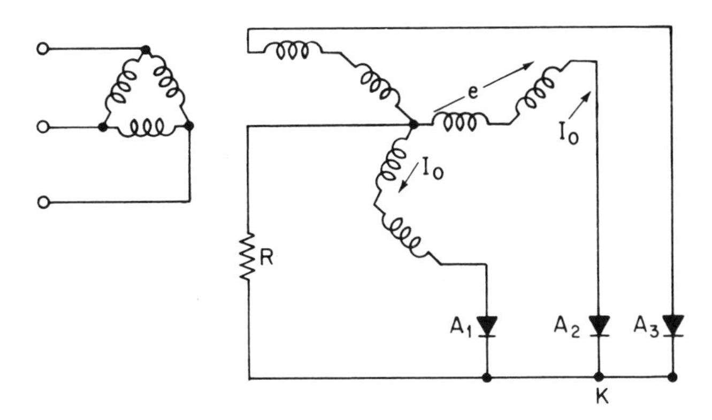

Figure 21-5. Three-phase zig-zag circuit.

21-3 m-PHASE RECTIFIER CIRCUITS

Since many rectifiers are built with 6, 12, 18, or more phases, using star-connected and branched or zigzag windings, it is appropriate to undertake a general study of rectifier circuits using m phases, with m diodes, each of which conducts for $2\pi/m$ radians per cycle, and with the load current of Fig. 21-6. This definition of m requires continuous load current, and applies also to the single-phase full-wave circuit but not to the single-phase half-wave application.

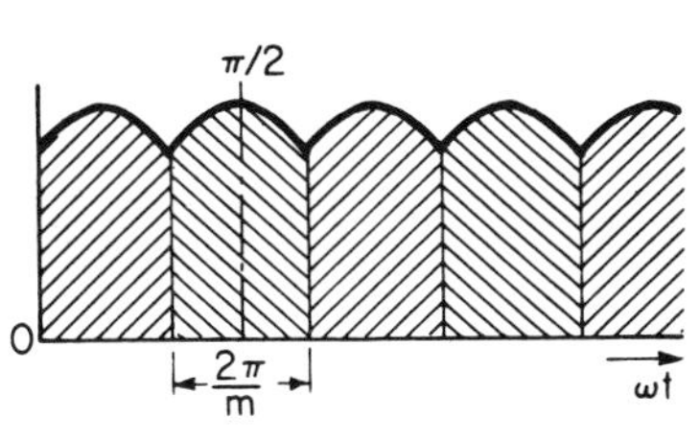

Figure 21-6. Current pulses in the m-phase rectifier circuit.

Neglecting losses, the dc load current of an m-phase rectifier is

$$I_{dc} = \frac{m}{2\pi}\int_{\pi/2-\pi/m}^{\pi/2+\pi/m} \frac{E_m \sin \omega t}{R}\, d\omega t = \frac{E_m}{R}\frac{m}{\pi}\sin\frac{\pi}{m} \tag{21-8}$$

Since $E_{dc} = I_{dc}R$, it is possible to write

$$\frac{E_{dc}}{E_m} = \frac{m}{\pi}\sin\frac{\pi}{m} \tag{21-9}$$

showing the effect of variation of m on the dc voltage. Figure 21-7 shows that there is little gain in dc voltage above six phases.

The dc power output is

$$P_{dc} = \frac{E_m^2}{R}\left(\frac{m}{\pi}\sin\frac{\pi}{m}\right)^2$$

The rms value of current through each diode can be found as

$$I_{rms} = \sqrt{\frac{1}{2\pi}\int_{\pi/2-\pi/m}^{\pi/2+\pi/m} \frac{E_m^2 \sin^2 \omega t}{R^2}\, d\omega t} = \frac{E_m}{R}\sqrt{\frac{1}{2\pi}\left(\frac{\pi}{m} + \sin\frac{\pi}{m}\cos\frac{\pi}{m}\right)}$$

and the ac power input from the transformer is obtainable as mI_{rms}^2R:

$$P_{in} = \frac{m}{2\pi}\frac{E_m^2}{R}\left(\frac{\pi}{m} + \sin\frac{\pi}{m}\cos\frac{\pi}{m}\right) \tag{21-10}$$

The maximum theoretical rectification efficiency becomes

$$\max \eta_R = \frac{(2m/\pi)\sin^2(\pi/m)}{\pi/m + \sin(\pi/m)\cos(\pi/m)} \times 100\% \tag{21-11}$$

The theoretical efficiency is 81.2 per cent for $m = 2$, and reaches 99.8 per cent for $m = 6$.

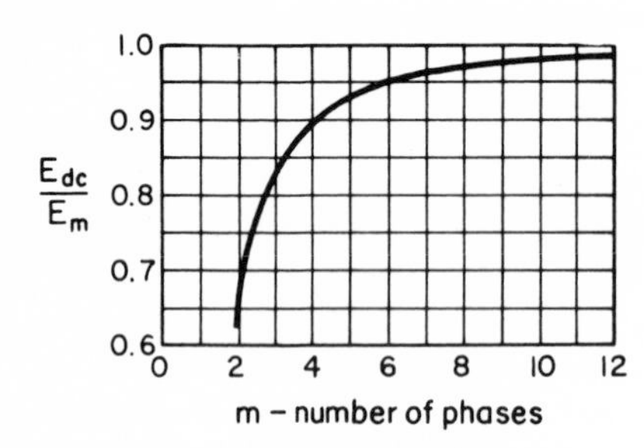

Figure 21-7. dc output voltage as a function of m.

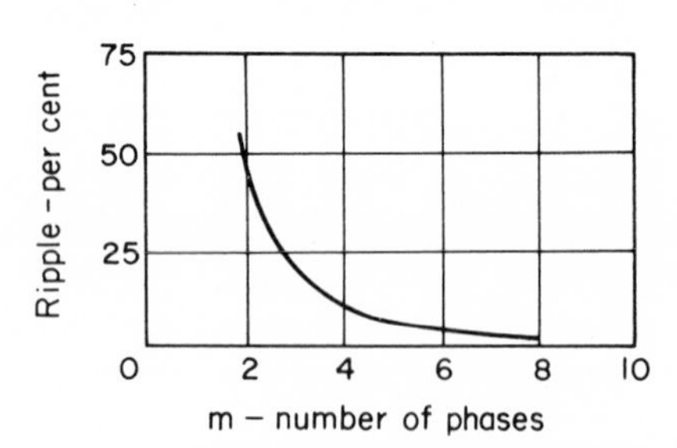

Figure 21-8. Ripple percentage vs. m.

The ripple percentage can be calculated by use of I_{dc} and I_{rms}. The curve of Fig. 21-8 shows that the ripple decreases rapidly with an increase in the number of phases. This fact makes the distortion of the primary-current wave form less in rectifier circuits having m large. This is important because harmonics introduced into the primary power circuit may cause telephone interference by induction.

21-4 EFFECT OF TRANSFORMER LEAKAGE REACTANCE

A rectifier circuit with leakage inductance L_s for each phase of the transformer shown in the anode leads, is drawn in Fig. 21-9. With leakage reactance neglected, it has been assumed that the current in one diode ceases before that in the next one begins. Actually, owing to the effect of the leakage L_s in maintaining the current, the second diode may conduct before the first ceases, with an overlap of the currents, as indicated by the cross-hatched areas of Fig. 21-9(b). The transfer of current from one diode to another is called *commutation.*

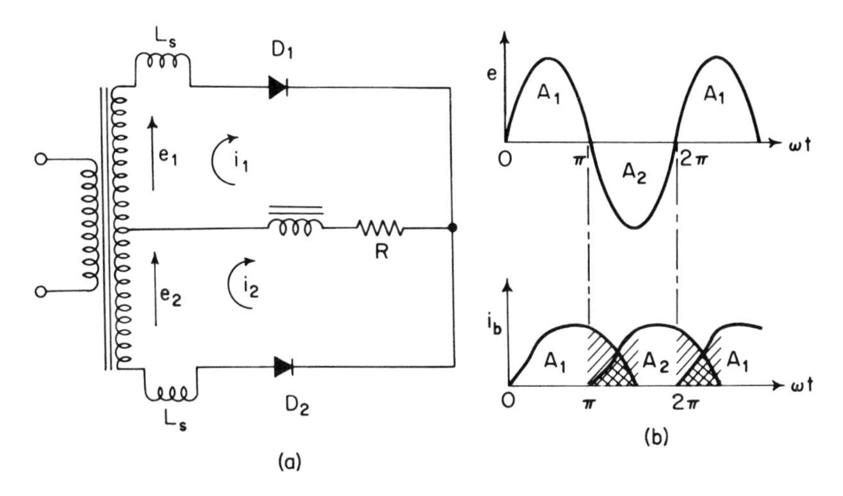

Figure 21-9. Full-wave circuit and current commutation.

Assuming that D_2 has started conducting before D_1 has ceased, an emf equation written through the transformer, the diodes, and the load gives

$$e_i - L_s\frac{di_1}{dt} = e_2 - L_s\frac{di_2}{dt} = e_L$$

Since L_s tends to maintain constant current during commutation, then $i_1 + i_2 = I_{dc}$, from which

$$\frac{di_1}{dt} + \frac{di_2}{dt} = 0$$

and

$$e_L = \frac{e_1 + e_2}{2}$$

so that the load voltage during commutation is the average of the phase voltages in that interval.

Using the general m-phase relations, it is possible to obtain

$$\frac{di_1}{dt} = -\frac{E_m}{L_s} \sin \frac{\pi}{m} \sin \omega t \tag{21-12}$$

as the commutation equation. Introduction of the m-phase dc voltage expression allows a solution to be written as

$$E_{dc} = E_m \frac{\pi}{m} \sin \frac{\pi}{m} - \frac{m}{\pi} \frac{\omega L_s I_{dc}}{2} \tag{21-13}$$

showing the effect of the leakage reactance in reducing the dc output voltage of a rectifier circuit.

The overlap of currents may be overlooked as far as dc current and ripple are concerned, and the more positive anode considered as conducting at any instant.

21-5 TRANSFORMER UTILIZATION FACTOR

Transformers for power use are designed for sinusoidal current, but transformers supplying rectifier circuits do not have sinusoidal currents. The larger the number of star-connected phases, the shorter is the time of current conduction per cycle in a given winding. The distorted current waves indicate the presence of harmonics, and the harmonic currents cause transformer heating, yet produce no useful dc output. A transformer rated at more than the dc power output is then required. The ability of a transformer to supply dc power in a rectifier circuit is called the *utilization factor* of the transformer, defined as

$$\text{U.F.} = \frac{P_{dc}}{\text{volt-amperes}} \tag{21-14}$$

The utilization factor is always less than unity.

The volt-ampere ratings of primary and secondary may differ, due to wave form variations. The total secondary volt-amperes are $mE_m/\sqrt{2}$ times the rms current, so that

$$\text{total secondary volt-amperes} = m\frac{E_m}{\sqrt{2}} \frac{E_m}{R} \sqrt{\frac{1}{2\pi}\left(\frac{\pi}{m} + \sin \frac{\pi}{m} \cos \frac{\pi}{m}\right)}$$

The dc power output is

$$P_{dc} = \frac{E_m^2}{R}\left(\frac{m}{\pi} \sin \frac{\pi}{m}\right)^2$$

and the secondary utilization factor for an m-phase rectifier transformer is

$$\text{secondary U.F.} = \frac{(2m/\pi) \sin^2 (\pi/m)}{\sqrt{\pi[(\pi/m) + \sin (\pi/m) \cos (\pi/m)]}} \tag{21-15}$$

Table 21-1

THEORETICAL† DESIGN DATA—RECTIFIERS WITH RESISTANCE LOAD

	Single-Phase		Three-Phase		Six-Phase Half-Wave
	Full-Wave	Bridge	Half-Wave	Full-Wave	
Value of m	2	2	3	6	6
Avg load current, I_{dc}	$0.636E_m$‡$/R$	$0.636E_m/R$	$0.827E_m/R$	$1.65E_m/R$	$0.955E_m/R$
Avg current/diode	$I_{dc}/2$§	$I_{dc}/2$	$I_{dc}/3$	$I_{dc}/3$	$I_{dc}/6$
Peak current/diode	$1.571I_{dc}$	$1.57I_{dc}$	$1.21I_{dc}$	$1.21I_{dc}$	$1.047I_{dc}$
E_{dc}	$0.636E_m$	$0.636E_m$	$0.827E_m$	$1.65E_m$	$0.955E_m$
Lowest ripple frequency	$2f_1$	$2f_1$	$3f_1$	$6f_1$	$6f_1$
Ripple	47%	47%	17%	4%	4%
Transformer V.A. rating:					
Secondary	$1.75P_{dc}$	$1.23P_{dc}$	$1.50P_{dc}$	$1.047P_{dc}$	$1.82P_{dc}$
Primary	$1.23P_{dc}$	$1.23P_{dc}$	$1.23P_{dc}$	$1.047P_{dc}$	$1.28P_{dc}$
Maximum theoretical efficiency	81.2%	81.2%	96.5%	99.8%	99.8%
P.I.V.	$3.14E_{dc}$	$1.57E_{dc}$	$2.09E_{dc}$	$1.045E_{dc}$	$2.09E_{dc}$

†Assuming diode drop and transformer leakage reactance both negligible.
‡E_m is the peak value of voltage to neutral of one phase of the transformer (see circuit diagrams).
§I_{dc} is defined as the average current in the load.

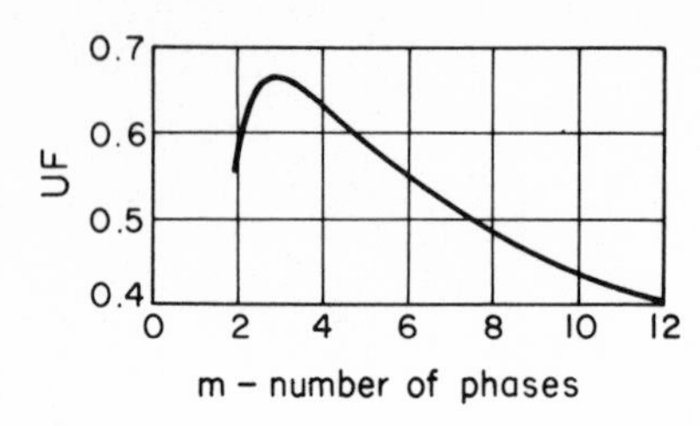

Figure 21-10. Secondary utilization factor of the m-phase circuit.

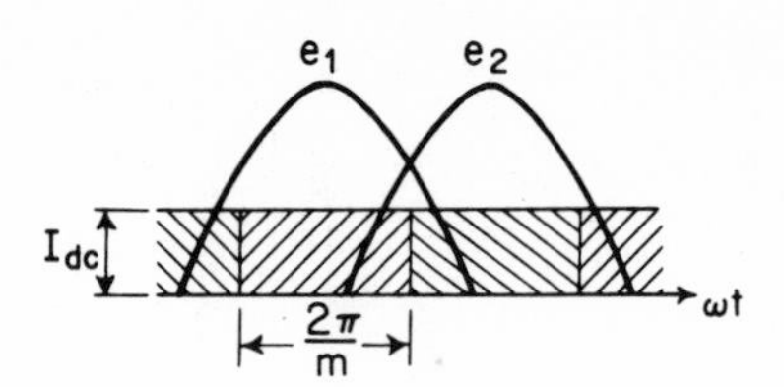

Figure 21-11. Current in m-phase circuit with series inductor.

A curve of this function appears in Fig. 21-10. A maximum occurs near $m = 3$, which implies that the most economic conduction angle is near 120°.

The opposite star-connected secondary windings are supplied by a single primary phase. The primary current is the effective value of two current pulses, so that

$$\text{primary } I_{\text{rms}} = \sqrt{2}\frac{E_m}{R}\sqrt{\frac{1}{2\pi}\left(\frac{\pi}{m} + \sin\frac{\pi}{m}\cos\frac{\pi}{m}\right)}$$

assuming 1:1 voltage transformation. If the circuit has p primary phases

$$\text{primary volt-amperes} = \frac{pE_m^2}{R}\sqrt{\frac{1}{2\pi}\left(\frac{\pi}{m} + \sin\frac{\pi}{m}\cos\frac{\pi}{m}\right)} \tag{21-16}$$

The primary utilization factor for the transformer is

$$\text{primary U.F.} = \frac{m}{p\sqrt{2}} \times \text{secondary U.F.} \tag{21-17}$$

Computations for circuits in which the dc components neutralize in a given transformer secondary follow the method outlined. Circuits such as the three-phase half-wave, which produce an unbalanced dc component in a transformer secondary, cannot be analyzed without further knowledge of primary-current wave forms. This phase of the subject is beyond the scope of the treatment given here.

A summary of rectifier circuit relations appears in Table 21-1.

21-6 *POLYPHASE RECTIFIER CIRCUITS WITH INDUCTIVE LOADS*

Because of the small ripple magnitudes and high ripple frequencies, a nominal inductance in the load of a polyphase rectifier circuit may give substantially constant load current, as illustrated in Fig. 21-11 for an m-phase circuit.

By the previous methods

$$E_{\text{dc}} = \frac{m}{2\pi}\int_{\pi/2-\pi/m}^{\pi/2+\pi/m} E_m \sin\omega t\, d\omega t = E_m\frac{m}{\pi}\sin\frac{\pi}{m} \tag{21-18}$$

Since the diode current is $I_{dc} = I_{rms}$, the power input from the transformer secondaries is

$$P_{ac} = \frac{m}{2\pi} \int_{\pi/2-\pi/m}^{\pi/2+\pi/m} I_{dc} E_m \sin \omega t \, d\omega t = I_{dc} E_m \frac{m}{\pi} \sin \frac{\pi}{m} \tag{21-19}$$

This is also the value of $P_{dc} = E_{dc} I_{dc}$, and thus the theoretical rectification efficiency has been raised to 100 per cent; no harmonic currents are present.

The effective current per diode is

$$I_{rms} = \sqrt{\frac{1}{2\pi} \int_{\pi/2-\pi/m}^{\pi/2+\pi/m} I_{dc}^2 \, d\omega t} = \frac{I_{dc}}{\sqrt{m}} \tag{21-20}$$

The ac volt-ampere input for all secondaries is

$$\text{secondary volt-amperes} = \frac{E_m}{\sqrt{2}} \frac{I_{dc}}{\sqrt{m}} m = \sqrt{\frac{m}{2}} E_m I_{dc}$$

and the utilization factors may be obtained as before. They are tabulated in Table 21-2. The secondary utilization factor has a maximum at $m = 2.7$, or between the two-phase and three-phase conditions.

Table 21-2 shows that for $m = 3$ or higher, there is negligible difference in performance with or without the filter inductance, since the wave forms are close approximations to rectangular form without inductance present.

Table 21-2

RECTIFIERS WITH INDUCTIVE LOAD†

	Single-Phase		Three-Phase		
	Full-Wave	Bridge	Half-Wave	Full-Wave	Six-Phase
m	2	2	3	6	6
Primary U.F.	0.90	0.90	0.827	0.955	0.780
Primary V.A.‡ rating	$1.11P_{dc}$	$1.11P_{dc}$	$1.21P_{dc}$	$1.047P_{dc}$	$1.28P_{dc}$
Secondary U.F.	0.636	0.90	0.675	0.955	0.552
Secondary V.A.‡ rating	$1.57P_{dc}$	$1.11P_{dc}$	$1.48P_{dc}$	$1.047P_{dc}$	$1.82P_{dc}$

†Based on theoretically infinite inductance. ‡V.A.=volt-ampere.

21-7 RECTIFIER CIRCUIT SELECTION

The major factors in determining the circuit to be used are usually ripple and transformer utilization. Transformers are large items in the cost of a rectifier installation and avoidance of dc saturation through some form of zigzag winding increases the transformer cost. Large numbers of phases increase the cost but decrease the ripple. A complete analysis of various circuits is needed to provide the economic factors, which may then be balanced against engineering performance.

In general, three diode ratings must be checked to make certain that a given type of service exceeds none of them. These ratings are the average current, the peak current, and the peak reverse-voltage rating of the diode. Any one of the three may provide a limit to the amount of power obtainable.

The single-phase circuits suffer from high ripple percentage, low ripple frequency, and low secondary utilization. The three-phase half-wave circuit has poor transformer utilization, and must make use of the zigzag winding, thereby increasing transformer cost.

The six-phase circuit is not often used because of low utilization, but modifications of the circuit are available to improve the transformer utilization through lengthening the conduction angle. A circuit demonstrating this principle appears in Fig. 21-12, and employs six secondary windings connected in two stars, the neutrals interconnected by an interphase reactor. At any instant there are two anodes carrying current, one in each group, and each anode conducts for 120°, or near the optimum for best utilization. Point-by-point addition of the two current waves shows that the load current will have a ripple frequency which is six times that of the supply. In effect, the two rectifiers are connected in parallel by the interphase reactor.

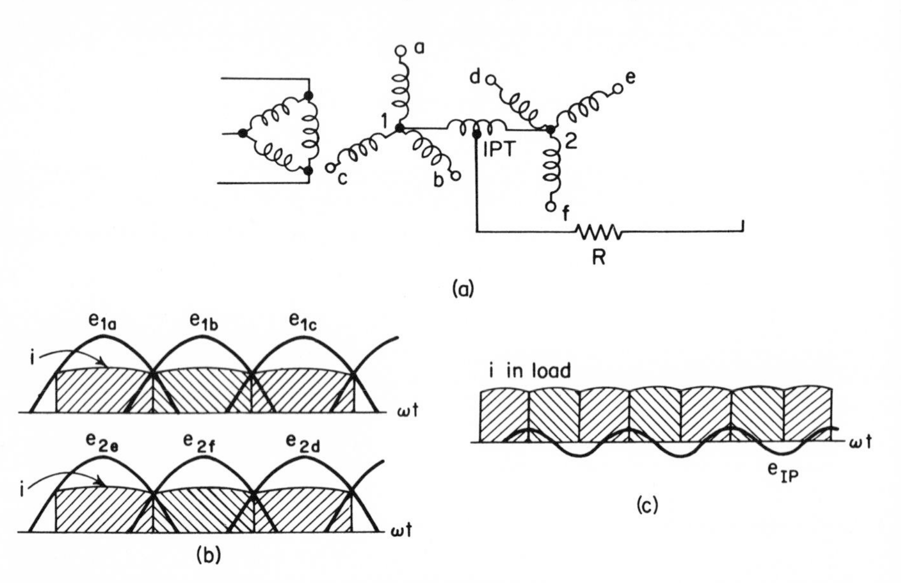

Figure 21-12. Three-phase double-Y circuit and wave forms.

21-8 CONTROL RECTIFIER DEVICES

Several high-current control devices are available, known variously as the *silicon-controlled rectifier* (SCR) or *thyristor*, the *triac*, and the *gate-controlled switch.* In general they show a trigger type of control, in contrast to the continuous control of current in the transistor, and are designed for power frequencies, currents, and voltages. The SCR can be triggered on by a current pulse, but can only be turned off by reducing the anode current below a minimum level. The triac is composed of two SCR units, in inverse-parallel, and can be triggered with positive or negative pulses. The gate-controlled switch may be triggered off as well as on.

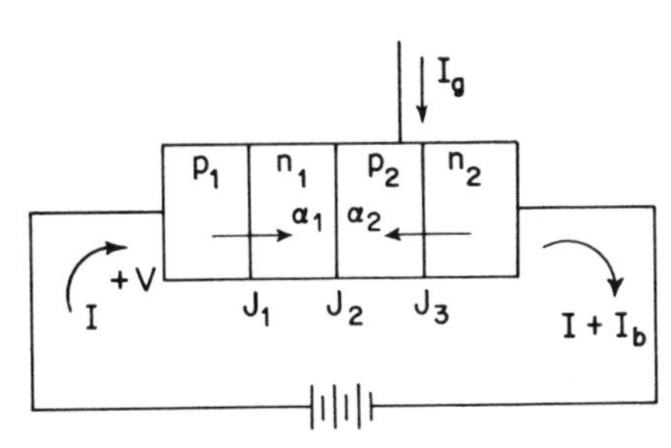

Figure 21-13. A silicon-controlled rectifier (SCR).

Built with four layer, pnpn or npnp configurations, as in Fig. 21-13, forward bias is placed on junctions J_1 and J_3 on the conducting half-cycle. Only the usual small forward voltages then appear across these junctions. This leaves junction J_2 with reverse bias, and the switch behavior is determined by the action of this junction.

Initially assume that I_g is zero. The total current across junction J_2 will be composed of three components, $\alpha_1 I$ of holes from the p_1 region, $\alpha_2 I$ of electrons from the n_2 region, and the leakage current, I_{CO}. Thus

$$I = \alpha_1 I + \alpha_2 I + I_{CO} = \frac{I_{CO}}{1 - (\alpha_1 + \alpha_2)} \qquad (21\text{-}21)$$

The leakage current in silicon is very small, and at low emitter currents the value of α in silicon is also small. Because of these conditions the current given by Eq. 21-21 is low. However, the value of α increases rapidly as the emitter current of a junction is increased, and as the sum of α_1 and α_2 approaches unity the current I becomes very large.

These factors may be correlated with the volt-ampere diagram of Fig. 21-14. At low currents in the region OA the value of $(\alpha_1 + \alpha_2) < 1$, and junction J_2 behaves as a pn junction with reverse bias. In the region AB the value of $(\alpha_1 + \alpha_2)$ increases and at B, $(\alpha_1 + \alpha_2) = 1$, with the so-called breakover or triggering value of voltage V_T, and current I_H. As the alphas further increase with current the junction J_2 is found to be forward-biased, and the whole device behaves as if it were a single pn junction biased in the forward direction. Operation is in the CD region of the volt-ampere curve, with high current and low voltage drop.

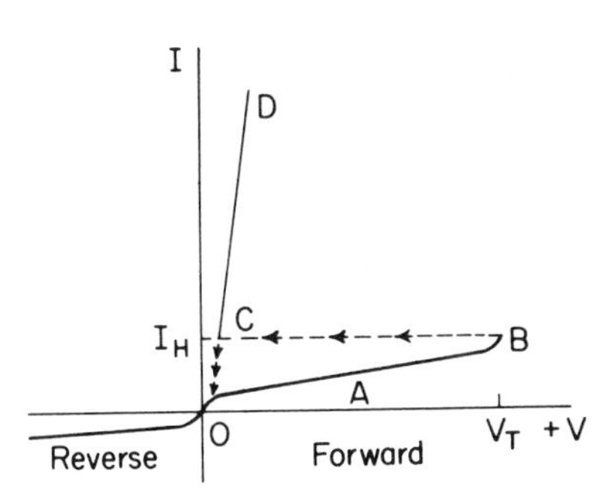

Figure 21-14. Volt-ampere curve of an SCR.

Because of the negative resistance properties of the BC region, conduction can be stopped only by reducing the current below the value I_H, the holding current; the operating point goes from C to O as indicated.

It is possible to increase α_2 independently by supplying a current I_g to p_2, and this is the principle of the silicon control rectifier (SCR). By use of an external circuit, the current I_g is driven in the same direction as I across junction J_3. This additional base current is effective in raising α_2 independently of the applied voltage V, and the condition of $(\alpha_1 + \alpha_2) = 1$ can be reached at lower values of breakover voltage, with small values of gate electrode current I_g.

In this mode of operation, using a triggering current to the gate electrode p_2, the volt-ampere curves become as in Fig. 21-15. The higher the triggering current, the lower is the value of applied voltage V at which breakover occurs. With sufficient values of I_g, as for 4, the device will turn on at small voltages and behave as a simple pn junction in the forward direction. When the current falls below the holding value I_H, conduction ceases and on the reverse half-cycle the device appears as two reverse-biased junctions in series.

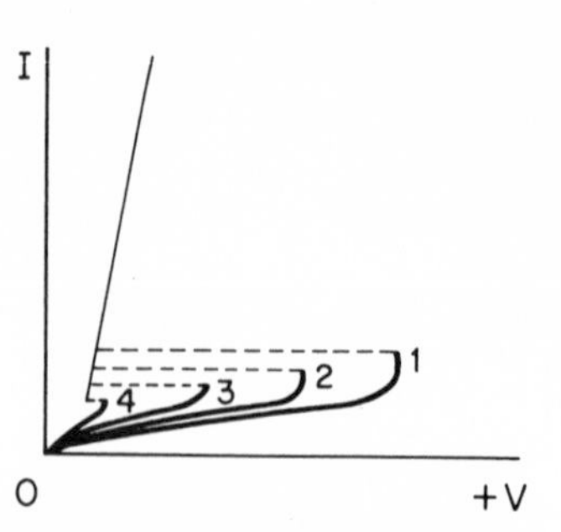

Figure 21-15. Volt-ampere curves for values of I_g; $I_{g4} > I_{g1}$.

In operation with ac, the control rectifier is operated with a peak voltage well below the breakover value, and triggering is accomplished by supplying a trigger current pulse to the gate electrode. Once the device has been triggered into conduction, the gate has no further control. Conduction can be started at any desired time in the positive half-cycle but will continue to the following zero crossing. The average value of the current can then be varied by variation of the triggering time. The only method of stopping conduction is to reduce the main current below the holding current I_H.

The difference between the SCR family and a power transistor might be considered. A power transistor achieves a high value of α or h_{FE} by using a very thin base layer, but a thin base between low resistivity emitter and collector regions is not compatible with the higher voltage desired for power application. By using a four-layer construction, the SCR family has been designed with a low α and wide base suited to high voltages, but also capable of developing a high effective α at high currents.

Mounting, cooling, and temperature considerations for the SCR will follow those discussed for the semiconductor diode. Ratings available have gate currents of 50 mA controlling operating currents to 50 A or more. Turn-on time approximates 1 μs and turn-off is accomplished in 5 to 10 μs.

21-9 CONTROLLED RECTIFICATION

With SCR or thyristor the dc output of a rectifier circuit can be varied through control of the triggering time in the ac cycle. If a pulse is applied to the gate at an angle θ_1, the current in the half-wave rectifier circuit of Fig. 21-16 is

$$i_b = \frac{E_m \sin \omega t}{R}, \qquad \theta_1 < \omega t < \pi \tag{21-22}$$

again neglecting diode drop. The direct current output will be the average of the wave form of Fig. 21-17, given by

$$I_{\text{dc}} = \frac{1}{2\pi}\int_0^{2\pi} i_b \, d\omega t = \frac{1}{2\pi}\int_{\theta_1}^{\pi} \frac{E_m \sin \omega t}{R} \, d\omega t = \frac{E_m}{2\pi R}(1 + \cos \theta_1) \tag{21-23}$$

Thus the direct current may be controlled by variation of the turn-on angle in each cycle. Figure 21-18(b) is a dimensionless plot of Eq. 21-23.

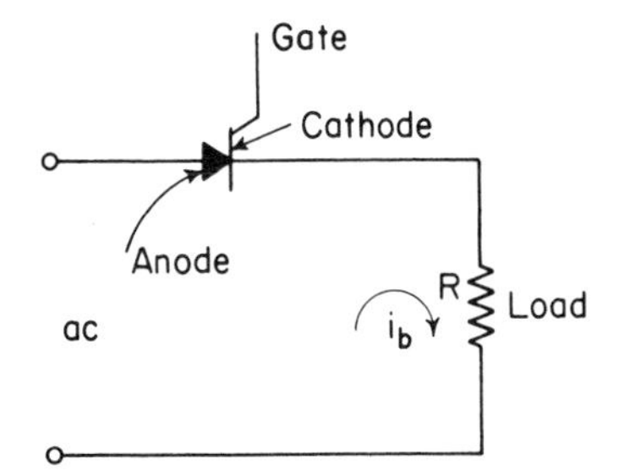

Figure 21-16. Silicon control rectifier circuit.

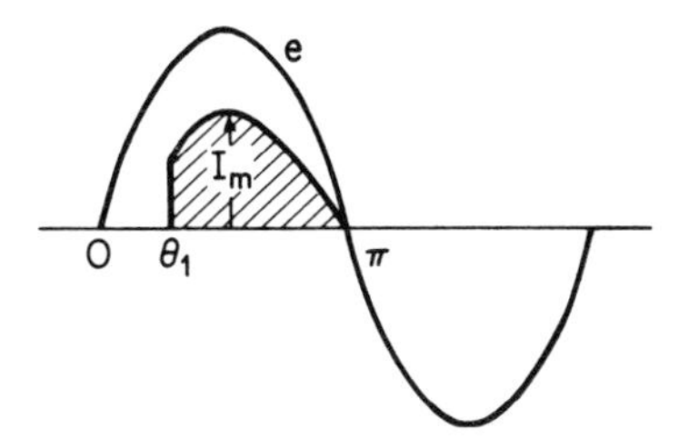

Figure 21-17. Control rectifier current wave form.

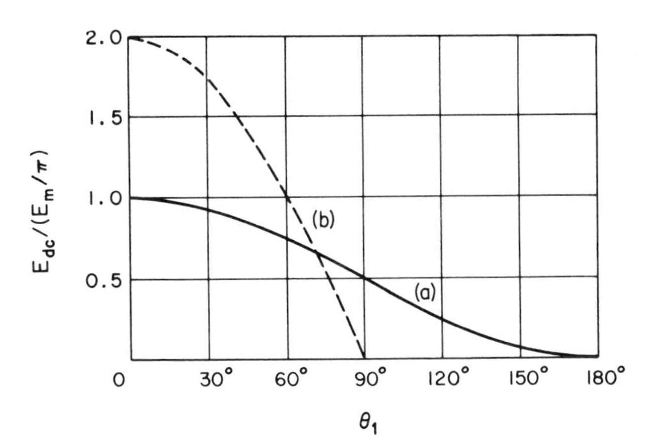

Figure 21-18. E_{dc} vs. triggering angle: (a) half-wave resistance load; (b) full-wave inductive load.

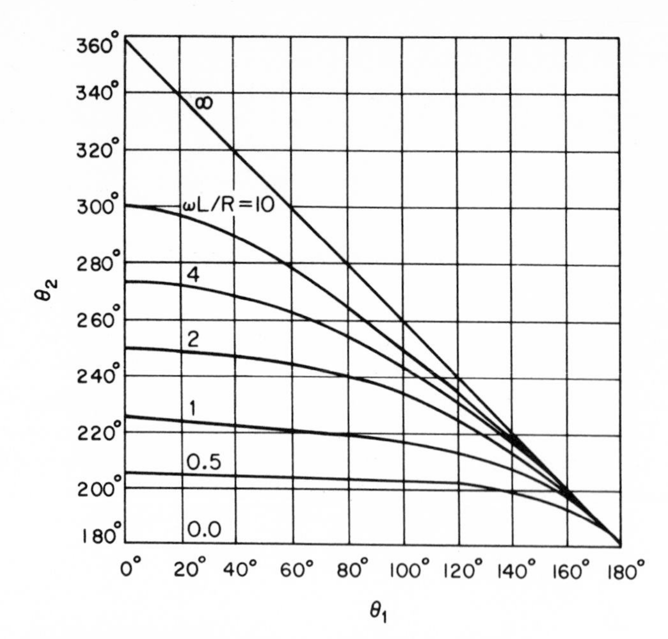

Figure 21-19. Cutoff angle vs. delay angle; $\omega L/R$ as a parameter.

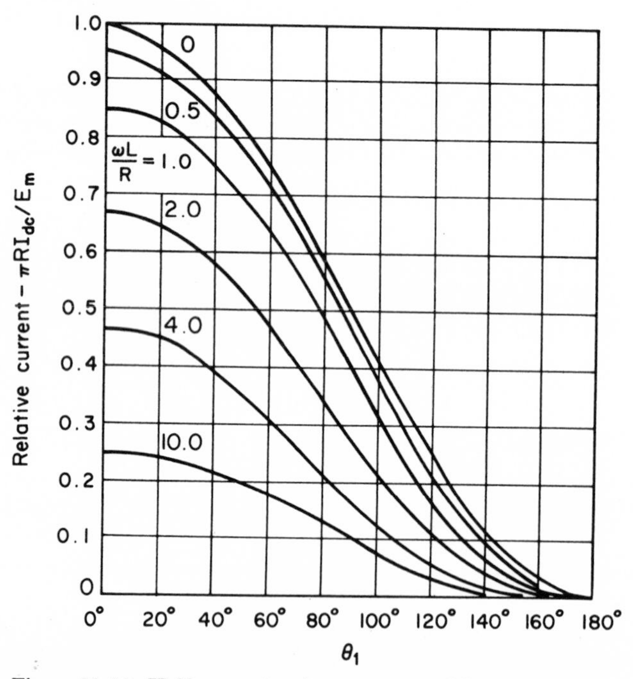

Figure 21-20. Half-wave circuit: current vs. delay angle.

When used with inductive loads an analysis similar to that for the R-L filter on the half-wave rectifier is required. It is possible to determine the cutoff angle, θ_2, for each value of θ_1 and ratio $\omega L/R$ of the load by graphical means, after deriving the usual transcendental equation. The values obtained for θ_2 are plotted in Fig. 21-19. It can be found that the average voltage will be

$$E_{\text{dc}} = I_{\text{dc}}R = \frac{E_m}{2\pi}(\cos\theta_1 - \cos\theta_2) \tag{21-24}$$

The average voltage is reduced by later triggering angles, as well as being a function of the load impedance angle. A plot of Eq. 21-24 appears in Fig. 21-20, in terms of the load parameter $\omega L/R$ and the delay angle θ_1.

When a full-wave circuit is used with an inductive load, the current is continuous. The results may again be derived, with attention to limits and the fact that the current at angle θ_1 is equal to that at angle $\theta_1 + \pi$. The direct voltage will be

$$E_{\text{dc}} = I_{\text{dc}}R = \frac{2E_m}{\pi}\cos\theta_1 \tag{21-25}$$

for full-wave operation, and continuous current. A plot of this equation appears as the solid curve of Fig. 21-18(a), showing a good control characteristic.

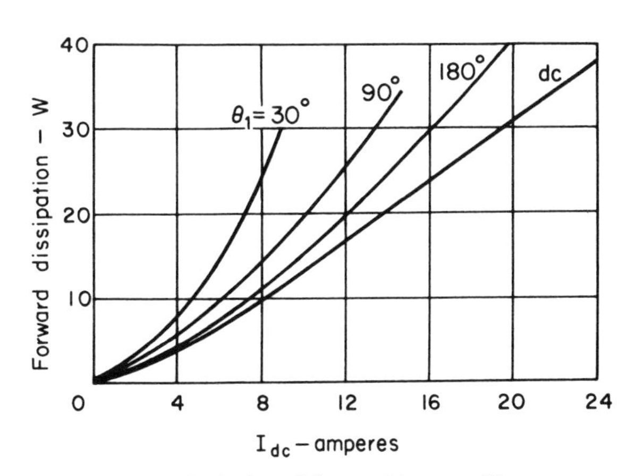

Figure 21-21. Variation of forward losses with conduction angle.

The forward power loss can be obtained by integration of the instantaneous current wave form of Fig. 21-17 and the forward voltage drop, and the results are plotted against the triggering angle in Fig. 21-21, for one typical unit. Gating losses are usually negligible.

21-10 TRIGGERING THE SILICON CONTROL RECTIFIER

In Fig. 21-22(a) a simple switch is used with gate circuit power taken from the ac supply. If S is closed, the rectifier D will conduct when the upper supply lead is positive, triggering the SCR. When the SCR conducts, its voltage drops to about one volt, and the gate current goes to essentially zero. Resistor R_g limits the gate current as $I_g = E_m/R_g$. Diode D prevents the high reverse voltage from being applied between the cathode and gate electrode on the reverse half-cycle. Opening switch S will terminate conduction. In Fig. 21-22(b) a phase-shift bridge of Section 21-11 is used to alter the time θ_1 in a similar circuit.

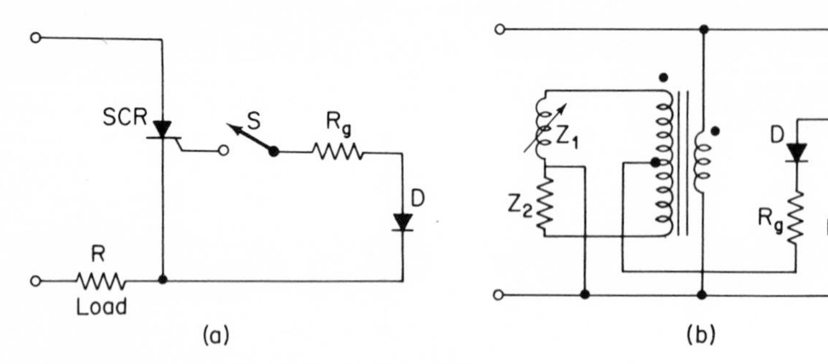

Figure 21-22. Firing circuits for the SCR.

The use of a unijunction transistor as an RC oscillator to develop a trigger input for a phase-controlled silicon control rectifier is shown in Fig. 21-23. Diode D is incorporated to synchronize the gate signal with the rectifier anode voltage. When terminal A and the anode of SCR are positive, diode D blocks current and transistor T is cut off. This allows capacitor C to charge through variable resistor R_x, introducing a time delay before the emitter of the unijunction transistor U reaches its peak voltage and triggers. The current through U to base 1 discharges C through R_2 in a sharp pulse which triggers the rectifier. Phase control, giving delay in triggering

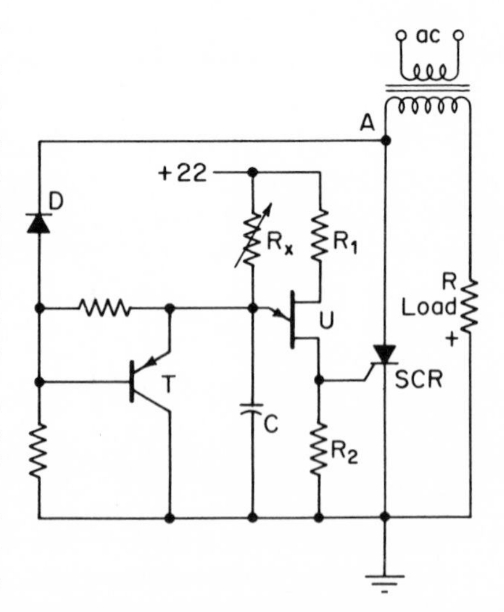

Figure 21-23. Phase control with a unijunction trigger circuit.

of the SCR, is possible by variation of the R_xC time constant. This basic principle is applied in many control functions.

The gate lead actually contacts only a small area of the gate electrode, and a finite time is needed for the trigger pulse to activate the whole cathode area of an SCR. Under low driving voltage only a small area of high gain is turned on, while under high driving voltage all the cathode area adjacent to the gate is turned on. For fast turn-on the SCR should be triggered by low-impedance circuits capable of large voltage input. The effect is illustrated in Fig. 21-24, and this explains the desirability of using the unijunction transistor and a low-impedance discharge type of trigger circuit. High drive also tends to eliminate differences in turn-on time between different SCR's, as shown by the narrowing of the operating range, at the higher voltages of the figure.

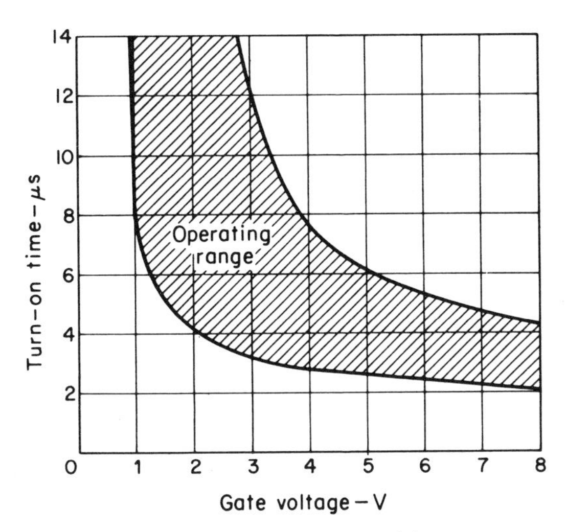

Figure 21-24. Turn-on time vs. gate drive.

21-11 THE PHASE-SHIFT CIRCUIT FOR CONTROL

In the circuits of Figs. 21-22(b) and 21-25(a), either $\mathbf{Z}_1$ or $\mathbf{Z}_2$ may be varied to shift the phase of the voltage between B and D. If A is in phase with the voltage at 1, then CA may be drawn as a phasor with B as the midpoint, as in (b). Voltage drops $\mathbf{IZ}_1 = \mathbf{I}R$ and $\mathbf{IZ}_2 = j\mathbf{I}X_c$ must add to voltage CA, and must also be mutually at right angles so that the locus of D becomes a semicircle with CA as the diameter. The phasor BD is then a variable-phase, constant-amplitude voltage, and the angle θ is its angle of lag behind the voltage at 1, which connects to the anode of the control rectifier.

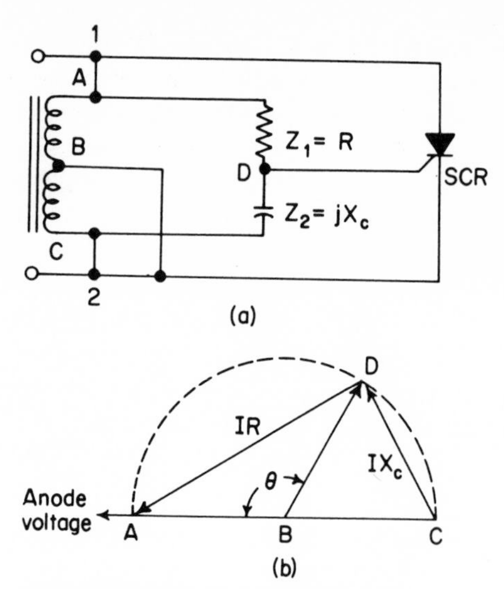

Figure 21-25. (a) Phase-shift bridge; (b) circle diagram.

The lag angle may be found from circuit constants and geometry as

$$\tan\frac{\theta}{2} = \frac{Z_1}{Z_2} = \omega CR \tag{21-26}$$

Variable inductors may also be used, making Z_1 inductive and Z_2 resistive. The angle then is

$$\tan\frac{\theta}{2} = \frac{Z_1}{Z_2} = \frac{\omega L}{R} \tag{21-27}$$

21-12 CONTROL RECTIFIER APPLICATIONS

In addition to dc voltage control, control rectifiers are adaptable to other services. One important use is for switching ac power loads at high speeds, where mechanical contactors are too slow. A typical circuit for switching a load R is shown in Fig. 21-26. The circuit is an *inverse-parallel* connection, so that one rectifier handles each half-wave, and alternating current is transmitted when the rectifiers are turned on by closure of switch S. Resistor R_g and the diodes provide protection of the gate electrode, as previously described. Switch S may be replaced with circuits

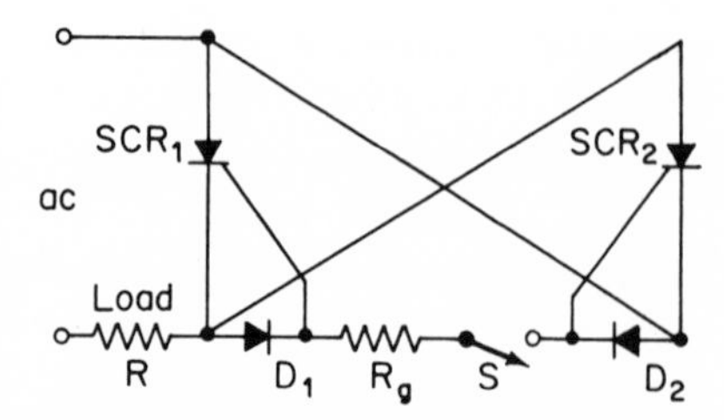

Figure 21-26. ac switch using silicon control rectifiers.

deriving signals from light, voltage, current, temperature, or other inputs.

Variation of the dc output voltage of a polyphase rectifier circuit can be accomplished by SCR's in the three-phase bridge circuit of Fig. 21-27. Since two rectifiers always conduct in series in a bridge circuit, it is unnecessary to control both, and so diodes D_1, D_2, D_3 are used instead of more expensive SCR units. The control circuit might be a three-phase adaptation of the unijunction transistor circuit of Fig. 21-23.

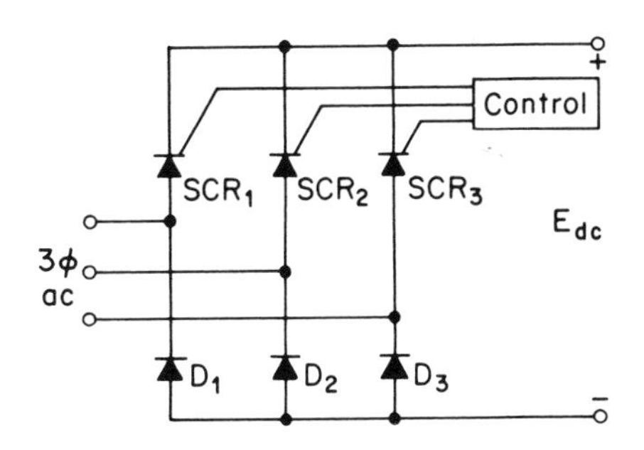

Figure 21-27. Three-phase controlled bridge rectifier.

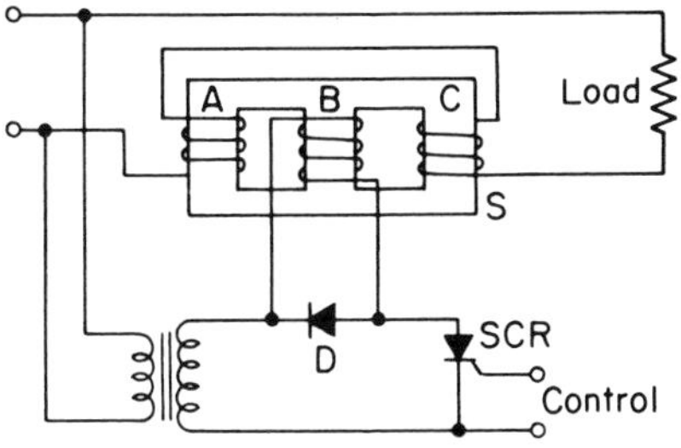

Figure 21-28. Control of ac loads by saturable reactor.

The control rectifier is also used to supply the direct current for saturable reactors, as in the temperature control of large electric furnaces, as in Fig. 21-28. The load current passes through coils A and C of the reactor, so arranged that their reactance is high with no direct current flowing and the load voltage is low. As the rectifier triggering voltage is made less lagging, direct current from SCR passes through coil B, saturating the iron, and reducing the reactance of the ac coils. With low reactance, the load current reaches the rated value.

The diode D is used to approximate full-wave action without a second control rectifier. In each cycle, while the current in coil B is increasing, the anode of SCR is maintained negative by the $L\, di/dt$ voltage of the reactor; when the current decreases, the sign of the inductive voltage reverses, and D conducts during the negative half-cycle when SCR is inoperative, until the magnetic energy in coil B is dissipated.

21-13 INVERTERS

A circuit that will change direct current to alternating current is called an *inverter*. A commonly used form of dc to ac or dc to dc inverter is shown in Fig. 21-29, employing switching transistors. The circuit is a feedback oscillator, operating at a frequency dependent on coils 1 and 3, the maximum flux density, and V_{CC}. Low-voltage dc, as from a storage battery, is supplied

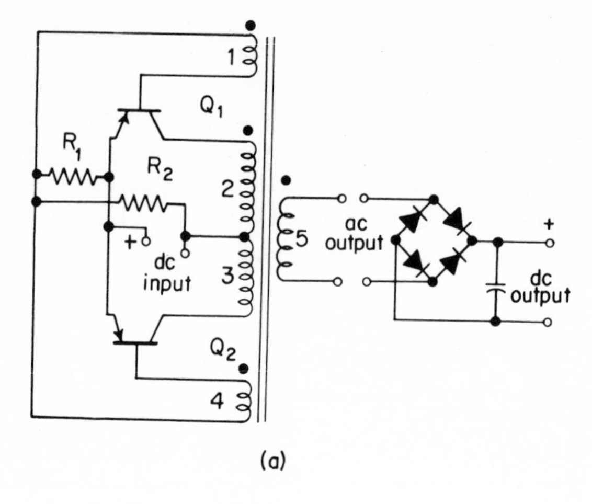

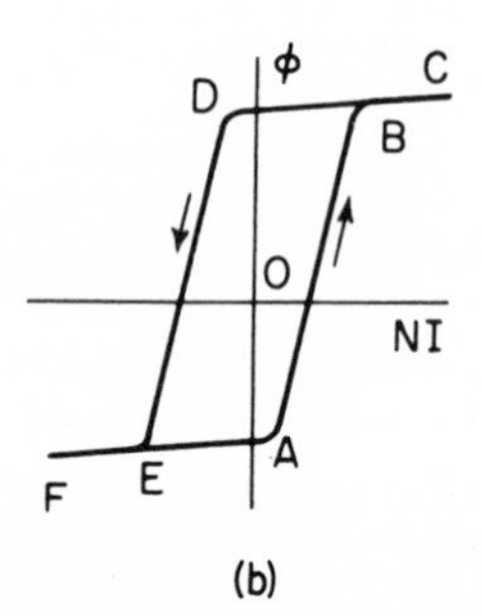

Figure 21-29. Transistor dc inverter.

to the transistors, and feedback is provided by coils 1 and 4. Output is taken from 5 as ac, or may be rectified for dc output at the desired voltage.

The transistors operate from cutoff, with a positive bias voltage, to saturation with negative base and almost the entire V_{CC} voltage appearing across the inductive load. The power dissipated is small in saturation, and while appreciable instantaneous power may be expended in the transistor during the switching interval, the switching time is so short that there is low average dissipation, and efficiencies exceed 90 per cent.

To consider the operation, let the core flux be at A on the hysteresis loop for the iron in Fig. 21-29(b), with the current in Q_1 exceeding that in Q_2. The flux will then begin to change toward B, inducing voltages in coils 1 and 4 which make the base of Q_1 negative and that of Q_2 positive, as indicated by the polarity dots. Then Q_1 saturates and Q_2 cuts off. The flux in the core continues toward positive saturation, almost the entire dc supply voltage appearing across coil 2. At B, the rate of change of flux decreases, the voltage on Q_1 rises and the increased current drives the core to C. However, $d\varphi/dt$ has fallen, the voltage in coil 1 drops, and transistor Q_1 goes toward cutoff, taking the operating point on the hysteresis loop toward zero NI near D. This change in flux induces an opposite voltage in coils 1 and 4, cutting off Q_1 and making the base of Q_2 negative and Q_2 conducts. There is current in coil 3 and the core flux begins to build toward negative saturation at E. When this point is reached, the induced voltages again reverse and the switching process repeats.

The secondary ac output is of square form, since the cycle time is largely occupied in flux change from A to B, or D to E on the flux loop, switching time of the wave being small. Thus the output is easily rectified and filtered if dc is desired.

PROBLEMS

21-1. A silicon diode having an average voltage drop of 0.5 V is used to supply a dc load of 12 V and 5.0 A. Find:
(a) The input ac volts.
(b) The peak anode current.
(c) The power loss in the diode.
(d) The non-dc power loss in the load.

21-2. A silicon diode is used in a half-wave circuit to supply direct current to a 15-ohm load. The source has a voltage of 12 rms. If diode drop is negligible, find:
(a) The peak diode current.
(b) The average diode current.
(c) The ac load current (rms).

21-3. A single-phase full-wave rectifier uses two silicon diodes rated at 1.5 A peak current and 0.225 A average current. The transformer voltage to center tap is 250 V rms and the load is 3000 ohms. If the diode drop is 1.0 V, find:
(a) dc load current.
(b) dc power output.
(c) Total diode loss.
(d) Peak inverse voltage.
(e) ac power input from the transformer.

21-4. Starting with Eq. 21-12, develop Eq. 21-13 for the dc output voltage.

21-5. A rectifier using six silicon diodes rated 35 A average, 150 A peak each, has a 230-V delta-connected primary and star-connected secondary. The load is 200 A at 600 V dc. Assuming negligible leakage reactance, find:
(a) Current and voltage ratings of all transformer windings.
(b) If the transformer is 97 per cent efficient, what is the overall power efficiency? Assume 0.5 V drop in each diode.

21-6. Design a 500-kW, 1500-V dc rectifier system using the three-phase half-wave, zigzag connection. Calculate:
(a) Transformer secondary voltage for all windings.
(b) Direct current through each diode.

(c) The kVA rating of the transformer primary and secondary. Assume the load has a large inductance for a filter.

21-7. A transformer having a 3000-V center-tapped secondary is used to supply either a single-phase full-wave or bridge rectifier. Find the ratio of the dc power output in full-wave to that in the bridge circuit for equal volt-ampere loading of the transformer.

21-8. For the m-phase rectifier without inductance, derive the general expression for the ripple.

21-9. You have a number of silicon power diodes rated as follows:

$$250 \text{ P.I.V.} \qquad \text{drop} = \text{negligible}$$
$$I_{\text{peak}} = 15 \text{ A} \qquad I_{\text{av}} = 5 \text{ A}$$

(a) Find the maximum dc-power available in the following service to a resistance load: one-phase, full-wave; three-phase, half-wave; six-phase, half-wave.
(b) The diodes cost $125 each and transformers cost $20 per kVA. Find the total cost per dc kW output for each of the three circuits.
(c) The transformers are 94 per cent efficient. If a return of 12 per cent must be paid for depreciation and interest on investment in equipment, and the rectifiers are used 5000 hours per year, find the cost of dc power per kilowatt-hour from each rectifier, if the input ac power costs 1.5¢ per kilowatt-hour.

21-10. A silicon control rectifier with anode supply of 200 V rms, 60 hertz, applied through a 100-ohm resistor, is fired at 60° of the anode cycle. If internal drop is zero:
(a) Find the average current flowing.
(b) Draw the current wave form.
(c) Draw the voltage wave form across the SCR.

21-11. The drop in a control rectifier is zero. If the anode supply is 120 V rms and the load a resistance of 10 ohms, find:
(a) The average current when the diode fires at 135°.
(b) The average current when the diode fires at 0°.

21-12. Find the needed value of series resistor to limit the average current to 0.6 A, if 120 V rms is applied to an SCR, with firing angle at 75° of the anode cycle.

21-13. An SCR is rated at 75 A peak, 20 A average. What is the greatest possible delay in the trigger angle, if direct current is at the rated value?

REFERENCES

1. Ebers, J. J., "Four-Terminal *pnpn* Transistors," *Proc. IRE*, Vol. 40, p. 1361 (1952).

2. Moll, J. L., M. Tanenbaum, and N. Holonyak, "*pnpn* Transistor Switches," *Proc. IRE*, Vol. 44, p. 1174 (1956).

3. Aldrich, R. W., and N. Holonyak, "Multiterminal *pnpn* Switches," *Proc. IRE,* Vol. 46, p. 1236 (1958).

4. Schmidt, P. L., "Voltage Conversion with Transistor Switches," *Bell Lab. Record,* p. 36 (1958).

5. Henkels, H. W., "Germanium and Silicon Rectifiers," *Proc. IRE,* Vol. 46, p. 1086 (1958).

6. Gentry, F. E., et *al.*, *Semiconductor Controlled Rectifiers.* Prentice-Hall, Inc., Englewood Cliffs, N. J., 1964.

7. Murray, R., Jr., editor, *Silicon Controlled Rectifier Designers' Handbook.* Westinghouse Electric Corp., Youngwood, Penna., 1964.

8. Gutzwiller, F. W., editor, *Silicon Controlled Rectifier Manual,* 4th ed. General Electric Co., Syracuse, N. Y., 1967.

22

LIGHT-SENSITIVE DEVICES

The first photosensitive cells employed selenium, which varied its electrical resistance with incident light. There followed cells employing electron emission induced by radiant energy, but with the development of new semiconductor materials, we have returned to photoconduction devices because of their small size and high sensitivity. Applications in the counting of objects, control of position, color measurement, or electrical generation in space are now common.

22-1 PHOTOSENSITIVE DEVICE CLASSIFICATION

Photosensitive devices convert radiant energy into electrical energy or control electrical energy. Various physical phenomena are employed, leading to the following classification of devices:

Photojunction devices, which utilize radiant energy to transfer charges across the forbidden energy gap of semiconductors into the conduction band.

Photoconductive cells, which vary their electrical resistivity in accordance with the radiant energy received, and employ semiconductor materials.

Photoemissive cells, in which the energy of the light beam causes emission of electrons from a metal cathode, enclosed in an evacuated or gas-filled glass or quartz envelope.

Photovoltaic cells, which generate an emf in proportion to the radiant energy received, as in the so-called *solar cell*.

22-2 SENSITIVITY DEFINITIONS

The sensitivity of a photoelectric device is stated in terms of current output per unit of total radiant power striking the cell surface for a specified wavelength range from a given source. The latter specification must be added because of sensitivity variations at different wavelengths of incident light. A source often used as a standard is a tungsten filament lamp operating at a temperature such that its spectrum of output radiant energy is that of a black body operating at 2870°K; the relative energy distribution from such a source appears dashed in Fig. 22-1(a).

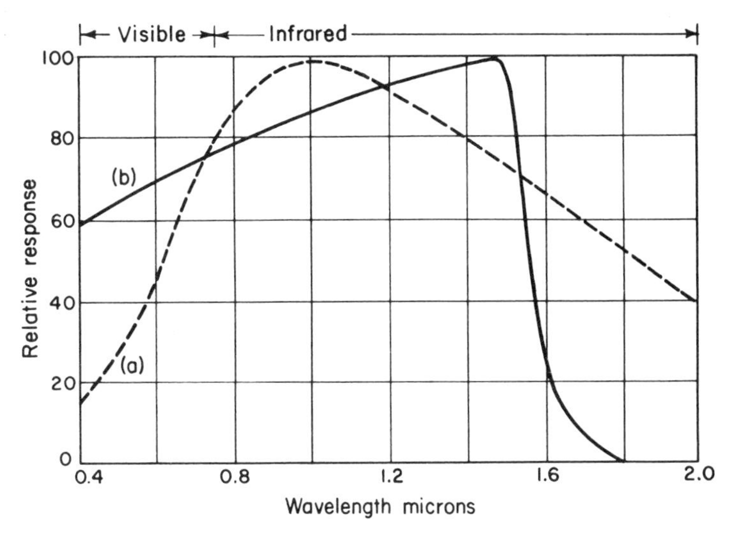

Figure 22-1. (a) Relative radiation from tungsten at a color temperature of 2870°K; (b) relative response of thin slab of germanium in photoconduction.

The *visible range* is usually taken as lying between the red at 0.7 micron (μ) and the violet at 0.4 micron wavelength. The *candle* is the standard source of luminous flux, giving a total light flux of 4π lumens. The solid angle surrounding a point source is 4π steradians, and the *lumen* is the light flux emitted per unit solid angle by a standard candle. More recently the *candela* was defined as the standard of luminous intensity; it is one-sixtieth of the luminous intensity of one sq cm of a black platinum source at the temperature of solidification. The difference between a candle and a candela is negligible.

Sensitivity may also be given in terms of visible light only, when it is called *luminous sensitivity*, and stated as current per lumen of incident visible light.

A point source radiates constant lumens per unit solid angle, but the flux incident per unit area decreases inversely as the square of the distance. To convert from the candlepower rating of a source to lumens, use:

$$L = \frac{CA}{d^2} \tag{22-1}$$

where L is the intensity in lumens from a point source of C candlepower at a distance of d m, on an area A m^2.

22-3 PHOTOJUNCTION DEVICES

Intrinsic conduction can be created in semiconductors by absorption of light quanta. This phenomenon assumes the absorption of radiant energy to raise an electron from the top of the filled band, across the unallowed gap, and into the conduction band. The light falling on the back-biased pn junction creates electron-hole pairs which are swept out as a photocurrent.

The minimum energy which must be supplied by the radiation must be equal to the gap energy, and this is about 0.7 eV for germanium. Since a photon carries a quantum of $W = hf$ joules and

$$\lambda_0 = \frac{c}{f_0} = \frac{hc}{eE_G} \quad \text{m} \tag{22-2}$$

the longest exciting wavelength is

$$\lambda_0 = \frac{6.62 \times 10^{-34} \times 3 \times 10^8}{1.60 \times 10^{-19} \times 0.7} = 17.7\ \mu = 17{,}700\ \text{Å}$$

This wavelength is in the infrared. A curve of relative output vs. wavelength for germanium is given in Fig. 22-1(b).

A small *photodiode* is illustrated in Fig. 22-2, with its characteristics in Fig. 22-3. Good linearity with light is obtained. With the threshold at 17.7μ such a cell has excellent infrared sensitivity. Some electrons from the impurity atoms will exist in the conduction band at normal temperatures, and in darkness. Produced by thermal energies, these charges produce dark conductivity known as "dark current."

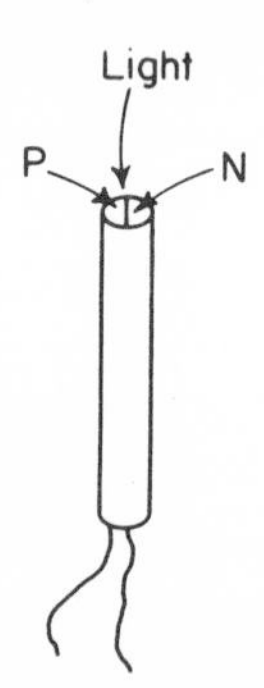

Figure 22-2. Typical junction photodiode.

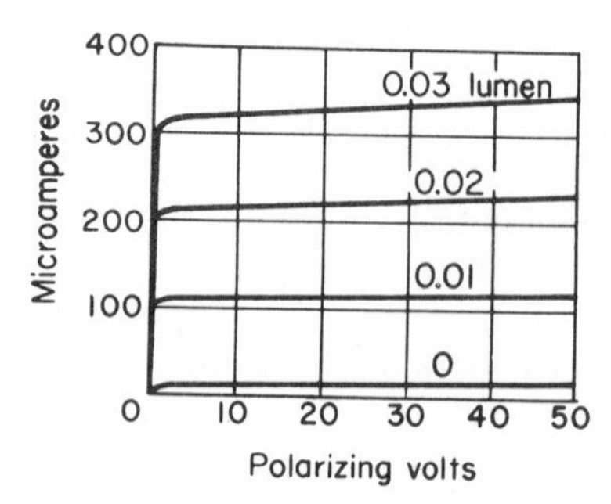

Figure 22-3. Output characteristics for 7223 photodiode.

Large area junctions are made with the junction very close to the surface. These cells are useful for direct conversion of solar energy to electrical energy, as in space satellite service. Very thin cells must be made, to have the photons absorbed by transfer of their energy to overcome the gap energy, rather than to have them absorbed as heat. Efficiencies of 10 to 15 per cent have been achieved, with voltages of 0.6 V and currents as high as 2.7 A per sq m.

Phototransistors follow logically from the photodiode; the combination adds transistor gain of 50 to 100 to the output of the diode. The collector is normally back-biased, and the base is floating. The radiant energy develops charges in the base region, as would normally be done by the base current. The construction is indicated in Fig. 22-4, with spectral response and collector characteristics in Fig. 22-5.

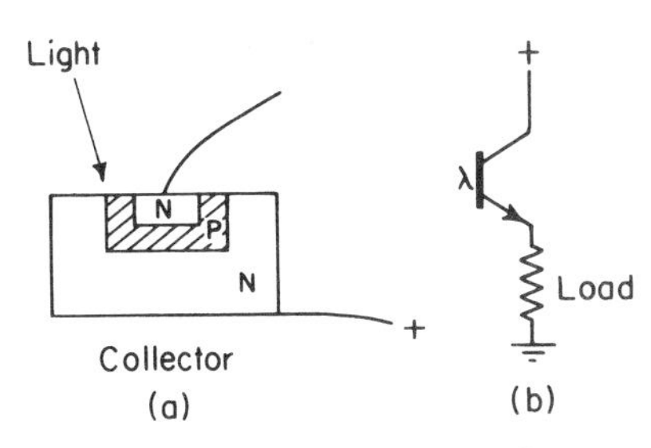

Figure 22-4. Phototransistor and symbol.

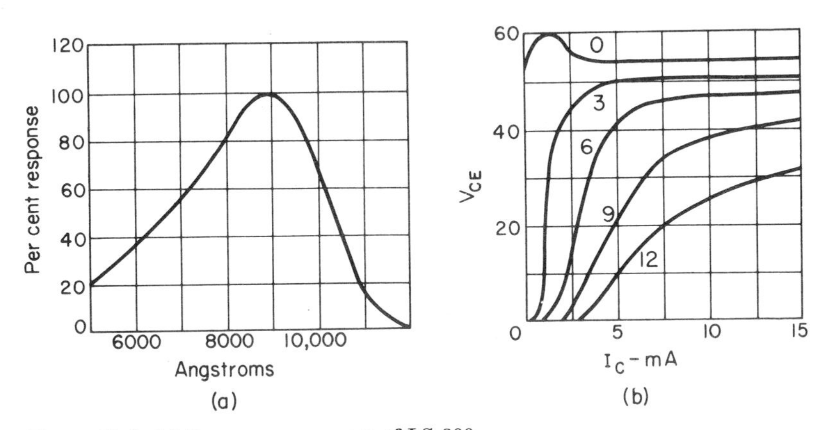

Figure 22-5. (a) Frequency response of LS-800 phototransistor; (b) collector characteristics, radiation as mW/cm².

22-4 PHOTOCONDUCTIVE CELLS

The very common cadmium-sulfide *photoconductive* cell is illustrated in Fig. 22-6(a). The electrodes are extended in an interdigital pattern to increase the contact area with the active material; corresponding volt-ampere curves as a function of light are shown in Fig. 22-6(b).

The speed of response of the CdS cell is rather slow, but the spectral response rather closely matches the human eye. Because of their large output

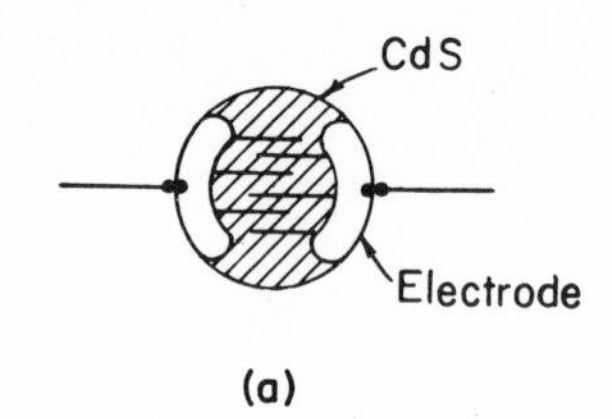

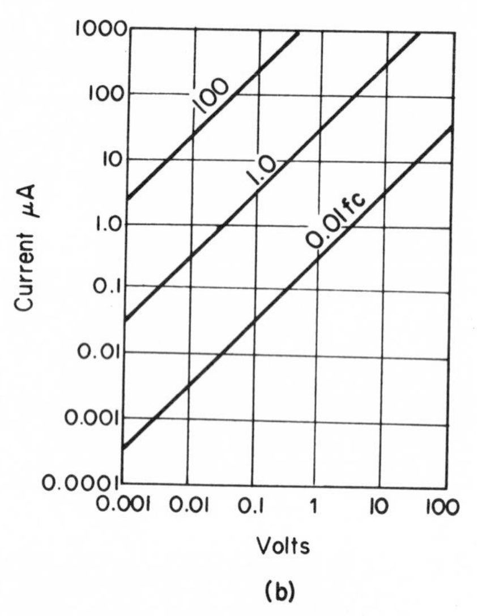

Figure 22-6. Response of cadmium-sulfide photoconductive cell.

and their simplicity they are used for automatic control of exposure for cameras, with the current supplied by a small battery.

22-5 PHOTOEMISSION

The Einstein equation for photoemission

$$hf = eE_W + \frac{mv^2}{2} \tag{22-3}$$

shows that emission will occur if the incident photon contributes an energy greater than the surface work function, E_W. Desirable cathode materials are the low work-function alkali metals having threshold emission frequencies in or below the visible region of the spectrum. Cesium, rubidium, potassium, and sodium are deposited in monoatomic films on the cathode bases and spectral sensitivity curves for a variety of surfaces are given in Fig. 22-7.

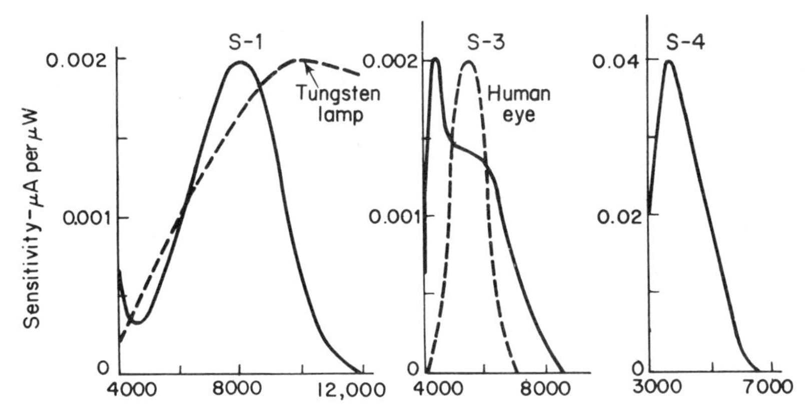

Figure 22-7. Color sensitivity of several photoemissive surfaces.

A comparison of the cesium S-1 curve with that of the radiation from a tungsten filament at 2870°K is made, and it can be seen that the S-1 surface is suitable for such a source. The S-3 surface of rubidium has its greatest sensitivity in the blue end of the spectrum. The S-4 surface, also of cesium, has a peak response 20 times greater than the peak of the S-3 surface. With the peak at 0.375 micron in the near ultraviolet, such a surface is useful with blue light sources.

22-6 THE PHOTOEMISSIVE CELL

A photoemissive cell consists of a radiation-sensitive cathode and an anode to collect the emitted electrons, assembled in a glass or quartz bulb, as in Fig. 22-8.

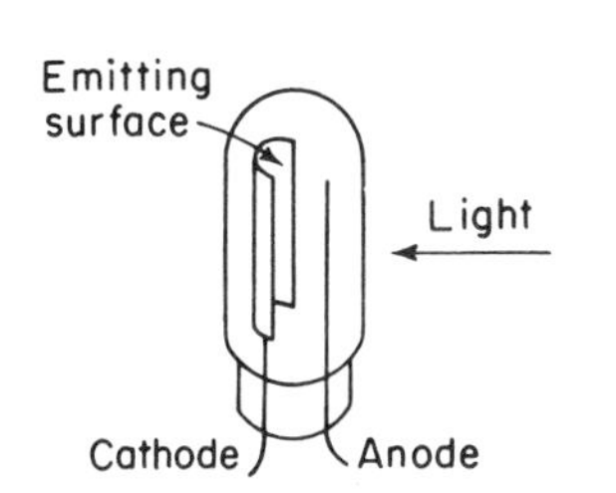

Figure 22-8. Photoemissive cell.

With light causing emission, a small applied voltage causes a small current. The current rises as the voltage is increased, but further increase in potential causes the current value to level off, as shown in Fig. 22-9. This indicates that all emitted electrons are being attracted to the anode. Owing to this saturation condition, the current is almost entirely independent of applied emf above values of 20 V. When operated in saturation, the output of a vacuum photocell is linear with light intensity.

The low slope of the volt-ampere curve indicates a very high internal resistance, of the order of hundreds of megohms. The form of equivalent circuit is that of a constant current generator connected to a load resistor.

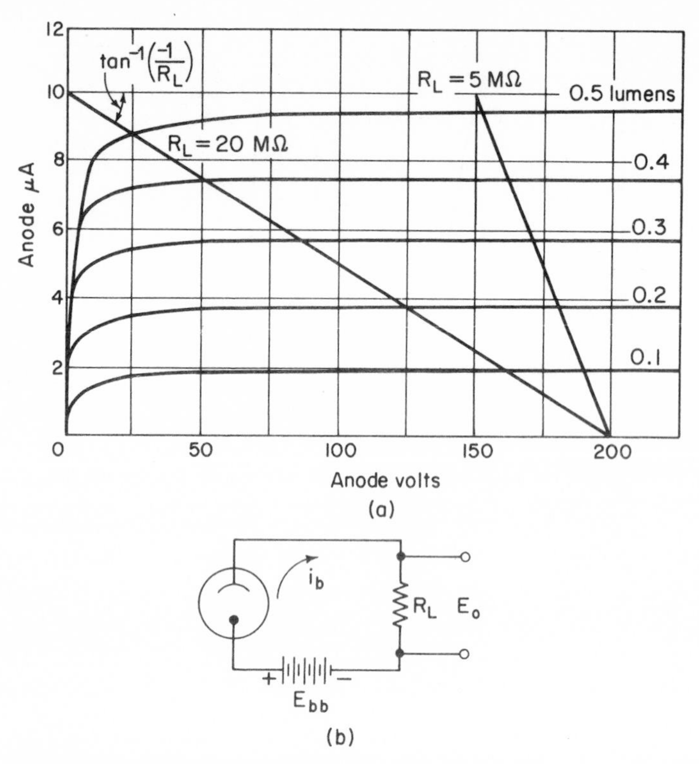

Figure 22-9. (a) Use of load line; (b) photoemissive cell circuit.

It would seem that a larger load should result in a higher output voltage, but leakage currents in the insulation serve as a limit. In practice the upper limit for the load resistance is usually 10 to 20 megohms.

By introduction of argon or other gas into a photoemissive tube, an increase in current output can be achieved. The electrons, in moving to the anode, collide with gas atoms and produce ionization. The electrons thus freed from the atoms join in the charge movement to the anode. The current may be increased four to seven times over that in an equivalent vacuum cell; this ratio is known as the *gas amplification factor*. Attempts to increase the ratio above ten usually result in bombardment of the cathode by positive gas ions and these cause damage to the surface.

The characteristics of the gas-filled cell are drawn in Fig. 22-10. Above potentials of 20 V, the gas-filled cell is more sensitive than the vacuum type. The curves show that the current output is not linear with light intensity, and an additional fault is a loss of sensitivity for high-frequency light variations, caused by the slow speeds of the positive ions.

Sinusoidal or other variations of intensity from a quiescent value may

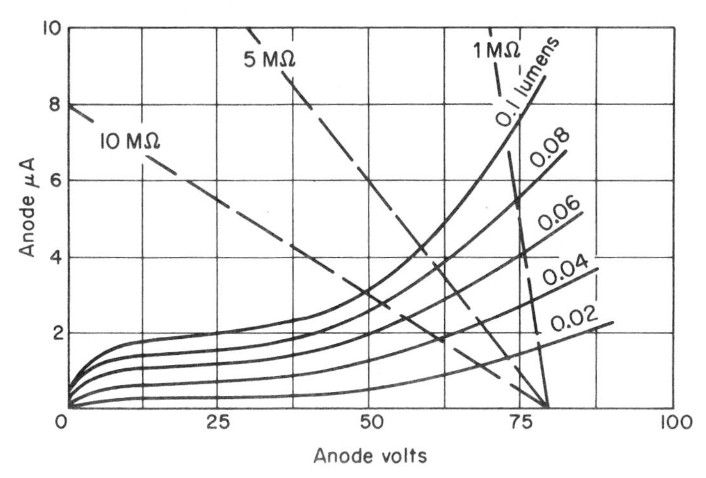

Figure 22-10. Characteristics for gas-filled 921 photoemissive cell.

be analyzed by use of the load line. For instance, with the cell of Fig. 22-9, with a 20 megohm load and a Q point at 0.3 lumen, a sinusoidal variation with peak-to-peak value of 0.2 lumen will produce 140 V peak-to-peak across the load.

A vacuum tube is a suitable relay and impedance transformer for a photoemissive cell, as shown in Fig. 22-11. Medium μ triodes are satisfactory and permit R_L values of 15 to 20 megohms. The potentiometer is used to adjust the plate current to a minimum, with zero light. When light strikes the cell, the grid becomes positive owing to the drop across R_L. As a result, the plate current increases and operates the relay. For applications in which the light is modulated, the relay may be replaced by a conventional RC-coupled ac amplifier.

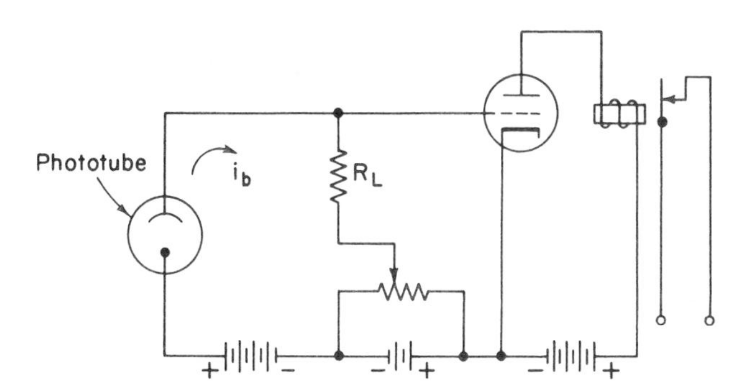

Figure 22-11. Circuit of photoelectric relay.

22-7 ELECTRON MULTIPLIERS

The currents in photoemissive cells under low light conditions are extremely small, but can be amplified by an application of secondary emission principles in an *electron multiplier*. Figure 22-12 shows the basic multiplier design in which a series of electrodes, all treated with low work-function emitting materials, are operated at successively higher positive potentials. For each electron leaving S, δ electrons are emitted from target T_1. These are attracted to target T_2 by the positive potential, and since each electron reaching T_2 causes δ electrons to be emitted, then δ^2 electrons leave T_2. The number of electrons leaving T_3 is δ^3, and so for n electrodes, the gain is

$$\text{gain} = \delta^n \qquad (22\text{-}4)$$

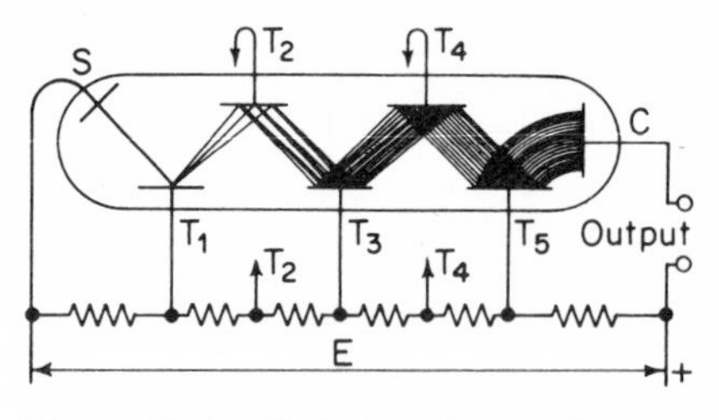

Figure 22-12. Photoelectric secondary emission multiplier.

where δ is frequently as high as 10, and as many as 10 stages may be used, for overall gains of several million.

Usually the targets, called dynodes, are shaped to focus the electrons successively on the targets, so as to prevent spreading of the beam and loss of part of the electrons.

22-8 PHOTOVOLTAIC CELLS

Several types of self-generating or *photovoltaic cells* employing semiconductor contacts against metals had much use prior to the advent of the photoconduction units. Electrons are raised to the conduction band by light energy, and these electrons flow over to the metal when the materials are placed in contact, as in Fig. 22-13. The copper-copper oxide cell and the iron-selenium cells are usual types.

The cells generate a few hundred millivolts in bright light, and under low resistance load, the current is almost linear with illumination. Because of the linearity and stability, these cells have been extensively used for photo-electric exposure meters, but are now being superseded by cadmium-sulfide photoconductive cells.

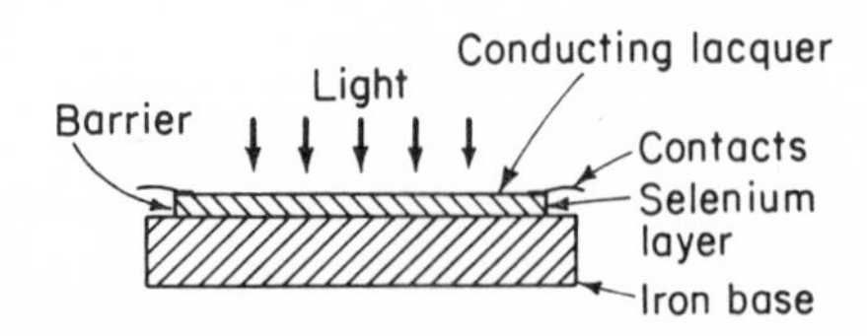

Figure 22-13. Iron-selenium photovoltaic cell.

PROBLEMS

22-1. A photocell receives a change in intensity from 0.12 to 0.32 lumen. If the tube is a 922, as in Fig. 22-9, and has a load of 10 megohms and applied voltage of 150, find the change in voltage across the load.

22-2. An illumination control for a schoolroom employs the circuit of Fig. 22-11 with a 922 photocell and a triode of $g_m = 2000$ μmho. The photocell operates at 150 V with $R_L = 5$ megohms; the tube has the grid bias adjusted for a plate current of 0.2 mA at $E_{bb} = 250$ V, with the resistance of the relay being 10,000 ohms. If the relay opens its contacts and turns off certain lamps at 4 mA coil current and closes the contacts at 1.8 mA, find the limits between which the room illumination will be allowed to vary. (*Hint*: The 922 cell is linear, so its characteristics may be interpolated.)

22-3. A 922 cell is used with a 10-megohm load and applied voltage of 100. Light striking the photocell changes sinusoidally from 0.1 to 0.32 lumen. What is the rms voltage across the load?

22-4. A tetrode having g_m of 1500 μmho is followed by four stages of current multiplication having $\delta = 3.4$. Find the output-current change expected because of 1-V rms signal on the tetrode grid.

22-5. A conventional photoemissive cell produces a change in output of 1 μA for a given change in light intensity. If a nine-stage multiplier having $\delta = 8$ follows this photocell, what is the change in output current of the multiplier?

22-6. A 50-cp lamp is used as a point source for a 921 cell in a circuit with a 5-megohm load and 80 V applied. How far away can the source be and still produce a current of 4 μA in the load? The area of the cell surface is 2.58 cm^2.

22-7. For the photodiode of Fig. 22-3, determine the load which will give maximum power output.

22-8. An 8000-ohm relay is used with the phototransistor of Fig. 22-5. It drops out at 0.1 mA and picks up at 0.3 mA. If 40 V is applied, find the light levels at which the relay will operate.

22-9. Determine the light flux in lumens on a photocathode illuminated by a point-source lamp of 32 candlepower at a distance of 50 cm. The cell area is 6.5 cm^2.

REFERENCES

1. Metcalf, G. F., "Operating Characteristics of Photoelectric Tubes," *Proc. IRE*, Vol. 17, p. 2064 (1929).

2. Zworykin, V. K., G. A. Morton, and L. Malter, "The Secondary Emission Multiplier," *Proc. IRE*, Vol. 24, p. 351 (1936).

3. Pierce, J. R., "Electron-Multiplier Design," *Bell Lab. Record*, Vol. 16, p. 305 (1938).

4. Shive, J. N., "The Phototransistor," *Bell Lab. Record*, Vol. 28, p. 337 (1950).

5. Chapin, D. M., C. S. Fuller, and G. L. Pearson, "A New Silicon *p-n* Junction Photocell for Converting Solar Radiation into Electrical Energy," *J. Appl. Phys.*, Vol. 25, p. 676 (1954).

6. Sawyer, D. E., and R. H. Rediker, "Narrow Base Germanium Photodiodes," *Proc. IRE*, Vol. 46, p. 1122 (1958).

23

MICROWAVE DEVICES AND THE LASER

The vacuum tube is inherently limited in upper frequency by the transit time of an electron between cathode and anode, and when this time becomes an appreciable part of a cycle, the possible gain becomes small.

Various efforts have led to a series of devices which overcome the transit-time limitation, and to other totally new forms using solid-state and quantum techniques, by which the frequency spectrum of electrical wave generation has now been expanded into the optical region.

23-1 VELOCITY MODULATION OF AN ELECTRON BEAM

In the first major step beyond the triode, the time of transit was utilized to allow electrons with cyclically varying velocities to catch up with preceding slower electrons, or to form *bunches*. These bunches then excited an output circuit as current pulses. This principle has become known as *velocity modulation*, and the device as the *klystron*.

The general arrangement of a klystron is shown in Fig. 23-1, with a beam of electrons from the cathode passing through holes in the *buncher*

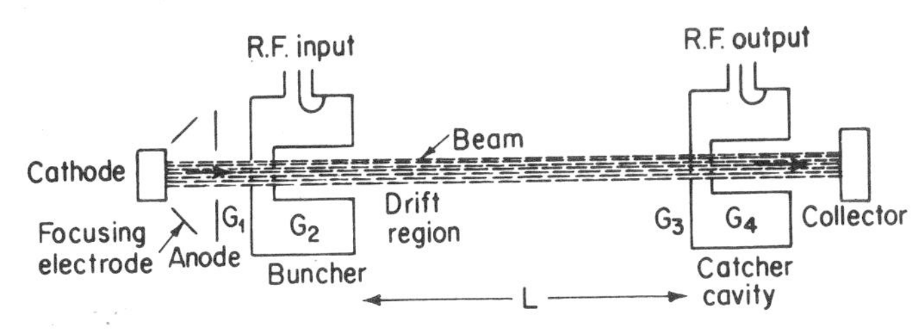

Figure 23-1. Two-cavity klystron amplifier.

into a drift region of length L, free of electric fields. The beam then passes through a *catcher* to reach a positive collector. The beam is assumed to remain well focused, or the mutual repulsion between electrons is neglected.

The buncher and catcher are resonant cavities which will sustain an alternating electric field between their left and right faces in the figure, and have properties similar to a parallel resonant circuit at a wavelength commensurate with their mechanical dimensions. The cyclic voltage existing between the G_1 and G_2 electrodes in the first cavity gives the electrons a small cyclic velocity variation

$$v = v_o(1 + \alpha \sin \omega t)$$

as they leave G_2 and enter the drift space.

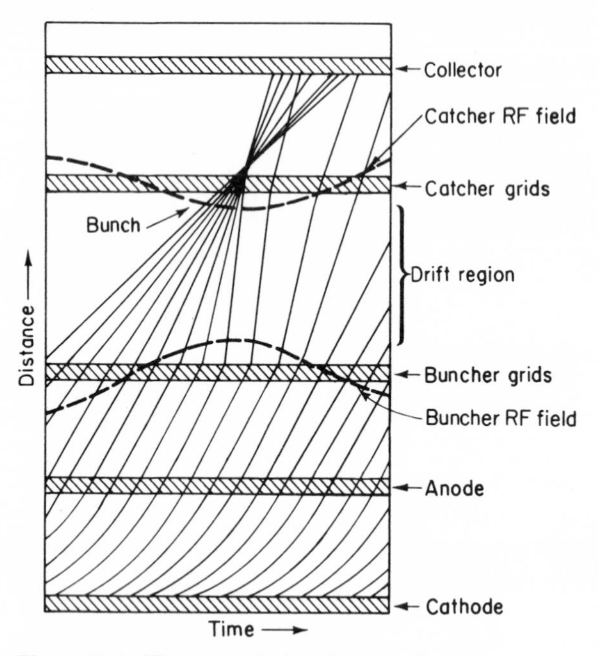

Figure 23-2. Klystron velocity–distance diagram.

In Fig. 23-2 the slopes of the electron paths represent the individual electron velocities. These velocities are uniform up to the buncher grids, and the figure shows the effects of the cyclic variation of velocity contributed by the buncher grids. Electrons passing the buncher grids at a time of accelerating field, or those given a velocity component of $+\alpha v_o \sin \omega t$ travel faster, and at the catcher grids have caught up with earlier electrons which were decelerated at the buncher, or given a velocity increment $-\alpha v_o \sin \omega t$.

As a result, the electron density passing the catcher grids varies with time, and this induces a current variation in the catcher cavity. By proper feedback the voltage variation in the cavity can be introduced into the buncher and made to excite that cavity, thus producing the original buncher-cavity variation, and the device becomes an oscillator. Since the cavities can have dimensions of centimeter order, the klystron frequency range can be carried down to such wavelengths.

23-2 THE KLYSTRON

One form of klystron is shown in Fig. 23-3(a). An electrical equivalent at (b) shows resonant circuits replacing the resonant cavities of (a). The cavities resonate electrically when a beam of electrons passes through them, much as a bottle develops an acoustical tone when air is blown across its mouth.

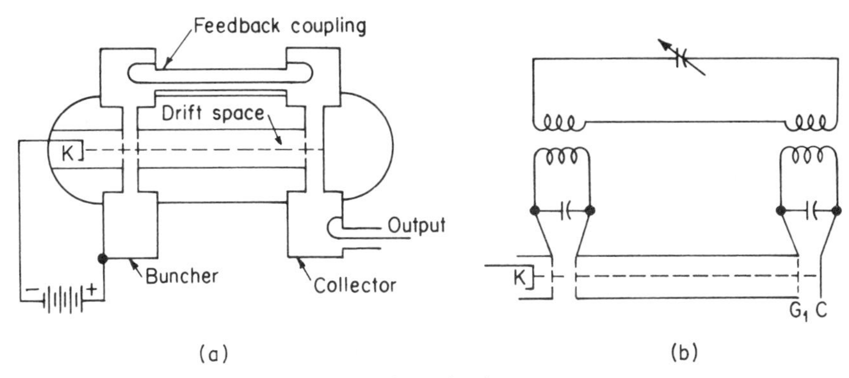

Figure 23-3. A klystron and its equivalent circuit.

In the buncher gap, the flow of electrons is uniform throughout the cycle. As many electrons are speeded up as are slowed down, and the net energy interchange is zero. However, as the electrons pass through the catcher cavity, a strong ac field is induced whose polarity is such as to slow down the bunch, or to extract energy from it. Half a cycle later the direction of the electric field across the gap has reversed, but this aiding field then acts on only the low-density or nonbunched portion of the beam. Thus more electrons are slowed down than are speeded up, and there is a net exchange of energy from the beam to the radio-frequency circuit in the catcher.

Since transit time is employed to aid in bunching of the electrons in the drift space, it is not a deterrent in high-frequency operation. Klystron tubes of this type are useful to 30 gigahertz, with power outputs from kilowatts at the lower frequencies to a few milliwatts at the higher end of the range.

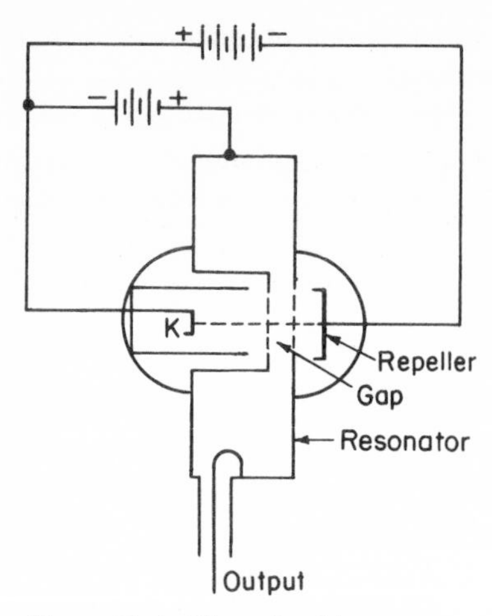

Figure 23-4. The reflex klystron.

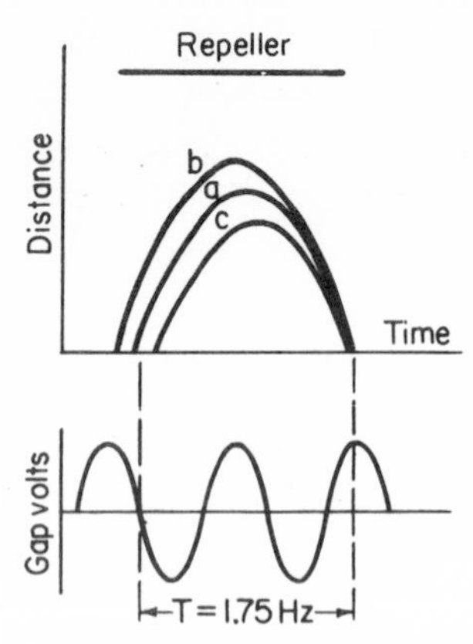

Figure 23-5. Production of bunches in the reflex klystron.

A more usual form of high-frequency klystron is the *reflex* type, of Fig. 23-4. The output resonant circuit is eliminated and the collector is replaced with a *repeller*, maintained at a negative potential. The reflex klystron utilizes an electron feedback which causes one resonant circuit to serve for both input and output. In passing through the two grids of the resonant cavity, the electron beam is given a velocity variation, as before. Faced by the negative repeller, the electrons are repelled back into the region of the two grids. Sufficient transit time has been available during this reversal action for the electrons to have become bunched, so that in passing through the resonant cavity in the reverse direction, they pass as bunches, inducing current in the resonant circuits as discussed previously.

The bunching action is illustrated in Fig. 23-5. Path a, plotted in distance and time, is that of an electron which passed through the input circuit at a time the potential was zero and becoming negative. It proceeds toward the repeller, but is slowed down, and returned to the gap. A second electron, at b, passes through the gap at the time of a positive voltage and is given a velocity higher than that of a. It travels further against the repelling field than the first electron, but it travels at a higher average velocity, and so it returns to the gap at the same instant as the first electron. A third electron, at c, goes through the gap at a negative potential time, and is slowed down. It does not travel as far against the repelling field but travels more slowly, so that c is also able to return to the gap at the same time as a.

As these bunches return through the gap, the phase conditions are such that energy is delivered from the bunched beam to the resonant cavity. The phasing of the bunches is controlled by the repeller voltage, while the frequency of operation is determined by the dimensions of the resonant circuit or cavity. The time of travel in the repeller or drift space must be $T = n + \frac{3}{4}$ cycles and is usually set at $1\frac{3}{4}$ or $2\frac{3}{4}$ cycles, with the cavity fixing the operation at one frequency. Broad tuning is possible through changing the dimensions of the resonant cavity, and fine tuning over about a one per cent range is possible with the repeller voltage.

23-3 BALLISTICS OF THE MAGNETRON

A magnetic field has been combined with a radial electric field to produce the basic *magnetron*, another efficient oscillator at microwave frequencies. As in Fig. 23-6, a cylindrical anode A has a cathode at K. The magnetic field is established longitudinally by a permanent magnet.

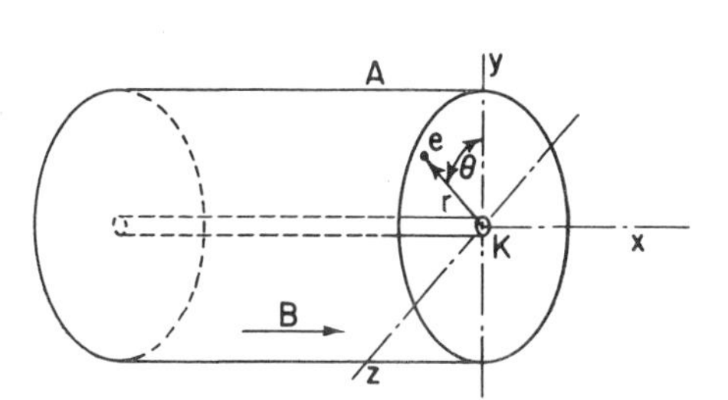

Figure 23-6. Basic geometry of the magnetron.

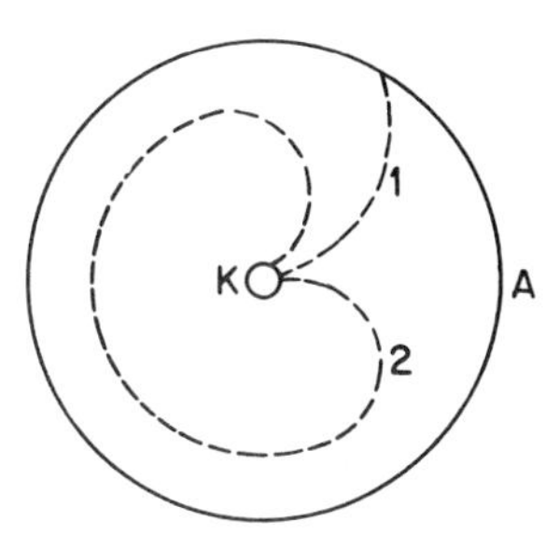

Figure 23-7. Possible electron paths in the magnetron.

In polar coordinates the position vector $\bar{r}$ of the electron e is

$$\bar{r} = r(\cos\theta + j\sin\theta)$$

By differentiation it can be found that

$$\frac{d^2\bar{r}}{dt^2} = \left[\frac{d^2r}{dt^2} - r\left(\frac{d\theta}{dt}\right)^2 + j\frac{1}{r}\frac{d}{dt}\left(r^2\frac{d\theta}{dt}\right)\right](\cos\theta + j\sin\theta) \qquad (23\text{-}1)$$

The motion may be described by two accelerations, a_r and a_θ, as

$$a_r = \frac{d^2r}{dt^2} - r\left(\frac{d\theta}{dt}\right)^2, \qquad a_\theta = \frac{1}{r}\frac{d}{dt}\left(r^2\frac{d\theta}{dt}\right) \qquad (23\text{-}2)$$

These are analogous to a_x and a_y in rectangular coordinates. The accelerations will produce two electron velocity components, radial or v_r, and tangential or ω_θ. The total energy of the electron at any point will be due to the radial electric field $\mathscr{E}_r$, since no work can be done by a fixed magnetic field.

The energy relations give

$$\frac{m}{2}(v_r^2 + r^2\omega_\theta^2) = \int_{rk}^{r} e\mathscr{E}_r\,dr \qquad (23\text{-}3)$$

Evaluation in terms of the parameters of the problem, with r_k as the cathode radius, and with radial electron emission assumed or $d\theta/dt = 0$ at the cathode surface, leads to

$$\frac{m}{2}\left[v_r^2 + \left(\frac{rBe}{2m}\right)^2\left(1 - \frac{r_k^2}{r^2}\right)^2\right] = -e\int_{rk}^{r} \mathscr{E}_r\, dr = V_{kr} \tag{23-4}$$

where V_{kr} is the potential between cathode and a point in the space at radius r. Consequently,

$$v_r = \sqrt{\frac{2eV_{kr}}{m} - \left(\frac{rBe}{2m}\right)^2\left(1 - \frac{r_k^2}{r^2}\right)^2} \tag{23-5}$$

For a fixed value of B, a large value of V on the anode will give a large radial velocity, and the electron paths will be as at 1 in Fig. 23-7. For smaller V, the radial velocity will go to zero at some r, and the path becomes that at 2, with the tube cut off.

23-4 *MAGNETRON OSCILLATORS*

The modern magnetron tube was developed during World War II, from the above basic principles; it became a major source of power for radar in the centimeter wavelength range.

The anodes are usually of the multicavity type, as in Figs. 23-8 and 23-9, although the exact cavity shape varies. The magnetic field is perpendicular to the plane of the page in Fig. 23-9(b), and a radial electric field is established between anode and cathode. The magnetic field is sufficient for the tube to be in the cutoff condition with no ac excitation present.

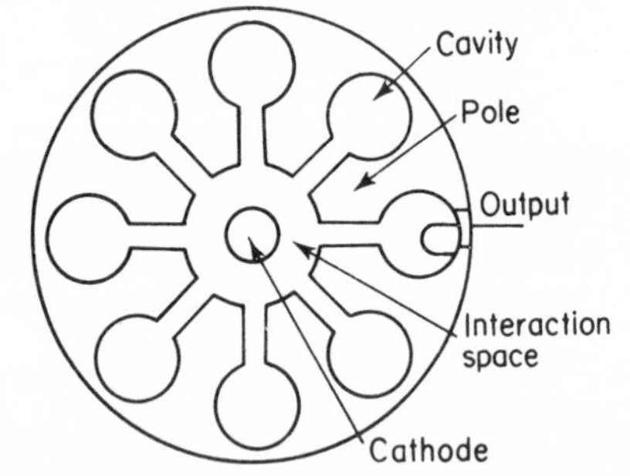

Figure 23-8. One form of multicavity anode.

The cavities serve as resonant circuits, and alternate poles become instantaneously positive and negative because of high-frequency voltages, although the whole anode is at positive dc potential to the cathode. The effect is that of a number of resonant circuits in series around the periphery for which the circuit of Fig. 23-9(a) is an approximate equivalent. The capacitance (electric field) is essentially concentrated across the gap, and the inductance (magnetic field) is formed by the inner part of the cavity. The capacitances C_2 constitute the gap capacitances, whereas those marked C_1 are the capacitances between pole faces and cathode.

Since the circuit returns upon itself, it can be seen that for a steady state to exist,

$$\beta = \frac{2\pi n}{N} \tag{23-6}$$

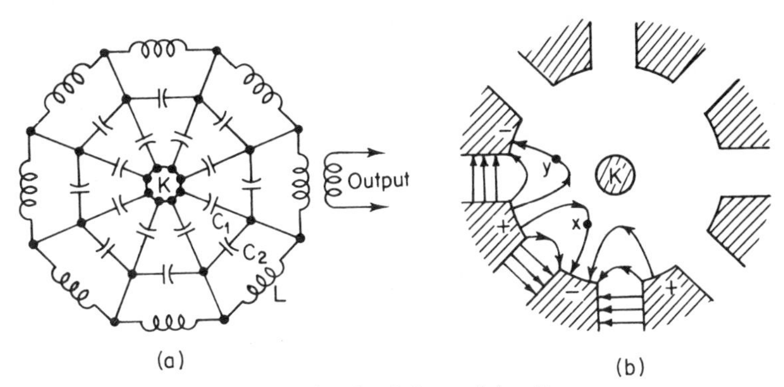

Figure 23-9. (a) Equivalent circuit of the multicavity magnetron; (b) the electric fields.

where

β = phase shift per section, in radians,
N = number of sections or pole faces,
n = number of cycles around cathode.

When $n = N/2$, the phase shift is π radians per section, and the pole faces alternate in polarity. This state, called the *π mode*, results in electric fields as in Fig. 23-9(b), and these fields rotate at a velocity such that two poles are passed per cycle.

With ac fields established between the poles, consider a single electron following a curved trajectory in the steady dc electric and magnetic fields. If it has a clockwise tangential component of velocity and is at X, the electric field due to the ac poles will be such as to increase its energy and its tangential velocity. An increase in tangential velocity will increase the magnetic force ($f_m = $ BeV) tending to accelerate the electron toward the cathode. This increase in magnetic force will tend to neutralize the radial force due to the positive potential on the anode and cause the electron to be quickly returned to the cathode.

An electron at y is in a tangential electric field which will slow it down, or extract energy from it. This process reduces the inward-directed magnetic force and allows the outward radial force due to the positive anode potential to accelerate the electron. Motion in an outward direction causes the magnetic force to redirect the electron with a tangential component. But tangential motion is opposed by the ac electric field, which extracts energy and slows down the electron. Electrons in the region of point y eventually reach the anode, having transferred energy from the static electric field to the dynamic electric field during their transit.

Electrons in positions such as x are quickly removed; electrons in positions near y are slowed down, redirected and accelerated, and again slowed down, with the process repeated until they reach the anode, so that a set of electron spokes are formed which rotate in synchronism with the field produced by the poles, as shown in Fig. 23-10.

The electrons rotating in the spoke efficiently convert energy from the static electric field to the dynamic electric fields set up by the resonant circuits. The all-electronic translation of energy in the magnetron accounts for the high efficiencies reached, which are in the range of 30 to 60 per cent. The frequencies reached are above 30,000 megahertz.

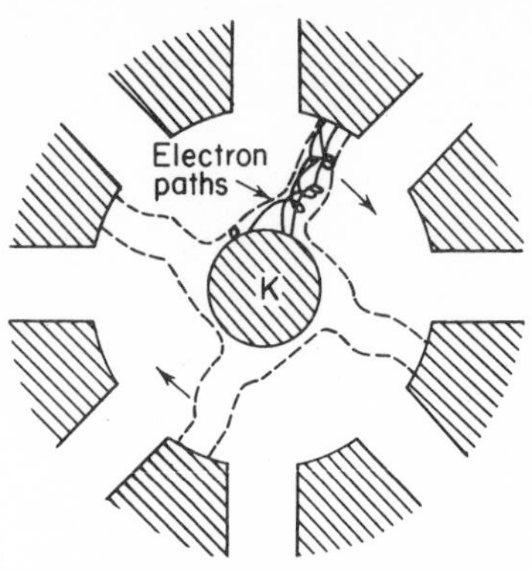

Figure 23-10. Electron paths and rotating electron "spokes."

Magnetrons are tunable over limited ranges by changing cavity volumes by inserting plungers near the back of the slots, thereby lowering inductance and raising frequency; or by inserting plungers near the forward ends of the slots, these increasing the capacity and lowering frequency.

23-5 TRAVELING-WAVE TUBES

As in the several preceding types, the *traveling-wave tube* (TWT) employs a process which converts the dc energy supplied to an electron beam into ac energy by interaction of the beam and an electric or magnetic field; the result is amplification of an applied signal.

An electron beam is formed and focused, caused to pass down a long interaction region, and finally arrives at a collector, as in Fig. 23-11(a). In the interaction region the electrons are prevented from dispersing in transverse directions, by a longitudinal magnetic field supplied by a coaxially located focusing coil.

A common form of TWT applies the signal to a small diameter helix which closely surrounds the electron beam, and the wave travels essentially at the velocity of light along the wire of the helix. Because of the helical path, the actual forward progress of the wave along the axis of the helix is a fraction of the velocity of light, and is called the *phase velocity*. This velocity is comparable to the velocities achieved by the electrons in the beam under moderate accelerating voltages, and adjustment of the accelerating voltage permits reaching a condition where the phase velocity of the wave and the electron velocity in the tube approach equality.

The radio-frequency field of the helix exerts a continuing force on the electrons of the beam; even though the signal fields are weak, the interaction distance is many centimeters. The electrons which are in a retarding field region are slowed down, and the electrons in an accelerating field region are speeded

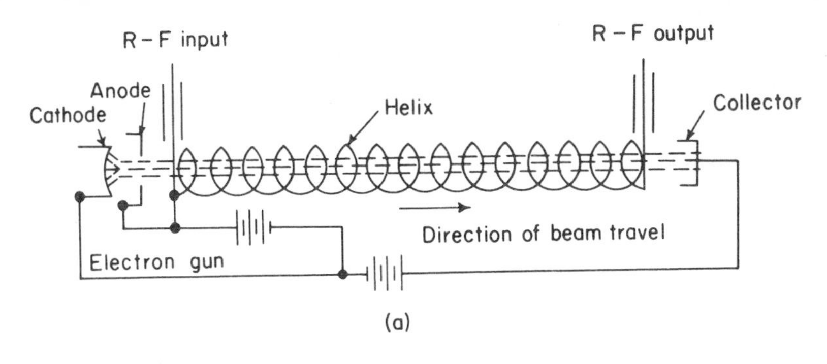

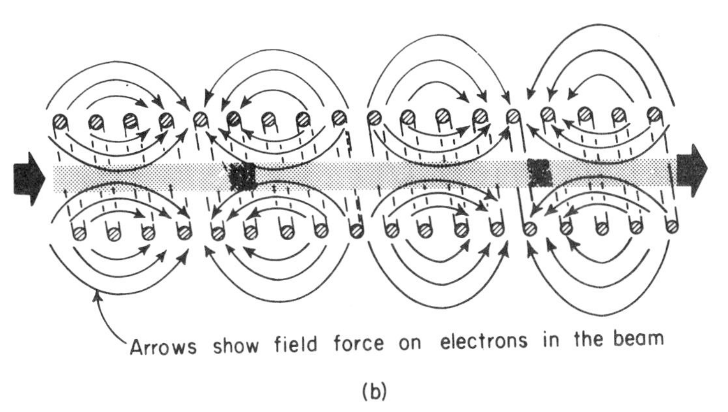

Figure 23-11. Principle of the travelling-wave tube.

up. The result is velocity modulation, and due to the length of the interaction region the beam will become bunched. As the beam travels down the helix, the electrons that are slowed down add energy to the circuit at the expense of beam energy, or induce a radio-frequency current in the helix. Both the density of the electrons in the bunch and the amplitude of the signal wave progressively increase as they travel together down the tube. At the collector end, the amplifier signal is transferred to an output circuit waveguide.

If the beam voltage is so adjusted that the velocity of the beam is slightly higher than the phase velocity of the circuit wave, then more electrons are decelerated than are accelerated, and a net transfer of energy from the beam to the circuit occurs. There will be a maximum in the gain vs. beam-voltage curve as a result.

Actually, the bunches induce a backward wave as well as the forward wave. Under the conditions described above, the phase relations and velocity of the backward wave components are such as to lead to ineffective coupling to the beam, and the effect may be neglected. In some forms of TWT's the backward wave is developed, and the forward wave then becomes negligible in effect.

In the klystron, the interaction between beam and circuit occurs over only a short distance, and to obtain a high unit field and transfer of energy a sharply-resonant, high-impedance circuit is needed, but this implies narrow-band frequency response. In the TWT the helix is a nonresonant circuit and the beam-circuit interaction occurs over a considerable distance at low field intensity. As a result, the TWT will amplify all signals which produce essentially the same phase velocity, and bandwidths may approximate an octave of frequency.

At the long-wave limit of gain, the helix becomes electrically short, and there is insufficient interaction distance to transfer the required energy; at the short-wave limit the interaction becomes poor because the circuit wave concentrates close to the helix wires and does not enter the beam region.

23-6 THE MASER

In discussion of the parametric amplifier in Chapter 17, it was mentioned that usual amplifier noise was due to the inherent particle nature of current, and in devices not employing current phenomena the inherent noise could be low, allowing amplification of very weak signals. The *maser* (microwave amplification by stimulated emission of radiation) and its optical offspring, the *laser*, operate by reason of quantum mechanical release of energy from excited atoms or molecules, and do provide very small noise levels in amplification.

The ammonia maser was the first such device. Its operation is dependent on the fact that the nitrogen atom in the ammonia molecule may occupy either of two energy levels with respect to the three hydrogen atoms, and that there is a certain probability of the atom being at either level at a given time. The difference in energy between the two levels is small and equal to the energy of a photon with a frequency of 24 gigahertz ($f = W/h$).

In ammonia molecules the two levels can be separated by an electric field, the low energy molecules being dispersed and those in the higher level being concentrated and passed to a resonant cavity, tuned to the critical frequency. If a microwave signal is introduced having a frequency exactly equal to the photon frequency which will be emitted by the molecule in its jump to the lower level, it can trigger or stimulate this jump. The ammonia molecules fall back to their lower energy states, and emit the energy difference as photons. Where we introduced one photon, we now have two. If these photons each trigger other molecules, the chain reaction can be self-sustaining. The amplifier turns into an oscillator, generating an output at the frequency of the photons emitted by the ammonia energy jump, as long as a supply of excited ammonia molecules is maintained.

The ammonia maser is at its best as an oscillator; to use it as an amplifier requires that the input signal be an almost exact frequency. The bandwidth accepted is so narrow as to make this difficult, whereas for oscillator purposes the narrow bandwidth or precision of frequency emitted is highly desirable. Such devices are being employed as very stable frequency standards, expected to maintain frequency within 1 part in 10^{11} over long time periods.

In the ammonia maser the over-population of high-energy states was achieved somewhat mechanically by electric-field sorting. In the next step, which led to development of solid-state masers, the desired high population density in a particular energy state was obtained by *pumping* or raising atoms in the normal energy state, to a higher energy level by supply of energy.

The various energy levels used depend on differences in the electron spin in the atom, rather than on atomic energy jumps in molecules. Most atoms are nonmagnetic, and the electrons are paired off with their spins or magnetism cancelled. There are a few atoms, however, in which some electrons have unpaired spins, and the material is then affected by a magnetic field or is paramagnetic; in particular those materials having more than one unpaired electron are used. In such atoms there will be one more energy level than there are unpaired electrons; chromium has three unpaired electrons and four energy levels, and ruby with chromium as one of its constituents is a particularly suitable maser material.

The value of using electrons having spin as the level differentiating factor is that the energy difference between these levels can be varied somewhat by the magnetic field to which the crystal is exposed. Also, these electrons are to some extent affected by the magnetic effects of neighboring atoms, so that not all the electrons will be subjected to exactly the same field. As a result the energy levels differ very slightly, and jumps from these levels will yield slightly differing frequencies, which gives the solid devices a greater bandwidth than the ammonia maser, and the response frequency can be tuned over a range of frequencies by variation of the magnetic field.

Two basic requirements must be met for amplification. First, radiation by spontaneous transition of electrons between the two working levels must be very much less probable than the emission of radiation when the jump between the same levels is triggered or stimulated by a small input of microwave energy of the correct frequency. This is a situation inherent in the metastable levels of atoms. Second, some mechanism must be provided to change the normal population density of the several energy levels employed.

The probability that a transition from a state n to a state m of lower energy will take place is of the form

$$P_{n,m} = A_{n,m} + uB_{n,m}$$

where $A_{n,m}$ is the spontaneous transition probability, u is the introduced radiation (photon) density at the position of the atomic system, and $B_{n,m}$

is the probability of stimulated emission. Einstein showed that

$$A_{n,m} = \frac{8\pi h f^3}{c^3} B_{n,m} \tag{23-7}$$

so that spontaneous emission increases as the cube of the frequency. Fortunately, in the microwave region, the probability of spontaneous transition for the energy levels under consideration is small, so that stimulated emission by input signal energy makes microwave amplification possible.

For the second requirement, we note that if there are equal numbers of atoms in states n and m, the probability of jumps upward, absorbing photons, is equal to the probability of jumps downward which emit photons. A net gain in outward radiation, or amplification, is possible only if there are more electrons in the upper energy state than in the lower one. Normally the ratio of population densities of atoms in the several states is such that $N_1 > N_2 > N_3$. If we can pump or excite atoms by introducing energy of frequency f_{13}, it may be possible to make $N_3 > N_1$. Amplification might then be possible by stimulated transitions from level 3 to level 2, at a frequency f_{32}. Another scheme, called a three-level maser, raises atoms to the 3 level with rapid spontaneous transitions following to the 2 level. If spontaneous transition from 2 to 1 is not very probable, a high density of electrons can be maintained at the 2 level, for stimulation by an input signal. This system allows continuous supply of energy or pumping of electrons from level 1 to 3, at frequency f_{13}, without interfering with the amplified and stimulated emission at frequency f_{21}.

A number of materials having the required transitory electron levels are available. One is ruby, with chromium the active impurity which also gives ruby its color; another is lanthanum ethyl sulfate with gadolinium as the active impurity; still another is iron-doped titanium dioxide.

Masers are thus suited to microwave amplification, even to the millimeter region. Inherent noise is a result of spontaneous emission, and this is related to a noise temperature as $T \geq hf/k = 5 \times 10^{-11} f$. This is equivalent to a noise temperature of 5°K at 100 gigahertz. Another limit is imposed by the pumping energy, which must be supplied at a frequency higher than signal frequency, since the energy input per electron must be in one jump instead of in two as for the emission. Bandwidth is dependent on the variation in energy levels of the individual atoms, due to slight magnetic intensity differences, or to line width from the spectral viewpoint, and approximates one per cent. Noise figures at liquid nitrogen temperatures are near zero.

Applications have been made as first-stage amplifiers for radio astronomy, space communications, and long-distance radar reception.

23-7 THE OPTICAL LASER

In the laser (light amplification by stimulated emission of radiation) the maser principle has been extended to optical light frequencies of 10^{14} or 10^{15}

hertz. In the past, light sources of high intensity utilized heated filaments or arcs as an aggregate of tiny light sources operating independently of each other. The emission was a continuous spectrum of incoherent light. The laser provides the first opportunity to generate high intensity coherent or monochromatic light; the radiation will produce interference patterns and can be readily focused in a small neighborhood.

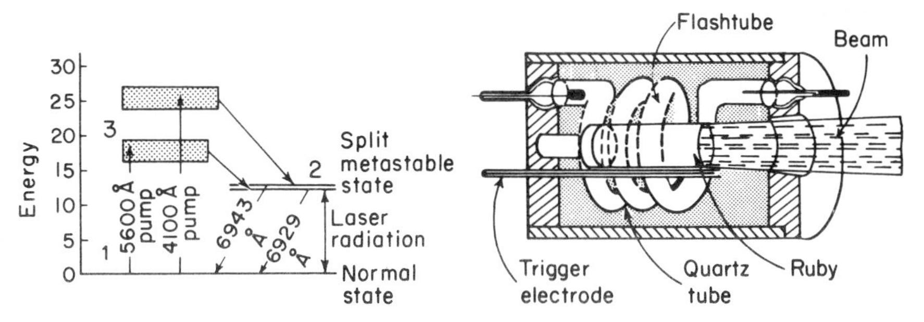

Figure 23-12. Energy levels in the ruby laser.

Figure 23-13. Ruby laser design.

The principle has been applied to generate light pulses in ruby, in which certain energy levels of the chromium atoms have been found to provide suitable output frequencies. The situation is illustrated in Fig. 23-12 where the pumping transitions occur in the green and yellow spectral region at 5600 Å and 4100 Å. This energy is supplied by repeated flashes of intense white light from a xenon flash tube wrapped around the ruby rod, as in Fig. 23-13. Spontaneous transition with emanation of heat then occurs in the crystal, and electrons drop from level 3 to the split metastable state 2. The population density of electrons at this level grows because of the long life time of about 10^{-3} seconds in this state.

When the population of electrons in state 2 exceeds that in 1 due to the pumping process, it is possible for laser action to begin. A triggering coherent beam of light at 6943 Å will stimulate other atoms to emit coherent radiation. This stimulated emission builds up very rapidly in an avalanche effect, until the stimulated downward transitions exceed the rate of production of excited atoms at 2 by the pumping process. The population excess is reduced below the maintaining level and the radiation dies, until pumping again establishes an excess population in the upper state. The result is a pulsed radiation, with repetition rate dependent on the amount of energy being supplied by the pump.

The ends of the ruby rod are optically flat and coated as partially reflecting mirrors. The mirrors turn part of the coherent beam back into the active volume, and the resulting feedback makes the ruby a regenerative oscillator. Only the modes which are reflected normal to the end mirrors will

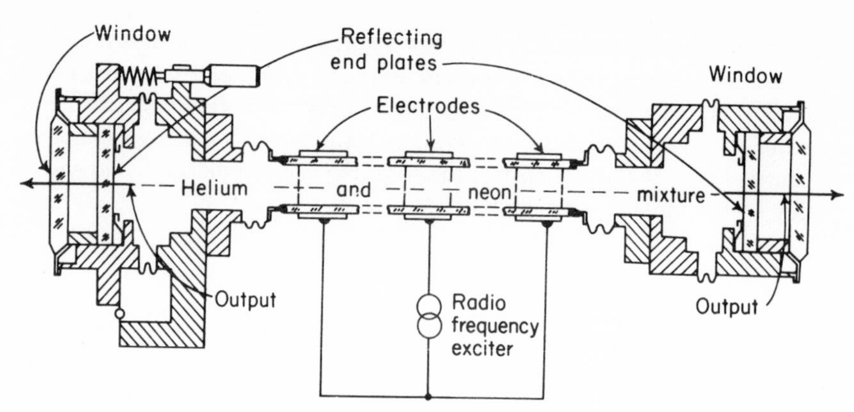

Figure 23-14. Gas continuous laser.

stimulate radiation, other modes ultimately being lost through the side surfaces of the rod, and the result is a very sharply parallel output beam through the semireflecting mirror surfaces at the rod ends.

It has been found possible to generate continuous output in a gas-phase laser, as indicated in Fig. 23-14, using a helium-neon mixture. The two electrons of helium can have parallel or antiparallel spins; in the first case the energy level is higher than in the second or net zero spin condition. Unexcited helium will have electrons in both levels, although the normal or ground state is that of the lowest spin energy. The higher energy level for the parallel-spin case is considerably above the normal state and is metastable, and a certain number of electrons are trapped in these levels.

Neon has an energy level very close to that of the helium metastable energy level. Collisions between metastable helium atoms and unexcited neon atoms have a high probability of energy transfer to the neon, and the neon population density in the excited level will increase. When these excited neon atoms return to their normal or ground state, they do so by several transitions which lead to output in the optical range.

Pumping is accomplished by establishing an electrical discharge in the gas by radio-frequency excitation. When the ionized helium atoms recombine, a fair percentage recombine as parallel-spin atoms and so the supply of metastable electrons is continually maintained.

While the process leads to continuous radiation, the gas-phase device is a low output device, due to the continuous output, whereas the ruby laser is capable of large peak power output, as the summation of the pumping energy supplied continuously.

23-8 THE GUNN OSCILLATOR

A solid-state generator of frequencies in the gigahertz range is the *Gunn oscillator*. Operation is dependent on a semiconductor material with two conduction bands, separated by an energy gap, and gallium arsenide satisfies this requirement.

The normal conduction band is at a lower energy level, and in the higher energy conduction band the electrons have a much lower mobility. In normal operation, conduction occurs by electrons present in the lower energy band, and the upper band is unfilled. As the applied electric field is increased from zero, the current increases at a normal rate as conduction electrons in the lower band are accelerated. With a further increase in applied field, more electrons are raised to the upper energy band with lower electron mobility. The lowered mobility means lowered current with increasing potential; this indicates a negative resistance. The resultant volt-ampere curve is similar to that of the tunnel diode.

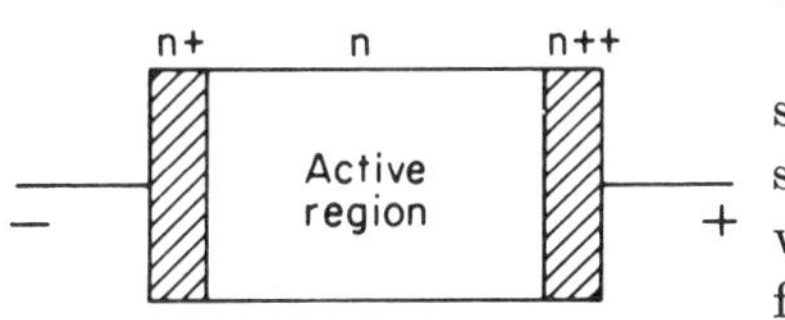

Figure 23-15. Sandwich Gunn oscillator diode.

Construction may take the form of a sandwich as in Fig. 23-15, the central section being n gallium arsenide, and with the outer sections epitaxially grown from gallium arsenide with increased impurity doping and higher conductivity. In passing from the n^+ region to the active n region a charge discontinuity occurs, causing a higher electric field and transfer of some of the electrons to the lower mobility band, where they move more slowly. Electrons from the n^+ region accumulate behind this barrier or depletion region, as in a diode. The region of high charge density moves across the active region at the average electron drift velocity. As the layer reaches the positive electrode, a new accumulation-depletion region builds up at the left face. Therefore the frequency generated is a function of drift velocity and of active region thickness.

By use of the n^+, lightly doped, and n^{++}, more heavily-doped regions (as contacting electrodes and mechanical supports), it is possible to obtain structures with active regions as thin as 4 microns. Frequencies in the range from 5 GHz and 100 milliwatts output up to frequencies of 35 GHz and 1 milliwatt output have been observed. Efficiencies of one to three per cent are presently possible. Tuning over a small range is possible by control of current and voltage.

REFERENCES

1. Varian, R., and S. Varian, "A High-Frequency Oscillator and Amplifier," *J. Appl. Phys.*, Vol. 10, p. 321 (1939).
2. Pierce, J. R., "Reflex Oscillators," *Proc. IRE*, Vol. 33, p. 112 (1945).
3. Fisk, J. B., H. O. Hagstrum, and P. L. Hartman, "The Magnetron as a Generator of Centimeter Waves," *Bell Syst. Tech. Jour.*, Vol. 25, p. 167 (1946).

4. Kompfner, R., "The Traveling-Wave Tube as an Amplifier at Microwaves," *Proc. IRE*, Vol. 35, p. 124 (1947).

5. Collins, G. B., *Microwave Magnetrons*. McGraw-Hill Book Company, New York, 1948.

6. Pierce, J. R., *Traveling-Wave Tubes*. D. Van Nostrand Co., Inc., Princeton, N. J., 1950.

7. Gordon, J. P., H. J. Ziegler, and C. H. Townes, "The Maser—New Type of Microwave Amplifier, Frequency Standard, and Spectrometer," *Phys. Rev.*, Vol. 99, p. 1264 (1955).

8. Bloembergen, N., "Proposal for a New Type of Solid-State Maser," *Phys. Rev.*, Vol. 104, p. 324 (1956).

9. Thomson, J., and E. B. Callick, *Electron Physics and Technology*. The Macmillan Company, New York, 1959.

10. Maiman, T. H., "Optical Action in Ruby," *Nature*, Vol. 187, p. 493 (1960).

11. *Proceedings of the IEEE*, special issue on Quantum Electronics, Vol. 51, pp. 1–294 (1963).

12. Gunn, J. B., "Instabilities of Current in III-V Semiconductors," *IBM Jour. Res. Devel.*, Vol. 8, p. 141 (1964).

13. *IEEE Transactions on Electron Devices*, ED-13, special issue on avalanche diodes and Gunn-effect devices (Jan. 1966).

APPENDIX

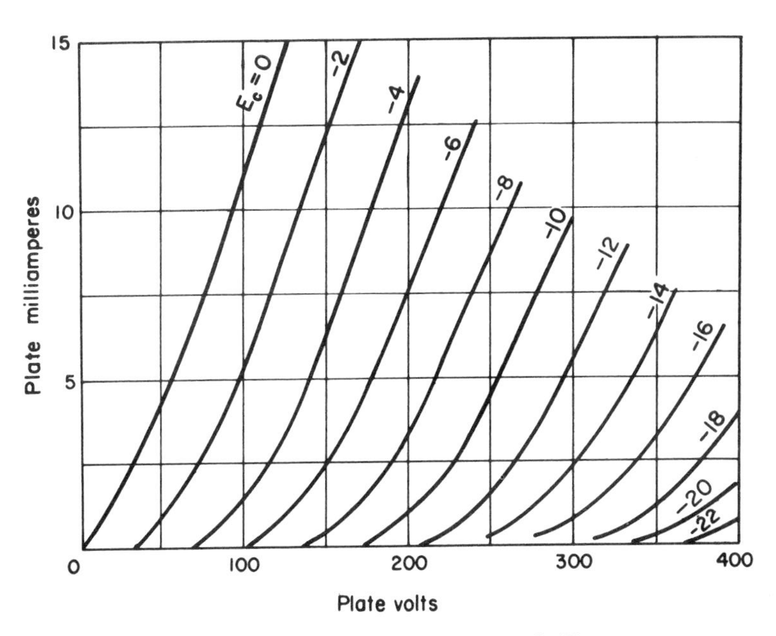

Figure A-1. 12AU7 triode. $C_{gp} = 3.8$ pF; $C_{gk} = 4.2$ pF; $C_{pk} = 5.0$ pF; Maximum anode dissipation = 2.5 W. At recommended Q point: $E_b = 250$ V; $\mu = 20$; $E_c = -8$ V; $r_p = 7700$ ohms; $I_b = 9$ mA; $g_m = 2600$ μmho.

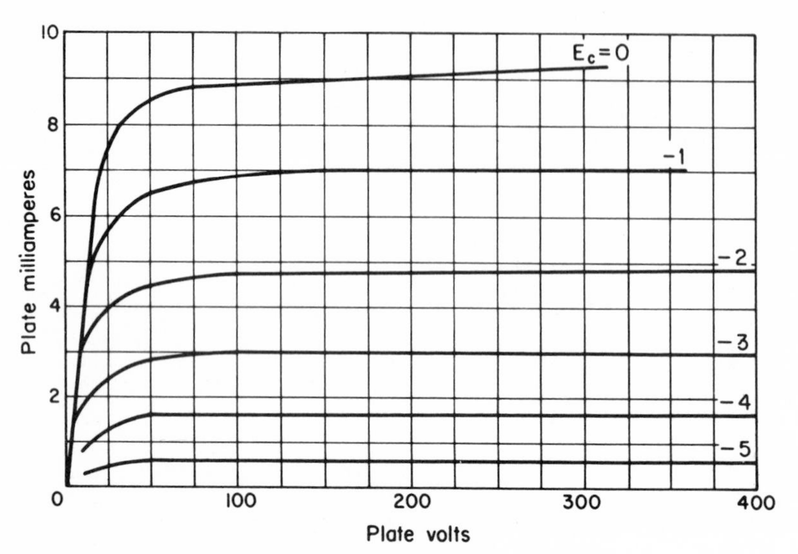

Figure A-2. 6AU6 pentode. $C_{gp} = 0.0005$ pF; $C_{gk} = 6$ pF; $C_{pk} = 7$ pF. Maximum anode dissipation = 2.5 W; maximum screen dissipation = 0.3 W. At recommended Q point: $E_b = 250$ V; $I_b = 3$ mA; $E_{c1} = -3$ V; $I_{c2} = 0.8$ mA; $E_{c2} = 100$ V; $E_{c3} = 0$ V; $r_p = 1$ megohm; $g_m = 1650$ μmho.

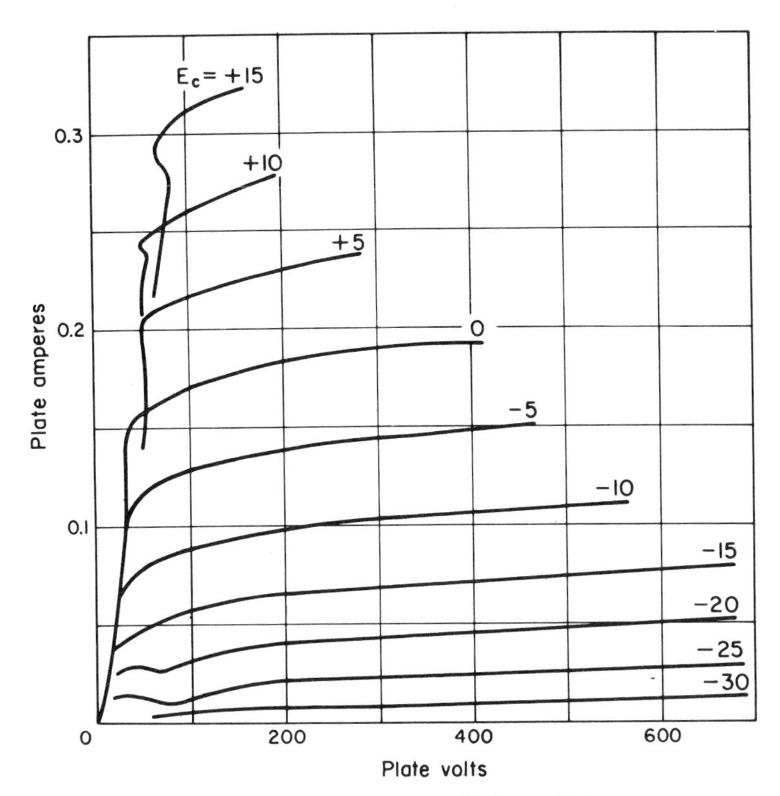

Figure A-3. 6L6 beam tetrode. $C_{gp} = 0.9$ pF; $C_{gk} = 11.5$ pF; $C_{pk} = 9.5$ pF. Maximum anode dissipation = 19 W; maximum screen dissipation = 2.5 W. At recommended Q point (Class A): $E_b = 350$ V; $I_b = 54$ mA; $E_{c1} = -18$ V; $I_{c2} = 2.5$ mA; $E_{c2} = 250$ V; $r_p = 33{,}000$ ohms; $g_m = 5200\ \mu$mho.

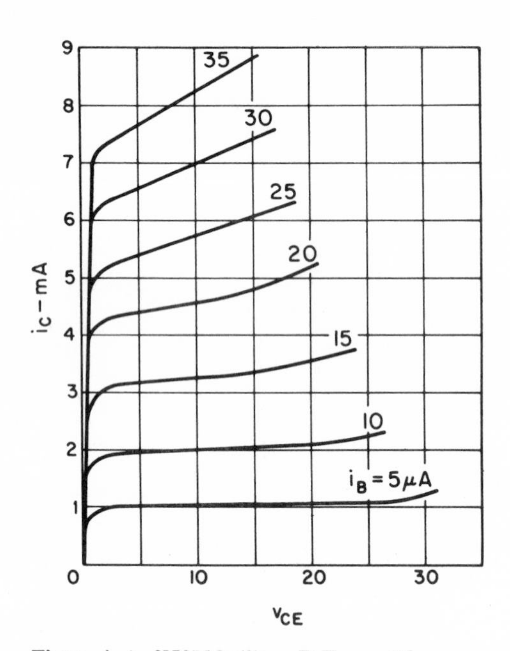

Figure A-4. 2N2712 silicon R.F. amplifier transistor. $V_{CE\ \max} = 18$ V; $h_{fe} = 150$; $h_{ie} = 220$ ohms. At 25°C, $P_d = 200$ mW and $I_{CO} = 0.5\ \mu$A.

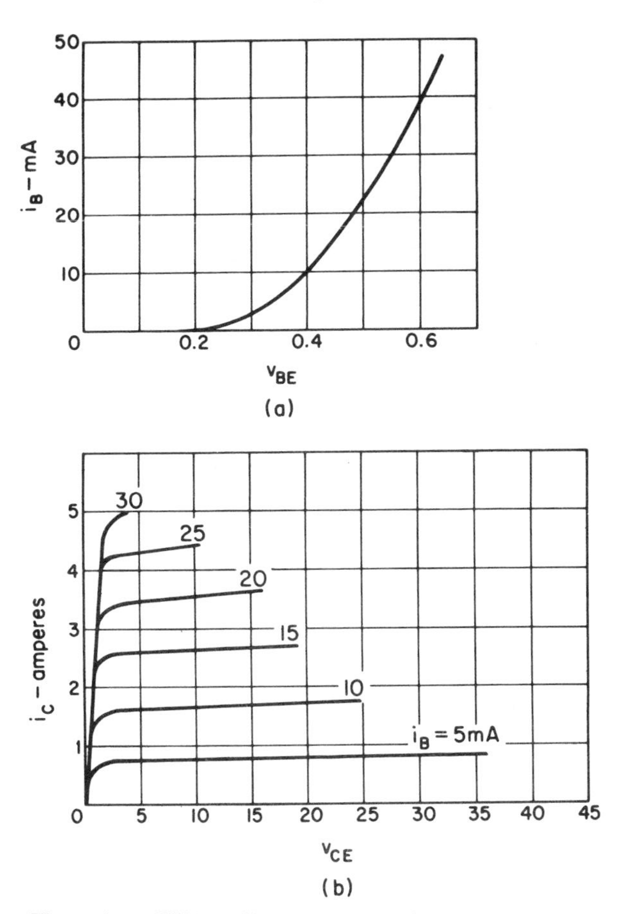

Figure A-5. 2N2147 silicon power transistor.
$V_{CE} = 50$ V max; $V_{EB} = 1.5$ V max; $I_C = 5$ A max;
$I_B = 1$ A max; $T_{J\,\max} = 100°C$; $\theta_{JC} = 1.5°C/W$;
$h_{FE} = 150$; GBW = 4 MHz.

Table A-1

SELECTED BESSEL-FUNCTION VALUES

x	$J_0(x)$	$J_1(x)$	$J_2(x)$	$J_3(x)$	$J_4(x)$	$J_5(x)$	$J_6(x)$	$J_7(x)$	$J_8(x)$	$J_9(x)$	$J_{10}(x)$
0	1.0000	0.0000	0.0000	0.0000	0.0000	0.0000	0.0000	0.0000	0.0000	0.0000	0.0000
1	0.7652	0.4401	0.1149	0.0196	0.0025	0.0003	0.0000	0.0000	0.0000	0.0000	0.0000
2	0.2239	0.5767	0.3528	0.1289	0.0340	0.0070	0.0012	0.0002	0.0000	0.0000	0.0000
3	−0.2601	0.3391	0.4861	0.3091	0.1320	0.0430	0.0114	0.0026	0.0005	0.0000	0.0000
4	−0.3971	−0.0660	0.3641	0.4302	0.2811	0.1321	0.0491	0.0152	0.0040	0.0009	0.0002
5	−0.1776	−0.3276	0.0466	0.3648	0.3912	0.2611	0.1310	0.0534	0.0184	0.0055	0.0015
6	0.1506	−0.2767	−0.2429	0.1148	0.3576	0.3621	0.2458	0.1296	0.565	0.0212	0.0070
7	0.3001	−0.0047	−0.3014	−0.1676	0.1578	0.3479	0.3392	0.2336	0.1280	0.0589	0.0235
8	0.1717	0.2346	−0.1130	−0.2911	−0.1054	0.1858	0.3376	0.3206	0.2235	0.1263	0.0608
9	−0.0903	0.2453	0.1448	−0.1809	−0.2655	−0.0550	0.2043	0.3275	0.3051	0.2149	0.1247
10	−0.2459	0.0453	0.2546	−0.0584	−0.2196	−0.2341	−0.0145	0.2167	0.3179	0.2919	0.2075
11	−0.1712	−0.1768	0.1390	0.2273	−0.0150	−0.2383	−0.2016	0.0184	0.2250	0.3089	0.2804
12	0.0477	−0.2234	−0.0849	0.1951	0.1825	−0.0735	−0.2437	−0.1703	0.0451	0.2304	0.3005
13	0.2069	−0.0703	−0.2177	0.0023	0.2193	0.1316	−0.1180	−0.2406	−0.1410	0.0670	0.2338
14	0.1711	0.1334	−0.1520	−0.1768	0.0762	0.2204	0.0812	−0.1508	−0.2320	−0.1143	0.0850

ANSWERS TO ODD-NUMBERED PROBLEMS

CHAPTER 1

1-1. (a) 7.33×10^6 m/s; (b) 1.07×10^{-9} s; (c) 2.44×10^{-17} J.

1-3. (a) 0.0333 m; (b) 14.2×10^{-9} s. Return to bottom plate.

1-5. (a) 94.9 V/m; (b) 1.90×10^{-4} W/m^2.

1-7. 45°; $x = 0.40$ m.

1-9. $v = 1.60 \times 10^6$ m/s upward; $W = 1.17 \times 10^{-20}$ J; $t = 5.62 \times 10^{-9}$ s.

1-11. 8.42×10^{-4} weber/m^2 out of page.

1-13. (a) 4.69 cm; (b) 6.6 cm.

1-15. $x = 27.6$ mm; $y = 0.0$ mm.

CHAPTER 2

2-3. 2.53×10^2 mho-m.

2-5. 5.22 ohm-m.

2-7. Ge: $\lambda = 1.66\ \mu$; Si: $\lambda = 1.08\ \mu$.

2-9. 332.

2-11. 0.049 ohm-m.

2-13. Falls 0.186 eV.

2-15. 0.179 V; 0.0179 ohms.

2-17. $n = 4.41 \times 10^{20}$/m^3; $p = 1.42 \times 10^{18}$/m^3.

CHAPTER 3

3-1. 1.05×10^6 m/s.

3-3. $\lambda_0 = 0.688\ \mu$; $v = 4.3 \times 10^5$ m/s.

3-5. Pt: 4.82 V; K: 1.78 V; Cd: 3.95 V; Mg: 3.62 V.

3-7. 30 mA/W.

3-9. (a) 0.0562 m^2; (b) 6.0×10^{11} m^2.

3-13. 167 V.

3-15. Diameter = 0.906 cm; length = 7.11 cm.

CHAPTER 4

4-1. 25 V; 16.7 W; 0.578.

4-3. 0.135 ohms (1 V diode drop).

4-5. 85 ohms.

4-9. (a) 535 V; (b) 6.2 A; (c) $\gamma = 0.0336$.

4-13. $\gamma = 0.266$; $E_{dc}/E_m = 425/621 = 0.685$. From Fig. 4-14, $\omega CR = 1.05$; $C = 1.5\ \mu$F.

4-15. (a) 0.12; (b) 0.174; (c) 0.0171.

4-17. $L = 2.65$ H; $C = 63\ \mu$F.

4-19. Capacitor input: 375 V rms to CT. Inductor input: 435 V rms to CT.

CHAPTER 5

5-1. $z_i = 7.08 + j2.47$ ohms; $z_0 = 1.43 + j5.80$ ohms; $z_r = z_f = 0.40 - j0.20$ ohms.

5-3.

$$y_i = y_0 = \frac{(Z_a + Z_c)(Z_c + Z_d)}{MN}$$

$$y_r = y_f = \frac{-(Z_a Z_d - Z_c^2)}{MN}$$

where

$$M = Z_a + Z_d + 2Z_c$$

$$N = (Z_a + Z_c)^2(Z_c + Z_d)^2 - (Z_a Z_d - Z_c^2)^2$$

$$h_i = \frac{(Z_a + Z_c)(Z_c + Z_d) - (Z_a Z_d - Z_c^2)^2}{Z_a + Z_d + 2Z_c}$$

$$h_0 = \frac{(Z_a + 2Z_c)Z_d}{(Z_a + Z_c)(Z_c + Z_d)}$$

$$h_r = h_f = \frac{Z_aZ_d - Z_c^2}{(Z_a + Z_c)(Z_c + Z_d)}$$

5-5. See Problem 5-3 (y parameters).

5-7. Since $Z_b = Z_c$,

$$Z_1 = \frac{Z_B(Z_A + Z_D) + 2Z_AZ_D}{Z_A + 2Z_B + Z_D}$$

$$Z_2 = \frac{Z_B(Z_A + Z_D) + 2Z_AZ_D}{Z_A + 2Z_B + Z_D}$$

$$Z_3 = \frac{Z_B^2 - Z_AZ_D}{Z_A + 2Z_B + Z_D}$$

5-9. $Z_{i1} = 7.1 + j2.47$ ohms; $Z_{o2} = 1.42 + j5.80$ ohms; $Z_{T12} = 10 + j17$ ohms.

5-11. $V_2 = 14.6$ V.

5-13. 13.01 dB.

CHAPTER 6

6-1. $i_C = 95\ \mu$A; $i_E = -102\ \mu$A; $i_B = -7\ \mu$A.

6-3. 960 μA.

6-5. $W = 9.35 \times 10^{-6}$ m.

CHAPTER 7

7-1. Q point at: $I_B = 200\ \mu$A, $I_C = 22$ mA. Then $i_{max} = 39$ mA; $i_{min} = 1$ mA.

7-3. $R_i = 1250$ ohms; $R_o = 3.07 \times 10^5$ ohms; $A_i = 5.81$; $A_v = -14.6$.

7-5. 16.27 dB.

7-9. C–E: 44.9 dB; C–B: 16.3 dB; C–C: 13.8 dB.

7-11. $R_{ie} = 8030$ ohms; $R_{oe} = 0.376 \times 10^6$ ohms; $A'_{ie} = 19$; $A'_{ve} = -23.7$.

7-13. $A_i = 0.66$.

7-15. $h_{ie} = 1600$ ohms $\quad h_{ib} = 31$ ohms
$h_{oe} = 0.42 \times 10^{-4}$ mho $\quad h_{ob} = 0.83 \times 10^{-6}$ mho
$h_{fe} = 49 \quad h_{fb} = -0.98$
$h_{re} = 12.5 \times 10^{-4} \quad h_{rb} = 93 \times 10^{-6}$
(b) 21,200; (c) 25.2×10^4 ohms.

CHAPTER 8

8-1. $S = (R_E + R_1)/[R_E + R_1(1 - \alpha)]$.

8-3. 96 μA.

8-7. From curves $h_{fe} = 60$; $S = 1$ if $R_E = R_L/\alpha$.

8-9. $R_B = 590{,}000$ ohms with $v_{EB} = 0.15$ V; $R_{ie} = 51{,}800$ ohms.

8-11. For 40°C rise: $\Delta I_C = 44\ \Delta I_{CBO}$; $\Delta I_C = 10\ \mu$A due to Δv_{EB}.

CHAPTER 9

9-3. $i_b = 10.8$ mA; $r_p = 6{,}450$ ohms; $g_m = 1775\ \mu$mho.

9-5. $I_b = 10.5$ mA; $i_{\text{max}} = 12.2$ mA; $i_{\text{min}} = 8.9$ mA.

9-9. $\dfrac{E_o}{E} = \dfrac{\mu R_L}{r_p + R_L + R_1 + \mu^2 R_1 R_2 (r_p + R_2)}$.

9-11. $E_o = 2.47$ V; $I_p = 0.89$ mA.

9-13. $E_s = 0.054$ V; $R_L = 146{,}000$ ohms.

9-17. 13.13 dB; 915 ohms.

9-19. See Fig. 9-28(a): $R_{k1} = 3000$ ohms, no bypass; $R_k = 5650$ ohms.

9-21. $R_L = 3500$ ohms; $E_o = 0.175$ V.

CHAPTER 10

10-1. $C_E = 2.03\ \mu$F.

10-3. $f_1 = 15$ Hz; $C = 0.042\ \mu$F.

10-5. $f_\beta = 54$ MHz; $f_T = 4.32$ GHz.

10-7. $f_2 = 126$ kHz; $f_1 = 19$ Hz.

10-9. $A_i = 7.0$ but $h_{fe} = 19$.

10-11. 9900 Hz.

10-13. $\alpha = 0.54$ magnitude.

10-15. Overall $f_2 = 43{,}600$ Hz; individual stage $f_2 = 85{,}500$ Hz.

10-17. $f_1 = 153$ Hz; $f_2 = 170$ kHz.

10-19. $C_d = 1186$ pF.

10-21. $A_i = 18.8$.

CHAPTER 11

11-1. BW = 1.27 MHz; 423rd harmonic.

11-3. (a) R_L = 5,170 ohms; (b) C = 595 pF; (c) E = 23.3 V.

11-5. With $\omega L \gg R_L$ at mid-frequency, and $R_L = R$ and $1/h_{oe}$ in parallel; $R_L = (2L/C)^{1/2}$.

11-7. E_2 = 2.76 V.

11-9. L = 500 μH; C = 5070 pF; k = 0.0109.

11-11. (a) ω_o = 9999; f_o = 1580 Hz; BW = 47 Hz; Q = 33.3.
(b) R = 3330 ohms; C = 1 μF; L = 0.01 H.

11-13. R_p = 2000 ohms; L = 3.98 μH.

11-15. Q = 75; L = 189 μH; C = 662 pF; M = 3.78 μH.

CHAPTER 12

12-1. $A = +70$; $\beta = -0.093$.

12-5. A/stage = 11.45; $\beta = -0.019$.

12-7. f_1 = 25.4 Hz; f_1' = 5.95 Hz.

12-11. R_F = 1.18 megohms.

12-13. Stable.

CHAPTER 13

13-1. 1.14 mA/V.

13-3. $e_2 = -0.75t^2$.

CHAPTER 14

14-3. (a) 8.19 W; (b) 5.03 per cent; (c) 14.8 W, 35.5 per cent.

14-5. P_d = 19.6 W.

14-7. S = 8.85; stable.

14-11. E_1/E_2 = 287.

14-13. (a) P_o = 8.13 W, P_{in} = 135 W, Efficiency = 6.0 per cent; (b) 333 ohms.

CHAPTER 15

15-1. $\omega = \sqrt{\dfrac{1}{C(L_1 + L_2 - L_1L_2h_{oe}/h_{ie})}}$

$h_{fe} \geqq \dfrac{L_1}{L_2}\left(1 - \dfrac{h_{oe}}{h_{ie}} L_2\right)$

15-3. $f = \dfrac{1}{2\pi}\sqrt{\dfrac{1}{L[C_1C_2/(C_1 + C_2)]}}\sqrt{1 + \dfrac{R}{r_p}\left(\dfrac{C_1}{C_1 + C_2}\right)}$

$f = 2.14$ MHz; $\mu = C_1/C_2$.

15-5. $A \geqq 11; f = 1300$ Hz.

CHAPTER 16

16-1. $P_c = 1$; each side band $= 0.031\ P_c$.

16-3. $e_o = 0.5E_1[\sin 2\pi \times 60t + 0.25 \sin 2\pi(60 + 6)t + 0.25 \sin 2\pi(60 - 6)t]$.

16-5. (a) $P_{\text{mod}} = 405$ W; (b) $a = 1.22$; (c) 520 W.

16-7. $\omega_o = 0.236$; $\omega_o \pm \omega_m = 0.115$; $\omega_o \pm 2\omega_m = -0.207$; $\omega_o \pm 3\omega_m = -0.229$; $\omega_o \pm 4\omega_m = 0.026$; $\omega_o \pm 5\omega_m = 0.267$; $\omega_o \pm 6\omega_m = 0.338$.

16-9. (a) 43 kHz; $\delta = 2$; (b) 20 kHz; $\delta = 50$.

16-13. $R = 620$ ohms.

16-15. $f_2/f_1 = 1.0125$.

16-17. 20.4 channels; $Q = 4.9$.

CHAPTER 17

17-1. $e = 0.60(e_s + 0.35e_s^2)$; $\omega_c = 1.5$ V; $\omega_m = 1.05$ V; $2\omega_c = 0.865$ V; $2\omega_m = 0.210$ V; $\omega_c \pm \omega_m = 0.60$ V; $2(\omega_c \pm \omega_m) = 0.105$ V; $2\omega_c \pm \omega_m = 0.523$ V; $2\omega_m/\omega_m = 20$ per cent distortion.

17-3. (a) 53 per cent; (b) $R_i = 263{,}000$ ohms; (c) $E_o = 7.5$ V.

17-7. $m_{a\ \max} = 0.555$.

17-9. 9.95 MHz, $E = 1.40$ V; 9.85 MHz, $E = 0.87$ V.

17-11. 1860 kHz.

CHAPTER 18

18-7. $f = 12{,}000$ Hz.

18-11. Error < 0.1 per cent.

CHAPTER 19

19-1.

A	B	C	$AB + AC$	$A(B + C)$
0	0	0	0	0
0	0	1	0	0
0	1	0	0	0
1	0	0	0	0
0	1	1	0	0
1	1	0	1	1
1	0	1	1	1
1	1	1	1	1

19-5. 563; 975; 852; 924.

19-7. 0, 1, 2, 3, 4, 5, 6, 7, 10, 11, 12, 13, 14, 15, 16, 17, 20, 21, 22, 23, 24, 173, 2077, 1707.

19-9. (a) 1010001 = 81; (b) 1010100 = 84; (c) 1000100110 = 550; (d) 1001001100 = 588.

19-11. $AB(C + D) = S$.

19-13.

Octal	135	313	343, 262	57	154, 333	225	254,
Decimal	93	203	227 178	47	108 219	149	172

CHAPTER 20

20-1. 01111, 10011, 10100, 10011, 01111, 01010, 00101, 00001, 00000, 00001, 00101.

CHAPTER 21

21-1. (a) 27.8 V rms; (b) 15.7 A; (c) 2.5 W; (d) 88 W.

21-3. (a) 0.074 A; (b) 16.4 W; (c) 0.148 W; (d) 7.3 V; (e) 20.8 W.

21-5. (a) P = 230 V, 386 A; S = 444 V, 81.5 A. (b) 96.9 per cent.

21-7. FW/bridge = 0.57.

21-9.

	1 phase	*3 phase*	*6 phase*
(a)	3.98 Kw	5.97 Kw	5.97 Kw
(b)	$98.80	$93.70	$143.80
(c)	1.89 c	1.91 c	2.11 c

21-11. (a) 0.79 A; (b) 5.4 A.

21-13. 47.6°.

CHAPTER 22

22-1. 38 V.

22-3. 14.1 V rms.

22-7. Pick-up: 0.028 l; drop-out: 0.009 l.

INDEX

A

D

E

F

G

H

I

S

T

U

V

W

Z